Lecture Notes in Earth Sciences 131

For further volumes:
http://www.springer.com/series/772

Joachim Reitner • Nadia-Valérie Quéric •
Gernot Arp

Advances in Stromatolite Geobiology

Springer

Prof. Dr. Joachim Reitner
Universität Göttingen
Geowissenschaftliches Zentrum
Abt. Geobiologie
Goldschmidtstr. 3
37077 Göttingen
Germany
jreitne@gwdg.de

Dr. Nadia-Valérie Quéric
Universität Göttingen
Geowissenschaftliches Zentrum
Abt. Geobiologie
Goldschmidtstr. 3
37077 Göttingen
Germany
nqueric@gwdg.de

Dr. Gernot Arp
Universität Göttingen
Geowissenschaftliches Zentrum
Abt. Geobiologie
Goldschmidtstr. 3
37077 Göttingen
Germany
garp@gwdg.de

ISSN 0930-0317
ISBN 978-3-642-10414-5 e-ISBN 978-3-642-10415-2
DOI 10.1007/978-3-642-10415-2
Springer Heidelberg Dordrecht London New York

Library of Congress Control Number: 2010937235

Cover design: SPi Publisher Services

Printed on acid-free paper

Springer is part of Springer Science+Business Media (www.springer.com)

Preface

Stromatolites are the most intriguing geobiological structures of the entire history of the earth since the early beginning of the fossil record in the Archaean. Traditionally, stromatolites and related microbial sediments are interpreted as biosedimentological remains of biofilms and microbial mats.

Stromatolites are important environmental and evolutionary archives that give us plenty of information about ancient habitats, biodiversity, evolution of complex benthic biosystems, and generally of Global Change. However, many aspects of the formation, biology, and geobiology of these structures are still cryptic and poorly understood.

The Geobiology Group in Göttingen has successfully been granted a large international research project to solve many of these open questions. Therefore, we organised a symposium under the auspices of the DFG-Research Unit FOR 571 "Geobiology of Organo- and Biofilms": Coupling Geosphere and Biosphere via Microbial Processes and the Courant Research Centre Geobiology, which is part of the German Federal Excellence Initiative.

The symposium was dedicated to Ernst Louis Kalkowsky (1851–1938), who has introduced the terms "Stromatolith" and "Oolith" to the earth science community in 1908. 2008 was the 100th anniversary of his remarkable publication published in *Zeitschrift der Deutschen geologischen Gesellschaft: "Oolith und Stromatolith im norddeutschen Buntsandstein"*.

However, one group on their own cannot answer all open questions, and therefore we have organised the stromatolite symposium in Göttingen together with our international colleagues and friends. This meeting somewhat stands in the tradition of the "Death Valley International Stromatolite Symposium", which was very successfully organised by Stanley M. Awramik and Robert Riding in 1994. This meeting has given us new and exceptional ideas and information on the formation and environmental setting of stromatolites, and we hope that the symposium in Göttingen has delivered us new insights into the scientific progress of this topic, which has taken place during the past 16 years.

More than 120 scientists from various interdisciplinary fields, e.g. biology, microbiology, biogeochemistry, geology, sedimentology, from 19 countries world-wide have joined the meeting and presented their most recent research on stromatolites and related topics.

The proceedings volume with more than 30 contributions is a most recent contribution to the geobiology of stromatolites, related microbial sediments, and microbial metabolic processes and covers further wide range of geomicrobiological topics. The editors hope that this publication will close some gaps of knowledge and will give new inspirations of future research dealing with stromatolites.

Göttingen, Germany
18 August 2010

Joachim Reitner
Nadia-Valérie Quéric
Gernot Arp

Acknowledgments

The Kalkowsky symposium was funded by the German Research Foundation (Deutsche Forschungsgemeinschaft DFG) via the Research Unit FOR 571 "Geobiology of Organo- and Biofilms – Coupling Geosphere and Biosphere via Microbial Processes" and the Courant Research Centre Geobiology which is part of the German Federal Excellence Initiative (Förderlinie 3) at the University of Göttingen. This enabled us to invite experts in the field of stromatolites and related systems worldwide to give keynote lectures and special laboratory and field courses. We are very grateful for this support. The compilation of this book would not have been possible without the kind help of the reviewers and the staff of the Geobiology department in Göttingen.

Finally we would like to thank the authors of this book who showed great patience during the publication of this book and for help in the reviewing process.

Contents

Part I
General Aspects and Research History

Founding of the Term ‘Stromatolite’: Ernst Louis Kalkowsky (1851–1938) and his Early Stromatolite Research

Alexander Gehler and Mike Reich

1 Biographical Notes

Ernst Louis Kalkowsky (1851–1938) (Figs. 1 and 2) was born September 9th, 1851, in Tilsit, East Prussia (now Sowetsk, Oblast Kaliningrad, Russia). He was the only child of Caroline Amalie and Friedrich Kalkowsky (Erler et al. 2008). After achieving his baccalaureate in June 1870 in his hometown, he matriculated at the University of Leipzig in natural sciences on October 17th, 1870, and received his doctorate in 1874 with a petrographic work on felsites and pitchstones. Kalkowsky’s doctoral advisor was Ferdinand Zirkel (1838–1912), a pioneer in the microscopic research on rocks and minerals.

In 1875, he accepted an employment as a field geologist (“Sektionsgeologe”) at the Royal Geological Survey of Saxony (“Königlich Sächsische Geologische Landesuntersuchung”), which he had to resign after only 7 months due to pulmonary disease (Erler et al. 2008; Kühne et al. 2006).

His habilitation thesis on the gneissic formation of the Silesian Eulengebirge (now Góry Sowie, south-western Poland), the oldest part of the Central European mountain range of the Sudetes, was published in 1878.

After an employment as assistant professor under F. Zirkel in Leipzig in 1881–1882, he worked self-employed in Königsberg (now Kaliningrad, Russia) for a short time, then in Gotha, amongst others for the publishing house “Justus Perthes Geographische Anstalt” (Anonymous 1938; Rimann 1940).

A. Gehler (✉)
Geoscience Centre, University of Göttingen, Isotope Geology Division, Goldschmidtstr. 1-3, 37077 Göttingen, Germany
e-mail: agehler@gwdg.de

M. Reich
Geoscience Centre, University of Göttingen, Museum, Collections and Geopark, Goldschmidtstr. 1-5, 37077 Göttingen, Germany
e-mail: mreich@gwdg.de

J. Reitner et al., *Advances in Stromatolite Geobiology*,
Lecture Notes in Earth Sciences 131, DOI 10.1007/978-3-642-10415-2_1,

Fig. 1 Ernst Louis Kalkowsky, photograph taken around 1928. [© Staatliche Naturhistorische Sammlungen Dresden, Museum für Mineralogie und Geologie, Archiv]

Fig. 2 Ernst Louis Kalkowsky [2nd row, 3rd from left] with colleagues during a meeting in Dresden on December 19th, 1895. [© SLUB Dresden / Deutsche Fotothek]

In 1886, he followed an appointment to the chair of mineralogy and geology in Jena, first employed as associate professor, from 1887 as full professor. He also obtained the head office of the "Grand Duke's Museum of Mineralogy". Eight years later, in April 1894, he moved to Dresden and succeeded Hanns Bruno Geinitz (1814–1900) as chair of mineralogy and geology, where in April 1898 he additionally assumed the direction of the "State Museum of Mineralogy and Geology", including the department of prehistory in Dresden, which was still occupied by Geinitz until then. For the next 25 years, Kalkowsky enriched and advanced the research at his faculty and the exhibition of the museum in an extraordinary way. In 1907, he was promoted to privy counsellor ("Geheimer Hofrat"). Kalkowsky retired and received emeritus status October 1st, 1919 (Anonymous 1938 and the personnel file, Technical University of Dresden).

Kalkowsky died on February 13th, 1938, in Dresden. His obituary was published in the "Zeitschrift der Deutschen geologischen Gesellschaft", of which he was a member for more than 64 years (Anonymous 1938). After cremation, his urn was entombed at the cemetery Paulinzella-Rudolstadt (Thuringia, Germany).

Even back in his school days, Kalkowsky was very passionate of mineralogy and geology. In those early times, the desire to become a professor of mineralogy emerged. Knowing this, his subsequent enthusiasm and meticulousness in this research topic is not surprising.

In the first 20 years of his scientific career after his doctor's degree, his research was mostly dedicated to optical mineralogy following the successful research of his doctoral advisor F. Zirkel in this new field of geology. In that period, Kalkowsky dealt with this subject in more than 25 papers, especially regarding the regional rock formations of Saxony (Rimann 1940; Mathé 1993).

His first and only comprehensive textbook "Elemente der Lithologie" (1886) was also published in this phase of his life.

After attaining the full professorship (Ordinary Professor) of mineralogy and geology at Dresden University, the demands of teaching and administration caused a certain decrease in his publication practice (Rimann 1940). Nevertheless, he developed a strong interest in nephrite, an amphibole mineral of the solid solution series of tremolite and actinolite. This is expressed in four substantial papers on this matter.

Beside the scientific research, Kalkowsky was a distinguished teacher and a brilliant museum scientist. Having already gained fundamental experience in Jena, he successfully restructured the exhibition of the State Museum for Mineralogy, Geology and Prehistory under modern aspects in his Dresden times (Rimann 1940; Mathé 1993).

Beyond this, Kalkowsky's dedication for the "Naturwissenschaftliche Gesellschaft Isis in Dresden" should be acknowledged. Shortly after his appointment in Dresden, he assumed the position of the vice-chairman of the mineralogical division of Isis; in December 1896, he became chairman. Kalkowsky held this position until 1919. Additionally, he presided over the entire society in the years 1899–1900 and 1907–1908. Anticipatory to his 70th birthday, he was appointed an honorary member of the society in November 1920 (Rimann 1940). Kalkowsky was

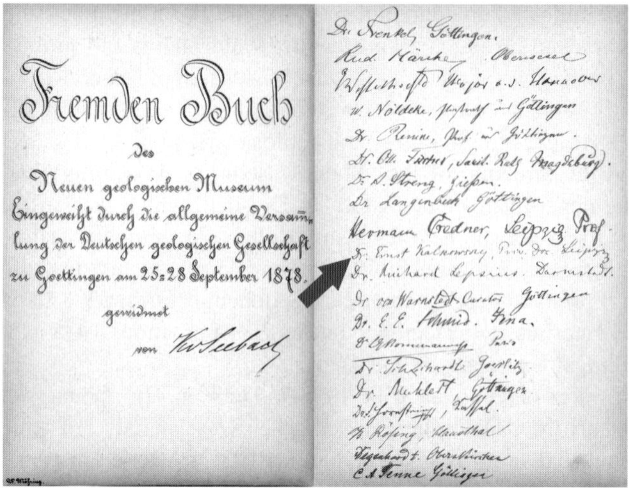

Fig. 3 Signature of Ernst Kalkowsky during the Annual meeting of the German Geological Society in Göttingen, September 1878 [guest book of the Geological Museum of Göttingen] [GZG.A.00010]

also a founding member of the "Paläontologische Gesellschaft" in 1912 (Jaekel 1914) as well as an active member of the German Geological Society ("Deutsche Geologische Gesellschaft") (Fig. 3).

2 Kalkowsky's Research on Stromatolites

Considering Palaeontology an ancillary science – belonging more to biology than to geology – Kalkowsky was strongly reserved in any research on this topic (Rimann 1940). However, with his paper "Oolith und Stromatolith im norddeutschen Buntsandstein" (1908; Figs. 4 and 5), in which he investigates oolitic and stromatolitic structures of the Lower Triassic in the Subhercynian Basin (Lower Saxony & Saxony-Anhalt, Germany), he provides an innovative contribution to all later research concerning stromatolitic structures.

In this paper, Kalkowsky proposed and defined the term "Stromatolith"; furthermore, he followed and strengthened the suggestion that these oolitic and stromatolitic structures are organosedimentary deposits. This was first mentioned by Ludwig and Theobald (1852) regarding the thermal waters of Bad Nauheim (Hesse, Germany) (e. g. Krumbein 2008).

Initially, his conclusions were doubted by most geologists in Germany (e.g. Reis 1908; Voss 1928); it was not until several years after his death that they were finally accepted in general. Today, 100 years after Kalkowsky's pioneering work, his view of stromatolites is still not far from the modern interpretation of this term.

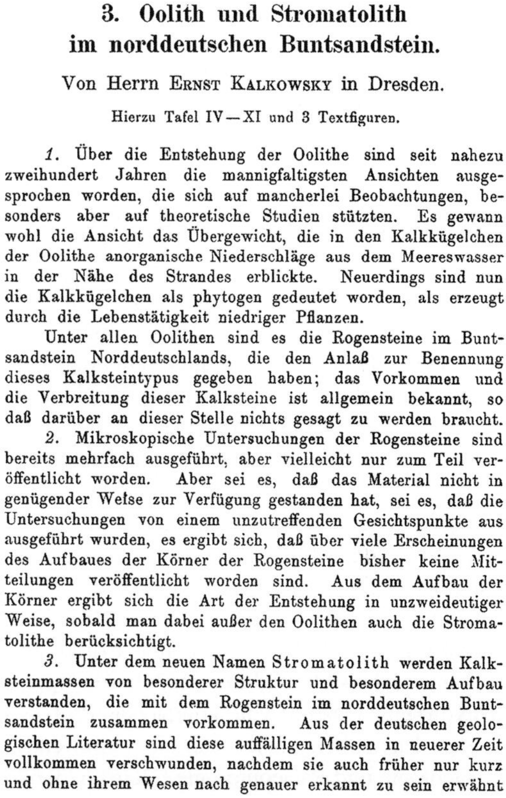

3. Oolith und Stromatolith im norddeutschen Buntsandstein.

Von Herrn Ernst Kalkowsky in Dresden.

Hierzu Tafel IV—XI und 3 Textfiguren.

1. Über die Entstehung der Oolithe sind seit nahezu zweihundert Jahren die mannigfaltigsten Ansichten ausgesprochen worden, die sich auf mancherlei Beobachtungen, besonders aber auf theoretische Studien stützten. Es gewann wohl die Ansicht das Übergewicht, die in den Kalkkügelchen der Oolithe anorganische Niederschläge aus dem Meereswasser in der Nähe des Strandes erblickte. Neuerdings sind nun die Kalkkügelchen als phytogen gedeutet worden, als erzeugt durch die Lebenstätigkeit niedriger Pflanzen.

Unter allen Oolithen sind es die Rogensteine im Buntsandstein Norddeutschlands, die den Anlaß zur Benennung dieses Kalksteintypus gegeben haben; das Vorkommen und die Verbreitung dieser Kalksteine ist allgemein bekannt, so daß darüber an dieser Stelle nichts gesagt zu werden braucht.

2. Mikroskopische Untersuchungen der Rogensteine sind bereits mehrfach ausgeführt, aber vielleicht nur zum Teil veröffentlicht worden. Aber sei es, daß das Material nicht in genügender Weise zur Verfügung gestanden hat, sei es, daß die Untersuchungen von einem unzutreffenden Gesichtspunkte aus ausgeführt wurden, es ergibt sich, daß über viele Erscheinungen des Aufbaues der Körner der Rogensteine bisher keine Mitteilungen veröffentlicht worden sind. Aus dem Aufbau der Körner ergibt sich die Art der Entstehung in unzweideutiger Weise, sobald man dabei außer den Oolithen auch die Stromatolithe berücksichtigt.

3. Unter dem neuen Namen Stromatolith werden Kalksteinmassen von besonderer Struktur und besonderem Aufbau verstanden, die mit dem Rogenstein im norddeutschen Buntsandstein zusammen vorkommen. Aus der deutschen geologischen Literatur sind diese auffälligen Massen in neuerer Zeit vollkommen verschwunden, nachdem sie auch früher nur kurz und ohne ihrem Wesen nach genauer erkannt zu sein erwähnt

Fig. 4 Title page of E. Kalkowsky's paper "Oolith und Stromatolith im norddeutschen Buntsandstein" in 1908

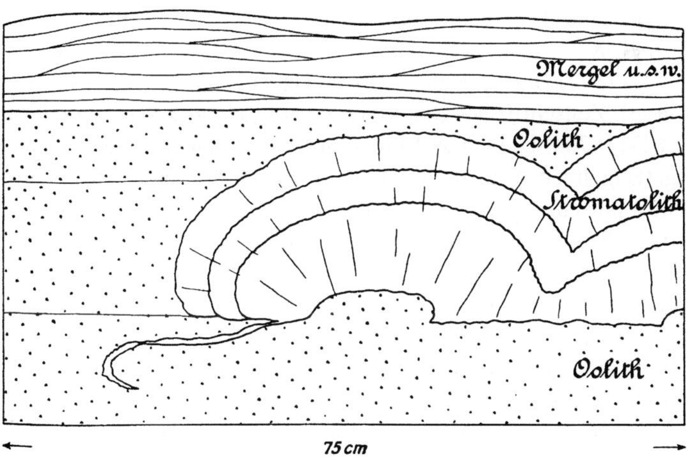

Fig. 5 Geological sketch of the stromatolitic and oolitic structures from E. Kalkowsky's paper "Oolith und Stromatolith im norddeutschen Buntsandstein" in 1908

Therefore, the knowledge of analogous structures he observed during various field trips, e.g. in Russia (Devonian), Germany (Permian and Upper Jurassic), England (Permian and Upper Triassic), and France (Neogene), was very helpful for his studies (Kalkowsky 1908; Rimann 1940).

3 Publications of E. Kalkowsky

Kalkowsky E (1874a) Mikroskopische Untersuchungen von Felsiten und Pechsteinen Sachsens. Mineralogische Mittheilungen [1874] (I):31–58. [Doctoral thesis]

Kalkowsky E (1874b) Die augithaltenden Felsitporphyre bei Leipzig. Zeitschrift der Deutschen geologischen Gesellschaft 26:586–599

Kalkowsky E (1875a) Mikroskopische Untersuchung des Glimmertrapps von Metzdorf. Neues Jahrbuch für Mineralogie, Geologie und Palaeontologie [1875]:488–505

Kalkowsky E (1875b) Ueber den Salit als Gesteinsgemengtheil. Mineralogische Mittheilungen [1875] (II):45–50

Kalkowsky E (1875c) Rother Gneiss und Kalkstein im Wilischthal im Erzgebirge. Zeitschrift der Deutschen geologischen Gesellschaft 27:623–630, 1 fig

Kalkowsky E (1876a) Über einige Eruptivgesteine des sächsischen Erzgebirges. Neues Jahrbuch für Mineralogie, Geologie und Palaeontologie [1876]:136–161

Kalkowsky E (1876b) Mittheilungen an Prof. G. Leonhard [Erwiderung an G. v. Rath das Tannenbergsthaler Gestein betreffend]. Neues Jahrbuch für Mineralogie, Geologie und Palaeontologie [1876]:623–626

Kalkowsky E (1876c) Ueber grüne Schiefer Niederschlesiens. Mineralogische Mittheilungen [1876] (II):87–116, (1 table), pl. VIII

Kalkowsky E (1876d) Das Glimmerschiefergebiet von Zschopau im sächsischen Erzgebirge. Zeitschrift der Deutschen geologischen Gesellschaft 28:682–749, (1 table), pl. X

Kalkowsky E (1877) Herr E. Kalkowsky an Herrn K. A. Lossen [Rother Gneiss im Erzgebirge]. Zeitschrift der Deutschen geologischen Gesellschaft 29:837–840

Kalkowsky E (1878a) Die Gneissformation des Eulengebirges. VI + 75 pp, 3 pls.; Leipzig (Engelmann). [„Habilitationsschrift durch welche unter Zustimmung der Philosophischen Facultät der Universität Leipzig zu seiner Sonnabend den 2. Februar 1878 Nachmittags 3 Uhr im Auditorium N$^{ro.}$ 6 im Bornerianum zu haltenden Probevorlesung über Mineralquellen einladet...]

Kalkowsky E (1878b) Der Granitporphyr von Beucha bei Leipzig. Neues Jahrbuch für Mineralogie, Geologie und Palaeontologie [1878]:276–286

Kalkowsky E (1878c) Der Leucitophyr vom Averner See. Neues Jahrbuch für Mineralogie, Geologie und Palaeontologie [1878]:727–729

Kalkowsky E (1878d) Ueber den Piperno. Zeitschrift der Deutschen geologischen Gesellschaft 30: 663–677, (1 table)

Kalkowsky E (1879a) Über die Thonschiefernädelchen. Neues Jahrbuch für Mineralogie, Geologie und Palaeontologie [1879]:382–387

Kalkowsky E (1879b) Ueber Krystallsystem und Zwillingsbildung des Tenorites. Zeitschrift für Krystallographie und Mineralogie 3 (III):279–287, pl. VI (9–10)

Kalkowsky E (1880a) Ueber die Erforschung der archäischen Formationen. Neues Jahrbuch für Mineralogie, Geologie und Palaeontologie [1880] (1):1–28

Kalkowsky E (1880b) Ueber Gneiss und Granit des bojischen Gneissstockwerkes im Oberpfälzer Waldgebirge. Neues Jahrbuch für Mineralogie, Geologie und Palaeontologie [1880] (1):29–42, pl. I

Kalkowsky E (1881a) Ueber Hercynit im sächsischen Granulit. Zeitschrift der Deutschen geologischen Gesellschaft 33:533–539

Kalkowsky E (1881b) Ueber den Ursprung der granitischen Gänge im Granulit in Sachsen. Ein Beitrag zur Kenntniss des Granites. Zeitschrift der Deutschen geologischen Gesellschaft 33:629–653, (1 table)

Kalkowsky E (1882) Einige Beobachtungen im sächsischen Granulitgebirge. Neues Jahrbuch für Mineralogie, Geologie und Palaeontologie [1882] (1):231–233

Kalkowsky E (1884a) Ueber die Polarisationsverhältnisse von senkrecht gegen eine optische Axe geschnittenen zweiaxigen Krystallplatten. Zeitschrift für Krystallographie und Mineralogie 9 (V–VI):486–497, pl. XIII

Kalkowsky E (1884b) Die Anwendung des Mikroskops in der Mineralogie und Geologie. Kapitel XI. In: Vogel J (ed.) Das Mikroskop und die wissenschaftlichen Methoden der mikroskopischen Untersuchung in ihrer verschiedenen Anwendung. Neu bearbeitet von Otto Zacharias. 4. Ed.; Leipzig (Denicke), pp 213–241

Kalkowsky E (1885) Ueber Olivinzwillinge in Gesteinen. Zeitschrift für Krystallographie und Mineralogie 10 (I):17–24, pl. II

Kalkowsky E (1886a) Elemente der Lithologie. Für Studierende bearbeitet. VIII + 316 pp, (31 tables); Heidelberg (Winter)

Kalkowsky E (1886b) Ueber Struvit von Homburg. Zeitschrift für Krystallographie und Mineralogie 11 (I):1–4, (1 table). pl. I

Kalkowsky E (1893) Ueber Geröll-Thonschiefer glacialen Ursprungs im Kulm des Frankenwaldes. Zeitschrift der Deutschen geologischen Gesellschaft 45:69–86

Kalkowsky E (1894) Mineralogie und Geologie an der Technischen Hochschule. Antrittsrede von Dr. Ernst Kalkowsky, Professor an der Technischen Hochschule zu Dresden. Der Civilingenieur (N. F.) 40 (4):301–312

Kalkowsky, E (1897) Das neue min.-geologische Institut der Kgl. Techn. Hochschule zu Dresden. Sitzungsberichte und Abhandlungen der Naturwissenschaftlichen Gesellschaft Isis in Dresden [1896] (2; July–December): 27–29

Kalkowsky E (1897) Ueber einen oligocänen Sandsteingang an der Lausitzer Ueberschiebung bei Weinböhla in Sachsen. Sitzungsberichte und Abhandlungen der Naturwissenschaftlichen Gesellschaft Isis in Dresden [1897] (1; January–June):80–89, pl. III

Kalkowsky E (1900) Hanns Bruno Geinitz. Die Arbeit seines Lebens. Rede in der öffentlichen Sitzung der Isis am 22. Februar 1900. Sitzungsberichte und Abhandlungen der Naturwissenschaftlichen Gesellschaft Isis in Dresden [1900] (1; January–June):V–XIII

Kalkowsky E (1902) Die Verkieselung der Gesteine in der nördlichen Kalahari. Sitzungsberichte und Abhandlungen der Naturwissenschaftlichen Gesellschaft Isis in Dresden [1901] (2; July–December):55–107, pls. II–IV

Kalkowsky E (1905) Die Markasit-Patina der Pfahlbau-Nephrite. Sitzungsberichte und Abhandlungen der Naturwissenschaftlichen Gesellschaft "Isis" in Dresden [1904] (2; July–December):51–60, 1 fig

Kalkowsky E (1906a) Der Nephrit des Bodensees. Sitzungsberichte und Abhandlungen der Naturwissenschaftlichen Gesellschaft "Isis" in Dresden [1906] (1; January–June):28–44, 1 fig

Kalkowsky E (1906b) Geologie des Nephrites im südlichen Ligurien. Zeitschrift der Deutschen geologischen Gesellschaft 58:307–378, (7 tables), pl. XVIII

Kalkowsky E (1907a) Die natürlichen Verhältnisse Dresdens. 1. Geologischer Aufbau, Wasserhorizonte, Bodenbeschaffenheit. In: Schäfer, F. (ed.): Wissenschaftlicher Führer durch Dresden. [79. Versammlung Deutscher Natur-Forscher und Ärzte]. pp 1–3; Dresden (v. Zahn & Jaensch)

Kalkowsky E (1907b) Die naturwissenschaftlichen Anstalten Dresdens. 3. Das Mineralogisch-Geologische Institut der Technischen Hochschule. In: Schäfer, F. (ed.): Wissenschaftlicher Führer durch Dresden. [79. Versammlung Deutscher Natur-Forscher und Ärzte]. pp 53–55; Dresden (v. Zahn & Jaensch)

Kalkowsky E (1907c) Technische Behörden und Anstalten und naturwissenschaftliche Museen Dresdens. 6. Das Königliche Mineralogisch-Geologische Museum, nebst der Prähistorischen Sammlung im Königlichen Zwinger. In: Schäfer, F. (ed.): Wissenschaftlicher Führer durch Dresden. [79. Versammlung Deutscher Natur-Forscher und Ärzte]. pp 112–114; Dresden (v. Zahn & Jaensch)

Kalkowsky E (1907d) Geologische Deutung des Nephrites von Gulbashen. In: Bauer, M.; Koken, E. & Liebisch, T. (eds.): Festband zur Feier des 100jährigen Bestehens. Neues Jahrbuch für Mineralogie, Geologie und Palaeontologie [1907]:159–168

Kalkowsky E (1908a) Der Korundgranulit von Waldheim in Sachsen. Sitzungsberichte und Abhandlungen der Naturwissenschaftlichen Gesellschaft "Isis" in Dresden [1907] (2; July–December):47–65

Kalkowsky E (1908b) Oolith und Stromatolith im norddeutschen Buntsandstein. Zeitschrift der Deutschen geologischen Gesellschaft 60:68–125, 3 figs., pls. IV–XI

Kalkowsky E (1908c) 53. Hauptversammlung der Deutschen geologischen Gesellschaft. Begrüßungsworte. Zeitschrift der Deutschen geologischen Gesellschaft, Monatsberichte 60 (8/10):191–195

Beck R, Credner H, Gäbert C, Hibsch JE, Kalkowsky E (1908) Exkursions-Pläne für die 53. allgemeine Versammlung der Deutschen geologischen Gesellschaft in Dresden. – 20 S., 12 figs., pls. I–V; Berlin (Schade)

Kalkowsky E (1909a) Europäische Entfernungen. Sitzungsberichte und Abhandlungen der Naturwissenschaftlichen Gesellschaft "Isis" in Dresden [1908] (2; July–December):33–40, (6 tables)

Kalkowsky E (1909b) Bericht über die Exkursionen an den Versammlungstagen. Zeitschrift der Deutschen geologischen Gesellschaft, Monatsberichte 61 (2):90–93

Kalkowsky E (1909c) Geologische Grundlagen der Entwicklungslehre. Sitzungsberichte und Abhandlungen der Naturwissenschaftlichen Gesellschaft "Isis" in Dresden [1909] (1; January–June):3–10

Kalkowsky E (1910) Geologie und Phantasie. Vortrag bei der Feier des 75jährigen Bestehens der Naturwissenschaftlichen Gesellschaft "Isis" am 26. Mai 1910. Sitzungsberichte und Abhandlungen der Naturwissenschaftlichen Gesellschaft "Isis" in Dresden [1910] (1; January–June):10–19

Kalkowsky E (1915a) Aluminokrate Schlieren im Frankensteiner Gabbro im Odenwald. Sitzungsberichte und Abhandlungen der Naturwissenschaftlichen Gesellschaft "Isis" in Dresden [1914] (2; July–December):33–42

Kalkowsky E (1915b) Opaleszierender Quarz. Zeitschrift für Krystallographie und Mineralogie 55 (I):23–50, pl. III

Kalkowsky E (1921) Mikroskopischer Coelestin im Röt von Jena als geologische Erscheinung. Zeitschrift der Deutschen geologischen Gesellschaft 73:1–23

Acknowledgements We like to thank V. J. Roden (Geoscience Centre, University of Göttingen) for her helpful comments on the manuscript.

References

Anonymous (1938) [obituary to E. L. Kalkowsky]. Zeitschrift der Deutschen geologischen Gesellschaft 90:173

Erler D, Krause R, Lange J-M (2008) Ernst Kalkowsky 1851-1938. Miniaturen zur Geologie Sachsens. GeoSzene 4:1–24

Jaekel O (1914) Bericht über die Gründung und Jahresversammlung der palaeontologischen Gesellschaft. Palaeontologische Zeitschrift 1:58–73

Kalkowsky E (1908) Oolith und Stromatolith im norddeutschen Buntsandstein. Zeitschrift der Deutschen geologischen Gesellschaft 60:68–125, 3 figs., pls. IV-XI.

Krumbein WE (2008) Biogenerated rock structures. Space Science Reviews 135:81–94

Kühne E, Lange J-M, Erler D (2006) Die Geschichte des Museums für Mineralogie und Geologie Dresden. Geologica Saxonica 50/51:13-95

Ludwig R, Theobald G (1852) Ueber die Mitwirkung der Pflanzen bei der Ablagerung des kohlensauren Kalkes. Annalen für Physik und Chemie 87:91-107

Mathé G (1993) Ernst Kalkowsky (1851-1938). Geologe, Hochschullehrer und Museumsdirektor. Abhandlungen des Staatlichen Museums für Mineralogie und Geologie zu Dresden 39:7–20

Reis OM (1908) [Referat zu:] Kalkowsky: Ueber Oolith und Stromatolith im norddeutschen Buntsandstein. Neues Jahrbuch für Mineralogie, Geologie und Paläontologie [1908] (2): 114–138

Rimann E (1940) Ernst Kalkowsky, sein Leben und sein Werk. Sitzungsberichte und Abhandlungen der Naturwissenschaftlichen Gesellschaft Isis in Dresden [1938/39]:69–95

Voss R (1928) Die paläogeographische Verbreitung des Rogensteins im deutschen Unteren Buntsandstein. Abhandlungen der Preußischen Geologischen Landesanstalt (N. F.) 107:1–66

Kalkowsky's Stromatolites and Oolites (Lower Buntsandstein, Northern Germany)

Josef Paul, Tadeusz M. Peryt, and Robert V. Burne

1 Introduction

Three world-wide used geological terms derived from the area around the German Harz Mountains: "Oolithi" (Brückmann 1721) and nearly 200 years later "Ooid" and "Stromatolith" (Kalkowsky 1908). All these three terms deal with carbonate grains and rocks of the Lower Triassic Buntsandstein Group around the Harz Mts. Therefore, this area is in a way the type area of stromatolites. But Kalkowsky did not only coin the name, but also deciphered their microbial origin. Therefore, it may be of interest to have a nearer look at these stromatolites, oolites and the environment in which they grew and to compare Kalkowsky's 100 years old observations and interpretations with recent ones.

2 Studied Area and Stratigraphic Frame

The lithostratigraphic unit hosting Kalkowsky's stromatolites is the Lower Buntsandstein Subgroup which is of Lower Triassic age and time-equivalent to the Induan and lower Olenekian of the International Chronostratigraphic Standard (Table 1). In the studied area, the Lower Buntsandstein Group is about 250 m

J. Paul (✉)
Geoscience Centre, University of Göttingen, Sedimentology Division, Goldschmidtstraße 3, 37077 Göttingen, Germany
e-mail: jpaul@gwdg.de

T.M. Peryt
Polish Geological Institute – National Research Institute, ul. Rakowiecka 4, 00-975 Warszawa, Poland

R.V. Burne
Research School of Earth Sciences, The Australian National University, Canberra, ACT, Australia

J. Reitner et al., *Advances in Stromatolite Geobiology*,
Lecture Notes in Earth Sciences 131, DOI 10.1007/978-3-642-10415-2_2,

Table 1 Stratigraphic scheme of the Lower Triassic in Germany

Ma	Chronostratigraphy		Lithostratigraphy: Subgroup	Lithostratigraphy: Formation
	Anisian	Aegean	Upper Buntsandstein	
245				
246	Olenekian	Spathian	Middle Buntsandstein	
247	Olenekian	Spathian	Middle Buntsandstein	
248	Olenekian	Spathian	Middle Buntsandstein	
249	Olenekian	Smithian	Lower Buntsandstein	Bernburg
250	Induan	Dienerian	Lower Buntsandstein	Calvörde
	Induan	Griesbachian	Lower Buntsandstein	Calvörde
251				
	Permian		Zechstein	Fulda

thick and consists of sandstones, mudstones and oolites. It is divided into two formations, the lower Calvörde Fm. and the upper Bernburg Fm. These sediments are organized in about 20 fining upward cycles starting with cross bedded sandstones, followed by oolites and gray mudstones. The cycle is terminated by structureless red mudstones containing small anhydrite nodules. The about 100 m thick Bernburg Fm. contains stromatolites and has more or less the same composition as the underlying Calvörde Fm. Only the percentage of oolites is higher and the Calvörde Fm. contains no stromatolites.

The Buntsandstein Group was deposited in the Central European Basin which was called the Germanic Basin in the past. It was an inland basin with only restricted access to the ocean (Fig. 1). The fining upward cycles are traceable from the Netherlands to Poland and can be used as markers (Geluk and Roehling 1997; Becker 2005). The common occurrence of oolite beds indicates the centre of the basin in contrast to the siliciclastic margins. An individual oolite bed may

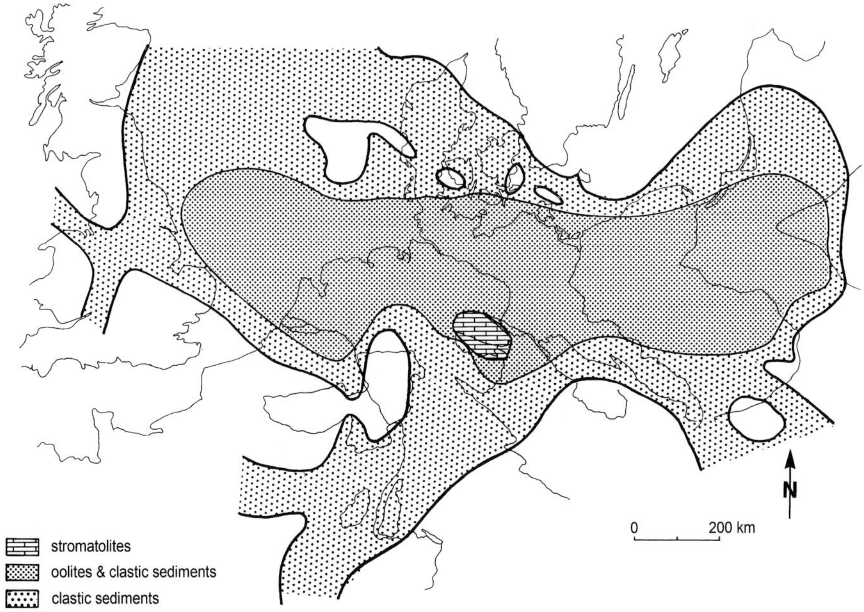

Fig. 1 The Lower Buntsandstein of the Central European Basin. After Paul (1982)

wedge out in some distance, but a series of oolite beds can be traced over some 100 km.

The stromatolites of the Bernburg Fm. are restricted to a relatively small area around the Harz Mts., the Eichsfeld Palaeohigh, which was situated between the Hessian Depression in the west and the Thuringian Subbasin in the east (Fig. 1). It persisted from the Permian until the end of the Triassic. Reduced thicknesses and condensed sequences are characteristic for this palaeohigh. Here, stromatolites are found quite frequently in quarries, road-sections and drilled cores (Dorn 1953; Paul and Klarr 1988; Paul and Peryt 2000). Investigations of Lower Buntsandstein oolites in various wells in other areas of Lower Saxony ended so far without success in finding stromatolites or even stromatolitic crusts.

3 Stromatolites

Ernst Kalkowsky (1908) described stocks or heads of layered or laminated carbonates in oolite beds of the Lower Buntsandstein in the Harz Mountains and defined them in this way (Kalkowsky 1908, p. 68, § 3):

> The new term "Stromatolite" is proposed for limestones with unique organisation and structures that occur associated with "roe-stone" (oolites).
>
> Stromatolites have a fine, more or less even layered fabric that contrasts with the concentric fabric of oolite grains.

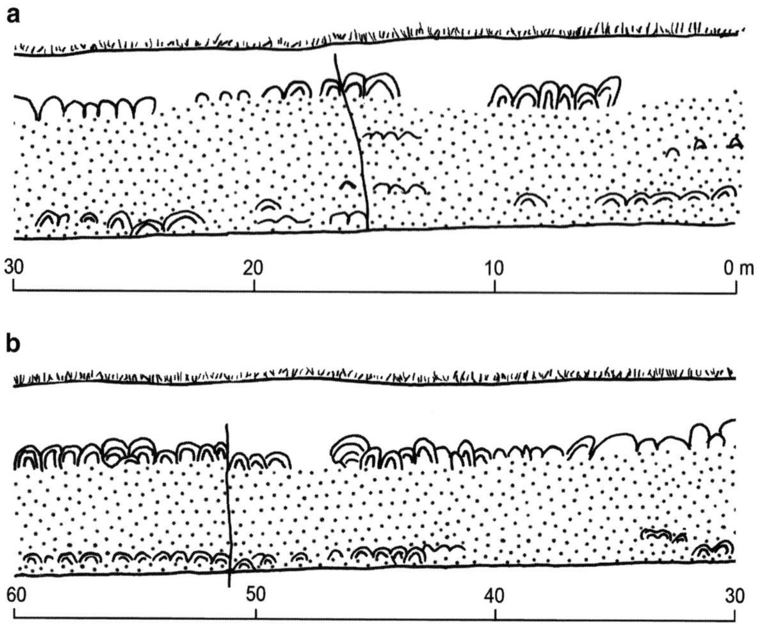

Fig. 2 (**a, b**) Oolites and stromatolites at a disused quarry at the Heeseberg

This even layered fabric, the lamination, is according to Kalkowsky (1908) the distinguishing main mark of stromatolites.

Stromatolites are always associated with to oolites. In most cases, they occur at the surface of oolite beds terminating the sedimentation of ooids. The thicker the oolite bed, the higher the probability of occurrence of stromatolites and the larger the stromatolites are. In Heeseberg in an abandoned quarry, a section of 7 m thick oolite beds with stromatolites is exposed for a length of more than 100 m (Figs. 2 and 3). At this 700 m^2 large outcrop, there are nearly no mud layers which normally occupy more than half of the Buntsandstein succession. Stromatolites are not randomly distributed, but occur preferentially in two horizons (Fig. 2). These succession of various types of ooids and stromatolites can be traced a distance of about 1 km.

3.1 Macroscopic Observations

Stromatolites occur as thin, some millimetres thick crusts or as up to 2 m high compound domes or smaller columnar forms (Figs. 4–7). Alternations of thin stromatolitic crusts and oolite layers are quite common (Fig. 8). The largest domes are at the centre of the occurrence of stromatolites. Towards the margins, their size and number decrease. Thin laminated encrustations veneer oolite beds. In some cases, encrustations outline ripples consisting of ooids and make them

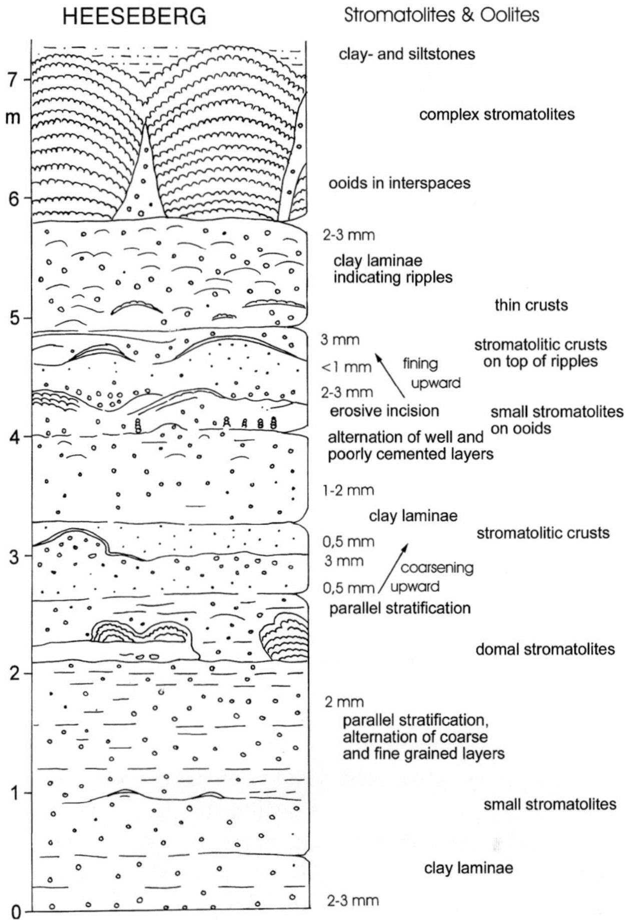

Fig. 3 Section of the oolite bed and domal stromatolites at the Heeseberg. After Paul and Peryt (2000)

visible. Crusts are thick on top of the ripple-crests and wedge out towards the troughs (Fig. 3). In some places, thin isolated columns of some millimetres diameter rise above individual ooids (Fig. 9). Columnar forms may be closed or wide spaced (Figs. 6 and 7). The most frequent stromatolitic forms are domes of various sizes. The larger domes are compound structures consisting of even or inclined layers and up to some cm high columnar forms branching like trees (Fig. 4). Ooids are in the interstices between the branches, but also sometimes incorporated into the stromatolitic tissue. Most likely, the height of the domes was only some centimeters above the sediment surface. Occasionally, the top of stromatolites domes as well as columnar forms is characterized by dissolved and etched surfaces. The size of columns varies between some centimetres and some millimetres (Figs. 4, 6 and 7). Thin stromatolitic crusts occur above oolites (Fig. 8). Very rare is a type of

Fig. 4 Compound domal stromatolite. Scale bar: 10 cm. Salzgitter

Fig. 5 Top view at small columnar stromatolitic crusts. Scale bar: 5 cm. Eisleben

Fig. 6 Close spaced columnar stromatolite. Scale bar: 2 cm. Salzgitter

Fig. 7 Alternation of wide spaced columns and stromatolitic crusts. At the centre shells, most likely of conchostracans. Scale bar: 5 cm. Eisleben

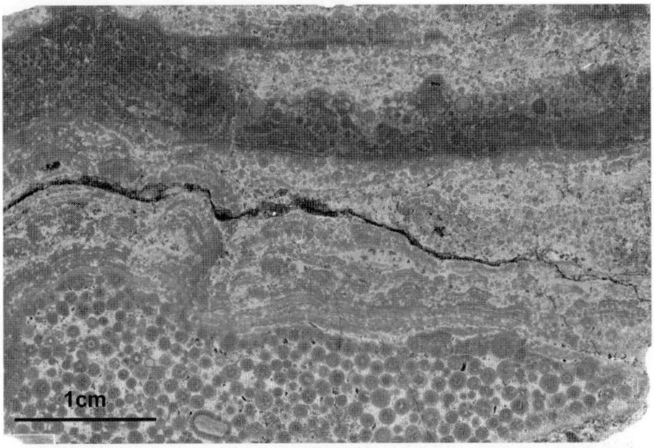

Fig. 8 Thin stromatolitic crust above an uneven surface of an oolite layer. Scale bar: 1 cm. Eisleben

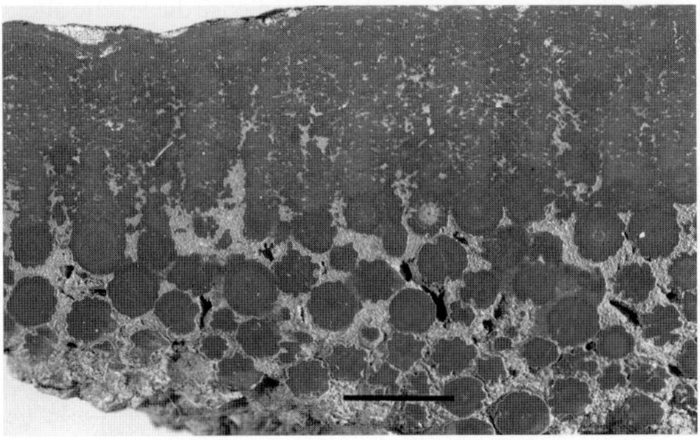

Fig. 9 Small columnar stromatolites growing above single ooids. Scale: 1 cm. Salzgitter

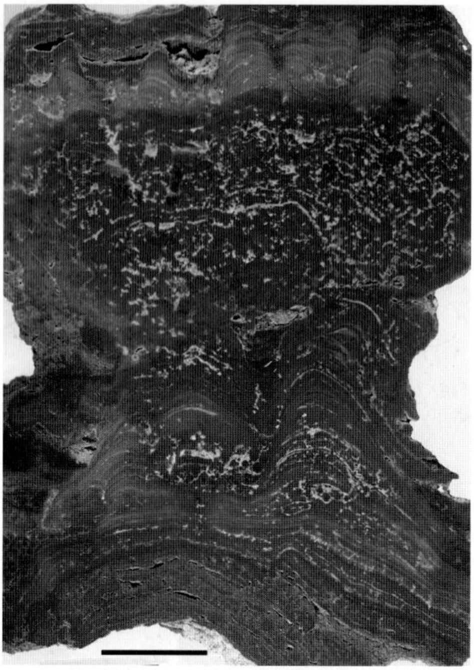

Fig. 10 Closely laminated stromatolitic crusts at the bottom and the top, thrombolite-like structures in the centre. Scale bar: 2 cm. Heeseberg

lamination resembling thrombolitic structure (Fig. 10). Smaller stromatolitic structures and ooids are red in colour due to an admixture (some percent) of haematite which is diagenetically incorporated into the calcite (Scherer 1975).

3.2 Microscopic Observations

The carbonates of the Bernburg Fm. underwent intense diagenetic alterations which destroyed the original tissues. In particular organic compounds have been extinguished by oxidizing conditions which dominated early and late diagenetic processes. Therefore, it is not possible to prove directly the microbial origin of Kalkowsky's stromatolites. But the autophototrophic growth forms on top of ripples, for instance, support this assumption (Fig. 7). Two types of stromatolitic microstructure can be distinguished, a spongy fenestrate and a fan-like which occasionally is a radial-filament (Paul and Peryt 2000) (Fig. 11).

Kalkowsky (1908) distinguished between stromatoids and stromatolites. There have been considerable discussions on what Kalkowsky meant by the term "stromatoid" (Hofmann 1973; Monty 1977; Krumbein 1983). In Kalkowsky's words (1908, p. 65):

> Oolites are composed of ooids and the cement that lithifies them. Stromatolites are composed of thin, more or less flat laminae of calcite with a specific texture. These thin

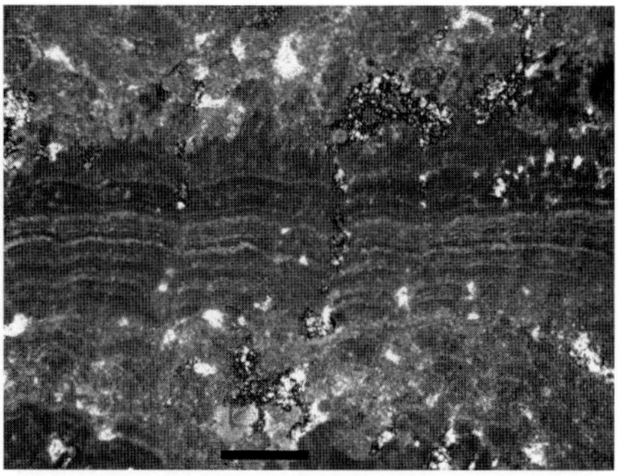

Fig. 11 Stromatolite layer. Thin laminated crusts within the oolites and at the base of stromatolites are formed by radial calcite crystals, whereas columnar stromatolites have spongy-fenestrate fabric Scale bar: 1 mm

> laminae are termed "Stromatoids". Stromatolites, unlike oolites, are not formed by limited individual colonies of constructing organisms; rather layers or mats of constructing organisms form them. There is a trend from ooid through poly-ooid and ooid bag to stromatolite in which the accreting layers show decreasing influence of a central nucleus in determining the form of growth.

From this it is clear that Kalkowsky considered ooids being produced by the growth of limited individual microbial colonies around a nucleus, and that the lithified accumulation of ooids together with the binding cement constitutes an "Oolite". By comparison he considered the individual colonies of microbes grown as non-nucleated layers producing a thin flat lamina of calcite which he termed "Stromatoid", and that the accumulation of cemented laminae, or "Stromatoids", together with their lithifying cements constitutes a "Stromatolite".

He therefore regarded "Ooids" and "Stromatoids" as end members of a continuum, with "poly-ooids" and "Ooid bags" as intermediate stages. However there is a hierarchical shortcoming to this concept in that ooids are individual sediment grains and an oolite is a lithology. A Stromatoid is often composed of grains (some of which may be ooids), and a Stromatolite is a structure within a lithology, though individual Stromatolites may coalesce together to form a lithology. Stromatoid in this sense is thin laminated organic material.

4 Oolites

The 100 m thick Bernburg Fm. at the Eichsfeld Palaeohigh contains up to 20% of oolites. They were investigated petrographically in detail by Usdowski (1962) and Richter (1983). Their sedimentology is worked out by Paul and Peryt (2000) and Palermo et al. (2008). In the centre of the Eichsfeld Palaeohigh, oolite beds are up to 7 m thick and ooids can reach a size of up to 10 mm, but usually they are between 0.2 and 3.5 mm in diameter (Figs. 12 and 13). Ooid size and thickness of the oolite beds decrease with increasing distance from the palaeohighs. Normally,

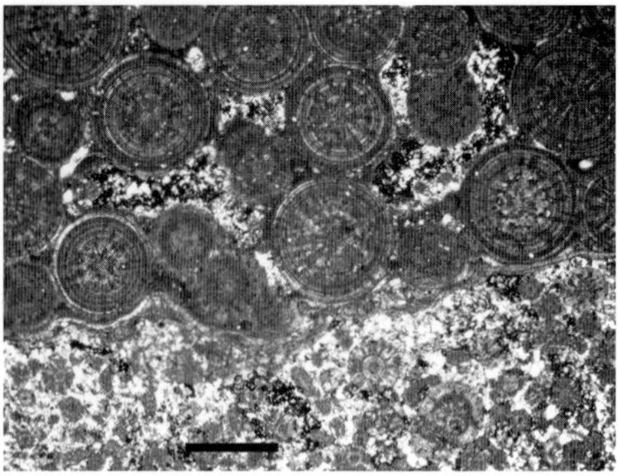

Fig 12 Well sorted ooids with a thin stromatolitic crust. Scale bar: 1 mm. Heeseberg

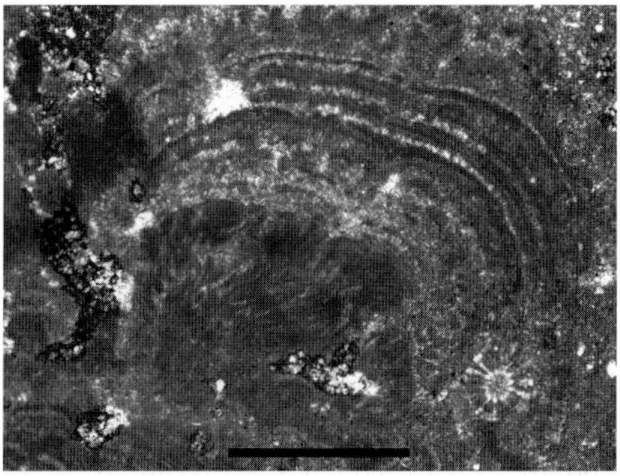

Fig. 13 Stromatolitic tissue with microbial tubes, most likely cyanobacterial sheaths. Scale bar: 1 mm. Heeseberg

the ooids consist of calcite and have a core of small quartz or mica grain or a calcite crystal. Richter (1983) assumed that the ooids originally consisted of magnesium calcite.

Under the microscope, there are two types of ooids recognizable. The majority of ooids consists of thin concentric layers of calcite and the other type has fine, radially oriented calcite fibres (Usdowski 1962). Both types are found side by side and can merge into each other. Often, the cores are more radially oriented, whereas the outer layers are still intact. This phenomen is independent of grain size or other parameters like position in the oolite bed. According to Scherer (1975), the ooids have an iron content of some percent. This is a clear indication of late diagenetic recrystallization during burial. Occasionally, ooids are meshed as consequence of pressure solution. Two or even more ooids are merged, which means that they have common outer layers. Clasts of ooids are wrapped up by thin stromatolitic layers (Fig. 14).

The 7 m thick oolite bed of the above mentioned outcrop at the Heeseberg consists of amalgamated oolite sheets. Careful examination indicated that most of the dm-thick oolite beds can be traced all over the outcrop. There are only very

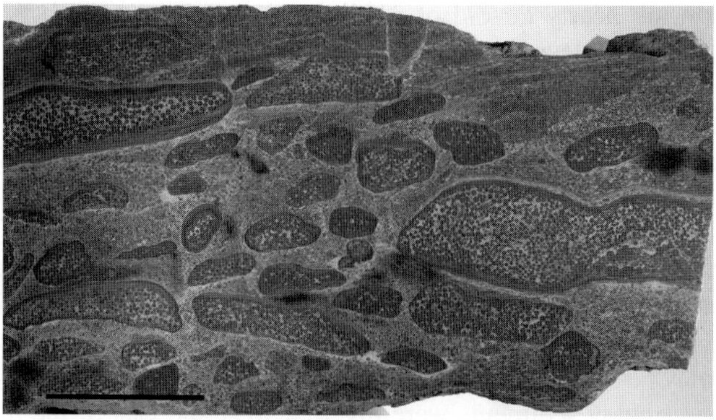

Fig. 14 Ooid intraclasts coated with thin stromatolitic crust. Scale bar: 5 cm. Salzgitter-Osterlinde

few erosional features like channels or cross-bedding. The oolites are well to very well sorted. Thicker oolite beds consist of several units which may have different grain sizes (Fig. 3). Individual layers of oolite beds correspond to storms or heavy current events. The units can be separated by traces of clayey laminae or thin stromatolitic crusts.

5 Interpretation and Discussion

The widespread Buntsandstein Group consists of red and variegated sandstones and mudstones which are present from Lorraine in France to the Holy Cross Mountains in eastern Poland (Fig. 1). It is a typical terrestrial redbed facies. These red beds are deposited in a large inland basin, the Central European Basin. Braided and meandering rivers brought siliceous grains and mud from the hinterland towards the centre of the basin (Weber 2000). The influx of material fluctuated during time as the climate was regularly changing between arid and less arid or even sub-humid due to orbital movements (Paul and Puff 2010). In certain positions of the orbit, the Triassic mega-monsoon could overlap to central Europe and bring some rain to the drainage area (Parrish 1993). The inflow of water led to large but shallow lakes in the centre of the basin. When the rainy seasons vanished, the water of the lake evaporated, leaving back dissolved ions like calcium, magnesium, sodium, sulfate, and chloride. Microbial calcite producing communities flourished as higher organisms were nearly absent due to rapidly changing environmental conditions. The calcite ooids formed in the shallow environment under the impact of wind and wave action bars or shoals. These several metres thick oolite deposits are present at the margin of the basin and palaeohighs from the Netherlands to Poland (Palermo et al. 2008). Finally, the lakes vanished and small ephemeral rivers reached the centre and brought only mud with them. Frequent desiccation cracks in the mudstones above and below the oolites indicate the shallow water regime. The sediments of the Lower

Buntsandstein are nearly devoid of fossils, body fossils as well as trace fossils. The reason of this scarceness may be not abnormal salinity, but rapidly changing environmental conditions as the shallow playa lake has no buffering capacity against fluctuations of various environmental parameters. The etched surfaces of stromatolites result from the decay of organic matter under a cover of clay or living biofilm led to formation of CO_2, lowering the pH and consequently leading to acidification of the water and dissolution of carbonate.

Some prerequisites of stromatolitic growth can be deduced from observations in the field. Muddy water or mud layers excluded stromatolites or terminated their growth. The microbial community did not survive a mud coverage or muddy water. This effect may be the reason for their restriction to the Eichsfeld Palaeohigh. Here, the sandy and muddy sediments were diverted west and east of the high on their way towards the basin. The more extended areas of oolites indicate that the ooid producing microbes are not so sensitive. The position of stromatolites at top of the oolite beds seems to reflect a directional evolution, most likely of the water chemistry, e.g. alkalinity or supersaturation in respect of calcium carbonate. Perhaps, the stromatolite producing microbial community needs a higher alkalinity than the ooid producing organisms. The lowermost millimeters of mudstones which have direct contact to stromatolites are grayish, not like normally reddish. The decay of the microbial mats after mud coverage led to reducing conditions. The iron(III) of clay flakes was reduced and removed.

The observed photoautotrophy points to cyanobacteria, at least forming a component of the microbial community (Fig. 13). There is a high potential of preservation through the absence of grazers and browsers and an early lithification, although the latter cannot be proved. In several passages of his article, Kalkowsky (1908, p. 112, 115, 119) mentioned and drew sketches of stromatolitic roots. After field observations and his drawings, we believe that thin stromatolitic crusts covering the uneven surface of oolites, served as attachments of larger domes.

Peryt (1975) described stromatolites from the Middle Buntsandstein in the Polish part of the Central European Basin. They are comparable with Kalkowsky's stromatolites, consisting of layered and columnar forms. They are also bound to oolites and oncolites, but grew most likely in a marine or brackish environment, as Peryt deduced from the frequent occurrence of gastropods, foraminifera and spirobid polychaetes.

Regarding the environment of Lower Buntsandstein in northern Germany, there is a long lasting discussion (see Becker 2005, with references herein). Up to now there is no proof that marine incursions or transgressions reached the western part of the Central European Basin during the Lower Buntsandstein.

Stromatolite producing organisms flourished when certain requirements were fulfilled. They needed certain physico-chemical environments, most likely stable substrate, clear not muddy waters, sunlight, high alkalinity, high calcium and carbonate contents of the water, and protection against grazers and burrowers.

The organic nature of stromatolites was generally accepted about 10 or 20 years after the appearance of Kalkowsky's publication. In contrast to that, there was a long-lasting discussion within the scientific community about the formation of ooids. During the nineteenth and twentieth century, most scientists thought of an inorganic origin, such as the precipitation of calcium and carbonate due to supersaturation (Usdowski 1962; Bathurst 1978; Simone 1980; Füchtbauer 1988). Calcite or aragonite may precipitate around a nucleus of a quartz or carbonate grain. Only Kalkowsky (1908) thought of an organic origin produced by colonies of lime secreting phyto-organisms. However, during the last 20 years an increasing number of indications were found that organisms are involved in the formation of ooids (Dahanayake et al. 1985; Davaud and Girardelos 2001; Plee et al. 2006). The in vitro production of ooid-like structures has been observed in cultures of spherical microbial communities (Brehm et al. 2004).

To summarize, Kalkowsky (1908) stated that

p. 100 § 64 Regarding the environment of the oolites in the north German Bunter Sandstone, it is generally assumed that they have formed in a shore facies. One could easily be tempted to think already now of salt lakes as area of their formation.

p. 118 § 88 Stromatolites were always associated with oolites.

p. 123 § 94 Ooids resemble growing bacterial colonies as observed in a Petri dish. Ooids are therefore probably produced by colonies of lime secreting phyto-organisms.

p.124 § 96 We have to assume that simple plants gave rise to limestone precipitation. My aim has been to show that the oolites and stromatolites of the north German Bunter Sandstone are inherently of organic origin.

Taking these statements of Kalkowsky into account, we may say that scientific progress during the last 100 years has not radically changed our understanding of the formation of the stromatolites and oolites of the Buntsandstein, other discoveries have been made that demonstrate that this association has recurred at several Geological Horizons, and is even found forming in modern seas.

Fifty years after Kalkowsky published this classic paper, Richard Chase recognized the first convincing modern analogues of "stromatoliths" around the shores of Hamelin Pool, Western Australia [R. Chase, personal communication to RVB]. Recent investigations of both localities reveal a number of interesting parallels between the environment of Hamelin Pool and that of the Basin with the association described by Kalkowsky. In both cases, stromatolites grow on stable or firm ground in turbulent environments characterized by low sedimentation rates, little fine grained sediment, virtually no terrigenous input, rapid cementation and abnormal or fluctuating salinity.

Kalkowsky's stromatolites occur on the surface of oolite beds. Laminated crusts (called stromatoid by Kalkowsky and interpreted as being formed by syndepositional cementation) also occur in these rocks. Both stromatolites and laminated crusts are concentrated in specific layers traceable throughout quarry faces, where the stromatolites are clearly syndepositional with rippled ooid sand (Fig. 15 in Paul and Peryt 2000). Spongy-fenestrate and fan-like stromatolitic microstructures can be distinguished, and both have undergone intense sparitization. The upper surfaces of some stromatolites are pitted due to syndepositional dissolution. The stromatolites may incorporate variable amounts of ooids, quartz grains and other material.

Hamelin Pool stromatolites also occur in association with ooid sands (Davies 1970; Logan et al. 1974). Subtidal stromatolites grow on rock substrate or crusts formed by penecontemporaneous cementation of marine sands, and are surrounded by mobile oolitic rippled sands and sand waves. The subtidal stromatolites have a laminoid fenestral fabric consisting of ooid and other carbonate sand grains cemented by micritic cements (Logan et al. 1974). Micritisation of sand grains begins soon after deposition and gradually destroys the original structure of the incorporated ooids and other grains (Monty 1976; Reid et al. 2003). Stromatolites in the intertidal zone are thought to be subtidal forms stranded by sea-level fall and modified by intertidal microbial communities (Burne 1991–1992). While the Buntsandstein stromatolites originated in a hyposaline and alkaline lake environment during the high stand of water level, and the Hamelin Pool stromatolites a forming in a hypersaline marine embayment during a period of regression, there are many environmental similarities. In both cases conditions favourable for ooid formation precedes the initiation of stromatolite growth, but the stromatolites co-exist with ooid sands, and incorporate ooid grains into their structures. The morphology of many of the subtidal Shark Bay stromatolites is clearly influenced by the erosive effects of ooid sand waves migrating around them. Once formed, early diagenesis progressively obliterates the structure of ooid grains incorporated into the stromatolites. The association of stromatolites and ooid sands is of considerable geological significance. In another present-day environment the stromatolites of Lee Stocking Island in the Bahamas show a similar association with migrating ooid sand waves to that found in Hamelin Pool (Dill 1991). The association of stromatolites and oolites dates back to the Archean. One of the oldest occurrences of the association is known from the 2.72 Ga Tumbiana Fm., Fortescue Gr. Pilbara Block in Western Australia.

Even the first stromatolites known to science are associated with oolitic limestones, 25 years before Kalkowsky's work was published. James Hall had formally named them *Cryptozoon proliferum*, and they occur in the oolitic Cambrian Hoyt Formation of Saratoga Springs, New York State (Hall 1883).

Acknowledgments We have to thank Michael Sosnitza for preparing and polishing the oolite and stromatolite samples. Max Hundertmark photographed the specimen. Cornelia Kaubisch prepared and digitalized the drawings. Dr. Ulrike Troitzsch, Canberra, helped to translate the sometimes difficult German of Kalkowsky.

References

Bathurst R (1978) Carbonate sediments and their Diagenesis. Developments in Sedimentology 12:1–658

Becker A (2005) Sequenzstratigraphie und Fazies des Unteren und Mittleren Buntsandsteins im östlichen Teil des Germanischen Beckens (Deutschland, Polen). Hallesches Jahrbuch für Geowissenschaften. Reihe B: Geologie, Paläontologie, Mineralogie. Beihefte 21:1–117

Brehm U, Palinska KA, Krumbein WE (2004) Laboratory cultures of calcifying biomicrospheres generate ooids – A contribution to the origin of oolites. – Carnets de Géologie/Notebooks on Geology, Letter 2004/03 (CG2003-L03):1–6

Brückmann FE (1721) Specimen physicum exhibens historiam naturalem, oolithi seu ovariorum piscium & concharum in Saxa. Mutatorum, Helmestadii, Salomoni & Schnorrii, 21 p

Burne RV (1991–1992) Lilliput's castles: Stromatolites of hamelin pool. Landscope 7:34–40

Dahanayake K, Gerdes G, Krumbein W (1985) Stromatolites, oncolites and oolites biogenically formed in situ. Naturwissenschaften 72:513–518

Davaud E, Girardelos S (2001) Recent freshwater ooids and oncoids from western lake Geneva (Switzerland): Indications of a common organically mediated origin. Journal of Sedimentary Research 71:423–429

Davies GR (1970) Carbonate Bank Sedimentation, Eastern Shark Bay, Western Australia. In: Logan BW, Davies GR, Read JF, Cebulski DE (eds) Carbonate Sedimentation and Environments, Shark Bay, Western Australia. The American Association of Petroleum Geologists, Memoir 13: 85–168

Dill RF (1991) Subtidal stromatolites, ooids and lime encrusted muds at the Great Bahama bank margin. Contributions in marine geology in honour of Farncis Parker Shephard. In: Osborne RH (ed) From Shoreline to Abyss. SEPM Special Publication 46:147–171

Dorn P (1953) Die Stromatolithen des Unteren Buntsandstein im südlichen Harzvorland. Neues Jahrbuch für Geologie und Paläontontologie, Abhandlungen 97:20–38

Füchtbauer H (1988) Sedimente und Sedimentgesteine. Schweitzerbart, Stuttgart, 1141 p

Geluk MC, Roehling HG (1997) High-resolution sequence stratigraphy of the Lower Triassic Buntsandstein in the Netherlands and Northwestern Germany. Geologie en Mijnbouw 76: 227–245

Hall J (1883) Plate VI and explanation: Cryptozoon, N. G., Cryptozoon proliferum n. sp. In: Pierson HR (ed) Thirtysixth annual report of the trustees of the State Museum of Natural History to the legislature. New York Senate paper 1883/53, Albany

Hofmann HJ (1973) Stromatolites: Characteristics and utility. Earth-Science Reviews 9: 339–373

Kalkowsky E (1908) Oolith und Stromatolith im norddeutschen Buntsandstein. Zeitschrift der deutschen geologischen Gesellschaft 60:68–125

Krumbein WE (1983) Stromatolites – The challenge of a term in space and time. Precambrian Research 20:493–915

Logan BW, Hoffman P, Gebelein CD (1974) Algal Mats, Cryptalgal Fabrics, and structures, Hamelin Pool, Western Australia. In: Logan BW, Read JF, Hagan GM, Hoffman P, Brown RG, Woods PJ, Gebelein CD (eds) Evolution and Diagenesis of Quaternary Carbonate Sequences, Sharc Bay, Western Australia. American Association of Petroleum Geologists, Memoir 22: 140–194

Monty CLV (1976) The origin and development of cryptalgal fabrics. In: Walter MR (ed) Stromatolites. Developments in Sedimentology 20:193–249

Monty CLV (1977) Evolving concepts on the nature and the ecological significance of stromatolites. In: Flügel E (ed) Fossil Algae. Springer, Berlin, pp 15–35

Palermo D, Aigner T, Geluk M, Poeppelreiter M, Pipping K (2008) Reservoir potential of a lacustrine mixed carbonate/ siliciclastic gas reservoir: The lower triassic rogenstein in the Netherlands. Journal of Petroleum Geology 31:61–96

Parrish JT (1993) Climate of the supercontinent Pangea. Journal of Geology 101:215–233

Paul J (1982) Der Untere Buntsandstein des germanischen Beckens. Geologische Rundschau 71:795–811

Paul J, Klarr K (1988) Feinstratigraphie und Fazies des Unteren und Mittleren Buntsandsteins in der Bohrung Remlingen 5. GSF-Bericht 8/87: 148 p

Paul J, Peryt TM (2000) Kalkowsky's stromatolites revisited (Lower Triassic Buntsandstein, Harz Mountains, Germany). Palaeogeography, Palaeoclimatology, Palaeoecology 161:435–459

Paul J, Puff P (2010) Das Klima des Buntsandsteins. Schriftenreihe der deutschen Gesellschaft für Geowissenschaften. 69 (in press)

Peryt TM (1975) Significance of stromatolites for the environmental interpretation of the Buntsandstein (Lower Triassic) rocks. Geologische Rundschau 64:143–158

Plee K, Ariztegui D, Sahan F, Martini R, Davaud E (2006) Microbes caught in the act: Disentangling the role of biofilms in the formation of low Mg calcite ooids in a freshwater lake. American Geophysical Union. Fall Meeting 2006. Abstract #B11A-1000

Reid RP, James NP, Macintyre IG, Dupraz CP, Burne RV (2003) Shark Bay stromatolites: Microfabrics and reinterpretation of origins. Facies 49:45–53

Richter DK (1983) Calcareous ooids: A synopsis. In Peryt TM (ed) Coated grains. Springer, Heidelberg, pp 71–99

Scherer M (1975) Fe-Anreicherung der Rogensteinzone des norddeutschen Unteren Buntsandsteins (Trias): Ein Hinweis auf die diagenetische Geschichte. Neues Jahrbuch Geologie und Paläontologie, Monatshefte 1975:568–576

Simone L (1980) Ooids: A review. Earth Sciences Review 16:319–355

Usdowski E (1962) Die Entstehung der kalkoolithischen Fazies des norddeutschen Unteren Buntsandsteins. Beiträge zur Mineralogie und Petrographie 8:141–179

Weber J (2000) Kieselsäurediagenese und gekoppelte Sedimentarchitektur – eine Beckenanalyse des Reinhardswald-Troges (Norddeutsches Becken, Solling-Folge, Mittlerer Buntsandstein). Kölner Forum Geologie und Paläontologie 7:1–165

The Nature of Stromatolites: 3,500 Million Years of History and a Century of Research

Robert Riding

1 Introduction

Stromatolites are widely regarded as layered, early lithified, authigenic microbial structures – often domical or columnar in form – that developed at the sediment water interface in freshwater, marine and evaporitic environments (Fig. 1). In addition to this unusually wide environmental distribution, the exceptionally long geological record of stromatolites spans at least 3,500 million years (Ma) (Vologdin 1962; Hofmann 1969, 1973; Walter 1976a; Grotzinger and Knoll 1999; Riding and Awramik 2000). Most of these examples, together with those described by Kalkowsky (1908), are essentially originally carbonate in composition. More than a century of research has revealed many details of their diverse fabrics and complex history, but much still remains to be understood about stromatolites. This is not surprising considering their wide distribution in time and space; and it helps to account for a continuing problem with their definition. Kalkowsky (1908) considered stromatolites to be microbial sediments, but it has become increasingly difficult to maintain this view for all ancient examples, especially those more than ~1,000 Ma old. The aim of this article is to evaluate progress in understanding what stromatolites are, since they were first described in the 1800s. This makes it impossible to avoid the thorny problem of how they should be defined.

The nature and definition of stromatolites have been persistent difficulties ever since Kalkowsky introduced the name. At first, the main question posed was "are stromatolites biogenic or abiogenic?" With time, the focus has shifted to whether all stromatolites are biogenic, or whether some are biogenic while others are abiogenic. Kalkowsky's (1908) microbial interpretation of stromatolites was

R. Riding
Department of Earth and Planetary Sciences, University of Tennessee, Knoxville, TN 37996, USA
e-mail: riding@cardiff.ac.uk

J. Reitner et al., *Advances in Stromatolite Geobiology*,
Lecture Notes in Earth Sciences 131, DOI 10.1007/978-3-642-10415-2_3,

Fig. 1 Loaf-shaped stromatolite in oolite. Early Triassic, Bernburg Fm, Heeseberg Quarry, Jerxheim, 50 km west of Magdeburg, Germany. Width of view 1.6 m

immediately challenged by the suggestion that they are abiogenic precipitates (Reis 1908). The subsequent century of research provided support from present-day and ancient examples for both of these views. In particular, persuasive evidence that some Precambrian stromatolites are essentially abiotic seafloor crusts grew out of pioneering studies of early Proterozoic examples in Canada (e.g., Kerans 1982; Grotzinger and Read 1983).

The challenge of defining stromatolites reflects the diversity and complexity of the structures they represent. The scarcity of present-day marine analogues for abiogenic seafloor crusts (Grotzinger and James 2000a, p. 9), such as those that occur in the Palaeoproterozoic, has hindered appreciation of the inorganic processes that can produce marine structures that have been described as stromatolites (Pope et al. 2000, p. 1149; Corsetti and Storrie-Lombardi 2003, p. 649; Perry et al. 2007, p. 169). The marine stromatolite record can be read as long-term change from less to more biogenic (Grotzinger and Kasting 1993, p. 235; Kah and Knoll 1996, p. 81; James et al. 1998). As a result, the time period from which stromatolites are viewed is critically important. It is not difficult to regard most Phanerozoic examples as essentially lithified microbial mats. In contrast many Precambrian examples regarded as stromatolites, especially those older than ~1,000 Ma, appear to contain, and in some cases entirely consist of, precipitated abiogenic crust. Furthermore, there is evidence that many late Archaean and early Proterozoic stromatolites consist of intimate interlayering of both lithified microbial mat and essentially abiogenic precipitated crust (Bertrand-Sarfati 1972, p. 155; Sami and James 1996, p. 217; Petrov and Semikhatov 2001, fig. 5a, b; Riding 2008, p. 95) that has been termed Hybrid Crust (Riding 2008). Interpretation of these deposits is hindered – especially in very old deposits – by recrystallization, but even in the Neoarchaean, relatively well-preserved examples retain clear indications of even and laterally very continuous layers that appear to consist of thin alternations of sparry and microcrystalline fabrics (Sumner and Grotzinger 2004, fig. 3). These "Boetsap laminae", named after a locality on the Boetsap River in South Africa, could be Hybrid Crust stromatolites (Riding 2008, p. 84) (Fig. 2). If so, Hybrid Crust is likely to be a major component of late Archaean and early Proterozoic

Fig. 2 Broad flat-topped stromatolite domes separated by a narrow shallow steep-sided depression. The well-developed layering is even and smooth, and can be traced from dome to dome across the intervening depression. Late Archaean Campbellrand–Malmani platform, Groot Boetsap River. South Africa. Width of view ~1.5 m

stromatolites. Boetsap laminae stromatolites are widespread in the extensive Neoarchaean Campbellrand-Malmani carbonate platform (Sumner and Grotzinger 2004, pp. 14–15).

Taking the long history of stromatolites as a whole, this suggests that some stromatolites are biogenic (e.g., lithified microbial carbonate), others are abiogenic precipitated crust, and that some are hybrid mixtures of the two. The concept of abiogenic stromatolites is not new. Logan et al. (1964, pp. 68–69), for example, recommended recognition of "inorganic stromatolite". The problem it presents is that it broadens the term stromatolite to potentially include speleothem and hot spring sinters, deposits that have not generally been regarded as stromatolites. As Walter (1976b, p. 1) recognized, if stromatolite definition is "so broad as to include a wide-range of non-biogenic structures" ... "the term would cease to be useful". Here layered authigenic microbial and hybrid crusts – but not abiogenic crusts – are regarded as stromatolites. A definition is proposed: *Stromatolites are macroscopically layered authigenic microbial sediments with or without interlayered abiogenic precipitates.*

This definition avoids the difficulty of encompassing abiogenic sinter within stromatolite. However, Hydra-like, and characteristic of stromatolite studies, it raises new challenges. If stromatolites are not biogenic, then how are ancient examples to be distinguished from abiogenic crusts and, specifically, what criteria can be used to confidently establish biogenicity? These are the perennial problems that led Logan et al. (1964), for example, to recognize abiogenic stromatolites. I suggest that it is now possible to use macro- and microfabric details to distinguish microbial and Hybrid Crust stromatolites from abiogenic crusts, where they are sufficiently well-preserved.

2 Stromatolites and Spongiostromids

2.1 *Stromatolith*

Kalkowsky (1908) introduced the term *Stromatolith* – layered stone – to describe columns and domes of well layered carbonate within beds of Early Triassic lacustrine oolite that occur near the Harz Mountains of northern Germany (see Paul and Peryt 2000) (Fig. 3). Bowl-shaped weathering products of these near Winnrode (probably Wienrode – near Blankenburg on the northern edge of the Harz) had earlier been called *Napfstein* (bowl-stone) (Naumann 1862, p. 741; Kalkowsky 1908, p. 69). Kalkowsky (1908, p. 125) suggested that *Stromatolithe* were formed by "*niedrig organisierte pflanzliche Organismen*" (simply organized plant-like organisms). In essence, he regarded stromatolites as laminated microbial structures (Riding 1999, p. 323), and he held a similar view for the ooids with which they are associated (Kalkowsky 1908, p. 68). But he was not the first to propose a general name for the structures that came to be termed stromatolites. Examples with well-preserved spongy microstructures in the Mississippian (Viséan) of Belgium had been named *spongiostromides* by Gürich (1906), who had placed them in new genera such as *Pycnostroma* and *Spongiostroma*. Gürich (1906) thought they were protozoans. Heim (1916, p. 566) introduced the term oncoid (*onkos* – nodule) for grains in the Jurassic of Switzerland. Pia (1927, pp. 36–37) may have intended to reflect the priority of Gürich's work over Kalkowsky's when he classified "Stromatolithi" and "Oncolithi" as sub-groups within the Spongiostromata, but this usage did not gain support. Instead, stromatolite became widely adopted as the general term, whereas spongiostrome is now (and more rarely) used to refer to the distinctive clotted fabrics found in many Phanerozoic stromatolites.

2.2 *Eozoön, Cryptozoon, Archaeozoon*

Furthermore, neither Kalkowsky nor Gürich was the first to recognize the structures that came to be called stromatolites. Similar, and also geologically older, examples

Fig. 3 Contact of oolite/pisolite (*left*) and stromatolite dome (*right*). Early Triassic, Bernburg Fm, Heeseberg Quarry, Jerxheim, 50 km west of Magdeburg, Germany

had long been recorded in North America. Steele (1825, pp. 17–18, pl. 2) described specimens in the Late Cambrian of New York State as "calcareous concretions" (Fig. 4). These were later named *Cryptozoon* by Hall (1883) (Fig. 5) who regarded them as the skeletons of simple animals. This opinion reflected Dawson's (1865) interpretation of *Eozoön*, from the Proterozoic of Québec, as a giant foraminifer. Dawson's claim of evidence for life in such ancient rocks attracted both interest and controversy. The "most instructive specimens of *Eozoön*", noted by Dawson (1876), are alternating laminae of serpentine or pyroxene and calcite in Grenvillian contact metamorphosed limestone from near Ottawa (Dawson 1865) (Fig. 6). *Eozoön* was keenly debated for many years and by the late 1900s most specimens were generally thought to be inorganic (O'Brien 1970; Hofmann 1971, p. 12; Adelman 2007).

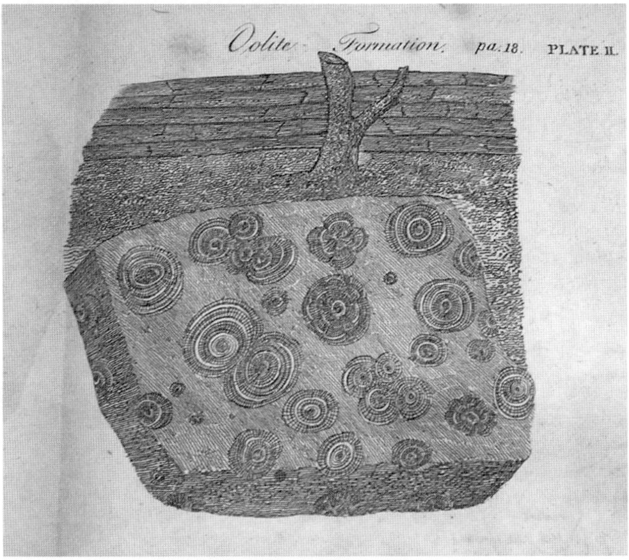

Fig. 4 Engraving of Late Cambrian stromatolites at Saratoga Springs, New York State, described as "calcareous concretions" (Steele 1825, p. 18, pl. 2). Compare Fig. 5

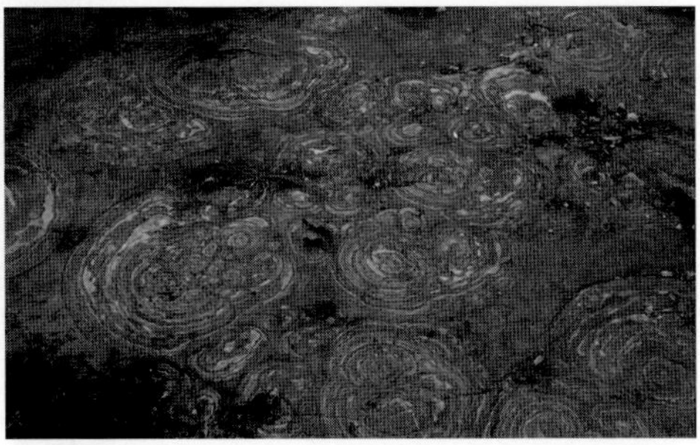

Fig. 5 Horizontal outcrop section of the stromatolite described by Steele (1825) and named *Cryptozoon proliferum* by Hall (1883), showing irregular and impersistent layering. Late Cambrian, Hoyt Limestone, Petrified Gardens, 5 km W. Saratoga Springs, New York State. Width of view ~2 m. Compare Fig. 4

Nonetheless, Dawson's idea that layered domes could be animal remains was perpetuated in the *zoon* suffix of the names given to other broadly similar specimens, such as *Cryptozoon* (Hall 1883). At the time that Dawson was studying *Eozoön*, Bell (1870, p. 324) was describing Palaeoproterozoic Gunflint stromatolites (Fig. 7) in western Ontario as "small coral-like siliceous concretions" that "show fine concentric rings" (see Hofmann 1969, p. 5). Similar "concentric nodular masses" were discovered by Bailey and Matthew (1872) in the Proterozoic of New Brunswick. Matthew (1890a) named them *Eozoön acadiense*, but immediately changed this to *Archaeozoon* (Matthew 1890b). These specimens resemble *Baicalia* and similar taxa

Fig. 6 *Eozoön canadense*, showing domical layering consisting of alternations of serpentine (*dark*) and calcite (*light*). Mesoproterozoic, Côte St. Pierre, north of Papineauville, 55 km ENE of Ottawa, Canada. GSC specimen 1992-234B collected by T.C. Weston (see Dawson 1876). Width of view ~11 cm. Photograph courtesy of Brian Chatterton

Fig. 7 Contact of silicified stromatolite dome (*right*) and adjacent *red-brown* carbonates. Basal Gunflint Formation, Palaeoproterozoic (1,880 Ma), 1 km. of Kakabeka Falls, Ontario, Canada

(Hofmann 1971, p. 58). Thus, Kalkowsky (1908) was neither the first to describe stromatolites nor to propose a general name for them, although of course he did create the name stromatolite that has been universally adopted. He may have been the first to regard them as microbial, although not by very long, because 6 years later Walcott (1914) compared them with cyanobacterial tufa.

3 Stromatolites as Lithified Microbial Mats

Understandably, interpretation of stromatolites has been closely linked to present-day deposits that could shed light on their origins. Although it is not clear exactly what reasons led Kalkowsky (1908) to regard stromatolites as microbial, they may include the close association of stromatolites with ooliths (which Kalkowsky named Ooide). At the time of Kalkowsky's (1908) article there was already an extensive literature suggesting that ooliths and similar grains might be organic (e.g., Linck 1903). This view was pioneered by Rothpletz (1892) who had found coccoid cyanobacteria in Great Salt Lake ooliths. A role for cyanobacteria in carbonate precipitation was further supported by Walcott's (1914) comparisons of *Collenia* and *Cryptozoon* with present-day lacustrine tufas. During the half-century or more since Kalkowsky's (1908) and Walcott's (1914) articles, the search for present-day analogues led from cool-water calcareous freshwater lakes and streams to subtropical marshes and shorelines, tidal flats and, eventually, deeper marine environments. It provided irrefutable evidence linking many ancient stromatolites, especially those in marine Phanerozoic environments, to lithified microbial mats.

3.1 Cyanobacterial Lacustrine Tufas

C.D. Walcott, studying Proterozoic sediments in the western United States, had found Neoproterozoic specimens at Nankoweap Butte in the Grand Canyon (Walcott 1895, p. 319) that he regarded as stromatoporoids, but which Dawson (1896, p. 208) thought were *Cryptozoon* and compared to foraminifers (Dawson 1896, pp. 211–212). Walcott (1906) encountered more examples as he extended his studies to the Mesoproterozoic of Montana, and he described them as new genera, including *Collenia*, named after a rancher in the Big Belt Mountains (Walcott 1914, p. 111). Walcott had seen *Cryptozoon* much earlier at Saratoga Springs in 1878 (Schopf 1999, p. 25; Lindemann and Yochelson 2005) but did not adopt the view that it was of animal origin. Instead, he compared it with present-day freshwater tufas in lakes in New York State (Walcott 1914) (Fig. 8). He argued that Proterozoic and Cambrian *Collenia* and *Cryptozoon* were "deposited through the agency of algae similar in type and activity to the (Cyanophyceae) Blue–green Algae" (Walcott 1914, p. 100). Therefore, although he appears to have been unaware of Kalkowsky's work and of the name *Stromatolithe*, Walcott's (1914) comparisons with tufa provided the first present-day support that stromatolites could be microbial. This was also the first

Fig. 8 Freshwater lacustrine microbial tufa domes, broadly similar to deposits in Green Lake, New York State, that C.D. Walcott (1914) compared with Mesoproterozoic stromatolites. Poza Azul, 8 km south-west of Cuatro Ciénegas, Coahuila, northern Mexico. Domes are ~0.5 m high

link between stromatolites and cyanobacteria: a connection confirmed by many subsequent researchers. It stimulated further studies, such as Roddy's (1915) of tufas in Little Conestoga Creek, Pennsylvania. Immediately, Wiman (1915) described late Proterozoic *Collenia* in Sweden as "cyanophycean", although Hadding (1927) subsequently regarded this specimen as inorganic (Vidal 1972).

3.2 Andros Marsh

Nearly 20 years after Walcott's comparison with tufa, the stromatolite connection with cyanobacteria was strongly reinforced by Black's (1933) study of laminated "blue-green algal" deposits at the margins of lakes and tidal creeks on Andros Island in the Bahamas. These both supported and differed from Walcott's approach. Although Black did not refer to Walcott, he new Roddy's work and pointed out that Andros mats do not closely resemble freshwater tufas, which generally show much stronger early lithification. Black's (1933) examples are relatively poorly lithified and he commented that they are "best developed in regions of low salinity", adding that Andros examples have little in common with the "hard, stony cyanophyceous limestones" that form in hardwater lakes and streams such as those described by Roddy (1915) (Black 1933, p. 185, p. 191). He wrote, perhaps with some understatement, that "Bahamian sediments show a certain resemblance" to "Precambrian and early Palaeozoic stromatolites" (Black 1933, p. 186). Some Andros mats are much more similar to Walcott's (1914) Mesoproterozoic Belt stromatolites than are freshwater tufas. This can be seen by comparing Black (1933, fig. 3) and Monty (1972, fig. 22) with *Collenia undosa* (Walcott 1914, pl. 13). Black's (1933) work provided support for a microbial interpretation at a time when there was considerable doubt about stromatolite biogenicity (e.g., Seward 1931). As a result, Fenton (1943, p. 95) was able to summarize "several lines of evidence which indicate that stromatolites are both organic and algal". Further, and unexpected, support for a cyanobacterial interpretation of stromatolites later came from some very old examples, when silicified cyanobacteria were reported in Palaeoproterozoic Gunflint stromatolites (Tyler and Barghoorn 1954; Barghoorn and Tyler 1965) (Fig. 7).

3.3 Coarse-Grained Thrombolitic Stromatolites

Black's specimens came from ponds and tidal creeks in the interior of northern Andros Island and, at most, are only marginal marine (Black 1933, p. 191). The search for marine stromatolites and oncoids was pursued in back-reef environments in Florida (Ginsburg et al. 1954; Ginsburg 1960) and Andros (Monty 1965), but these specimens too were mainly poorly lithified (Monty 1972, p. 745). Breakthrough came with the discovery of large (generally up to ~0.5 m high) well lithified microbial domes and columns along the shoreline of seasonally hypersaline Shark Bay in Western Australia (Fig. 9). These now famous stromatolites were noted by geologists in 1954 (Playford and Cockbain 1976, p. 389), and described by Logan (1961) who emphasized their cyanobacterial mats and resemblance to *Cryptozoon*. The columns on the beach at Shark Bay are thought to have formed subtidally and then been exposed by recent relative sea-level fall (Playford and Cockbain 1976, p. 399). Externally these columnar microbial carbonates can closely resemble some

Fig. 9 Individual and laterally amalgamated thrombolitic stromatolite columns, ~50 cm high. Intertidal zone, Carbla Point, Hamelin Pool, Shark Bay, Western Australia. Photograph courtesy of Eric Mountjoy

very ancient examples such as 1,800 Ma Pethei elongate columns (Hoffman 1989, fig. 9b), but internally they are often less well layered and coarser grained. Crudely layered columns at Shark Bay, illustrated by Logan (1961, pl. 1, fig. 4) and described by Aitken (1967, p. 1171) as "thrombolitic stromatolites", are largely composed of fine sand (Logan et al. 1974).

Large columnar and domical stromatolites are very rare in present-day marine environments. Apart from Shark Bay, the only other region in which they have been reported are the Bahama Banks (Dravis 1983; Reid et al. 1995, 2000), where they are especially well-developed in the tidal channel between Lee Stocking Island and Norman's Pond Cay in the Exuma Cays (Dill et al. 1986; Riding et al. 1991a) (Fig. 10). Shark Bay and Lee Stocking columns occur in wave- and current-swept environments. Their formation can be attributed to factors that include the mat communities, water movement, and grainy conditions. Strong waves and currents raise sand to the accreting mat surface, and the grainy environment deters overgrowth by reefal encrusters (Dill et al. 1989, p. 10). Seasonal hypersalinity at Shark Bay also limits competitors. Accretion rates are high due to the combined presence of thick soft microbial mats that contain abundant EPS (extracellular polymeric substances) (Decho et al. 2005), and rapid grain supply. In addition to cyanobacteria, the mats contain diatoms and filamentous green algae that enhance trapping (Awramik and Riding 1988; Dill et al. 1989; Riding et al. 1991a). The upper mat

Fig. 10 "Molar tooth" thrombolitic stromatolite domes surrounded, and half-buried, by rippled ooid sand. Domes have exposed heights of ~0.5–1 m. Depth ~10 m in tidal channel between Lee Stocking Island and Norman's Pond Cay, Exuma Islands, Bahamas

remains soft and sticky because it is largely uncalcified, and the early lithification necessary to support these large columns mainly occurs in, or below, the lower part of the mat. Microbial lithification by sulphate reduction (Visscher et al. 2000) is limited to very thin micritic crusts (Reid et al. 2000), cyanobacterial sheaths are uncalcified (Reid et al. 2000, p. 992), and calcification of algal filaments mainly occurs in cavities (Dravis 1983; Whittle et al. 1993, p. 224). Thus, grainy current swept conditions, thick soft surface mats, and early subsurface lithification allow decimetric, and locally metric, columns to develop. Similar coarse-grained thrombolitic stromatolite domes are well developed in the late Miocene of SE Spain (Riding et al. 1991b; Braga et al. 1995; Feldmann and McKenzie 1997) but have not been reported from older rocks.

3.4 Tidal Flats

In contrast to the scarcity of large subtidal domes and columns, low relief – often poorly lithified – microbial mats are often widespread in many muddy-sandy intertidal and supratidal habitats, in both carbonate and siliciclastic environments, as well as in natural and artificial saline lakes (Fig. 11). Complex layered algal-bacterial communities develop on these intermittently illuminated and wetted sediment surfaces, and microbial energy cycling and physicochemical gradients generate vertical zonation (e.g., Revsbech et al. 1983; Cohen and Rosenberg 1989; van Gemerden 1993; Des Marais 2003; Stal 2000). This can be seen in distinctive millimetric colour bands produced by the microbes and their mineral products. Examples from Danish tidal flats were illustrated in Flora Danica (Hornemann 1813, pl. 1485) and reported from Trindelen in Odense Fjord (Ørsted 1842). Similar deposits in northern Germany were described as "Farbstreifen-Sandwatt" (colour-banded sand-flat) (Schulz 1936; Kremer et al. 2008, fig. 2c), and varicoloured mats

Fig. 11 Thick soft layered microbial mat on sediment. Coloured layers from the top down are: *yellow* EPS-rich layer with scytonemin; *green* cyanobacterial layer; *red* sulphur bacteria layer; *grey* sulphate reduction layer (although there is likely to be highest activity of sulphate reducers in the *green* layer). Salt pan the western side of Lagoa Pitanguinha, 85 km east of Rio de Janeiro, Brazil. Width of view 10 cm

are also well known at Great Sippewisset saltmarsh on Buzzard's Bay, Massachusetts (Nicholson et al. 1987).

These environments are stressed by physical and chemical factors (Cohen and Rosenberg 1989; Decho 2000) such as desiccation, ultra-violet radiation, temperature, and salinity that deter invertebrate competitors. As a result, cohesive microbial mats are generally much more extensive in present-day marine intertidal than shallow subtidal environments (Browne et al. 2000, p. 236). Following description of Shark Bay columns (Logan 1961), the accessibility of intertidal mats made them a focus of the geological search for marine stromatolites, e.g., in Andros (Monty 1967), the southern Persian Gulf (Kendall and Skipwith 1968), at Shark Bay (Davies 1970; Logan et al. 1974), and near the Coorong Lagoon in South Australia (Walter et al. 1973). This coincided with increased geomicrobiological research on microbial mats in hypersaline lagoons such as Solar Lake, adjacent to the Gulf of Aqaba (Por 1967; Jørgensen and Cohen 1977; Krumbein et al. 1977) and adjacent to the Pacific coast near Lázaro Cárdenas in northern Baja California, Mexico (Horodyski and Vonder Haar 1975; Horodyski 1977; Horodyski and Bloeser 1977; Javor and Castenholz 1981).

3.5 Wrinkle Marks

Present-day tidal flat mats typically form poorly lithified low relief structures that contrast markedly with Shark Bay columns (Kendall and Skipwith 1968). Aitken (1967, pp. 1163–1164) gave stratiform mats a different name – "cryptalgal laminite" – to describe "planar-laminated carbonate bodies bearing evidence of algal-mat activity". Whereas Aitken (1967) nonetheless continued to recognized these stratiform deposits as stromatolites, some authors have distinguished them from stromatolites because they lack characteristic "columnar, branched, and hemispheroidal structures that accrete from a single point" (Ginsburg and Planavsky 2008, p. 177); presumably because this makes it more difficult to interpret them as biogenic (e.g., Schieber 1998, p. 106). Nonetheless, cohesive mats can preserve the effects of synsedimentary deformation to which they are particularly prone on tidal flats, and these can impart distinctive features that aid their recognition in the rock record. In arid conditions, desiccation causes cracking and fragmentation (Gerdes et al. 1993, 2000, p. 205) (Fig. 12), whereas wet flexible mats are bent and folded by waves and currents, as well as gas bubbles (Cameron et al. 1985). Early lithification can preserve such synsedimentary mat deformation structures (Kendall and Skipwith 1968, p. 1056), but research has shown that mats can also leaves traces in sediments that are not strongly affected by early lithification, such as siliciclastic sand. As a result, deformed and imprinted mats have an extensive although often subtle geological record. From near White Sulphur Springs in central Montana, where he had collected Mesoproterozoic *Collenia*, Walcott (1914, p. 107, pl. 11, fig. 3) described *Kinneyia* and *Newlandia*, which have patterned

Fig. 12 Desiccated cohesive microbial mat, patterned by broad polygonal shrinkage cracks. Salt pan on the western side of Lagoa Pitanguinha, 85 km east of Rio de Janeiro, Brazil

ribbed or ripple-like surfaces. This original *Kinneyia* is in carbonate (the Newland Limestone) but was compared with somewhat similar structures in siliciclastic sediments (Martinsson 1965; Bloos 1976; Seilacher 1982). In the past, both *Newlandia* (Holtedahl 1921) and *Kinneyia* (Fenton and Fenton 1936, pp. 612–615; Fenton 1943, p. 86) have been regarded as inorganic, but more recently *Kinneyia*, in particular, has been linked to "Runzelmarken" (wrinkle marks), originally described from Jurassic siltstones of northern Germany (Häntzschel and Reineck 1968).

Horodyski (1982) drew attention to bedding plane markings in silty siliciclastic mudstones of the Mesoproterozoic Snowslip Formation in Glacier National Park, Montana, interpreting them as "impressions of thin, wrinkled algal mats" by comparing them with present-day mats in Florida Bay. Detailed studies of present-day intertidal siliciclastic mats (Cameron et al. 1985; Gerdes et al. 1985) revealed numerous delicate but distinctive structures (Gerdes and Krumbein 1994) that have subsequently been recognized in ancient sediments and interpreted as evidence for ancient mats. For example, Schieber (1986) interpreted carbonaceous silty shales in the Mesoproterozoic Newland Formation as remnants of benthic microbial mats, and found further evidence for microbial mats in shales and sandstones of similar age in the Belt Supergroup (Schieber 1998). Thus, "wrinkle marks" and other distinctive folds and elevated and depressed patterns on bedding planes, together with *Kinneyia*-like structures, are now widely linked to synsedimentary deformation of flexible cohesive sediment binding mats (Hagadorn and Bottjer 1997). These "microbially induced" (Noffke et al. 1996) or "microbially mediated" (Hagadorn and Bottjer 1997) "sedimentary structures" (MISS) (Noffke et al. 2001) have also been reported from Archaean rocks (Noffke et al. 2003, 2008). They can therefore be important indicators of mats in very old rocks where other indications of microbial life are uncertain. In addition to deformation structures, organic matter and filamentous fabrics can be preserved (Noffke et al. 2006), although these interpretations are not without the difficulties of distinguishing mat-related from other sedimentary structures (Porada and Bouougri 2007; McLoughlin et al. 2008; Porada et al. 2008).

3.6 Reef Crusts

Late Cenozoic reefal microbial crusts are much less well known than intertidal and hypersaline stromatolites and mats, but are locally important deposits in reef cavities and on fore-reef slopes. They typically occur as cryptic veneers on corals and other skeletons in framework cavities and closely resemble some of the thick clotted, peloidal stromatolitic crusts that are locally common in reefs in the Palaeozoic and Mesozoic. In a series of important papers, Macintyre (1977, 1984, 1985) drew attention to fine-grained peloidal crusts in Holocene Caribbean reefs and reef caves. Their non-skeletal fabrics are evidently precipitated. Together with other examples in the Great Barrier Reef (Marshall and Davies 1981; Marshall 1983), they attracted attention as examples of submarine lithification (James and Ginsburg 1979; Land and Moore 1980; Macintyre and Marshall 1988). Similar crusts occur in Messinian coral reefs in the western Mediterranean (Pedley 1979) and in some cases crust volume exceeds that of coral (Dabrio et al. 1981). Their clotted-peloidal fabrics were interpreted as calcified bacterial organic matter containing grains trapped in adhesive biofilm (Riding et al. 1991b). Similar crusts were found in reef caves at St Croix (Zankl 1993) and Lizard Island (Reitner 1993), and in late Quaternary reefs in the Pacific (Montaggioni and Camoin 1993; Cabioch et al. 1999; Webb et al. 1998). They have been studied in detail by Gilbert Camoin and colleagues at Tahiti (Camoin and Montaggioni 1994; Cabioch et al. 1999; Camoin et al. 1999, 2006, 2007).

Reefal microbial crusts are millimetric to decimetric in thickness (Cabioch et al. 2006, p. 304) with irregular, domical or dendritic surfaces (e.g., Land 1971; Macintyre 1977, pp. 507–508; Marshall 1986; Sherman et al. 1999; Camoin et al. 1999, 2006) (Fig. 13). Smooth domes usually show better internal layering than the columns. Silt-size peloids, generally $<50\ \mu m$ across, are characteristic crust components (Macintyre and Marshall 1988) (Fig. 14) and appear to "float" in fenestral microspar. Chafetz (1986) proposed that such peloids form in semi-isolated cavities in present-day reefs by bacterially-induced precipitation around suspended bacterial colonies, and many researchers have inferred a generally bacterial origin for fine-grained reef crusts (Pedley 1979; Brachert and Dullo 1991; Jones and Hunter 1991; Riding et al. 1991b; Montaggioni and Camoin 1993; Zankl 1993; Webb et al. 1998). More specifically, bacterial sulphate reduction (BSR) is suggested by the typically magnesian calcite composition, stable isotope values, and biomarkers. Crust values of magnesian calcite in the range 12–18 mole % Mg (Macintyre et al. 1968) are difficult to obtain inorganically (Morse and Mucci 1984, p. 287) and suggest bacterial activity (Pigott and Land 1986, pp. 355–356) such as BSR (Malone et al. 2001, p. 891, and fig. 10). BSR has also been inferred from carbon and oxygen isotope values of peloidal crusts and fills (Land and Goreau 1970; Pigott and Land 1986, figs. 9–11; Reitner et al. 2000, p. 153). Acidic macromolecules including diaminopimelic acids (Reitner et al. 1995) and biomarker evidence for anaerobic heterotrophs (Reitner et al. 2000, pp. 158–159) were found in Lizard Island and St Croix peloidal crusts. In Tahiti reef crusts, Camoin et al. (1999,

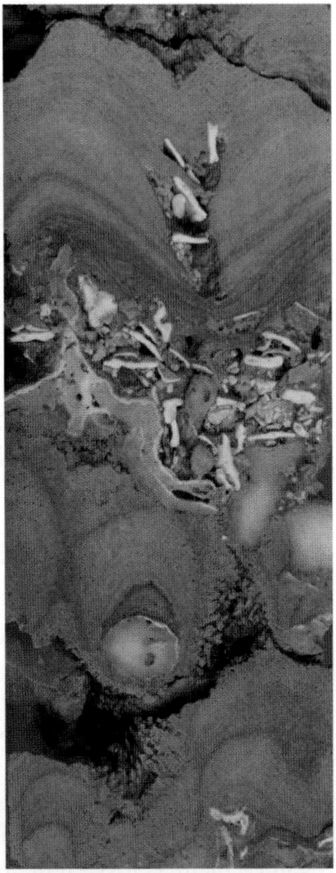

Fig 13 Reefal microbial stromatolite crusts (laminated *grey-brown*) on lighter coloured coral skeletons and bioclasts. Core through late Pleistocene-early Holocene reef, SW Tahiti, IODP 310, Maraa eastern transect, Hole M0015A, Last Deglacial Sequence, Subunit 1C, interval 310-M0015A-21R-01, 19–31 cm (Expedition 310 Scientists 2007, fig. 44). Width of view, 5.5 cm. © IODP/ECORD

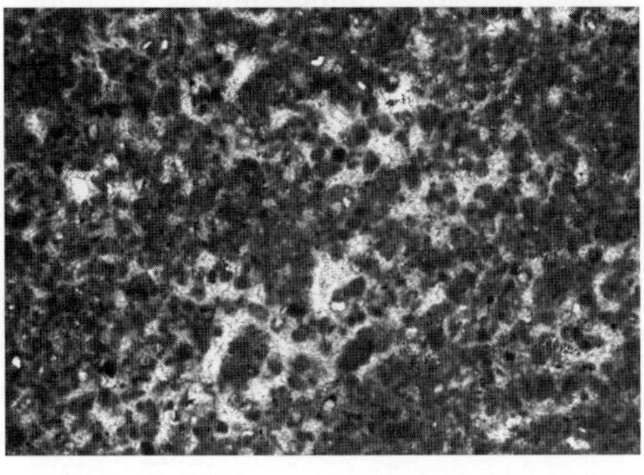

Fig. 14 Microfabric of reefal microbial crust. Peloid microspar with irregularly amalgamated dark peloidal masses in a light microspar, locally fenestral, matrix. Pleistocene −988 m drowned reef, Kohala, NW Hawaii; approx. age 375,000–400,000 years. Sample T302-R34, provided by Jody Webster. Width of view 1.4 mm

p. 297) found muramic and diaminopimelic acids characteristic of bacterial cell walls, and Heindel et al. (2009) found fatty acids typical of sulphate reducers.

In Late Pleistocene–Holocene reefs, crusts occur on wave-swept reef margins in dark enclosed framework cavities and on deep dark fore-reef slope surfaces

(Land and Goreau 1970; James et al. 1976; Macintyre 1977, p. 513; James and Ginsburg 1979, fig. 6-2e, fig. 6–17d; Marshall and Davies 1981; Lighty 1985; Marshall 1986; Camoin and Montaggioni 1994; Webb et al. 1998; Camoin et al. 2006). Their position between sciaphilic encrusters (e.g., corallines, foraminifers) and overlying reef or pelagic sediment reflects their formation in the closing stage of framework growth. As light diminishes, reef surfaces are colonized by a succession of increasingly sciaphilic (shade-loving) skeletal organisms (Garrett 1969; Martindale 1992, fig. 7): coralline algae, bryozoans, sclerosponges, foraminifers, and serpulids. The ability of the crusts to form in dark environments is consistent with them as products of essentially heterotrophic bacterial communities. Preference for wave-swept reef margins suggests the effect of seawater flushing on precipitation (e.g., James et al. 1976, p. 541; Macintyre 1977; Marshall 1986, p. 23). This too is consistent with a microbial origin since bacterial calcification is strongly dependent on environmental factors that promote precipitation (Riding 2000).

4 Stromatolites as Abiogenic Structures

Whereas biogenic interpretations of stromatolites, from Kalkowsky (1908) onwards, focused entirely on microbial sediments, abiogenic interpretations have been much more wide-ranging. In addition to sub-aqueous precipitates, such as hot spring sinters and, eventually, seafloor crusts, they involved post-depositional structures such as folds and diagenetic concretions. Organic interpretations of stromatolite-like structures were not new in the early 1900s. Most descriptions of *Cryptozoon*, *Spongiostroma* and similar deposits in the late 1800s and early 1900s regarded them as fossil remains. But whereas controversy enveloped *Eozoön*, there appears to have been little initial resistance to interpretations of *Cryptozoon* and *Spongiostroma* as simple animals. In contrast, the views of Kalkowsky (1908) and Walcott (1912, 1914) aroused some strong opposition. Possibly this was because Kalkowsky (1908) interpreted not only stromatolites but also ooids as microbial. In the case of Walcott (1912, 1914) criticism centred on his algal-cyanobacterial interpretations of Archaean and Proterozoic specimens that could convincingly be compared with inorganic structures such as crystal fans and diagenetic concretions. These questions created widespread uncertainty about the nature of stromatolites that continued for 20 years (Seward 1931, pp. 83–89) until the tide turned with Black's (1933) recognition of present-day stromatolitic microbial mats on Andros Island. Fold structures have also been drawn into the debate, especially for some Archaean stromatolite-like structures (Hofmann 1971, p. 58; Lowe 1994, p. 389).

4.1 Sinters

Kalkowsky's (1908) suggestion that both ooids and stromatolites are microbial was immediately challenged by Reis (1908), who argued that they are inorganic precipitates. Bucher (1913) supported Reis, observing that both stromatoliths and oolites had "a structure of delicate layers". Later he stated: "stromatoliths are the sedimentary equivalent of the calcareous and siliceous 'sinter' of the hot springs" and he contrasted them with "the coarse calcareous crusts which are formed by thick, felted masses of fresh-water algae and mosses" (Bucher 1918, p. 608). This was echoed by Bradley (1928) who, in marginal facies of the lacustrine Eocene Green River Formation in Wyoming, interpreted coarse fabrics similar to those of spongy Green Lake tufa fabrics as algal, contrasting them with finely banded and radial fibrous layers which he regarded as inorganic (Bradley 1928, p. 209, p. 217, pl. 34a,c; pl. 45a).

4.2 Concretions

Walcott (in Lawson 1912, pp. 16–23) gave the name *Atikokania* to Archaean structures found by A.C. Lawson in Archaean Steep Rock carbonates, and also identified a specimen as *Cryptozoan*? (NB, not *Cryptozoon*). Walcott suggested that *Atikokania* might be a sponge, but retracted this when Abbott (1914) compared it with Permian concretionary carbonates in north-east England (Hofmann 1971, p. 25). The idea that *Atikokania* might be a sponge persisted (de Laubenfels 1955) but it has generally been regarded as inorganic (Glaessner 1962, p. 472), for example as crystal fans (Hofmann 1971) precipitated on the seafloor (Sumner and Grotzinger 2000, p. 134).

Following Abbott's (1914) example, Holtedahl (1921) compared some of Walcott's (1914) Mesoproterozoic specimens from the Belt Series with concretions in the Permian Magnesian Limestone of the Zechstein evaporite basin in north-east England. Holtedahl (1921, p. 202) agreed with Sedgwick (1829) that these Permian structures were diagenetic and he particularly compared them with Walcott's (1914) *Greysonia* and *Newlandia* from the Newland Limestone of Montana. Holtedahl (1921) considered that they "must have had a similar origin" (p. 201) and were "inorganic and secondary" (p. 203). However, he concluded: "The discovery of algae and bacteria in pre-Cambrian strata, reported by Walcott, has therefore lost none of its importance, even if it should be found that these organisms are not responsible for the many curious structures found in the Algonkian Newland limestone" (p. 206). Nonetheless, as a result of these comparisons with Zechstein concretions, Seward (1931, pp. 83–89) was sceptical that *Cryptozoon* and similar structures were organic. Seward recognized that blue-green algae were associated with nodules, such as those described by Mawson (1929) from South Australia, but he was hesitant to infer "that their presence is essential" to nodule formation (Seward 1931, p. 83), and he also doubted that they could have produced "reefs of *Cryptozoon*" (Seward 1931, pp. 86–87). But Goldring (1938, p. 21) pointed out that the irregularity of *Cryptozoon* contrasted with the regularity to be expected from inorganic precipitates, and in a footnote Seward (1931, p. 86) himself recognized that bacterial deposits could closely resemble *Cryptozoon*. Although Fenton

(1943, p. 93, fig. 7) considered that it does not closely resemble *Newlandia*, some of Holtedahl's Zechstein specimens are stromatolitic in appearance. Fenton (1943, p. 86) too, considered that some of Walcott's (1914) Newland samples were inorganic products of "segregation of calcite and dolomite during deformation".

4.3 Folds

At Steep Rock Lake in Ontario, Jolliffe (1955, pp. 380) suggested that Walcott's (in Lawson 1912, pp. 16–23) "*Cryptozoan*? fragment probably came from what are much the most common organic-like forms in the Steeprock carbonate – the so-called 'algal' structures". These include large stromatolitic domes (Wilks and Nisbet 1985). However, Hofmann (1971, p. 58 and pl. 22, fig. 1) noted that "The sample appears to be deformed, and it is probable that the lamination of this particular specimen is a tectonic foliation and not stromatolitic". Lowe (1994, fig. 3d, p. 389) considered that a domical structure interpreted as a ~3,450 Ma stromatolite at Pilbara, Australia (Walter et al. 1980), had formed by soft-sediment deformation. Furthermore, he suggested that all three of the well-documented reports of stromatolites older than 3,200 Ma known at that time "probably formed through non-biological processes" such as deformation or evaporitic inorganic precipitation (Lowe 1994, p. 390).

4.4 Seafloor Crusts

Reliance on non-marine deposits for examples of abiogenic stromatolite-like structures continued into the 1970s (e.g., Read 1976; Thrailkill 1976; Walter et al. 1976). In the 1980s, a significant new development occurred when petrographic studies suggested that well-preserved Palaeoproterozoic stromatolites in Arctic Canada could be interpreted as essentially abiogenic precipitates (Kerans 1982; Grotzinger and Read 1983). Although these deposits were often at first described as "fibrous marine cements" (Grotzinger 1989b, p. 10), they were evidently seafloor crusts. Grotzinger (1986a) considered the possibility that microdigitate stromatolites are "entirely abiotic", and Grotzinger and Rothman (1996, p. 424) proposed that "abiotic mechanisms" could account for the growth of large Palaeoproterozoic stromatolites such as those described by Jackson (1989, fig. 13) (Fig. 15). Thus, some Proterozoic deposits described as stromatolites appear, from petrofabric evidence, to be essentially abiotic seafloor precipitates. Literature survey indicates at least four general categories of subaqueous Sparry Crust based on Precambrian examples (Riding 2008, table 1); three of which include stromatolitic deposits: (1) Botryoidal fans and crystal pseudomorphs include small radial fibrous millimetric microbotryoids that build *Tarioufetia* and *Tungussia* (Bertrand-Sarfati 1972); (2) "microdigitate stromatolites": small laminated columns of radial crystals (Grotzinger and Read 1983; Hofmann and Jackson 1987); (3) isopachous laminite (Jackson 1989; Pope and Grotzinger 2000, fig. 8e; Sumner and Grotzinger 2004).

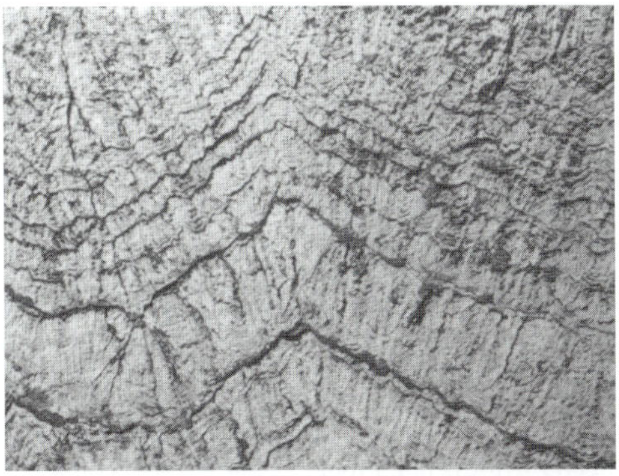

Fig. 15 Isopachous laminite forming "peaked stromatolitic carbonate", interpreted as potentially abiogenic by Grotzinger and Rothman (1996, p. 424). Cowles Lake Formation (~1,900 Ma). Width of view 45 cm. Jackson (1989, fig. 13), CSPG© 1989, reprinted by permission of the CSPG whose permission is required for further use

Fig. 16 Small, partly silicified, domical to cuspate stromatolites showing even and laterally continuous, smooth lamination. Late Archaean Campbellrand–Malmani platform, near Groot Boetsap River. South Africa

This made it clear that essentially abiogenic laminated crusts had precipitated on the floors of seas in the geological past, as well as forming in non-marine lakes and pools at the present-day. Recognition of these abiogenic marine deposits justified Hoffman's (1973) apprehension that at least some "ancient stromatolites" ... "might not be biogenic at all". They emphasized the importance of clearly discriminating between these crusts and lithified microbial sediments. Appropriately, it was the petrographic descriptions of these seafloor crusts themselves that contributed significantly to resolving the dilemmas of stromatolite definition, by providing fabric criteria that can be used to distinguish biogenic and abiogenic layered authigenic carbonates.

5 Stromatolite Fabrics

Regular, as opposed to uneven, layering is a distinct feature of some stromatolites (Fig. 16). Pope et al. (2000, p. 1139) interpreted "isopachous stromatolites to have been dominated by chemogenic precipitation in the absence of microbial mats, and the growth of peloidal stromatolites to have been controlled by sedimentation in the presence of microbial mats" and considered "thinly laminated, isopachous stromatolites" ... "to have a largely abiotic origin" (p. 1149). These insights benefited from decades of research that had helped elucidate the structure of present-day

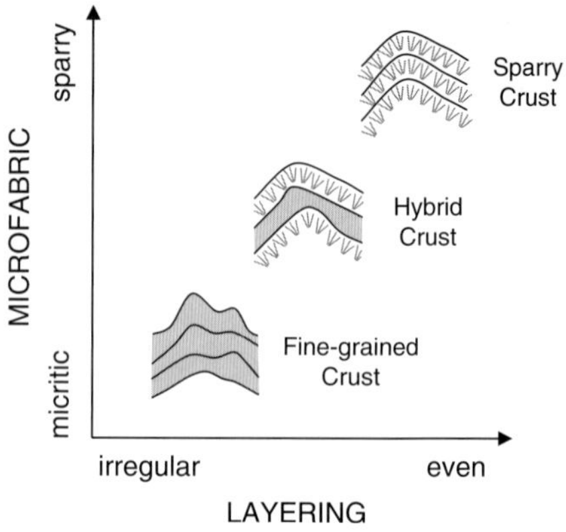

Fig. 17 Conceptual representation of Fine-grained, Hybrid and Sparry Crust, comparing their layering and microfabric. Fine-grained and Sparry Crust are interpreted as essentially microbial and abiogenic, respectively; Hybrid Crust is a mixture of the two. Fine-grained and Hybrid crust are here considered to be stromatolite; Sparry Crust is not. From Riding (2008, fig. 14)

sinters and lithified microbial mats, and of similar ancient examples that included abiogenic seafloor crusts that appeared to lack modern marine equivalents. Although these deposits as a whole are complex, it is possible to identify a few principal components. Two distinct fabrics are Fine-grained Crust with uneven layering, and Sparry Crust with regular layering, which can be interpreted as lithified microbial carbonate and abiogenic carbonate respectively (Riding 2008, pp. 77–80, p. 90). These two fabrics occur separately, but also in intimate association as Hybrid Crust (Fig. 17).

5.1 *Fine-Grained Biogenic Crust*

Fine-grained biogenic crust is characterized by fine-grained microfabric and irregular or poorly layered macrofabric. The microfabrics can be dense, clotted, peloidal, and/or filamentous (Riding 2000, figs. 5–7). They are dominated by micrite and microspar, and may contain fenestrae and allochthonous grains. Macroscopically, Fine-grained Crust stromatolites typically exhibit uneven to discontinuous – sometimes poorly defined – layers, usually with poor inheritance (Figs. 5 and 18). In some cases the incorporated grains are sand size, as in Shark Bay and Lee Stocking columns, but the matrix remains fine. In these cases, the larger grains can make the macrofabric correspondingly more uneven. Fine-grained crust stromatolites are widespread in the Neoproterozoic and Phanerozoic (Riding 2008, p. 73).

5.2 *Sparry Crust*

Sparry Crust typically has coarsely crystalline, often radial-fibrous, microfabric. The macrofabrics may consist of even, approaching isopachous, laterally persistent layers with good inheritance. There are diverse varieties, two of which are

Fig. 18 Small steep-sided stromatolite dome, showing uneven and discontinuous layering. Early Cambrian, Série Lie de Vin, Tiout, near Taroudant, Anti-Atlas Mountains, Morocco

microdigitate "tufa" (Hoffman 1975, p. 262; Grotzinger and Read 1983, p. 712, fig. 1f; Hofmann and Jackson 1987, p. 964) and isopachous laminite (Jackson 1989, figs 6, 13; Grotzinger and Knoll 1999, fig. 6a, b). Sparry Crust stromatolites were widespread in the Palaeoproterozoic and Mesoproterozoic, with microdigitate forms typically in shallower (Hoffman 1975, p. 262) and isopachous laminite in deeper water environments (Jackson 1989). In the Phanerozoic, Sparry Crust deposits locally occur in evaporite basins (Pope et al. 2000, p. 1139).

5.3 *Hybrid Crust*

Hybrid crust consists of interlayered Fine-grained and Sparry Crust (Riding 2008, p. 74). Detailed descriptions (e.g., Vologdin 1962; Komar et al. 1965; Hofmann 1969, fig. 13; Walter 1972, pls 5, 6, 10, 12; Bertrand-Sarfati et al. 1994; Sami and James 1996, p. 217; Petrov and Semikhatov 2001, fig. 6, p. 269) suggest that many Proterozoic stromatolites consist of intimate alternations of both microbial and Sparry Crust fabrics (Riding 2008, p. 95). For example, in large *Conophyton*, Kerans (1982) noted that "bladed cement crusts were precipitated on microbial laminae while stromatolites were growing" (see Grotzinger 1986b, p. 840). These do not appear to be minor occurrences. Examples with Hybrid-like "Boetsap lamination" are locally widespread in the Neoarchaean (Fig. 2). The subsequent very long interval represented by the Palaeo- and Mesoproterozoic,

~2,500–1,000 Ma ago, was the "Golden Age" of large and abundant stromatolites (Awramik and Riding 1986). An extensive literature (see references in Riding 2000) suggests that many of these Palaeo- and Mesoproterozoic stromatolites consist of Hybrid Crust and may exceed other coeval stromatolites in abundance (Riding 2008, p. 91). The often delicate alternations in Hybrid Crust (Petrov and Semikhatov 2001, fig. 6a; Riding 2008, fig. 9) may reflect seasonal changes (Bertrand-Sarfati 1972, p. 155). It could be that combination of microbial growth and abiogenic precipitation enhanced the growth rates of these deposits, some of which exhibit exceptional size and relief (Grotzinger and Knoll 1999, p. 352; Sumner and Grotzinger 2004, fig. 10; Riding 2008, pp. 91–92).

6 Stromatolite Definition

Stromatolites are regarded here as layered benthic microbial deposits (Kalkowsky 1908; Burne and Moore 1987; Riding 1999). The three types of authigenic carbonate crust outlined above can all form layered deposits (both biogenic and abiogenic) that have at various times been described as stromatolites (Riding 2008, fig. 12). Fine-grained Crust (typically unevenly layered and with complex clotted and peloidal microfabrics) and Hybrid Crust (thin alternations of Sparry and Fine-grained crust) are here regarded as essentially biogenic and therefore as stromatolites. Sparry Crust (typically with regular, even layering and sparry microfabric) is here regarded as essentially abiogenic and is not regarded as stromatolite.

Proposed definition: *Stromatolites are macroscopically layered authigenic microbial sediments with or without interlayered abiogenic precipitates.*

Stromatolites therefore include microbial and hybrid types; they can form domes and columns, but also commonly occur as sheet-like masses. In the definition, *with* abiogenic precipitates can refer to Hybrid Crust, and *without* abiogenic precipitates can refer to Fine-grained Crust. In contrast, thrombolites, dendrolites, and leiolites are categories of non-layered microbial deposits characterized by clotted, dendritic and aphanitic macrofabrics, respectively.

Definition of stromatolites as *microbial* is consistent with Kalkowsky's (1908) view which has been widely supported (e.g., Awramik and Margulis 1974; Walter 1976b, p. 1; Monty 1977, p. 23; Burne and Moore 1987; Riding 1999). It is also consistent with stromatolites being regarded as microbialite, as Burne and Moore (1987) intended, which has been widely adopted. Furthermore, this definition emphasizes genetic nature, which is a major interest in the study of stromatolites. Much of the interest in very old stromatolites, for example, centres on whether or not they are biogenic (e.g., Lowe 1994; Hofmann et al. 1999; Allwood et al. 2006). A definition that does not address this misses a crucial point. No abiogenic present-day deposits have generally been regarded as stromatolites, and although strong arguments have been expressed in favour of a descriptive definition that recognizes abiogenic as well as biogenic stromatolites (Logan et al. 1964; Hofmann 1973, p. 342, p. 346; Semikhatov et al. 1979) these appear to be essentially designed to

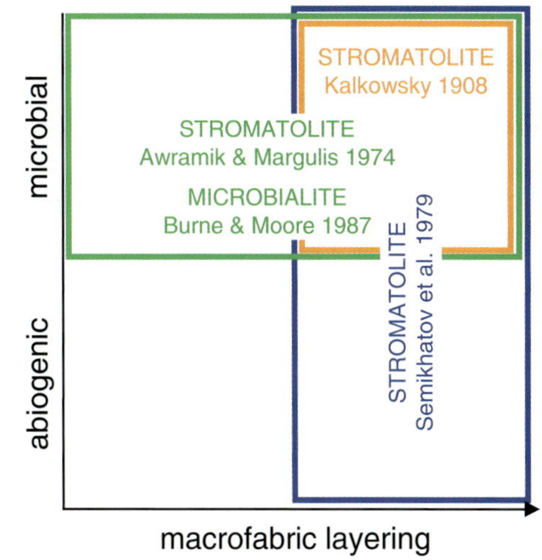

Fig. 19 Diagrammatic summary of contrasting stromatolite definitions, as they relate to degree of macrolayering and microbial/abiogenic origins. Circumscription of microbialite (Burne and Moore 1987) corresponds with Awramik and Margulis's (1974) definition of stromatolite. Kalkowsky's (1908) definition (supported by Burne and Moore 1987; Riding 1999) is supported here

avoid the difficulty of discriminating between abiogenic and biogenic examples in ancient deposits. Such a broad definition has the disadvantage that it could "cease to be useful" (Walter 1976b, p. 1). For example, "abiogenic stromatolites" could encompass deposits such as speleothem, travertine, and hot spring sinter.

Definition of stromatolites as *layered* is also consistent with Kalkowsky's (1908) definition, but is more restricted than either Awramik and Margulis's (1974) or Semikhatov et al.'s (1979) definitions which also permitted stromatolites to be unlayered or abiogenic respectively (Fig. 19).

Since this definition reaffirms that of Kalkowsky (1908), it raises the same question that has been at the centre of the stromatolite debate for over a century: if stromatolites are biogenic, how are ancient examples to be distinguished from abiogenic crusts? I propose to base this on the macrofabric and microfabric criteria outlined above (Stromatolite Fabrics). Macrofabric mainly involves evenness, regularity, and persistence of the layers. Microfabric ranges from sparry to fine-grained (clotted and peloidal) and filamentous. Hybrid Crust consists of thin alternations of Sparry and Fine-grained crust. Application of these criteria, especially microfabric, necessarily relies on good preservation. The extent to which this proposal will be useful remains to be seen.

7 Discussion

Timelines show how concept development tracked fieldwork and research progress discoveries among ancient examples and potential present-day analogues (Fig. 20). In this context, the "definition difficulty" that has hindered stromatolite studies reflects several factors. One of these is the availability of present-day analogues. Stromatolite studies could not have progressed without the comparative

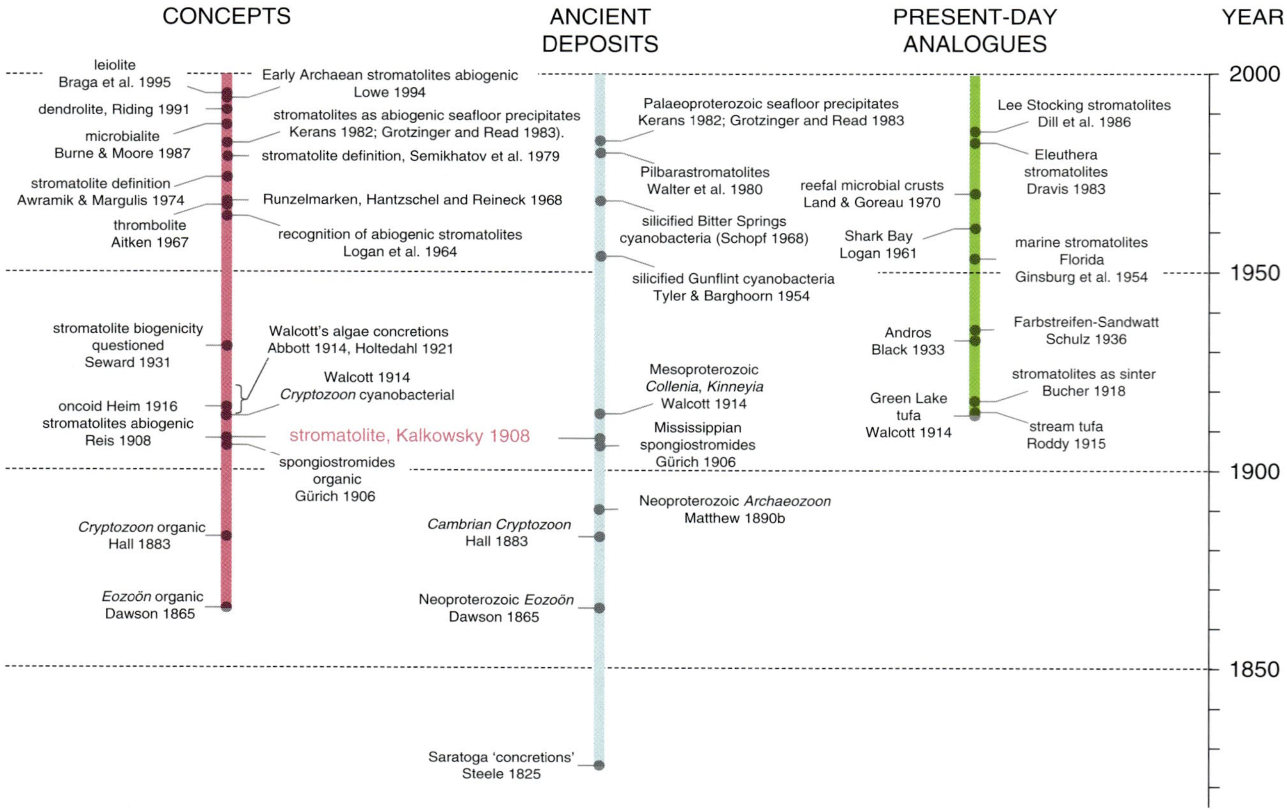

Fig. 20 Time-lines of stromatolite concept development and related ancient and present-day deposits mentioned in the text, from 1825 to 2000, beginning with recognition of *Cryptozoon* and *Eozoön*

information provided by the wide variety of present-day abiogenic and microbial mat deposits briefly summarized above. But despite this extensive research, efforts to base stromatolite interpretation entirely on present-day analogues have not been altogether successful. This is because not all ancient stromatolite-like deposits are well-represented today; most notably the abiogenic seafloor crusts first clearly recognized in the Proterozoic by Kerans (1982), Grotzinger and Read (1983), and Grotzinger (1989a, b) and others. Stromatolites have had such a long geological history, influenced by major changes in both biology and seawater chemistry, that finding good examples of all types of stromatolite today might hardly be expected. Another factor is the geological origin of the stromatolite concept, rooted in specimens that are hundreds of million years old (Kalkowsky 1908; Hall 1883). Stromatolites have been reported to be most abundant 2,000 Ma ago (Grotzinger 1990), and the term has been widely applied to specimens as old as 3,500 Ma (Allwood et al. 2006). Because stromatolites were first described and defined in the rock record, their genesis was not immediately demonstrable and demanded resourceful interpretation. This question of origins has become the motif of stromatolite studies, focused on the difficulty of distinguishing abiogenic crusts from microbial carbonates.

The paradox presented by stromatolites is conveyed by Ginsburg's (1991, p. 25) impish comment that "few observers have any difficulty identifying archetypical stromatolites . . . yet defining stromatolites is controversial". This is like saying that everyone knows what stromatolites look like, but no one can agree what they are. It was this daunting prospect, which threatened to leave stromatolite definition in a state of perpetual limbo, that led Semikhatov et al. (1979) to propose a descriptive definition. Yet, broad definitions encompassing both biogenic and abiogenic structures avoid the central question of how stromatolites have formed (Walter 1976b, p. 1). This brings the debate full circle, back to trying to find ways to make a microbial definition work (Burne and Moore 1987).

Nonetheless, there are clear pointers to be drawn from the considerable advances derived from studies of present-day deposits. One is that the term stromatolite is only very seldom applied to present-day deposits known to be inorganic (an exception is Maliva et al.'s (2000, p. 934) description of modern stromatolites "formed largely by inorganic precipitation"). This suggests that researchers have tended to view stromatolites as essentially biogenic, and that the only reason for grouping abiogenic and microbial crusts in ancient deposits is inability to distinguish them (Semikhatov et al. 1979). Another is that there are distinct fabric differences between abiogenic and microbial crusts, as was recognized very early (Bucher 1913). The pace of development of limestone petrography accelerated in the 1960s, but focused more on young and grainy carbonates than on old and authigenic ones. But substantial progress has been made in distinguishing laminated authigenic crusts of microbial and abiogenic origin. The argument here is that fabric criteria can resolve these questions, and that it is important to develop them inorder to overcome issues that impede stromatolite studies. It should be possible not only to formulate a microbial definition of stromatolite, but also to confidently apply it sufficiently widely in the geological record.

7.1 Abiogenic and/or Biogenic Stromatolites?

Despite all the difficulties, in many respects the sometimes contentious half-century of debate vindicated the views of both sets of early contenders; Kalkowsky (1908) and Walcott (1914), on the one hand, and of Reis (1908) and Bucher (1913) on the other. By the 1960s there was growing acknowledgement of the existence of both organic and inorganic stromatolites. A decade or more prior to Awramik and Margulis (1974), Logan et al. (1964, p. 68) succinctly stated: "Stromatolites are laminated structures that have been previously termed fossil algae. It is now recognized that such structures may be formed by a number of different processes and organisms." They did not offer a general definition of stromatolite, but their recommendation was clear: "To be useful, the term stromatolite should be preceded whenever possible by an adjective signifying the kind of stromatolite under consideration, for example, algal stromatolite, foraminiferal stromatolite, inorganic stromatolite, and so forth" (Logan et al. 1964, p. 69). This was echoed by Hofmann (1973), who emphasized that stromatolites need not be biogenic (p. 342) and recognized chemogenic stromatolites (p. 346) and the need to distinguish "biogenic stromatolites from chemical and mechanical ones" (p. 350).

These calls by well-known researchers did not lead to peaceful reform of stromatolite definition. Instead they precipitated a schism between those who regarded stromatolites as essentially biogenic, and those who believed it was necessary for the term stromatolite to encompass both biogenic and abiogenic laminated structures. In addition to discovery of cyanobacteria in ~1880 Gunflint stromatolites (Barghoorn and Tyler 1965) (Fig. 7), similar – and impressive – fossils had been found in the ~850 Ma Bitter Springs silicified stromatolitic mats (Schopf 1968; Knoll and Golubic 1979), emphasizing the biogenicity of Proterozoic stromatolites. Against this background, two definitions were published expressing these contrary convictions. Awramik and Margulis (1974) adhered to Kalkowsky's (1908) view that stromatolites are microbial, whereas Semikhatov et al. (1979) followed Logan et al.'s (1964) lead and recognized both biogenic and abiogenic stromatolites (Fig. 19).

Awramik and Margulis (1974) defined stromatolites as "megascopic organosedimentary structures produced by sediment trapping, binding and/or precipitation as a result of growth and metabolic activity of organisms, primarily blue-green algae". This genetic approach was endorsed in the introduction to Walter's (1976b, p. 1) seminal stromatolite volume. Its advantage was its straightforward restatement of Kalkowsky's (1908) microbial view. But it required biogenicity to be demonstrated, and this was difficult for ancient examples. Hoffman's (1973) had frankly admitted that "Something that haunts geologists working on ancient stromatolites is the thought that they might not be biogenic at all". This question of confidently establishing biogenicity in ancient examples is the enduring central problem of stromatolite studies. Reis (1908) used it to challenge Kalkowsky (1908). It was clearly expressed by Seward (1931, pp. 83–89) and Logan et al. (1964), and preoccupied Semikhatov et al. (1979), Ginsburg (1991, pp. 25–27), Grotzinger and Knoll (1999), and Brasier et al. (2006, p. 894). It led Buick et al. (1981) to suggest that "structures of uncertain origin that resemble stromatolites should be called 'stromatoloids'" and Awramik and Grey (2005) to term abiogenic stromatolite-like structures *pseudostromatolites*.

Semikhatov et al. (1979, p. 992) (but with S.M. Awramik dissenting), proposed to define stromatolites as "laminated, lithified, sedimentary growth structures that accrete away from a point or limited surface of attachment. They are commonly, but not necessarily, of microbial origin and calcareous composition". The operative words here are "commonly, but not necessarily", that permit some stromatolites to be abiogenic. They supported this descriptive, rather than genetic, approach to stromatolite definition by stating that "it was the lamination that Kalkowsky stressed, not its origin" (Semikhatov et al. 1979, p. 994). This was misleading; Kalkowsky (1908) emphasized organic origin as well as layering in describing stromatolites. But Semikhatov et al. (1979) were correct to point out the difficulties of confidently applying a genetic definition: "if a biological origin had to be demonstrated before a geological object could be called a stromatolite the term would in most instances be inapplicable (or at best provisional)" (Semikhatov et al. 1979, p. 994). Their approach was bold and realistic, a clear departure from Kalkowsky (1908), and an attempt to express a dual view of stromatolites in which many were microbial, but some could be abiogenic. It aroused criticism for exactly this reason. It appeared to be an all-encompassing definition that failed to either readily separate different structures or to circumscribe related ones. Walter (1976b, p. 1) had already perceptively foreseen that if non-genetic definitions "are so broad as to include a wide-range of non-biogenic structures", "the term would cease to be useful". Subsequently, Ginsburg's (1991, p. 27) opinion was that the Semikhatov et al.'s (1979) definition "includes structures of a variety of origins ranging from tufa domes ... to laminated structures of mineralized organisms ... and even some of the laminated zones of caliches and calcretes as well as certain speleothems." It could perhaps even be considered to include diagenetic concretions. A disadvantage with its specific wording is that it appears designed to exclude laterally extensive deposits such as Aitken's (1967) cryptalgalaminates. But the main difficulty can be summed up by the question, if stromatolites are "laminated, lithified, sedimentary growth structures that accrete away from a point or limited surface of attachment" (Semikhatov et al. 1979), then what exactly is *not* a stromatolite?

7.2 Microbialite

The name stromatolite purposely emphasized layered structure; *Stroma* and *stromat* – indicate a bed coverlet in Greek and Latin. Kalkowsky (1908, p. 102) stated: "*Alle Stromatolithe zeigen im vertikalen Schnitt deutliche Lagenstruktur*" "All stromatolites show distinct layering in vertical section". As geological studies progressed it eventually became clear that there are deposits that resemble stromatolites in external form but which appear to lack internal layering. Aitken (1967, p. 1164) introduced *thrombolite* "for cryptalgal structures related to stromatolites, but lacking lamination and characterized by a macroscopic clotted fabric" to describe structures that were common in the Late Cambrian and Early Ordovician. During the 1970s, growing awareness of the local importance of thrombolites led to doubts about the value of requiring layering as an integral part of stromatolite description. For example, Awramik and Margulis (1974) defined stromatolites as "megascopic organosedimentary structures produced by sediment trapping, binding and/or precipitation as a result of growth and metabolic activity of organisms, primarily blue-green algae". This required stromatolites to be microbial, but not necessarily layered, and therefore permitted thrombolite to be regarded as a type of stromatolite. But this left no specific term for laminated stromatolites. This deficiency was resolved when Burne and Moore (1987) applied the essence of Awramik and Margulis' (1974) definition of stromatolite to a new term: *microbialite*. "Microbialites are organosedimentary deposits that have accreted as a result of a benthic microbial community trapping and binding detrital sediment and/or forming the locus of mineral precipitation" (Burne and Moore 1987, pp. 241–242). This then allowed stromatolites to be regarded as macro-laminated microbialites, and thrombolites as macro-clotted microbialites. It also encouraged subsequent additions to the microbialite family, such as dendrolite (dendritic; Riding 1991, p. 34) and leiolite (aphanitic; Braga et al. 1995, p. 347). Since then, macrofabric has been a fundamental descriptor for these structures and stromatolites (Riding 2000, pp. 189–195).

Burne and Moore (1987) carefully evaluated the processes that might contribute to the formation of microbialites and associated deposits. As Hofmann (1973, fig. 5) and Riding (1977, fig. 1) had done, they noted the importance of sediment trapping, inorganic calcification and biologically influenced calcification (Burne and Moore 1987, figs. 1 and 2, pp. 243–244). Accordingly, they recognized three types of microbialite to reflect these processes respectively: microbial boundstone, tufa and framestone (Burne and Moore 1987, pp. 242–243). Microbialite was therefore designed as a broad term to encompass deposits formed by both grain trapping and mineral precipitation associated with benthic microbes, and this specifically included "microbial tufa – formed when microorganic material is incorporated during inorganic precipitation of carbonate" (p. 243). In detail, this could be considered to differ from Awramik and Margulis (1974) who defined stromatolites as "megascopic organosedimentary structures produced by sediment trapping, binding and/or precipitation as a result of growth and metabolic activity of organisms, primarily blue-green algae". The reasoning is that "precipitation as a result of growth and metabolic activity of organisms" emphasizes organic processes whereas "incorporated during inorganic precipitation of carbonate" emphasizes inorganic processes. However, these nuances are not evident in Burne and Moore's (1987, pp. 241–242) actual definition of microbialite, "organosedimentary deposits that have accreted as a result of a benthic microbial community trapping and binding detrital sediment and/or forming the locus of mineral precipitation", which is closer to that of Awramik and Margulis (1974). Furthermore, Burne and Moore (1987,

p. 249) advocated "a return to Kalkowsky's original meaning for 'stromatolite'" and used it "to refer to one possible internal structure of a microbialite".

The point of this rather protracted discussion is that – whatever the intention – in practice microbialite provided a necessary umbrella term for stromatolites, thrombolites and similar deposits and it became widely used. As the name obviously indicates, all these deposits are considered microbial. Thus, the train of events that led Burne and Moore (1987) to base microbialite on Awramik and Margulis's (1974) definition of stromatolite, strengthened the view that stromatolites are essentially microbial. This in turn tended to undermine the conviction, expressed by Logan et al. (1964) and Hofmann (1973) Semikhatov et al. (1979), that it was necessary to also recognize abiogenic stromatolites. But not everything was going in the direction of a microbial definition. Already, in the 1980s, studies of Proterozoic carbonates were interpreting some stromatolites as abiogenic seafloor crusts. This looked set to rebalance the debate, because it presented stromatolites as genetically heterogeneous structures.

7.3 *Seafloor Crusts*

The realization that some Proterozoic stromatolites are essentially abiogenic seafloor precipitates, set in train by Kerans (1982) and Grotzinger and Read (1983), was a break-through; arguably comparable in significance to the discovery of Shark Bay stromatolites 20 years previously. It revealed fundamental differences between Precambrian and present-day carbonate sedimentation (Grotzinger 1990; Grotzinger and Kasting 1993), and it provided good reason to recognize abiogenic stromatolites. But appreciation of its importance took time to spread. One reason for this may have been a lack of terminological clarity. Grotzinger and Read (1983, p. 712, fig. 1f) described the components of Rocknest microdigitate stromatolites as "cement laminae" and "cement crusts". Use of "seafloor cement" to describe microdigitate stromatolites and isopachous Sparry Crust continued for more than a decade (Kah and Knoll 1996, p. 79; Pope et al. 2000, p. 1145), even after Grotzinger and Knoll (1995, p. 579) pointed out that seafloor crusts/encrustations should be distinguished from "true cements which bind sediment particles and line voids". There was also an element of restraint. Grotzinger described microdigitate stromatolites as "in essence, evaporites" (Grotzinger 1986b, p. 842) but cautiously interpreted them as "microbially influenced inorganic calcification (although it is possible that they are entirely abiotic in origin)" (Grotzinger 1986a). Whatever the reasons, it was 15 years before more outspoken statements were expressed about the origins of these seafloor crusts. At the end of the century, Grotzinger and James (2000a, p. 7) included microdigitate stromatolites and isopachous millimetric laminites (i.e., Isopachous Sparry Crust) in their summary of Precambrian marine "abiotic precipitates", and Pope et al. (2000, p. 1149) considered "thinly laminated isopachous stromatolites" to be "largely abiotic". Thus, this research revealed a wide

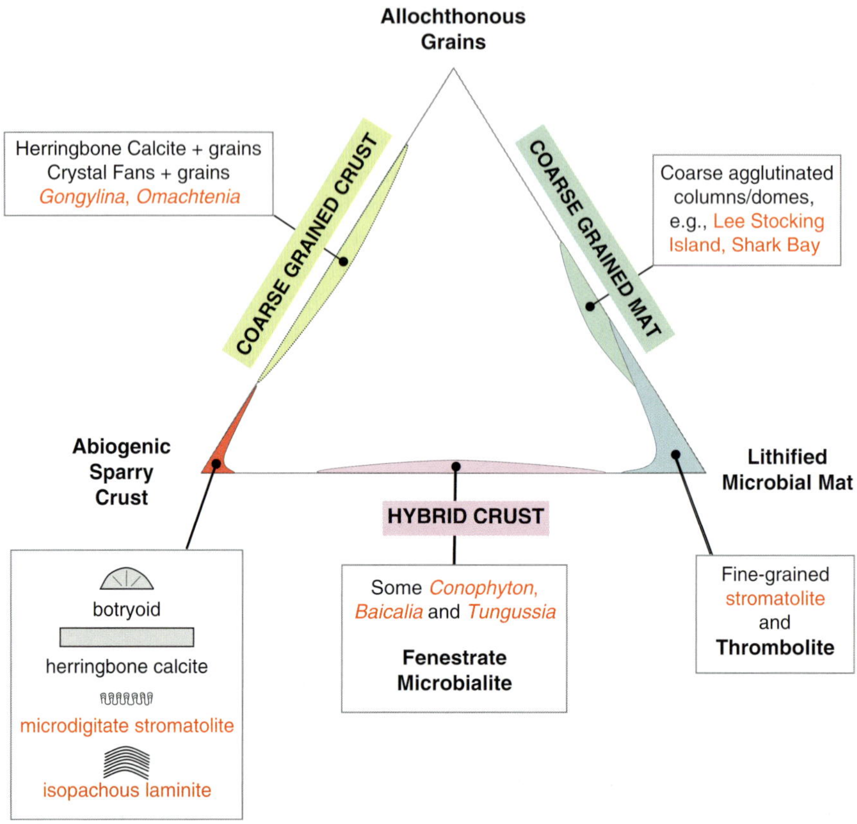

Fig. 21 Interpretive summary of Precambrian authigenic crusts, with layered deposits (both biogenic and abiogenic) that have been described as stromatolites indicated in *red*. Principal components are Sparry Crust (essentially abiogenic precipitate), Fine-grained Crust (lithified microbial mat), and allochthonous grains. Intermediate categories are Hybrid Crust, Coarse Grained Crust, and Coarse Grained Mat. From Riding (2008, fig. 12)

range of authigenic seafloor crusts (e.g., Sumner and Grotzinger 2004), and layered varieties have often, at one time or another, been termed stromatolite (Fig. 21).

7.4 Biogenicity Criteria

Recognition of abiogenic seafloor crusts reinforced the key question first pinpointed by Reis (1908): if stromatolites are biogenic, as Kalkowsky (1908) believed, then how are ancient examples to be distinguished from abiogenic crusts and, specifically, what criteria can be used to confidently establish stromatolite biogenicity. Progress in addressing these questions had been fitful. Following studies of microbial fabrics and fossils in stromatolites (Gürich 1906; Garwood 1913, Pia

1927, pp. 36–40) was able to distinguish filamentous (porostromate) and clotted (spongiostromate) microfabrics, and Black (1933) began to relate microbes to sedimentary structures in present-day stromatolites. But research on these two fronts advanced at different rates and cross-fertilization tended to be limited. For example, Johnson (1946) documented the microfabrics of Late Palaeozoic oncoids and *Cryptozoon* long before Shark Bay columns were discovered and described as *Cryptozoon* by Logan (1961). Considerable efforts were made to relate microbes to microfabric development in the 1970s (e.g., Golubic 1976; Monty 1976) and to document abiogenic fabrics (e.g., Gebelein 1974; Walter 1976c) and Semikhatov et al. (1979, pp. 1004–1005) were able to summarize microbial effects on carbonate grain/crystal relationships, lamina thickness and relief, and early diagenesis. But applying these to the interpretation of ancient examples was not straightforward. Buick et al. (1981, pp. 165–167) suggested eight biogenicity criteria for ancient stromatolites. Yet Lowe (1994, p. 389) was able to argue that five of these "are common not only to three-dimensional stromatolites but also to many if not most inorganic precipitates" and that the remaining three features "characterize <5% of stromatolites of any age". These final three all involve the presence of microfossils. By the early 1990s this approach showed surprisingly little advance on the work of Bucher (1913). In the Archaean, for example, Buick (1992, p. 255) used complex branching, tufted microfabric, and trapped grains to argue that Tumbiana stromatolites "are clearly biogenic", and Lowe (1994, p. 388) used fine, smooth, continuous lamination to question a biological origin for Barberton stromatolites.

It was at this stage that petrographic studies began to add important new microfabric detail to augment biogenicity criteria. Well-preserved Proterozoic microfabrics (e.g., Sami and James 1996; Knoll and Semikhatov 1998) allowed Grotzinger and Knoll (1999, fig. 3, pp. 320–323) to distinguish microbial mat from sea-floor precipitated crusts. In spite of this, Grotzinger and Knoll (1999, p. 316) were pessimistic about applying a genetic definition of stromatolite. Their focus was on numerical simulations that could reflect stromatolite growth processes, but they recognized the difficulty of distinguishing between mat growth and "the growth of abiotic marine crusts", and concluded with Grotzinger and Rothman (1996) that "morphology may well be a non-unique parameter" (Grotzinger and Knoll 1999, p. 343). Nonetheless, their insights into fabrics, together with those of Pope et al. (2000), put biogenicity into a broader perspective: the requirement now was for criteria not simply to establish the nature of biogenic deposits, but of abiogenic deposits too. This was advanced by articles in Grotzinger and James (2000b). Riding (2000, pp. 186–188) summarized the peloidal, clotted and filamentous fabrics of Phanerozoic microbial carbonates, and Corsetti and Storrie-Lombardi (2003, p. 652, fig. 1c–f) emphasized isopachous lamination and radiating crystal fan fabrics as indications of abiotic precipitation in non-marine stromatolites. At this point, criteria were available to distinguish two distinct types of layered authigenic crust: essentially microbial and essentially abiogenic (Pope et al. 2000; Perry et al. 2007, p. 169).

The presence of Hybrid Crusts complicates the otherwise simple view of stromatolites based on two basic fabric types that do not generally co-occur. Hybrid

Crusts are composed of intimate alternations of Fine-grained and Sparry Crust and are therefore partly biogenic and partly abiogenic. But here again, petrofabric studies enable these to be deciphered. Already, Bertrand-Sarfati (1972, pl. 11(4), pl. 22(2)) had suggested that layered alternations in stromatolites might reflect seasonal changes in accretion, and Kerans (1982) (see Grotzinger 1989b, p. 10) had pointed out that "cement crusts were precipitated on microbial laminae while stromatolites were growing". Locally, Hybrid Crusts appear to be extensively developed in Proterozoic stromatolite reefs such as ~1,000 Ma *Baicalia* reefs in Siberia (Petrov and Semikhatov 2001) (see Riding 2008, p. 82). Similarly, in the Pethei Group (~1,850 Ma) of northern Canada, Sami and James (1994, p. 120) suggested that millimetric spar-micrite couplets reflect alternation of "cement precipitation and microbial mat growth".

7.5 Fabric Criteria Through the Geological Record

Macro- and microfabrics are the keys to deciphering stromatolites (Riding 2000, p. 206). The three principal fabric types are interpreted to be essentially biogenic, mixed and abiogenic (Fig. 22), and have created a variety of layered and unlayered deposits (Fig. 23). These criteria, proposed here to discriminate between these diverse deposits, build fundamentally on the observations first emphasized by Reis (1908) and subsequent researchers when they compared the delicate layering and radial structures of sinters with the coarse spongy fabrics of microbial deposits (Bucher 1913, 1918; Bradley 1928). These have since been examined in detail. See, for example, the following publications and their contained references on

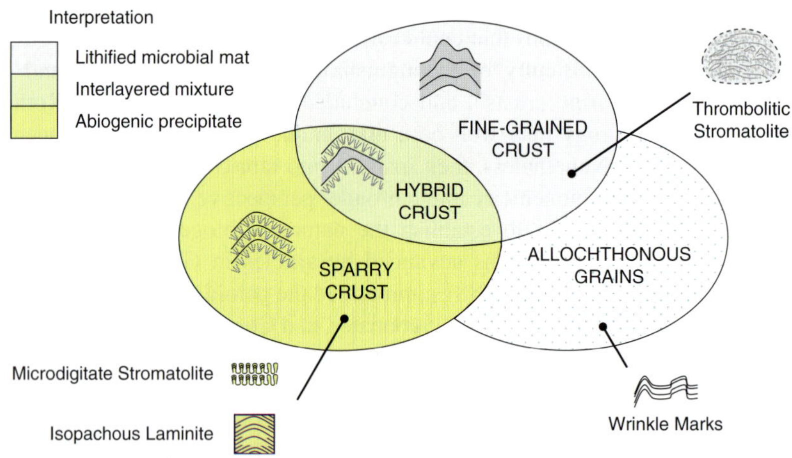

Fig. 22 Representation to show how Sparry and Fine-grained Crust, as well as incorporated allochthonous grains, contribute to the formation of Hybrid Crust, thrombolitic stromatolite, other macrolayered authigenic carbonate crusts, and wrinkle marks

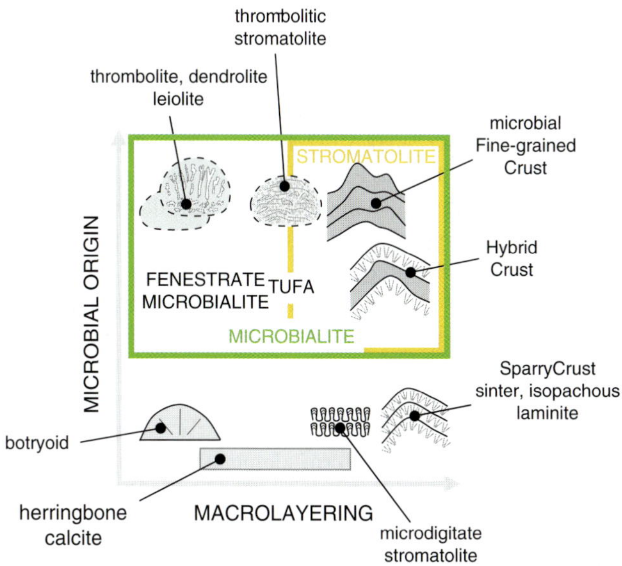

Fig. 23 Major categories of authigenic carbonate crust deposits compared with regard to degree of microbial origin and macrolayering. Principal microbial groupings are stromatolite [as defined by Kalkowsky (1908), and also here] and thrombolite (Aitken 1967), together with related non-layered microbial carbonates (dendrolite, leiolite). These are circumscribed by microbialite (Burne and Moore 1987). Examples of abiogenic crust, shown in the lower part of the diagram, include botryoids, herringbone calcite, microdigitate stromatolites and Sparry Crust deposits [for summary details see Sumner and Grotzinger (2000, 2004) and Riding (2008)]

Phanerozoic (e.g., Monty 1981; Bertrand-Sarfati and Monty 1994) and Precambrian (Grotzinger and Knoll 1999; Grotzinger and James 2000b; Pope et al. 2000; Sumner and Grotzinger 2004; Riding 2008) marine stromatolites and crusts, present-day microbial mats (e.g., Monty 1976; Arp et al. 2003, 2010; Dupraz et al. 2009) and hot spring carbonates (Pentecost 2005). Sparry Crust, Fine-grained Crust, and Hybrid Crust are recognizable in sufficiently well-preserved deposits. Proterozoic and especially Archaean deposits are especially challenging due to often poor preservation, but in these too fabric details allow cautious interpretation:

Sparry Crust may be present in the Palaeoarchaean, e.g., in the Warrawoona Group (~3,450 Ma) of north-western Australia, in stromatolites exhibiting continuous laminae (Lowe 1980, 1983; Hofmann 2000) that have sparry microfabrics (Hofmann et al. 1999, fig. 3), but these could well be secondary in origin (Hofmann et al. 1999, p. 1259). Thinly layered isopachous Sparry Crust appears to be present as "crinkly laminite facies" in the ~2,600 Ma Cheshire Formation of Zimbabwe (Sumner and Grotzinger 2000, p. 128). It is definitely present in the ~1,800 Ma Pethei Group of north-west Canada (Jackson 1989; Pope and Grotzinger 2000, p. 112). Isopachous Sparry Crust remained common in marine environments until the late Mesoproterozoic (Kah and Knoll 1996, fig. 4; see Riding 2008, fig. 5). It is associated with Hybrid Crust into the Neoproterozoic (see Hybrid Crust, below), and is probably present in evaporite basin sequences during the Phanerozoic (Pope et al. 2000).

Fine-grained Crust appears to be present in 2,724 Ma Tumbiana stromatolites from north-western Australia (Buick 1992, fig. 3e; Lepot et al. 2008, fig. 1b), and is definitely present in the ~1,800 Ma Pethei Group as peloidal clotted micrite (Sami and James 1996, fig. 7). It remains common in present-day marine stromatolites.

Hybrid Crust may be present as "Boetsap lamination" in the latest Archaean (2,550 Ma) Campbellrand-Malmani platform of South Africa (Sumner and Grotzinger 2004; see Riding 2008, p. 84), and again is definitely present as "spar-micrite couplets" in the ~1,800 Ma Pethei Group (Sami and James 1996, p. 217). Hybrid Crust remained common in marine environments to the latest Mesoproterozoic, e.g., in *Baicalia lacera* of the ~1,020 Ma Burovaya Fm of Turukhansk, Siberia (Petrov and Semikhatov 2001, fig. 6a). In the early Neoproterozoic it may be represented by some of the "lamelliform elements" described by Aitken (1989, pp. 15–16) in the ~800 Ma Little Dal reef (see Turner et al. 2000, fig. 10b).

This documentation needs to be extensively developed, fabric criteria require detailed refinement, and many aspects of biogenic and abiogenic influences on mineral precipitation in stromatolites require elucidation (e.g. Arp et al. 2010). These will also benefit by being augmented by analytical and modelling approaches to stromatolite macrofabric that build on previous work (e.g., Grotzinger and Rothman 1996; Batchelor et al. 2000, 2003; Corsetti and Storrie-Lombardi 2003; Storrie-Lombardi et al. 2004; Dupraz et al. 2006). In addition to providing important new insights into long-lived and wide-ranging deposits, fabric recognition provides the keys required to stabilize stromatolite definition.

8 Conclusions

1. Kalkowsky's (1908) microbial interpretation of ancient stromatolites was at first vigorously challenged due to their resemblance to hot spring crusts and other present-day laminated precipitates that were regarded as essentially abiogenic, and due to confusion between stromatolites and diagenetic concretions. During the following decades, criticism diminished as microbial stromatolites were recognized in present-day lake, marsh and shallow-marine environments. Logan et al. (1964) suggested recognition of both microbial and abiogenic stromatolites. This was opposed by support for the original microbial view. Awramik and Margulis (1974) proposed a modified definition that required stromatolites to be microbial, but not necessarily thinly layered. When Burne and Moore (1987) used this concept to define microbialite, stromatolites became a sub-group within microbialites, and reverted to being laminated microbial deposits. Meanwhile, Semikhatov et al. (1979) had proposed a descriptive definition of stromatolite that included both microbial and abiogenic deposits,

so long as they were thinly layered. This history of concept development resulted in at least three alternative views of stromatolites: (a) laminated and microbial (Kalkowsky 1908; Burne and Moore 1987; Riding 1999); (b) just microbial (Awramik and Margulis 1974); (c) laminated and microbial or abiogenic (Semikhatov et al. 1979). None of these approaches has gained unanimous support and there is currently no generally accepted definition of stromatolite.

2. Problems of stromatolite definition arise from the difficulties of distinguishing laminated authigenic crusts of microbial and abiogenic origin in ancient examples. Similarities in layering and external morphology between these deposits raise doubts about their interpretation, and confuse stromatolite definition by encouraging inclusion of abiogenic as well as microbial structures. Present-day crusts that have been regarded as essentially abiogenic, e.g., in cave and hot-spring sinters, have not normally been regarded as stromatolites. The only justification for grouping abiogenic and microbial crusts is the difficulty of distinguishing them in ancient deposits.
3. The geological record of layered marine authigenic carbonate crusts includes three distinctive types of deposit that have all been regarded as stromatolites: (a) Isopachous Sparry Crust – abiotically precipitated well-layered crust with sparry microfabric; (b) Fine-grained Crust – lithified irregularly, sometimes poorly, layered microbial carbonate, generally with fine-grained, (e.g., clotted, peloidal) but also sometimes coarse-grained microfabrics; (c) Hybrid Crust – thinly interlayered combinations of both preceding fabrics, with generally well-developed layering. Further work is required, but these fabric criteria can be used to broadly distinguish essentially abiogenic from essentially microbial crusts (Pope et al. 2000), provided that these are sufficiently well preserved.
4. The definition proposed here is: *Stromatolites are macroscopically layered authigenic microbial sediments with or without interlayered abiogenic precipitates.* This recognizes at least two types of stromatolite: (a) microbial, characterized by Fine-grained, locally coarse, Crust and (b) hybrid, composed of Hybrid Crust (i.e., alternations of Fine-grained and Sparry Crust). In contrast, thrombolite, dendrolite and leiolite are microbial carbonates distinguished from each other, and from stromatolites, by clotted, dendritic and aphanitic macrofabrics, respectively.
5. Thus, it is possible to use carbonate fabric criteria to distinguish essentially abiogenic crusts from essentially microbial ones. This ability permits practical application of a microbial definition of stromatolite, as Kalkowsky (1908) intended.

Acknowledgements I am grateful to Joachim Reitner and the Organizing Committee for supporting my participation in the 2008 Kalkowsky Symposium. Brian Chatterton and Eric Mountjoy generously provided photographs of *Eozoön* and Shark Bay columns, respectively, and Jody Webster a thin-section of reef crust from Kohala, Hawaii. Canadian Society of Petroleum Geologists kindly gave permission to reproduce Fig. 15. I am indebted to Phil Fralick, Dawn Sumner and Pieter Visscher, respectively, for showing me the Gunflint, Campbellrand-Malmani, and Lagoa Pitanguinha localities. Gernot Arp and Joachim Reitner made helpful suggestions that improved the final manuscript.

References

Abbott G (1914) Is '*Atikokania lawsoni*' a concretion? Nature 94: 477–478

Adelman J (2007) *Eozoön*: debunking the dawn animal. Endeavour 31: 94–98

Aitken JD (1967) Classification and environmental significance of cryptalgal limestones and dolomites, with illustrations from the Cambrian and Ordovician of southwestern Alberta. Journal of Sedimentary Petrology 37: 1163–1178

Aitken JD (1989) Giant "algal" reefs, Middle/Upper Proterozoic Little Dal Group (>770, <1200 Ma), Mackenzie Mountains, N.W.T., Canada. In: Geldsetzer HHJ, James NP, Tebbutt GE (eds) Reefs, Canada and Adjacent Area. Canadian Society of Petroleum Geologists, Memoir 13: 13–23

Allwood AC, Walter MR, Kamber BS, Marshall CP, Burch IW (2006) Stromatolite reef from the Early Archaean era of Australia. Nature 441: 714–718

Arp G, Reimer A, Reitner J (2003) Microbialite formation in seawater of increased alkalinity, Satonda Crater Lake, Indonesia. Journal of Sedimentary Research 73: 105–127

Arp G, Bissett A, Brinkmann N, Cousin S, de Beer D, Friedl T, Mohr KI, Neu TR, Reimer A, Shiraishi F, Stackebrandt E, Zippel B (2010) Tufa-forming biofilms of German karstwater streams: microorganisms, exopolymers, hydrochemistry and calcification. In: Pedley M, Rogerson M (eds) Tufas and Speleothems. Geological Society of London Special Publication 336: 83–118

Awramik SM, Grey K (2005) Stromatolites: biogenicity, biosignatures, and bioconfusion. Proceedings of SPIE 5906: 5906P-1-5906P-9

Awramik SM, Margulis L (1974) Stromatolite Newsletter 2: 5

Awramik M, Riding R (1986) "Golden Age" stromatolites and modern analogs. Society of Economic Paleontologists and Mineralogists Annual Midyear Meeting, Raleigh, North Carolina, September 1986, Abstracts III: 3–4

Awramik SM, Riding R (1988) Role of algal eukaryotes in subtidal columnar stromatolite formation. Proceedings of the National Academy of Sciences of the United States of America 85: 1327–1329

Bailey L, Matthew GF (1872) Preliminary report on the geology of New Brunswick. Geological Survey of Canada, Reports of Progress 1870–1871, 15–57

Barghoorn ES, Tyler SA (1965) Microorganisms from the Gunflint Chert. Science 147: 563–575

Batchelor MT, Burne RV, Henry BI, Watt SD (2000) Deterministic KPZ model for stromatolite laminae. Physica A 282: 123–136

Batchelor MT, Burne RV, Henry BI, Watt SD (2003) Mathematical and image analysis of stromatolite morphogenesis. Mathematical Geology 35: 789–803

Bell R (1870) Report on lakes Superior and Nipigon. Geological Survey of Canada, Reports of Progress 1866–1869, 313–364

Bertrand-Sarfati J (1972) Stromatolites columnaires du Pré-cambrien supérieur du Sahara Nord-Occidental. CNRS, Paris, Centre de Recherches sur les Zones Arides, Géologie: 14, xxxvii+ 245 pp

Bertrand-Sarfati J, Monty C (eds) (1994) Phanerozoic Stromatolites II. Kluwer, Dordrecht, 471 pp

Bertrand-Sarfati J, Freytet P, Plaziat JC (1994) Microstructures in Tertiary nonmarine stromatolites (France). Comparison with Proterozoic. In: Bertrand-Sarfati J, Monty C (eds) Phanerozoic Stromatolites II. Kluwer, Dordrecht, pp 155–191

Black M (1933) The algal sedimentation of Andros Island Bahamas. Philosophical Transactions of the Royal Society (London) Series B: Biological Science 222: 165–192

Bloos G (1976) Untersuchungen über Bau und Entstehung der feinkörnigen Sandsteine des Schwarzen Jura (α) (Hettangium und tiefstes Sinemurium) im schwäbischen Sedimentationsbereich. Arbeiten aus dem Institut für Geologie und Paläontologie an der Universität Stuttgart 71: 1–277

Brachert TC, Dullo W-C (1991) Laminar crusts and associated foreslope processes, Red Sea. Journal of Sedimentary Research 61: 354–363

Bradley WH (1928) Algae reefs and oolites of the Green River Formation. US Geological Survey Professional Paper 154: 203–233
Braga JC, Martín JM, Riding R (1995) Controls on microbial dome fabric development along a carbonate-siliciclastic shelf-basin transect, Miocene, SE Spain. Palaios 10: 347–361
Brasier M, McLoughlin N, Green O, Wacey D (2006) A fresh look at the fossil evidence for early Archaean cellular life. Philosophical Transactions of the Royal Society (London) B Biological Sciences 361: 887–902
Browne KM, Golubic S, Seong-Joo L (2000) Shallow marine microbial carbonate deposits. In: Riding R, Awramik SM (eds) Microbial Sediments. Springer, Berlin, pp 233–249
Bucher WH (1913) Über einige Fossilien und über Stromatolithbildung im Tertiär der bayerischen Rheinpfalz. München Geognostische Jahreshefte, Jahrgang 26: 76–102
Bucher WH (1918) On oölites and spherulites. Journal of Geology 26: 593–609
Buick R (1992) The antiquity of oxygenic photosynthesis: evidence from stromatolites in sulphate-deficient Archaean lakes. Science 255: 74–77
Buick R, Dunlop JSR, Groves DI (1981) Stromatolite recognition in ancient rocks: an appraisal of irregularly laminated structures in an Early Archaean chert-barite unit from North Pole, Western Australia. Alcheringa 5: 161–181
Burne RV, Moore L (1987) Microbialites; organosedimentary deposits of benthic microbial communities. Palaios 2: 241–254
Cabioch G, Camoin GF, Montaggioni LF (1999) Postglacial growth history of a French Polynesian barrier reef tract, Tahiti, central Pacific. Sedimentology 46: 985–1000
Cabioch G, Camoin G, Webb GE, Le Cornec F, Garcia Molina M, Pierre C, Joachimski MM (2006) Contribution of microbialites to the development of coral reefs during the last deglacial period: case study from Vanuatu (South-West Pacific). Sedimentary Geology 185: 297–318
Cameron BW, Cameron D, Jones JR (1985) Modern algal mats in intertidal and supratidal quartz sands, northeastern Massachusetts, U.S.A. In Curran HA (ed) Biogenic structures: their use in interpreting depositional environments. SEPM Special Publication 35: 211–224
Camoin GF, Montaggioni LF (1994) High energy coralgal-stromatolite frameworks from Holocene reefs (Tahiti, French Polynesia). Sedimentology 41: 655–676
Camoin GF, Gautret P, Montaggioni LF, Cabioch G (1999) Nature and environmental significance of microbialites in Quaternary reefs: The Tahiti paradox. Sedimentary Geology 126: 271–304
Camoin G, Cabioch G, Eisenhauer A, Braga J-C, Hamelin B, Lericolais G (2006) Environmental significance of microbialites in reef environments during the last deglaciation. Sedimentary Geology 185: 277–295
Camoin GF, Iryu Y, McInroy DB, Expedition 310 Scientists (2007) Proceedings of the Integrated Ocean Drilling Program, volume 310, 83 pp. Washington, DC, 310, doi:10.2204/iodp.proc.310.2007
Chafetz HS (1986) Marine peloids; a product of bacterially induced precipitation of calcite. Journal of Sedimentary Petrology 56: 812–817
Cohen Y, Rosenberg E (eds) (1989) Microbial Mats; Physiological Ecology of Benthic Microbial Communities. American Society for Microbiology, Washington, DC
Corsetti FA, Storrie-Lombardi MC (2003) Lossless compression of stromatolite images: a biogenicity index? Astrobiology 3: 649–655
Dabrio CJ, Esteban M, Martin JM (1981) The coral reef of Nijar, Messinian (uppermost Miocene), Almeria Province, S.E. Spain. Journal of Sedimentary Petrology 51: 521–539
Davies G R (1970) Carbonate bank sedimentation, eastern. Shark Bay, Western Australia. American Association of Petroleum Geologists Memoirs 75: 85–168
Dawson JW (1865) On the structure of certain organic remains in the Laurentian limestones of Canada. Quarterly Journal of the Geological Society London 21: 51–59
Dawson JW (1876) Notes on the occurrence of *Eozoön canadense* at Côte St. Pierre. Quarterly Journal of the Geological Society London 32: 66–75
Dawson W (1896) Note on *Cryptozoon* and other ancient fossils. The Canadian Record of Science 7: 203–219

de Laubenfels MW (1955) Porifera. In: Moore RC (ed) Treatise on Invertebrate Paleontology, Part E, Archaeocyatha and Porifera. Geological Society of America and University of Kansas Press, Lawrence, pp E21–E112

Decho AW (2000) Microbial biofilms in intertidal systems: an overview. Continental Shelf Research 20: 1257–1273

Decho AW, Visscher PT, Reid RP (2005) Production and cycling of natural microbial exopolymers (EPS) within a marine stromatolite. Palaeogeography, Palaeoclimatology, Palaeoecology 219: 71–86

Des Marais DJ (2003) Biogeochemistry of hypersaline microbial mats illustrates the dynamics of modern microbial ecosystems and the early evolution of the biosphere. Biological Bulletin 204: 160–167

Dill RF, Shinn EA, Jones AT, Kelly K, Steinen RP (1986) Giant subtidal stromatolites forming in normal salinity waters. Nature 324: 55–58

Dill RF, Kendall CGStC, Shinn EA (1989) Giant subtidal stromatolites and related sedimentary features. 28th International Geological Congress, American Geophysical Union, Washington, DC, Field Trip Guidebook T373, 33 pp

Dravis, JJ (1983) Hardened subtidal stromatolites, Bahamas. Science 219: 385–386

Dupraz C, Pattisina R, Verrecchia EP. (2006) Translation of energy into morphology: simulation of stromatolite morphospace using a stochastic model. Sedimentary Geology 185: 185–203

Dupraz C, Reid RP, Braissant O, Decho AW, Norman RS, Visscher PT (2009) Processes of carbonate precipitation in modern microbial mats. Earth-Science Reviews 96: 141–162

Expedition 310 Scientists (2007) Maraa eastern transect: sites M0015–M0018. In: Camoin GF, Iryu Y, McInroy DB, Expedition 310 Scientists (eds). Proceedings of the IODP, 310: Washington, DC (Integrated Ocean Drilling Program Management International, Inc.), 83 pp. doi:10.2204/iodp.proc.310.106.2007

Feldmann M, McKenzie JA (1997) Messinian stromatolite-thrombolite associations, Santa Pola, SE Spain: an analogue for the Palaeozoic? Sedimentology 44: 893–914

Fenton CL (1943) Pre-Cambrian and early Paleozoic algae. American Midland Naturalist 30: 83–111

Fenton CL, Fenton MA (1936) Walcott's 'Pre-Cambrian Algonkian algal flora' and associated animals. Geological Society of America Bulletin 47: 609–620

Garrett P (1969) The geology and biology of large cavities in Bermuda reefs. In: Ginsburg RN, Garrett P (eds), Reports of research 1968 seminar on organism-sediment relationships. Bermuda Biological Field Station Research Special Publication 6: 77–88

Garwood EJ (1913) On the important part played by calcareous algae at certain geological horizons, with special reference to the Palaeozoic rocks. Geological Magazine Decade 5, 10: 440–446, 490–498, 545–553

Gebelein CD (1974) Biologic control of stromatolite microstructure: implications for precambrian time stratigraphy. American Journal of Science 274: 575–598

Gerdes G, Krumbein WE (1994) Peritidal potential stromatolites – a synopsis. In: Bertrand-Sarfati J, Monty C (eds) Phanerozoic Stromatolites II. Kluwer, Dordrecht, pp 101–129

Gerdes G, Krumbein WE, Reineck H-E (1985) The depositional record of sandy, versicolored tidal flats (Mellum Island, southern North Sea). Journal of Sedimentary Petrology 55: 265–278

Gerdes G, Claes M, Dunajtschik-Piewak K, Riege H, Krumbein WE, Reineck H-E (1993) Contribution of microbial mats to sedimentary surface structures. Facies 29: 61–74

Gerdes G, Krumbein WE, Noffke N (2000) Evaporite microbial sediments. In: Riding R, Awramik SM (eds) Microbial Sediments. Springer, Berlin, pp 196–208

Ginsburg RN (1960) Ancient analogues of recent stromatolites. International Geological Congress, 21st, Copenhagen, part 22, 26–35

Ginsburg RN (1991) Controversies about stromatolites: vices and virtues. In: Muller DW, McKenzie JA, Weissert H (eds), Controversies in Modern Geology; Evolution of Geological Theories in Sedimentology, Earth History and Tectonics, Academic Press, London, pp 25–36

Ginsburg RN, Planavsky NJ (2008) Diversity of Bahamian stromatolite substrates. In: Dilek Y, Furnes H, Muehlenbachs K (eds) Links between geological processes, microbial activities and evolution of life. Modern Approaches in Solid Earth Sciences 4: 177–195

Ginsburg RN, Isham LB, Bein SJ, Kuperberg J (1954) Laminated Algal Sediments of South Florida and their Recognition in the Fossil Record. Marine Laboratory, University of Miami, Coral Gables, Florida, Unpublished Report, 54–21, 33 pp

Glaessner MF (1962) Pre-cambrian fossils. Biological Reviews 37: 467–493

Goldring W (1938) Algal barrier reefs in the Lower Ozarkian of New York with a chapter on the importance of coralline algae as reef builders through the ages. Bulletin of the New York State Museum 315: 5–75

Golubic S (1976) Organisms that build stromatolites. In: Walter MR (ed) Stromatolites. Developments in Sedimentology 20. Elsevier, Amsterdam, pp 113–140

Grotzinger JP (1986a) Cyclicity and paleoenvironmental dynamics, Rocknest platform, northwest Canada. Geological Society of America Bulletin 97: 1208–1231

Grotzinger JP (1986b) Evolution of Early Proterozoic passive-margin carbonate platform, rocknest formation, wopmay orogen, Northwest Territories, Canada. Journal of Sedimentary Petrology 56: 831–847

Grotzinger JP (1989a) Facies and evolution of Precambrian carbonate depositional systems: emergence of the modern platform archetype. In: Crevello PD, Wilson JL, Sarg JF, Read JF (eds) Controls on carbonate platform and basin development. SEPM Special Publication Number 44: 79–106

Grotzinger JP (1989b) Introduction to Precambrian reefs. In: Geldsetzer HHJ, James NP, Tebbutt GE (eds) Reefs, Canada and adjacent areas. Canadian Society of Petroleum Geologists Memoir 13: 9–12

Grotzinger JP (1990) Geochemical model for Proterozoic stromatolite decline. American Journal of Science 290: 80–103

Grotzinger JP, James NP (2000a) Precambrian carbonates: evolution of understanding. In: Grotzinger JP, James NP (eds) Carbonate sedimentation and diagenesis in the evolving Precambrian world. SEPM Special Publication Number 67: 3–20

Grotzinger JP, James NP (eds) (2000b) Carbonate sedimentation and diagenesis in the evolving Precambrian World. SEPM Special Publication Number 67: 364

Grotzinger JP, Kasting JF (1993) New constraints on Precambrian ocean composition. Journal of Geology 101: 235–243

Grotzinger JP, Knoll AH (1995) Anomalous carbonate precipitates: is the Precambrian the key to the Permian? Palaios 10: 578–596

Grotzinger JP, Knoll AH (1999) Stromatolites in Precambrian carbonates: evolutionary mileposts or environmental dipsticks? Annual Reviews of Earth and Planetary Sciences 27: 313–358

Grotzinger JP, Read JF (1983) Evidence for primary aragonite precipitation, lower Proterozoic (1.9-Ga) Rocknest Dolomite, Wopmay Orogen, Northwest Canada. Geology 11: 710–713

Grotzinger JP, Rothman DR (1996) An abiotic model for stromatolite morphogenesis. Nature 383: 423–425

Gürich G (1906) Les spongiostromides du Viséen de la Province de Namur. Musée Royal d'Histoire Naturelle de Belgique, Mémoires 3(4): 1–55

Hadding A (1927) The pre-Quaternary sedimentary rocks of Sweden. I. A survey of the pre-Quaternary rocks of Sweden. II. The Paleozoic and. Mesozoic conglomerates of Sweden. Lunds Universitets Årsskrift, Nya Förhandlingar (2) 23: 41–171

Hagadorn JW, Bottjer DJ (1997) Wrinkle structures: microbially mediated sedimentary structures common in subtidal siliciclastic settings at the Proterozoic-Phanerozoic transition. Geology 25: 1047–1050

Hall J (1883) *Cryptozoön*, n.g.; *Cryptozoön proliferum*, nsp. New York State Museum of Natural History, 36th Annual Report of the Trustees, plate 6

Häntzschel W, Reineck H-E (1968) Fazies-Untersuchungen im Hettangium von Helmstedt (Niedersachsen). Mitteilungen des Geologischen Staatsinstuts Hamburg 37: 5–39

Heim A (1916) Monographie der Churfirsten-Mattstock-Gruppe, III. Stratigraphie der Unteren Kreide und des Jura. Zur Lithogenesis. Beiträge zur geologischen Karte der Schweiz NF 20: 369–662

Heindel K, Birgel D, Peckmann J, Kuhnert H, Westphal H (2009) Sulfate-reducing bacteria as major players in the formation of reef-microbialites during the last sea-level rise (Tahiti, IODP 310). Geochimica et Cosmochimica Acta 73 (13), Goldschmidt Conference Abstracts, p A514

Hoffman PF (1973) Recent and ancient algal stromatolites: seventy years of pedagogic cross-pollination. In: Ginsburg RN (ed) Evolving Concepts in Sedimentology. The Johns Hopkins University Studies in Geology 21. The Johns Hopkins University Press, Baltimore, London, pp 178–191

Hoffman PF (1975) Shoaling-upward shale-to-dolomite cycles in the Rocknest Formation (lower Proterozoic), Northwest Territories, Canada. In: Ginsburg RN (ed) Tidal Deposits. Springer, Berlin, pp 257–265

Hoffman PF (1989) Pethei reef complex (1.9 Ga), Great Slave Lake, N.W.T. In: Geldsetzer HHJ, James NP, Tebbutt GE (eds), Reefs, Canada and adjacent area. Canadian Society of Petroleum Geologists, Memoir 13: 38–48

Hofmann HJ (1969) Attributes of stromatolites. Geological Survey of Canada Paper 69-39: 58 pp

Hofmann HJ (1971) Precambrian fossils, pseudofossils, and problematica in Canada. Geological Survey of Canada Bulletin 189: 146

Hofmann HJ (1973) Stromatolites: characteristics and utility. Earth-Science Reviews 9: 339–373

Hofmann HJ (2000) Archean stromatolites as microbial archives. In: Riding RE, Awramik SM (eds) Microbial Sediments. Springer, Berlin, pp 315–327

Hofmann HJ, Jackson JD (1987) Proterozoic ministromatolites with radial fibrous fabric. Sedimentology 34: 963–971

Hofmann HJ, Grey K, Hickman AH, Thorpe RI (1999) Origin of 3.45 Ga coniform stromatolites in Warrawoona Group, Western Australia. Geological Society of America Bulletin 111: 1256–1262

Holtedahl O (1921) Occurrence of structures like Walcott's Algonkian algae in the Permian of England. American Journal of Science 1: 195–206

Hornemann JW (ed) (1813) Flora Danica, vol. 9, fasc. 25. Hof-Bogtrykker, Nicolaus Miller, Copenhagen

Horodyski RJ (1977) *Lyngbya* mats at Laguna Mormona, Baja California, Mexico; comparison with Proterozoic stromatolites. Journal of Sedimentary Petrology 47: 1305–1320

Horodyski RJ (1982) Impressions of algal mats from the Middle Proterozoic Belt Supergroup, northwestern Montana, USA. Sedimentology 29: 285–289

Horodyski RJ, Bloeser B (1977) Laminated algal mats from a coastal lagoon, Laguna Mormona, Baja California, Mexico. Journal of Sedimentary Petrology 47: 680–696

Horodyski RJ, Vonder Haar SP (1975) Recent calcareous stromatolites from Laguna Mormona (Baja California), Mexico. Journal of Sedimentary Petrology 45: 894–906

Jackson MJ (1989) Lower Proterozoic Cowles Lake foredeep reef, N.W.T., Canada. In: Geldsetzer HHJ, James NP, Tebbutt GE (eds), Reefs, Canada and adjacent area. Canadian Society of Petroleum Geologists, Memoir 13: 64–71

James NP, Ginsburg RN (1979) Petrography of limestones from the wall and fore-reef. In: James NP, Ginsburg RN (eds) The Seaward Margin of Belize Barrier and Atoll Reefs. IAS Special Publication Number 3. Blackwell, Oxford, pp 111–152

James NP, Ginsburg RN, Marszalek DS, Choquette PW (1976) Facies and fabric specificity of early subsea cements in shallow Belize (British Honduras) reefs. Journal of Sedimentary Petrology 46: 523–544

James NP, Narbonne GM, Sherman AG (1998) Molartooth carbonates: shallow subtidal facies of the mid- to late Proterozoic. Journal of Sedimentary Research 68(5): 716–722

Javor BJ, Castenholz RW (1981) Laminated microbial mats, Laguna Guerrero Negro, Mexico. Geomicrobiology Journal 2: 237–273

Johnson JH (1946) Lime-secreting algae from the Pennsylvanian and Permian of Kansas. Geological Society of America Bulletin 57: 1087–1120

Jolliffe AW (1955) Geology and iron ores of Steep Rock Lake (Ontario). Economic Geology 50: 373–398

Jones B, Hunter IG (1991) Corals to rhodolites to microbialites; a community replacement sequence indicative of regressive conditions. Palaios 6: 54–66

Jørgensen BB, Cohen Y (1977) Solar Lake (Sinai). 5. The sulfur cycle of the benthic cyanobacterial mats. Limnology and Oceanography 22: 657–666

Kah LC, Knoll AH (1996) Microbenthic distribution of Proterozoic tidal flats: environmental and taphonomic considerations. Geology 24: 79–82

Kalkowsky E (1908) Oölith und Stromatolith im norddeutschen Buntsandstein. Zeitschrift der Deutschen geologischen Gesellschaft 60: 68–125, pls 4–11

Kendall CGStC, Skipwith PAd'E (1968) Recent algal mats of a Persian Gulf lagoon. Journal of Sedimentary Petrology 38: 1040–1058

Kerans C (1982) Sedimentology and stratigraphy of the Dismal Lakes Group, Proterozoic, Northwest Territories. PhD thesis, Carleton University, Ottawa, Canada. Unpublished

Knoll AH, Golubic S (1979) Anatomy and taphonomy of a Precambrian algal stromatolite. Precambrian Research 10: 115–151

Knoll AH, Semikhatov MA (1998) The genesis and time distribution of two distinctive Proterozoic stromatolite microstructures. Palaios 13: 408–422

Komar VA, Raaben ME, Semikhatov MA (1965) *Conophyton* in the Riphean of the USSR and their stratigraphic importance. Trudy Geological Institute, Leningrad 131: 72 pp, in Russian

Kremer B, Kazmierczak J, Stal JL (2008) Calcium carbonate precipitation in cyanobacterial mats from sandy tidal flats of the North Sea. Geobiology 6: 46–56

Krumbein WE, Cohen Y, Shilo M (1977) Solar Lake (Sinai). 4. Stromatolitic cyanobacterial mats. Limnology and Oceanography 22: 635–656

Land LS (1971) Submarine lithification of Jamaican reefs. In: Bricker OP (ed) Carbonate Cements. Johns Hopkins University Press, Baltimore, pp 59–60

Land LS, Goreau TF (1970) Submarine lithification of Jamaican reefs. Journal of Sedimentary Petrology 40: 457–462

Land LS, Moore CH (1980) Lithification, micritization and syndepositional diagenesis of biolithites on the Jamaican island slope. Journal of Sedimentary Petrology 50: 357–369

Lawson AC (1912) The geology of Steep Rock Lake, Ontario. Geological Survey of Canada Memoir 28: 7–15

Lepot K, Benzerara K, Brown GE, Philippot P (2008) Microbially influenced formation of 2,724-million-year-old stromatolites. Nature Geoscience 1: 118–121

Lighty RG (1985) Preservation of internal reef porosity and diagenetic sealing of submerged early Holocene barrier reef, southeast Florida shelf. In: Schneidermann N, Harris PM (eds) Carbonate cements. SEPM Special Publication 36: 123–152

Linck G (1903) Die Bildung der Oolithe und Rogensteine. Neues Jahrbuch für Mineralogie, Geologie, und Paläontologie 16: 495–513

Lindemann RH, Yochelson EL (2005) C.D. Walcott and the Hoyt Limestone: an historic encounter at Saratoga Springs, New York. Northeastern Geology & Environmental Sciences 27: 177–186

Logan BW (1961) *Cryptozoon* and associated stromatolites from the Recent, Shark Bay, Western Australia. Journal of Geology 69: 517–533

Logan BW, Rezak R, Ginsburg RN (1964) Classification and environmental significance of algal stromatolites. Journal of Geology 72: 68–83

Logan BW, Hoffman P, Gebelein CD (1974) Algal mats, cryptalgal fabrics, and structures, Hamelin Pool, Western Australia. In: Logan BW, Read JF, Hagan GM, Hoffman P, Brown RG, Woods PJ, Gebelein CD (eds) Evolution and diagenesis of quaternary carbonate sequences, Shark Bay, Western Australia. American Association of Petroleum Gelogists Memoir 22: 140–191

Lowe DR (1980) Stromatolites 3,400–3,500 Myr old from the Archean of Western Australia. Nature 284: 441–443
Lowe DR (1983) Restricted shallow-water sedimentation of early Archean stromatolitic and avaporitic strata of the Strelley Pool Chert, Pilbara Block, Western Australia. Precambrian Research 19: 239–283
Lowe DR (1994) Abiological origin of described stromatolites older than 3.2 Ga. Geology 22: 387–390
Macintyre IG (1977) Distribution of submarine cements in a modern Caribbean fringing reef, Galeta Point, Panama. Journal of Sedimentary Petrology 47: 503–516
Macintyre IG (1984) Extensive submarine lithification in a cave in the Belize Barrier Reef Platform. Journal of Sedimentary Petrology 54: 221–235
Macintyre IG (1985) Submarine cements – the peloidal question. In: Schneidermann N, Harris PM (eds) Carbonate cements. SEPM Special Publication 36: 109–116. Tulsa, Oklahoma, USA
Macintyre IG, Marshall JF (1988) Submarine lithification in coral reefs: some facts and misconceptions. Proceedings 6th International Coral Reef Symposium, Townsville, Australia, 8–12 August 1988, 1: 263–272
Macintyre IG, Mountjoy EW, d'Anglejan BF (1968) An occurrence of submarine cementation of carbonate sediments off the west coast of Barbados, W.I. Journal of Sedimentary Petrology 38: 660–664
Maliva RG, Missima TM, Leo CL, Statom RA, Dupraz C, Lynn M, Dickson JAD (2000) Unusual calcite stromatolites and pisoids from a landfill leachate collection system. Geology 28: 931–934
Malone MJ, Slowey NC, Henderson GM (2001) Early diagenesis of shallow-water periplatform carbonate sediments, leeward margin, Great Bahama Bank (Ocean Drilling Program Leg 166). Geological Society of America Bulletin 113: 881–894
Marshall JF (1983) Submarine cementation in a high-energy platform reef; One Tree Reef, southern Great Barrier Reef. Journal of Sedimentary Petrology 53: 1133–1149
Marshall, J.F. 1986. Regional distribution of submarine cements within an epicontinental reef system: central Great Barrier Reef, Australia. In: Schroeder JH, Purser BH (eds), Reef Diagenesis. Springer, Berlin, pp 8–26
Marshall JF, Davies PJ (1981) Submarine lithification on windward reef slopes; Capricorn-Bunker Group, southern Great Barrier Reef. Journal of Sedimentary Petrology 51: 953–960
Martindale W (1992) Calcified epibionts as palaeoecological tools: examples from the Recent and Pleistocene reefs of Barbados. Coral Reefs 11: 167–177
Martinsson A (1965) Aspects of a Middle Cambrian thanatotope on Öland. Geologiska Föreningens i Stockholm Förhandlingar 87:181–230
Matthew GF (1890a) On the existence of organisms in the pre-Cambrian rocks. Natural History Society New Brunswick, Bulletin 2(9): 28–33
Matthew GF (1890b) Eozoön and other low organisms in Laurentian rocks at St. John. Natural History Society New Brunswick, Bulletin 2(9): 36–41, 67
Mawson D (1929) Some South Australian algal limestones in process of formation. Quarterly Journal of the Geological Society, London 85: 613–620
McLoughlin N, Wilson LA, Brasier MD (2008) Growth of synthetic stromatolites and wrinkle structures in the absence of microbes – implications for the early fossil record. Geobiology 6: 95–105
Montaggioni LF, Camoin GF (1993) Stromatolites associated with coralgal communities in Holocene high-energy reefs. Geology 21: 149–152
Monty C (1965) Recent algal stromatolites in the Windward lagoon, Andros Island, Bahamas. Annales de la Société Géologique de Belgique 88: 269–276
Monty C (1967) Distribution and structure of recent stromatolitic algal mats, eastern Andros Island, Bahamas. Société Géologique de Belgique, Annales 90: 55–99
Monty C (1972) Recent algal stromatolitic deposits, Andros Island Bahamas. Preliminary report. Geologische Rundschau 61: 742–783

Monty C (1976) The origin and development of cryptalgal fabrics. In: Walter MR (ed) Stromatolites. Developments in Sedimentology 20. Elsevier, Amsterdam, pp 193–249
Monty C (1977) Evolving concepts on the nature and the ecological significance of stromatolites. In: Flügel E (ed) Fossil Algae, Recent Results and Developments. Springer, Berlin, pp 15–35
Monty C (ed) (1981) Phanerozoic Stromatolites. Springer-Verlag, Berlin, 249 pp
Morse JW, Mucci A (1984) Composition of carbonate overgrowths produced on Iceland spar calcite crystals buried in Bahamian carbonate-rich sediments. Sedimentary Geology 40: 287–291
Naumann CF (1862) Lehrbuch der Geognosie, Band 2. Engelmann, Leipzig, 1092 pp
Nicholson JAM, Stolz JF, Pierson BK (1987) Structure of a microbial mat at Great Sippewissett Marsh, Cape Cod, Massachusetts. FEMS Microbiology Letters 45: 343–364
Noffke N, Gerdes G, Klenke T, Krumbein, WE (1996) Microbially induced sedimentary structures – examples from modern sediments of siliciclastic tidal flats. Zentralblatt Geologie Paläontologie I (1995) 1/2: 307–316
Noffke N, Gerdes G, Klenke T, Krumbein WE (2001) Microbially induced sedimentary structures: a new category within the classification of primary sedimentary structures. Journal of Sedimentary Research 71: 649–656
Noffke N, Hazen R, Nhleko N (2003) Earth's earliest microbial mats in a siliciclastic marine environment (2.9 Ga Mozaan Group, South Africa). Geology 31: 673–676
Noffke N, Eriksson KA, Hazen RM, Simpson EL (2006) A new window into Early Archean life: microbial mats in Earth's oldest siliciclastic tidal deposits (3.2 Ga Moodies Group, South Africa). Geology 34: 253–256
Noffke N, Beukes N, Bower D, Hazen RM, Swift DJP (2008) An actualistic perspective into Archean worlds – (cyano-)bacterially induced sedimentary structures in the siliciclastic Nhlazatse Section, 2.9 Ga Pongola Supergroup, South Africa. Geobiology 6: 5–20
O'Brien C F (1970) *Eozoön canadense* 'The dawn animal of Canada'. Isis 61: 206–223
Ørsted AS (1842) Beretning om en Excursion til Trindelen, en alluvialdannelse i Odensefjord. Krøyer, Naturhistorisk Tidsskrift 3: 552–569
Paul J, Peryt TM (2000) Kalkowsky's stromatolites revisited (Lower Triassic Buntsandstein, Harz Mountains, Germany). Palaeogeography, Palaeoclimatology, Palaeoecology 161: 435–458
Pedley HM (1979) Miocene bioherms and associated structures in the Upper Coralline limestone of the Maltese Islands: their lithification and palaeoenvironment. Sedimentology 26: 577–591
Pentecost A (2005) Travertine. Springer, Berlin, 445 pp
Perry RS, McLoughlin N, Lynne BY, Sephton MA, Oliver JD, Perry CC, Campbell K, Engel H, Farmer JD, Brasier MD, Staley JT (2007) Defining biominerals and organominerals: direct and indirect indicators of life. Sedimentary Geology 201: 157–179
Petrov PYu, Semikhatov MA (2001) Sequence organization and growth patterns of late Mesoproterozoic stromatolite reefs: an example from the Burovaya Formation, Turukhansk Uplift, Siberia. Precambrian Research 111: 257–281
Pia J (1927) Thallophyta. In: Hirmer M (ed) Handbuch der Paläobotanik 1. Oldenburg, Munich, pp 31–136
Pigott JD, Land LS (1986) Interstitial water chemistry of Jamaican reef sediment: sulfate reduction and submarine cementation. Marine Chemistry 19: 355–378
Playford PE, Cockbain AE (1976) Modern algal stromatolites at Hamelin Pool, a hypersaline barred basin in Shark Bay, Western Australia. In: Walter MR (ed) Stromatolites. Elsevier, Amsterdam, pp 389–411
Pope MC, Grotzinger JP (2000) Controls on fabric development and morphology of tufas and stromatolites, uppermost Pethei Group (1.8 Ga), Great Slave Lake, northwest Canada. In: Grotzinger JP, James NP (eds) Carbonate sedimentation and diagenesis in the evolving Precambrian world. SEPM Special Publication Number 67: 103–121
Pope MC, Grotzinger JP, Schreiber BC (2000) Evaporitic subtidal stromatolites produced by in situ precipitation: textures, facies associations, and temporal significance. Journal of Sedimentary Research 70: 1139–1151

Por FD (1967) Solar Lake on the shores of the Red Sea. Nature 218: 860–861

Porada H, Bouougri EH (2007) Wrinkle structures – a critical review. Earth Science Reviews 81: 199–215

Porada H, Ghergut J, Bouougri EH (2008) *Kinneyia*-type wrinkle structures – critical review and model of formation. Palaios 23: 65–77

Read JF (1976) Calcretes and their distinction from stromatolites. In: Walter MR (ed), Stromatolites. Developments in Sedimentology 20. Elsevier, Amsterdam, pp 55–71

Reid RP, Macintyre IG, Steneck RS, Browne KM, Miller TE (1995) Stromatolites in the Exuma Cays, Bahamas: uncommonly common. Facies 33: 1–18

Reid RP, Visscher PT, Decho AW, Stolz JF, Bebout BM, Dupraz C, Macintyre IG, Paerl HW, Pinckney JL, Prufert-Bebout L, Steppe TF, DesMarais DJ (2000) The role of microbes in accretion, lamination and early lithification of modern marine stromatolites. Nature 406: 989–992

Reis OM (1908) Kalkowsky: Ueber Oölith und Stromatolith im norddeutschen Buntsandstein. Neues Jahrbuch fiir Mineralogie, Geologie und Paläontologie 2: 114–138

Reitner J (1993) Modern cryptic microbialite/metazoan facies from Lizard Island (Great Barrier Reef, Australia). Formation and concepts. Facies 29: 3–39

Reitner J, Gautret P, Marin F, Neuweiler F (1995) Automicrites in a modern marine microbialite. Formation model via organic matrices (Lizard Island, Great Barrier Reef, Australia). Bulletin de l'Institut d'Océanographique de Monaco Special Number 14: 237–263

Reitner J, Thiel V, Zankl H, Michaelis W, Wörheide G, Gautret P (2000) Organic and biogeochemical patterns in cryptic microbialites. In: Riding RE, Awramik SM (eds) Microbial Sediments, Springer, Berlin, pp 149–160

Revsbech NP, Jørgensen BB, Blackburn TH, Cohen Y (1983) Microelectrode studies of photosynthesis and O_2, H_2S and pH profiles of a microbial mat. Limnology and Oceanography 28: 1062–1074

Riding R (1977) Skeletal stromatolites. In: Flügel E (ed) Fossil Algae, Recent Results and Developments. Springer, Berlin, pp 57–60

Riding R (1991) Classification of microbial carbonates. In: Riding R (ed) Calcareous Algae and Stromatolites. Springer-Verlag, Berlin, pp 21–51

Riding R (1999) The term stromatolite: towards an essential definition. Lethaia 32: 321–330

Riding R (2000) Microbial carbonates: the geological record of calcified bacterial-algal mats and biofilms. Sedimentology 47(Suppl 1): 179–214

Riding R (2008) Abiogenic, microbial and hybrid authigenic carbonate crusts: components of Precambrian stromatolites. Geologia Croatica 61(2–3): 73–103

Riding R, Awramik SM (eds) (2000) Microbial Sediments. Springer, Berlin, 331 pp

Riding R, Awramik SM, Winsborough BM, Griffin KM, Dill RF (1991a) Bahamian giant stromatolites: microbial composition of surface mats. Geological Magazine 128: 227–234

Riding R, Martín JM, Braga JC (1991b) Coral stromatolite reef framework, Upper Miocene, Almería, Spain. Sedimentology 38: 799–818

Roddy HJ (1915) Concretions in streams formed by the agency of blue-green algae and related plants. Proceedings American Philosophical Society 54: 246–258

Rothpletz A (1892) Über die Bildung der Oolithe. Botanisches Centralblatt 51: 265–268

Sami TT, James NP (1994) Peritidal carbonate platform growth and cyclicity in an early Proterozoic foreland basin, upper Pethei Group, northwest Canada. Journal of Sedimentary Research B64: 111–131

Sami TT, James NP (1996) Synsedimentary cements as Paleoproterozoic platform building blocks, Pethei Group, northwestern Canada. Journal of Sedimentary Research 66: 209–222

Schieber J (1986) The possible role of benthic microbial mats during the formation of carbonaceous shales in shallow Proterozoic basins. Sedimentology 33: 521–536

Schieber J (1998) Possible indicators of microbial mat deposits in shales and sandstones: examples from the mid-Proterozoic Belt Supergroup, Montana, U.S.A. Sedimentary Geology 120: 105–124

Schopf JW (1968) Microflora of the Bitter Springs Formation, Late Precambrian, central Australia. Journal of Paleontology 42: 651–688

Schopf JW (1999) Cradle of Life: The Discovery of Earth's Earliest Fossils. Princeton University Press, Princton, New Jersey, USA, 336 pp

Schulz E (1936) Das Farbstreifen-Sandwatt und seine Fauna, eine ökologische biozönotische Untersuchung an der Nordsee. Kieler Meeresforschung 1: 359–378

Sedgwick A (1829) On the geological relations and internal structure of the Magnesian Limestone, and the lower portions of the New Red Sandstone Series in their range through Nottinghamshire, Derbyshire, Yorkshire, and Durham, to the southern extremity of Northumberland. Transactions of the Geological Society of London, Second Series 3: 37–124

Seilacher A (1982) Distinctive features of sandy tempestites. In: Einsele G, Seilacher A (eds) Cyclic and Event Stratification. Springer, Berlin, pp 333–349

Semikhatov MA, Gebelein CD, Cloud P, Awramik SM, Benmore WC (1979) Stromatolite morphogenesis – progress and problems. Canadian Journal of Earth Science 16: 992–1015

Seward AC (1931) Plant Life Through the Ages. Cambridge University Press, Cambridge, 601 pp

Sherman CE, Fletcher CH, Rubin KH (1999) Marine and meteoric diagenesis of Pleistocene carbonates from a nearshore submarine terrace, Oahu, Hawaii. Journal of Sedimentary Research 69: 1083–1097

Stal LJ (2000) Cyanobacterial mats and stromatolites. In: Whitton BA, Potts M (eds) The Ecology of Cyanobacteria. Their Diversity in Time and Space. Kluwer, Dordrecht, The Netherlands, pp 61–120

Steele JH (1825) A description of the Oolitic Formation lately discovered in the county of Saratoga, and state of New-York. American Journal of Science 9: 16–19, part of pl. 2

Storrie-Lombardi MC, Corsetti FA, Grigolini P, Ignaccolo M, Allegrini P, Galatolo S, Tinetti G (2004) Complexity analysis to explore the structure of ancient stromatolites. Chaos, Solitons and Fractals 20: 139–144

Sumner DY, Grotzinger JP (2000) Late Archean aragonite precipitation: petrography, facies associations, and environmental significance. In: Grotzinger JP, James NP (eds) Carbonate sedimentation and diagenesis in the evolving Precambrian world. SEPM Special Publication Number 67: 123–144

Sumner DY, Grotzinger JP (2004) Implications for Neoarchaean ocean chemistry from primary carbonate mineralogy of the Campbellrand-Malmani platform, South Africa. Sedimentology 51: 1–27

Thrailkill J (1976) Speleothems. In: Walter MR (ed) Stromatolites, Developments in Sedimentology 20. Elsevier, Amsterdam, pp 73–86

Turner EC, Narbonne GM, James NP (2000) Framework composition of early Neoproterozoic calcimicrobial reefs and associated microbialites, Mackenzie Mountains, N.W.T., Canada. In: Grotzinger JP, James NP (eds) Carbonate sedimentation and diagenesis in the evolving Precambrian world. SEPM Special Publication Number 67: 179–205

Tyler SA, Barghoorn ES (1954) Occurrence of structurally preserved plants in Pre-Cambrian rocks of the Canadian Shield. Science 119: 606–608

van Gemerden H (1993) Microbial mats: a joint venture. Marine Geology 113: 3–25

Vidal G (1972) Algal stromatolites from the Late Precambrian of Sweden. Lethaia 5: 353–367

Visscher PT, Reid RP, Bebout BM (2000) Microscale observation of sulphate reduction: correlation of microbial activity with lithified micritic laminae in modern marine stromatolites. Geology 28: 919–922

Vologdin AG (1962) The Oldest Algae of the USSR. Academy of Sciences of the USSR, Moscow, 657 pp, in Russian

Walcott CD (1895) Algonkian rocks of the Grand Canyon. Journal of Geology 3: 312–330

Walcott CD (1906) Algonkian formations of northwestern Montana. Geological Society of America Bulletin 17: 1–28

Walcott CD (1912) Notes on fossils from limestone of Steeprock series, Ontario. Geological Survey Canada Memoir 28: 16–23

Walcott CD (1914) Cambrian geology and paleontology III: Precambrian Algonkian algal flora. Smithsonian Miscellaneous Collection 64: 77–156
Walter MR (1972) Stromatolites and the biostratigraphy of the Australian Precambrian and Cambrian. Special Papers in Palaeontology 11: 190
Walter MR (ed) (1976a) Stromatolites. Developments in Sedimentology 20. Elsevier, Amsterdam, 790 pp
Walter MR (1976b) Introduction. In: Walter MR (ed) Stromatolites. Developments in Sedimentology 20. Elsevier, Amsterdam, pp 1–3
Walter MR (1976c) Geyserites of Yellowstone National Park: an example of abiogenic "stromatolites". In: Walter MR (ed) Stromatolites. Developments in Sedimentology 20. Elsevier, Amsterdam, pp 87–112
Walter MR, Golubic S, Preiss WV (1973) Recent stromatolites from hydromagnesite and aragonite depositing lakes near the Coorong Lagoon, South Australia. Journal of Sedimentary Petrology 43: 1021–1030
Walter MR, Bauld J, Brock TD (1976) Microbiology and morphogenesis of columnar stromatolites (Conophyton, Vacerrilla) from hot springs in Yellowstone National Park. In M.R. Walter (ed) Stromatolites. Developments in Sedimentology 20. Elsevier, Amsterdam, pp 273–310
Walter MR, Buick R,Dunlop JSR (1980) Stromatolites 3,400–3,500 Myr old from the North Pole area, Western Australia. Nature 284: 443–445
Webb GE, Baker JC, Jell JS (1998) Inferred syngenetic textural evolution in Holocene cryptic reefal microbialites, Heron Reef, Great Barrier Reef, Australia. Geology 26: 355–358
Whittle GL, Kendall CGStC, Dill RF, Rouch L (1993) Carbonate cement fabrics displayed: a traverse across the margin of the Bahamas platform near Lee Stocking Island in the Exuma Cays. Marine Geology 110: 213–243
Wilks ME, Nisbet EG (1985) Archaean stromatolites from the Steep Rock Group, northwestern Ontario, Canada. Canadian Journal of Earth Sciences 22: 792–799
Wiman C (1915) Om Visingsö -kalkstenen vid Gränna. Geol Foren i Stockholm Förh 37: 367–375
Zankl H (1993) The origin of high-Mg-calcite microbialites in cryptic habitats of Caribbean coral reefs – their dependence on light and turbulence. Facies 29: 55–60

Part II
Stromatolite Formation and Microbial Biomineralisation

Modern Marine Stromatolites of Little Darby Island, Exuma Archipelago, Bahamas: Environmental Setting, Accretion Mechanisms and Role of Euendoliths

R. Pamela Reid, Jamie S. Foster, Gudrun Radtke, and Stjepko Golubic

1 Introduction

The search for modern stromatolites was initiated by geologists in an attempt to understand the processes that governed the Earth for about five sixths of the history of Life. Entire land- and seascapes dominated by stromatolites are rare in the modern world of plants and animals. Modern stromatolites were first encountered under extreme environmental conditions, such as the hypersaline ponds of Hamelin Pool, Shark Bay (Logan 1961; Reid et al. 2003). Discovery of stromatolite formation under normal sea water conditions in Schooner Cays and at Lee Stocking Island in the Bahamas (Dravis 1983; Dill et al. 1986) was therefore, of particular significance. These stromatolites grow under peculiar conditions of sand transportation along tidal channels, and accrete primarily through the trapping and binding of sand grains. Similar stromatolites were later encountered at a variety of locations throughout the Exuma Cays (Reid et al. 1995); those in the shallows of a wave-exposed coast at Highborne Cay (Reid et al. 1999, 2000) generated a number of publications dealing with production and decomposition of organic matter, lithification of microbial mats, and successions, interactions and seasonal distribution of microbial communities [reviewed by Stolz et al. (2009)].

R.P. Reid (✉)
Marine Geology and Geophysics, Rosenstiel School of Marine and Atmospheric Science, University of Miami, 4600 Rickenbacker Cswy, Miami, FL 33149, USA
e-mail: preid@rsmas.miami.edu

J.S. Foster
Department of Microbiology and Cell Science, University of Florida, Gainesville, FL 32611-0700, USA, Space Life Sciences Lab, Kennedy Space Center, Merritt Island, FL 32899, USA

G. Radtke
Hessisches Landesamt für Umwelt und Geologie, Rheingaustr. 186, 65203 Wiesbaden, Germany

S. Golubic
Biological Science Center, Boston University, 5 Cummington Street, Boston, MA 02215, USA

J. Reitner et al., *Advances in Stromatolite Geobiology*,
Lecture Notes in Earth Sciences 131, DOI 10.1007/978-3-642-10415-2_4,

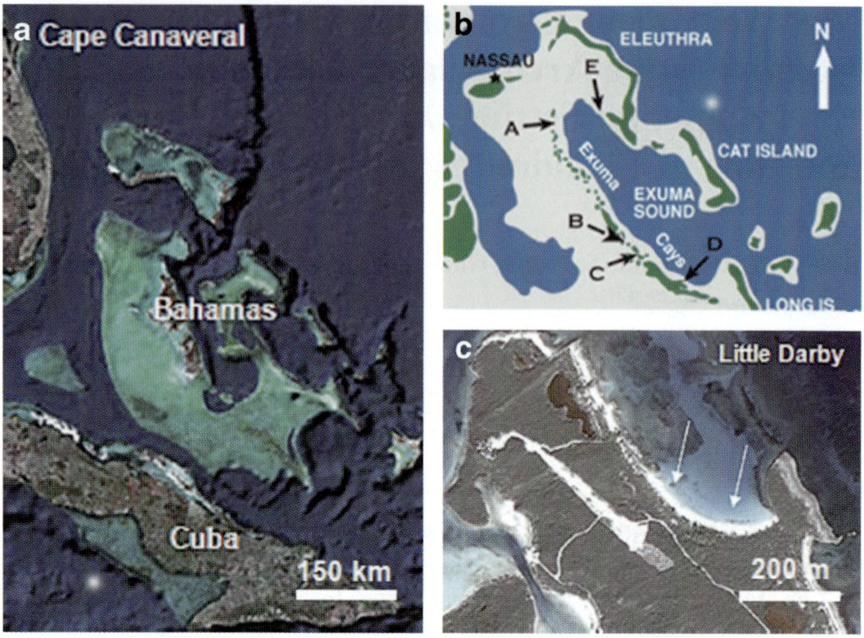

Fig. 1 Geographical location of Little Darby Island and stromatolites. (**a**) Google Earth image of Bahamas platform. (**b**) Diagrammatic map of the location of stromatolites found within the Exuma archipelago. Highborne Cay (A); Little Darby (B); Lee Stocking Island (C); Stocking Island (D); and Schooner Cays (E). (**c**) Google Earth image of the stromatolite beach on the north east coast of Little Darby Island. *Arrows* point to the major clusters of actively accreting stromatolites

In this paper, we present new observations on subtidal stromatolites, in a North-Northeast-exposed sandy bay of Little Darby Island, Exuma Archipelago, Bahamas (Fig. 1). The data, which represent initial results from the first science expedition at the new Little Darby Research Station, focus on the dynamics of stromatolite accretion and the role of euendolithic microorganisms in stromatolite formation.

2 Environmental Setting of Little Darby Stromatolites

Large fields of stromatolitic structures are distributed, parallel to the beach, at a depth ranging from 1 to 2 m (Fig. 2a). Some of these structures are standing alone in the sand; others coalesce into reef complexes (Fig. 2b–f). Sand is transported by the waves, whereby the stromatolites acted as obstacles, which modify the turbulence and sand transport. During the recent study in September 2009, the upper surfaces of many stromatolites, about 30–50 cm above ground were within the range of periodic sand supply during windy days. They were covered by active microbial mats and showed different degrees of induration. Soft microbial mats

Fig. 2 Stromatolitic structures of the Little Darby Cay, Exuma archipelago, Bahamas. (**a**) The field of subtidal, shallow water stromatolites (*dark areas* in front of the beach). (**b**) Two stromatolites monitored for their accretion rates, note the upward expansion of the structures and zonal coloration of their lateral walls. The nail in upper right is 7 cm long. (**c**) Upper surface of the stromatolites with 5–10 cm diameter domes where active accretion takes place. Stromatolite upper left is 60 cm wide. (**d**) Vertical section through the mat on the upper surface of a stromatolite,

were comprised of sheathed filaments of oscillatoriacean cyanobacteria of the genus *Schizothrix* with incorporated sand grains. These mats, studied earlier at Stocking Island (Reid and Browne 1991), Lee Stocking Island (Browne 1993; Feldmann and Mackenzie 1998) and Highborne Cay (Reid et al. 2000) formed small domal protuberances 1–5 cm in diameter (Fig. 2b, c) on top of Little Darby stromatolites by trapping and binding of sediment particles combined with upward and radial movement of *Schizothrix* trichomes. A vertical section through such protuberances (Fig. 2d) indicated that the vertical accretion of the mat shifts laterally, on a centimeter scale, depending on the combination of local sand supply and cyanobacterial response in movement and growth. During periods of calm weather the accumulation of cyanobacterial biomass produced translucent horizons, which in deeper layers became obscured by sand grain compaction. Lateral shifts in accretion include and slightly favor stromatolite margins, resulting in the gradual widening of the active upper surface, the flat appearance of the stromatolite tops, and the conical outline of stromatolite profiles with overhang side walls. The side walls did not appear to trap significant amounts of sediment; they were covered by different microbial populations than the stromatolite upper surfaces and accordingly were colored differently (Fig. 2e). The bases of these stromatolites were in frequent contact with moving sand ripples, thus exposed to abrasion by moving sands. The surfaces of stromatolites on the more easterly side of the beach were overgrown by macro-algae, predominantly by *Batophora* and *Acetabularia* (Fig. 2f). Other stromatolitic structures in the studied fields were buried in the sand, and some were exposed and colonized by dark colored coating, the black coloration contrasting with the white of the carbonate sand (Fig. 2f, lower right).

3 Accretional Processes in Little Darby Stromatolites

In stromatolites of Little Darby Island, which are similar to those studied at Lee Stocking Island and on Highborne Cay, the accretion takes place within a soft mat along the upward turned surfaces of the stromatolite structures (Fig. 2c). Two components are essential in the process of stromatolite accretion: (a) the presence of the *Schizothrix* mat and (b) the flux of carbonate sand grains supplied by waves or currents. The cyclicity of wind and wave turbulence is marked by pulses in sand supply to the stromatolites and is recorded in fine, mm scale lamination (Fig. 3a).

Fig. 2 (Continued) showing domes of active accretion with a lateral displacement over time. Darker layers comprised of exopolymers of *Schizothrix* mat mark the pauses in sediment supply during calm weather conditions. Scale bar is 1 cm long. (**e**) Groups of stromatolites coalescing into coherent reefs (*background*) with two units showing color zonation on lateral walls. Note the irregular rippling of sand due to interference of wave action around stromatolitic mounds. (**f**) Stromatolites sheltered from sand supply are overgrown by algae. The unit in the center is overgrown mainly by *Batophora* on the top and by *Acetabularia* on its sides. The stromatolitic structure on lower right is re-emerging after having been buried by sand

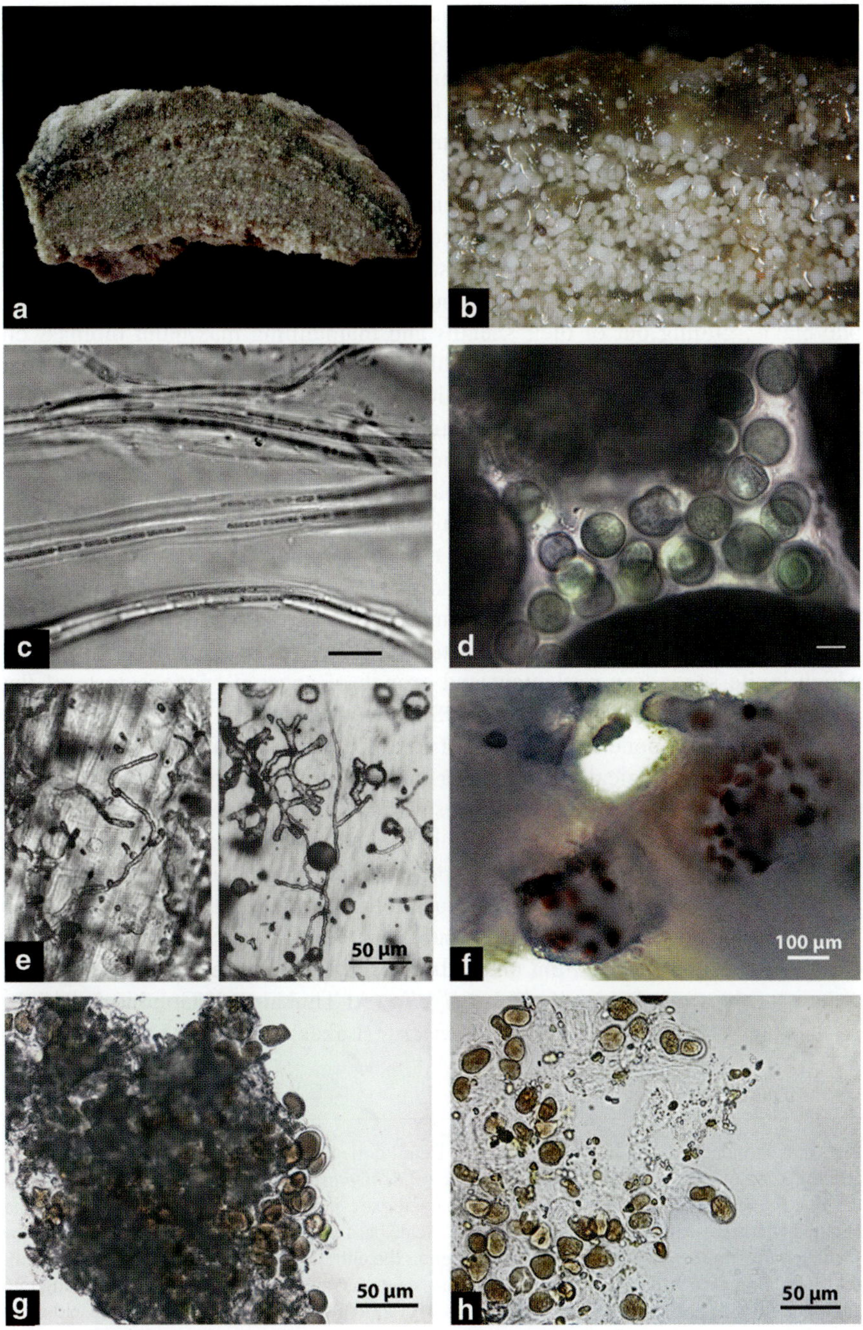

Fig. 3 Early stages in action and solidification of the *Schizothrix* mat responsible for the accretion of stromatolites on Little Darby Cay. (**a**) Section of a domal unit from the upper surface of stromatolites, note the change in the size of entrapped sand grains from layer to layer reflecting the

Gliding trichomes of *Schizothrix*, the primary producer, remain on top of the structure (Fig. 3b) and move upward with stromatolite accretion.

The predominant mat-forming organism is *Schizothrix gebeleinii* (Fig. 3c); with highly motile gliding trichomes bundled inside exopolymer sheaths (see Golubic and Browne 1996). Minor components consist of other rapidly gliding organisms belonging to cyanobacterial genera *Schizothrix*, *Oscillatoria*, *Spirulina* as well as gliding anoxygenic bacteria of the *Chloroflexus*-type. Molecular signatures of this group has been noted in the stromatolites of Highborne Cay (Foster and Green 2010). During times of wave turbulence, sand particles stick to the mat, gradually burying the cyanobacteria. The latter respond by positive phototaxis, gliding out of their sheaths and maintaining maximum biomass on the top of the structure. These young filaments contain single trichomes, which later duplicate by fragmentation and glide next to each other forming bundles in the common sheath (Fig. 3c). The "mature", i.e. multitrichomous filaments form during calm periods when cyanobacterial biomass accumulates without incorporation of sand (Fig. 3b). The principal primary product contributions are exopolymer sheaths of *Schizothrix*, which are left behind several mm deeper in the mat, and become subject to degradation by organotrophic bacteria as shown for the Highborne stromatolites (Decho et al. 2005) which includes anaerobic respiration via sulfate reduction (Baumgartner et al. 2006). Additional primary production occurs in sheltered interstitial spaces between the entrapped grains, which invite secondary settlement and growth of coccoid cyanobacteria of the genus *Aphanocapsa* producing soft EPS embedded colonies (Fig. 3d).

The second component, consisting of entrapped and thus stabilized sand grains have their own biological history. Prior to incorporation into stromatolites, they were part of mobile, shoaling sand dunes with their own interstitial microflora and fauna, and with a substantial contribution by phototrophic euendoliths (Wild et al. 2006; Tribollet et al. 2010). Shoaling ooids and other carbonate sand particles are bored by a specialized assemblage of euendolithic cyanobacteria, which are able to settle and persist in a highly mobile substrate, experiencing profound and rapid changes in availability of light and nutrients. That assemblage is dominated by species of the cyanobacterial genera *Hyella* (Al-Thukair and Golubic 1991a, b; Al-Thukair et al. 1994) and *Cyanosaccus* (Lukas and Golubic 1981). This

Fig. 3 (Continued) coeval wave energies. (**b**) Detail collected during calm weather resulting in accumulation of sand-free Schizothrix biomass. (**c**) *Schizothrix gebeleinii*, dominant organism in the mat. Note the elongated cells and the characteristic presence of more than one trichome bundled within the common sheath of filaments. Scale bar is 10 μm. (**d**) Soft mats of *Aphanocapsa* sp. developing in the interstices between sand grain the entrapped in the mat. Scale bar is 10 μm. (**e**) Microborings found in particles of calcareous sands surrounding stromatolites. Sand particles are bored prior to entrapment into stromatolites. (**f**) Growth burst of the euendolithic cyanobacterium *Solentia sanguinea* in sand grains following entrapment and stabilization. Note that some cells are in focus on the surface of the grains. (**g**) Crushed sand grain from **f** releasing *Solentia*. (**h**) Colony of *Solentia* upon dissolution of entrapped sand grain by dilute HCl assemblage is very different from euendoliths inhabiting solid substrates, such as coastal limestone and coral reefs (Radtke et al. 1996, 1997b; Radtke and Golubic 2010). Thus, sand particles are pre-bored by the time they are incorporated in the mats (Fig. 3e), but only a minute proportion of those borings are inhabited by live and active euendoliths.

4 Grain Stabilization Results in a Shift in Euendolith Composition

Ecological conditions for euendoliths undergo a major change after the grains are entrapped and stabilized in stromatolitic structures, resulting in a population explosion of *Solentia sanguinea* (Fig. 3f). *Solentia sanguinea* has been earlier isolated in culture from ooid sand grains off the Lee Stocking Island and from the Arabian Gulf, but is exceedingly rare in free sand grains (Golubic et al. 1996). The growth of the organism is apparently stimulated when the substrate is stabilized as by entrapment in stromatolites or in culture on agar. Similar mass development of *Solentia* has been observed in Highborne Cay stromatolites (Reid et al. 2000; Stolz et al. 2009: fig. 4d). In the stromatolites of Little Darby Island, *Solentia* occurred both on the surfaces of the entrapped sand grains, and bored into the grain's interior (Fig. 3g), with cells penetrating at the end of long gelatinous stalks (Fig. 3h). Development of endolithic blooms of *Solentia* coincides with the process of mat lithification.

5 Lithification of Bahamian Stromatolites

Lithification of modern stromatolites is one of the processes that link the genesis of these structures with their historic record. The timing of the process appears essential even for the definition of the stromatolite concept (e.g. Krumbein et al. 2003). Stromatolite-forming microbial mats experience early lithification. Two closely coupled components drive carbonate precipitation within microbial mats: (1) changes in the ion activity product $Ca^{2+} \times CO_3^{2-}$ and (2) the organic matrix in which the minerals nucleate (e.g. Pentecost and Riding 1986; Reitner 1993; Dupraz and Visscher 2005). Activity of CO_3^{2-}, which determines the saturation index of $CaCO_3$ minerals in many natural systems, is affected via pH and alkalinity changes by microbial metabolism, such as photosynthesis or sulfate reduction, and environmental parameters. The organic matrix of extracellular polymeric substances (EPS), which embed the microbial communities, is the physical location where carbonate minerals nucleate and grow. Fresh EPS typically binds cations (e.g. Ca^{2+}), thereby inhibiting CaCO3 precipitation. Precipitation results when degradation of the EPS reduces this inhibition and releases previously bound Ca^{2+}, or by oversaturation of the cation-binding capacity (Braissant et al. 2007). Microbial activity is ultimately responsible for EPS production and degradation leading to mineral precipitation.

In Bahamian marine stromatolites, active accretion by *Schizothrix* mats is punctuated by thin horizons of carbonate precipitates and cemented grain layers, as documented on vertical sections through stromatolites (e.g. Reid et al. 2000: fig. 4). Studies at Highborne Cay indicate that thin micritic crusts form in surface biofilms with abundant heterotrophic bacteria and high rates of sulfate reduction (Visscher et al. 2000; Andres et al. 2006). Precipitation of aragonite within the biofilm forms a micritic lamina, which caps the underlying ooids. The high level of carbonate saturation of the interstitial waters in and around Bahamian stromatolites allows minor changes in pH to trigger precipitation, which makes it difficult to distinguish among, abiotic, biogenic and organogenic causes of calcification. In theory, several microbial metabolic activities operating in actively accreting stromatolites may act as triggers of precipitation. Highborne Cay studies suggest, however, that heterotrophs play a more direct role in crust precipitation than previously assumed (Visscher et al. 2000; Andres et al. 2006).

The cemented grain layers that punctuate *Schizothrix* trapping and binding are characterized by an abundance of *Solentia sanguinea*. *Solentia* tunnels through grains, leaving bore holes filled with EPS. Aragonite precipitation within the EPS results in micritization of the grains and, when tunnels cross between grains, welds the grains together. This process forms a well-cemented layer (Macintyre et al. 2000; Reid and Macintyre 2000; Reid et al. 2000). The crystallization pattern of carbonate precipitate observed in the *Solentia* borings complies with the texture of the EPS stalks of this organism (Reid and Macintyre 2000). The compliance of mineral arrangement with the organic template offered by specific EPS product illustrates the principle of organo-mineralization (Trichet and Défarge 1995; Reitner et al. 1995). Precipitation within the *Solentia* layers indicates that carbonate precipitation may be an integral part of the activity of euendoliths, which dissolve carbonate in the process of boring and re-deposit it elsewhere as suggested by Kobluk and Risk (1977a, b), Schneider and Le Campion-Alsumard (1999) and Garcia-Pichel (2006).

6 Post-accretional Transformations of Stromatolites

The existence of stromatolites as actively growing submarine structures in modern environments is a rarity, and attention typically focuses on accretion. However, as noted above, stromatolitic structures that became sheltered from sand supply enter extended non-accretional periods. At Little Darby Island, these sheltered structures were overgrown by algae, mainly by *Acetabularia* and *Batophora* (Figs. 2f and 4a), becoming practically indistinguishable from any carbonate hardground under the same setting. This observation confirms the importance of sand supply as one of the essential components of stromatolite growth. Although some entrapment of sand grains among the thalli of macroalgae is inevitable, there is no evidence of stromatolite accretion by macroalgae as proposed by Awramik and Riding (1988). Other processes contributing to the diagenetic history of stromatolitic structures include micro- and macroborings and secondary encrustations that may ultimately

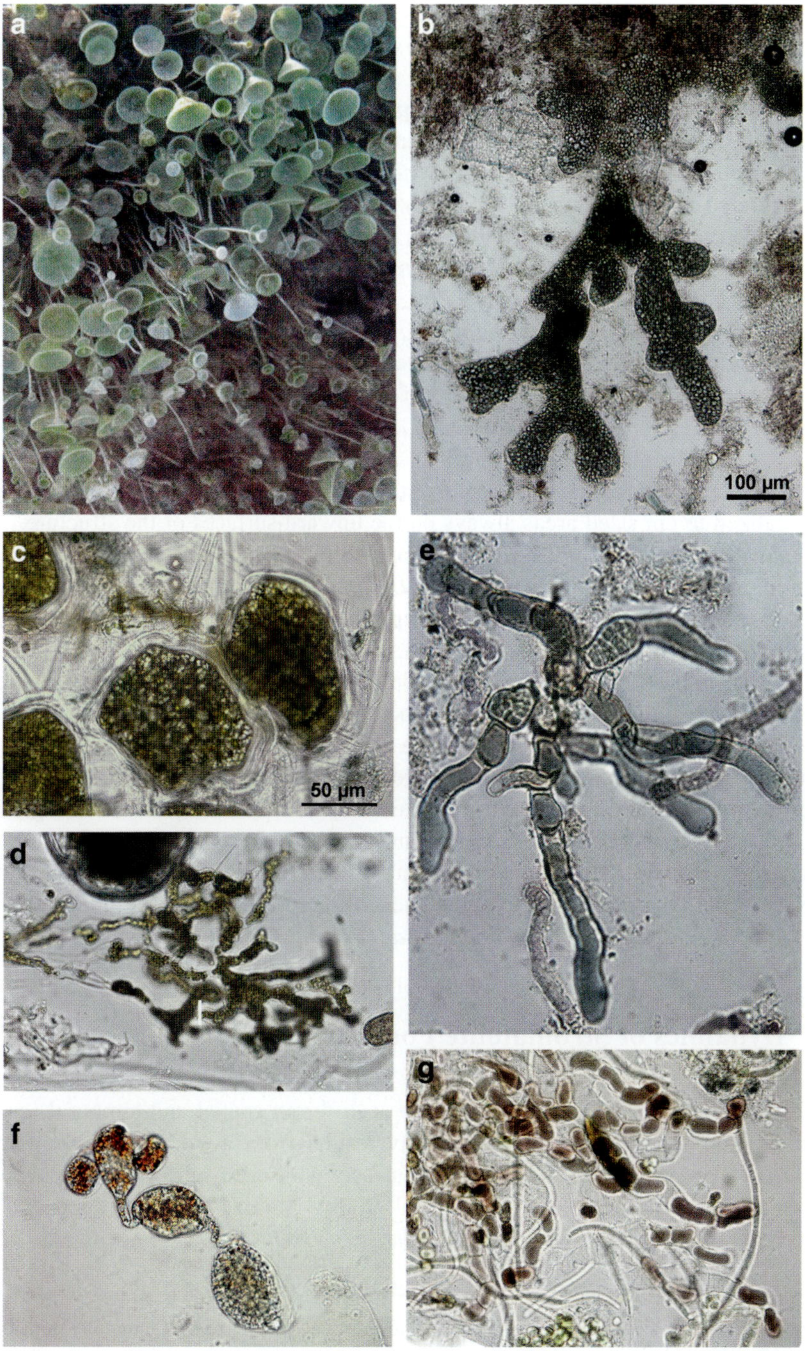

Fig. 4 Assemblages of epi- and endoliths following stases in stromatolite accretion. (**a**) A population of *Acetabularia* overgrowing the side of a stromatolite (detail of Fig. 2f). (**b**) The euendolithic rhizoid of *Acetabularia* in growth position, extracted by acid treatment of the crust.

obscure the original accretional features. Such structural modifications may lead to formation of thrombolitic textures when viewed in section or in outcrop (Planavsky and Ginsburg 2009).

The extended stasis provides new opportunities for the settlements of epi- and endoliths alike, resulting in an increase in diversity and complexity of the resident microboring community apparently at the expense of *Solentia,* which is rarely encountered in these samples. Prominent are algae, including eukaryotic euendoliths, especially *Acetabularia* (Fig. 4a) and its euendolithic rhizoids (Fig. 4b) formerly misidentified as *Ostreobium brabantium* (see Radtke et al. 1997a). The hardened surfaces are often paved by large polygonal chlorophyte cells with chaetae (Fig. 4c), rhizoids of *Batophora oerstedtii*, euendolithic networks of the siphonal chlorophyte *Ostreobium quekettii* (Fig. 4d). Among cyanobacterial euendoliths dominant are various species of *Hyella,* which have not been observed anywhere else (Fig. 4e). In addition, Conchocelis stages of a local marine red alga are occasionally observed (Fig. 4f).

The accretion of every stromatolitic structure observed in the Little Darby field is localized to cm sized areas always on the upward surfaces of the structure. The sides of the stromatolites are stationary and firmly cemented. They are externally coated by different mat types (including *Leptolyngbya* spp.) and, under these mats, harbor a highly diverse community of epilithic and euendolithic phototrophs, an assemblage comparable to that found on structures with extended stasis.

The lower parts of the stromatolite side walls are covered by dark, almost black veneer which is locally abraded, and show patches of white carbonate (Fig. 1f, bottom right). These lower parts of the stromatolites and sometimes the entire structure are occasionally completely covered by sand. The stress imposed by light deprivation and exposure to repeated abrasion by sand prevents the growth of most algae and other epilithic organisms, and exerts selective pressure on euendoliths as well. Under the conditions of sand-erosion pressure and repeated covering and uncovering by sand, the euendolithic assemblage has reduced diversity and may be dominated by *Solentia sanguinea* (Fig. 4g).

7 Dynamics of Stromatolite Formation

The dynamics of microbial mats operate at different spatial and temporal scales. Stromatolite accretion at Highborne Cay results from a cycling between three different mat types on the stromatolite surface, with each subsurface layer representing a former surface mat (Reid et al. 2000). Similar dynamics appear to be important in the stromatolites of Little Darby Island.

Fig. 4 (Continued) (**c**) Puzzle shaped chlorophyte cells with tubular bristles lining the surfaces of the walls. The scale in **c** is valid for **d–g**. (**d**) *Ostreobium quekettii* the most common euendolith under low light conditions. (**e**) *Hyella* sp., probably a new species. (**f**) Conchocelis stage of a bangialean rhodophyte. (**g**) *Solentia sanguinea* in erosion stressed base of a stromatolite

The soft mat is dominated by *Schizothrix*, which is responsible for trapping of sand and a vertical growth of local domes (=Type 1 mat of Reid et al. 2000). In the process, trichomes of *Schizothrix* move to the mat surface leaving significant amounts of their photosynthetic product, their empty EPS sheath behind to bacterial consumers. Both layers moved upward in the process of the accretion of the stromatolitic structure, interrupted by short pauses in sedimentation. These pauses are marked by biofilm development and corresponding precipitation of thin micritic crusts (=Type 2 mats). Extended stasis in stromatolite accretion is accompanied by changes in microbial euendolith composition. The changes feature a population increase of *Solentia* – first in the entrapped sand particles, and later from grain to grain over calcified bridges (=Type 3 mat). During these extended pauses in accretion, the stromatolite hardens. The endolith diversity also increases over time (Stolz et al. 2009) and may involve removal and truncation of bored grains by grazers, together with the resident euendoliths. The resumption of stromatolite accretion starts another cycle in mat development, incorporating the truncated grains as a horizon within the stromatolite (Reid et al. 2000). This cyclicity involves alternation of the accretionary phase with mat types 1, with sorter or longer stationary phases resulting in mat types 2 and 3 respectively. A vertical sequence of this cyclicity is historical and in vertical petrographic section it is evident as a stratigraphic record.

Accretional periods may be interrupted by much longer periods of stasis, which permit profound changes in microbial composition and an overgrowth by algae and animals. Similar interruptions may be caused by burial of stromatolites in the sand. In the deep tidal channels of the Lee Stocking Island, where sand moves passing stromatolites in massive dunes, both burial and deprivation of sediment supply tend to occur cyclically, leaving accretion as limited "windows of opportunity" (Golubic and Browne 1996; Seong-Joo Lee et al. 2000). In the case of Little Darby, local sheltering of some stromatolites from sediment supply can result in long periods of stasis with algal overgrowth, whereas burial of stromatolites may take place during major storms.

Acknowledgements This is Research Initiative on Bahamian Stromatolites (RIBS) contribution #59. The international collaboration was supported by the Alexander-von-Humboldt Foundation, Bad Godesberg, and Hanse Wissenschaftskolleg, Delmenhorst, Germany. We appreciate the critical comments of Joachim Reitner and Gernot Arp.

References

Al-Thukair AA, Golubic S (1991a) Five new *Hyella* species from the Arabian Gulf. Algological Studies 64:167–197

Al-Thukair AA, Golubic S (1991b) New endolithic cyanobacteria from the Arabian Gulf. I. *Hyella immanis* sp. nov. Journal of Phycology 27:766–780

Al-Thukair AA, Golubic S, Rosen G (1994) New euendolithic cyanobacteria from the Bahama Bank and the Arabian Gulf: *Hyella racemus* sp. nov. Journal of Phycology 30:764–769

Andres MS, Sumner DY, Reid RP, Swart PK (2006) Isotopic fingerprints of microbial respiration in aragonite from Bahamian stromatolites. Geology 34:973–976

Awramik SM, Riding R (1988) Role of algal eukaryotes in subtidal columnar stromatolite formation. Proceedings of the National Academy of Sciences of the United States of America 85:1327–1329

Baumgartner LK, Reid RP, Dupraz C, Decho AW, Buckley DH, Spear JR, Przekop KM, Visscher PT (2006) Sulfate reducing bacteria in microbial mats: Changing paradigms, new discoveries. Sedimentary Geology 185:131–145

Braissant O, Decho AW, Dupraz C, Glunk C, Przekop KM, Visscher PT (2007) Exopolymeric substances of sulfate-reducing bacteria: Interactions with calcium at alkaline pH and implication for formation of carbonate minerals. Geobiology 5:401–411

Browne KM (1993) Lamination in Recent Bahamian Subtidal Stromatolites: Origin and Lithification. Unpublished PhD dissertation, University of Miami, Coral Gables, pp 296

Decho AW, Visscher PT, Reid RP (2005) Production and cycling of natural microbial exopolymers (EPS) within a marine stromatolite. Palaeogeography, Palaeoclimatology, Palaeoecology 219:71–86

Dill RF, Shinn EA, Jones AT, Kelly K, Steinin RP (1986) Giant subtidal stromatolites forming in normal saline waters. Nature 324:55–58

Dravis JJ (1983) Hardened subtidal stromatolites, Bahamas. Science 219:385–386

Dupraz C, Visscher PT (2005) Microbial lithification in marine stromatolites and hypersaline mats. Trends in Microbiology 13:429–438

Feldmann M, Mackenzie J (1998) Stromatolite thrombolite association in a modern environment, Lee Stocking Island, Bahamas. Palaios 13:201–212

Foster JS, Green SJ (2010) Microbial diversity in modern marine stromatolites. In: Tewari V (ed) Cellular Origin, Life in Extreme Habitats and Astrobiology: Stromatolites. Springer, Berlin, in press

Garcia-Pichel F (2006) Plausible mechanisms for the boring on carbonates by microbial autotrophs. Sedimentary Geology 185:205–213

Golubic S, Browne KM (1996) *Schizothrix gebeleinii* sp. nov. builds subtidal stromatolites, Lee Stocking Island, Bahamas. Algological Studies 83:273–290

Golubic S, Al-Thukair AA, Gektidis M (1996) New endolithic cyanobacteria from the Arabian Gulf and the Bahama Bank: *Solentia sanguinea* sp. nov. Algological Studies 83:291–301

Kobluk DR, Risk MJ (1977a) Micritization and carbonate-grain binding by endolithic algae. American Association of Petroleum Geologist Bulletin 61:1069–1082

Kobluk DR, Risk MJ (1977b) Calcification of exposed filaments of endolithic algae, micrite envelope formation and sediment production. Journal of Sedimentary Petrology 47:517–528

Krumbein WE, Brehm U, Gorbushina AA, Levit G, Palinska KA (2003) Biofilm, biodictyon and biomat – biolaminites, ooolites, stromatolithes – geophysiology, global mechanism and parahistology. In: Krumbein WE, Paterson DM, Zavarzin GA (eds) Fossil and Recent Biofilms. Kluwer Acad Publ, Dordrecht, pp 1–27

Logan BW (1961) Cryptozoon and associate stromatolites from the Recent, Shark Bay, Western Australia. Journal of Geology 69:517–533

Lukas KJ, Golubic S (1981) New endolithic cyanophytes from the North Atlantic Ocean: I. *Cyanosaccus piriformis* gen. et sp. nov. Jounral of Phycology 17:224–229

Macintyre IG, Prufert-Bebout L, Reid RP (2000) The role of endolithic cyanobacteria in the formation of lithified laminae in Bahamian stromatolites. Sedimentology 47:915–921

Pentecost A, Riding R (1986) Calcification in cyanobacteria. In: Leadbeater BSC, Riding R (eds) Biomineralization of lower plants and animals. SAS 30, Clarendon, Oxford, pp 73–90

Planavsky N, Ginsburg RN (2009) Taphonomy of modern marine Bahamian microbialites. Palaios 24:5–17

Radtke G, Golubic S (2010) Microbial euendolithic assemblages and microborings in intertidal and shallow marine habitats: insight in cyanobacterial speciation. In: Reitner J, Queric, N-V, Arp G (eds) Advances in Stromatolite Geobiology. Lecture Notes in Earth Sciences 131, Springer, Berlin, pp 213–241

Radtke G, Le Campion-Alsumard T, Golubic S (1996) Microbial assemblages of the bioerosional "notch" along tropical limestone coasts. Algological Studies 83:469–482

Radtke G, Gektidis M, Golubic S, Hofmann K, Kiene WE, Le Campion-Alsumard T (1997a) The identity of an endolithic alga: *Ostreobium brabantium* Weber-van Bosse is recognized as carbonate-penetrating rhizoids of *Acetabularia* (Chlorophyta, Dasycladales). Courier Forschungsinstitut Senckenberg 201:341–347

Radtke G, Le Campion-Alsumard T, Golubic S (1997b) Microbial assemblages involved in tropical coastal bioerosion: An Atlantic-Pacific comparison. Proceedings of the 8th International Coral Reef Symposium 2:1825–1830

Reid RP, Browne KM (1991) Intertidal stromatolites in a fringing Holocene reef complex, Bahamas. Geology 19:15–18

Reid RP, Macintyre (2000) Microboring versus recrystallization: Further insight into the micritization process. Journal of Sedimentary Research 70:24–28

Reid RP, Macintyre IG, Steneck RS, Browne KM, Miller TE (1995) Stromatolites in the Exuma Cays, Bahamas: Uncommonly common. Facies 33:1–18

Reid RP, Macintyre IG, Steneck RS (1999) A microbialite/algal ridge fringing reef complex, Highborne Cay, Bahamas. Atoll Research Bulletin 466:1–18

Reid RP, Visscher PT, Decho AW, Stolz JF, Bebout BM, Dupraz CP, Macintyre IG, Paerl HW, Pinckney JL, Prufert-Bebout L, Steppe TF, DesMarais DJ (2000) The role of microbes in accretion, lamination and early lithification of modern marine stromatolites. Nature 406:989–992

Reid RP, James NP, Macintyre IG, Dupraz CP, Burne RV (2003) Shark Bay stromatolites: Microfabrics and reinterpretation of origins. Facies 49:45–53

Reitner J (1993) Modern cryptic microbialite/metazoan facies from Lizard Island (Great Barrier Reef, Australia) – Formation and concepts. Facies 29:3–40

Reitner J, Gautret P, Marin F, Neuweiler F (1995) Automicrites in a modern microbialite – Formation model via organic matrices (Lizard Island, Great Barrier Reef, Australia). Bulletin de l´Institut océanographique Monaco 14(2):237–263

Schneider J, Le Campion-Alsumard T (1999) Construction and destruction of carbonates by marine and freshwater cyanobacteria. European Journal of Phycology 34:417–426

Seong-Joo Lee, Browne KM, Golubic S (2000) On stromatolite lamination. In: Riding R, Awramik SM (eds) Microbial sediments. Springer, Berlin, pp 16–24

Stolz JF, Reid RP, Visscher PT, Decho AW, Norman RS, Aspden RJ, Bowlin EM, Franks J, Foster JS, Paterson DM, Przekop KM, Underwood GJC, Prufert-Bebout L (2009) The microbial communities of the modern marine stromatolites at Highborn Cay, Bahamas. Atoll Research Bulletin 567:1–29

Tribollet A, Golubic S, Radtke G, Reitner J (2010) On Microbiocorrosion. In: Reitner J, Queric, N-V, Arp G (eds) Advances in Stromatolite Geobiology. Lecture Notes in Earth Sciences 131, Springer, Berlin, pp 243–253

Trichet J, Défarge C (1995) Non-biologically supported organomineralization. Bulletin de l´ Institut océanographique Monaco Spec 14:203–236

Visscher PT, Reid RP, Bebout BM (2000) Microscale observations of sulfate reduction: Correlation of microbial activity with lithified micritic laminae in modern marine stromatolites. Geology 28: 919–922

Wild C, Laforsch C, Huettel M. (2006) Detection and enumeration of microbial cells within highly porous calcareous reef sands. Marine and Freshwater Research 57:415–420

Molecular Approaches to Studying Living Stromatolites

Brendan P. Burns, Nithya Baburajendran, and Joannita Dharmawan

1 Introduction

The presence of stromatolites dates back to some 3.5–3.8 billion years in earth's history, a time period that witnessed the appearance of the first forms of life (Schopf 2006). The microbial communities that constituted these complex structures played a significant role in the global biogeochemical cycles and sedimentation (Dupraz and Visscher 2005). The impetus behind stromatolite genesis is not clearly known and the modes of accretion, and lithification have long been debated; both biotic and abiotic mechanisms have been suggested for stromatolite formation (Grotzinger and Knoll 1999; Grotzinger and Rothman 1996). Irrespective of their origin, the contributions of stromatolites to the primeval environmental conditions are undisputed. Owing to their importance, stromatolites have formed the core of a growing area of microbial ecology and evolution-based research. Modern stromatolites, the living counterparts of fossilized stromatolites, offer a glimpse as to how microbial communities may have prevailed on the early Earth, and allow unprecedented access to the functional complexity of these systems. Microbial mats that are also considered as living analogues of ancient stromatolites have been studied to interpret ancient stromatolites and have been particularly resourceful in comprehending the formation of extant stromatolites. Layering of lithified microbial mats is in fact attributed to stromatolite

B.P. Burns (✉), N. Baburajendran, and J. Dharmawan
School of Biotechnology and Biomolecular Sciences, University of New South Wales, Sydney, NSW 2052, Australia
Australian Centre for Astrobiology, University of New South Wales, Sydney, NSW 2052, Australia
e-mail: brendan.burns@unsw.edu.au

J. Reitner et al., *Advances in Stromatolite Geobiology*,
Lecture Notes in Earth Sciences 131, DOI 10.1007/978-3-642-10415-2_5,

Fig. 1 Living stromatolites of Shark Bay, Western Australia

development (Dupraz and Visscher 2005). Over the years great progress has been made in unraveling the mysteries of these organosedimentary structures with much focus on the microbial diversity and the process of lithification.

The most extensively studied groups of modern stromatolites belong to the open marine waters of Exuma Sound, Bahamas, and the Hamelin Pool at Shark Bay, Western Australia (Fig. 1). The studies on these groups of stromatolites have uncovered significant data that constitute our present understanding of stromatolite formation and ecology. The initial techniques used for the study of stromatolites have ranged from microscopy to traditional culturing. Progressively RNA and DNA based methods are being utilized. Not more than 1% of microorganisms found in many different environments can be cultured and hence, characterized (Handelsman 2004). Most of the microbial communities such as those associated with stromatolites are highly complex and contain large numbers of microbes; studying such communities by means of traditional methods is not feasible. The advent of culture independent techniques has helped shed light on the "uncultured world" (Handelsman 2004). Molecular techniques that utilize DNA extracted directly from environmental samples have the advantage of overcoming the bias associated with culture-based methodologies, and are hence being preferentially applied in stromatolite research to facilitate numerous downstream analyses.

2 DNA Isolation from Stromatolites: Pros and Cons

The recent research on modern stromatolites vividly exhibits the potential of DNA based techniques in stromatolite biology (Burns et al. 2004, 2005, 2009; Papineau et al. 2005; Green et al. 2008; Baumgartner et al. 2009; Goh et al. 2009;

Allen et al. 2009; Breitbart et al. 2009; Foster et al. 2009). However, the success of such analyses often rest on the first and fundamental rate-limiting step of direct DNA isolation from stromatolites. Extraction of stromatolite community DNA is in fact quite challenging, owing to the presence of cemented calcium carbonate matrix that hinders access to the associated microbial communities. As there are no universal protocols available for direct DNA isolation from stromatolites, some of the various methods that have been employed in recent studies are compared and discussed below.

Most of the studies begin with a physical step that involves pulverisation of the stromatolite sample, usually with a mortar and pestle (Burns et al. 2004; Goh et al. 2009) or a bead-beating procedure (Papineau et al. 2005; Green et al. 2008; Baumgartner et al. 2009; Foster et al. 2009). This physical step is essential to release as much cellular material as possible that is in physical association with sediments/minerals, which will then allow for optimal cell lysis in subsequent chemical steps. As a modification on these initial physical disruption methods, the need to obtain high quality high molecular weight DNA for downstream metagenomic analyses has led some groups to adopt DNA extraction based on combined chemical methods. The presence of exopolymeric substances (EPS) in microbial mats and stromatolites (Decho 1990) in particular is a significant challenge in extracting good quality DNA, and the use of freeze-thaw and enzymatic methods have helped overcome this issue. These have included an extended lysozyme/Proteinase K step of 1 h and having one freeze-thaw cycle (Burns et al. 2004), and up to four and five freeze-thaw cycles, respectively (Goh et al. 2009; Allen et al. 2009) for extraction from Shark Bay stromatolites. These freeze-thaw steps are also thought to improve cell lysis. A further chemical step, the addition of the EPS-binding reagent cetyl trimethyl ammonium bromide (CTAB), is also employed in many protocols to ensure all EPS is removed (Breitbart et al. 2009; Allen et al. 2009; Foster et al. 2009; Goh et al. 2009). In addition to EPS removal, many stromatolite samples have associated calcium carbonate material that needs to be removed before effective cell lysis. A common method for carbonate removal is incubation of samples in an EDTA slurry (0.5 M) for varying times depending on the amount of carbonate in a given sample. Gentle shaking overnight at 37°C has proven successful in dissolving carbonate from stromatolite samples (Wade and Garcia-Pichel 2003), however for particularly carbonate-rich samples from Shark Bay a second overnight incubation with EDTA was required to remove all the carbonate (Allen et al. 2008).

Subsequent to EPS and carbonate removal, many protocols then employ a number of incubation steps designed to both lyse cells and clean up the DNA from associated impurities. A range of buffers have and are being employed in stromatolite DNA extractions, including potassium ethyl xanthogenate (Tillet and Neilan 2000), various combinations of phosphate buffers, TNE buffers and SDS, as well as a range of phenol: chloroform extractions (Burns et al. 2004, 2005, 2009; Papineau et al. 2005; Goh et al. 2009; Allen et al. 2009; Breitbart et al. 2009; Foster et al. 2009). In comparison, some studies have also employed kit-based methods for extraction of total genomic DNA from environmental stromatolite samples (Green et al. 2008; Breitbart et al. 2009; Baumgartner et al. 2009). Examples of these include the FastDNA Spin Kit for Soil (QBiogene), UltraClean Soil DNA Kit (Mo Bio), and the Mo Bio SOIL DNA extraction kit (Mo Bio). As these studies involved examining different kinds of stromatolites from different geographical locations it is difficult to assess the efficacy of the various kits available, however the published data suggest they are all viable options for stromatolite DNA extractions. However, recent work on the metagenomics of microbialites in Mexico (Breitbart et al. 2009) found that these kits were unsuccessful on their samples, possibly due to excess EPS. To overcome this these researchers employed modifications of the freeze-thaw, CTAB, and phenol: chloroform extraction protocols (Breitbart et al. 2009).

All of these methods are aiming for the same outcome: to increase DNA purity and yield. Downstream molecular methods such as PCR are extremely sensitive and require pure forms of DNA; contaminations can often inhibit DNA amplification (Burns et al. 2004). For construction of libraries and analyzing community structures large amounts of DNA are desired with minimal shearing. Thus an ideal protocol should aim at achieving high purity and yield. As described above, the procedures available for the isolation of DNA from stromatolite samples generally consists of combinations of physical, chemical and enzymatic steps. Use of freezing and thawing can result in loss of DNA, however, the amount of contamination and shearing is often less when compared to bead mill homogenization. Proteinase K and lysozyme treatments are the most frequently used enzymatic techniques, and the enzyme concentrations used will depend on the sample size; increasing the enzyme concentration and decreasing sample mass promotes cell lysis resulting in higher DNA yields.

Regardless of the type of cell lysis method used, it is essential to subject the DNA extracts to repeated purification to remove contaminating substances such as humics. Contamination is a common problem in environmental samples and it is necessary to minimize the contaminants as they can eventually affect the processing of the DNA samples. Single purification steps however, often do not completely remove the contaminants; multiple purification cycles can increase the purity but can also lead to loss of DNA. Thus it is often desirable to start with multiple samples of the stromatolite material, and the DNA extracted from these is ultimately pooled to allow for both optimal purity and yield for the purpose accessing the genetic and functional diversity of stromatolites.

3 Molecular Tools: Accessing the Genetic Diversity and Beyond

Once DNA has been successfully extracted from stromatolite material, two molecular approaches in particular have been employed that have revolutionized stromatolite biology. These molecular approaches when compared to the traditional techniques have proved more insightful in establishing the genetic diversity of stromatolites and are currently being utilized to further enhance our knowledge on stromatolite formation, the microbial community structure, and their function.

3.1 PCR-Based rDNA Analysis

Ribosomal RNA gene sequence was first used by Pace et al. (1986) to establish the genetic diversity of an environmental sample without culturing. 16S and 5S rRNA sequences were used for the study. This analysis was based on the knowledge that ribosomal RNA genes served as evolutionary chronometers allowing for the phylogentic classification of microorganisms (Woese 1987). rRNA gene sequences are highly conserved amongst evolutionarily related species. Before the advent of PCR, rRNA analysis was carried out by means of direct sequencing or filter hybridization of rRNA extracts (Stahl et al. 1988), screening of clones from DNA libraries for rDNA inserts and in-vitro amplification of rDNA (obtained by reverse transcription of rRNA) followed by separation on DGGE or TGGE (Wintzingerode et al. 1997). RFLP and ARDA (amplified ribosomal DNA restriction analysis) have also been used to determine the genetic diversity of different microbial communities (Weidner et al. 1996; Köhler et al. 1991).

With the power to directly amplify the DNA extracts from samples, PCR has been used to amplify rDNA genes with the help of specific primers. Such amplicons are used for constructing 16S rDNA libraries, and the clones randomly screened and eventually sequenced. These sequences can then be probed against nucleotide sequences of well-characterized organisms by means of bioinformatic tools thus, providing a culture independent method for ascertaining the microbial diversity in various environments. This technique has been efficiently used to study the microbial ecology of modern stromatolites in various geographical locations, and members of cyanobacteria, bacteria and archaea have been identified (Burns et al. 2004, 2005, 2009; Papineau et al. 2005; Baumgartner et al. 2009; Goh et al. 2009; Allen et al. 2009; Breitbart et al. 2009; Foster et al. 2009).

The early work by Burns et al. (2004) on stromatolites in Hamelin Pool revealed the presence of various cyanobacterial genera including *Synechococcus*, *Xenococcus*, *Microcoleus*, *Leptolyngbya*, *Plectonema*, *Symploca*, *Cyanothece*, *Pleurocapsa* and *Nostoc*. The study also reported the presence of potentially free- living *Prochloron* – a genus commonly found as a symbiont. Eubacteria including well-classified groups like Proteobacteria, low G+C Gram-positive Planctomycetes and Acidobacteria and few unknown groups OP9, OP109 and Marine A group were identified. Halophilic archaea belonging to the divisions Euryarchaeota and Crenarchaeota and methonogenic archaea from the order Methanosarcinales were also detected. In this investigation both culture based and culture independent techniques were used for analysing the genetic diversity of Hamelin Pool stromatolites, however DNA based methods were shown to be more efficient in identifying the stromatolites microbiota (Burns et al. 2004).

Furthermore, the difference in the microbial populations between the different morphological classes, the surface and the interior of Hamelin Pool stromatolites have also been described (Burns et al. 2004; Papineau et al. 2005). The relative abundance of each group of microbes has also been established in all the previous works; approximately 90% are bacterial representatives while 10% are archaeal. *Schizothrix* and *Solentia* that are predominant in Bahamian stromatolites have not been observed in the Hamelin pool stromatolites. Papineau et al. (2005) demonstrated the relationship between stromatolite morphology and the microbial communities. Substantial difference was observed amongst the communities of domal and irregular stromatolites, and also between the surface and endolithic communities. No new phylogenetic groups were described in the study, however, the cyanobacterial representatives which have long been believed to dominate the stromatolite communities where shown to be less abundant. Such low abundance has also been seen in case of the Bahamian stromatolites. Other microorganisms thus contribute to the biomass. Though sometimes present in small numbers Papineau et al. (2005), cyanobacteria are considered as the primary producers while some anoxygenic phototrophs have also been deemed as contributors of the overall primary productivity. The mass and turnover rate of the primary producers control the size of other microbes required to maintain a community and ultimately influence stromatolite morphogenesis.

3.2 *Metagenomics*

Metagenomics, also known as environmental/community/population genomics involves the direct DNA isolation from a community of microorganisms and construction of a genomic DNA library, which can be screened for various aspects. There are two fundamental but complementary approaches to a metagenomic analysis: sequence-based analyses and functional analyses. In sequence-based analysis, the clones from a metagenomic library are sequenced randomly and genes of interests are identified. The phylogenetic markers flanking the desired gene are then used to assign a taxonomic group to the gene. Thus the organism and its characteristic feature can be identified. Another methodology involves the sequencing of DNA segments from a specific taxon. Though this technique is mainly based on phylogenetic markers, due to the lack of significant numbers of such markers, a variation in this approach involving the random sequencing of the clones on a large scale has been devised. With the vast amount of sequence data obtained the dynamics of the community can be determined (Pedros-Alio 2006).

In functional metagenomcis, the metagenomic library is expressed and the functions of the clones analysed. Thus by screening the metagenome, specific genes can be identified without prior knowledge of the sequence. Such functional screens can also be used to assign functions to uncharacterized proteins. Once the clones carrying the genes of interest are identified in the metagenomic library, the clones can be sequenced (Ferrer et al. 2009). A further interesting aspect towards this approach is that clones that have similar functions can be grouped and sequenced and phylogenetic anchors can be identified; thus the diversity of genomes sharing the same functions can be characterized (Handelsman 2004).

In a recent study by Desnues et al. (2008) of the modern marine stromatolites from HighBorne Cay, Bahamas and freshwater stromatolites and thrombolites from Pozes Azules II and Rio Mesquites, Mexico, sequence based metagenomic analysis of the viral communities was carried out and the biogeography of the phages was established. The sequences of the three metagenome were compared with each other and the distribution of the viral population was determined. The lack of sequence overlaps (cross-contigs) between the metagenomes indicated the presence of genetically unique viral particles in each microbialite. This difference in the viral community between geographically separated microbialites has been associated with the evolutionary origin of these structures.

4 The Future of Stromatolite Research

The 16S rRNA based PCR techniques have been particularly useful in establishing the genetic diversity of extant stromatolites. Analysis of the microbial diversity is one of the first steps towards comprehension of the complex communities that constitute the evolutionarily and ecologically significant stromatolites. Understanding the relationships between the microbial communities, the metabolic activities that sustain these communities and ultimately influence stromatolite formation will form the premise for future research. PCR based methods present certain limitations with such aspects as they do not allow for functional analyses, and a powerful tool such as metagenomics will have a significant role in such endeavours. Nevertheless, 16S rRNA analyses can complement metagenomics; the clones of a stromatolite metagenome can be screened for 16S phylogenetic markers using PCR and the functions can be correlated with specific organisms. Furthermore, characterisation of both novel organisms and genes, as well as developing new molecular techniques are providing an excellent platform for future research on stromatolites (Pomati et al. 2004; Goh et al. 2006; Marshall et al. 2007; Leuko et al. 2007, 2008, 2009; D'Elia et al. 2007; Münchoff et al. 2007; Allen et al. 2008).

Various aspects of microbial ecology of complex communities have been uncovered by means of metagenomics. Symbiotic relationships, competition, communication, accumulation of nutrients and energy production in diverse environments have been studied with the help of metagenomics. Thus there is significant potential in utilizing metagenomics in the study of stromatolite ecology. Bioactive compounds such as antibiotics, secondary metabolites and enzymes have also been discovered by means of metagenomics. Secondary metabolite production in stromatolites has been documented; a putative anticancer agent has been reported in Hamelin pool stromatolites (Burns et al. 2005). Thus screening the stromatolite metagenomes for the synthesis of substances that have possible biotechnological applications is another fascinating aspect of stromatolite research. The knowledge garnered through such potential studies can shed light on the past and will also help appreciate the present ecological roles played by modern stromatolites.

References

Allen MA, Goh F, Leuko S, Echigo A, Mizuki T, Usami R, Kamekura M, Neilan BA, Burns BP (2008) *Haloferax elongans* sp. nov. and *Haloferax mucosum* sp. nov., isolated from microbial mats from Hamelin Pool, Shark Bay. International Journal of Systematic and Evolutionary Microbiol 58:798–802

Allen MA, Goh F, Burns BP, Neilan BA (2009) Bacterial, archaeal and eukaryotic diversity of microbial mat communities in the hypersaline lagoon of Shark Bay. Geobiology **7**:82–96

Baumgartner LK, Spear JR, Buckley DH, Pace NR, Reid RP, Dupraz C, Vissche PT (2009) Microbial diversity in modern marine stromatolites, Highborne Cay, Bahamas. Environmental Microbiology 11(10): 2710–2719

Breitbart M, Hoare A, Nitti A, Siefert J, Haynes M, Dinsdale E, Edwards R, Souza V, Rohwer F, Hollander D (2009) Metagenomic and stable isotopic analyses of modern freshwater microbialites in Cuatro Cienegas, Mexico. Environmental Microbiology 11:16–34

Burns BP, Goh F, Allen M, Neilan BA (2004) Microbial diversity of extant stromatolites in the hypersaline marine environment of Shark Bay, Australia. Environmental Microbiology 6:1096–1101

Burns BP, Seifert A, Goh F, Pomati F, Neilan BA (2005) Genetic potential for secondary metabolite production in stromatolite communities. FEMS Microbiology Letters 243:293–301

Burns BP, Anitori R, Butterworth P, Henneberger R, Goh F, Allen MA, Ibanez-Peral R, Bergquist PL, Neilan BA, Walter MR (2009) Modern analogues and the early history of microbial life. Precambrian Research 173:10–18

D'Elia M, Blanco A, Liocchelli D, Orofino V, Fonti S, Pomati F, Burns BP (2007) Physical characterisation of biotic and abiotic minerals of exobiological interest. Il Nuovo Cimento 121:833–842

Decho A (1990) Microbial exopolymer secretions in ocean environments: their role(s) in food webs and marine processes. Oceanography & Marine Biology Annual Review 28:73–154

Desnues C, Rodriguez-Brito B, Rayhawk S, Kelley S, Tran T, Haynes M, Liu H, Furlan M, Wegley L, Chau B, Ruan Y, Hall D, Angly FE, Edwards RA, Li L, Thurber RV, Reid RP, Siefert J, Souza V, Valentine DL, Swan BK, Breitbart M, Rohwer F (2008) Biodiversity and biogeography of phages in modern stromatolites and thrombolites. Nature 452:340–343

Dupraz C, Visscher PT (2005) Microbial lithification in marine stromatolites and hypersaline mats. Trends in Microbiology 13:429–438

Ferrer M, Beloqui A, Timms KN, Golyshin PN (2009) Metagenomics for mining new genetic resources of microbial communities. Journal of Molecular Microbiology and Biotechnology 16:109–123

Foster JS, Green SJ, Ahrendt SR, GolubiC S, Reid RP, Hetherington KL, Bebout L (2009) Molecular and morphological characterization of cyanobacterial diversity in the stromatolites of Highborne Cay, Bahamas. ISME Journal 3:573–587

Goh F, Leuko S, Allen MA, Bowman JP, Kamekura M, Neilan BA, Burns BP (2006) *Halococcus* hamelinensis sp. nov., a novel halophilic archaeon isolated from stromatolites in Shark Bay, Australia. International Journal of Systematic and Evolutionary Microbiology 56:1323–1329

Goh F, Allen MA, Kawaguchi T, Decho AW, Neilan BA, Burns BP (2009) Determining the specific microbial populations and their spatial distribution within the stromatolite ecosystem of Shark Bay. ISME Journal 3:383–396

Green SJ, Blackford C, Bucki P, Jahnke LL, Prufert-Bebout L (2008) A salinity and sulfate manipulation of hypersaline microbial mats reveals stasis in the cyanobacterial community structure. ISME Journal 2:457–470

Grotzinger JP, Knoll AH (1999) Stromatolites in Precambrian carbonates: evolutionary mileposts or environmental dipsticks? Annual Review of Earth and Planetary Sciences 27:313–358

Grotzinger JP, Rothman DH (1996) An abiotic model for stromatolite morphogenesis. Nature 383:423–425

Handelsman J (2004) Metagenomics: application of genomics to uncultured microorganisms. Microbiology and Molecular Biology Reviews 68:669–685

Köhler G, Ludwig W, Schleifer KH (1991) Differentiation of lactococci by rRNA gene restriction analysis. FEMS Microbiology Letters 84:307–312

Leuko S, Allen MA, Goh F, Burns BP, Walter, MR, Neilan BA (2007) Analysis of intergenic spacer region length polymorphisms to investigate the halophilic archaeal diversity of stromatolites and microbial mat. Extremophiles 11:203–210

Leuko S, Goh FR, Ibáñez-Peral R, Burns BP, Walter MR, Neilan BA (2008) Lysis efficiency of standard DNA extraction methods for *Halococcus* sp. in an organic rich environment. Extremophiles 12:301–308

Leuko S, Raftery M, Burns BP, Walter MR, Neilan BA (2009) Global protein-level responses of *Halobacterium salinarum* NRC-1 to prolonged changes in external sodium chloride concentrations. Journal of Proteome Research 8:2218–2225

Marshall CP, Leuko S, Coyle CM, Walter MR, Burns BP, Neilan BA (2007) Carotenoid analysis of halophilic archaea by resonance Raman spectroscopy. Astrobiology 7:631–643

Münchoff, J, Hirose H, Maruyama T, Burns BP, Neilan BA (2007) Phylogeography of *Prochloron* and its didemnid ascidian host. Environmental Microbiology 9:890–899

Pace NR, Stahl DA, Lane DJ, Olsen GJ (1986) The analysis of natural microbial populations by ribosomal RNA sequences. Advances in Microbial Ecology 9:1–55

Papineau D, Walker JJ, Mojzsis SJ, Pace NR (2005) Composition and structure of microbial communities from stromatolites of Hamelin pool in Shark Bay, Western Australia. Applied and Environmental Microbiology 71:4822–4832

Pedros-Alio C (2006) Genomics and marine microbial ecology. International Microbiology 9:191–197

Pomati F, Burns BP, Neilan BA (2004) Identification of a Na^+ Dependent transporter associated with saxitoxin producing strains of the cyanobacterium *Anabaena circinalis*. Applied and Environmental Microbiology 70:4711–4719

Schopf JW (2006) Fossil evidence of Archaean life. Philosophical Transactions of the Royal Society of London B 361:869–885

Stahl DA, Flesher B, Mansfield HR, Montgomery L (1988) Use of phylogenetically based hybridization probes for studies of ruminal microbial ecology. Applied and Environmental Microbiology 54:1079–1084

Tillet T, Neilan BA (2000) Xanthogenate nucleic acid isolation from cultured and environmental cyanobacteria. Journal of Phycology 36:251–258

Wade BD, Garcia-Pichel F (2003) Evaluation of DNA extraction methods for molecular analyses of microbial communities in modern calcareous microbialites. Geomicrobiology Journal 20:549–561

Weidner S, Arnold W, Puhler A (1996) Diversity of uncultured microorganisms associated with the seagrass *Halophila stipulacea* estimated by restriction fragment length polymorphism analysis of PCR-amplified 16S rRNA genes. Applied and Environmental Microbiology 62:766–771

Wintzingerode FV, Göbel UB, Stackebrandt E (1997) Determination of microbial diversity in environmental samples: pitfalls of PCR-based rRNA analysis. FEMS Microbiology Reviews 21:213–229

Woese CR (1987) Bacterial evolution. Microbiological Reviews 51:221–271

Magnesium Inhibition Controls Spherical Carbonate Precipitation in Ultrabasic Springwater (Cedars, California) and Culture Experiments

Patrick Meister, Orion Johnson, Frank Corsetti, and Kenneth H. Nealson

1 Introduction

The formation of spherical mineral aggregates is commonly observed in nature and well-known from carbonates formed in marine and hypersaline environments and based on their internal structure called ooids, pellets, pisolites, etc. (e.g. Bathurst 1967; Donahue 1969; Davies et al. 1978; Tucker 1984; Tucker and Wright 1990). Since spherical carbonates are often associated with microbial precipitation, many studies have suggested that the spherical crystal shape is a result of microbial carbonate formation (e.g. Gerdes et al. 1994; Vasconcelos et al. 1995; Verrecchia et al. 1995; Rivadeneyra et al. 2006). Recently, laboratory experiments using the pure Ca-carbonate system were successful in artificially reproducing the spherical crystal formation (e.g. Cölfen and Qi 2001; Dupraz et al. 2004; Braissant et al. 2003) by adding different polymeric sugars and amino acids as analogue for microbial extracellular polymeric substances (EPS). Also, Mitterer (1972), Suess and Fütterer (1972), Davies et al. (1978), Gonzáles-Muñoz et al. (2000), and Meldrum and Hyde (2001) showed in experiments a formation of spherical carbonates by interaction of organic compounds and dissolved Mg^{2+}. Spherical crystals at nanometer scales have first been interpreted by Folk (1999) as nanobacterial fossils. According to results from culture experiments (Warthmann et al. 2000; Bontognali et al. 2008) and field studies (Bontognali et al. 2008; Sánchez-Román et al. 2008), however, nanocrystals are associated with, but not entombing microbial cells. Instead, the spherical shape was suggested to be influenced by EPS within the microbial biofilm. Considering a pure Ca-carbonate system, Bosak and

P. Meister (✉)
Max-Planck-Institute for Marine Microbiology, Celsiusstr. 1, 28359 Bremen, Germany
e-mail: patrick.meister@alumni.ethz.ch

O. Johnson, F. Corsetti, and K.H. Nealson
Department of Earth Sciences, University of Southern California, 3651 Trousdale Parkway, Los Angeles, CA 90089-0740, USA

J. Reitner et al., *Advances in Stromatolite Geobiology*,
Lecture Notes in Earth Sciences 131, DOI 10.1007/978-3-642-10415-2_6,

Newman (2005) experimentally demonstrated the growth of star shape aggregated crystals in the presence of sulphate reducing bacteria vs. rhombohedral crystals under abiotic conditions. Even if the outer shape appeared somewhat rounded, no radial aggregation was obvious from the extinction under cross-polarized light in the interior of the crystals. Despite these different observations, the molecular mechanism and kinetic control of spherical crystal formation remains poorly understood and some of the conclusions remain contradictory. Furthermore it is still unclear how the spherical vs. rhombohedral crystal formation is controlled in the natural environment and in the geological past.

In this study, spherical and rhombohedral calcite and aragonite precipitation are described from Cedars Springs (California), where carbonate sinter forms from highly alkaline water (pH $>$ 11) derived from continental serpentinization (Barnes et al. 1967; Barnes and O'Neil 1969) of unaltered peridotite within the Coast Range Ophiolite of California. The carbonates, which are clearly abiotically induced, are petrographically described and mineralogically analyzed by X-ray diffractometry. Carbonate $\delta^{13}C$ and $\delta^{18}O$ data are presented and the findings are discussed in relation to the mineralogy, carbonate morphology and water chemistry. The field observations were simulated in precipitation experiments using cultures of different strains of the bacterial genus *Shewanella*, a heterotrophic bacterium isolated from aquatic environments and commonly used for microbial culture experiments. As at Cedars Springs, the Mg/Ca-carbonate system was considered with variable Mg/Ca ratios, and the same experiments were performed with abiotic controls, where bicarbonate was added. Processes leading to carbonate precipitation are discussed, based on simple morphological observations of carbonate crystals grown at Cedars Springs and the precipitation experiments under different conditions in order to better understand the biotic vs. abiotic control of spherical and rhombohedral crystal formation. In particular, it is discussed how these processes are controlled in the environment such as alkaline systems, but also under marine conditions. An alternative model to the often discussed influence of EPS on spherical crystal formation is suggested, and Cedars spring carbonates are discussed as a model for the influence of ultrabasic vents on secular variation of ocean chemistry and on the precipitation of spherical vs. rhombohedral carbonate. This study aims at a better understanding of carbonate crystal structures occurring throughout the geological record.

2 Study Area

The Cedars Springs (Fig. 1) occur as a series of highly alkaline springs at the bottom of a deeply incised valley within the ultrabasic Cedars block, located 15 km north of Cazadero, Sonoma County, California. The Cedars block is one of several remnants of the Coast Range Ophiolite of California (Geological Map of California, USGS), which were underthrusted during the Cretaceous by the

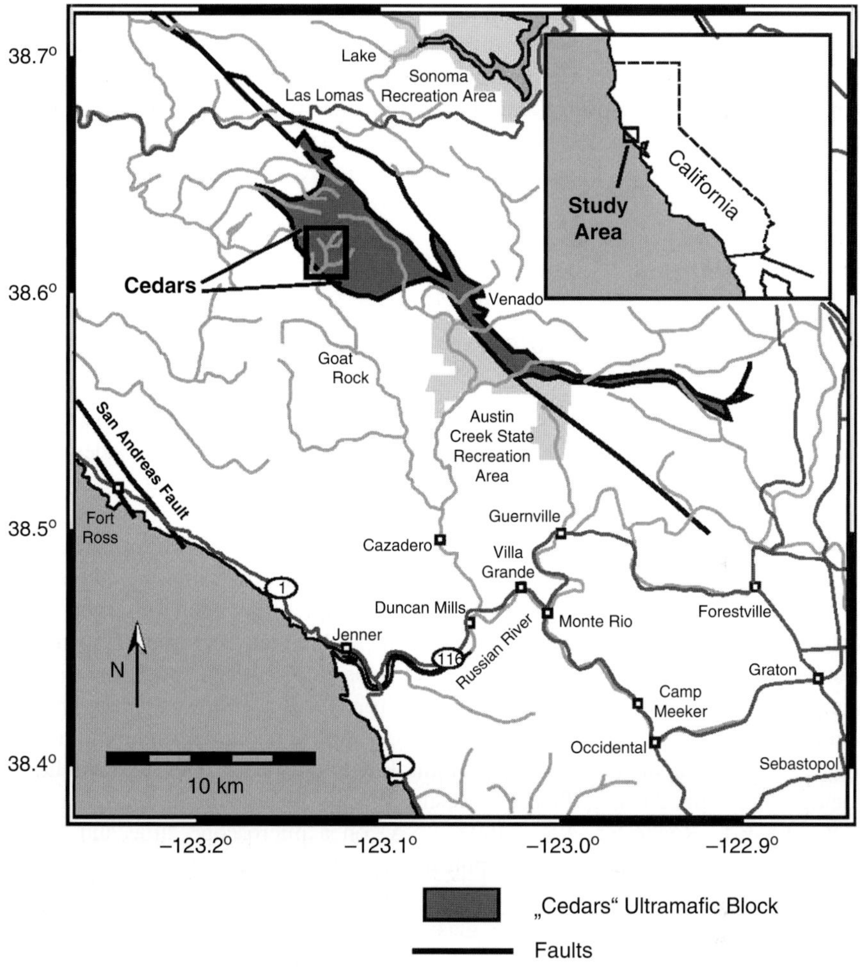

Fig. 1 Map and satellite image of the Cedars Alkaline Spring complex, California

Franciscan accretionary prism and juxtaposed on high-pressure, low-temperature metamorphic rocks during later normal faulting (Platt 1986). The unaltered state of the peridotite due to its tectonic history makes it prone for serpentinization. Analysis of the water chemistry shows that the spring water is meteoric water, which has interacted with the ultrabasic rock during terrestrial serpentinization (Barnes and O'Neil 1969). The waters derived from serpentinization are commonly depleted in Mg^{2+}, but contain Ca^{2+} due to the leaching of clinopyroxene. Cedars spring waters reach Ca^{2+} concentrations of up to 1.5 mM, but contain essentially no Mg^{2+} (Barnes et al. 1967). The alkaline spring waters with a pH > 11 strongly contrast the surface water flowing in the nearby creek. The latter shows a pH of 8 and is enriched in Mg^{2+} (up to 1.5 mM), whereas the Ca^{2+} concentration is relatively low (~100 μM; Barnes et al. 1967). The two different water types, however, are commonly mixed, since the pools are located nearby or within the creek. Preliminary water analysis shows the presence of H_2 and H_2S (cf. Sleep et al. 2004), which also commonly results from the serpentinization reaction. The water chemistry is variable between different springs and there is some site to site variation of carbonate precipitation.

3 Methods

3.1 Petrography and Isotopes

Ten carbonate samples were collected at various alkaline springs of the Cedars Ultrabasic Complex in July 2006. Small pieces were scraped off the sides of the sinter terraces. More friable carbonate from the inside of the different pools could be easily collected with the sample tubes. The tubes were filled with water of the pools and the samples were stored refrigerated until microscopy. Microphotographs of the fresh sample material were taken within one week after sampling. No change in shape and amount of precipitates of the samples has been observed since collection. Large specimens were taken from the laminated riverbed cement and from fragments of older terrace walls, and slabs were cut by a rock saw for the preparation of thin sections. Brittle material from the field site was gently broken, mounted on a glass slide in the wet state and viewed under a petrographic microscope with normal and cross-polarized light. Precipitates from both solid and liquid media were embedded in Piccolyte resin (Ward's Natural Science, 5100 West Henrietta Rd., Rochester, NY, 14692, USA) on a microscope slide and subsequently inspected under the petrographic microscope. Samples for X-ray diffraction (XRD) were suspended in ethanol, finely ground in an agar mortar and pestle and mounted on the frosted side of a glass slide. The samples were continuously scanned from 5 to 65° using Cu Kα radiation on a Rigaku Geigerflex D/max-A XRD at the Earth Science Department, University of Southern California.

Carbon and oxygen isotopes in powdered bulk and micro-drilled sub-samples were analysed with a Finnigan MAT 251 mass spectrometer coupled to an automated acidification device (type "Kiel") at the Geological Institute, University of Bremen. The analytical precision of the mass spectrometer is +0.05‰ for $\delta^{13}C$, and +0.07‰ for $\delta^{18}O$. Powdered Solnhofner Plattenkalk calibrated against the NBS standard was used as a working standard, and the $\delta^{13}C$ and $\delta^{18}O$ values of the carbonates are reported relative to the Vienna Pedee Belemnite Standard (VPDB).

3.2 Precipitation Experiments

Precipitation experiments were performed in solutions with 10 mM $CaCl_2$, and 10 mM $CaCl_2$ + 50 mM $MgCl_2$, respectively (Table 1), which is the concentration

Table 1 Composition of the solutions and media used for precipitation experiments

	Control solution	Medium
For carbonate supersaturation		
$MgCl_2$	50.00	50.00
$CaCl_2$	10.00	10.00
$NaHCO_3$	10.00	10.00
pH	8.50	8.50
Mineral salts		
NaCl	–	30.00
NH_4Cl	–	28.40
KCl	–	1.34
Yeast extract (PO_4^{3-} source) (g/l)	–	0.10
Nutrients		
Trace elements[a] (100× stock)	–	10 ml
Vitamine mix[b] (100× stock)	–	10 ml
Na-acetate	–	10.00
For solid media		
Agar, noble (g/l)	–	10.00
Supplements		
SO_4^{2-}	28.00	28.00
Yeast extract (g/l)		
Microbial cells		
NaCl	30.00	
NH_4Cl	28.40	
KCl	1.34	

[a]Wolfe mineral solution (Wolin et al. 1963)
[b]Myers and Nealson (1988)

of these ions in modern seawater. De-ionized water from a Millipore filter apparatus was used for the experiments. The pH was adjusted to 8.5 by adding NaOH. $NaHCO_3$ was prepared as a 100× stock solution and added to reach a final concentration of 10 mM. The same amount of $NaHCO_3$ was added to pure spring water. Different experiments were performed, where 0.01 g/l yeast extract, 28 mM SO_4^{2-}, 10 ml/l amino acid solution, 50 mM Na-lactate, and different salt concentrations were added, which may influence the precipitation kinetics and, hence, mineralogy and crystal structure.

In order to test for biological effects on mineralogy and crystal structure, precipitation experiments were performed with aerobic heterotrophic bacteria of the species *Shewanella*. Liquid and solid minimal media for *Shewanella* were used [Table 1; modified after Myers and Nealson (1988)], supplemented with the same constituents as in the abiotic experiments. Strains of *Shewanella* were chosen, because they commonly occur in natural aquatic environments and can also be easily grown under aerobic conditions. No bicarbonate was added to the cultures since alkalinity was produced bacterially when acetate was added to the medium. Both cultures and controls were incubated at 30°C from 24 h to several days. The pH was monitored using a pH electrode for the liquid media, and pH indicator paper for the colonies formed on the agar plates. Crystal growth was monitored under a dissecting microscope, and crystals were further analysed by microscopy and XRD.

4 Results

4.1 Petrography and Mineralogy of Spring Carbonates

Several carbonate sinter terraces with m^2 sized pools are forming downstream of the different springs (Fig. 2a). The walls of the terraces were observed to form rapidly within a single season (Johnson, pers. comm.) and they consist of hard cemented, often laminated calcite and sometimes aragonite (Fig. 2b). Aragonite appears to be present where ultrabasic spring water is mixed with river water, whereas terraces that form exclusively from spring water consist of calcite. Further downstream, carbonate precipitates occur throughout the riverbed (Fig. 2c), where

Fig. 2 Photograph of the Cedars ultrabasic spring complex showing (**a**) the different pools and terraces, at Barnes Spring. (**b**) Close up of (**a**). (**c**) Photo of riverbed with (**d**) close up showing regular lamination in the rock

blocks and gravel of ultrabasic rock are cemented by meter thick layers of laminated calcite and aragonite cement (Fig. 2d). In contrast to the terrace walls and massive precipitates in the riverbed, the bottoms of the pools are covered with slower growing fragile and delicately branched precipitates of carbonate. Often, a thin layer of carbonate "ice", a swimming layer of carbonate at the air/water interface, forms especially at the pools nearest to the source.

Microscopic analysis of the spring carbonates reveals that calcite cements in proximity to the sources consist mainly of perfectly developed, around 50 μm sized rhombohedral crystals (Fig. 3a). These rhombohedral calcites seem to predominantly form the hard cemented terrace walls, whereas finer crystal needles and bundles of needles were recovered as friable carbonate in the stagnant pools. In smear slides of scraped off material from the terraces, perfectly developed carbonate rhombs were observed. Further downstream, where river water is significantly contributing to the mixed water in the pools, up to 100 μm sized spherical aggregates or bundles of crystal needles are formed (Fig. 3b, c). In general, the relative amount of spherical crystals in the sample correlates with the amount of aragonite

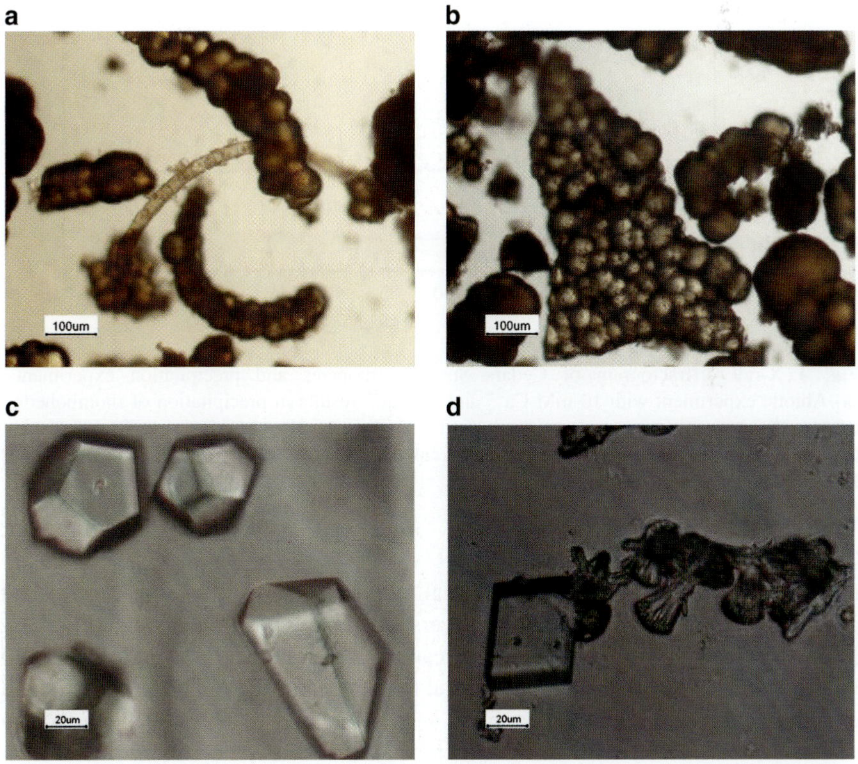

Fig. 3 Photomicrographs of carbonates collected from Cedars ultrabasic spring complex: (**a**) and (**b**) show spherical and bothryoidal carbonate (aragonite) forming friable precipitates in the pool; (**c**) shows the rhombohedral calcite from hard encrusted terraces. (**d**) Showing the two distinctly different morphologies of crystal fans and rhombohedral crystals

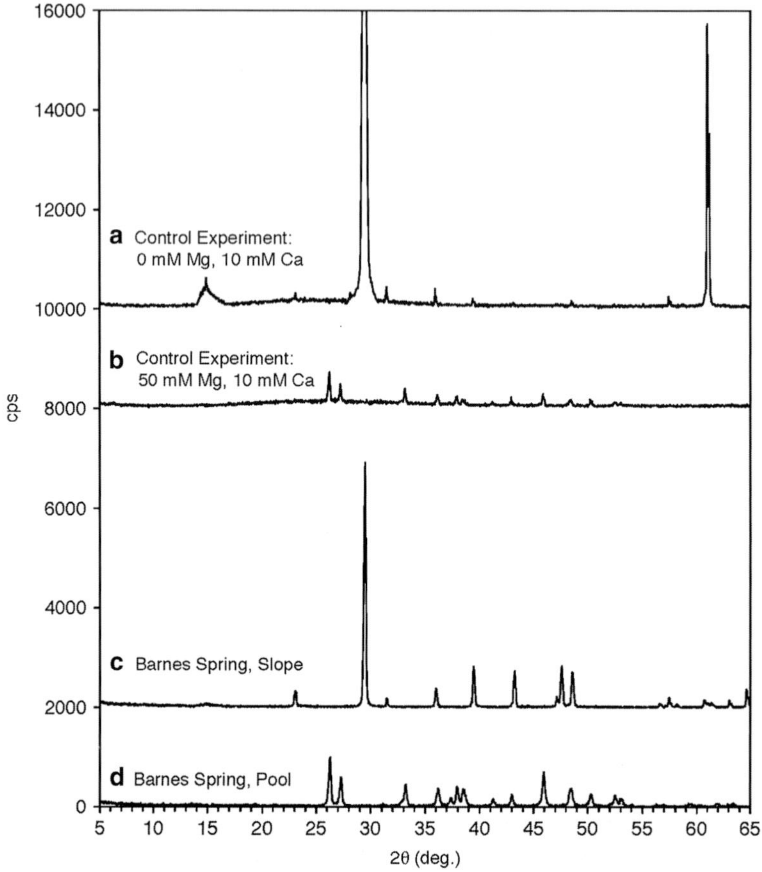

Fig. 4 X-ray diffractogram of Cedars Spring carbonates and precipitation experiments: (**a**) Abiotic experiment with 10 mM Ca^{2+} and no Mg^{2+} results in precipitation of rhombohedral calcite. (**b**) Aragonitic mineralogy results after addition of 50 mM Mg^{2+}. (**c**) Rhombohedral carbonate from hard encrusted terraces form calcite; (**d**) spherical carbonates from the pools show aragonitic mineralogy

detected by XRD (Fig. 4). The spherical carbonates show a smooth surface and often grow together in sheaths or layers, incrusting algal filaments (Fig. 3c). Microscopic analysis of the laminated carbonates reveals an alternation between laminae with spherical crystals and crystal fans consisting of aragonite, and laminae of blocky calcite cement (Fig. 5a, b).

For comparison, it is to mention here that also carbonate chimneys found in submarine ultrabasic springs, such as the Lost City Hydrothermal Field (Kelley et al. 2001, 2005; Von Damm 2001; Früh-Green et al. 2003, 2007) consist predominantly of aragonite forming delicately branched structures, and it seems from Ludwig et al. (2006; Fig. 4a therein) that branches consist of crystal fans.

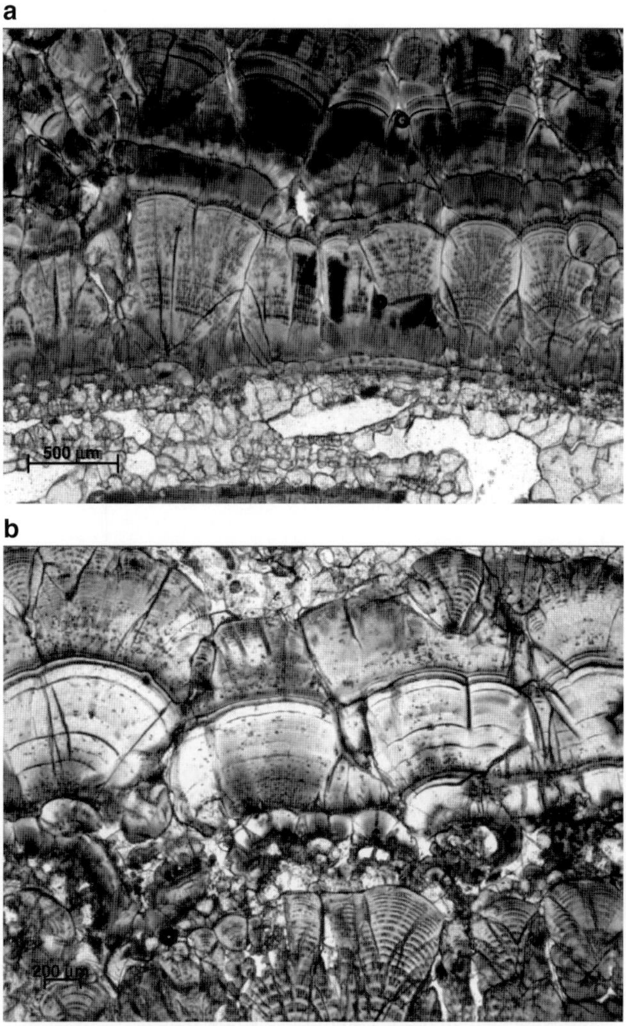

Fig. 5 Thin section photomicrograph of laminated spring carbonate showing alternation between spherical crystal fans and blocky cement. (**a**) The crystal fans are growing out of point sources of truncation surfaces. (**b**) Crystal fans showing variable radius in different laminae

4.2 Carbon and Oxygen Isotopes

Stable C and O isotope signatures were measured for ten samples collected from separate locations in order to represent the whole range of isotope values that occur in the spring carbonates (Table 2). In a crossplot (Fig. 6), values are scattered between −3‰ and −20‰ for $\delta^{18}O$ and between −10‰ and −25‰ for $\delta^{13}C$ with

Table 2 Mineralogy, $\delta^{13}C$ and $\delta^{18}O$ values of Cedars Springs carbonates

Site	Sample	Aragonite (% of carbonate)	$\delta^{13}C$ (% VPDB)	$\delta^{18}O$ (% VPDB)
NS06	Hand specimen, white lamina	31.65	−12.03	−4.16
NS06	Hand specimen, beige lamina	3.27	−12.92	−5.11
NS06	Hand specimen, white hard lamina	14.91	−16.21	−8.72
NS06	Hand specimen, beige, fine lamina	0.00	−20.21	−14.11
NS06-4	Pool	100.00	−14.18	−3.47
BS06-1	Pool	17.01	−23.96	−18.24
BS06-1/2	Slide	0.00	−25.03	−17.59
BS06-2	Terrace		−17.52	−10.30
BS06-2	Pool	100.00	−12.29	−4.02
BS06-7	Crust		−24.15	−17.62
BS06-8	Pool		−23.84	−15.60

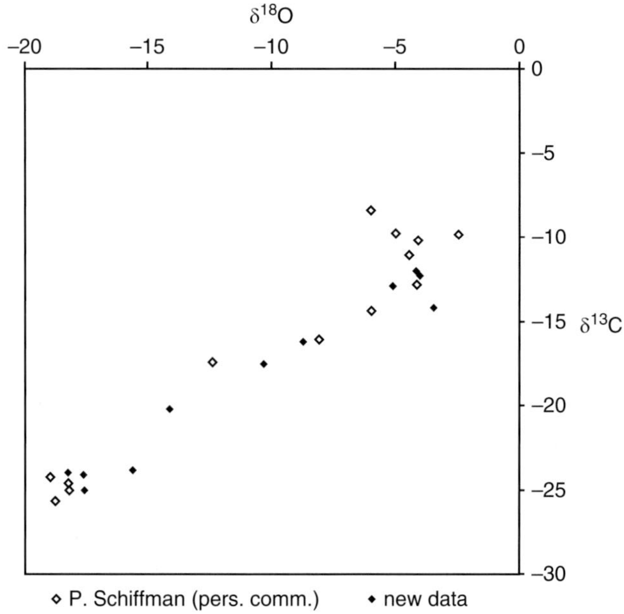

Fig. 6 Cross-plot of isotopic signature from aragonite (*open diamonds*) and calcite (*closed diamonds*), and superimposed data from P. Schiffman (pers. comm.; crosses). The data align between $\delta^{18}O = -4‰$, $\delta^{13}C = -10‰$ and $\delta^{18}O = -20‰$, $\delta^{13}C = -25‰$, with reflects more or less the distance from the source. Now direct correlation with the mineralogy was observed

all values aligned between these two end-member concentrations. The values agree well with the scatter observed in an independent study (P. Schiffman, pers. comm.), supporting that the selection of samples represents the range of values present in the spring carbonates. However, if the data are plotted versus the relative amount of aragonite in the samples, no clear trend is observed. If the distribution of $\delta^{18}O$ and $\delta^{13}C$ values at the field sites is considered, the major trend indicates progressively less negative values downstream from the sources. A minor trend is that carbonates within the pools are slightly less negative than values from the fast growing terrace walls of the same pools.

The $\delta^{18}O$ value of the water, according to Barnes and O'Neil (1969) is −8.2‰ for Mg–HCO_3 water (Adobe Canyon, Red Mountain, California) and −8.7‰ for the Ca–OH water (at Blackbird Valley, Red Mountain). P. Schiffman (pers. comm.) measured values of ~6.5‰ for the water at Cedars Springs, in general. These values indicate a meteoric origin of the serpentinizing fluid, which, according to Barnes and O'Neil (1969) is not strongly altered by interaction with the ultrabasic rock, probably because of high flow rates compared to the rates of serpentinization. The isotopic composition of the water is in equilibrium with the positive end-member of the range in $\delta^{18}O$ values of the spring carbonates under surface temperature. Due to extremely low dissolved inorganic carbon (DIC) concentrations of the spring water (most DIC is taken up by the $pH > 11$ water as CO_2 from the atmosphere), the $\delta^{13}C$ of DIC was not determined.

4.3 *Precipitation Experiments*

Spherical aggregates of aragonite were precipitated from a solution of pure water with a normal marine concentration of 50 mM Mg^{2+} and 10 mM Ca^{2+} (Fig. 7a). The crystals show the typical extinction cross under polarized light (Fig. 7b) due to the radial aggregation of the needle shaped crystals. The size of the aggregates varies between 10 and 100 μm and always two generations of aggregates occurred, which differed in crystal size, shape, and time (usually over night) of precipitation. In general, the spheres are not perfectly spherical and the somewhat fuzzy rims are due to the different lengths of the crystal needles. When no Mg^{2+} was added, small well-defined calcite rhombs and no radial aggregates developed in experiment solutions (Fig. 7c, d). When supplements were added to 10 mM $CaCl_2$ solution without the addition of Mg^{2+}, in general no significant change in crystal structure was observed, except, when 28 mM SO_4^{2-} was added, clearly elongate crystals developed. The elongated crystals frequently form twins, orienting themselves to each other nearly at 90°. No crystals formed in a solution with 10 mM $CaCl_2$ and 0.01 g/l yeast extract.

Carbonate spheres were precipitated on agar plates, when supplemented with 50 mM Mg^{2+} and 10 mM Ca^{2+} and inoculated with strains of *Shewanella* (Fig. 8a–f). In liquid media, microbially induced precipitation only occurred, when a bacterial biofilm (pellicle) formed at the surface of the solution. The crystals that formed on the agar plates commonly occurred within the bacterial colonies, although in a few cases crystal formation was observed inside of the agar, remote from the colonies. Crystals on the agar plates only formed when the bacteria were fully grown, which usually was the case after 48 h. The pH in the different colonies commonly reached 8.5–9, independent of the initial pH. The carbonate spheres were perfectly developed with a smooth surface and extremely fine, radially aggregated crystal needles showing a typical cross of extinction under

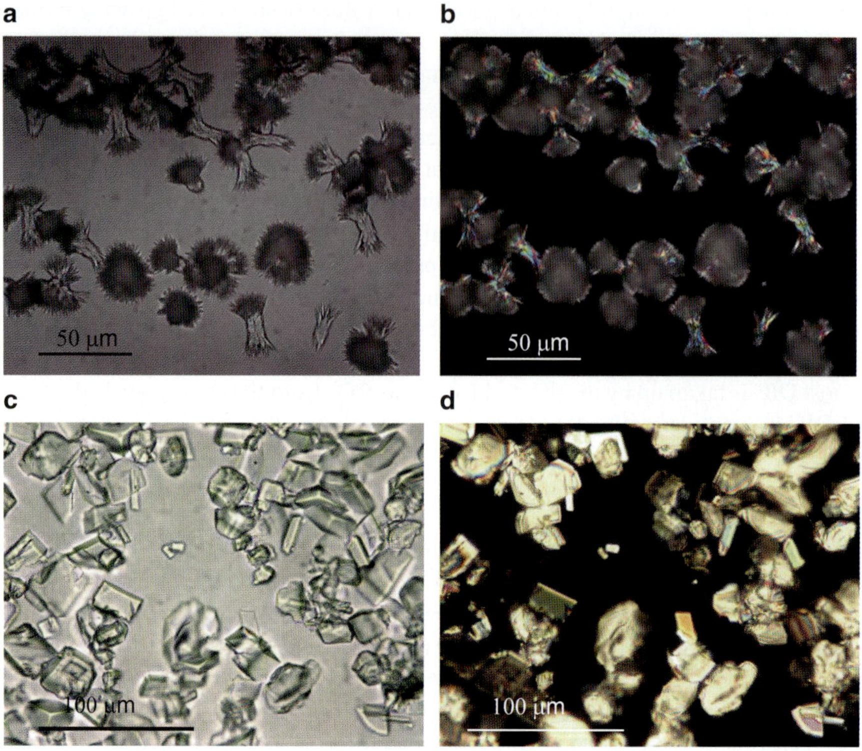

Fig. 7 Photomicrographs of nearby radially aggregated aragonite crystals forming from pure H_2O with 10 mM $CaCl_2$ and 50 mM $MgCl_2$, 10 mM $NaHCO_3$ and pH adjusted to 8.5 (**a**) under transmitted light, (**b**) under cross-polarized light. (**c**) Calcite crystals with rhombohedral morphology from solutions with 10 mM $CaCl_2$ but no $MgCl_2$, (**d**) under cross-polarized light

cross-polarized light. This perfectly developed spherical shape differs from the somewhat fuzzy spheres precipitated from the abiotic experiments. Also, the spheres from culture experiments show, in contrast to the aragonitic aggregates from the abiotic experiment, a poorly ordered crystal structure. The absence of aragonite could, however, result from an inhibition effect by the agar. Again in the cultures, crystals of two different generations occurred, which differed in crystal size, shape, and time of precipitation. The larger crystals often reached 100 μm or more in diameter with some spheres growing together to form double or multiple sphere clusters (Fig. 8a, b). The smaller generation was usually in the less than 10 μm range with the crystals forming dumbbell shapes (Fig. 8c, d), identical to the shapes observed in the microbially mediated dolomite from precipitation experiments by Vasconcelos et al. (1995) and Warthmann et al. (2000). When no Mg^{2+} was added, small well-defined calcite rhombs rather than spheres developed (Fig. 8e, f). Crystals further developed in the culture during storage of the plates at 4°C and after 3 months, radially grown, concentric geometries were observed, which were similar to the crystal aggregates observed by Bosak and Newman

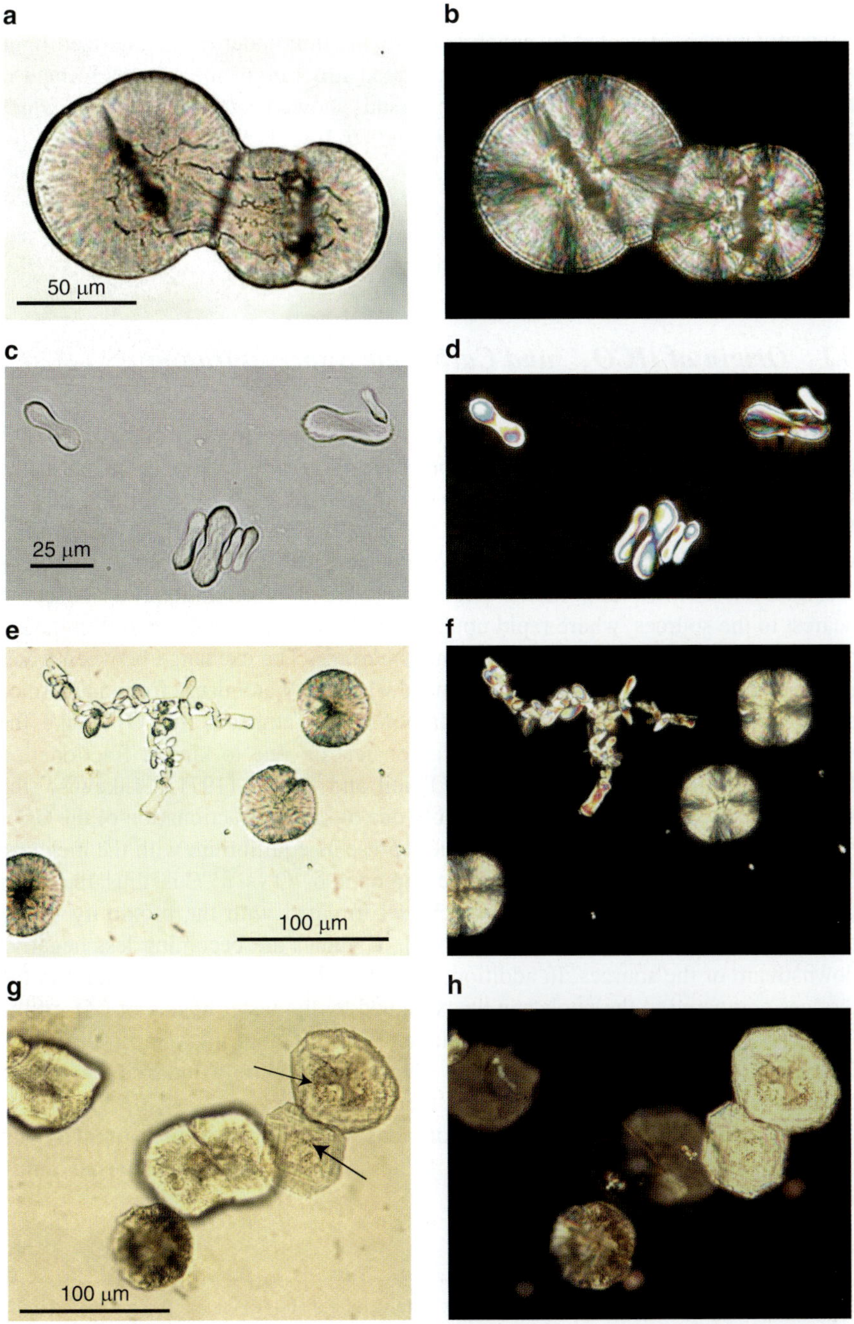

Fig. 8 (**a**) Carbonate crystals with spherical morphology from solid minimal media with 10 mM $CaCl_2$ and 50 mM $MgCl_2$ and inoculated with *Shewanella* under transmitted light, and (**b**) under cross-polarized light. (**c**) Dumbbell shaped smaller generation of carbonate crystals from cultures

(2005). Although somewhat round in outer shape, still no radial aggregation of the inner structure is indicated by synchronous extinction under cross-polarized light, and differed clearly from a radially aggregated structure of the spheres formed in the experiments with Mg^{2+}. Some of the crystals showed entombed bacteria, which are randomly oriented within the crystal (Fig. 8g, h).

5 Discussion

5.1 Origin of HCO_3^- and Carbonate Supersaturation

The source water of the Cedars ultrabasic springs has an extremely high pH, but it is nearly free of DIC. If uptake of hydrothermal CO_2 would have occurred in the subsurface, under high pH, carbonate would immediately have formed in situ. Instead, after emerging at the surface, CO_2 is rapidly absorbed from the atmosphere, which progressively neutralizes the pH of the source water and results in carbonate precipitation downstream. In the pools, most rapid precipitation was observed nearest to the sources, where rapid uptake of CO_2 leads to formation of thin layers of carbonate "ice" directly at the air/water interface. The exchange between water and atmospheric CO_2 is especially enhanced when the water drops from one terrace to the next. As essentially all inorganic carbon is from atmospheric CO_2 uptake, the strongly negative $\delta^{13}C$ values in the carbonates are due to kinetic fractionation during CO_2 uptake [as concluded by O'Neil and Barnes (1971)]. Likewise, the strongly negative $\delta^{18}O$ values are probably due to kinetic fractionation of the CO_2-oxygen, and only the most positive value represents equilibrium with the meteoric $\delta^{18}O$ values of the spring water. The linear trend in $\delta^{18}O$ vs. $\delta^{13}C$ is, thus, the result of different rates of CO_2 uptake, which is correlated with the proximity to the ultrabasic sources and both the $\delta^{18}O$ and $\delta^{13}C$ values are becoming less negative downstream of the sources. In addition, the somewhat more negative values at the terraces compared to the pools can be explained by the faster uptake of CO_2 of the flowing water compared to the slow uptake of CO_2 of stagnant water in the pools. Overall, precipitation of Ca-carbonate seems to be entirely induced by abiotic conditions due to its super-saturation in the spring water.

In the precipitation experiments, solutions were clearly supersaturated with respect to Ca-carbonate, except in the bacterial culture media, where alkalinity was produced by microbial metabolism.

Fig. 8 (Continued) with 50 mM $MgCl_2$ under transmitted light and (**d**) under cross-polarized light. Spheres and dumbbells often occur simultaneously (**e**) under transmitted light and (**f**) under cross-polarized light. (**g**) Microbially mediated carbonate crystals from *Shewanella* cultures without $MgCl_2$ under transmitted light, (**h**) under cross-polarized light

5.2 *Kinetic Effects on Crystal Structure*

Despite super-saturation, often minerals do not precipitate because their formation is kinetically inhibited. Aragonite is known to form in marine environments when calcite precipitation is inhibited due to high Mg/Ca ratios (Berner 1975). Also in our precipitation experiments, the addition of Mg^{2+} led to the formation of aragonite. As shown in the abiotic control experiments, the inhibition of crystal growth by Mg^{2+} seems to be expressed in the spherical crystal shape of the carbonate crystals. According to Kralj et al. (2004) and Arvidson et al. (2005), elongate (needle shaped) crystal growth is due to differential inhibition of growth in different directions and it seems that spherical growth is the logical consequence if needle shaped crystals are free to grow in all directions out of a nucleation point. Similarly, SO_4^{2-}, which was suggested as a kinetic inhibitor for dolomite (Mg, $Ca(CO_3)_2$) formation at high temperatures (Baker and Kastner 1981), was shown to cause elongated crystal growth, but not as efficiently as Mg^{2+}. Other additives had no significant effect on crystal structure except yeast extract (used for the culture experiments), which completely inhibited precipitation from the solutions. Yeast extract contains large amounts of carbohydrate polymers, which may inhibit carbonate precipitation. These observations suggest that spherical calcite growth in precipitation experiments with different carbohydrates (Cölfen and Qi, 2001; Braissant et al. 2003) is similarly caused by an inhibitory effect.

Considering the carbonate precipitates at Cedars Springs, we suggest that the formation of spherical aragonite crystals is due to significant mixing of the spring water with Mg-rich river water. Mg–HCO_3^- water originates from surface weathering of the ultrabasic rock, whereas Ca–OH^- water from the ultrabasic springs have low Mg^{2+} concentrations. Thus, the distribution of spherical aragonite and rhombohedral calcite in the Cedars ultrabasic spring complex should reflect the chemistry of the water at the different sites of precipitation. Indeed, carbonates in pools, where spring water is frequently mixed with surface water from the nearby river, as it is the case at Nipple Spring, aragonite tends to be more abundant. Moreover, aragonite generally forms in the pools as delicate friable and slow growing carbonate whereas the fast growing, massively incrusted terraces mainly consist of calcite, which is reflected by slightly more negative $\delta^{13}C$ and $\delta^{18}O$ isotope values in the terrace. However, because the pH and Mg^{2+} concentration of the two mixing waters are not necessarily coupled – e.g. pH and Mg^{2+} concentration can both be high when surface water is mixed with fresh source water – the isotopic composition may not show direct correlation with the aragonite/calcite ratio of the carbonate. The isotope composition rather seems to be dependent of the distance from the springs downstream, while morphology depends on the relative amounts in which the two water types are mixing. Alternating cement structures in laminated carbonates are also likely controlled by mixing of the two water types downstream. In addition, temperature may influence the effect of Mg^{2+} on crystal morphology (Morse et al. 1997).

5.3 *Biological Effects on Crystal Structure*

The aragonite spherules found in the Cedars spring carbonates commonly are aggregated to form sheaths or layers, and often incrust algal filaments. According to the model of Braissant et al. (2003), spherical growth is induced in the presence of carbohydrates and amino acids, as an analogue for the effects of extracellular polymeric substances (EPS) in microbial biofilms. At the Cedars, the precipitation of spherical crystals may also be microbially mediated, and algal filaments or EPS layers may act as a substrate for crystal growth. Nevertheless, the concentration of organic compounds in the Cedars Spring waters is in all probability too low to be mainly responsible for spherical crystal structure. In the presented experiments, we show that bacterially mediated carbonates from Mg-free solutions show only minor differences to the abiotic controls. The outer shapes are somewhat more rounded, but the interior shows an uniform crystallographic orientation, similar to calcite crystals reported from cultures with sulphate-reducing bacteria (Bosak and Newman 2005). Meldrum and Hyde (2001) showed that organic matter contributes to the spherical crystal structure, and Gonzáles-Muñoz et al. (2000) suggested that a microbial factor is involved besides the effect of Mg^{2+}. In our experiments, spherical crystals were only formed in the presence of Mg^{2+}, but are more smoothly and regularly developed in the presence of bacteria than in the abiotic control. Over all, a biological mediation effect on crystal structure at the Cedars spring carbonates may not be completely excluded. However, this effect would be minor, and not the primary reason for the formation of spherical aragonite.

In nature, spherical crystal structures are commonly ascribed to microbially mediated precipitation, maybe because the round shapes are reminiscent of something "living" (e.g. Gerdes et al. 1994; Vasconcelos et al. 1995; Verrecchia et al. 1995; Folk 1999; Van Lith et al. 2003; Rivadeneyra et al. 2006; Sánchez-Román et al. 2008).

In fact, round crystal shapes can be purely crystallographic features due to kinetic inhibition, rather than biologic induction of crystal growth. The spherical structure of many bio-minerals may thus not be the direct consequence of biological "mediation", but bio-mediation may become relevant, when precipitation is otherwise kinetically inhibited. This effect may play an important role and may even be precisely regulated within living tissues and biofilms through the concentration levels of Mg^{2+} or other (organic?) inhibitors. For the precipitation of spring carbonates at the Cedars, a biological factor may then play a role, when the pH is moderate and the precipitation of calcite is partially inhibited through the presence of Mg^{2+} from mixing with adjacent river water. Under these (less alkaline) conditions, algal filaments can grow and may serve as a substrate for crystal growth.

The absorption of Ca^{2+} by EPS, an effect discussed by Braissant et al. (2007) may locally increase the Mg/Ca ratio and indirectly affect the precipitation kinetics, in particular, during times of low marine Mg^{2+} concentrations. Nevertheless, spherical aragonite formation in Cedars Springs and precipitation experiments do not support the EPS model for spherical carbonate formation under natural, in particular marine, conditions and Mg-inhibition may impose a much larger effect on crystal growth than EPS.

5.4 Spherical Carbonates Through Earth's History

Varying contribution of Mg-rich water from surface weathering vs. Ca-rich hydrothermal fluid from serpentinization, as observed at a small scale at the Cedars alkaline springs, can be observed on a global scale in the secular variation of the Mg/Ca ratio of seawater during times of fast and slow seafloor spreading. Variation in Mg-content led to periods with predominantly aragonite or calcite precipitation (aragonite seas and calcite seas: Given and Wilkinson 1985; Hardie 1996, 2003; Stanley and Hardie 1998; Burns et al. 2000). The Mg content also affected ooid morphology – aragonitic ooids with concentric structure were predominantly formed in aragonite seas, calcitic radial ooids formed in calcite seas (Sandberg 1983; Wilkinson et al. 1985; Wilkinson and Given 1986).

Aragonite ooids show micritic and concentric fibrous lamination at micrometer scale, whereby the crystallites are randomly oriented. In this case, the round shape may not be a crystallographic feature and a microbial factor was considered for their formation (e.g. Folk and Lynch 2001). Micritic ooids were similarly described from freshwater Lake Geneva (Plee et al. 2008), from a system where carbonate is barely supersaturated. Micritic ooids, thus, may be the result of microbial metabolism, locally increasing supersaturation, rather than a kinetic effect. Indeed, micritic ooids formed typically in aragonite seas, in which carbonate saturation was probably low (as a result of slow spreading rates producing less CO_2 in the atmosphere and less continental weathering during lower tectonic activity). In contrast, calcitic ooids formed in calcite seas at the same time with increased carbonate supersaturation (Riding and Liang 2005). These radial calcitic ooids are more comparable with the spherical crystals at Cedars Springs. The interplay of Mg/Ca ratio and carbonate supersaturation provides, thus, a mechanism that can explain the varying type of ooid formation as the result of thermodynamic vs. kinetic limitation.

High seawater carbonate supersaturation particularly occurred during several times in the Precambrian, so that carbonate precipitation would have been kinetically controlled. Indeed, some Precambrian precipitates, often considered as stromatolites, show round structures and crystal fans (Grotzinger and Knoll 1999), analogous to crystal fans observed in the laminated cements from the Cedars spring carbonates. These crystal fans may have been formed completely abiotically and their shape can be explained by Mg^{2+} inhibition. Also during the Precambrian, calcite and aragonite ooids are reported (Tucker 1984; Ries et al. 2008), while secular variation in Mg/Ca ratio was reconstructed (Hardie 2003). Still more research is required to understand the biotic and abiotic kinetic controls of carbonate precipitation, as well as better constraints on chemistry of past oceans, in order to interpret the processes that govern carbonate precipitation throughout Earth's history.

6 Conclusions

Carbonate precipitates abiotically due to high carbonate supersaturation from highly alkaline groundwater at Cedars Springs, whereby CO_2 is absorbed from the atmosphere by the spring water. Alternating precipitation of rhombohedral calcite and spherical aragonite crystals is most likely controlled by variable Mg/Ca ratios due to mixing of ultrabasic Ca–OH dominated spring water with Mg–HCO_3^- – rich river water derived from surface weathering. Precipitation experiments show that spherical crystal structure is mainly a kinetic effect due to Mg^{2+} inhibition, but the presence of microbial biofilms may contribute to a finer surface structure of the carbonate spherules. These observations suggest that, particularly in the marine environment, spherical carbonates may not result from interaction with microbial EPS, but simply due to an inhibiting effect by the Mg^{2+}. The frequently observed association of spherical crystals with microbial biofilms can be explained by microbial mediation that becomes the controlling factor of carbonate precipitation only when precipitation is otherwise inhibited. Accordingly, spherical crystallographic features, such as crystal fans and radial ooids became then prominent in Earth's history, when precipitation was kinetically limited. In contrast, micritic ooids formed when precipitation was thermodynamically limited. Secular variation in the Mg/Ca ratio and carbonate supersaturation may thus explain the distribution of crystal fans and spherical carbonate crystal structure through Earth's history.

Acknowledgements We thank Monika Segl (University of Bremen) and Gaute Lavik (MPI, Bremen) for isotope analysis, as well as Peter Schiffman (UC Davis) for allowing us to present his unpublished isotope data. Gail Arnold (MPI, Bremen) and Carolina Reyes (UC Santa Cruz) provided useful comments and improved the clarity of the manuscript. Discussions with Gijs Kuenen, Ana Obraztsova, John Platt, and Rolf Warthmann contributed significantly to the study. We thank Damon Rogers for giving us the opportunity to visit the Cedars Springs and for allowing us to use his camping facility. This study was sponsored by the Agouron Institute and the Swiss National Science Foundation (SNF).

References

Arvidson RS, Collier M, Davis KJ, Vinson MD, Amonette JE, Lüttge A (2005) Magnesium inhibition of calcite dissolution kinetics. Geochimica et Cosmochimica Acta 70:583–594

Baker P, Kastner M (1981) Constraints on the formation of sedimentary dolomite. Science 213:214–216

Barnes I, O'Neil JR (1969) The relationship between fluids in some fresh alpine-type ultramafics and possible modern serpentinization, Western United States. The Geological Society of America Bulletin 80:1947–1960

Barnes I, LaMarche VC, Himmelberg GR (1967) Geochemical evidence of present-day serpentinization. Science 56:830–832
Bathurst RGC (1967) Oolitic films on low energy carbonate sand grains, Bimini Lagoon, Bahamas. Marine Geology 5:89–109
Berner RA (1975) The role of magnesium in the crystal growth of calcite and aragonite from sea water. Geochimica et Cosmochimica Acta 39:489–504
Bontognali TRR, Vasconcelos C, Warthmann RJ, Dupraz C, Bernasconi SM, McKenzie JA (2008) Microbes produce nanobacteria-like structures, avoiding cell entombment. Geology 36:663-666
Bosak T, Newman DK (2005) Microbial kinetic controls on calcite morphology in supersaturated solutions. Journal of Sedimentary Research 75:190–199
Braissant O, Cailleau G, Dupraz C, Verrecchia EP (2003) Bacterially induced mineralization of calcium carbonate in terrestrial environments: the role of exopolysaccharides and amino acids. Journal of Sedimentary Research 73:485–490
Braissant O, Decho AW, Dupraz C, Glunk C, Przekop KM, Visscher PT (2007) Exopolymeric substances of sulfate-reducing bacteria: interactions with calcium at alkaline pH and implication for formation of carbonate minerals. Geobiology 5:401–411
Burns SJ, McKenzie JA, Vasconcelos C (2000) Dolomite formation and biogeochemical cycles in the Phanerozoic. Sedimentology 47:49–61
Cölfen H, Qi L (2001) A systematic examination of the morphogenesis of calcium carbonate in the presence of a double-hydrophilic block copolymer. Chemistry – A European Journal 1:106–116
Davies PJ, Bubela B, Ferguson J (1978) The formation of ooids. Sedimentology 25:703–729
Donahue J (1969) Genesis of oolite and pisolite grains: an energy index. Journal of Sedimentary Petrology 39:1399–1411
Dupraz C, Visscher PT, Baumgartner LK, Reid RP (2004) Microbe–mineral interactions: early carbonate precipitation in a hypersaline lake (Eleuthera Island, Bahamas). Sedimentology 51:745–765
Folk RL (1999) Nannobacteria and the precipitation of carbonate in unusual environments. Sedimentary Geology 126:47–55
Folk RL, Lynch FL (2001) Organic matter, putative nannobacteria and the formation of ooids and hardgrounds. Sedimentology 48:215–299
Früh-Green GL, Kelley DS, Bernasconi SM, Karson JA, Ludwig KA, Butterfield DA, Boschi C, Proskurowski G (2003) 30,000 years of hydrothermal activity at the Lost City Vent Field. Science 301:495–498
Früh-Green GL, Delacour A, Boschi C, Bernasconi SM, Butterfield DA, Kelley DS, Proskurowski G (2007) Building Lost City: serpentinization, mass transfer and life in a peridotite-hosted hydrothermal system. Geochimica et Cosmochimica Acta 71/15(Suppl. 1):A298
Gerdes G, Dunajtschik-Piewak K, Riege H, Taher AG, Krumbein WE, Reineck H-E (1994) Structural diversity of biogenic carbonate particles in microbial mats. Sedimentology 41:1273–1294
Given RK, Wilkinson BH (1985) Kinetic control of morphology, composition, and mineralogy of abiotic sedimentary carbonates. Journal of Sedimentary Petrology 55:109–119
Gonzáles-Muñoz MT, Ben Chekroun K, Ben Aboud A, Arias J, Rodriguez-Gallego M (2000) Bacterially induced Mg-calcite formation: role of Mg^{2+} in development of crystal morphology. Journal of Sedimentary Research 70:559–564
Grotzinger JP, Knoll AH (1999) Stromatolites in Precambrian carbonates: evolutionary mileposts or environmental dipsticks? Annual Review of Earth and Planetary Sciences 27:313–358
Hardie LA (1996) Secular variation in seawater chemistry: an explanation for the coupled secular variation in the mineralogies of marine limestones and potash evaporates over the past 600 m.y. Geology 24:279–283
Hardie LA (2003) Secular variation in Precambrian seawater chemistry and the timing of Precambrian aragonite seas and calcite seas. Geology 31:785–788

Kelley DS, Karson JA, Blackman DK, Früh-Green GL, Butterfield DA, Lilley MD, Olson EJ, Schrenk MO, Roe KK, Lebon GT, Rivizzigno P, AT3-60 Shipboard Party (2001) An off-axis hydrothermal vent field near the Mid-Atlantic Ridge at 30°N. Nature 412:145–149

Kelley DS, Karson JA, Früh-Green GL, Yoerger DR, Shank TM, Butterfield DA, Hayes JM, Schrenk MO, Olson EJ, Proskurowski G, Jakuba M, Bradley A, Larson B, Ludwig K, Glickson D, Buckman K, Bradley AS, Brazelton WJ, Roe K, Elend MJ, Delacour A, Bernasconi SM, Lilley MD, Baross JA, Summons RE, Sylva SP (2005) A serpentinite-hosted ecosystem: the Lost City hydrothermal field. Science 307:1428–1434

Kralj D, Kontrec J, Brecevic L, Falini G, Nöthig-Laslo V (2004) Effect of inorganic anions on the morphology and structure of magnesium calcite. Chemistry – A European Journal 10:1647–1656

Ludwig KA, Kelley DS, Butterfield DA, Nelson BK, Früh-Green G (2006) Formation and evolution of carbonate chimneys at the Lost City Hydrothermal Field. Geochimica et Cosmochimica Acta 70:3625–3645

Meldrum FC, Hyde ST (2001) Morphological influence of magnesium and organic additives on the precipitation of calcite. Journal of Crystal Growth 231:544–558

Mitterer RM (1972) Biogeochemistry of aragonitic muds and oolites. Geochimica et Cosmochimica Acta 36:1407–1422

Morse JW, Wang Q, Tsio MY (1997) Influences of temperature and Mg:Ca ratio on $CaCO_3$ precipitates from seawater. Geology 25:85–87

Myers CR, Nealson KH (1988) Bacterial manganese reduction and growth with manganese oxide as the sole electron acceptor. Science 240:1319–1321

O'Neil JR, Barnes I (1971) C13 and O18 compositions in some fresh-water carbonates associated with ultramafic rocks and serpentinites: western United States. Geochimica et Cosmochimica Acta 35: 687–697

Platt JP (1986) Dynamics of orogenic wedges and the uplift of high-pressure metamorphic rocks. The Geological Society of America Bulletin 97:1037–1053

Plee K, Ariztegui D, Martini R, Davaud E (2008) Unravelling the microbial role in ooid formation-results of an in situ experiment in modern freshwater Lake Geneva in Switzerland. Geobiology DOI: 10.1111/j.1472-4669.2007.00140.x

Riding R, Liang L (2005) Geobiology of microbial carbonates: metazoans and seawater saturation state influence on secular trends during the Phanerozoic. Palaegeography, Palaeoclimatology, Palaeoecology 219:101–115

Ries JB, Anderson MA, Hill RT (2008) Seawater Mg/Ca controls polymorph mineralogy of microbial $CaCO_3$: a potential proxy for calcite-aragonite seas in Precambrian time. Geobiology 6:106–119

Rivadeneyra MA, Martín-Algarra A, Sánchez-Navas A, Martín-Ramos D (2006) Carbonate and phosphate precipitation by Chromohalobactern marismortui. Geomicrobiology Journal 23:89–101

Sánchez-Román M, Vasconcelos C, Schmid T, Dittrich M, McKenzie JA, Zenobi R, Rivadeneyra MA (2008) Aerobic microbial dolomite at the nanometer scale: implications for the geologic record. Geology 36:879–882

Sandberg PA (1983) An oscillating trend in Phanerozoic non-skeletal carbonate mineralogy. Nature 305:19–22

Sleep NH, Meibom A, Fridriksson Th, Coleman RG, Bird DK (2004) H_2-rich fluids from serpentinization: geochemical and biotic implications. Proceedings of the National Academy of Sciences of the United States of America 101:12818–12823

Stanley SM, Hardie LA (1998) Secular oscillations in the carbonate mineralogy if reef-building and sediment-producing organisms driven by tectonically forced shifts in seawater chemistry. Palaeogeography, Palaeoclimatology, Palaeoecology 144:3–9

Suess E, Fütterer D (1972) Aragonitic ooids: experimental precipitation from seawater in the presence of humic acid. Sedimentology 19:129–139

Tucker ME (1984) Calcitic, aragonitic, and mixed calcitic-aragonitic ooids from the mid-Proterozoic Belt Supergroup, Montana. Sedimentology 31:627–644

Tucker ME, Wright VP (1990) Carbonate sedimentology. Blackwell Science, Oxford

Van Lith Y, Warthmann R, Vasconcelos C, McKenzie JA (2003) Microbial fossilization in carbonate sediments: a result of the bacterial surface involvement in dolomite precipitation. Sedimentology 50:237–245

Vasconcelos C, McKenzie JA, Bernasconi S, Grujic D, Tien AJ (1995) Microbial mediation as a possible mechanism for natural dolomite formation at low temperatures. Nature 377:220–222.

Verrecchia E, Freytet P, Verrecchia KE, Dumont JL (1995) Spherulites in calcrete laminar crusts: biogenic $CaCO_3$ precipitation as a major contributor to crust formation. Journal of Sedimentary Research A65:690–700

Von Damm KL (2001) Lost city found. Nature 412:127–128

Warthmann R, Van Lith Y, Vasconcelos C, McKenzie JA, Karpoff AM (2000) Bacterially induced dolomite precipitation in anoxic culture experiments. Geology 28:1091–1094

Wilkinson BH, Given KR (1986) Secular variation in abiotic marine carbonates: constraints on Phanerozoic atmospheric carbon dioxide contents and oceanic Mg/Ca ratios. Journal of Geology 94:321–333

Wilkinson BH, Owen RM, Carrol AR (1985) Submarine hydrothermal weathering, global eustasy, and carbonate polymorphism in Phanerozoic marine oolites. Journal of Sedimentary Petrology 55:171–183

Wolin EA, Wolin MJ, Wolfe RS (1963) Formation of Methane by Bacterial Extracts. The Journal of Biological Chemistry 238:2882–2886

Microbial Control on Lamina Formation in a Travertine of Crystal Geyser, Utah

Chiduru Takashima, Tomoyo Okumura, Shin Nishida, Toshihiko Shimamoto, Hiroko Koike, and Akihiro Kano

1 Introduction

Travertines – carbonates mostly precipitated in situ from calcareous hot-spring water – often show stromatolitic lamination which has been interpreted as daily banding (Folk et al. 1985; Guo and Riding 1992; Pentecost 1994). Prevailing interpretation for the daily rhythm has tied up with photosynthesis of cyanobacteria and algae living on the travertine surface. Textures of the laminated travertines are various, and have been ascribed to different processes such as the rise of supersaturation by daytime photosynthesis (Guo and Riding 1992), trapping of fine-grained particles on glutinous extracellular polymeric substances (EPS) (Takashima and Kano 2008), and inhibition of crystal growth by cation-binding EPS. All these studies underline the importance of the microbial activity in terms of developing the daily lamination.

Regardless of the microbial control on the texture, it is widely accepted that carbonates in hot-spring environments are precipitated principally by inorganic processes, such as CO_2 degassing and elevated supersaturation (Herman and Lorah 1988; Pentecost 2005). Because CO_2 partial pressure in calcareous hot springs often reaches 1 atm and sometimes even more (Kitano 1963), the degassing on the ground is so active and unlikely influenced by a subtle effect from the microbial photosynthesis. However, such inorganic degassing could be affected by a change in hydrology. For instance, the flow rate of springwater controls part of the CO_2 degassing and heat exchange with an ambient environment, both of which

C. Takashima (✉), T. Okumura, S. Nishida, H. Koike, and A. Kano
Environment Course, Faculty of Culture and Education, Saga University, Saga 840-8502, Japan
e-mail: chizurut@cc.saga-u.ac.jp

T. Shimamoto
Department of Earth and Planetary Systems Science, Hiroshima University, Higashi-hiroshima 739-8526, Japan

J. Reitner et al., *Advances in Stromatolite Geobiology*,
Lecture Notes in Earth Sciences 131, DOI 10.1007/978-3-642-10415-2_7,

influence the degree of supersaturation with respect to carbonate minerals. Therefore, laminated textures could be formed inorganically where the hydrological processes are fluctuating.

This expectation can be tested in geyser settings where the growing CO_2 pressure in the underground is released by springwater discharge at regular temporal intervals. We examined travertine deposits in the Crystal Geyser, central Utah, located along the Little Grand Wash fault at the bank of the Green River (Shipton et al. 2004). This geyser discharges calcareous water since 1935 when a well for oil exploration was drilled through a 21.5-m-thick travertine mound to the base of the Triassic section (Baer and Rigby 1978). Thus, the modern spring is artificial, although the pre-existing travertine mound indicates that the water used to discharge naturally from the fault. Since the well was drilled, the mound has been growing into an oval of 80×100 m^2 (Fig. 1) by precipitation of calcite from the springwater.

2 Hydrochemistry of the Crystal Geyser

The Crystal Geyser discharges water intermittently as a result of releasing CO_2 pressure gradually growing in underground. A single discharge event appeared a quite unique sequence consisting of "bubbling" (intermittent weak discharge; Fig. 2a) for 2–5 h and the following "explosion" (continuous and strong discharge up to 25 m high; Fig. 2b) for 0.5–1.5 h. From 27 to 31 August 2007, we have studied the geyser and observed six discharge events. By considering that we missed one event in the morning of 29 August, the interval between the events was approximately 16 h (Fig. 2c). During the explosion, the travertine mound was entirely covered with water, but dried out during the time between the discharges. We evaluated the amount of water by monitoring the flow rate and the concentration of chemical tracer during the discharge event on 31 August, which had lasted for a relatively long time. The whole sequence of this event discharged 1,040 m^3 of water (96 by bubbling and 944 by explosion).

Water from the geyser vent is cold (~18°C), weakly acid pH (~6.3) and rich in cations and anions (Table 1). A substantial concentration was recorded for sodium (4.54 g/L), chlorite (3.99 g/L), sulfate (2.10 g/L) and calcium (1.05 g/L; Table 1). A distinct feature is its extraordinary high equilibrated CO_2 partial pressure (~2.2 atm; Table 1) that provides motive force for the geyser and also enlarges the content of dissolved inorganic carbon (207 mM; 9.12 gL^{-1} as CO_2). Assuming that these chemical compositions were stable, we can evaluate that the discharge event on 31 August emitted 9.5 tons of carbon dioxide.

Due to CO_2 degassing as well as to O_2 absorption on the ground, supersaturation induces precipitation of calcite and iron oxide. This was demonstrated by change in concentration of Ca^{2+} and Fe^{2+} along the water passage from the vent to site 6 (Fig. 1). Concentration of Fe^{2+} decreases by precipitation of iron oxide in a proximal part of the mound, which is consistent with the observation of reddish

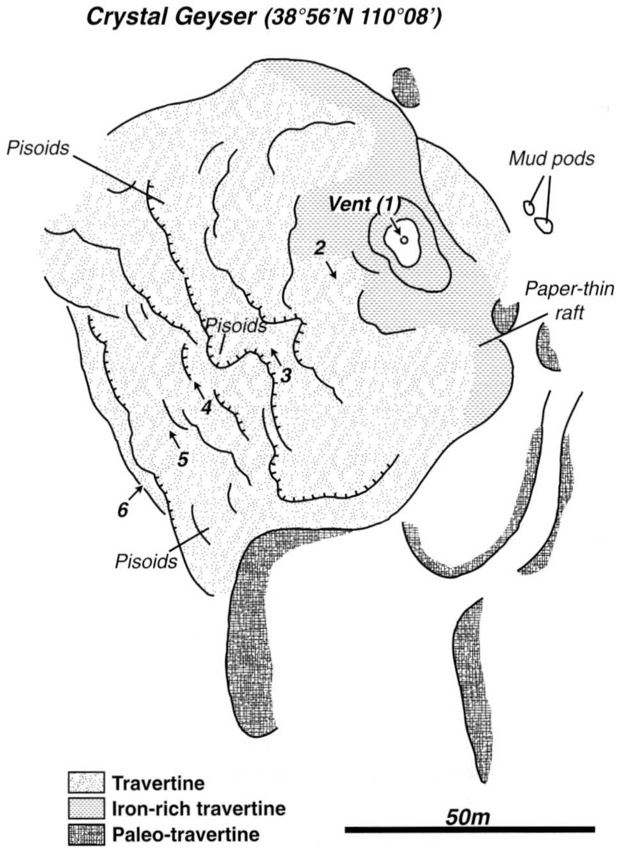

Fig. 1 Map of Crystal Geyser in central Utah, United States, showing sites of water chemistry analysis (1–6). The analyzed sediment was collected from site 4

deposits near the vent (Fig. 1). A series of downstepping travertine mounds radiate out downslope from the vent to sw direction. The mound surface shows several terraces where calcite precipitation is likely active because quick water flow enhances the CO_2 degassing. The active calcite precipitation reflected on a substantial decreasing of the Ca^{2+} concentration between sites 3 and 6 (Table 1). Concentration of the other ions at the vent (Mg^{2+}, Na^+, K^+, SO_4^{2-}, and Cl^-) slightly increased at the mound (Table 1). This enigmatic phenomenon resulted from dissolution of salts (e.g., gypsum and halite) that were formed by evaporation of water remaining from the prior explosion event.

The travertine on these downslope terraces consists of calcite and often exhibits laminated internal structures (Fig. 3a). Microbial life is common on the mound. Some travertine faces on the terraces are faintly greenish indicating the presence of photosynthetic microbes. Green colored mats by filamentous algae or cyanobacteria was observed in small-scale pools, and often accompanied with formation of pisoids and paper-thin calcite rafts (Fig. 1).

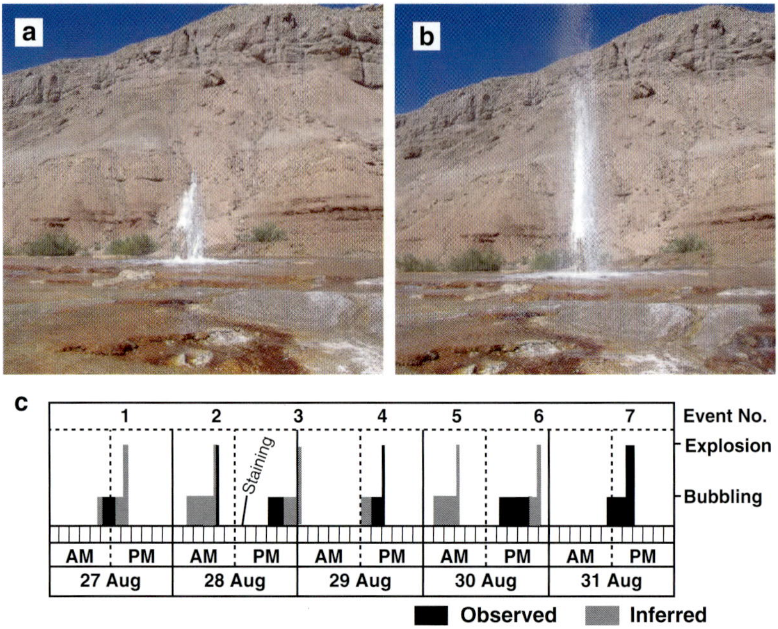

Fig. 2 Discharge event at Crystal Geyser. Each event consists of bubbling (**a**) and explosion (**b**). Seven discharge events happened during the observed five days (**c**; 27–31 August, 2007). The analyzed sample was stained at mid-day of 28 August

Table 1 Chemical composition, saturation index of calcite (SIc), equilibrated partial pressure of CO_2 of water from the vent (site 1) and of the travertine mound (site 2–6; Fig. 1) at the Crystal Geyser, central Utah. Ion concentrations are given in mg L^{-1}, partial pressure in atmospheres (atm)

Site and date	Temp	pH	Ca^{2+}	Mg^{2+}	Fe^{2+}	Na^{+}	K^{+}	SO_4^{2-}	Cl^{-}	SI_c[a]	pCO_2[a]
1 – Aug 29 (bub)	18.1	6.33	1,045	247	4.9	4,540	309	2,100	3,990	0.94	2.17
2 – Aug 29 (bub)	19.2	6.67	1,108	255	2.7	4,630	330	2,260	4,025	1.31	1.03
3 – Aug 29 (bub)	20.2	6.77	1,119	266	2.0	4,660	333	2,250	4,100	1.42	0.83
4 – Aug 29 (bub)	23.2	7.25	954	255	1.3	4,650	297	2,200	4,030	1.84	0.27
5 – Aug 29 (bub)	26.1	7.40	691	233	0.9	4,634	313	2,220	4,040	1.86	0.18
6 – Aug 29 (bub)	29.6	7.45	532	229	1.0	4,650	324	2,310	4,050	1.81	0.15

[a]Calculated by following methods of Plummer and Bosenberg (1982) and assuming the charge balance with bicarbonate

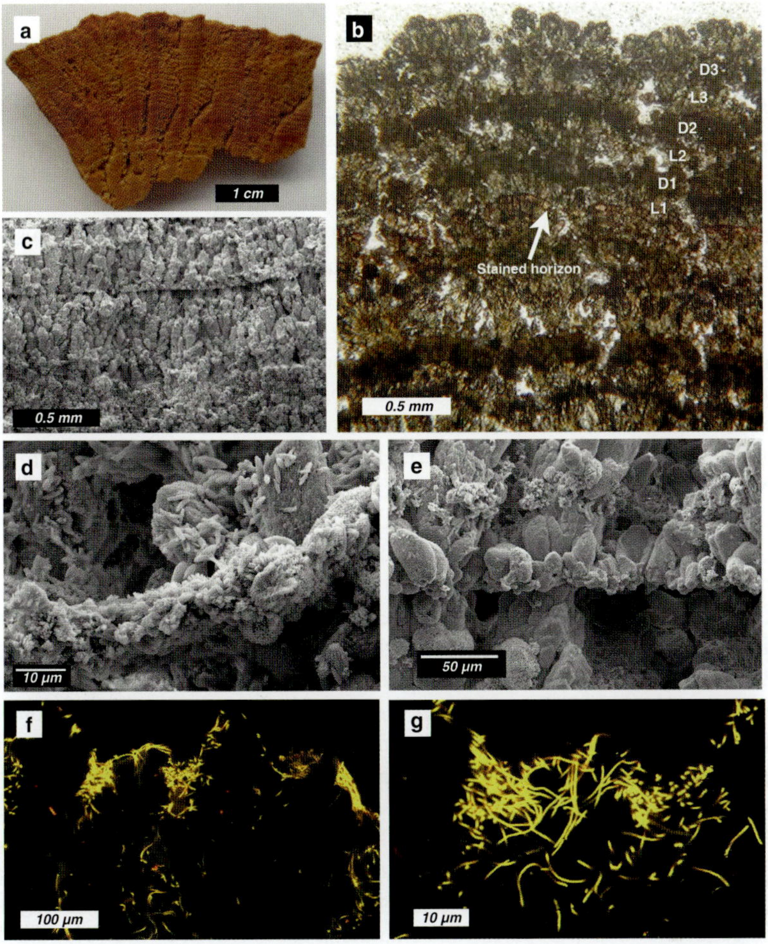

Fig. 3 A laminated travertine collected at 17:00 on 31 August from site 4 in Crystal Geyser. (**a**) Cross section of the specimen. (**b**) Thin section image showing the lamination that consists of lighter-colored (L1–3) and darker-colored layers (micritic bands; D1–3). The *arrow* indicates the horizon stained at mid-day of 28 August. (**c**) A SEM image of the laminated travertine. (**d**) A SEM image of a micritic band showing a filament encrusted by a fine-grained calcite. (**e**) A SEM image of a lighter-colored layer. (**f**) A fluorescence image of the travertine surface showing cyanobacterial biofilm. (**g**) An enlarged image of Fig. 3f showing clusters of filamentous cyanobacteria

3 Methods

This study focuses on the origin of the travertine lamination in the setting of the Crystal Geyser. The samples were collected from site 4 (Fig. 1) on the route of water chemistry analysis, where water flow had been continuously kept during a discharge event. We have investigated textural characteristics by optical, fluorescence, and scanning electron microscopes (SEM), and examined temporal development of the lamination, and microbial composition on the travertine.

Dried travertine surface in the meantime between the discharges events provide an occasion to stain the surface to mark an exact time line for monitoring the travertine growth. We stained the travertine surface in mid-day of 28 August by Alizarin Red S that dyes calcite red and collected the specimen after the discharge event on 31 August.

In order to specify the microbial community on the travertine surface, a sample was collected and immediately fixed in ethanol. The surface of sediment sample was cleaned by DNA contamination removal solution (DNA-OFF™, TaKaRa) and 206 mg of the sample were scraped using grinder. This powdered sample was decalcified by 0.5 M EDTA (pH 8.0) at room temperature for over 16 h and centrifuged to collect cell pellet. After the pellet was washed by super distilled water, total DNA were extracted with a DNeasy Blood and Tissue kit (QIAGEN) following to the instructions for bacteria. Bacterial 16S rRNA genes were amplified by PCR with the Bacteria-specific 27F forward primer (5′-AGAGTTTGATCCTG GCTCAG-3′) and the universal 1492R reversal primer (5′-GGTTACCTTGTT ACGACTT′) (DeLong 1992). The amplification was performed with a T-Gradient thermocycler (Biometra) with total of 40 cycles; 94°C for 30 s, 55°C for 45 s, and 72°C for 60 s. PCR products were subcloned using the TOPO TA Cloning Kit (Invitrogen Corp.) into the PCR 2.1 TOPO-vector and then transformed into the TOP 10F′ strain of *E. coli*. Positive colonies were selected and screened for the correct product insert. After the initial screen and the enzymatic purification (PCR Product Pre-Sequencing kit, USB Corporation), the PCR products were sequenced using a Dye Terminator Cycle Sequencing Kit (Beckman Coulter Inc.), according to the manufacturer's protocol. The sequencing consisted of 30 cycles of 96°C for 20 s, 50°C for 20 s, 60°C for 150 s. Capillary sequencing was conducted using the CEQ2000XL DNA sequencer (Beckman Coulter Inc.) and data were analyzed using CEQ Software, Sequence Analysis ver. 2 (Beckman Coulter Inc.). Alignment of the sequence data was performed manually using MEGA 4.02 (Tamura et al. 2007). Sequences of the 16S rRNA gene from this study were grouped into operational taxonomic units (phylotypes) on the basis of a nucleotide similarity of >97%. Sequence similarity was determined by the Basic Local Alignment Search Tool (BLAST) on the NCBI website (http://www.ncbi.nlm.nih.gov/blast). The 16S rRNA gene sequences were registered in the DDBJ (GenBank accession Nos. CG01:AB491699 to CG25:AB491723).

4 Textures and Mineralogy of the Laminated Travertine

The travertine at site 4 (Fig. 1) shows the laminated texture (Fig. 3a) that appears as an alternation of lighter-colored and darker-colored layers in thin section when using transmitted light. The lighter layers consist of aggregates of calcite crystals, and their thickness ranges from 0.2 to 0.7 mm (Fig. 3b). The darker layers are 0.1–0.2 mm thick, composed finer-grained carbonate crystals (Fig. 3b), and here called micritic bands. These crystals form a tightly-packed framework of the well-lithified travertine, but leave some sub-mm-scale void spaces between the crystals (Fig. 3b). The calcite normally occurs as subhedral-anhedral crystals of 50–200 μm in diameter, which are slightly expanded upwards (Fig. 3c). Some crystals consist of rhombic crystallites that have been often reported in calcite travertines of other localities (Folk et al. 1985; Jones et al. 2005; Takashima and Kano 2008). The larger calcite crystals have grown from smaller particles that are aligned certain horizons (Fig. 3c), likely corresponding to micritic bands in optical microscopic views. The enlarged SEM view of such horizon shows filamentous structures encrusted by rod-shaped crystals of a few μm in diameter. The structure is likely an encrusted photosynthetic microbe, as recognized by autofluorescence on the thin sections (Fig. 3f, g). They are 1–2 μm in diameter (Fig. 3g) and occur on the travertine surfaces as well in void spaces between the calcite crystals (Fig. 3f).

Alizarin Red S successfully stained the specimen, and the horizon dyed in red marks as the time-line of mid-day of 28 August. The dyed horizon is observed ~1 mm below the surface in the middle of a lighter-colored layer (Fig. 3b). The travertine above the horizon includes three micritic bands (D1-3; Fig. 3b).

The XRD analysis indicated that the travertine are mainly calcite and includes a minor fraction of aragonite. This result is consistent with lack of needle crystals that are the main constituents of the aragonite travertines (Kano et al. 2006).

5 Phylogenetic Analysis

30 sequences extracted from the travertine at site 4 are spread into 25 phylotypes that contain four metabolic groups, cyanobacteria, anoxygenic phototrophs, aerobic and anaerobic heterotrophic bacteria (Table 2). Four phylotypes (CG12, CG13, CG14 and CG15) are cyanobacteria, and the closest taxa by FASTA are rod-shaped *Synechococcus* sp. that has been often reported from hot spring environments (Allewalt et al. 2006). While similarity search using BLAST identified the closest identity (97%) for CG13 to filamentous *Phormidium* sp. 10 phylotypes (11 clones) belong to the alpha-proteobacteria. CG01–CG03, CG04 and CG05 are related to genera *Loktanella*, *Sulfitobacter* and *Thalassobaculum,* respectively (Table 2), all of which are rod-shaped methophilic aerobic heterotrophic bacteria (Trappen et al. 2004; Pukall et al. 1999; Ivanova et al. 2004; Zhang et al. 2008). CG07 is close to a rod-shaped genus *Rubrimonas* that can grow on aerobic respiration in aerobic condition and on anoxygenic photosynthesis in anoxic condition (Suzuki et al. 1999). Although the identity is low, one phylotype (CG22) was related to a gamma-proteobacterium *Ectothiorhodospira* sp. that is phototrophic sulfur bacteria (Zahr et al. 1992). Six phylotypes (CG16–21) are delta-proteobacteria that includes sulfate reducers, such as *Desuluromusa kysingii*, *Desulfobacterium anilini*, and *Desulfobacterium corrodens* (Table 2). *D. kysingii* grow at 4–35°C and pH

Table 2 Bacterial 16S rDNA phylotypes identified in the travertine at Crystal Geyser. 25 phylotypes were detected from 30 gene sequences. The closest match and identity are based on BLAST similarity search

Phylotype (No. of clones)	Phylogenetic relationship			
	Taxonomic categorie	Closest database match	Accession No.	Identity (%)
CG12 (4)	Cyanobacteria	*Synechococcus* sp. JA-3-3Ab	(AY884052)	90
CG14 (1)	Cyanobacteria	*Synechococcus* sp. JA-3-3Ab	(AY884052)	88
CG15 (1)	Cyanobacteria	*Synechococcus* sp. JA-3-3Ab	(AY884052)	91
CG13 (1)	Cyanobacteria	*Synechococcus* sp. T71	(AF448073)	92
CG01 (1)	α-proteobacteria	*Loktanella vestfoldensis*	(AJ582226)	99
CG02 (1)	α-proteobacteria	*Loktanella* sp. WM1	(EF421433)	99
CG03 (1)	α-proteobacteria	*Loktanella vestfoldensis*	(AJ582226)	95
CG04 (2)	α-proteobacteria	*Sulfitobacter* sp. P76	(EU823294)	99
CG05 (1)	α-proteobacteria	*Thalassobaculum* sp. P24	(EU656113)	99
CG06 (1)	α-proteobacteria	*Pseudoruegeria aquimaris*	(DQ675021)	94
CG07 (1)	α-proteobacteria	*Rubrimonas cliftonensis*	(D85834)	94
CG08 (1)	α-proteobacteria	*Ochrobactrum* sp. LJJS1-2	(DQ133574)	86
CG09 (1)	α-proteobacteria	*Micavibrio* sp. EPC4	(DQ186615)	86
CG10 (1)	α-proteobacteria	*Parvibaculum* sp. P31	(FJ182044)	94
CG11 (1)	Chloroflexi	*Dehalococcoides* sp. BHI80-15	(AJ431246)	84
CG16 (1)	δ-proteobacteria	*Desulfuromusa kysingii*	(X79414)	93
CG19 (1)	δ-proteobacteria	*Desulfuromusa kysingii*	(X79414)	96
CG17 (1)	δ-proteobacteria	*Desulfobacterium anilini*	(EU020016)	94
CG20 (1)	δ-proteobacteria	*Desulfobacterium corrodens*	(AY274450)	99
CG18 (1)	δ-proteobacteria	*Polyangium thaxteri*	(AJ233943)	90
CG21 (2)	δ-proteobacteria	*Bacteriovorax* sp. PNEc1	(AY294221)	87
CG22 (1)	γ-proteobacteria	*Ectothiorhodospira shaposhnikovii*	(EU700082)	88
CG23 (1)	γ-proteobacteria	*Alteromonadales* bacterium JK46	(EF554897)	91
CG24 (1)	Bacteroidetes	*Aequorivita ferruginea*	(AY027803)	92
CG25 (1)	Firmicutes	*Ruminococcus* sp. NML 00-0124	(EU815223)	87

6.5–7.9 (Liesack and Finster 1994). *D. anilini* is rod shape and its optimum pH and temperature are around 6.9–7.5 and 35°C (Schnell et al. 1989).

6 Discussion

The SEM and fluorescence views indicate the close association between filamentous phototropic microbes and micritic bands (Fig. 3d, f, g). The phototrophs that were dominated by cyanobacteria but including anoxygenic taxa (Table 2) supported highly diverse heterotrophic community living on the surface or inside of the travertine in Crystal Geyser. The filamentous cyanobacteria are known to secrete extracellular polymeric substances (EPS) that form the glutinous biofilm on the travertine surface (Fig. 3f).

The EPS-bearing biofilm plays a central role in forming the lamination of stromatolitic carbonates (Riding 2000), but the specific process has been interpreted differently. A traditional interpretation for calcified filaments is biologically-induced precipitation associated with photosynthetic activity (Pentecost and Riding 1986) that rise supersaturation of calcite. Recent researches using microelectrodes confirmed that cyanobacterial biofilm photosynthesis drives calcification within the microenvironment of tufa biofilms (Bissett et al. 2008; Shiraishi et al. 2008). However, this process is effective only with relatively low alkalinity and pH around 8 (Arp et al. 2001), which are inconsistent with the water chemistry of the Crystal Geyser. Second, EPS can induce calcite nucleation and precipitation by binding calcium ions to their negatively charged sites (e.g., Pentecost 1985). This process may have happened occurred on the biofilm of the Crystal Geyser travertine, and is consistent with the calcified filamentous textures observed in the micritic bands (Fig. 3d). However, the cation-binding EPS can also inhibit carbonate precipitation by decreasing activity of Ca^{2+} in the vicinity of the biofilm (Arp et al. 1999), and the inhibition associated with EPS was demonstrated for an aragonite travertine in SW Japan.

The third potential process is trapping of suspended particles that were previously suggested for the modern marine stromatolites by Monty (1976). This process is still valid although the later researches of the Bahamian stromatolites suggested that the lamination is formed in over a longer term period (typically months) by temporal replacement of the microbial community and lithification associated with sulfate reduction (Reid et al. 2000; Andres et al. 2006). Our previous study on a calcite travertine at the Shionoha Hot Spring in Japan concluded that this process was responsible for the daily rhythm of the micritic bands (Takashima and Kano 2008).

Evidences in this study not specify which process was important, and it is quite possible that all three processes contributed to the band formation. Whatever the case, we observed the spatial association between the micritic bands and the microbes, and demonstrated that the microbes are dominated by photoautotrophs (Table 2). It is therefore likely that the micritic bands were formed in daytime when the microbes can photosynthesize.

If so, the observed laminated pattern (Fig. 3b) can be correlated to the actual discharge pattern from the Geyser (Fig. 2c). The travertine above the stained horizon consists of three darker-colored micritic band (D1–3) alternated with three lighter colored layers of larger calcite crystals (L1–3; Fig. 2b). This part should have been formed in a period from mid day of 28 August to evening of 31 August, and during the period five discharge events happened (No. 3–7; Fig. 2c). Just above the stained horizon, the first layer L1 (Fig. 3b) should have grown with the discharge event 3 that happened after the sunset. The next layer, the micritic band D1 was formed with the event No. 4 during the early afternoon of 29 August (Fig. 2c). Then, the event No. 5 corresponds to L2. The long-lasting event No. 6 likely formed both a micritic band (N2) and a lighter layer (L3), which corresponding to the daytime bubbling and the subsequent discharge at night, respectively. Finally, the last event No. 7 precipitated the uppermost layer, D3 (Fig. 3c). We can conclude that the darker bands (D1–3) were formed when both water and sunlight were available on the travertine surface. Interpretation of water supply did not reflect on the lamination pattern, but could reflect on a discontinuous surface that could be observed in SEM images (e.g., Fig. 3e).

This study confirmed importance of photosynthetic bacteria in forming the daily lamination of the stromatolitic carbonate. The bacteria are not necessary being cyanobacteria, and can be anoxygenic photosynthetic bacteria, such as green sulfur of purple non-sulfur bacteria. Stromatolites developed in the anoxic Earth surface before 2.3 Ga (Kopp et al. 2005) could have formed by microbial community including anoxygenic photosynthetic bacteria.

References

Allewalt JP, Bateson MM, Revsbech NP, Slack K, Ward DM (2006) Effect of temperature and light on growth of and photosynthesis by *Synechococcus* isolates typical of those predominating in the Octopus Spring microbial mat community of Yellowstone National Park. Applied and Environmental Microbiology 72:544–550

Andres MS, Sumner DY, Reid RP, Swart PK (2006) Isotopic fingerprints of microbial respiration in aragonite from Bahamian stromatolites. Geology 34:973–976

Arp G, Thiel V, Reimer A, Michaelis W, Reitner J (1999) Biofilm exopolymers control microbialite formation at thermal springs discharging into the alkaline Pyramid Lake, Nevada, USA. Sedimentary Geology 126:159–176

Arp G, Reimer A, Reitner J (2001) Photosynthesis-induced biofilm calcification and calcium concentrations in Phanerozoic ocean. Science 292:1701–1704

Baer JL, Rigby JK (1978) Geology of the Crystal Geyser and the environmental implications of its effluent, Grand County, Utah. Utah Geology 5:125–130

Bissett A, de Beer D, Schoon R, Shiraishi F, Reimer A, Arp G (2008) Microbial mediation of stromatolite formation in karst-water creeks. Limnology and Oceanography 53:1159–1168

DeLong EF (1992) Archaea in coastal marine environments. Proceedings of the National Academy of Sciences of the United States of America 89:5685–5689

Folk RL, Chafetz HS, Tiezzi PA (1985) Bizarre forms of depositional and diagenetic calcite in hot-spring travertines, central Italy. In: Scheiderman N, Harris PM (eds.) Carbonate Cements, SEPM Special Publications 36 pp 349–369

Guo L, Riding R (1992) Aragonite laminae in hot water travertine crusts, Rapolano Terme, Italy. Sedimentology 39:1067–1079

Herman MM, Lorah SJ (1988) The chemical evolution of a travertine-deposition stream: Geochmeical processes and mass transfer reactions. Water Resources Research 24:1541–1552

Ivanova EP, Gorshkova NM, Sawabe T, Zhukova NV, Hayashi K, Kurilenko VV, Alexeeva Y, Buljan V, Nicolau DV, Michailov VV, Christen R (2004) *Sulfitobacter delicatus* sp. nov. and *Sulfitobacter dubius* sp. nov., respectively from a starfish (*Stellaster equestris*) and sea grass (*Zostera marina*). International Journal of Systematic and Evolutionary Microbiology 54:475–480

Jones B, Renaut WR, Owen BR, Torfason H (2005) Growth patterns and implications of complex dendrites in calcite travertines from Lysuhóll, Snæfellsnes, Iceland. Sedimentology 52:1277–1301

Kano A, Takashima C, Ohtsuka S (2006) Hot-spring in eastern Kyushu and their related sedimentation and microbial processes. ISC 2006 Field Excursion Guidebook FE-A10:10 pp

Kitano Y (1963) Geochemistry of calcareous deposits found in hot springs. Journal of Earth Science Nagoya University 1:68–100

Kopp EK, Kirschvink JL, Hilburn IA, Nash CZ (2005) The Paleoproterozoic snowball Earth: A climate disaster triggered by the evolution of oxygenic photosynthesis. Proceedings of the National Academy of Sciences of the United States of America 102:1131–1136

Liesack W, Finster K (1994) Phylogenetic analysis of five strains of gram-negative, obligately anaerobic, sulfur-reducing bacteria and description of *Desulfirornusa* gen. nov., including *Desulfuromusa kysingii* sp. nov., *Desulfurornusa bakii* sp. nov., and *Desulfuromusa succinoxidans* sp. nov. International Journal of Systematic Bacteriology 44:753–758

Monty CLV (1976) The origin and development of cryptalgal fabric. Developments in Sedimentology 20:193–249

Pentecost A (1985) Association of cyanobacteria with tufa deposits: Identity, enumeration and nature of the sheath material revealed by histochemistry. Geomicrobiology Journal 4:285–298

Pentecost A (1994) Formation of laminate travertines at Bagno Vignone, Italy. Geomicrobiology Journal 12:239–251

Pentecost A (2005) Travertine. Springer-Verlag, Berlin, 445 pp

Pentecost A, Riding R (1986) Calcification in cyanobacteria. In: Leadbeater BSC, Riding R (eds.) Biomineralization in Lower Plants and Animals. Systematic Association Special Publication, 30. Clarendon Press, Oxford, pp 73–90

Plummer LN, Bosenberg E (1982) The solubilities of calcite, aragonite and vaterite in CO_2-H_2O solutions between 0 and 90°C, and an evaluation of the aqueous model for the system $CaCO_3^-CO_2$-H_2O. Geochimica et Cosmochimica Acta 46:1011–1040

Pukall R, Bunterfuß D, Frühling A, Rohde M, Kroppenstedt RM, Burghardt J, Lebaron P, Bernard L, Stackebrandt E (1999) *Sulfitobacter rnediterraneus* sp. nov., a new sulfite-oxidizing member of the α-Proteobacteria. International Journal of Systematic Bacteriology 49:513–519

Reid RP, Visscher PT, Decho AW, Stolz JF, Bebout BM, Dupraz C, Macintyre IG, Paerl HW, Pinckney JL, Prufert-Bebout L, Steppe TF, DesMarais DJ (2000) The role of microbes in accretion, lamination and early lithification of modern marine stromatolites. Nature 406:989–992

Riding R (2000) Microbial carbonates: the geological record of calcified bacterial-algal mats and biofilms. Sedimentology 47:179–214

Schnell S, Bak F, Pfennig N (1989) Anaerobic degradation of aniline and dihydroxybenzenes by newly isolated sulfate-reducing bacteria and description of *Desuifobacterium aniline*. Archives of Microbiology 152:556–563

Shipton ZK, Evans JP, Kirschner D, Kolesar PT, Williams AP, Heath J (2004) Analysis of CO_2 leakage through 'low-permeability' faults from natural reservoirs in the Colorado Plateau, east-central Utah. Special Publications, Geological Society London 233:43–58

Shiraishi F, Bissett A, de Beer D, Reimer A, Arp G (2008) Photosynthesis, respiration and exopolymer calcium-binding in biofilm calcification (Westerhöfer and Deinschwanger Creek, Germany). Geomicrobiology Journal 25:83–94

Suzuki T, Muroga Y, Takahama M, Shiba T, Nishimura Y (1999) *Rubrimonas cliftonensis* gen. nov., sp. nov., an aerobic bacteriochlorophyl II-containing bacterium isolated from a saline lake. International Journal of Systematic Bacteriology 49:201–205

Takashima C, Kano A (2008) Microbial processes forming daily lamination in a stromatolitic travertine. Sedimentary Geology 208:114–119

Tamura K, Dudley J, Nei M, Kumar S (2007) MEGA4: Molecular Evolutionary Genetics Analysis (MEGA) software version 4.0. Molecular Biology and Evolution 24:1596–1599

Trappen SV, Mergaert J, Swings J (2004) *Loktanella salsilacus* gen. nov., sp. nov., *Loktanella fryxellensis* sp. nov. and *Loktanella vestfoldensis* sp. nov., new members of the Rhodobacter group, isolated from microbial mats in Antarctic lakes. International Journal of Systematic and Evolutionary Microbiology 54:1263–1269

Zahr M, Fobel B, Mayer H, Imboff JF, Campos PV, Weckesser J (1992) Chemical composition of the lipopolysaccharides of *Ectothiorhodospira shaposhnikovii*, *Ectothiorhodospira mobilis*, and *Ectothiorhodospira halophila*. Archives of Microbiology 157:499–504

Zhang GI, Hwang CY, Cho BC (2008) *Thalassobaculum litoreum* gen. nov., sp. nov., a member of the family Rhodospirillaceae isolated from coastal seawater. International Journal of Systematic and Evolutionary Microbiology 58:479–485

Photosynthesis-Induced Stromatolite Formation in the Freshwater Creeks

Fumito Shiraishi

1 Introduction

Stromatolites, the oldest macrofossil in the Earth history, appeared about 3.5 billion years ago and considered to be formed by the interaction between the activities of benthic microorganisms and the surrounding environment, and hence are expected to provide important information for the evolution of life and Earth's environment. However, fossil stromatolites usually lack microfossils (e.g., Riding 1994), which makes it difficult to interpret their formation mechanisms. One approach to overcome this problem is the investigation of recent analogs in detail to understand how microorganisms produce stromatolitic constructions. Promising analogs were thought to be the recent stromatolites developing in seawater settings, and thus, they have been keenly studied last several decades. However, some researchers pointed out their critical disadvantage: the recent marine stromatolites are composed mainly of agglutinated sand grains, and are different from ancient ones composed mainly of fine minerals precipitated *in situ* (e.g., Awramik 1984; Riding 2008). Therefore, it is necessary to seek and investigate more suitable recent analogs to understand stromatolite formation.

The DFG research unit 571 (Geobiology of organo- and biofilms: Coupling of the geosphere and the biosphere by microbial processes) focused on recent freshwater stromatolites (tufa stromatolites) developing at karst creeks as a possibly more suitable analog because they are composed of fine carbonate minerals, resembling to the fabric of ancient stromatolites. Our 3-year investigation revealed that the photosynthetic activities of investigated biofilms have strong impact on ambient water chemistry, induce $CaCO_3$ precipitation, and result in the stromatolite

F. Shiraishi
Department of Environmental Change, Graduate School of Social and Cultural Studies, Kyushu University, 744 Motooka, Nishi-ku, Fukuoka 819-0395, Japan
Geoscience Centre, University of Göttingen, Goldschmidtstraße 3, 37077 Göttingen, Germany
e-mail: fshirai@scs.kyushu-u.ac.jp

J. Reitner et al., *Advances in Stromatolite Geobiology*,
Lecture Notes in Earth Sciences 131, DOI 10.1007/978-3-642-10415-2_8,

formation. The present note briefly summarizes the results of our project (Arp et al. 2010; Bissett et al. 2008a, b; Shiraishi et al. 2008a, b).

2 Materials and Methods

Freshwater stromatolites developing at two German karst creeks, Deinschwanger Bach and Westerhöfer Bach, were investigated [for detail field settings, see Arp et al. (2001) and Shiraishi et al. (2008b), respectively]. Visualization of microorganisms in the mineralized stromatolite biofilms was achieved by applying improved FISH (fluorescence *in situ* hybridization) and CARD (catalyzed reporter deposition) – FISH together with tape-stabilized sectioning technique [for technical detail, see Shiraishi et al. (2008c)]. *In situ* and *ex situ* microelectrode measurements (pH, O_2, Ca^{2+} and CO_3^{2-}) were conducted to evaluate the metabolic impact of stromatolite biofilms on $CaCO_3$ precipitation [for technical detail, see Bissett et al. (2008a, b) and Shiraishi et al. (2008a, b)]. In order to trace calcium, stromatolite samples were immersed in the creek water containing ^{45}Ca, incubated for 24 h either under light or in the dark, and labeled calcium was visualized by β-imaging [for technical detail, see Bissett et al. (2008a)]. Finally, required ambient chemical conditions for photosynthesis-induced $CaCO_3$ precipitation (see below) was examined by PHREEQC simulations. For this examination, 800 μmol L^{-1} of CO_2 was subtracted from given water concentration, and the calcite saturation states before and after subtraction were compared as $\Delta\Omega_{act}$.

3 Results and Discussion

Mineralized biofilms such as stromatolites are usually not suitable for FISH investigation because (1) positively charged mineral surfaces bind negatively charged FISH probes to cause severe false binding, and (2) it is very difficult to make thin sections of hard textures. However, improved hybridization and sectioning techniques applied in our study could successfully avoid these inherent difficulties of calcified biofilms. The results showed that the filamentous cyanobacteria densely colonized in the uppermost 200–500 μm of stromatolite biofilms, while heterotrophic bacteria were abundantly distributed from the surface to the deeper part (Shiraishi et al. 2008c, fig. 5B).

Microelectrode measurements conducted by Bissett et al. (2008a, b) and Shiraishi et al. (2008a, b) showed that the water chemistry near the biofilm surface (diffusive boundary layer; ca. 200–500 μm from the biofilm surface) is significantly different from that of bulk water: pH, O_2 and CO_3^{2-} increases coupled with Ca^{2+} decrease occurred during illumination, while pH and O_2 decreases coupled with slight Ca^{2+} increase occurred in the dark (Fig. 1). Saturation state of calcite at the biofilm surface calculated from Ca^{2+} and CO_3^{2-} profiles was higher than that

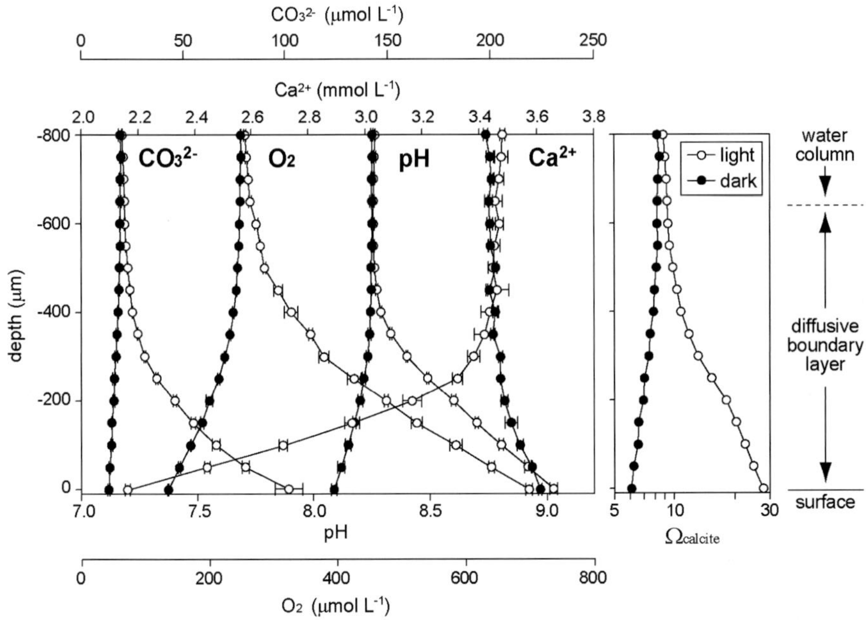

Fig. 1 Microelectrode profiles of tufa stromatolite (pH, O_2, Ca^{2+} and CO_3^{2-}) from the Westerhöfer Bach. Open circles indicate light profiles, and closed circles indicate dark profiles. Saturation state ($\Omega_{calcite}$) calculated from Ca^{2+} and CO_3^{2-} profiles is also shown. Modified after Shiraishi et al. (2008a)

of bulk water under illumination, and lower under the dark. These results are considered to be indicating massive $CaCO_3$ precipitation induced by photosynthetic activity of stromatolite biofilms due to illumination, while respiration inhibits precipitation in the dark (Bissett et al. 2008a; Shiraishi et al. 2008a, b). Indeed, the actual depositional rate (-1.85 to -2.86×10^{-6} mol m^{-2} s^{-1}) roughly coincides with Ca^{2+} flux measured by microelectrode [-3.81×10^{-6} mol m^{-2} s^{-1}; see Shiraishi et al. (2008a)]. This interpretation was further confirmed by ^{45}Ca trace experiments, showing that calcium accumulation occurred only under illumination (Bissett et al. 2008a). Thus, it is concluded that the investigated stromatolites are formed by photosynthesis. No $CaCO_3$ precipitation was detected by microelectrodes at the surface of biofilm-free limestone plate, regardless of surrounding $CaCO_3$ supersaturation (ca. 10-fold). This fact indicates that photosynthesis has significant influence on $CaCO_3$ precipitation kinetics (Shiraishi et al. 2008a).

Despite the critical role of cyanobacterial photosynthetic activity in context to the formation of investigated stromatolites, carbon isotopic record of stromatolites reflected the change in the creek water, and no ^{13}C enrichment in the precipitated $CaCO_3$, that is considered as signature for photosynthetic carbonate precipitation (e.g., Pentecost and Spiro 1990), was recognized (Shiraishi et al. 2008b, fig. 10).

One possible explanation for this observation is that the photosynthetic CO_2 assimilation is not sufficient enough for establishing a steep ^{13}C gradient in the diffusive boundary layer of investigated stromatolites. To confirm this view, it is necessary to study detail fluxes of various carbonate species and their isotopic equilibrium at the stromatolite surface. In any case, it should be noted that the absence of photosynthetic signatures in carbon stable isotopes does not necessarily indicate a lack of photosynthetic contribution for $CaCO_3$ precipitation (Shiraishi et al. 2008b).

Cyanobacterial calcification fabric is also considered to be useful for distinguishing photosynthetic calcification [sheath impregnation for photosynthetically-induced, and sheath encrustation for inorganically-induced calcification; see Merz-Preiß and Riding (1999)]. The investigated stromatolites, however, contain only encrusted sheaths (Shiraishi et al. 2008b, fig. 3a, b), pointing to the fact that sheath impregnation or encrustation alone cannot be used as indicators for photosynthetic-induced $CaCO_3$ precipitation (Shiraishi et al. 2008b).

Numerical simulations conducted by Shiraishi (submitted) suggest that photosynthesis-induced $CaCO_3$ precipitation, as recognized in the investigated freshwater stromatolites, can even occur in saline settings such as seawater, but only within a defined range of $CO_{2(aq)}$, $CO_3{}^{2-}$ and Ca^{2+} concentrations: the most suitable $CO_{2(aq)}$ and $CO_3{}^{2-}$ concentrations for increasing $\Delta\Omega_{act}$ are attained around a pH of 8.5 and a dissolved inorganic carbon (DIC) concentration of 3–10 mmol L^{-1}. Too high and too low concentrations of $CO_{2(aq)}$ and $CO_3{}^{2-}$ (which points to relatively high and low pH and DIC concentrations) inhibit increasing $\Delta\Omega_{act}$ (Fig. 2). Ca^{2+} concentration positively correlates with $\Delta\Omega_{act}$, and thus, a higher Ca^{2+} concentration increases the chemical suitability for photosynthesis-induced $CaCO_3$ precipitation ($\Delta\Omega_{act}$) and vice versa.

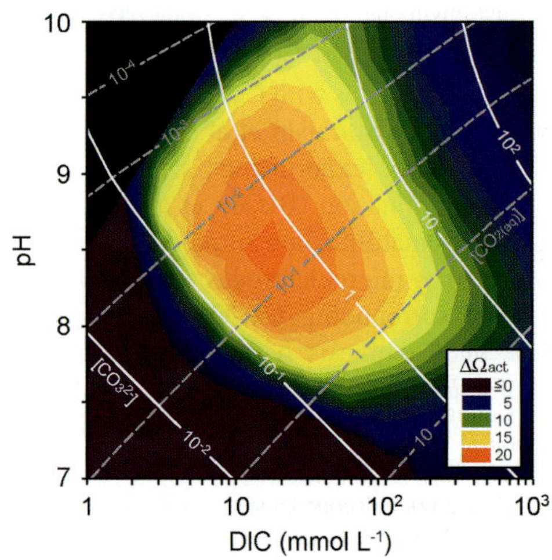

Fig. 2 Contour plot of CO_2 removal effect on calcite precipitation ($\Delta\Omega_{act}$) for a wide range of pH–DIC conditions calculated using standard seawater. $[CO_{2(aq)}]$ and $[CO_3^{2-}]$ contour lines (mmol L^{-1}) are shown with gray dashed lines and white solid lines, respectively

4 Conclusions

The investigation of freshwater stromatolites developing at the karst creeks revealed that they are formed by photosynthetic activity of cyanobacteria. Despite of critical role of cyanobacterial photosynthesis, investigated stromatolites did not contain previously suggested indicators for photosynthesis-induced $CaCO_3$ precipitation. Numerical simulations revealed that the photosynthetic $CaCO_3$ precipitation can theoretically occur even in the seawater settings, but requires optimum pH, DIC and Ca^{2+} concentrations.

Acknowledgements Dr. Gernot Arp kindly provided valuable advices for improving the manuscript. This is the part of the Research Unit 571 "Geobiology of organo- and biofilms", funded by the German Research Foundation (DFG–FOR 571; AR 335/5; publication #49).

References

Arp G, Wedemeyer N, Reitner J (2001) Fluvial tufa formation in a hard-water creek (Deinschwanger Bach, Franconian Alb, Germany). Facies 44:1–22

Arp G, Bissett A, Brinkmann N, Cousin S, de Beer D, Friedl T, Mohr K, Neu TR, Reimer A, Shiraishi F, Stackebrandt E, Zippel B (2010) Tufa-forming biofilms of German karstwater creeks: Microorganisms, exopolymers, hydrochemistry and calcification. In: Rogerson M, Pedley MM (eds), Tufas and Speleothems: Unravelling the Microbial and Physical Controls. Geological Society London Special Publication 336:83–118

Awramik SM (1984) Ancient stromatolites and microbial mats. In: Cohen Y, Castenholtz RW, Halvorson HO (eds) Microbial Mats: Stromatolites. New York, Alan R Liss Inc, pp 1–22

Bissett A, de Beer D, Schoon R, Shiraishi F, Reimer A, Arp G (2008a) Microbial mediation of stromatolite formation in karst-water creeks. Limnology and Oceanography 53:1159–1168

Bissett A, Reimer A, de Beer D, Shiraishi F, Arp G (2008b) Metabolic microenvironmental control by photosynthetic biofilms under changing macroenvironmental temperature and pH. Applied and Environmental Microbiology 74:6306–6312

Merz-Preiß M, Riding R (1999) Cyanobacterial tufa calcification in two freshwater streams: ambient environment, chemical thresholds and biological processes. Sedimentary Geology 126:103–124

Pentecost A, Spiro B (1990) Stable carbon and oxygen isotope composition of calcites associated with modern freshwater cyanobacteria and algae. Geomicrobiology Journal 8:17–26

Riding R (1994) Evolution of algal and cyanobacterial calcification. In: Bengtson S (ed) Early Life on Earth. New York, Columbia Univ. Press, pp 426–438

Riding R (2008) Marrying stromatolite perspectives: 3500 million year of history and a century of research. In: Reitner J, Quéric NV, Reich M (eds) Abstract volume of International Kalkowsky-Symposium. Universitätsdrucke Göttingen, Göttingen, pp 29–30

Shiraishi F, Bissett A, de Beer D, Reimer A, Arp G (2008a) Photosynthesis, respiration and exopolymer calcium-binding in biofilm calcification (Westerhöfer and Deinschwanger Bach, Germany). Geomicrobiology Journal 25:83–94

Shiraishi F, Reimer A, Bissett A, de Beer D, Arp G (2008b) Microbial effects on biofilm calcification, ambient water chemistry and stable isotope records (Westerhöfer Bach, Germany). Palaeogeography, Palaeoclimatology, Palaeoecology 262:91–106

Shiraishi F, Zippel B, Neu TR, Arp G (2008c) *In situ* detection of bacteria in calcified biofilms using FISH and CARD–FISH. Journal of Microbiological Methods 74:6306–6312

The Role of Purple Sulphur Bacteria in Carbonate Precipitation of Modern and Possibly Early Precambrian Stromatolites

Rolf Warthmann, Crisógono Vasconcelos, Anne Greet Bittermann, and Judith A. McKenzie

1 Introduction

Since the description of stromatolites a century ago by Kalkowsky (1908) as products of ancient microbial activity, the exact understanding of its formation remained unclear and is still matter of debate. Stromatolites are definitely the first visible fossil record of life in Earth history, and are dated back as far as 3.45 Ga (Grotzinger and Knoll 1999; Riding 2000), which implicates very early activities of life, possibly different than those today. The environmental conditions in the Early Precambrian, such as ocean chemistry or surface temperature are not fully understood today. Several studies tried to explain the formation of Precambrian stromatolites by analyzing modern analogues such a the Highborne cay stromatolites of the Bahamas (Dupraz and Visscher 2005) or the Shark Bay (Australia) stromatolites (Burns et al. 2004; Allen et al. 2009). Studies of ancient stromatolites apparently showed fossil traces of filamentous or coccoid cells (Westall et al. 2006). Ancient microbial activity is supported by geochemical indications, which show isotope discriminations of carbon and sulphur isotopes (Grassineau et al. 2001; Schidlowski 2001). However, often these geochemical data were not unambiguous. An essential feature of some Precambrian stromatolites is lamination caused by alternating layers of carbonate (calcite, micrite, dolomite), intermitted with other mineral layers, mostly silica minerals in ancient stromatolites (Allwood et al. 2006).

R. Warthmann (✉)
Institute of Biotechnology, Zurich University of Applied Sciences, 8820 Wdenswil, Switzerland
e-mail: R.Warthmann@gmail.com

C. Vasconcelos and J.A. McKenzie
Geological Institute, ETH Zürich, 8092 Zürich, Switzerland

A.G. Bittermann
Center for Microscopy and Image Analysis, University of Zurich, 8006 Zurich, Switzerland

J. Reitner et al., *Advances in Stromatolite Geobiology*,
Lecture Notes in Earth Sciences 131, DOI 10.1007/978-3-642-10415-2_9,

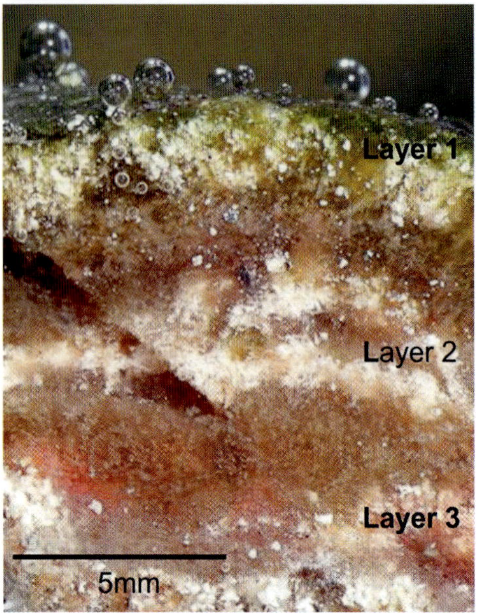

Fig. 1 Cross-section through a microbial mat of a stromatolite from the hypersaline lagoon *Lagoa Vermelha*. Note the sequence of colors indicating the dominant organisms, especially the *pinkish layer* containing purple sulphur bacteria. Top *green layer*, cyanobacteria; *brown layers*, heterotrophic bacteria; *red layer*, purple sulphur bacteria. The *white layers* are in situ formed carbonate layers

Whereas in modern stromatolites carbonate accretion is in the majority of cases a result of trapping, binding and cementing of detrital particles by activity of cyanobacteria and sulphate-reducing bacteria (Reid et al. 2000; Dupraz and Visscher 2005), the ancient counterparts do not show detrital carbonate accumulation, but rather an in situ precipitation of carbonates. Phototrophic sulphur bacteria are candidates which were presumably dominant on Early Earth (Blankenship 1992) and a possible role of anoxygenic phototrophic bacteria in carbonate formation has been discussed recently (Bosak et al. 2007).

In this paper, the study of living stromatolites and microbial mats of a hypersaline lagoon in Brazil (Vasconcelos et al. 2006) give new insights into the role of purple sulphur bacteria and their sulphur metabolism (Fig. 1). The mode of in situ carbonate laminations, which was studied in modern biofilms and in artificially grown stromatolites, possibly applies to the Early Precambrian type of stromatolites as well.

2 Methods

Artificial stromatolites: Anaerobic chambers (~5 l) were filled with an anoxic gas mixture of 90% N_2, 5% CO_2 and 5% H_2. A catalyst pack (Oxoid Ltd.) in presence of H_2 provides strictly anoxic conditions. Furthermore, H_2 can serve as electron donor for initial sulphate reducing-bacteria and anoxygenic phototrophic bacteria. A basin was placed in the chamber containing 0.7 l artificial seawater (Coolen and Overmann 2000) with increased carbonate alkalinity (20 mM). To prevent growth of cyanobacteria, 50 μm DCMU [3-(3′,4′-Dichlorphenyl)-1,1-Dimethylurea] was provided to the liquid. A block (9 × 9 × 3 cm) of 3% triple washed agar served as a gelatinous growth surface, similar to natural stromatolites and microbial mats. The experiment was inoculated by adding 1 cm^3 homogenized microbial mat from Lagoa Vermelha (Brazil), and was inoculated in a 12 h day – night cycle provided by a 400 W HQI reflector lamp (6,500 K) in a distance of 1 m, which gives a similar illumination as a tropical light regime. Incubation was run under room temperature (~22°C).

Microsensor measurements: Needle-type microsensors (O_2, sulphide, pH) were obtained from Unisense (Aarhus DK), equipped with computer-controlled micromanipulators and were used according to the manual. Illumination was provided by a 150 W HQI reflector lamp, producing typical full-spectrum tropical light conditions.

Phylogenetic analysis: 16S rDNA was extracted and amplified from pure cultures of microorganisms using standard methods. PCR products were partially sequenced and aligned using the ARB software. Phylogenetic affiliations were established using the Ribosomal Database Project (RDP) with standard parameters, see http://rdp.cme.msu.edu/.

Scanning electron microscopy (SEM) and element dispersive x-ray microanalysis (EDX): Samples were fixed with 2.5% glutaraldehyde in artificial seawater, dehydrated in a graded series of ethanol, dried over the critical point of CO_2, mounted and sputter-coated with 5 nm platinum. Topographic images were acquired as secondary electron-signal in a Zeiss Supra 50 VP SEM at 4 kV. Element analysis was performed at 10 kV using an EDAX-detector and EDAX Genesis-software.

Samples of Abu Dhabi microbial mats: Sampling area of Abu Dhabi sabkha microbial mats is described in Bontognali et al (2009).

3 Results

3.1 Morphology of Hypersaline Microbial Mats

Stromatolites from Lagoa Vermelha and layered microbial mats from the Abu Dhabi sabkha showed the typical sequence of green, brown and pinkish- to red layers embedded in a matrix of exopolymeric substances (EPS). White carbonate layers appear in association with distinct microbial layers in the studied mats (Fig. 1). No trapping and binding, but chemical in situ precipitation is assumed to be responsible for the carbonate accretion, as no (or very few) detrital carbonate particles were observed. The mode of formation of these layered carbonates will be discussed later on.

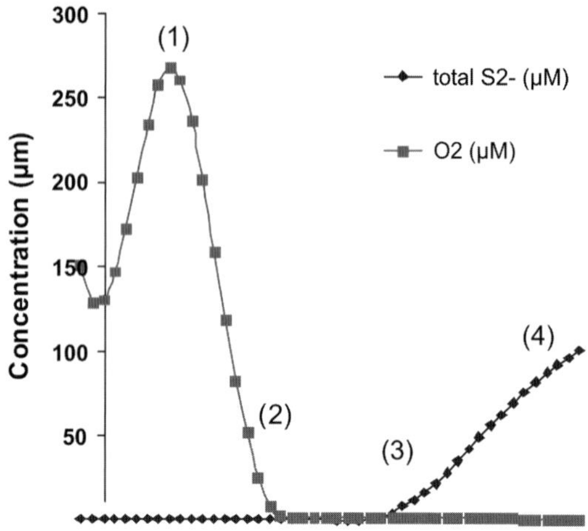

Fig. 2 Profiles of oxygen and sulfide in a stromatolite-building microbial mat of *Lagoa Vermelha*. Obviously, the profiles of oxygen and sulfide do not overlap, indicating anaerobic sulfide oxidation. Similar profiles were obtained with a microbial mat from the Abu Dhabi sabkha. Annotations: (1) Oxygen peak produced by cyanobacteria, (2) oxygen decrease by heterotrophs, (3) Sulfide consumption by anaerobic sulfide-oxidizing purple bacteria, and (4), sulphide supply by sulfide-producing sulfate- and sulphur reducers deeper in the sediment

3.2 *Micro-scale Profiles of Oxygen and Sulphide*

Repeated measurements of microbial mats from *Lagoa Vermelha* and the Abu Dhabi sabkha showed under illumination that the vertical profiles of oxygen and sulphide did not overlap. Oxygen produced by cyanobacteria in the top layer decreased within 3 mm to zero, while sulphide appeared at about 4.8 mm, implying that there is a gap between O_2 and sulphide (Fig. 2). This finding was confirmed in multiple measurements. The profile shows the activity of four types of organisms: (1) Oxygen-producing cyanobacteria, (2) oxygen consuming heterotrophs, (3) anaerobic sulphide-oxidizing purple bacteria, and (4), sulphide-producing sulphate- and sulphur reducers deeper in the sediment.

3.3 *Artificial Anoxic Microbial Mats*

Artificial microbial mats were grown in anoxic chambers filled with artificial seawater, inoculated with *Lagoa Vermelha* mat microbes and incubated for several months in a diurnal light cycle (for details see Sect. 2). The result was a red microbial mat of about 0.5 mm in size at the surface of the semi-solid substratum (Fig. 3), which appeared after 4–6 weeks incubation time. Carbonate precipitated in the red microbial mat as 20–40 μm globular structures, similar as observed before in other microbial experiments (van Lith et al. 2003; Warthmann et al. 2005; Bontognali et al. 2007, 2008). The produced carbonate was a mixture of calcite and hydromagnesite, as analyzed by SEM-EDX (Fig. 4). Microscopic observations showed that the dominant microorganisms in the microbial community were purple sulphur bacteria, with typical sulphur inside the cells. One dominant strain was isolated and identified as *Ectothiorhodospira marismortui* (belonging to modern proteobacteria) as closest neighbor by partial 16S rDNA analysis. This fits well, because these organisms are marine, anoxic sulphide-oxidizing bacteria, and are typical microbial mat inhabitants. Purple sulphur bacteria in general play an important role in the intensive sulphur cycling in microbial mats (Franks and Stolz in press).

The primary metabolism of purple sulphur bacteria is the assimilation CO_2 by means of H_2S oxidation according to $2HS^- + 3H^+ + HCO_3^- \rightarrow \langle CH_2O \rangle + 2S^\circ + 2H_2O$. Two weak acids ($H_2S/HS^-$ and CO_2/HCO_3^-) are converted to neutral compounds (biomass, S° and H_2O) by that reaction, resulting in a shift of pH to alkaline, and thereby promoting carbonate precipitation. Purple sulphur bacteria can produce large amounts of elemental sulphur (S°) (Pfennig and Trüper 1989),

Fig. 3 Artificial microbial mat grown under anoxic conditions in the light. A *red* microbial mat developed after 2 months of incubation, containing purple sulphur bacteria

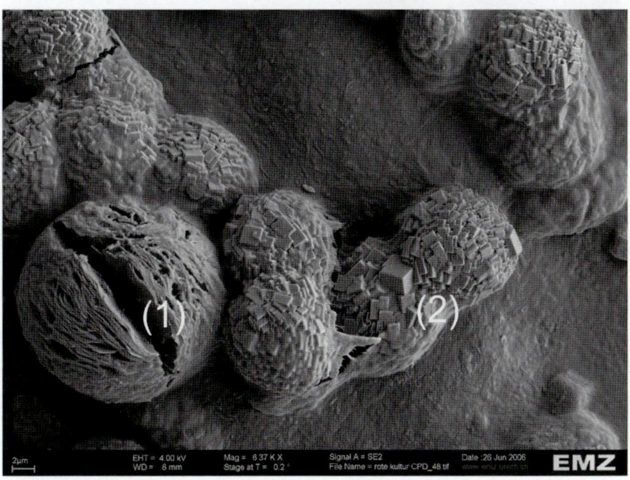

Fig. 4 Scanning electron micrograph showing minerals from the artificial microbial mat. Carbonate minerals, such as calcite (2) and hydromagnesite (1) were formed within the *red layer* of purple sulphur bacteria

which is typically either stored in the cells or is excreted, and thus producing a pH imbalance in the microbial mat according the equation above. The pH shift of anaerobic sulphide oxidation of is even stronger than by oxygenic photosynthesis. The overall reaction including carbonate precipitation would lead to a maximum of 3 calcite formed by 1 carbon fixed into biomass, according to:

$$\mathbf{2HS^- + 3Ca^{2+} + 4HCO_3^- \rightarrow 3CaCO_3 + <CH_2O> + 2S^\circ + 2H_2O}$$

The pinkish layer with purple sulphur bacteria is a site in the mat where carbonates nucleate and grow, as observed frequently in the natural microbial mats from *Lagoa Vermelha* (van Lith et al. 2003; Vasconcelos et al. 2006) and in artificial microbial mats (Warthmann et al. 2006). As a proof that elemental sulphur is formed inside the cells, a pure culture of the isolated bacteria was investigated by SEM-EDAX, which showed clearly that the inclusions in the cells were pure sulphur (Fig. 5).

4 Discussion

4.1 Photosynthesis and Carbonate Precipitation in the Archean

Environmental conditions of the Archean were probably very different from those today. An anoxic reducing atmosphere, high UV radiation and elevated Earth surface temperatures were the main characteristics. Nevertheless, the first traces

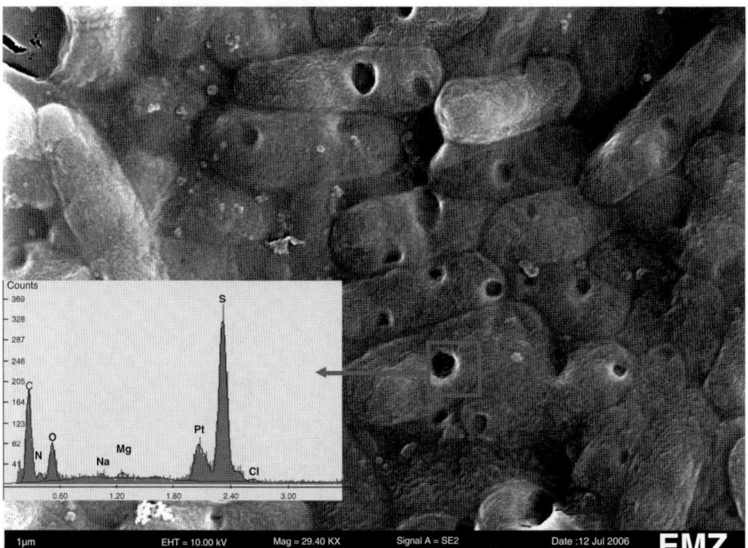

Fig. 5 Element analysis on artificial microbial mat. The topographic SEM-image shows a compact mass of *Ectothiorhodospira* sp. bacteria, containing elemental sulphur (S°) globules inside the cells. The EDX spectrum (box insert), showing sulphur besides Pt from the coating and C, O, N from biomass

of life as the 3.5-Ga Warrawoona sequence in Australia were attributed to the existence of microbial mats (Tice and Lowe 2004). Cyanobacteria were most likely absent at that time, so H_2S and H_2 served as major electron donor for an early anaerobic photosynthesis (Olson 2006). These electron donors, which were probably abundant at that time, enabled phototrophic biomass formation without oxygen production in the euphotic zones of shallow ocean seafloors or lagoons protected of strong wave energy. By their metabolism, carbonate can be formed as layers, because purple sulphur bacteria are typically organized in distinct layers in a microbial mat (Krumbein 1983). A model for the Precambrian carbonate formation by purple sulphur bacteria is outlined in Fig. 6. As no competition by higher organisms or grazers occurred, mats could develop and proliferate to form stable structures over some time. The Precambrian H2S-oxidizers forming the early microbial mats were probably phylogenetically different than the modern bacteria doing the same reactions today, but the process in principle did not change, according to Robert N. Ginsburg saying "The players may have changed, but the game is still the same".

5 Conclusions

The findings that purple sulphur bacteria are able to form microbial mats, which precipitate carbonate in distinct layers, expand our understanding of carbonate biomineralization and stromatolite formation in early anoxic ecosystems. In summary, we can conclude that even before the evolution of oxygenic photosynthesis, anaerobic sulphur bacteria may have contributed to the formation of

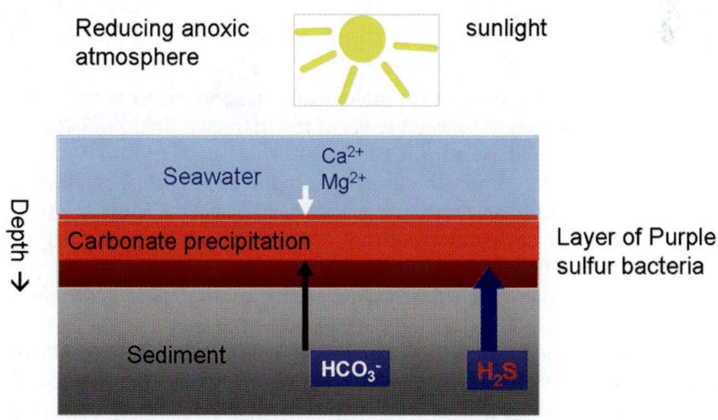

Fig. 6 Model of microbial carbonate precipitation in the Early Precambrian. Anoxygenic purple sulphur bacteria (*red layer*) producing a pH shift in the light, inducing precipitation of in situ carbonate. Ca^{2+} is provided by seawater, and sulfide is produced by sulphur- and sulphate-reducing bacteria

biologically-mediated carbonates resulting in the well-known carbonate lamination of fossil microbial mats and stromatolites.

References

Allen MA, Goh F, Burns BP, Neilan BA (2009) Bacterial, archaeal and eukaryotic diversity of smooth and pustular microbial mat communities in the hypersaline lagoon of Shark Bay. Geobiology 7:82–96

Allwood AC, Walter MR, Kamber BS, Marshall CP, Burch IW (2006) Stromatolite reef from the Early Archaean era of Australia. Nature 441:714–718

Blankenship RE (1992) Origin and early evolution of photosynthesis Photosynthesis Research 33:91–111

Bontognali T, Vasconcelos C, Warthmann R, McKenzie JA (2007) Dolomite nucleation on extracellular polymeric substances. In: 17th Annual V M Goldschmidt Conference, Cologne, pp A108–A108

Bontognali TRR, Vasconcelos C, Warthmann RJ, Dupraz C, Bernasconi SM, McKenzie JA (2008) Microbes produce nanobacteria-like structures, avoiding cell entombment. Geology 36:663–666

Bontognali TRR, Vasconcelos C, Warthmann RJ, Bernasconi SM, Dupraz C, Strohmenger CJ, McKenzie JA (2009) Dolomite formation within microbial mats in the coastal sabkha of Abu Dhabi (United Arab Emirates). Sedimentology, published online: Dec. 8 2009

Bosak T, Greene SE, Newman DK (2007) A likely role for anoxygenic photosynthetic microbes in the formation of ancient stromatolites. Geobiology 5:119–126

Burns BP, Goh F, Allen M, Neilan BA (2004) Microbial diversity of extant stromatolites in the hypersaline marine environment of Shark Bay, Australia. Environmental Microbiology 6:1096–1101

Coolen MJ, Overmann J (2000) Functional exoenzymes as indicators of metabolically active bacteria in 124,000-year-Old sapropel layers of the eastern Mediterranean Sea. Applied and Environmental Microbiology 66:2589–2598

Dupraz C, Visscher PT (2005) Microbial lithification in marine stromatolites and hypersaline mats. Trends in Microbiology 13:429–438

Franks J, Stolz JF (2009) Microbial Mats in Earth's Fossil Record of Life: Geobiology. Earth-Science Reviews 96:163–172

Grassineau NV et al. (2001) Antiquity of the biological sulphur cycle: evidence from sulphur and carbon isotopes in 2700 million-year-old rocks of the Belingwe Belt, Zimbabwe. Proceedings of the Royal Society of London Series B-Biological Sciences 268:113–119

Grotzinger JP, Knoll AH (1999) Stromatolites in Precambrian carbonates: evolutionary mileposts or environmental dipsticks? Annual review of earth and planetary sciences 27:313–358

Kalkowsky E (1908) Oolith und Stromatolith im norddeutschen Buntsandstein. Zeitschrift der Deutschen geologischen Gesellschaft 60:68–125

Krumbein WE (1983) Stromatolites – the challenge of a term in space and time. Precambrian Research 20:493–531

Olson JM (2006) Photosynthesis in the Archean Era. Photosynthesis Research 88:109–117

Pfennig N, Trüper HG (1989) Anoxygenic phototrophic bacteria. In: Staley JT, Bryant, MP, Pfennig N, Holt JG (ed) Bergey's Manual of Systematic Bacteriology. Williams and Wilkins, Baltimore/Hong Kong/London/Sydney, pp 1635–1639

Reid RP et al. (2000) The role of microbes in accretion, lamination and early lithification of modern marine stromatolites. Nature 406:989–992

Riding R (2000) Microbial carbonates: the geological record of calcified bacterial-algal mats and biofilms. Sedimentology 47:179–214

Schidlowski M (2001) Carbon isotopes as biogeochemical recorders of life over 3.8 Ga of Earth history: evolution of a concept. Precambrian Research 106:117–134

Tice MM, Lowe DR (2004) Photosynthetic microbial mats in the 3,416-Myr-old ocean. Nature 431:549–552

van Lith Y, Vasconcelos C, Warthmann R, McKenzie JA (2003) Sulphate-reducing bacteria induce low-temperature dolomite and high Mg-calcite formation. Geobiology 1:71–79

Vasconcelos C, Warthmann R, McKenzie JA, Visscher PT, Bittermann AG, van Lith Y (2006) Lithifying microbial mats in Lagoa Vermelha, Brazil: Modern Precambrian relics? Sedimentary Geology 185:175–183

Warthmann R, Vasconcelos C, Sass H, McKenzie JA (2005) Desulfovibrio brasiliensis sp. nov., a moderate halophilic sulfate-reducing bacterium from Lagoa Vermelha (Brazil) mediating dolomite formation. Extremophiles 9:255–261

Warthmann R, Vasconcelos C, McKenzie JA (2006) Anaerobic sulfur bacteria inducing lithification in modern- and possibly Precambrian stromatolites. Geochimica et Cosmochimica Acta 70:A689–A689

Westall F, de Ronde CEJ, Southam G, Grassineau N, Colas M, Cockell C, Lammer H (2006) Implications of a 3.472–3.333 Gyr-old subaerial microbial mat from the Barberton greenstone belt, South Africa for the UV environmental conditions on the early Earth. Philosophical Transactions of the Royal Society B Biological Sciences 361:1857–1875

Precipitation of $CaCO_3$ Under Sulphate-Reduction Conditions

Dorota Wolicka and Andrzej Borkowski

1 Introduction

The major part of geochemical processes in the lithosphere, including the formation of carbonate minerals, may be linked with the activity of living organisms. These processes are influenced by physical and chemical factors in the environment that significantly control the occurrence of particular mineral phases. Because the reactions of carbonate precipitation are controlled by, e.g., carbon dioxide content, biotic factors seem to play a significant role in this process. Carbon dioxide is utilised both as a donor and acceptor of electrons by various groups of microorganisms in their metabolic processes. The key role in CO_2 generation is played by microorganisms, particularly those occurring in environments favouring the precipitation of carbonate minerals. It should be noted that 10^8–10^9 tonnes of carbon are annually deposited as $CaCO_3$ in the oceans.

The possibility of $CaCO_3$ formation in favourable physical and chemical conditions is enormous because of the very low value of the $CaCO_3$ solubility constant [$K_{sol} = 10^{-8.32}$]. Due to this fact, even a low content of carbon dioxide introduced into an environment containing Ca^{2+} ions may cause concentrations to exceed the solubility-product limit and result in the precipitation of calcium carbonate. It should be remembered, however, that the formation of $CaCO_3$ is strictly linked with the physical and chemical conditions of the environment, mainly pH (Hammes and Verstraete 2002).

D. Wolicka (✉)
Institute of Geochemistry, Mineralogy and Petrology, University of Warsaw, ul. Żwirki i Wigury 93, 02-089 Warsaw, Poland
e-mail: d.wolicka@uw.edu.pl

A. Borkowski
Department of Soil Science, University of Life Sciences (SGGW), ul. Nowoursynowska 159, 02-776 Warsaw, Poland

J. Reitner et al., *Advances in Stromatolite Geobiology*,
Lecture Notes in Earth Sciences 131, DOI 10.1007/978-3-642-10415-2_10,

The solubility of calcite in water depends also on the temperature, but this relationship does not favour the precipitation of calcite at low temperatures. The equation showing the reaction of $CaCO_3$ dissolution is as follows:

$$CaCO_3 \leftrightarrow Ca^{2+} + CO_3{}^{2-} \quad \mathbf{\Delta H^0} = -13\text{kJ/mol} \tag{1}$$

Due to an enthalpy below zero, the reaction is exothermic and an increase in the solubility constant with a decrease of water temperature is expected. According to the van 't Hoff equation, it can be calculated that, e.g., at a water temperature of 5°C, the K_{sol} of $CaCO_3$ will increase to $10^{-8.14}$. Due to this fact, carbonate deposits, particularly calcite and aragonite, are not formed on the seabeds of deep oceans. The boundary depth is assumed to be at 4.5–5 km and is denoted as the carbonate compensation depth (CCD).

In the process of carbon dioxide generation, an important role is played by different groups of microorganisms that are responsible for the mobilisation of carbon and its biogeochemical cycle. Numerous groups of bacteria have the ability to alkalise the environment, transforming $Ca(HCO_3)_2$ into insoluble $CaCO_3$. Conversely, other bacteria, e.g., those occurring in marine deposits, have the ability to dissolve $CaCO_3$. It should be noted that carbon occurring in carbonates is available to living organisms only in a restricted range, e.g., due to the dissolution of $CaCO_3$ during sulphide oxidisation (2) or ethanol fermentation (3) (Dupraz and Visscher 2005).

$$3HS^- + 4O_2 + CaCO_3 + HCO_3{}^- \rightarrow 2[CH_2O] + Ca^{2+} + 3SO_4{}^{2-} \tag{2}$$

$$3[CH_2O] + CaCO_3 + H_2O \rightarrow HCO_3{}^- + Ca^{2+} + C_2H_5OH \tag{3}$$

Biological formation of carbonates is the result of the metabolic activities of many different groups of microorganisms, both aerobic and anaerobic. Carbonates can be formed as the by-products of various biological processes, including photosynthesis, e.g., in blue-green bacteria, as evidenced by the formation of stromatolites, as well as in photosynthetic bacteria or during anaerobic sulphate reduction (Borowitzka 1989; Castanier et al. 1999; Ehrlich 2001; Dittrich et al. 2004; Dupraz and Visscher 2005).

In ecosystems with a restricted oxygen supply, such as the bottom deposits of lakes, seas and oceans, are found bacteria that acquire energy for their metabolic processes through anaerobic transpiration (Postgate 1984). Very important among this group are the sulphate-reducing bacteria (SRB), which take part in almost all mineral- and rock-forming processes in anaerobic conditions (Lysnes et al. 2004). The SRB comprise a phylogenetically diverse group of anaerobic bacteria displaying physiological and ecological similarities. Their presence was noted for the first time in laboratory conditions by Beijerinck in 1985 (Gibson 1990). A characteristic environment for SRB occurrence is marine deposits, in which the sulphate concentration averages 28 mM (De Wit 1992), as well as oil fields and crude-oil magazines (Magot et al. 2000). The preferred carbon sources for SRB are compounds of low molecular weight, such as organic acids (i.e., propionate, formate and malonate), volatile organic acids (i.e., acetate), and alcohols (e.g., ethanol, propanol and butanol) (Fauque et al. 1991). Single SRB species are known to utilise amino acids as the sole carbon source: *Desulfovibrio aminophilus* (Baena et al. 1998), *Desulfobacterium vacuolatum* (Rees et al. 1997), and *Desulfovibrio mexicanus* (Hernandez-Eugenio et al. 2000). Some species, e.g., *Desulfotomaculum antarcticus*, may utilise glucose as the sole carbon source, but this is a very rare case among SRB (Fauque et al. 1991). All SRB are Gram-negative, excluding the *Desulfonema* species. This group of bacteria is extremely variable; it includes soil and water types, psychrophilic, meso- and thermophilic, halophilic and barophilic bacteria. The utilisation of low-molecular-weight compounds by SRB is linked to the fact that they do not possess hydrolytic enzymes and are so not able to biodegrade polymers. Thus, they play a key role in the final stage of organic-matter biodegradation in environments containing sulphates utilised as the last acceptors of electrons in the transpiration cycle. In the natural environment, the source of sulphate ions may be from easily soluble sulphates such as Na_2SO_4, or even hanebachite ($CaSO_4$), jarosite ***($KFe_3[OH]_6|(SO_4)_2$), anglesite ($PbSO_4$), barite ($BaSO_4$) (Karnachuk et al. 2002) or gypsum (Wolicka 2006).

Mixed populations of SRB have the ability to precipitate different minerals, e.g., sulphates (gypsum and anhydrite), carbonates (calcite, dolomite and siderite) (Perry and Taylor 2006), and phosphates (apatite) (Wolicka 2006). The formation of dolomite, calcite or aragonite in typical SRB environments and in conditions favouring the reduction of sulphates typically takes place in anoxic lagoons with a high salinity (Warthman et al. 2000), in gypsum and anhydrites (Peckmann et al. 1999), as well as in sulphate-rich natural lakes with a high salinity (Wright 2000).

In a different process, SRB may also play an indirect although important role in carbonate production. Some of these bacteria very often form close relationships with methane-producing anaerobic microorganisms, i.e., Archaea. These microorganisms, previously classified as bacteria, have the ability to produce methane from simple organic compounds or directly from carbon dioxide. Some may utilise carbon dioxide from the biodegradation of organic compounds conducted by SRB living in similar conditions. This relationship may develop into a very close relationship known as syntrophy, where SRB utilise the methane produced, oxidising it to carbon dioxide, which then may be used once again by the archaean syntrophic partner. The sulphates used by SRB in the transpiration process are supplied from the external environment. This process is known as AOM (anaerobic oxidation of methane) and proceeds according to the following equation (Valentine 2002):

$$CH_4 + SO_4^{2-} \rightarrow HCO_3^- + HS^- + H_2O \quad (4)$$

The process of anaerobic methane oxidation may be a crucial factor in the carbon cycle for two reasons. Firstly, it plays an important role in the reactions by which global methane production is transformed into carbon dioxide. Secondly, this process can be the key process in the formation of carbonate minerals (Moore et al. 2004), particularly within the continental shelves, where the AOM process is most intense (Miyajima and Wada 1999; Ogrinc et al. 2003).

The study was focused on testing the possibility of carbonate formation in conditions favouring SRB selection. The aim of study was to elucidate the role of SRB in the biogeochemistry of $CaCO_3$ precipitation under sulphate-reducing conditions.

2 Methodology

2.1 *Microorganisms*

The bacterial communities were isolated from Vistula River mud (Poland). In the selection of anaerobic bacterial communities, lactate and/or ethanol were used as the sole carbon sources (Postgate 1984). The microorganisms were multiplied using the method of *microcosms*: samples of the mud (10 g) were placed in transparent 100-mL containers. The containers were tightly closed and set aside for 6 weeks to allow the selection of anaerobic, sulphidogenic consortia capable of carrying out the reduction of sulphate. The cultures were incubated at room temperature (~20 °C).

2.2 *Media*

A modified Postgate medium (Postgate 1984), in which Na_2SO_4 was replaced with gypsum at a concentration equivalent to 4.5 g Na_2SO_4/L and yeast extract and sodium citrate were not added, was used to select SRB able to oxidise only one carbon source. The second medium was a minimal medium (1 g/L NH_4Cl). Both media were enriched with sodium lactate (3 mg/L) and/or ethanol (3 mg/L). All media contained gypsum (5 g/L) as the sole electron acceptor for SRB and the source of calcium. Resazurin was added to all cultures at 0.001 g/L as an indicator of redox conditions in the medium.

2.3 *Determinations*

The time from inoculation to the end of experiment was 14 days. Sulphates in the cultures were determined every day with the hot barium method, chemical oxygen demand (COD) was determined by the dichromate method, and pH determined using a pH meter; the pH of the cultures was corrected with 0.1 N HCl or 0.1 N NaOH as needed. Determinations involving postculture sediments (residual gypsum and sediments precipitated during bacterial activity) and fluids were made using X-ray analysis with a DRON- 2 X-ray diffractometer. The bacterial communities obtained were observed using a scanning electron microscope (SEM).

3 Results and Discussion

A total of 16 cultures were established: eight on Postgate medium (four with lactate and four with ethanol) and eight on minimal medium (four with lactate and four with ethanol). Generally, a higher activity of the isolated communities was observed on the Postgate medium than on the minimal medium. The highest sulphate reduction by the SRB, about 2,400 mg/L after incubation (corresponding to a calculated sulphate reduction of almost 95%), was noted in cultures on a medium with lactate as the sole carbon source. A distinctly lower activity was noted in communities isolated on the minimal medium, not exceeding 1,200 mg/L (Fig. 1).

In the SRB cultures obtained, a high degree of sulphate reduction was observed on the Postgate medium, i.e., 2,400 and 1,600 mg/L, the equivalent of ***860 mg HS^-/L on a medium with lactate and 590 mg HS^-/L on a medium with ethanol as the sole carbon source. According to Reis et al. (1992), the toxicity of sulphides and free H_2S produced during the microbiological reduction of sulphates, as well as the competition for the carbon source, represent two factors responsible for the inhibition of SRB growth and the accompanying microflora. In SRB populations growing on a lactate medium, the maximal concentration of sulphides was observed at pH = 6.7. The content of HS^-, at 547 mg/L (16.1 mM), was considered to be completely inhibitory to growth. Studies on the toxicity of hydrogen sulphide were carried out by Reiss on cultures of *Desulfovibrio* sp. In studies of mixed communities, the values of HS^- are lower, e.g., Hilton and Oleskiewicz (1989) reported 400 mg S/L and McCartney and Oleszkiewicz (1991) reported 448 mg S/L. In some cultures of these isolated communities the obtained HS^- concentrations were higher, showing the adaptation of the microorganisms to higher concentrations of H_2S.

In all cultures of sulphidogenic consortia, the reduction of sulphates was accompanied by the utilisation of organic compounds. The initial COD value reached 3,000 mg O_2/L, whereas by the end it fell to 300 mg O_2/L for the culture on Postgate medium with lactate, and it fell to 500 mg O_2/L on the medium with ethanol.

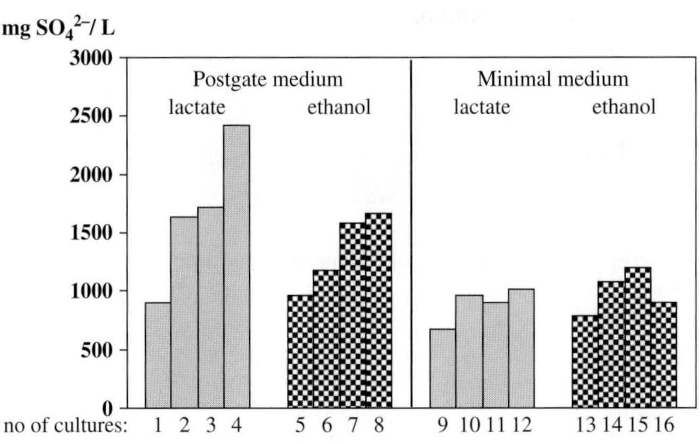

Fig. 1 Maximal reduction of sulphate in cultures of isolated bacterial communities

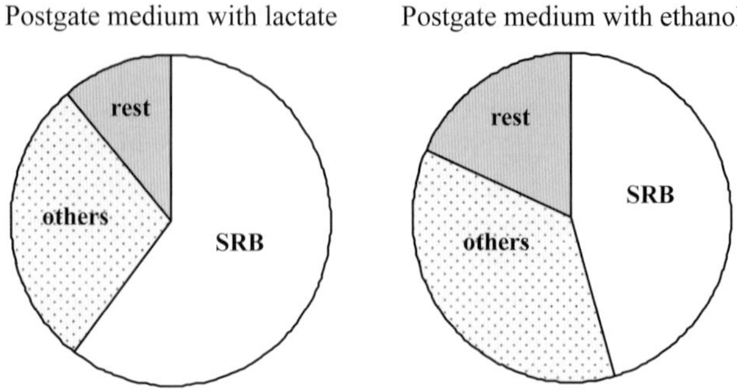

Fig. 2 Participation of SRB in COD reduction

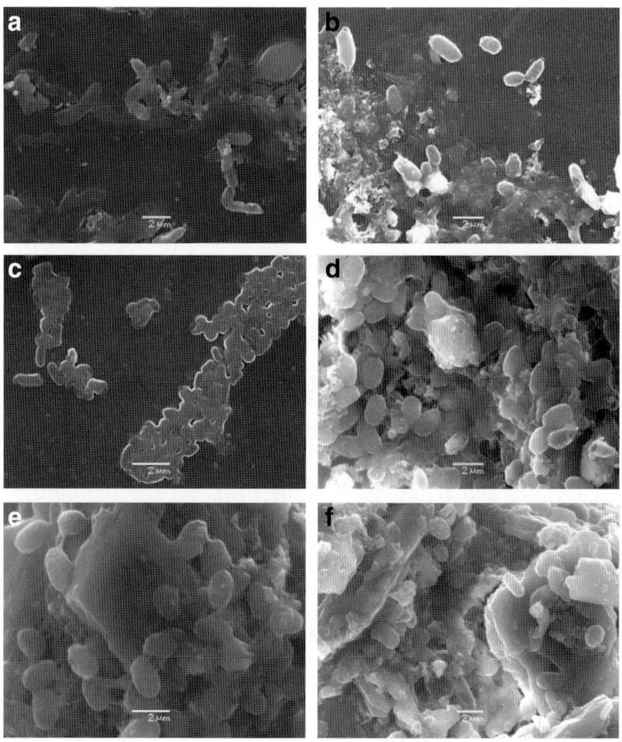

Fig. 3 Isolated sulphate-reducing bacterial communities in Postgate medium with lactate (**a–c**) and with ethanol (**d–f**) (SEM micrographs)

From the stoichiometry of the sulphate-reduction process (1 mole of HS^- per mole of SO_4), the ratio of the sulphides produced and the COD consumed by the entire bacterial community is known to be COD/SO_4 = 0.67 (Hao et al. 1996); thus, the participation of SRB in the reduction of organic compounds was calculated (Fig. 2).

During incubation, the culture supernatants were sampled for microscopic observations. They confirmed that the isolated communities were heterogeneous and contained various morphological forms. The communities included such forms as *vibrio*, *spirillum*, *bacterium* and *coccus*. The bacteria were accompanied by mineral phases in the form of both carbonates and sulphates (Fig. 3).

After incubation, diffractometric analysis of the postculture sediments was performed (Fig. 4). In the postculture sediments, the presence of secondary mineral phases was confirmed, e.g., calcite resulting from microbial activity. It is interesting to note that in the culture on Postgate medium with ethanol, which was less active than the same culture on a medium with lactate (Fig. 1), the postculture sediments contained only calcite, showing that the entire amount of gypsum introduced to the culture underwent biotransformation. Part of the sulphate ions may remain in the solution (Wolicka, unpublished data) and cannot be used by the isolated SRB community due to autoinhibition by HS^-. There are at least two reasons for this phenomenon. First, CO_2 formed in the cultures derived not only from the SRB but also from the accompanying microflora, thus increasing the content of carbon dioxide in the culture. The excess calcium ions in postculture solutions can easily bind with CO_2 forming $CaCO_3$ when the pH is not acidic. Additionally, the process of calcite formation could possibly be linked with the substitution of gypsum by carbonates (Fig. 5).

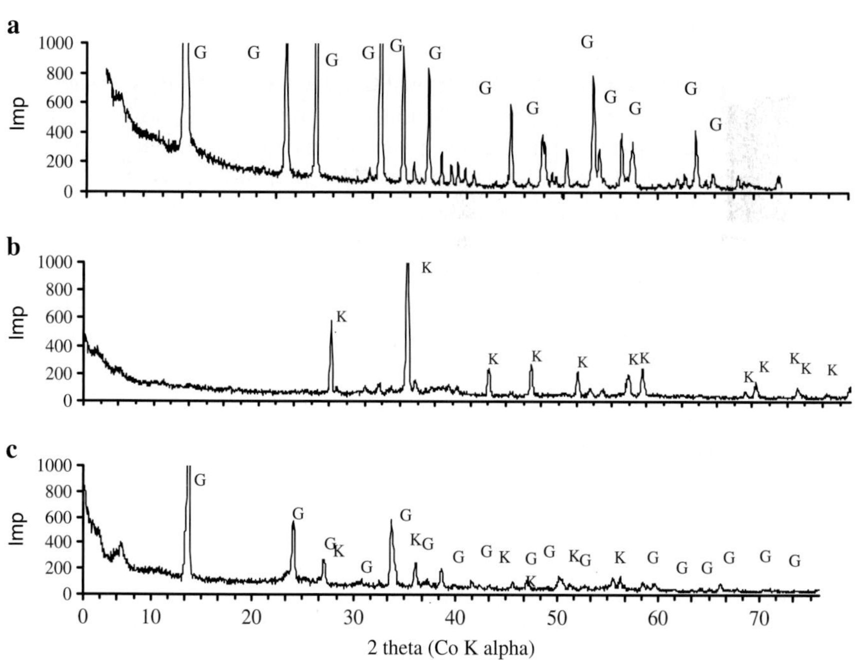

Fig. 4 Diffractograms: gypsum (**a**) and post-cultures sediments on Postgate medium with ethanol (**b**) and with lactate (**c**). The symbols indicate: *G* gypsum, *K* calcite

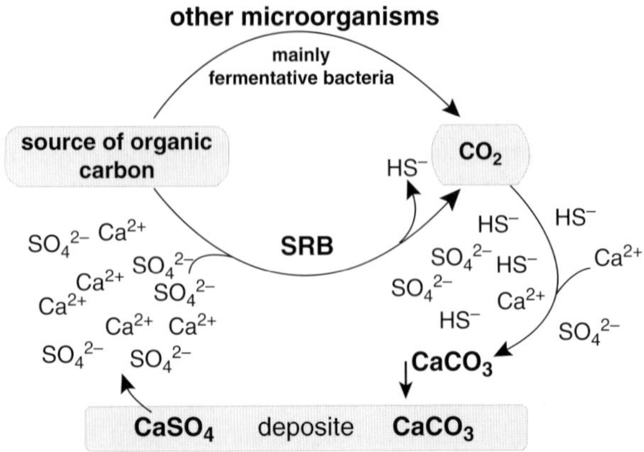

Fig. 5 Participation of sulphate-reducing bacteria in carbonate formation

4 Conclusions

The isolated SRB cultures showed a high activity for sulphate reduction with simultaneous effective biodegradation of the applied organic compounds (lactate and ethanol). In effect, biotransformation and the formation of secondary mineral phases in the form of calcite took place in the postculture sediments. Chemical analyses showed that SRB played a crucial role in the biodegradation of organic compounds, as reflected in the high fractional COD reductions by these bacteria. It thus seems likely that most of the calcite formed in the cultures was the result of SRB activity. The involvement of other groups of microorganisms accompanying SRB in environments favouring the formation of carbonates cannot be ruled out; however, it appears that the role of these microorganisms is much less significant, particularly in environments rich in sulphates, where, in the last phase of mineralisation, the SRB successfully compete for organic carbon with other groups of microorganisms. In light of this fact, the role of SRB in the formation of carbonates in anaerobic environments is extremely important, but is often underestimated.

References

Baena S, Fardeau ML, Labat M, Ollivier B, Garcia JL, Patel BKC 1998 *Desulfovibrio aminophilus* sp.nov., a novel amino acid degrading and sulfate reducing bacterium from an anaerobic dairy wastewater lagoon. Systematic and applied microbiology 21:498–504

Borowitzka MA 1989 Carbonate calcification in algae – initiation and control. In: Mann S, Webb J, Wiliams RJP (eds) Biomineralization, Weinheim, Germany: VCH Verlagsgesellschaft pp 116–135.

Castanier S, Metayer-Levrel GL, Perthuisot J-P 1999 Ca-carbonates precipitation and limestone genesis – the microbiogeologist point of view. Sedimentary Geology 126:9–23

De Wit R 1992 Sulphide-containing environments. In: Lederberg J (ed) Encyclopedia of Microbiology, San Diego, CA: Academic 4:105–121
Dittrich M, Kurz P, Wehrli B 2004 The role of autotrophic picocyanobacteria in calcite precipitation in an oligotrophic lake. Geomicrobiology Journal 21:45–53
Dupraz C, Visscher P 2005 Microbial lithification in marine stromatolites and hypersaline mats. Trends in Microbiology 13(9):429–438
Ehrlich H 2001 Microbial formation and degradation of carbonates. In: Geomicrobiology, New York: Mercel Dekker, Inc. pp 183–227
Fauque G, Legall J, Barton LL 1991 Sulfate-reducing and sulfur reducing bacteria. In: Shively JMI, Barton LL (eds) Variations in autotrophic life. New York: Academic
Gibson G 1990 Physiology and ecology of the sulphate-reducing bacteria. Journal of Applied Bacteriology 69:769–797
Hammes F, Verstraete W 2002 Key roles of pH and calcium metabolism in microbial carbonate precipitation. Re/Views in Environmental Science & Bio/Technology 1:3–7
Hao OJ, Chen JM, Huang L, Buglass RL 1996 Sulfate-reducing bacteria. Critical reviews in environmental science and technology 26:155–187
Hernandez-Eugenio G, Fardeau ML, Patel BKC, Garcia JL, Ollivier B 2000 *Desulfovibrio mexicanus* sp.nov., a sulfate-reducing bacterium isolated from an upflow anaerobic sludge blancet (UASB) reactor treating cheese wastewaters. Anaerobe 6:305–312
Hilton BL, Oleskiewicz JA 1989 Sulfide-induced inhibition of anaerobic digestion. Journal of Environmental Engineering 144:1377–1391
Karnachuk OV, Kurochkina SY, Tuovinen OH 2002 Growth of sulfate-reducing bacteria with solid-phase electron acceptors. Applied Microbiology Biotechnology 58:482–486
Lysnes K, Thorseth IH, Steinsbu BO, Øvreas L, Torsvik T, Pedersen K 2004 Microbial community diversity in seafloor basalt from the Artic spreading riges. FEMS Microbiological Ecology 50:213–230
Magot M, Ollivier B, Patel BKC 2000 Microbiology of petroleum reservoirs. Antonie van Leeuwenhoeck 77:103–116
McCartney DM, Oleszkiewicz JA 1991 Sulfide inhibition of anaerobic degradation of lactate and acetate. Water Research 25:203–209
Miyajima T, Wada E 1999 Sulfate-induced isotopic variation in biogenic methane from a tropical swamp without anaerobic methane oxidation. Hydrobiologia 382:113–118
Moore TS, Murray RW, Kurtz AC, Schrag DP 2004 Anaerobic methane oxidation and the formation of dolomite. Earth and Planetary Science Letters 229:141–154
Ogrinc N, Hintelmann H, Eckley C, Lojen S 2003 Biogeochemical influence on carbon isotope signature in boreal lake sediments. Hydrobiologia 494:207–213
Peckmann J, Thiel V, Michaelis W, Clari P, Gaillard C, Martire L, Reitner J 1999 Cold seep deposits of Beauvoisin (Oxfordian; southeastern France) and Marmorito (Miocene; northern Italy): microbially induced, authigenic carbonates. International Journal of Earth Sciences 88:60–75
Perry CT, Taylor KG 2006 Inhibition of dissolution within shallow water carbonate sediments: impacts of terrigenous sediment input on syn-depositional carbonate diagenesis. Sedimentology 53:495–513
Postgate JR 1984 The sulfate-reducing bacteria. 2nd edition. Cambridge University Press, Cambridge
Rees GN, CG Harfoot, AJ Sheehy 1997 Amino acid degradation by the mesophilic sulfate-reducing bacterium *Desulfobacterium vacuolatum*. Archives of Microbiology 169:76–80
Reis MAM, Almeida JS, Lemos PC, Carrondo MJT 1992 Effect of hydrogen sulfide on growth of sulfate reducing bacteria. Biotechnology and Bioengineering 40:593–600
Valentine DL 2002 Biogeochemistry and microbial ecology of methane oxidation in anoxic environments: a review. Antonie van Leeuwenhoek 81:271–282
Warthman R, Van Lith Y, Vasconcelos C, Mckenzie JA, Karpoff AM 2000 Bacterially induced dolomite precipitation in anoxic culture experiments. Geology 28:1091–1094

Wolicka D 2006 Biotransformation of phosphogypsum in cultures of bacteria selected from petroleum W: Proceedings of the International Conference Protection and Restoration of the Environmental VIII Chania, Greece, 3–7 July, pp 217–218
Wright DT 2000 Benthic microbial communities and dolomite formation in marine and lacustrine environments – a new dolomite model. In: Glenn CR, Lucas J, Prevot L (red). Marine Authigenesis from Global to microbial. SEPM Spec Publ 66:7–20

Myxococcus xanthus Colony Calcification: An Study to Better Understand the Processes Involved in the Formation of this Stromatolite-Like Structure

Concepcion Jimenez-Lopez, Kaoutar Ben Chekroun, F. Jroundi, Manuel Rodríguez-Gallego, Jose Maria Arias, and Maria Teresa González-Muñoz

1 Introduction

Calcium carbonate precipitation is a common phenomenon in nature and has been observed to be mediated by a number of microorganisms (for a review, see Castanier et al. 2000; Wright and Oren 2005). Bacterially induced carbonate mineralization is important in a wide range of processes including atmospheric CO_2 budgeting (Braissant et al. 2002; Ehrlich 2002), carbonate sediment and rock formation (Riding 2000; Ben Chekroun et al. 2004), biogeochemical cycling of elements (Banfield and Nealson 1997; Ehrlich 2002), human pathological concretions (Keefe 1976) and have also relevant technological applications for example in metal-contaminated groundwater bioremediation (Warren et al. 2001; Fujita et al. 2004; Mitchell and Ferris 2005; Köhler et al. 2007) and even in conservation of ornamental stone (Rodriguez-Navarro et al. 2003; Jimenez-Lopez et al. 2007).

Much research has been dedicated to the study of extracelullar precipitation of calcium carbonate by different microorganisms [i.e.: sulfate-reducing bacteria and cyanobacteria (i.e.: Wright 1999), *Bacillus* (i.e.: Castanier et al. 2000), Myxobacteria (i.e.: Ben Chekroun et al. 2004), *Halobacillus* (i.e.: Rivadeneyra et al. 2004) and *Pseudomonas* (i.e.: Baskar et al. 2006)]. Most of these studies are mainly dedicated to the understanding of the role of bacteria on the specific biomineralization phenomenon and to the characterization of the resulting biominerals. In this context, passive and active mechanisms have been proposed for the precipitation of bacterially induced calcium carbonate. While passive mechanisms for calcium

C. Jimenez-Lopez, K.B. Chekroun, F. Jroundi, J.M. Arias, and M.T. González-Muñoz (✉)
Departamento de Microbiología, Facultad de Ciencias, Universidad de Granada, Campus de Fuentenueva s/n, 18071, Granada, Spain
e-mail: mgonzale@ugr.es

M. Rodríguez-Gallego
Departamento de Mineralogía y Petrología, Facultad de Ciencias, Universidad de Granada, Campus de Fuentenueva s/n, 18071, Granada, Spain

J. Reitner et al., *Advances in Stromatolite Geobiology*,
Lecture Notes in Earth Sciences 131, DOI 10.1007/978-3-642-10415-2_11,

carbonate precipitation refer to the mineral formation as the result of changes on the chemistry of the environment caused by bacterial metabolic activity, active mechanisms for calcium carbonate precipitation refer to the formation of minerals mediated by bacterial wall, cell membranes and/or cell debris, which act as nuclei for crystallization (Wright and Oren 2005; Ercole et al. 2007).

With respect to passive calcium carbonate precipitation, there are a number of metabolic processes able to create favorable conditions for calcium carbonate precipitation. One of the most studied passive mechanisms is photosynthesis. In this regard, phototrophic bacteria like cyanobacteria have been associated with carbonate precipitation in the context of marine and freshwater sediments. However, there is controversy regarding the mechanisms leading to the alkalinization of the environment, crucial for the precipitation of calcium carbonate (Golubic 1973; Ehrlich 1996; McConnaughey and Whelan 1997; Altermann et al. 2006). Heterotrophic bacteria are also important producers of $CaCO_3$. Such a precipitation occurs as a consequence of chemical changes in the environments where those microorganisms live, from the release of different metabolites resulting from a variety of metabolic activities, including ammonification, dissimilatory nitrate reduction, degradation of urea or uric acid and dissimilatory sulfate reduction. All these metabolic activities lead to an increase in the pH of the environment and also to the production of CO_2, which results in the rise of the supersaturation with respect to the relevant mineral phase, thus triggering the precipitation of calcium carbonate (i.e.: Krumbein 1979; Wright 1999; Castanier et al. 2000; Riding 2000; Visscher et al. 2000; Wright 2000; Bachmeier et al. 2002; Hammes et al. 2003; Rodriguez-Navarro et al. 2003).

Regarding active precipitation of calcium carbonate by bacteria, there are a number of studies focused in the role of the cell wall, extracellular polymeric substances (EPS), cell membranes, cell debris and other organics in the precipitation of minerals. In this context, it has been demonstrated that exopolysaccharides, aminoacids and capsular polysaccharides mediate the precipitation of calcite and vaterite and even influence the resulting morphology of the crystals (Kawaguchi and Decho 2002; Braissant et al. 2003; Ercole et al. 2007). González-Muñoz et al. (1996) observed the precipitation of calcite and struvite in jellified culture medium inoculated with cellular membranes of *Myxococcus xanthus*. Also, oriented precipitation of calcium phosphate in the periplasm and inside the cells, associated with integrated membrane structures have been observed in other studies (Benzerara et al. 2004). The process of cell calcification and the role of cell wall and EPS in vaterite precipitation have been further analyzed by Rodriguez-Navarro et al. (2007).

In spite of the extensive literature regarding calcium carbonate precipitation by bacteria, EPS, cell membranes and bacterial debris, little is known about the calcification that would occur in a whole population, where all these factors are integrated. Such a study is, however, important, since it is a better approach to what would occur in natural systems where most microbial cells form communities inside microbial biofilms (Decho 1990, 2000; Sutherland 2001). The understanding of how colonies calcify and how passive and active mechanisms interact to give rise to an specific colony calcification pattern would provide hints regarding the environmental conditions where such a calcification occurred, the size of the population involved in the mineralization, how active was the metabolism of such a population, and potential gradients along the colony. All these data could provide valuable information regarding the mineralization processes resulting in the formation of biomineral structures in sediments (i.e. formation of lithified mats in nature) and even, as an approximation, in pathological concretions. It is, nevertheless, a first step and simplistic approximation to understand mechanisms leading to the formation of calcified structures, since it has been pointed out that microbial mats in nature are constituted by different populations which create a more complex ecosystem (Visscher and Stolz 2005). This work focus in investigating passive and active mechanisms involved in the calcification of colonies and which factors determine the calcification and formation of specific patterns observed.

2 Materials and Methods

2.1 Microorganism and Culture Media

The microorganism used was *Myxococcus xanthus* (strain number 422 provided by the Spanish Type Culture Collection, Burjasot, Valencia, Spain). *M. xanthus*, a Gram negative, common soil bacterium, belonging to the δ-subdivision of the Proteobacteria, has been chosen since it has been demonstrated that such bacteria is an effective producer of extracellular calcium carbonate (calcite and vaterite; Rodriguez-Navarro et al. 2003, 2007) and that calcium carbonate may even be also induced in the sole presence of cell membranes by a purely active mechanism (González-Muñoz et al. 1996).

For inoculum preparation, *M. xanthus* was cultured in liquid medium CT (Dworkin and Kaiser 1993) for 48 h at 28°C and at 0.455 × *g* (Gallenkamp rotary shaker) until it reached the exponential growth phase. The size of the population at this time was ~2×10^9 cells/ml (González-Muñoz et al. 2000).

Biomineralization tests were conducted in M-3 and CC solid culture media. Two different culture media were used in the experiments to allow for different initial supersaturation conditions with respect to the relevant phase of calcium carbonate and also to test conditions in which the growth rate of *M. xanthus* is different. Jellified culture media were used to simulate the biofilms in which most cells are embedded in natural environments (Costerton et al. 1987). To prepare M-3 and CC solid culture media, 1.8 % agar-agar (Difco) was added to the filtered liquid culture media M-3 and CC described, respectively, in Rodriguez-Navarro et al. (2003) and Ben Chekroun (2000), with the following composition (percentages represent wt/v): M-3 (1% Bacto Casitone, 1% $Ca(CH_3COO)_2{\cdot}4H_2O$, 0.2% $K_2CO_3{\cdot}1/2H_2O$ in distilled water, pH 8), and CC (0.3% Bacto Casitone, 0.4% $Ca(CH_3COO)_2{\cdot}4H_2O$, 0.1% $CaCl_2$, 0.3% $NaHCO_3$, 0.1% yeast extract in distilled water, pH 8). Those culture media were sterilized by autoclaving for 20 min at 120°C and distributed in sterile Petri dishes.

2.2 *Experimental Procedures, Analyses and Calculations*

2.2.1 Experimental Procedure and Study of the Mineralization

Thirty Petri dishes were prepared with each culture media (M-3 y CC). Out of these, twenty seven dishes were inoculated with eight drops/dish of 20 μL of the inoculum of *M. xanthus*, while three dishes were kept sterile, acting as control. The number of drops per plate was determined to allow the growth of enough colonies of *M. xanthus* needed for the study, while maintaining a considerable space between those drops. All dishes were sealed with Parafilm to avoid dehydration and incubated at 28°C for a maximum of 30 days. Borders and inners of the colonies were observed every day by optical microscopy (total magnification 100×) with the goal of detecting calcification.

At predetermined time intervals (1, 2, 4, 6, 7, 9, 12, 15 and 30 days), three dishes of each M-3 and CC runs were selected to perform analyses of the colonies and of the minerals produced within the colonies, following the sequence detailed below:

(a) *Mineralogical analyses by XRD of the solid precipitated within the colonies.* Half of the colonies of each Petri dish were selected and scratched with a metal spatula. The solid recovered was placed on a glass sample holder to be analyzed by X-ray Diffraction (Philips PW-1710, Cu Kα radiation).

(b) *Analyses by Electron Microscopy of the calcification of the colonies.* The other half of the colonies remaining in each Petri dish were fixed by a new method specially developed to be able to observe the colony by FESEM (Field Emission Scanning Electron Microscopy) and HRTEM (High Resolution Transmission Electron Microscopy) without disturbing the structure of such a colony. Colonies were fixed by adding to the Petri dish a solution of 2% (v/v) glutaraldehide in 0.1M PSB buffer for 2 h at 4°C and thereafter rinsed three times in 0.1M PSB buffer for 10 min.

The glutaraldehyde-fixed colonies were then harvested and post fixed with OsO_4, rinsed, and dehydrated by using an ethanol ramp (50–100%). Thereafter, each colony was cut in two (S-samples and T-samples). From now on, each samples (S and T) followed different treatments: S-samples were dried by the critical point method (Anderson 1951) and carbon coated for FESEM analyses (LEO Carl Zeiss, mod. GEMINI-1530). T-samples were embedded in Epon-812 resin for HRTEM analyses (Philips CM200, 200 kV).

T-samples corresponding to S-samples in which calcification was observed in the colony by FESEM, were selected to be analyzed by HRTEM. For those analyses, T-samples, embedded in resin, were cut to 250 and 500 Å thickness. Some thin sections were conventionally stained with uranyl acetate and lead citrate, while others were kept unstained. To determine whether calcium carbonate precipitated around the cells, and/or in cell membranes or cell debris, the presence of calcium at those locations was tested by HRTEM-EDAX (collection time was 200 s for a 200 × 1,000 Å window) and the crystallinity of such a Ca-bearing phase was determined by Selected Area Electron Diffraction (SAED).

2.2.2 Determination of the Trends of the Growth of the Bacterial Population

The evolution of bacterial population over time in both media was an important factor, since it has a crucial role understanding the chemical changes occurring in those media leading to the calcification of the colonies. Since measurements of cell density within a colony are very difficult to make and induce large errors, the size of the bacterial population growing in each culture media was calculated by measuring the optical density (OD) of cultures of *M. xanthus*. Those cultures were prepared by inoculating five milliliters of liquid M-3 and CC media with 0.5 mL of the *M. xanthus* inoculum prepared as described above. Cultures were incubated at 28°C at 0.455 × *g* (Gallenkamp rotary shaker). Optical density was measured by using the Spectrophotometer Genesys 10 UV Thermo Electron Corporation (USA). In M-3 culture medium, OD was measured from 0 to 180 h in intervals ranging from 3 to 5 h. In CC runs, OD was measured from 0 to 128 h every 3 h. Cell population (CFU/mL) was calculated from optical density by applying the correlation detailed in Benhamu et al. (1979). These authors determined curves of optical density of a culture of *Myxococcus* over time and also, over identical time interval, they measured the number of cells present in such a culture. Form those data, the correlation of OD and cell number was established, that being CFU/mL $= 10^9$ OD –311872, $R^2 = 0.9963$, $n = 10$.

2.2.3 Study of the Chemistry of the Culture Media

The initial supersaturation of M-3 and CC media with respect to calcite and vaterite was calculated from measured values of pH and $Ca_{T(aq)}$ in the nutritive solutions before agar-agar addition (then corrected to the final volume to take into account changes in the volume associated to the addition of the agar) and calculated values of alkalinity. The pH was measured by using a combination pH electrode (Crison). $Ca_{T(aq)}$ was determined by atomic absorption spectrophotometry (AAS, Perkin-Elmer 1100B). Experimental error for pH and $Ca_{T(aq)}$ measurements was ±0.05 and ±0.1 mM, respectively.

Carbonate alkalinity and acetate concentration was determined following the same methodology described in (Rodriguez-Navarro et al. 2007). Activities and activity coefficients for all aqueous species were calculated using the EQ3/6 program (Wolery 1992) from measured values of $Ca_{T(aq)}$, pH, and calculated values of alkalinity and acetate. Ion activity products (IAP) were calculated as the product of the activity of calcium and carbonate in solution ($a_{Ca^{2+}} \times a_{CO_3^{2-}}$). Saturation state

(Ω) with respect to the particular $CaCO_3$ phase was calculated from IAP and K_{sp} values ($pK_{sp,vaterite} = 7.913$ and $pK_{sp,calcite} = 8.48$; Plummer and Busenberg 1982).

3 Results

Chemical analyses of the culture media reveals that initial $Ca_{(aq)}$ was 52.5 mM for M-3 and 15.5 mM for CC culture media, respectively, while initial alkalinity was 15.6 mM for M-3 and 17.2 mM for CC culture media, respectively. The initial value of pH in both media was 8.00. Calculated saturation (Ω) with respect to the relevant mineral phase (vaterite in M-3 and calcite/vaterite in CC, as reported below) were the following (a) M-3, $\Omega_{vaterite} = 51.6$; (b) CC, $\Omega_{vaterite}$ was 4.0 and $\Omega_{calcite}$ was 14.8.

Optical density measurements of the *M. xanthus* cultured in liquid M-3 and CC (Fig. 1) show that, in the M-3 run, the lag phase extended up to 49 h (OD = 0.074, corresponding to CFU/mL = 7.4×10^7), leading to the exponential growth phase, which lasted until 6 days (OD = 0.63; CFU/mL = 6.3×10^8), giving then rise to the stationary phase. Conversely, in CC runs, the lag phase lasted only 9 h (OD = 0.078; CFU/mL = 7.8×10^7), giving then rise to the exponential growth phase, which extended only until 36 h (OD = 0.326; CFU/mL = 3.2×10^8).

Colonies of *M. xanthus* were evident at 48 h of incubation in both CC and M-3 culture media. They displayed the characteristic expanding border, as a result of the gliding motility of this bacterium (McBride et al. 1993). In M-3 medium, even though it was filtered prior to the experiment, few scarce crystals were still observed

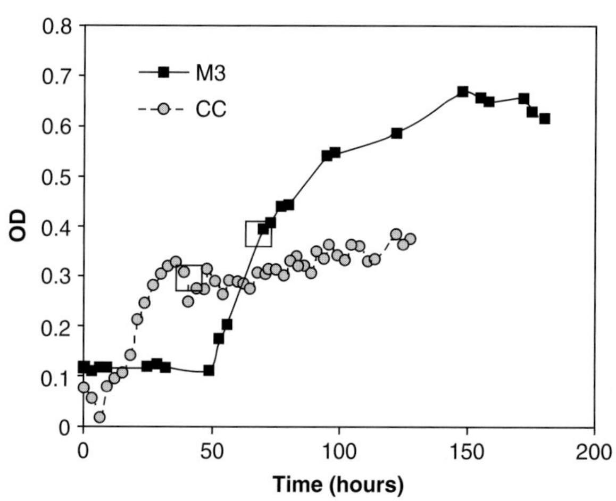

Fig. 1 Measurements of optical density (OD) over time of liquid culture media M-3 and CC inoculated with *M. xanthus*. Squares represent the time at which the first precipitation of calcium carbonate was observed in the colonies of *M. xanthus* growing in solid M-3 and CC

at the beginning of the experiment in the bulk medium, outside the colony (Fig. 2a). That was not the case in CC-runs (Fig. 3a). Nevertheless, new carbonate precipitation in the bulk medium (outside the colonies) were not observed in either M-3 or CC media during the time course experiment.

The evolution of the mineralization of the colonies in both M-3 and CC culture media is shown in Figs. 2 and 3, respectively. In M-3 culture medium, calcification of the colonies was observed with time, almost parallel to the growth of the population, firstly on the borders of the colonies (at 3 days) and later also in the inner part of the colonies. Few scarce crystals formed in the border of the colonies at 3 days. With

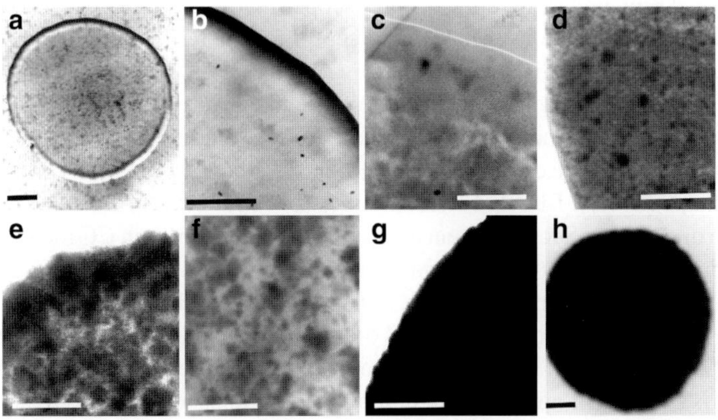

Fig. 2 Time evolution of the colony calcification in M-3 culture medium as observed by optical microscope. (**a**) Colony observed at 48 h; (**b**) Border of the colony at 3 days; (**c**) Border at 4 days; (**d**) Border at 9 days; (**e**) Border at 12 days; (**f**) Inner of the colony at 12 days; (**g**) Border at 30 days; (**h**) colony at 30 days. Scale bars are 1 mm

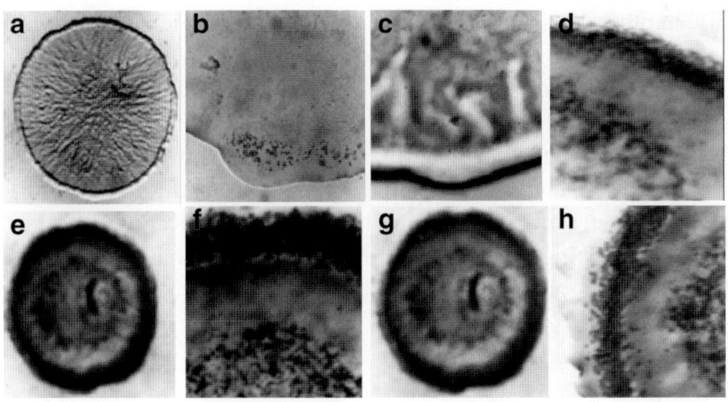

Fig. 3 Time evolution of the colony calcification in CC culture medium as observed by optical microscope. (**a**) Colony observed at 48 h; (**b**) Border of the colony at 48 h; (**c**) Border at 4 days; (**d**) Border at 9 days; (**e**) Colony at 12 days; (**f**) Border and inner at 12 days; (**g**) Colony at 30 days; (**h**) Border and inner of the colony at 30 days

time, clusters of calcium carbonate appeared, preferentially at the borders of the colonies (Figs. 2b–e) and a few scarce ones at the inner part of the colonies (Fig. 2f). The number of clusters gradually increased during the course of the experiment and ended up with the calcification of the whole colony, which become dense and opaque under the optical microscope after 12 days (Fig. 2g, h). According to XRD analyses, solid precipitated in M-3 culture medium was only vaterite.

In contrast, something very different occurred in CC runs. Crystals were first detected at the borders of the colony at 48 h (Fig. 3b). With time, calcified areas were also detected in the inner part of the colony (after 4 days), though restricted to a concentric ring (Figs. 3c–h). Such a calcification appeared, however, less dense than in M-3 runs under the optical microscope. Interestingly, no calcification was detected elsewhere within the colony. According to XRD analyses, solid precipitated in CC culture medium was a mixture of vaterite (~75%) and calcite (~25%).

FESEM images show the presence of cells embedded in a dense matrix that also contains calcium carbonate crystals formed extracellularly, which appear mainly as spherulites and micritic crystals. That holds true for both CC and M-3 runs (Figs. 4a and 5a, respectively). EDAX-FESEM analyses of the surface of the cells show the presence of calcium. Such a presence was first detected at 56 h in CC medium, and at 12 days in M-3 medium (Figs. 4b and 5b, respectively). However, the presence of Ca in the surface does not imply that cells are calcified. With the goal of testing whether or not cells were calcified, thin sections of the calcified colonies were observed with the HRTEM. Those HRTEM images show cells embedded in a matrix of EPS and cell debris (Figs. 4c–f and 5c–f). In non-stained samples, electron dense areas were observed in the cell walls, inside of some cells, in the EPS, and associated to fractions of cell membranes (Figs. 4d, e and 5c, d). EDAX analyses of those darker areas show the presence of calcium and SAED demonstrate that those Ca-containing phases were crystalline (Figs. 4c and 5c). Not all the cells show the same level of calcification, as it can be observed in Figs. 4c, e and 5c, d. While there are cells that appear totally calcified (cell wall and inside), in others, only the cell wall is calcified and, finally, in other cells, only a fraction of the inside is calcified. Images of stained samples also show dark areas corresponding to stained cell membranes and other dark areas that are calcified structures (Figs. 4f and 5f). Broken and/or very deteriorated membranes can be observed in these stained images.

At 30 days, SEM pictures, (Figs. 4g, h and 5g, h), show a calcified matrix that contains numerous holes with diameters ranging from 0.3 to 0.6 μm [which is similar to the diameter of *M. xanthus* cells (Dworkin and Kaiser 1993)], and empty calcium carbonate tubing of about 3–4 μm of length [which is similar to *M. xanthus* length (Dworkin and Kaiser 1993)].

4 Discussion

The calcification of a colony will be discuss in two levels: as a whole (Sects. 4.1 and 4.2) and at the micron/nano scale in the Sect. 4.3.

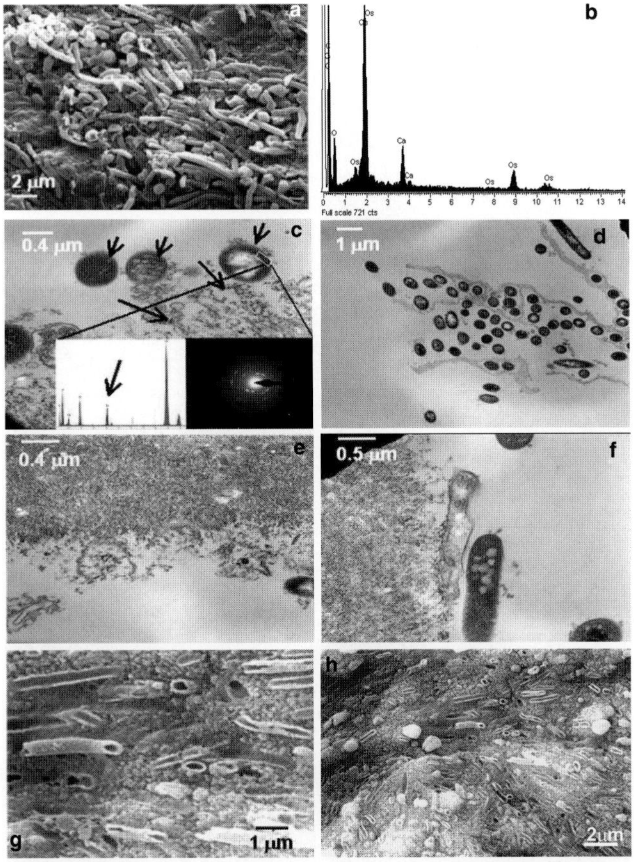

Fig. 4 SEM and HRTEM analyses of the colonies of *M. xanthus* cultured in CC medium (**a**) SEM image of part of the colony fixed following the fixation method described in the present paper; (**b**) EDAX analyses of the surface of the cells; (**c**)–(**f**) HRTEM images showing calcified cells and calcified EPS, (**g**) and (**h**) calcified cells embedded in calcified EPS that also contains rounded crystallites ((**h**) is a detail of (**g**)). (**a**)–(**f**) colonies cultured for 56 h. (**g**) and (**h**) SEM images of colonies incubated at 30 days. (**c**)–(**e**) non-stained samples. (**f**) stained sample. Details of (**c**) show electron diffractions and EDAX analyses of part of the coating observed around the cell. The presence of calcium in EDAX is pointed with an *arrow*

4.1 Patterns of Colony Calcification

It is remarkable the two different colony calcification patterns observed in our experiments: a whole colony calcification in M-3 runs versus a calcification of only concentric rings within the colony in CC runs [following a Liesegang-ring pattern (Henisch 1988), observed previously by González-Muñoz et al. (2000) while studying the production of Mg-calcite by *M. xanthus*]. Both the "whole-colony calcification" pattern and the "Liesegang-ring" pattern can be explained by

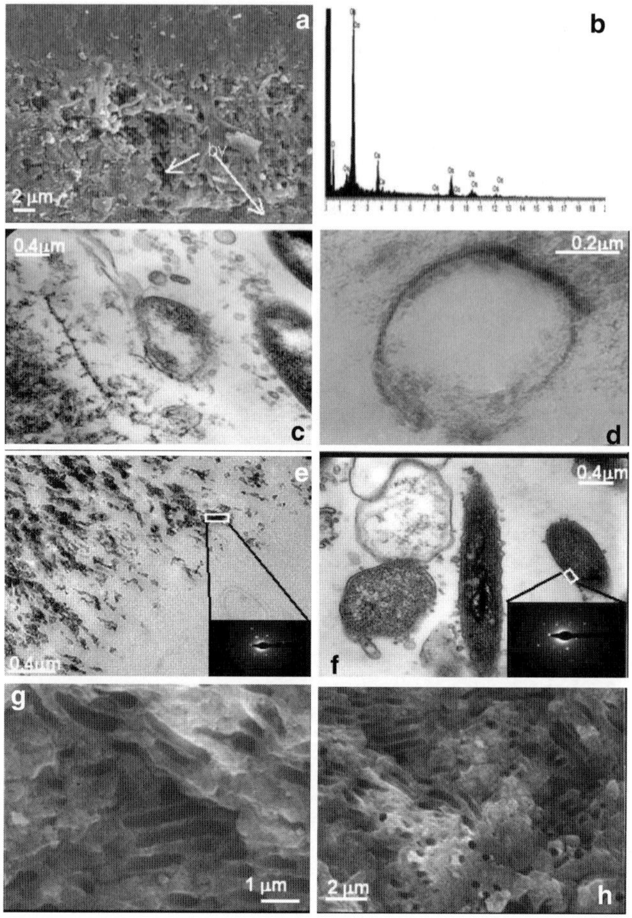

Fig. 5 SEM and HRTEM analyses of the colonies of *M. xanthus* cultured in M-3 medium (**a**) SEM image of part of the colony fixed following the fixation method described in the present paper; (**b**) EDAX analyses of the surface of the cells; (**c**)–(**f**) HRTEM images showing calcified cells and calcified EPS, (**g**) and (**h**) calcified cells embedded in calcified EPS. (**a**)–(**f**) colonies cultured for 12 days. (**g**), (**h**) SEM images of colonies incubated at 30 days. (**c**)–(**e**) non-stained samples. (**f**) stained sample. (**d**) slice of 500 Å thickness. Details of (**c**), (**f**) show electron diffractions of part of the coating observed around the cell and also of dark areas in EPS

means of the saturation conditions within the colony during the calcification process, which is the result of a number of processes and parameters discussed below.

The evolution of the supersaturation with respect to calcium carbonate within the colony over time is a function of the concentration of calcium, pH and alkalinity at each specific location. The metabolic activity of *M. xanthus* increases the pH and alkalinity in the environment where the bacterium live. While increases in the pH are caused by the NH_3 release as a consequence of the oxidative deamination of the aminoacids, increases in the alkalinity of the system are caused by the CO_2 release, both two processes occurring as a consequence of the use of the peptone as a source of carbon and nitrogen (Rodriguez-Navarro et al. 2003). Therefore, and since pH values and alkalinity are related with bacterial metabolism, maximum pH and alkalinity values are expected at those areas of maximum concentration of cells. The diffusion of NH_3 and CO_2 from those areas to the bulk culture medium will create a gradient of pH and alkalinity within the colony. With respect to calcium, and being that a macronutrient necessary for the growth of the cells, calcium is expected to diffuse from the bulk medium (outside the colony) to the colony. Therefore, diffusion of metabolites from the colony to the bulk culture medium and counter-diffusion of Ca^{2+} from the bulk culture medium to the colony probably results in the radial formation of gradients of saturation from the border to the inner of the colony. Those gradients create certain areas where the system is supersaturated with respect to a specific calcium carbonate phase, thus inducing the precipitation of such a phase (González-Muñoz et al. 2000).

The highest saturation values probably occur firstly at the borders, due to an accumulation of bacterial metabolites diffused from all cells in the colony to the bulk culture medium. Moreover, the concentration of calcium at the borders is also likely to reach the maximum value compared to other locations within the colony, since, once the calcium reach the cells, it is going to be used, thus decreasing the concentration of such a cation. As a result, maximum supersaturation levels are expected at the borders of the colony, being that consistent with our results which show that calcium carbonate precipitation in both M-3 and CC runs was first observed at the borders of the colony (Figs. 2b, c and 3b, c).

Moreover, the precipitation of calcium carbonate also modulates the saturation of the system with respect to calcium carbonate within the colony. Once the precipitation of calcium carbonate occurs, the saturation of the system decreases as a consequence of (1) the sequestering of Ca^{2+} and CO_3^{2-} from the culture medium to be incorporated into the crystal structure and also, (2) the acidification resulting from the release of protons. The faster and the more intense the precipitation occurs, the sharper is the decrease of the saturation. Different output can then happen depending of the scenario defined by the initial saturation of the environment:

(a) *Low saturated systems*. If the initial saturation of the system is not very high, once calcium carbonate precipitates, this saturation may decrease in the adjacent areas to a value where the system is not saturated anymore with respect to the precipitation of calcium carbonate, being, thus, forbidden such precipitation. This situation holds until the saturation of the system increases, due to bacterial metabolism, to a value that allows again the nucleation of calcium carbonate in another area within the colony. This new value of supersaturation may be different from that of the former, leading to the precipitation of other polymorphs of calcium carbonate that require lower saturation values. Therefore, in those low-saturation cases, both precipitation in a "Liesegang-ring pattern" and a zonation of polymorphic phases of calcium carbonate should be expected.

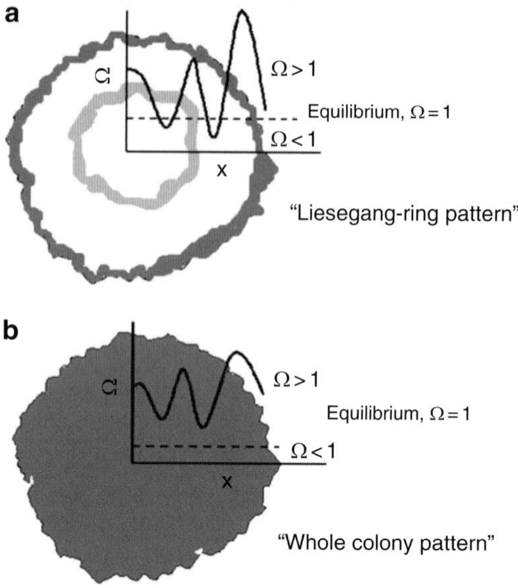

Fig. 6 Model of the evolution of the saturation of the system with respect to calcium carbonate from the border to the inner of the colony in (**a**) CC and (**b**) M-3 culture media. *Gray color* indicates calcification. Scales of *gray* represent the intensity of calcification

Such a behaviour, represented in Fig. 6a, is likely to occur in CC culture medium ($\Omega_{vaterite} = 4.0$ and $\Omega_{calcite} = 14.8$). It may happen that, in the borders, the first precipitate is vaterite and then, after this first precipitation event, the system became undersaturated, thus forbidding a new formation of calcium carbonate in the adjacent area. When saturation rises again, a new precipitation event occurs, leading to the calcification of the colony in a "Liesegang-ring pattern". However, this new saturation value may not be enough for vaterite to precipitate but for calcite, since the latter has a lower solubility product, thus allowing the precipitation of this other polymorph of calcium carbonate. This hypothesis is in agreement with the pattern observed in CC medium and with the mixture of calcite and vaterite detected. Nevertheless, although a certain polymorph may be dominant at each ring, a mixture of polymorphs may still exist, given that the supersaturation inside a ring is not uniform either, due to the existence of highly-saturated areas created by the prevalent absorbance of calcium (for instance, negatively charged areas of the cell wall, EPS, cell debris or by-products of cell metabolic activity).

(b) *Highly saturated systems*. If the initial saturation of the system is very high, decreases due to the precipitation of a $CaCO_3$ are probably not enough to reach values below saturation in the adjacent areas where such precipitation occurs. Therefore, the system is always supersaturated with respect to calcium carbonate, and the precipitation of such a phase occurs all over the colony (Fig. 6b). If the superaturation conditions along the colony are high enough, then no polymorphic zonation may occur, as the system is always saturated with respect to a specific polymorphic phase. This is the case of M-3 runs ($\Omega_{vaterite} = 51.6$), where the whole colony calcifies and the solid precipitate is 100% vaterite precipitated all over the colony.

(c) *Intermediate saturated systems*. In those systems where saturation value is not so high, calcium carbonate precipitation may still occur all over the colony, since the system may be always supersaturated with respect to calcium carbonate. However, a zonation of polymorphs of calcium carbonate may still occur, since the saturation value, after a precipitation event, may not be enough to allow for the precipitation of the polymorph that formed in the previous event. Therefore, whether the colony calcifies completely or only at certain areas and whether or not there is a polymorph zonation, may help identifying if the calcification process occurred at high, intermediate or low saturation values.

The "Liesegang-ring pattern" observed in our experiments resembles the concentric patterns of ooids. Although ooids in modern environments are composed of aragonite, in the past, calcitic, dolomitic and bimineralic ooids have also been found (Wright and Oren 2005). Results from this study show that a microorganism, *M. xanthus*, is able to induce the calcification of the population with this specific concentric pattern, and that such a pattern may provide insights of the geochemical conditions in which the calcification occurred.

4.2 Critical Size of the Population Necessary for the Calcification of the Colony

Our results show that even though the bulk culture media were supersaturated with respect to calcium carbonate, no precipitation was observed outside the colony, and, even within the colony, mineral formation did not occur until after 2 days in CC medium and until after 3 days in M-3 medium. Figure 1 shows that the exponential growth phase of the population of *M. xanthus* occurred earlier in CC than in M-3 medium. Therefore, the critical size of the population required to produce a critical supersaturation as a result of bacterial metabolism and to provide enough pre-existing surfaces (cell wall, cell debris, EPS by-products of bacterial metabolism), being both necessary for the calcification of the colony, would be reached earlier in CC culture medium than in M-3. This result is in agreement with the earlier calcification detected in CC (2 days) compared to M-3 culture medium (3 days).

Being then the critical size understood as the size of the population needed to reach a critical supersaturation in the environment, such a number should be dependent on other factors, like, for instance, the baseline saturation of the environment and the temperature. The existence of this critical size may provide insights of the size and how active was the bacterial population involved in the formation of calcified structures. Examples of such a calcified structures can be pathological calcified concretions [calcific periarthritis, Milwaukee shoulder syndrome, osteoarthritis (Molloy and McCarthy 2003), kidney stone formation (Kajander and Çiftçioglu 1998; Çiftçioglu et al. 1999), and dental pulp stone (Çiftçioglu et al. 1998) and calcified structures in nature, as lithified mats forming stromatolites in different environments [Precambian sediments (Walter 1983;

Grotzinger and Knoll 1999; Altermann 2004), Palaeozoic and Mesozoic marine environments (Riding 2000), modern calcareous streams, lakes and terrestrial sediments (Riding 2000), tropical marine and quasi-marine sites (e.g. Dravis 1983; Reid et al. 2003), and extreme environments (Kempe and Kazmierczak 1990)]. Nevertheless, and because calcified structures in nature represent a much more complicated process than the one here investigated, more research needs to be done to determine the critical number of cells required for calcification when different bacterial populations are present and are interacting, using other microorganisms and considering different biominerals.

4.3 Passive and Active Mechanisms of Calcium Carbonate Precipitation

As it will be discussed in detail below, calcification of the colonies is the result of a number of several precipitation events acting in different locations and include (1) homogeneous precipitation of calcium carbonate within the colony (resulting from passive mechanisms of precipitation) and (2) heterogeneous precipitation of calcium carbonate in negatively charged areas of the cell wall, EPS, cell membranes and cell debris (resulting from active mechanisms of precipitation).

4.4 Passive Mechanisms of Calcium Carbonate Precipitation Giving Rise to an Homogeneous Precipitation

Since mineral formation was only detected within the colonies of *M. xanthus* and not in the bulk culture medium (even though the culture medium was supersaturated with respect to vaterite and calcite), it is clear that the presence of bacterium is a prerequisite for the precipitation of calcium carbonate. Both passive and active biomineralization processes make the nucleation and growth of carbonate crystals both thermodynamically and kinetically possible (Dupraz and Visscher 2005). In the context of thermodynamics, the supersaturation of the system with respect to a particular mineral phase (Ω) is the driving force for the precipitation of such a phase, being possible when $\Omega > 1$ (Garside 1982). Such an increase in supersaturation, as a consequence of bacterial metabolic activity, resulted in the homogeneous precipitation of spherula and micritic crystals that were not associated with any cell structures and/or debris, as well as, in the heterogeneous precipitates associated with cell structures and/or debris that will be discussed below.

The fact that *M. xanthus*, depending on the culture media (which, in turn, set different saturation states), is able to produce different polymorphs (vaterite in M-3 and calcite-vaterite in CC), shows that polymorph selection is probably related to supersaturation rather than to the sterochemical or geometrical matching between the organic substrate and the mineral proposed by Mann et al (1988). These findings are in accordance with those of Rodriguez-Navarro et al. (2007). However, as suggested by those authors, the role of the organic matter in stabilizing the less stable polymorph due to its incorporation into the structure of the mineral should not be neglected. The precipitation of vaterite in the higher saturated medium (M-3) and vaterite-calcite in the lower saturated medium (CC) is explained by the higher solubility product of vaterite compared to that of calcite, which makes necessary higher supersaturation conditions for the precipitation of vaterite (Ogino et al. 1987). M-3 was more saturated than CC, not only at the beginning of the experiment ($\Omega_{\text{vaterite, M-3}} = 51.6$ vs $\Omega_{\text{vaterite, CC}} = 4.0$), but also probably during the time course experiment, given the fact that the size of the population of *M. xanthus* in M-3 was higher than that in CC (Fig. 1). Therefore, pH and alkalinity increases within the colony due to bacterial metabolism are expected to be higher in M-3 runs compared to that in CC runs. Those data support the precipitation in M-3 of the polymorphic phase that requires higher supersaturation conditions.

4.4.1 Active Mechanism of Calcium Carbonate Precipitation

Not only supersaturation makes nucleation of crystals possible, since it is also necessary to overcome some energy barriers determined by kinetic factors. One of these barriers is the energy required for the system to create a new surface (Garside 1982). In this regard, bacteria may favor the kinetics of calcium carbonate precipitation by serving, the cell, cell structures, cell debris and even by-product of the bacterial metabolic activity, as nucleation sites for mineral deposition, then acting as a pre-existing surface (Chafetz and Buczynski 1992; Ben Omar et al. 1994; Knorre and Krumbein 2000; Paerl et al. 2001; Dupraz and Visscher 2005; Rodriguez-Navarro et al. 2007). All these structures acting as pre-existing surfaces will be here referred to CFPS.

The heterogeneous precipitation of calcium carbonate on CFPS is the result of several processes including the creation of saturating areas at these locations and the interaction between the substrate and the mineral. Supersaturation in those areas should be higher than that of the culture medium, since the negatively charged areas in CFPS have the ability to concentrate calcium (Addadi and Weiner 1992). In the case of the cell wall, and being the saturation dependent on the extracellular release of CO_2 and NH_3, the highest saturations are expected in the surroundings of the cell wall. Those high saturations are favored by the negatively charged surface functional groups, such as carboxyl, hydroxyl and phosphate sites (Schultze-Lam et al. 1996). Therefore, the first calcium carbonate precipitation is expected to happen at these areas. Later, as CO_2 and NH_3 diffuse from the cells to the medium, other locations will also begin to saturate, reaching the highest supersaturations those in which calcium accumulates, like in specific areas of other CFPS. In particular, EPS has been shown to be a rich matrix of polymers, including polysaccharides, proteins, glycoproteins, nucleic acids, phospholipids, and humic acids, able to concentrate calcium (Beveridge and Graham 1991; McSwain et al. 2005). In fact, a number of bacterial macromolecules and structures (cell wall, cell membranes, EPS, aminoacids and capsular exopolisacharides) have been found to mediate the precipitation of calcium carbonate and even to control the morphology and the orientation of the crystals (González-Muñoz et al. 1996; Kawaguchi and Decho 2002; Braissant et al. 2003; Benzerara et al. 2004; Ercole et al. 2007; Rodriguez-Navarro et al. 2007). The mechanisms by which those substrates favor the precipitation of calcium carbonate are under discussion. Several ways have been proposed (1) the substrate templates the nucleation of $CaCO_3$ crystals (geometrical matching, Pentecost 1985; Mann et al 1988; Dupraz and Visscher 2005); (2) the substrate favor the oriented aggregation of homogenously nucleated calcium carbonate crystals by electrostatic affinity between negatively charged areas and specific atomic planes of the crystals (ionotopic effect; Rodriguez-Navarro et al. 2007), and (3) in the case of EPS, the substrate traps seed crystals which act as nuclei for heterogeneous calcium carbonate precipitation (Knorre and Krumbein 2000; Turner and Jones 2005).

Regarding calcium carbonate precipitation on cell membranes, an interesting remark is that such precipitation was observed in fractions of cell membranes, both in M-3 and CC media (Figs. 4f and 5f). Such an observation does not agree with that of Benzerara et al. (2004), who concluded that the precipitation of calcium phosphate induced by *Ramlibacter tataouinensis* requires integrated membrane structure. However, and given the fact that the biominerals describe in the paper from those authors (phosphates) and those described in the present study (carbonates) are different, other mechanisms may account for the differences observed, like, for instance, the role of the phospholipids of the membranes on the precipitation of calcium phosphate. This issue, which does not fall within the focus of the present work, needs to be further considered in future studies.

Not only CFPS, but also other organics (macromolecules) inside the cell can act as pre-existing surfaces, resulting in the calcification of the interior of the cells, as it can be observed in Figs. 4c, d and 5c, d. Such features have been also observed by Benzerara et al. (2004). HRTEM images of stained preparations show that membranes are not intact (Figs. 4f and 5f), thus, this calcification inside the cells probably occurs once the cell lyses. It is little plausible that such a calcification takes place while the cell is living or intact, since it has been described by Rosen (1987) that most live bacterium cells have extrusion mechanisms to maintain low concentrations of intracellular calcium, so, therefore, the protoplasm may be undersaturated with respect to the precipitation of calcium carbonate. The fact that not all the cells present the same level of calcification (from only the cell wall to also all the protoplasm with all the transient stages) is probably related to the existence of different generations in the colony, from the first to latter descendants, thus making it possible to observe different stages of biomineralization in the same sample.

With time, the calcium carbonate structure formed in the cell wall is thick enough to affect the permeability of the cell wall, thus triggering the lyses of the cells. As shown in Figs. 4g, h and 5g, h, once the cell lyses, only the calcium carbonate coating remains, forming the empty tubing and holes observed in the calcified matrix in the abovementioned figures. Such features have been observed in stromatolites in freshwater cements, most of them related to cyanobacteria (Krumbein 1979; Merz-Preiß and Riding 1999; Riding 2000). Here we extend this observation to a heterotrophic soil bacterium. Moreover, the possibility of having a porous calcium carbonate biogenic material with a certain porous size and length may be of interest in biotechnological application, although such an application is out of the focus of the present work.

5 Conclusions

The study of the processes involved in the calcification of colonies of *M. xanthus* shows that such a calcification is the result of a number of several precipitation events acting in different locations and include the homogeneous precipitation of calcium carbonate within the colony and the heterogeneous precipitation of calcium carbonate in negatively charged areas of the cell wall, EPS, cell membranes and cell debris. For that calcification to occur, a critical size of the population is needed. Such a critical size is not only understood just as size of the population, but it better refers to the size of the population needed to reach a critical metabolic activity able to change the supersaturation of the environment.

Finally, different patterns are observed in the calcified colonies: a "Liesegang-ring pattern" and a "whole colony pattern", depending on the initial saturation of the culture medium and the evolution over time of the saturation within the colony. Such an evolution of saturation depends on a number of factors, being the most relevant the size of the population, the metabolic activity of such a population and previous precipitation events. As a consequence of those factors, saturation gradients occur from the inner to the border of the colony that result in different calcification patterns and even in the precipitation of several polymorphs of calcium carbonate. The results of this study are important, since, when applied to environments where calcified structures are found (for instance, pathological concretions and lithified mats) they may provide valuable information regarding both the chemical conditions under which those structures formed, the size of the population that was involved in the formation of such a structure and the biotic versus inorganic origin of such a lithified mat.

From the methodological point of view, the method designed in this study to fix the colony allows the study of fractions of such a colony by Electron microscopy without disturbing the structure of this colony.

Acknowledgements This work was financed by grants CGL2007-61489/BOS and MAT 2006-05411 (Spanish Ministry of Culture) and RNM-03943 (Junta de Andalucia) and supported by Research Groups BIO 103 (Junta de Andalucía). CJL thanks also go to grant CGL2007-63859 from MEC. Thanks go to Dr. Alejandro Rodriguez-Navarro for his ideas in the discussion of the results, comments to the figures and revising the initial version of the manuscript and to CIC personal from the University of Granada for technical assistance.

References

Addadi L, Weiner S (1992) Control and design principles in biological mineralization. Angewandte Chemie, International Edition in English 31:153–169

Altermann W (2004) Evaluation of Life and Precambriam Bio. Geology. In: Altermann W, Nelson DR, Mueller W, Catuneanu O (eds) The Precambrian Earth Tempos and Events. Developments in Precambrian Geology. Elsevier, Amsterdam, pp 539–545

Altermann W, Kazmierczak J, Oren A, Wright DT (2006) Cyanobacterial calcification and its rock-building potential during 3.5 billion years of Earth history. Geobiology 4:147-166

Anderson TF (1951) Techniques for the preservation of three-dimensional structure in preparing specimens for the electron microscope. Transactions of the New York Academy of Sciences 13:130–133

Bachmeier KL, Williams AE, Warmington JR, Bang SS (2002) Urease activity in microbiologically-induced calcite precipitation. Journal of Biotechnology 93:171-181

Banfield JF, Nealson KH (1997) Geomicrobiology: Interactions Between Microbes and Minerals. Reviews in Mineralogy 35: 448 p Mineralogical Society of America, Washington, DC

Baskar S, Baskar R, Mauclaire L, McKenzie JA (2006) Microbially induced calcite precipitation in culture experiments: Possible origin for stalactites in Sahastradhara caves, Dehradun, India. Current Science 90(1):58–64

Ben Chekroun K (2000) Producción de diversos minerales por núcleos heterogéneos de cristalización aportados por *Myxococcus xanthus* y comparación con otros microorganismos. Ph.D. thesis, Universidad de Granada, Granada, Spain

Ben Chekroun K, Rodriguez-Navarro C, Gonzalez-Muñoz MT, Arias JM, Cultrone G, Rodriguez-Gallego M (2004) Precipitation and growth morphology of calcium carbonate induced by *Myxococcus xanthus*: Implications for recognition of bacterial carbonates. Journal of Sedimentary Research 74:868–876

Ben Omar N, Entrena M, González-Muñoz MT, Arias JM, Huertas F (1994) The effects of pH and phosphate on the production of struvite by *Myxococcus xanthus*. Geomicrobiology Journal 12:81–90

Benhamu C, Arias JM, Montoya E (1979) Efecto de la glucosa sobre el crecimiento de *Myxococcus coralloides* D. Ars Pharmaceutica 20(3):237–246

Benzerara K, Menguy N, Guyot F, Skouri F, de Luca G, Barakat M, Heulin T (2004) Biologically controlled precipitation of calcium phosphate by *Ramlibacter tataouiensis*. Earth and Planetary Science Letters 228:439–449

Beveridge TJ, Graham LL (1991) Surface-layers of bacteria. Microbiological Reviews 55(4): 684–705

Braissant O, Verrecchia EP, Aragno M (2002) Is the contribution of bacteria to terrestrial carbon budget greatly underestimated? Naturwissenschaften 89:366-370

Braissant O, Cailleau G, Dupraz C, Verrecchia APJ (2003) Bacterially induced mineralization of calcium carbonate in terrestrial environments: the role of exopolysaccharides and amino acids. Journal of Sedimentary Research 73:485–490

Castanier S, Le Métayer-Levrel G, Orial G, Loubière JF, Pethuisot JP (2000) In: Ciferri O et al. (eds) Of Microbes and Art: The Role of Microbial Communities in the Degradation and Protection of Cultural Heritage. Plenun, New York, pp 203–218

Chafetz HS, Buczynski C (1992) Bacterially induced lithification of microbial mats. Palaios 7:277–293

Çiftçioglu N, Çiftçioglu V, Vali H, Turcott E, Kajander EO (1998) Sedimentary rocks in our mouth: Dental pulp stones made by nanobacteria. In: Hoover RB (ed) Proceedings of the Society of Photo-Optical Instrumentation Engineers (Spie) 3441:130–136

Çiftçioglu N, Bjorklund M, Kuoriksoski K, Bergstrom K, Kajander E (1999) Nanobacteria: an infectious cause for kidney stone formation. Kidney International 56:1893–1898

Costerton JW, Cheng KJ, Geesey GG, Ladd TI, Nickel JC, Dasgupta M, Marrie TJ (1987) Bacterial biofilms in nature and disease. Annual Review of Microbiology 41:435–464

Decho AW (1990) Microbial exopolymer secretions in ocean environments: their role(s) in foodwebs and marine processes. Oceanography and Marine Biology: An Annual Review 28:73–154

Decho AW (2000) In: Riding RE, Awramik SM (eds) Microbial Sediments. Springer, Heidelberg, pp 1–9

Dravis JJ (1983) Hardened subtidal srtomatolites, Bahamas. Science 219:385–386

Dupraz C, Visscher PT (2005) Microbial lithification in marine stromatolites and hypersaline mats. Trends in Microbiology 13(9):429–436

Dworkin M, Kaiser D (1993) *Myxobacteria II*. American Society for Microbiology, Washington, DC

Ehrlich HL (1996) Geomicrobiology, Marcel Dekker Inc., New York

Ehrlich HL (2002) Geomicrobiology, Marcel Dekker Inc., New York

Ercole C, Cacchio P, Botta AL, Centi V, Lepidi A (2007) Bacterially induced mineralization of calcium carbonate: the role of exopolysaccharides and capsular polysaccharides. Microscopy and Microanalysis 13:42–50

Fujita Y, Redden GD, Ingram JC, Cortez MM, Ferris FG, Smith RW (2004) Strontium incorporation into calcite generated by bacterial ureolysis. Geochimica et Cosmochimica Acta 68:3261–3270

Garside J (1982) In: Nancollas GH (ed) Biological Mineralization and Demineralization. Springer, Berlin 23, pp 23–35

Golubic S (1973) In: Carr N, Whitton BA (eds) The Biology of Blue-Green Algae. Blackwell Scientific Publications, Oxford, pp 434–472

González-Muñoz MT, Ben Omar N, Martínez-Cañamero M, Rodríguez-Gallego M, López Galindo A, Arias JM (1996) Struvite and calcite crystallization induced by cellular membranes of *Myxococcus xanthus*. Journal of Crystal Growth 163:434-439

González-Muñoz MT, Ben Chekroun K, Ben Aboud A, Arias JM, Rodríguez-Gallego M (2000) Bacterially induced Mg-calcite formation: Role of Mg^{2+} in development of crystal morphology. Journal of Sedimentary Research 70:559–564

Grotzinger JP, Knoll AH (1999) Stromatolites in Precambriam carbonates: evolutionary mileposts or environmental dipsticks? Annual Review of Earth and Planetary Sciences 27:313–358

Hammes F, Boon N, de Villiers J, Verstratet W, Siciliano SD (2003) Strain-specific ureolytic microbial calcium carbonate precipitation. Applied and Environmental Microbiology 69: 4901–4909

Henisch HK (1988) Crystals in Gels and Liesegang Rings. Cambridge University Press, Cambridge

Jimenez-Lopez C, Rodriguez-Navarro C, Pinar G, Carrillo-Rosúa FJ, Rodriguez-Gallego M, Gonzalez-Muñoz MT (2007) Consolidation of degraded ornamental porous limestone stone by calcium carbonate precipitation induced by the microbiota inhabiting the stone. Chemosphere 68(10):1929–1936

Kajander EO, Çiftçioglu N (1998) Nanobacteria: an alternative mechanism for pathogenic intra- and extracellular calcification and stone formation. Proceedings of the National Academy of Sciences of the United States of America 95:8274–8282

Kawaguchi T, Decho AW (2002) A laboratory investigation of cyanobacterial extracellular polymeric secretions (EPS) in influencing $CaCO_3$ polymorphism. Journal of Crystal Growth 240:230–235

Keefe WE (1976) Formation of crystalline deposits by several genera of the family *Enterobacteriaceae*. Infection and Immunity 14:590-592

Kempe S, Kazmierczak J (1990) In: Ittekkot V, Kempe S, Michaelis W, Spitzy A (eds) Facets of Modern Biogeochemistry. Springer, Berlin, pp 255–278

Knorre HV, Krumbein WE (2000) In: Riding RA, Awramik SM (eds) Microbial Sediments. Springer, Berlin, pp 25–31

Köhler SJ, Cubillas P, Rodríguez-Blanco JD, Bauer C, Prieto M (2007) Removal of cadmium from wastewaters by aragonite shells and the influence of other divalent cations. Environmental Science & Technology 41:112–118

Krumbein WE (1979) In: Trudinger PA, Swaine DJ (eds) Biogeochemical Cycling of Mineral-Forming Elements. Elsevier, Amsterdam, pp 47–68

Mann S, Heywood BR, Rajam S, Birchall JD (1988) Controlled crystallization of $CaCO_3$ under stearic-acid monolayers. Nature 334:692–695

McBride MJ, Hartzell P, Zusman DR (1993) In: Dworkin M, Kaiser D (eds) Myxobacteria II. American Society for Microbiology, Washington, DC, pp 285–305

McConnaughey TA, Whelan JF (1997) Calcification generates protons for nutrient and bicarbonate uptake. Earth-Science Reviews 42:95–117

McSwain BS, Irvine RL, Hausner M, Wilderer PA (2005) Composition and distribution of extracellular polymeric substances in aerobic flocs and granular sludge. Applied and Environmental Microbiology 71(2):1051–1057

Merz-Preiß M, Riding R (1999) Cyanobacterial tufa calcification in freshwater streams: ambient environment, chemical thresholds and biological processes. Sedimentary Geology 126: 103–124

Mitchell AC, Ferris FG (2005) The coprecipitation of Sr into calcite precipitates induced by bacterial ureolysis in artificial groundwater: Temperature and kinetic dependence. Geochimica et Cosmochimica Acta 69 (17):4199–4210

Molloy ES, McCarthy GM (2003) Hydroxyapatite deposition disease of the joint. Current Rheumatology Reports 5:215–221

Ogino T, Suzuki T, Sawada K (1987) The formation and transformation mechanism of calcium carbonate in water. Geochimica et Cosmochimica Acta 51:2757–2767

Paerl HW, Steppe TF, Reid RP (2001) Bacterially mediated precipitation in marine stromatolites. Environmental Microbiology 3:123–130

Pentecost A (1985) Association of cyanobacteria with tufa deposits: identity, enumeration and nature of the sheath material revealed by histochemistry. Geomicrobiology Journal 4:285–298

Plummer LN, Busenberg E (1982) The solubilities of calcite, aragonite and vaterite in CO_2-H_2O solutions between 0 °C and 90 °C, and an evaluation of the aqueous model for the system $CaCO_3$-CO_2-H_2O. Geochimica et Cosmochimica Acta 46:1011–1040

Reid RP, James NP, Macintyre IG, Dupraz CB, Burne RV (2003) Shark Bay stromatolites: microfabrics and re-interpretation of origins. Facies 33:1-17

Riding R (2000) Microbial carbonates: the geological record of calcified bacterial-algal mats and biofilms. Sedimentology 47:179–214

Rivadeneyra MA, Párraga J, Delgado R, Ramos-Cormenzana A, Delgado G (2004) Biomineralization of carbonates by Halobacillus trueperi in solid and liquid media with different salinities. FEMS Microbiology Ecology 48: 39–46

Rodriguez-Navarro C, Rodriguez-Gallego M, Ben Chekroun K, Gonzalez-Muñoz MT (2003) Conservation of ornamental stone by *Myxococcus xanthus*-induced carbonate biomineralization. Applied and Environmental Microbiology 69:2182–2193

Rodriguez-Navarro C, Jimenez-Lopez C, Rodriguez-Navarro A, González-Muñoz MT, Rodriguez-Gallego M (2007) Complex biomineralized vaterite structures encapsulating bacterial cells. Geochimica et Cosmochimica Acta 71:1197–1213

Rosen BP (1987) Bacterial calcium transport. Biochimica et Biophysica Acta 906:101–110

Schultze-Lam S, Fortin D, Davis BS, Beveridge TJ (1996) Mineralization of bacterial surfaces. Chemical Geology 132:171–181

Sutherland IA (2001) The biofilm matrix – an immobilized but dynamic microbial environment. Trends in Microbiology 9:222–227

Turner EC, Jones B (2005) Microscopic calcite dendrites in cold-water tufa: implications for nucleation of micrite and cement. Sedimentology 52(5):1043–1066

Visscher PT, Stolz JF (2005) Microbial mats as bioreactors: populations, processes, and products. Palaeogeogryphy, Palaeoclimatology, Palaeoecology 219:87–100

Visscher PT, Reid RP, Bebout BM (2000) Microscale observation of sulphate reduction: correlation of microbial activity with lithified micritic laminae in modern marine stromatolites. Geology 28:919–922

Walter MR (1983) In: Schopf JW (ed) Earth's Earliest Biosphere: Its Origin and Evolution. Princeton University Press, Princeton, pp 187–213

Warren LA, Maurice PA, Parmar N, Ferris FG (2001) Microbially mediated calcium carbonate precipitation: Implications for interpreting calcite precipitation and for solid-phase capture of inorganic contaminants. Geomicrobiology Journal 18:93–115

Wolery TJ (1992) EQ3/6, A Software Package for Geochemical Modeling of Aqueous Systems. Version 7.0. University of California, Livermore.

Wright DT (1999) The role of sulphate-reducing bacteria and cyanobacteria in dolomite formation in distal ephemeral lakes of the Coorong region, South Australia. Sedimentary Geology 126: 147–157

Wright DT (2000) Benthic microbial communities and dolomite formation in marine and lacustrine environments – a new dolomite model. In: Glenn C, Prévot-Lucas L, Lucas J (eds) Marine Authogenesis: From Global to Microbial. SEPM special publication, Tulsa, OK, 66:7–14

Wright DT, Oren A (2005) Non-photosynthetic bacteria and the formation of carbonates and evaporates though time. Geomicrobiology Journal 22:27-53

Are Stromatolites the Most Ancient Skeletal Organisms?

Evgenia Sumina, Vladimir Orleansky, and Dmitry Sumin

Stromatolites have an ancient and contradictory research history. Initially, they were classified as non-organic fossils and either concretions or layered limestones. Over time, their organic nature was established, but details regarding the stromatolite-forming organisms were not known. Walcott (1914) proposed that stromatolites are produced by blue-green algae (cyanobacteria) and this was confirmed by the present-day observation of stromatolite-forming microbiota. Since that time stromatolite-forming organisms were considered as colonies of microorganisms with the stromatolite being the result of the interaction between the colonies and trapping sediment. This description of the nature of stromatolites was successfully used for the Precambrian. However, the stromatolite-forming microorganisms were never considered as an organism. Thus the following controversy arose: the cyanobacterial accumulation is not able to undergo morphogenesis, but is capable of forming a large morphogenically shaped mineral bodies, the stromatolites, which are stable in space and time (Fig. 1). To resolve this controversy, laboratory modelling was performed on the stromatolite forming biota (Орлеанский and Раабен 1997). To continue this experiment, separate research project on the morphogenetic reaction society (here "society" refers to a compact settlement of independent elements- filamentous cyanobacteria" with a non-determined degree of integrity) of contemporary filamentous cyanobacteria as a whole community was performed.

Ten types of structures (Fig. 2) were detected for filamentous cyanobacteria creation as a result of their regrouping (Sumina 2002). The forms of the structures are specific and grow as a behaviour response to various influences, sometimes growing up to 10 cm. So the community appears to possess differentiation and

E. Sumina (✉) and V. Orleansky
Paleontological Institute RAS, Profsoyuznaya st., 123, 117647 Moscow, Russia
e-mail: stromatolit@list.ru, orleanor@mail.ru

D. Sumin
Independent researcher, Zhebrunova st., 5, 44, 107014 Moscow, Russia

J. Reitner et al., *Advances in Stromatolite Geobiology*,
Lecture Notes in Earth Sciences 131, DOI 10.1007/978-3-642-10415-2_12,

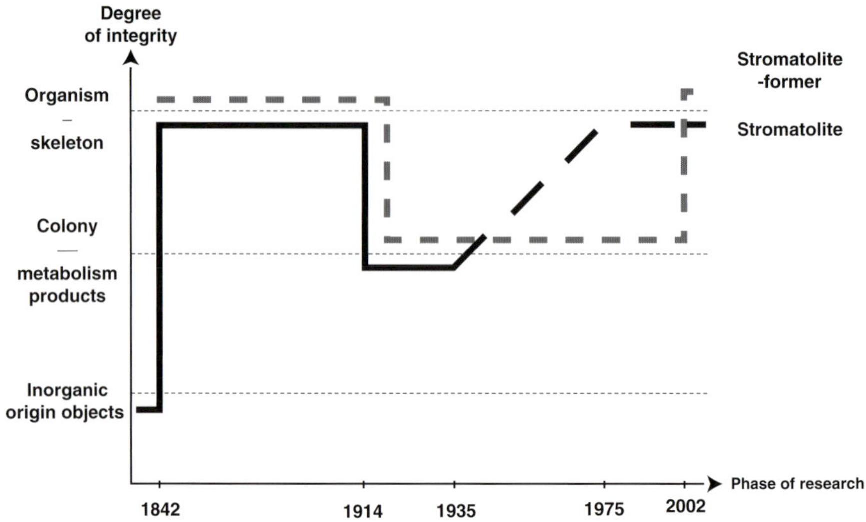

Fig. 1 Scheme illustrating changes in interpreting stromatolites and stromatolite-forming organisms

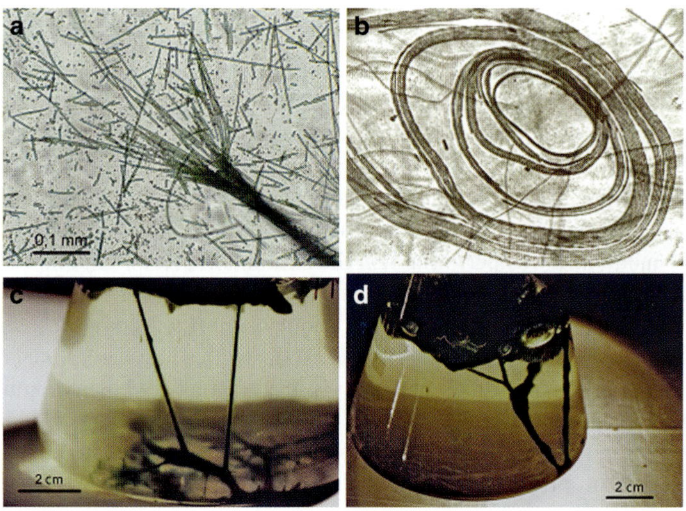

Fig. 2 The structures formed by filaments: (**a**) bundle; (**b**) rings; (**c**) and (**d**) contraction of bundles

morphogenesis abilities, not as a number of separated filaments, but as an entire unity structure. A stromatolite may be also regarded as community morphogenesis evidence under conditions of mineral sediment. So far as the stromatolite form and its evolution correspond to the needs of the stromatolite-former as a photosynthesising organism. We may consider a stromatolite to be a supporting mineral formation, which appears as a result of morphogenesis. The morphogenesis ability

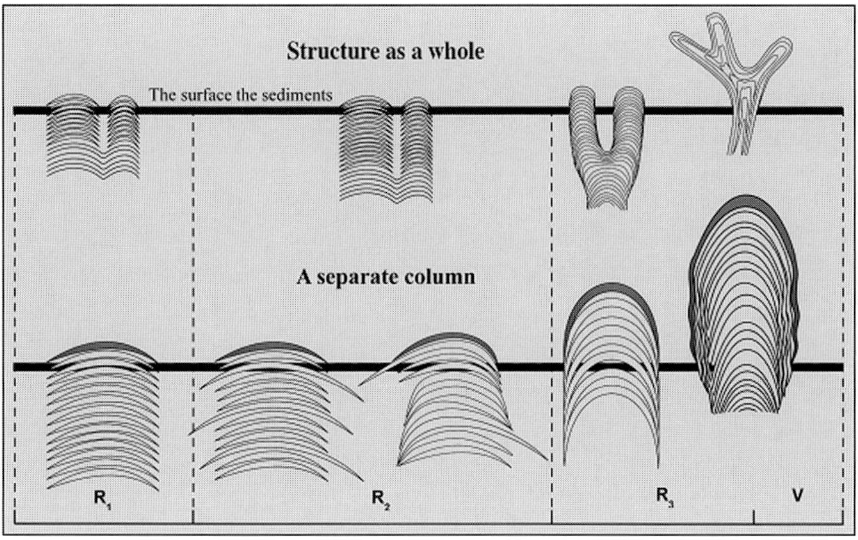

Fig. 3 Scheme of morphogenetic transformations of Upper Precambrian stromatolites

is inherent both to the stromatolite-former as a cyanobacterial community and also to the stromatolite itself. The latter ability occurs in the adaptive features of single parts and also for the building as a whole. The comparison of experimental results and evolutionary variations in fossil stromatolites allow us to consider the stromatolites as a peculiar analogue to the skeleton of eukaryotic organisms. The coordination, agreement and dependence of the variations in the evolution of morphological features up to Upper Precambrian is evident (Fig. 3). The comparison of stromatolite and eukaryote skeletons shows the following features of skeletons: morphological definition, hierarchical organization, definition and evolutionary stasis. However, the direct comparison of stromatolites and eukaryotic skeletons is not possible because of the prokaryotic nature of the producers. The term "skeleton" is currently utilized only for eukaryotic organisms. It is evident that the "skeleton" assumes a comparison between the supporting prokaryotes and eukaryotes mineral formations from the most general point of view.

Acknowledgements This work was supported by RFBR №080400484.

References

Орлеанский ВК, Раабен МЕ (1997) Живая лабораторная модель – аналог пассивно-ветвящихся столбчатых строматолитов. Киев, Альгология, Т 7 (№ 2): 185–188

Sumina EL (2002) Laboratory observation on culture of cyanobacteria and problem of stromatolites morphogenesis. Proceedings of SPIE, 4939: 101–102, Bellingham

Walcott CD (1914) Cambrian geology and paleontology III: Precambrian Algonkian algal flora. Smithsonian Miscellaneous Collection 64: 77–156

Microbial Mats and Microbialites in the Freshwater Laguna Bacalar, Yucatan Peninsula, Mexico

Eberhard Gischler, Stjepko Golubic, Michael A. Gibson, Wolfgang Oschmann, and J. Harold Hudson

1 Introduction

Laguna Bacalar, a karstic freshwater lake in the state of Quintana Roo, Yucatan Peninsula, Mexico (Fig. 1) features in its southern part possibly the largest freshwater microbialite structures known. These structures extend continuously for over 10 km (Gischler et al. 2008). The laguna is a narrow, elongated body of freshwater, about 40 km long and 1–2 km wide, extending NE to SW with an adjacent strip of periodically flooded lowland. The lagoon is tectonically rooted on the Bacalar fault, which is the northern continuation of the Rio Hondo fault zone (Kenkmann and Schönian 2006, their fig. 6). Laguna Bacalar gets as deep as 20 m with the deepest areas located between Bacalar and Zipline Point. The lake is located in the low Neogene karstic landscape and fed by a series of karstic springs originating from deep circular sinkholes locally called cenotes. Based on hydrogeochemical analyses of groundwater on the Yucatan peninsula, Perry et al. (2009) concluded that the cavern system that includes the Bacalar cenotes probably predates the Tertiary; strontium isotopes suggest that waters derive their ion load from late Cretaceous and/or Paleogene carbonates and evaporites. As in other karstic systems, the emerging groundwater is supersaturated with respect to carbonate and this condition and strong currents along the narrow passages of the lagoon are closely associated with carbonate precipitation and microbialite accretion. The currents

E. Gischler (✉) and W. Oschmann
Institut für Geowissenschaften, Goethe-Universität, 60438 Frankfurt am Main, Germany
e-mail: gischler@em.uni-frankfurt.de

S. Golubic
Biological Science Center, Boston University, Boston, MA 02215, USA

M.A. Gibson
Department of Geology, Geography and Physics, University of Tennessee, Martin, TN 38238, USA

J.H. Hudson
ReefTech Inc., 8325 SW 68th Street, Miami, FL 33143, USA

J. Reitner et al., *Advances in Stromatolite Geobiology*,
Lecture Notes in Earth Sciences 131, DOI 10.1007/978-3-642-10415-2_13,

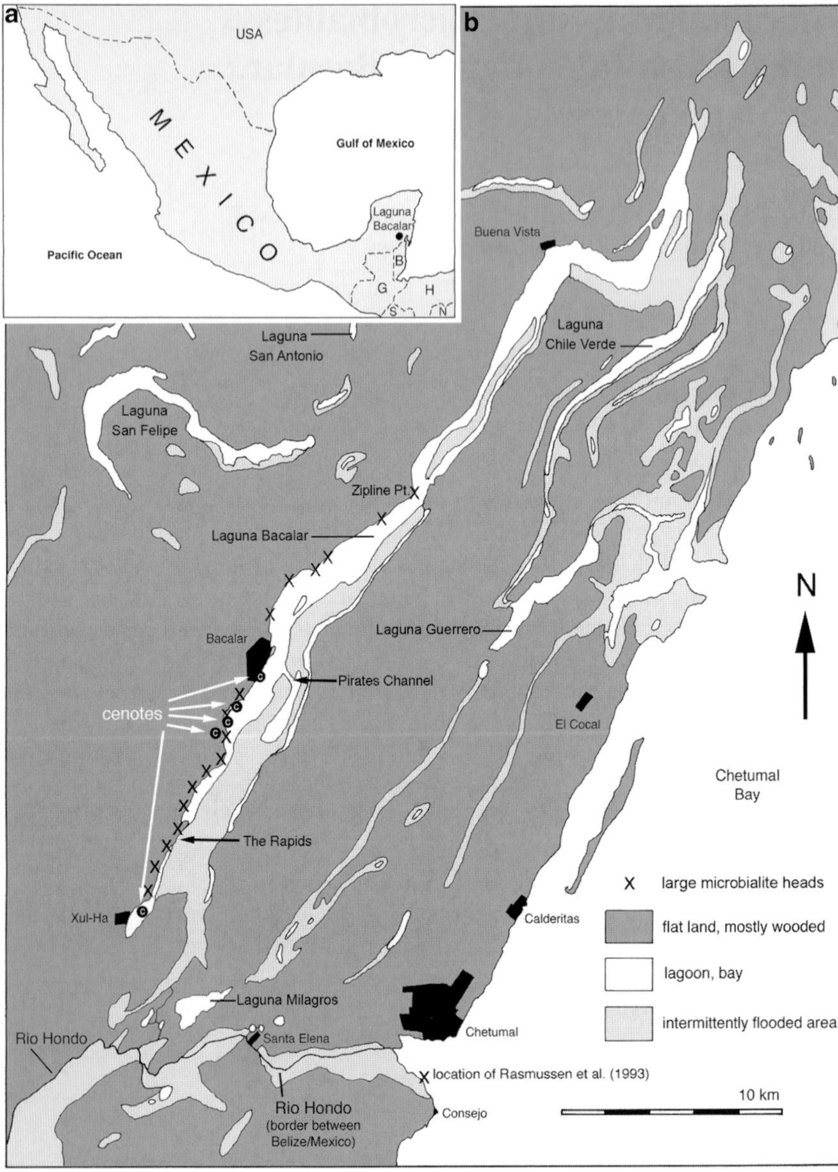

Fig. 1 Map of Laguna Bacalar west of Chetumal Bay, SE Yucatan peninsula, Mexico. Occurrences of large microbialite heads and cenotes are marked

are also modified by carbonate accretion and enhanced by narrowing of the passages. The outer shapes of microbialites observed in the lagoon include domes, ledges, low-relief crusts, and oncolites.

The internal texture of the stromatolites appears heterogeneous when viewed in section, comprised of finely laminated parts alternating in section with upward radiating filament arrangements; both interrupted by cavities and incorporated mytilid bivalves. The accretional history of the Laguna Bacalar stromatolites extends over a thousand years as suggested by radiometrically dated cores from microbialites in the southern part of the lagoon. These ages have to be handled with some reservation though, because of the strong hard water effect in Laguna Bacalar (Gischler et al. 2008). The current fauna in the lagoon shows low diversity and consists of recently immigrated mollusks including the mytilid zebra mussel (*Dreissena* sp.) and the apple snail (*Pomacea* sp.). *Dreissena* is especially abundant in the high current environment in the southwest part of the lake called the Rapids. Deeper in the cores, we detected layers with a small and thin-shelled lucinid bivalve and an unidentified small and high-spired gastropod.

In the present contribution we present new results from a study covering the entire lagoon and some of the adjacent bodies of water. We present data of water chemistry as it changes through the system and describe actively growing microbial mats associated with carbonate deposition and microbialite formation.

2 Materials and Methods

Surface sediment, living mat, core, and water samples were collected at 57 sample locations around Laguna Bacalar (Fig. 2). Analysis of water chemistry for 37 sampling sites along both sides of the lake was performed by atomic absorption spectrometry (AAS) for cations and by titration for anions in the Laboratory of Geochemistry of the Environment at Goethe-University, Frankfurt/Main and at the Analytical Laboratory Aachen (A.L.A), respectively. A total of 25 push cores up to 1.8 m long were taken from selected microbialite heads and from the adjacent lagoon floor to study the accretional history of these structures. Numerous surface samples of sediment and microbial mats were taken along Laguna Bacalar. Petrographic thin sections were prepared and internal texture of carbonate deposits identified and compared with the currently growing microbial mats. X-ray diffraction was performed to determine the mineralogy of the precipitates, and ^{14}C dating was carried out for the cores taken by Beta Analytic Inc. Miami, FL, USA.

Samples of actively growing microbial mats were preserved in 4% formaldehyde solution in environmental water and in 70% ethanol. Samples were prepared for light microscopy, sectioned and microscopically analyzed using Zeiss Universal Microscope (Zeiss, Oberkochen) and photomicrographically documented. Morphotypes of dominant mat-building cyanobacteria were measured from photomicrographs and in-scale projections, and then morphometrically analyzed using Sigma-Scan Image software (Jandel Scientific, Sausalito, CA). Determination of cyanobacteria was carried out using manuals of Komárek and Anagnostidis (2005), Geitler (1932)

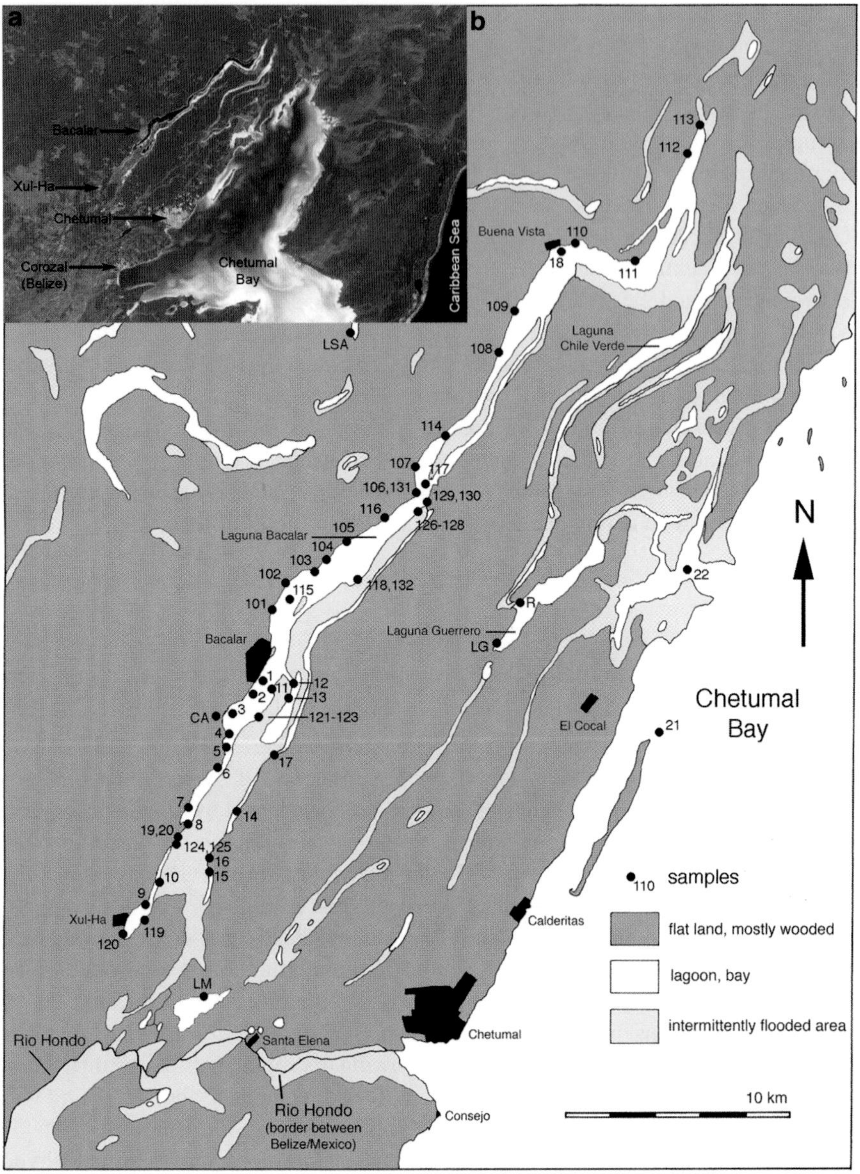

Fig. 2 (**a**) Satellite image of Laguna Bacalar. (**b**) Sampling sites in and around Laguna Bacalar. *LSA* Laguna San Antonio, *CA* Cenote Azul, *LM* Laguna Milagros, *LG* Laguna Guerrero, *R* Raudales

and Bergey's Manual of Systematic Bacteriology (Castenholz 2001). Selected parts of the mats were critical-point-dried and investigated by scanning electron microscopy (SEM, CAMSCAN S4).

3 Results

Water chemistry and carbonate deposition in the tropical karstic Laguna Bacalar and adjacent waters were studied in March 2005 and August 2006 (Gischler et al. 2008). In March 2009, the distribution and typology of microbialites in the entire system was recorded, the microbial composition of actively growing microbial mats analyzed and the study of water chemistry was extended to include surrounding lakes: Laguna Guerrero, Laguna Milagros, and Laguna San Antonio (Fig. 2).

3.1 Hydrology and Water Chemistry

Laguna Bacalar is fed by Xul-Ha cenote and four other cenotes located around the village of Bacalar (Fig. 1). Three of these four cenotes are located within the lagoon and an adjacent one (Cenote Azul; site CA) is separated from the lagoon without a surface flow connection. The cenotes are much deeper than the lake, (according to local fishermen as deep as 90 m). There is a strong northward current originating from Xul-Ha cenote at the southwest end of the lake. The flow velocity is strongest, reaching several m/s in the narrow channel called The Rapids. Towards Bacalar village, the current velocities are reduced and the water flow is divided towards east and south through the Pirates Channel, and to the north of the lagoon towards the village of Buena Vista. The southern flow connects to Chetumal Bay via Rio Hondo and an artificial aqueduct. The northern flow connects with Chetumal Bay through Laguna Chile Verde and Laguna Guerrero.

The southwestern part of Laguna Bacalar between the villages of Xul-Ha and Bacalar is characterized by carbonate concentrations that exceed those of the marine realm (Table 1). Calcium concentrations are only slightly lower as compared to the ocean. North of Bacalar, carbonate concentrations decrease, but they are still elevated as compared to the ocean at sites 104 and 115 north of Bacalar village. Carbonate concentrations also decrease towards the east. High carbonate concentrations outside Laguna Bacalar were also found in Cenote Azul and in the Laguna San Antonio (site LSA) to the west. Locations closer to the brackish Chetumal Bay (sites 14–17, 104–114, LG, R) showed elevated concentrations of chlorine and sodium. This finding suggests that the flow regime might be reversed, i.e., that at times the water from Chetumal might have flown westward. Such reversals might have occurred when the water levels in the system of karstic lagoons are extremely low. The annual variation in water levels in Laguna Bacalar is about 30 cm. Lagoonal water is apparently controlled by precipitation, which is high during the summer-fall and low during winter-spring seasons. However, at the end of the dry season, in March 2009, we have recorded eastward currents in the Pirates channel at location 12 and a flow from Laguna Chile Verde to Laguna Guerrero at our sampling site R at the village Raudales.

Table 1 Results of water analyses. Values for mean ocean water are given for comparison (Milliman 1974, his table 3). For locations of samples see Fig. 2. Values are mg/l. Note that cations in sample 1–22 were measured using AAS, and cations in samples 104 to LSA by titration

Sample	Na^{2+}	Mg^{2+}	Ca^{2+}	HCO_3^-	Cl^-	SO_4^{2-}
1	56	79	325	183	41	1,072
2	56	72	323	165	45	1,100
3	60	76	326	140	43	1,060
4	55	73	323	177	70	1,160
5	51	70	325	201	67	1,137
6	49	72	323	183	66	1,113
7	49	76	325	214	71	1,139
8	49	72	314	207	48	1,019
9	50	70	309	232	345	1,031
10	49	734	314	238	51	1,038
11	68	82	325	165	78	1,211
12	72	82	320	146	79	1,106
13	70	84	316	189	82	1,147
14	75	85	314	116	102	1,154
15	80	87	315	110	110	1,194
16	78	85	313	104	102	1,165
17	68	84	313	104	185	1,154
18	108	89	310	104	44	1,030
19	43	71	334	226	39	1,023
20	44	73	329	220	47	1,185
21	4,174	499	295	171	5,730	1,346
22	2,535	345	351	153	3,060	1,544
104	90	97	404	140	146	1,190
108	150	103	391	110	261	1,200
111	213	109	388	110	361	1,170
112	212	109	381	110	379	1,150
113	223	105	381	110	378	1,150
114	116	97	388	120	184	1,140
115	88	91	408	150	131	1,140
116	98	90	394	130	151	1,140
117	108	100	412	130	167	1,140
118	100	94	400	120	155	1,140
LM	122	114	422	84	181	1,300
LG	1,040	218	497	120	1,780	1,530
R	614	173	478	120	1,100	1,480
CA	37	95	497	300	45	1,240
LSA	33	10	62	180	32	70
Ocean water	10,760	1,295	415	140	19,350	2,700

3.2 *Distribution and Shape of Microbialites*

The presence of microbialites and degree of their induration correlates with the degree of supersaturation of ambient waters with respect to calcium carbonate. The most intensive calcification was observed in microbialite heads and ledges along the channel The Rapids (sampling sites 8, 19, 20, 124, 125). Oncolites were found abundantly between the Rapids and Bacalar village. Large microbialite heads over 1 m in diameter and height occurred along the western shore in the southern part of Laguna Bacalar (Fig. 2, sites 8). Microbialite heads tend locally to coalesce and form platforms (at sites 2 and 103). In 2009, we found large heads also north of the Bacalar toward Zipline Point, but less abundant than south of the village. North of Zipline Point and along the entire eastern shores of the lagoon, microbialites form only low-relief crusts (Fig. 1). E-W depth sounder traverses across the lagoon about 2 km north and south of the village of Bacalar revealed mostly flat bottom, indicating that no microbialite heads form in deeper water. No microbialites were found in the adjacent Laguna San Antonio, Laguna Milagros, Laguna Guerrero, and along the coast of Chetumal Bay.

The presence of microbialites and degree of induration correlate with water chemistry, their shapes, however, seem to depend largely on shoreline topography and relief so that microbialite crusts are found along low angle shorelines, heads occur at intermediate angle, and ledges at steep, high angle slopes (Fig. 3).

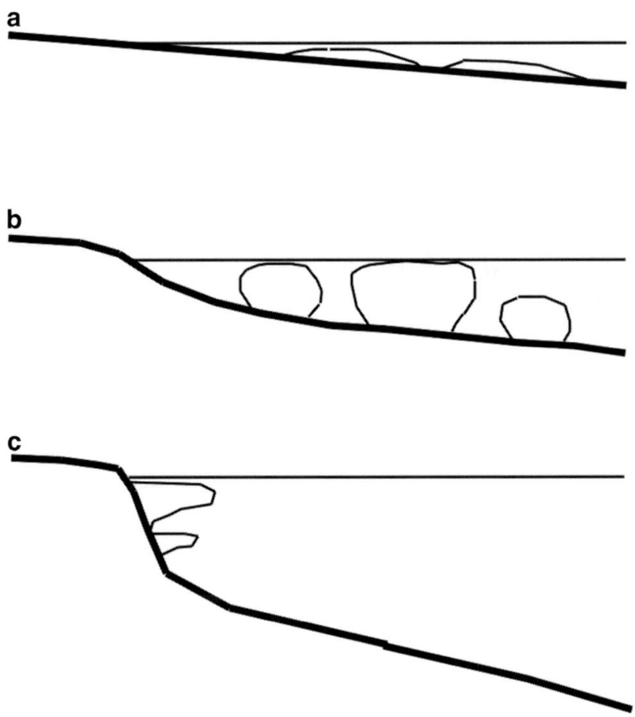

Fig. 3 Simplified schematic presentation showing microbialite shape as related to shoreline topography. (**a**) Low angle slope with crusts. (**b**) Intermediate angle slope with heads. (**c**) High angle slope with ledges

3.3 Microbial Mats

Microbial mats cover only mm-deep surface layer of stromatolitic structures, identifying the site of active accretionary process of the microbialite structure. The consistency of mats varies between soft velvety coatings to hard crusts or elastic-leathery consistency. They are mostly beige to brown in color, distributed over the upper surfaces of microbialite crusts and on tops and slopes of microbialite heads. Microbial crusts are the most common microbialite structures in the lake. They vary in degree of calcification and are usually prostrate. They occur along shores with mild slope and are widely distributed along the entire eastern shoreline and in northern parts of the lake. Lithified hemispheric, internally zoned stromatolitic structures, about 5 cm in diameter occur on calcareous heads and ledges in sections with high current velocity (The Rapids; sites 8, 19, 20). Bright orange colored microbial mats located on the flat tops of microbialite domes are the most conspicuous mat structures, however these appear to have a limited seasonal growth (sites 2, 103).

Microscopic analysis of the composition of different microbial mat types has established that each is dominated by a particular cyanobacterium.

3.3.1 Calothrix Mat

The most common microbial mat is constructed by upward oriented tapered filaments of the rivulariacean cyanobacterium *Calothrix* (Fig. 4). The mat forms expanded semi-indurated crusts along shores at low angles (Fig. 4a). The filaments of *Calothrix* are frequently falsely branched to form upward diverging bushes that preserve well and constitute a common sedimentary element of the stromatolite interior (Fig. 4b). Sheaths surrounding the trichomes are dark brown, pigmented by the extracellular sun-screening pigment scytonemin (Portwich and Garcia-Pichel 2003), which are also preserved in calcified parts of the stromatolite (Fig. 5). Trichomes are frequently curved at their base, distally tapered, unconstricted to slightly constricted at the crosswalls and oriented upright or upward-radiating (Fig. 4c). Filaments are variable in size, at the base 15.57 ± 2.51 (31) μm in diameter. Cells of the trichomes 6.07 ± 0.79 (45) μm wide and 3.68 ± 1.24 (45) μm long, short and wide at the base (Fig. 6), becoming distally isodiametric, tapering toward the tips into long hair-like whips comprised of long and narrow cells with pale vacuolized content. The heterocysts form at the base of the trichomes and false branches, separated at later stages of development by extracellular polymeric substance (EPS) that constitutes the sheath and is produced only by vegetative, carbon fixing trichome cells. Many filaments do not differentiate heterocysts and could be misidentified under the genus *Homoeothrix*.

Calothrix-dominated mats have the widest distribution in the lake and maintain also the highest diversity of cyanobacteria, including colonies of the other filamentous

Fig. 4 *Calothrix* mat. (**a**) Low angle shoreline with prostrate semiindurated microbialite crust. The western shore north of Zipline Point and the entire eastern lagoon shores are characterized by low-relied microbial crusts, at the location 121. (**b**) Tapered tips of *Calothrix* filaments in typical upright orientation. (**c**) Close up view of *Calothrix* filaments surrounded by diatoms. Note the club-shaped basis of a filament. Scale bar in **b** is 50 μm in **c** 10 μm long. (**d**) Microbialite domes close to cenotes, in forefront coalescing to form contiguous platforms

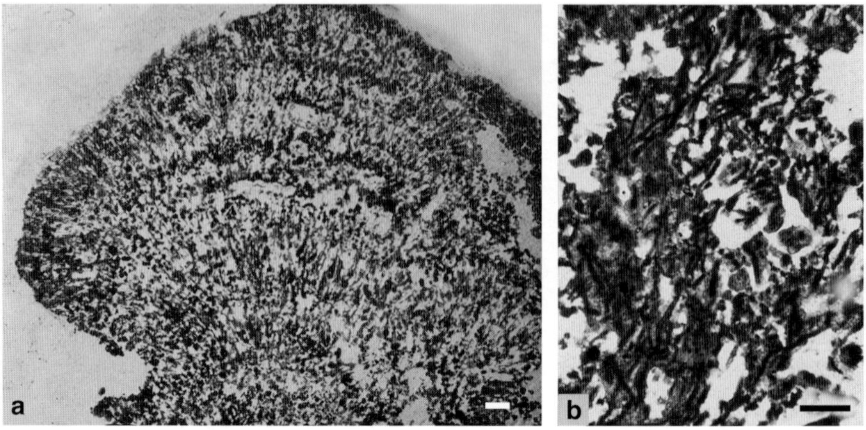

Fig. 5 Petrographic thin section of a fossil colony of *Calothrix*. (**a**) Upward radiating orientation of calcified filaments. (**b**) Detail showing preserved filaments. Scale bars are 100 μm long [Images from Gischler et al. (2008)]

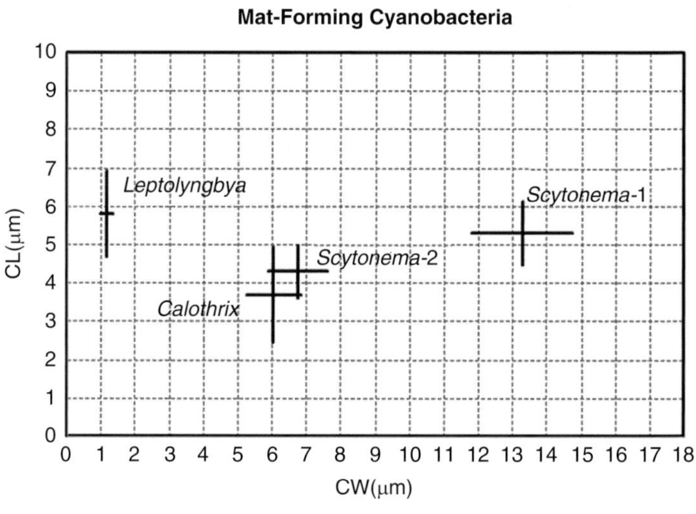

Fig. 6 Cross-diagram display of dimensions of *Leptolyngbya* sp., *Calothrix* sp., and two *Scytonema* spp. projected as means (intersection) and standard deviation (arms) for trichome cells

cyanobacteria: *Schizothrix*, two species of *Scytonema*, coccoid cyanobacteria *Entophysalis* and *Gloeothece* and mixed assemblages of diatoms and other eukaryotic algae. The *Calothrix* mat also forms meter-sized microbialite domes, often coalescing into platforms (Fig. 4d). They form along shores with intermediate angles, especially in the vicinity of cenotes.

3.3.2 Scytonema Mat

Distinct, firmly lithified hemispheroid structures 4–6 cm in diameter are formed in the habitat with strong currents called The Rapids (Fig. 7a). These microbialites are almost exclusively built by populations of a single cyanobacterial taxon, *Scytonema* sp. 1. The structures are dome-shaped, densely covered by cyanobacterial biofilm (Fig. 7b) internally permeated throughout by calcified sheaths (Fig. 7c) showing internal concentric lamination in sagittal section (stromatolitic morphology). The lamination indicates that the degree of calcification changes rhythmically over time, as the concentric laminae differ in mineral load. The filaments of this heterocystous cyanobacterium are 38.31 ± 5.49 µm wide, with occasional paired false branching (Fig. 7d). The cells of the trichomes are 14.87 ± 1.46 (50) µm wide and 5.28 ± 0.84 (80) µm long (Fig. 6, right). Filaments are curved and intertwined but, unlike *Calothrix,* are not oriented upward. Trichomes show predominant apical growth, i.e., they exhibit the highest growth rates at the trichome tips (Fig. 7e). Heterocysts, the sites of dinitrogen fixation are intercalary, connected by two opposing gates with vegetative carbon-fixing cells (Fig. 7f). In the sampled material we observed a large proportion of fragmented trichomes, commonly exiting from their sheaths, which opened by a funnel-shaped diverging tip (Fig. 7g).

The systematic position of this organism is close to *Scytonema mirabile*, but differs from that species by wide non-laminated sheaths, and frequent opening of sheaths at the tip of filaments. A much smaller *Scytonema* sp. 2 was present as a minor constituent in the *Calothrix* mat.

3.3.3 Leptolyngbya Mat

Bright orange colored mats grow on top of large microbialite heads, close to the water surface and partially exposed to the atmosphere. They frequently form a slightly elevated rim enclosing a water-filled depression on the microbialite tops making an appearance of a mini-atoll (Fig. 8a). These mats were produced by a sheathed filamentous cyanobacterium with narrow trichomes currently classified as *Leptolyngbya* Anagnostidis et Komárek (former *Phormidium* Kützing). The filaments consist of unbranched cellular trichomes, each surrounded by a thin but firm exopolymer sheath. Cells are cylindrical, significantly longer then wide unconstricted to barely constricted at the cross wall; the end cells are rounded. Cells are 1.20 ± 0.19 (26) µm wide and 5.77 ± 1.12 (63) µm long surrounded by exopolymer sheaths to form unbranched filaments 2.56 ± 0.32 (54) in diameter (Fig. 6, left). Trichomes move by gliding motility, often abandoning their sheaths and forming new ones, so that the trichomes are concentrated close to the surface of the mat, whereas deeper parts of the mat consist of predominantly empty sheaths. The bright orange coloration of these mats (Fig. 8b) reflects high concentrations of carotenoids in the cells, a pigmentation known to serve for protection from damage by high light intensity and UV (Ehling-Schulz and Scherer 1999; Seckbach and Oren 2007). Below this protecting layer the mat is blue-green in color.

Fig. 7 *Scytonema* mat. (**a**) Rapids, the site with high velocity water currents and development of microbiolitic ledges. (**b**) Hemispheric laminated and calcified colony of *Scytonema* sp.-1. (**c**) Vertical section through the surface layer of the hemispheroid. (**d**) Pair of false branches characteristic for the genus *Scytonema*. (**e**) Funnel-shaped diverging opening of the sheath and exiting trichomes. (**f**) *Scytonema* filament containing a trichome with intercalary heterocyst. (**g**) Massive abandoning of sheaths by trichome fragments. Scale bar in **d** is 50 μm valid for (**d–g**)

Vertical section (perpendicular to the surface) through the orange *Leptolyngbya* mat, when magnified, reveals layering comprised of lighter and darker laminae (Fig. 8c) marked by grains of calcium carbonate. The mat shows initial irregularities

Fig. 8 Orange *Leptolyngbya* mat (**a**) forming mini-atolls on large microbialite heads, at location 2, western shore. (**b**) Hand-held specimen of the *orange* mat with underlain *blue-green* layer.

in the lower part of the section, following or emphasizing the relief of the base it colonizes, but finally forms an even surface.

The base of the mat consists of heavily calcified upward diverging colonies of a different organism, and a frequent local discontinuity between the base and the finely laminated mat above (Fig. 8c, lower part). In the natural state, the interstitial spaces between filaments of the finely laminated orange mat are filled with capillary environmental water, which becomes periodically supersaturated with respect to carbonate, resulting in horizons with local carbonate precipitate (white lines in Fig. 8c). A closer look identifies the filament orientation (Fig. 8d): The filaments of *Leptolyngbya* sp. show rhythmic change between predominantly horizontal and predominantly vertical orientation. In horizontally prostrate position the filaments produce a dense opaque lamina. When upright, they are loosely arranged making a thicker, translucent lamina (Fig. 8e). Trapped sediment particles occurred interspersed within the laminated structure with higher concentrations associated with darker laminae. Individual filaments appear distinctly resolved when viewed by SEM of critical-point-dried specimens (Fig. 9). In vertical section through the mat, the filaments appear to form a network with filaments of the adjacent thinner and thicker laminae perpendicular to each other (Fig 9a). The SEM preparations document that the interstitial spaces between filaments do not contain exopolymer substances, the latter being confined to the sheaths. The surface view shows prostrate layers of filaments oriented parallel to each other (Fig. 9b), apparently aligned with the prevailing current.

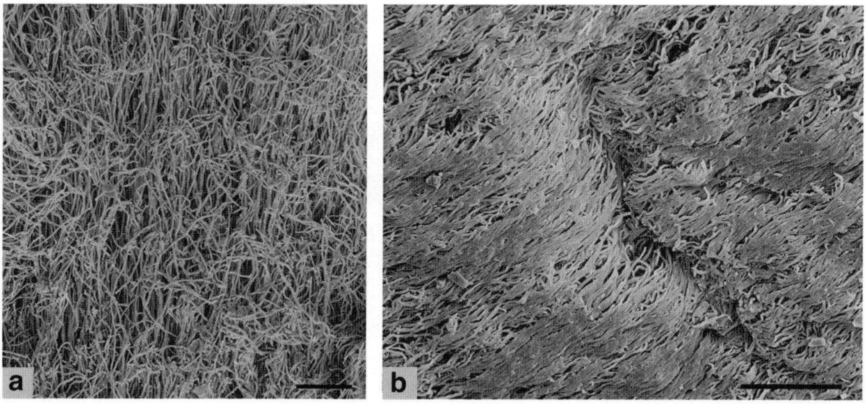

Fig. 9 Critical Point dried preparation of *Leptolyngbya* mat observed as SEM images. (**a**) *Leptolyngbya* filaments forming an interior mat network comparable with the section perpendicular to the surface. (**b**) View of prostrate filaments on the mat surface, aligned in the direction of the current. Scale bars are 10 μm long

Fig. 8 (Continued) (**c**) Vertical section through the mat, showing fine lamination. (**d**) Detail showing density difference between vertically oriented and prostrate filament arrangement. (**e**) Detail, showing alternation in filament orientation. Scale bar in **c** is 1 mm long, in **d** and **e** is 100 μm long

The studied mat showed a marked discontinuity about 1 cm below the surface (Fig. 8b, c). The mat beneath the laminated orange mat had a coarser texture and appeared calcified. It was comprised of upward radiating filaments, significantly wider than those forming the finely laminated mat. These features are consistent with the properties of the *Calothrix* mat, which apparently preceded the formation of *Leptolyngyba* on top of the microbialite head and was most likely the main architect of these large domal structures.

3.3.4 Internal Structures of Bacalar Microbialites

Calcium carbonate deposition in Laguna Bacalar is not uniform, neither are the carbonate saturation levels in different parts of the lake. Highest levels of dissolved carbonate have been consistently found along the western shore of the lake, in the vicinity of the cenotes, and in passages exhibiting higher current velocities. All precipitates in the lake consist of low magnesium calcite. Petrographic thin sections of the interior portions of the Bacalar microbialites prepared at different levels in the collected cores, show distinct areas, some with finely laminated, and others with upward radiating carbonate textures, the latter associated with larger diameter filaments (encountered mainly as sheath remains). In addition, the internal texture close to the surface is interrupted by cavities of various sizes and contains embedded *Dreissena* bivalve shells. Because of this inhomogeneous internal structure the microbialites of the Laguna Bacalar can be categorized as thrombolitic.

4 Discussion

The overall architecture of the microbialite deposits of the Laguna Bacalar suggests that the structures are constructed by more than one agent, which likely included the processes active today. The three studied microbial assemblages prevalent today represent three separate ecological niches occupied and dominated by different microorganisms and showing different biological responses.

The distribution of microbialites in Laguna Bacalar is controlled primarily by elevated carbonate concentrations of lagoonal waters, with possible contribution by photosynthetic activities of cyanobacteria in the structure and the presence of exopolymers produced by them (Merz-Preiss 2000). Microbialites are largest and most abundant in the southern parts of the lagoon where the karstic springs called cenotes are located and where high carbonate concentration is maintained by the influx of karstic groundwaters. The primary producers are comprised of mat-building cyanobacteria and benthic diatoms which support abundant grazing populations of *Pomacea* gastropods. Carbonate accretion rates are apparently high enough to counteract destructive effect of grazing activity and protect the sedimentary structures which have been formed over a millennium and recognized in cores of lithified stromatolites (Gischler et al. 2008). Rasmussen et al. (1993) described local occurrences of stromatolites at the mouth of Rio Hondo river in the nearby Chetumal Bay. These authors suggested that exclusion of grazers in brackish water was decisive for stromatolite buildup. We observed abundant grazers in Laguna Bacalar stromatolites and suggest that they are maintained by rapid calcification.

Abundant mytilid bivalves are incorporated into the outer, more recently deposited parts of the microbialite structure, which contributes to their porosity. This effect can be best observed in the occurrences of The Rapids, where the mytilids are particularly abundant. Currents and flushing of pore spaces by carbonate-saturated waters assures a continuation of carbonate precipitation and cementation of the structure by lining of pore spaces (Gischler et al. 2008; their fig. 4g).

In addition to water chemistry and flushing, shoreline topography clearly influences occurrences and shapes of microbialites in Laguna Bacalar. The angle of the slope which controls the accommodation space apparently controls the direction of stromatolite accretion. Low angle shorelines are characterized by low relief microbialite crusts as observed along the eastern shoreline. Steep angle shorelines such as in the Rapids channel lead to the growth of microbialite ledges. Large heads are found were slope angles are intermediate (Fig. 3). When the available accomodation space is filled up, microbialites tend to grow sideways, i.e., they prograde and in some cases even coalesce, as at locations 2 and 103 (Fig. 4d).

The interpretation of the studied actively growing mats in terms of their role in the construction of large lithified microbialites of Laguna Bacalar poses objective difficulties. All mats contain low magnesium calcite as mineral components, however, they differ in the degree they are calcified. Accordingly, the interpretation of their roles in the accretionary history of these microbialites remains based on structural similarities and is tenuous. The three types of microbial mats observed in the Laguna Bacalar respond differently to the process of carbonate deposition.

The upward oriented (and divergent) growth of *Calothrix* accommodates a wide range of habitats and moderate rates of calcification and microbialite accretion. The organism requires time for growth, which is generated from the base of the trichomes and differentiates by tapering upward and extruding into the water column. The calcification is predominantly encasing the sheaths whereas photosynthetically active trichomes are gradually dislocated upward by growth and false branching. Under too low calcification rates, the community may be affected or even removed by grazers. The success of the community is also limited by burial due to excessive calcification and accretion rates that exceed the growth rates of *Calothrix* trichomes. Many filaments do not differentiate heterocyst, which is also consistent with slow growth and availability of nitrogen as nutrient for growth (Fogg et al. 1973).

In cases of highest rates of calcification, as observed in the Rapids section of the lagoon, the colonization by *Scytonema* mats and formation of hemispheroid stromatolites is favored. *Scytonema* trichomes show maximum growth in distal rather than proximal parts as in *Calothrix*. In addition, we could observe another adaptation of this organism to respond to high calcification rates by fragmenting the trichomes (Fig. 7g), which are able to leave their cemented sheaths by gliding and producing new ones on top of the sediment. The mat is thereby maintained on the stromatolite surface on top of a firmly calcified base. The proportion of free trichomes and those in the process of leaving their sheaths observed upon decalcification of *Scytonema* mats (Fig. 7e, g) testified about the high rate at which the process takes place. Also the filaments and the abandoned sheaths of *Scytonema* were intertwined and oriented predominantly parallel to the substrate surface rather than perpendicular as in *Calothrix*, suggesting a mechanism in escaping the burial by motility rather than by growth of trichomes. *Scytonema*-constructed microbialites with different and changing growth orientation occur in waters of different and variable salinities (Monty 1967; Reitner et al. 1996).

The *Leptolyngbya* mats occupy a narrow zone defined by water level in an area where the calcification rates are moderate. The mats are characterized by a rapid combination of growth, motility and change in orientation of their trichomes, which leave a mark in positioning of sheaths within the predominantly organic structure. Carbonate particles originating from suspension become trapped in their laminae. Autochthonous carbonate occurred only in some horizons (white laminae in Fig. 8c) and was comprised by fine micritic precipitate. This is in contrast to the adjacent heavily calcified *Calothrix* mat, with filaments often completely encased in carbonate, although both mats were apparently exposed to the water of the same chemistry. A possible reason for the observed difference in the degree of their calcification may be explained by the difference in their growth rates.

The alternate change in orientation of filaments and the resulting fine lamination characteristic of *Leptolyngbya* is indistinguishable from those found in the marine *Leptolyngbya* (*Phormidium*) *hendersonii* (Howe) Anagnostidis et Komárek (Golubic and Focke 1978). The lamination in this organism was found to be nocti-diurnal: The upright orientation of the filaments was promoted by positive phototaxis of cyanobacterial trichomes, which subsequently assumed a prostrate position during the night. The timing has been shown by Monty (1965, 1967) using carborundum marking. The process of upward gliding motility of trichomes is accompanied by the production of exopolymer sheaths, which are left behind and constitute the main constructive element of the structure. This rhythmic change in filament orientation is consistent with the density difference between upright and prostrate layers. The intricate dense meshwork of tiny filaments accounts for the firmness of the structure, although the space between filaments is filled with water rather than exopolymer (observed as air space by SEM following critical-point-drying procedure) (Fig. 9a). The correlation of entrapped particles with the prostrate layers is also consistent with the timing of the mat accretion: They accumulate during the night when the trichomes rest and are diluted during the phototactic movement and growth of the microorganism by day. The rapid increase in thickness of these mats can be measured by counting of the laminae. The relatively short time of the growing season of these mats (calculated to be between 30 and 60 days), is commensurate with the low rate of sedimentation and calcification within the mat.

The rapid formation of the *Leptolyngbya* mat as described above is in contrast to the significantly slower growth of the *Calothrix* mat, which is produced by a nonmotile cyanobacterium. The orientation of its filaments rest on phototropism rather than on phototaxis, whereas their thick, scytonemin-enriched sheaths offer prolonged protection of the trichomes. In addition, the mat remains open to water circulation during its growth and is longer exposed to calcification. The organic remains of the *Calothrix* mat are present in the cores and correlate well with the upward radiating texture of the colonies, however, the filaments of the much more delicate *Leptolyngbya* did not preserve in the lithified interior of the structure. The finely laminated carbonate portions within the lithified cores may represent carbonate templates after *Leptolyngbya* mats, but direct evidence of the identity could not be established.

If the identity of the three currently active microbial mats, and the fossil carbonate texture in the cores were confirmed, it would suggest that historic changes in local water supply was sufficient to cause changes in dominant mat communities. The microbialite accretion alters the position of the mat in reference to water levels, and also modifies the water flow, which, in turn, may alter the conditions and rates of precipitation (e.g., by narrowing the channels and increasing the water flow). Such interactions and mutual adjustments between water access to biota and carbonate deposition may explain the irregular spatial distribution of textures during the process of stromatolite accretion as reflected in the textures found in the cores.

Acknowledgements We like to thank the Deutsche Forschungsgemeinschaft (DFG) for their support (Gi 222/15-2). International cooperation was supported by Alexander-von-Humboldt foundation, Bonn and Hanse Wissenschaftskolleg, Delmenhorst to S. Golubic. M. Gibson thanks the faculty and research development grant program at the University of Tennessee for financial support. Paul Blanchon (UNAM, Puerto Morelos) helped with logistics in Mexico.

References

Castenholz RW (2001) Oxygenic photosynthetic bacteria. In: Boon DR, Castenholz RW (eds) Bergey's Manual of Systematic Bacteriology, 2nd ed. Springer, New York, pp 473–600.

Ehling-Schulz M, Scherer S (1999) UV protection in cyanobacteria. Eur J Phycol 34: 329–338.

Fogg GE, Stewart WDP, Fay P, Walsby AE (1973) The Blue-green Algae. Acad Press, London, 459 pp.

Geitler L (1932) Cyanophyceae. Rabenhorst's Kryptogamenflora von Deutschland, Österreich und der Schweiz 14. Akademischer Verlag, Leipzig, 1196 pp.

Gischler E, Gibson MA, Oschmann W (2008) Giant Holocene freshwater microbialites, Laguna Bacalar, Quintana Roo, Mexico. Sedimentology 55: 1293–1309.

Golubic S, Focke JW (1978) *Phormidium hendersonii* Howe: identity and significance of a modern stromatolite building microorganism. J Sediment Petrol 48: 751–764.

Kenkmann T, Schönian F (2006) Ries and Chicxulub: impact craters on earth provide insights for Martian ejecta blankets. Meteorit Planet Sci 41: 1587–1603.

Komárek J, Anagnostidis K (2005) Cyanoprokaryota 2. Teil Oscillatoriales. In: Büdel B, Gärtner G, Krienitz L, Schlager M (eds) Süsswasserflora von Mitteleuropa. Fischer, Jena, 759 pp.

Merz-Preiss M (2000) Calcification in cyanobacteria. In: Riding RE, Awramik SM (eds) Microbial Sediments. Springer-Verlag, Berlin-Heidelberg, pp. 50–56.

Milliman JD (1974) Marine Carbonates. Springer, Berlin, 375 pp.

Monty CLV (1965) Recent algal stromatolites in the windward lagoon, Andros Island, Bahamas. Ann Soc Geol Belgique 88: 269–276.

Monty CLV (1967) Distribution and structure of recent stromatolitic algal mats, eastern Andros Island, Bahamas. Ann Soc Geol Belgique 90: 55–100.

Perry E, Paytan A, Pedersen B, Velasquez-Oliman G (2009) Groundwater geochemistry of the Yucatan peninsula, Mexico: constraints on stratigraphy and hydrogeology. J Hydrol 367: 27–40.
Portwich A, Garcia-Pichel F (2003) Biosynthetic pathway of mycosporines (mycosporine.-like amino acids) in the cyanobacterium *Chlorogloeopsis* sp. strain PCC 6912. Phycologia 42: 384–392.
Rasmussen KA, Macintyre IG, Prufert L (1993) Modern stromatolite reefs fringing a brackish coastline, Chetumal Bay, Belize. Geology 21: 199–202.
Reitner J, Paul J, Arp G, Hause-Reitner D (1996) Lake Thetis domal microbialites – a complex framework of calcified biofilms and organomicrites (Cervantes, Western Australia). In: Reitner J, Neuweiler F, Gunkel F (eds) Global and Regional Controls on Biogenic Sedimentation. Göttinger Arb Geol Paläont, Sb2, Göttingen, pp 85–89.
Seckbach J, Oren A (2007) Oxygenic photosynthetic microorganisms in extreme environments. In: Seckbach J, Oren A (eds) Algae and Cyanobacteria in Extreme Environments. Springer, Dordrecht, pp 5–25.

Part III
Microbial Ecology and Fossil Record

Geomicrobiology of Fluid Venting Structures at the Salse di Nirano Mud Volcano Area in the Northern Apennines (Italy)

Christina Heller, Martin Blumenberg, Sebastian Kokoschka, Christoph Wrede, Michael Hoppert, Marco Taviani, and Joachim Reitner

1 Introduction

Recent studies suggest that geological sources of methane like onshore and marine seeps, micro-seepage and mud volcanoes are an important source of this greenhouse gas (Etiope 2004). They represent the second most important natural emission after wetlands. New estimations indicate a global emission of methane from mud volcanoes in a range of 6-9 Tg year-1 (Etiope and Ciccioli 2009).

Mud volcanoes are present in many terrestrial and marine areas worldwide (Mazurenko et al. 2003; Scholte 2005), but their geographical distribution is strongly controlled by the geological settings. The majority is localized in areas of recent tectonic activity, particularly in zones of compression like accretionary complexes, thrust belts and e.g. in the forelands of Alpine orogenic structures (Dimitrov 2002; Kopf 2002). Furthermore, since the sediments in mud volcanoes are often hydrocarbon-rich, a relation between the formation of hydrocarbons and mud volcanoes was suggested (Dimitrov 2002).

The geographical occurrence of mud volcanoes in Italy can be divided into three main groups: (i) northern and (ii) central Italy and (iii) Sicily (Martinelli and Judd

C. Heller, M. Blumenberg, and J. Reitner (✉)
Geoscience Centre, University of Göttingen, Geobiology Division, Goldschmidtstraße 3, 37077, Göttingen, Germany
e-mail: jreitne@gwdg.de

S. Kokoschka, C. Wrede, and M. Hoppert
Institute of Microbiology and Genetics, University of Göttingen, Grisebachstr. 8, 37077, Göttingen, Germany

M. Taviani
ISMAR-CNR, Via Gobetti 101, 40129, Bologna, Italy

J. Reitner et al., *Advances in Stromatolite Geobiology*,
Lecture Notes in Earth Sciences 131, DOI 10.1007/978-3-642-10415-2_14,

2004). The study site is located near Modena in Northern Italy, at a mud volcano area named Salse di Nirano. The Salse di Nirano is situated in the Emilia Romagna, covers a surface of approximately 75.000 m^2 and represents one of the biggest mud volcano areas in Italy (Martinelli and Judd 2004).

Generally, mud volcanoes are formed by the expulsion of water, mud and gases (in particular methane and higher hydrocarbons), exhibit often anoxic niches and contain various electron acceptors. These preconditions potentially promote the activity of microorganisms performing the anaerobic oxidation of hydrocarbons, in particular methane. The anaerobic oxidation of methane (AOM) is suggested to be usually coupled to sulphate reduction (Reeburgh 1980) and is carried out by a symbiotic association of methanotrophic archaea and sulphate reducing bacteria (SRB), namely members of the *Methanosarcinales* and the *Desulfosarcina*/*Desulfococcus* group (DSS) (e.g., Hinrichs et al. 1999; Thiel et al. 1999; Boetius et al. 2000; Orphan et al. 2001a, 2001b, 2002; Elvert et al. 2003; Reitner et al. 2005a, 2005b; Treude et al. 2005), although many aspects are still insufficiently understood. Recent phylogenetic and biochemical studies have suggested that the anaerobic methanotrophic (ANME) archaea have supposedly reversed the methanogenic pathway (Hoehler et al. 1994; Hallam et al. 2003, 2004; Krüger et al. 2003). Although AOM has been mainly found in marine sediments and at marine methane seeps, AOM has been recently also found in a specific terrestrial mud volcano area near Paclele Micci in Romania (Alain 2006).

Our work aims to understand the functioning of the system of the Salse di Nirano mud volcanoes and the yet unknown microbial communities by organic-geochemical and molecular microbiological methods. The Salse di Nirano mud volcano area is located near Modena (Emilia Romagna, Northern Italy), and consists currently of four main conical vents and several satellite gryphons or "Salse". The main volcanoes reach 3 m in height and emit muddy fluids and gas. The mud volcano area is situated at the bottom of an oval depression near an anticline in the outcrop of Plio-Pleistocene clays (Martinelli and Rabbi 1998; Capozzi et al. 1994; Capozzi and Picotti 2002; Accaino et al. 2007; Castaldini 2008). The region belongs to an area of active thrusting along the Pede-Apennines margin of the Northern Apennines (Benedetti et al. 2003).

Over the time the actual number and the shape of the cones varied (Martinelli and Judd 2004). The surface colour of the mud volcanoes is grey. The emitted muddy fluids run down the cones; thereby, the clay deposits increase the size of the cones. During our observations in August 2008 and June 2009 several cones showed such an overflow, the other cones exhibited liquid mud breccias and gas bubbling deep in their craters. In the majority of the craters and smaller "Salse" gas bubbles from 1 cm up to 15 cm in diameter broken through the surface. At the surface of some mud pools an oily film of liquid hydrocarbons was visible.

Here we describe the environmental conditions and our first results of the geochemical and biochemical analyses of the phases seeping out at this exceptional geomicrobiological environment.

2 Materials and Methods

2.1 Study Site and Sampling

The samples were collected during two campaigns (August 2008 and June 2009) from several mud volcanoes in the Emilia Romagna area in Northern Italy. Mud and gases were collected from four active cones, which showed constant development of gas bubbling. This study is focussed on one of the active cones of the Salse di Nirano (Fig. 1).

Mud samples were collected in a depth of approximately 1.5 m with a hand-operated vacuum pump, connected to a flexible tube with a metal pipe at the open end (UniSampler, Bürkle, Bad Bellingen, Germany) and were transferred to the laboratory under cooled conditions. The samples for the lipid analyzes were freeze-dried and stored by −20°C. Gas bubbles were collected in the central part of the gryphon using a funnel and special gas vials (Labco Vials, Labco Limited, Buckinghamshire, United Kingdom).

The temperature, redox potential and the pH-values of the mud were measured directly in the pools.

2.2 Gas Analysis

The concentrations of CH_4, CO_2 and N_2O were measured with an automated gas chromatographic system (GC 14A, Shimadzu, Duisburg, Germany) equipped with

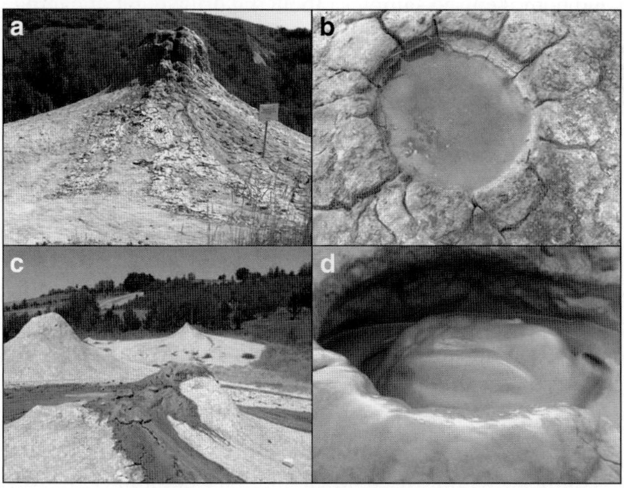

Fig. 1 Images of the Salse di Nirano mud volcanoes Northern Apennines, Italy. a) and c) mud volcano cones and craters of the Salse di Nirano. b) Liquid hydrocarbons on the surface of a pool. d) A bubbling salse

two detectors, a flame ionization detector (FID) for CH_4 analysis and an electron capture detector (ECD) for CO_2 and N_2O analyses (detailed description: Loftfield 1997).

The carbon isotope compositions ($\delta^{13}C$, $\delta^{2}H$) of the methane were measured with an isotope ratio mass spectrometer (IRMS; Delta plus XP, Thermo, Bremen, Germany; $\delta^{13}C$) coupled to a PreCon and a High Temperature Conversion/ Elemental Analyzer in conjunction with an IRMS (Delta V plus, Thermo, Bremen, Germany; $\delta^{2}H$).

2.3 Geochemical Analyses

The total carbon (TC), nitrogen (TN) and sulphur (TS) concentrations were measured by a CNS Elemental analyser (Euro Vector Instruments and Software, Milano, Italy). The total inorganic carbon (TIC) was determined after acidification with H_3PO_4. The total organic carbon (TOC) was calculated by the difference between TC and TIC.

The elemental content of the particle-free fluids were analyzed by an Inductively Coupled Plasma Optical Emission Spectrometry (ICP-OES; Perkin Elmer Optima 3300 DV) according to Thomson and Walsh (1983).

2.4 Lipid Biomarker Analyses

Lyophilised and homogenised samples (about 4 g dry weight of campaign 2008 and about 30 g of campaign 2009 sample, respectively) were hydrolysed using 6 % KOH in methanol (pH 14) in excess (2h at 80°C in ultrasonication bath) to extract free and release ester-bound lipids. The resulting alkaline reaction solution was extracted with *n*-hexane (5x) yielding the neutral lipid fraction. The neutral lipid fraction was further separated by column chromatography (Merck silica gel 60) and eluents of increasing polarity (n-hexane, dichloromethane, MeOH) providing a hydrocarbon and a polar fraction containing mainly alcohols. The alcohol fraction was treated with N,O-bis(trimethylsilyl)trifluoroacetamide (BSTFA) for 2 h at 80°C to silylate alcohols. Fatty acids (FA) were obtained by acidification of the residue of the alkaline reaction solution to a pH of 1-2 and subsequently extracted using *n*-hexane (5x). Prior to analysis, fatty acids were converted to their methyl esters by adding trimethylchlorosilane in methanol (1:9; v:v; 2 h, 80°C).

The above mentioned fractions were analyzed by gas chromatography-mass spectrometry (GC-MS) using a Varian 1200L MS (EI, 70 eV) coupled to a Varian CP 3800 GC equipped with a fused silica capillary column (ZB5-MS, 30 m, 0.32 mm inner diameter, 0.25 μm film thickness). Carrier gas was Helium. Temperature program: 3 min at 80°C; from 80°C to 310°C at 6°C min^{-1}; 20 min at 310°C. Lipids were identified by comparison of GC-retention times and published mass spectra of reference compounds (NIST library and own spectra). $\delta^{13}C$-values of lipids were analyzed (min. of three replicates) using a Thermo Trace GC coupled to a Thermo Delta Plus isotope-ratio mass spectrometer (GC-C-IRMS). Combustion of the components to CO_2 was performed with a CuO/Ni/Pt-furnace operated at 940°C. The stable carbon isotope compositions are reported in the delta notation ($\delta^{13}C$) vs. the V-PDB standard (standard deviation was usually less than 0.5 ‰). The typical GC-program used for GC-C-IRMS analyses was the same as described above for GC-MS.

2.5 Microbiological Methods

Cultures for isolation of sulphate reducing bacteria were prepared anaerobically. Media were composed as already described, with lactate or ethanol as carbon sources (Postgate 1951; Sakaguchi et al. 2002). Liquid media were inoculated with different volumes (10%–50%) of samples from mud volcano fluids. From enrichment cultures, grown at 20°C, pure cultures were obtained after separation of colonies by agar tube dilution series (Evans et al. 1977). The strains were further characterised by 16S rDNA-analysis according to established procedures (primers: 27F, 1525R, Lane 1991, E334F, E939R, Baker et al. 2003). Negative staining of cells was performed with 1% phosphotungstic acid (Hoppert 2003). Electron microscopy was performed with a Zeiss EM 902 transmission electron microscope.

3 Results and Discussion

3.1 Geochemistry

The largest part of the collected gas emitted by the mud volcanoes at the Salse di Nirano was methane (~ 99%), but small amounts of other hydrocarbons, carbon dioxide and nitrous gas were also found to be present (< 1%). Measurements of the $\delta^{13}C$-values of the released CH_4 gave a mean value of –50 ‰, and a $\delta^{2}H$ -value of −175 ‰. The $\delta^{13}C$-value indicates a thermogenic origin of the gas, potentially with minor contributions from a biogenic source (Whiticar 1999).

The material expelled is semi-liquid and is composed of 24% carbonates, 34% feldspar, 41% quartz and 1% halite analyzed by XRD (X-Ray diffraction). The content of the total inorganic carbon (TIC) in the mud is 2.64% dry weight and the total organic carbon is 0.41% dry weight (Table 1). The amount of total sulphur is 0.19% dry weight and for the total nitrogen the analyses revealed a content of 0.04% dry weight.

As became evident during the lipid biomarker analyses the mud samples contain high amounts of liquid hydrocarbons. The hydrocarbon fractions of both campaigns were dominated by oil hydrocarbons with carbon chains from C18 to C38, with a maximum at C28 and no odd over even predominance. This distribution clearly

Table 1 Geochemical composition and environmental conditions of a Salse di Nirano mud volcano (A: pore water: B: mud)

A											
		[mV]		[°C]	[ppm]						
	Sample	E_H (corr)	pH	T	Mg^{2+}	Ca^{2+}	Fe^{2+}	Ni^{2+}	SO_4^{2-}	Cl^-	Na^+
	NR7	−110	7,9	23	191	81	0,1	0,04	200*	7300*	4492
B		% [w/w]									
	Sample	TIC	TOC	TS	TN						
	NR7	2.49	0.43	0.19	0.04						

* from: Martinelli and Ferrari (1991)

shows the presence of mature organic matter in the underlying sediments of the mud volcanoes at the Salse di Nirano. A similar predominance of *n*-alkanes was also found in mud from Romanian mud volcanoes (Alain et al. 2006). Furthermore, fresh allochthonous organic matter, mainly from higher plants, was also identified in the polar fraction and may be transported into the mud via wind and/or by the flow of the mud through the underlying palaeosoils. These components mainly include even-chained wax ester-derived homologues of long-chain fatty acids and long chain alcohols, both maximizing at C_{26} ($\delta^{13}C$ about −30 to −33‰) and suite of plant derived sterols (e.g. Stigmastanol, Fig. 3).

A low redox potential of −110 mV, a pH-value of 7.9 and a temperature of about 23°C was measured for the mud pool analysed for this study (Table 1). Elemental analyses of the particle-free fluid water revealed a concentration of 191 ppm Mg^{2+}, 81 ppm Ca^{2+} and 4492 ppm Na^+. These data evidence that the waters are brackish and similar to spring and groundwaters from the same area (Conti et al 2000). The low redox potential, the pH-value, the temperature of about 23°C and the presence of various electron acceptors (e.g. SO_4^{2-} from underlying gypsum layers) allow microbial sulphate reduction and establishment of diverse sulphate reducing bacteria (SRB) as well as the growth of other microorganisms.

3.2 *Microbiology: Enrichment Cultures*

Conditions for enrichment of organisms aimed at the isolation of sulphate-reducing bacteria. The samples were taken from two mud volcanoes. In the original mud volcano fluids, occasionally, large vibroid cells could be observed. From the enrichment cultures, four different pure cultures of anaerobic bacteria were obtained. 16S rDNA analysis revealed 99 – 100 % identity of the isolates to already described organisms: The isolates include the sulphate-reducing Gram-negative *Desulfovibrio psychrotolerans*, which was, so far, only known from a salt water lake situated in the Himalaya (Jyothsna et al. 2008). It may be due to the presence of yeast extract in the enrichment media that also (facultative) fermenting organisms were isolated. In the mud volcano fluid, the *Clostridium thiosulfatireducens*-isolate may ferment proteins from other bacterial sources. The organisms are depicted in figure 2. The flagellated *Clostridium* cell exhibits a terminal spore (Fig. 2 A),

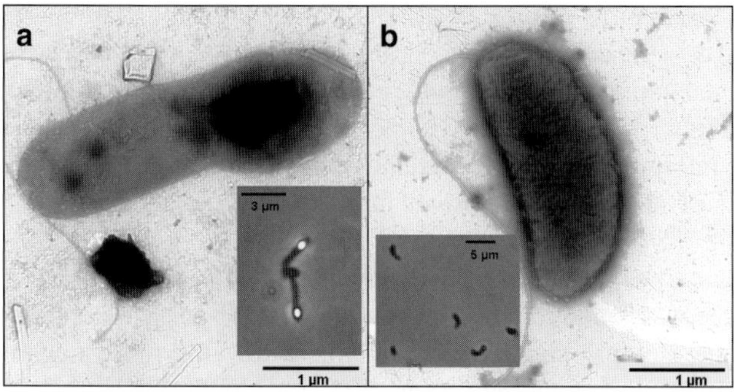

Fig. 2 Electron micrographs of the isolates *Clostridium thiosulfatireducens* (a) and *Desulfovibrio psychrotolerans* (b)-strains after negative staining. The insets show light micrographs of the respective strains

Desulfovibrio has a vibroid cell shape and shows one polar flagellum (Fig. 2 B). The occurrence of two bacteria from mammalian intestines (*Enterobacter aerogenes* and *Enterococcus faecalis*) may be explained by contamination of the setting with ground- or surface water from pastures situated around the sampling site. From the enrichment cultures and isolates as well as the outgasing of H_2S at the sampling site it is obvious that sulphate reduction in the fluids is an important process. Archaea have, so far, not been isolated from the mud.

3.3 Lipid Biomarkers

Eukaryotes, bacteria, and archaea are characterised by partially very specific lipid membrane components, so called lipid biomarkers. Moreover, ratios of stable carbon isotopes in these biomarkers contain information on the carbon fixation pathway used, the biosynthesis of lipids but especially on the carbon substrate used (Hayes 2001). In particular, processes where ^{13}C-depleted substrates are used by microorganisms can be easily identified by compound specific stable isotope analyses even in complex and heterogeneous microbial settings. An excellent example is the anaerobic oxidation of methane (AOM), where the usually isotopically light methane carbon is transferred into the lipids of the closely operating AOM performing consortia of sulphate-reducing bacteria and methane oxidizing archaea (e.g., Blumenberg et al. 2004; Hinrichs et al. 1999; Pancost et al. 2001b).

In addition to the above mentioned allochthonous biosignatures, we also found biomarkers with slightly ^{13}C-depleted stable isotope ratios suggesting microbial sources which fed on ^{13}C-depleted carbon substrates. These structures are several dialkyl glycerol diethers (DAGE) with the highest depletion of −51.7‰ (campaign 2008) found for a DAGE with two hexadecane chains (16/16-DAGE). Also depleted are *ai*15/*i*15-DAGE (−41.4‰) and the *ai*15/*ai*15-DAGE (−50.1‰) with the first suggesting contributions of sources with partially conventional $\delta^{13}C$ values. Respective DAGE have so far mainly described from thermophilic bacteria like *Aquifex pyrophilus* (Huber et al. 1992) and the sulphate-reducing bacterium *Thermodesulfobacterium* commune (Langworthy et al. 1983). However, at many marine, normal temperated sites where the AOM is the predominant microbial process, strongly ^{13}C-depleted DAGE occur in high amounts. Consequently, these DAGE were interpreted as been sourced by AOM-involved SRB (Blumenberg et al. 2004; Elvert et al. 2005; Pancost et al. 2001a). We therefore interpret a high proportion of *i*15/*ai*15-, *ai*15/*ai*15- and 16/16-DAGE present in the mud of the Salse di Nirano to be produced by SRB involved in AOM. However, since the $\delta^{13}C$ of the methane was found to be −50‰ and that AOM-performing microbes further fractionate the isotopic signal of the methane towards more negative values a high proportion of the tentatively AOM SRB-derived DAGE must be also sourced by additional SRB or other bacteria. For instance, glycerol ether lipids with *iso*-branched pentadecane chains were recently also found in the spore-forming and widespread soil bacterium Myxobacterium *Myxococcus xanthus* (Ring et al. 2006) which might be also partial sources of *i*15/*i*15-DAGE in the mud, explaining the mixed $\delta^{13}C$ signal found in the mud.

The majority of archaeal biomarkers present in the mud are only slightly or not ^{13}C-depleted (particularly archaeol from the 2008 campaign −36 and from the 2009 campaign −29.3‰ although in the latter sample co-elution with a C38:2 ethyl alkenone was observed). Isoprenoid hydrocarbons such as crocetane and

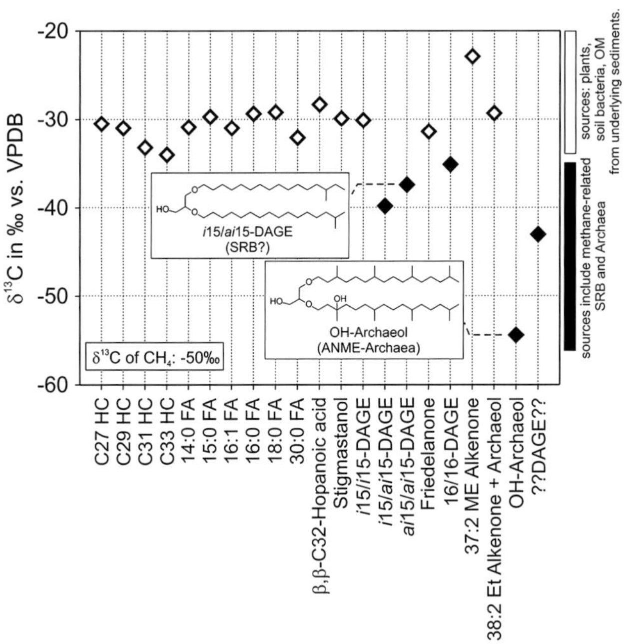

Fig. 3 Stable carbon isotopes of selected lipid biomarkers extracted from mud samples from the 2009 campaign and the potential sources (exemplified two structures). Note that the amount of OH-archaeol was very low

2,6,10,15,19-pentamethylicosane, all often present at AOM-sites (e.g. Michaelis et al. 2002; Thiel et al. 1999), are not present or, more likely, are superimposed by other co-eluting components. Nevertheless, the vast majority of archaea present in the mud is obviously neither performing the anaerobic oxidation of methane nor feeding on any other ^{13}C-depleted carbon source. Nevertheless, the presence of trace amounts of hydroxyarchaeol (OH-archaeol) with a δ^{13}C value of −51‰ demonstrates that archaea are also present which are actively involved in the anaerobic turnover of methane.

We also observed differences in the distribution and stable isotope signatures of biomarkers tentatively sourced by microorganisms performing the anaerobic oxidation of methane, between both sampling campaigns. Higher concentrations, accompanied by stronger ^{13}C-depletions were found in the sample from the 2008 campaign (strongest ^{13}C-depletion −51.4‰ for 16/16-DAGE) whereas the same compound in the mud sample from the 2009 campaign revealed a δ^{13}C value of −41‰ (Fig. 3). This demonstrates that the proportion of AOM-involved bacteria and archaea was different at both sampling times, suggesting the intensity of AOM to vary strongly with respect to time and/or space.

4 Conclusions

Terrestrial mud volcanoes of the Salse di Nirano were found to be excellent settings for divers microorganisms. The bacterial lipid biomarkers and the enrichment cultures support the presence of sulphate reducing bacteria in the mud volcanoes. Furthermore, lipid biomarkers demonstrate that methanotrophic archaea are present in the system of the Salse di Nirano mud volcanoes suggesting that AOM is taking place. The slightly ^{13}C-depleted stable isotope ratios of the bacterial biomarkers indicates that these microorganism feed on ^{13}C depleted carbon sources. However, the low content of the archaeal biomarkers in the samples like archaeol and hydroxyarchaeol and the fact that these biomarkers are only slightly or not depleted in ^{13}C shows that AOM in this fluid venting structures it is not very important *in situ*. Our data indicate that the majority of the yet unknown archaea and bacteria present in the mud are neither involved in AOM nor feeds on any other ^{13}C-depleted carbon source.

Acknowledgments We are grateful to the authorities of the Salse di Nirano Natural Reserve for granting permit to carry out field research and to the Guardie Ecologiche for their support, especially Augusta and Luciano Callegari. We thank the Competence Centre for Stable Isotopes of the University of Goettingen and Klaus Simon (Department of Geochemistry, Geoscience Centre of the University of Goettingen) for the analyses of our samples, This study received financial support by Deutsche Forschungsgemeinschaft (DFG grants Re 665/31-1, Ho 1830/2-1, Bl 971/1-1) and ISMAR-CNR Bologna scientific contribution n. 1641. Thanks are also due to Jens Dyckmans, Lars Swecz, and Reinhard Langel who carried out the stable isotope analyses.

References

Accaino F, Bratus A, Conti S, Fontana D, Tinivella U (2007) Fluid seepage in mud volcanoes of the northern Apennines: An integrated geophysical and geological study. Journal of Applied Geophysics 63:90–101

Alain K, Holler T, Musat F, Elvert M, Treude T, Krüger M (2006) Microbiological investigation of methane- and hydrocarbon-discharging mud volcanoes in the Carpathian Mountains, Romania. Environmental Microbiology 8:574–590

Baker GC, Smith JJ, Cowan DA (2003) Review and re-analysis of domain-specific 16S primers. Journal of Microbiological Methods 55:541–555

Benedetti LC, Tapponnier P, Gaudemer Y, Manighetti I, Van der Woerd J (2003) Geomorphic evidence for an emergent active thrust along the edge of the Po Plain: The Broni-Stradella fault. Journal of Geophysical Research 108:2238

Blumenberg M, Seifert R, Reitner J, Pape T, Michaelis W (2004) Membrane lipid patterns typify distinct anaerobic methanotrophic consortia. Proceedings of the National Academy of Sciences of the USA 101:11111–11116

Boetius A, Ravenschlag K, Schubert CJ, Rickert D, Widdel F, Gieseke A, Amann R, Jørgensen BB, Witte U, Pfannkuche O (2000) A marine microbial consortium apparently mediating anaerobic oxidation of methane. Nature 407:623–626

Capozzi R, Menato V, Rabbi E (1994) Manifestazioni superficiali di fluidi ed evoluzione tettonica recente del margine Appenninico Emiliano-Romagnolo: indagine preliminare. Atti Ticinensi Scienze Terra 1:247–254

Capozzi R, Picotti V (2002) Fluid migration and origin of a mud volcano in the Northern Apennines (Italy): The role of deeply rooted normal faults. Terra Nova 14:363–370

Castaldini D (2008) Maps and multimedia tool for the environmental tourism in protected areas of the Modena Apennines (northern Italy). GeoJournal of Tourism and Geosites 1(1):13–33

Conti A, Sacchi E, Chiarle M, Martinelli G, Zuppi GM (2000) Geochemistry of the formation waters in the Po Plain (Northern Italy): an overview. Applied Geochemistry 15:51–65

Dimitrov LI (2002) Mud volcanoes – the most important pathway for degassing deeply buried sediments. Earth Science Reviews 59:49–76

Elvert M, Boetius A, Knittel K, Jørgensen BB (2003) Characterization of specific membrane fatty acids as chemotaxonomic markers for sulphate-reducing bacteria involved in anaerobic oxidation of methane. Geomicrobiology Journal 20:403–419

Elvert M, Hopmans EC, Treude T, Boetius A, Suess E (2005) Spatial variations of methanotrophic consortia at cold methane seeps: implications from a high-resolution molecular and isotopic approach. Geobiology 3:195–209

Etiope G (2004) New directions: GEM – Geologic Emissions of Methane, the missing source in the atmospheric methane budget. Atmospheric Environment 38:3099

Etiope G, Ciccioli P (2009) Earth´s Degassing: A missing ethane and propane source. Science 323:478

Evans JB, Harrell LJ (1977) Agar shake tube technique for simultaneous determination of aerobic and anaerobic susceptibility to antibiotics. Antimicrobial Agents and Chemotherapy 12:534–536

Hallam SJ, Girguis PR, Preston CM, Richardson PM, DeLong EF (2003) Identification of methyl coenzyme M reductase A (mcrA) genes associated with methane-oxidizing archaea. Applied and Environmental Microbiology 69:5483–5491

Hallam SJ, Putnam CM, Preston JC, Detter D, Rokhsar PM, Richardson PM, DeLong EF (2004) Reverse methanogenesis: testing the hypothesis with environmental genomics. Science 305:1457–1462

Hayes JM (2001) Fractionation of the isotopes of carbon and hydrogen in biosynthetic processes. In: Valley JW, Cole DR (eds) Stable isotope geochemistry. Mineralogical Society of America, Washington D.C. Reviews in Mineralogy and Geochemistry. 43:225–278

Hinrichs K-U, Hayes JM, Sylva SP, Brewer PG, DeLong EF (1999) Methane-consuming archaebacteria in marine sediments. Nature 398:802–805

Hoehler TM, Alperin MJ, Albert DB, Martens CS (1994) Field and laboratory studies of methane oxidation in an anoxic marine sediment: Evidence for a methanogen-sulfate reducer consortium. Global Biogeochemical Cycles 8:451–464

Hoppert M (2003) Microscopic Techniques in Biotechnology. Wiley-VCH, Weinheim

Huber R, Wilharm T, Huber D, Trincone A, Burggraf S, König H, Rachel R, Rockinger I, Fricke H, Stetter KO (1992) Aquifex pyrophilus gen. nov. sp. nov., represents a novel group of marine hyperthermophilic hydrogen-oxidizing bacteria. Systematic and Applied Microbiology 15: 340–351

Jyothsna TSS, Sasikala C, Ramana CV (2008) Desulfovibrio psychrotolerans, sp. nov., a psychrotolerant and moderately alkaliphilic sulfate-reducing deltaproteobacterium from the Himalayas. International Journal of Systematic and Evolutionary Microbiology 58:821–825

Kopf AJ (2002) Significance of mud volcanism. Reviews of Geophysics 40:2.1–2.52

Krüger M, Meyerdierks A, Glöckner FO, Amann R, Widdel F, Kube M, Reinhardt R, Kahnt J, Böcher R, Thauer RK, Shima S (2003) A conspicuous nickel protein in microbial mats that oxidize methane anaerobically. Nature 426:878–881

Lane DJ (1991) 16S/23S rDNA sequencing. In: Stackebrandt E and Goodfellow M (eds) Nucleic Acid Techniques in Bacterial Systematics. Wiley, Chichester, pp 115–175

Langworthy TA, Holzer G, Zeikus JG, Tornabene TG (1983) Iso- and anteiso-branched glycerol diethers of the thermophilic anaerobe Thermodesulfotobacterium commune. Systematic and Applied Microbiology 4:1–17

Loftfield N, Flessa H, Augustin J, Beese F (1997) Automated gas chromatographic system for rapid analysis of the atmospheric trace gases methane, carbon dioxide, and nitrous oxide. Journal of Environmental Quality 26:560–564

Martinelli G, Ferrari G (1991) Earthquake forerunners in a selected area of Northern Italy: recent developments in automatic geochemical monitoring. Tectonophysics, 193, 397–410.

Martinelli G, Rabbi E (1998) The Nirano mud volcanoes. In: Curzi PV, Judd AG (eds) Abstracts and Guide Book, Vth International Conference on Gas in Marine Sediments, Bologna, Italy; September 1998. Grafiche A & B, Bologna, pp 202-206.

Martinelli G, Judd A (2004) Mud volcanoes of Italy. Geological Journal 39:49–61

Mazurenko LL, Soloviev VA, Gardner JM, Ivanov MK (2003) Gas hydrates in the Ginsburg and Yuma mud volcano sediments (Moroccan Margin): results of chemical and isotopic studies of pore water. Marine Geology 195:201–210

Michaelis W, Seifert R, Nauhaus K, Treude T, Thiel V, Blumenberg M, Knittel K, Gieseke A, Peterknecht K, Pape T, Boetius A, Amann R, Jørgensen BB, Widdel F, Peckmann J, Pimenov NV, Gulin MB (2002) Microbial reefs in the Black Sea fueled by anaerobic oxidation of methane. Science 297:1013–1015

OrphanVJ, House CH, Hinrichs K-U, McKeegan K, DeLong EF (2001a) Methane-consuming archaea revealed by directly coupled isotopic and phylogenetic analysis. Science 293, 484–487.

Orphan VJ, Hinrichs K-U, Ussler W, III, Paull CK, Taylor LT, Sylva SP, Hayes JM, DeLong FM (2001b) Comparative analysis of methane-oxidizing archaea and sulphate reducing bacteria in anoxic marine sediments. Applied and Environmental Microbiology 67:22–1934

OrphanVJ, House CH, Hinrichs K-U, McKeegan KD, DeLong EF (2002) Multiple archaeal groups mediate methane oxidation in anoxic cold seep sediments. Proceedings of the National Academy of Sciences 99:7663–7668

Pancost RD, Bouloubassi I, Aloisi G, Sinninghe Damsté JS, the Medinaut Shipboard Scientific Party (2001a) Three series of non-isoprenoidal dialkyl glycerol diethers in cold-seep carbonate crusts. Organic Geochemistry 32:695–707

Pancost RD, Hopmans EC, Sinninghe Damsté JS, The Medinaut Shipboard Scientific Party (2001b) Archaeal lipids in Mediterranean cold seeps: Molecular proxies for anaerobic methane oxidation. Geochimica et Cosmochimica Acta 65:1611–1627

Postgate JR (1951) On the nutrition of Desulfovibrio desulfuricans. Journal of General Microbiology 5:714–724

Reeburgh WS (1980) Anaerobic oxidation of methane; rate depth distributions in Skan Bay sediments. Earth and Planetary Science Letters 47:345–352

Reitner J, Peckmann J, Blumenberg M, Michaelis W, Reimer A, Thiel V (2005a) Concretionary methane-seep carbonates and associated microbial communities in Black Sea sediments. Palaeogeography, Palaeoclimatology, Palaeoecology 227:18–30

Reitner J, Peckmann J, Reimer A, Schumann G, Thiel V (2005b) Methane derived carbonate build-ups and associated microbial communities at cold seeps on the lower Crimean shelf (Black Sea). Facies 51:66–79

Ring MW, Schwar G, Thiel V, Dickschat JS, Kroppenstedt RM, Schulz S, Bode HB (2006) Novel iso-branched ether lipids as specific markers of developmental sporulation in the myxobacterium *Myxococcus xanthus*. Journal of Biological Chemistry 281:36691–36700

Sakaguchi T, Arakaki A, Matsunaga T (2002) Desulfovibrio magneticus sp. nov., a novel sulfate-reducing bacterium that produces intracellular single-domain-sized magnetite particles. International Journal of Systematic and Evolutionary Microbiology 52:215–221

Scholte KH (2005) Hyperspectral Remote Sensing and Mud Volcanism in Azerbaijan. PrintPartners Ipskamp B.V., The Netherlands

Thiel V, Peckmann J, Seifert R, Wehrung P, Reitner J, Michaelis W (1999) Highly isotopically depleted isoprenoids: Molecular markers for ancient methane venting. Geochimica et Cosmochimica Acta 63:3959–3966

Thompson M, Walsh J-N (1983) A Handbook of Inductively Coupled Plasma Spectrometry. – (Chapman & Hall, New York, NY, United States; Blackie & Son, Glasgow, United Kingdom

Treude T, Knittel K, Blumenberg M, Seifert R, Boetius A (2005) Subsurface microbial methanotrophic mats in the Black Sea. Applied and Environmental Microbiology 71:6375–6378

Whiticar MJ (1999) Carbon and hydrogen isotope systematics of bacterial formation and oxidation of methane. Chemical Geology 161:291–314

Trace Element and Biomarker Signatures in Iron-Precipitating Microbial Mats from the Tunnel of Äspö (Sweden)

Jens Kurz, Klaus Simon, Christine Heim, Joachim Reitner, Nadia -Valérie Quéric, and Volker Thiel

1 Introduction

The Äspö Hard Rock Laboratory (Äspö HRL) is a tunnel located near Oskarshamn in the southeast of Sweden, that serves as a testing environment for the disposal of nuclear waste. The Äspö HRL hosts and makes accessible a wide spectrum of microbially driven subsurface ecosystems (Pedersen 1997). 1,127 m from the tunnel entrance and 160 m below the Baltic Sea floor, a natural pond has established beneath a fluid spring from the diorite rock (Rönnback 2005). The pond contains microbial mats of several cm thickness, that can be separated into ochre-coloured surface mats and underlying mats showing an obvious pattern of dark, mm-thick and more or less horizontal streaks. Spatially separated below these mats, a third type comprising dense black mats occur. Previous studies showed that ochre mats, which abundantly occur in the Äspö HRL, are dominated by the iron-oxidizing bacterium *Gallionella ferruginea* (Ferris et al. 1999; Anderson and Pedersen 2003; Hallberg and Ferris 2004; Anderson et al. 2006). Based on these earlier studies, emphasis was placed here on the geochemical investigation of the pond habitat with its variety of different mat types. Specifically, trace element patterns were explored for inorganic biosignatures, arising from different biogeochemical pathways driving the distinct microbial systems. The lipid biomarker composition of these mats was investigated and linked to the trace element distributions. Additionally,

J. Kurz
Max Planck Institute for Biophysical Chemistry, Am Faßberg 11, 37077, Göttingen, Germany

K. Simon
Geoscience Centre, University of Göttingen, Geochemistry Division, Goldschmidtstraße 1, 37077, Göttingen, Germany

C. Heim, J. Reitner, N.-V. Quéric, and V. Thiel (✉)
Geoscience Centre, University of Göttingen, Geobiology Division, Goldschmidtstraße 3, 37077, Göttingen, Germany
e-mail: vthiel@gwdg.de

J. Reitner et al., *Advances in Stromatolite Geobiology*, 221
Lecture Notes in Earth Sciences 131, DOI 10.1007/978-3-642-10415-2_15,

the involved biominerals were examined to receive a better understanding of the microbial mat structures and explore inorganic patterns that may reveal potential biosignatures for these systems in ancient environments.

2 Material and Methods

2.1 Sample Collection

Samples were taken in April 2008 at the sample location 1127B Pond, which is situated 1,127 m from the tunnel entrance and 160 m below the Baltic Sea floor. 32 syringes were pressed into the microbial layers and taken out with the core inside after one hour. The cores were immediately frozen with liquid nitrogen and stored at −80°C until analysis.

2.2 Sample Preparation

Ten representative cores were chosen. Three cores completely consisted of ochre mats; two were black with a minor admixture of ochre mats only at the top layer, while the other five were ochre with dark streaks in between. The latter are referred to as streak mats. The cores were partitioned with a scalpel by their colour zonation in (1) ochre, (2) streak and (3) black segments. In total, 26 segments were used for trace element and 18 segments for biomarker analysis. The samples were lyophilized overnight. For trace element analysis 50 mg of the dry powder was dissolved in 1 mL of concentrated HNO_3 and 100 μL of an internal standard were added before the sample was oxidized with 5 mL of 30% H_2O_2. 45 mL of ultrapure water were added, leading to a dilution of 1:1,000. Lipid biomarkers were extracted from ca. 100 mg of the lyophilized sample with 5 mL of dichlormethane/methanol (1:1; v:v), 5 mL of dichlormethane/methanol (3:1; v:v) and 5 mL of dichlormethane respectively (ultrasonication, 20 min, 40°C). Each extraction was followed by a centrifugation step (15 min). The combined extracts were dried under a stream of nitrogen and re-dissolved in 500 μL of dichlormethane. Aliquots were (1) derivatized with N,O-Bis(trimethylsilyl)trifluoroacetamide (BSTFA) (80°C, 1.5 h) to produce trimethylsilane (TMS-)derivatives of alcohols and free fatty acids. A second aliquot was transesterified with trimethylchlorosilane/methanol (1:10; v:v; 80°C, 1.5 h) to produce methyl ester derivatives of bound fatty acids. The reaction mixture was extracted three times with *n*-hexane. The fatty acid methyl ester (FAME) fraction was purified using a glass column (ID 12 mm) filled with 0.6 g silica gel (Merck Kieselgel 60). The extract was re-dissolved in 7 mL *n*-hexane and placed on the column. The FAME fraction was eluted with 7 mL of dichlormethane, dried under a gentle stream of nitrogen, and re-dissolved in 20 μL of *n*-hexane for biomarker analysis (Sect. 2.4).

2.3 Trace Element Analysis

A PerkinElmer ELAN DRC II ICP-MS was used for the analysis of Li, Ti, Mn, Co, Ni, Cu, Zn, Zr, Nb, Mo, Sn, Rb, Sr, Cs, W, Pb, Th, U and REE. The RF power was 1,250 W; the MEINHARD nebulizer had a flow rate of 0.9 L/min, the auxiliary gas flow was 1 L/min and the cooling gas flow 14.5 L/min. For Al, Ca, Fe, Mg, Na and P, a PerkinElmer Optima 3,300 DV ICP-OES was used. For the REE plot (Fig. 2) the data was normalized on the Continental Earth's Crust (Wedepohl 1995).

2.4 Biomarker Analysis

The coupled gas chromatography and mass spectrometry (GC/MS) system used was a Varian 3800 GC interfaced to a Varian 1200 Quadrupole MS. The GC was equipped with a fused silica capillary column (Phenomenex ZB-1, 30 m 0.25 mm ID, 0.1 μm film thickness). Helium was used as carrier gas at a flow rate of 1 mL/min, and 1 μL sample was injected manually. The injector oven was programmed as follows: 70°C isothermal for 1 min, from 70°C to 320°C at 150°C/min, and 10 min isothermal. The GC-oven was programmed as follows: 80°C isothermal for 1 min, from 80° to 310°C at 10°C/min, and 30 min isothermal. Mass spectra were acquired in electron impact mode (70 eV) with an ion source temperature of 180°C, a scan range of *m/z* 50 to *m/z* 500 and a scan time of 0.75 s. The identification of the compounds was based on comparison of the mass spectra and of GC retention times with those of published data and of reference compounds (Supelco 37 FAME standard).

2.5 Factor Analysis

The software Statistica 6.0 (Statsoft Inc., Tulsa, U.S.A.) was applied to correlate the trace element analyses. The trace element data (concentrations in ppb) of 51 isotopes for 26 ochre and streak mats, six black mats and one mixed (A8-23-1) sample were taken into account for this analysis.

2.6 Structure Analysis

For X-ray diffraction (XRD) analyses, a Philips PW 1877 instrument was used with a PW 1729 X-ray generator at wave lengths of $CuK_{\alpha 1} = 1.5406$ Å and $CuK_{\alpha 2} = 1.54442$ Å. Raman spectroscopy was undertaken with a Melles Griot Argon Ion Laser IMA 106020B0S at wave length 488 nm, an Olympus BX41 microscope, a HORIBA Jobin Yvon LabRAM HR 800 UV spectrometer with 800 nm focal length, and a HORIBA Jobin Yvon CCD Symphony LN_2-Series detector. For electron microscopy and energy-dispersive X-ray spectroscopy (EDS) analysis, a FEI Quanta 200

FEG Environmental scanning electron microscope (SEM) was applied in the low vacuum mode.

3 Results

3.1 Structure of the Microbial Mats

Combined SEM/EDS analyses of the ochre and streak mats showed a dense network of *Gallionella* stalks (Fig. 1) that made it difficult to distinguish further structures present. EDS analyses revealed that O, Fe and C had the largest abundance, indicating that iron-oxides, strongly intertwined with organic material, are the main components. The presence of iron-oxides as the main inorganic component was further corroborated by XRD analyses of dry powder which revealed amorphous iron-oxides as the predominant mineral phase for the ochre and streak mats.

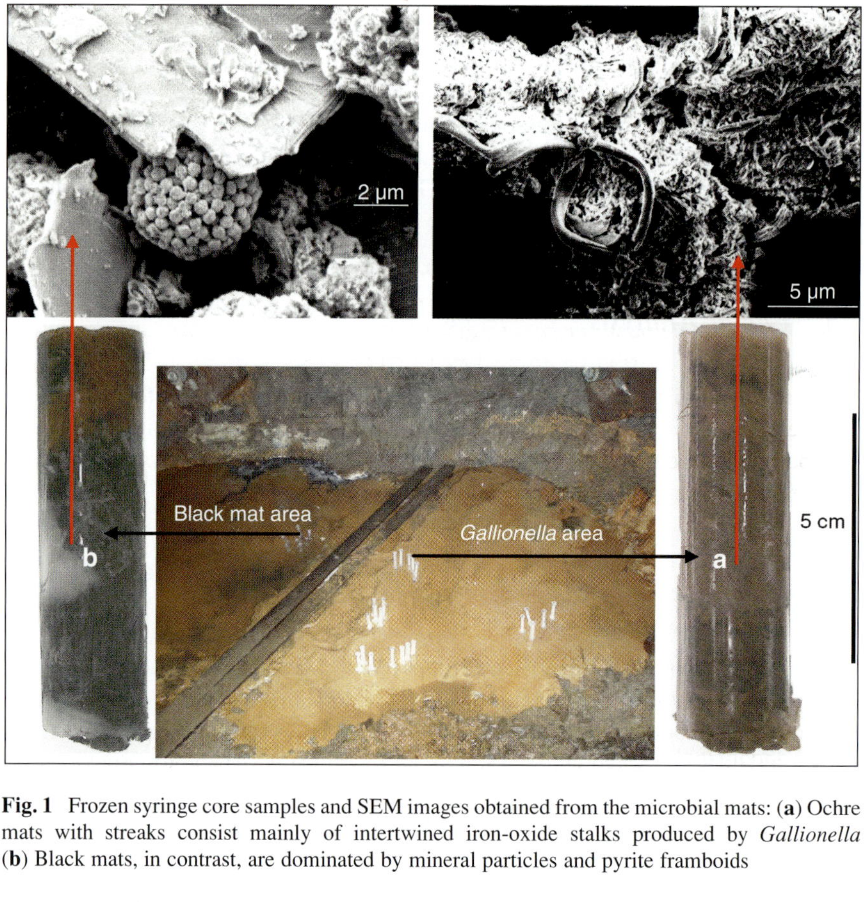

Fig. 1 Frozen syringe core samples and SEM images obtained from the microbial mats: (**a**) Ochre mats with streaks consist mainly of intertwined iron-oxide stalks produced by *Gallionella* (**b**) Black mats, in contrast, are dominated by mineral particles and pyrite framboids

Raman spectroscopy revealed hematite (Fe_2O_3) and magnetite (Fe_3O_4) as major iron-oxide phases. No differences were visible between the ochre and the streak mats, hence in the following they are summarized under the term "*Gallionella* mat". The black mats contain abundant diorite particles, which are covered by bacteria and associated with iron-sulfides. Abundant ca. 5 μm sized sphere-like structures (Fig. 1) were observed with SEM/EDS. The spheres show S and Fe as the main components in a ratio of 2:1, and were interpreted as pyrite framboids. The XRD spectrum of the black mat shows predominant diorite-derived signals for quartz, albite and calcite, whereas weak peaks characteristic for pyrite were tentatively identified. Using Raman spectroscopy, only the diorite minerals were identified in the black mat. Testing of the ochre, the streak and the black dry mat powder with a bar magnet revealed magnetism for all samples.

3.2 *Trace Elements*

The microbial mats are massively enriched in trace elements relative to the continental earth crust. The element distribution patterns were almost similar for the streak and ochre mats, however, Mn, Mo, W and U showing somewhat higher abundance in the further, and P, Ti, and Zn in the latter (Fig. 2). A very different element distribution was observed for the black mats. Compared to the ochre and streak mats, the black mats are lower in Na, P, Mn, Fe, Sr, W and heavy Rare Earth Elements (HREE), but significantly higher in Mg, Ca, Ti, Co, Ni, Cu, Zn, Rb, Cs, Pb and light Rare Earth Elements. The ochre and streak mats reveal, in addition to a strong enrichment of HREE, a marked W-tetrad effect. Thus, although the feeder fluid is the same, each mat type displayed a distinctive trace element signature. Factor analysis of all 51 measured elements unequivocally confirmed the finding of nearly identical element patterns in the ochre and streak mats and a distinctively different element distribution in the black mat (Fig. 3).

3.3 *Biomarkers*

Both the ochre and streak mats contained straight-chain saturated and monounsaturated C_{16} and C_{18} *n*-fatty acids and low concentrations of saturated C_{20}, C_{22} and C_{24} fatty acids. In addition, hopanoids were identified, namely hop-17(21)-ene, diplopterol, and 17β(H),21β(H)-bishomohopan-32-ol. The black mat samples revealed abundant methylated fatty acids, namely terminally branched iso (*i*-) and anteiso (*ai*-) C_{15}-, C_{17}-, and 10-methyl-C_{16} homologues (Fig. 4). In addition to C_{14}–C_{24} *n*-fatty acids, the black mat furthermore showed low but significant concentrations of non-isoprenoid dialkylglycerol diethers (DAGE). As for fatty acids, terminally branched (*i*- and *ai*-) C_{15} and C_{17} moieties comprise abundant alkyl groups in these compounds. Hopanoids were present in much lower concentrations as in the ochre and streak mat samples. Bishomohopanol was not observed in the black mats.

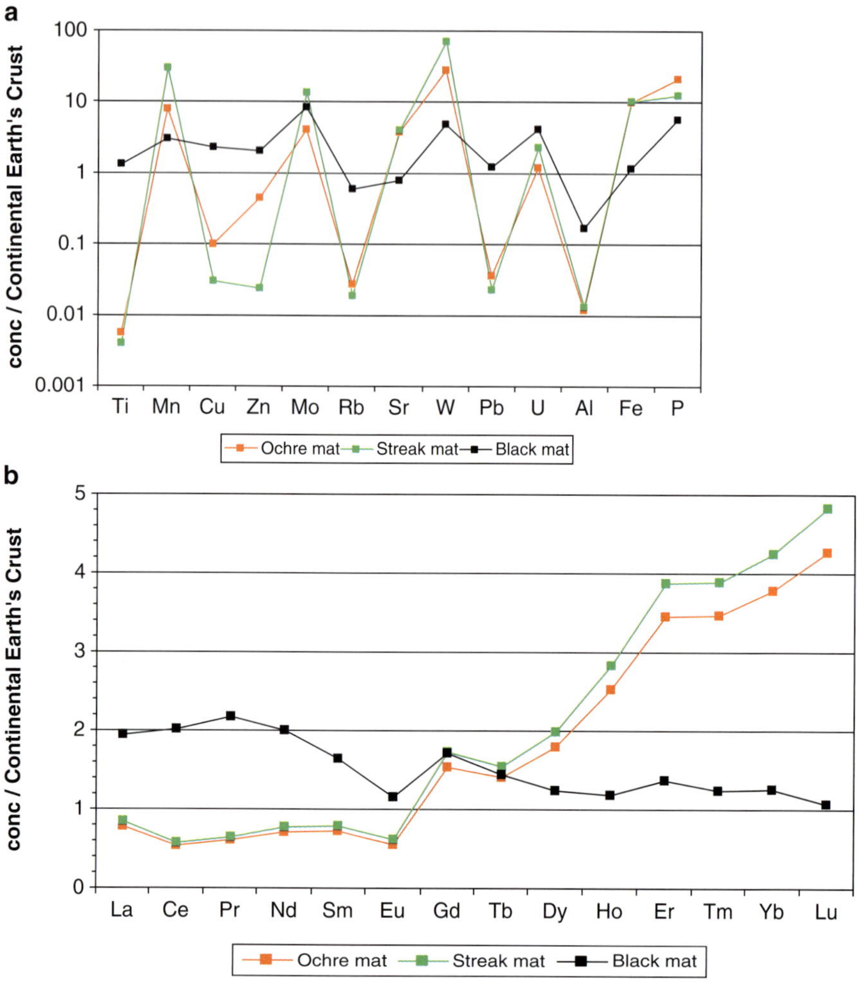

Fig. 2 ICP-MS and ICP-OES trace element analysis based on the averages of 10 ochre, 13 streak and 3 black samples. (**a**) Multielement analysis (**b**) REE plot normalized on the Continental Earth's Crust (Wedepohl 1995). A W-tetrad effect is visible for the ochre and streak mats

4 Discussion

4.1 Structure of the Microbial Mats

Fe is the dominant element in the ochre/streak *Gallionella* mats (around 40%) and occurs mainly as amorphous iron-oxides, as revealed by XRD and Raman spectroscopy (see also Ferris et al. 1999). The stoichiometric formula of these oxides has been calculated by Hallberg and Ferris (2004) as Fe_2O_3, thus equalling hematite.

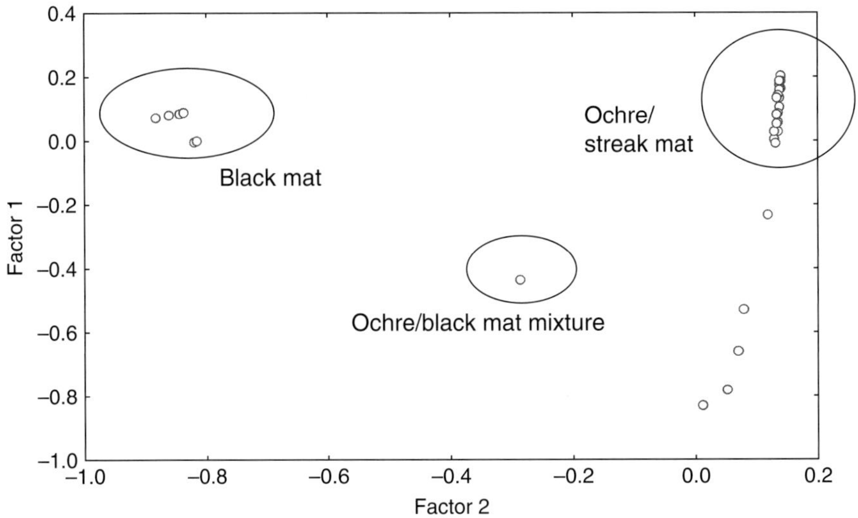

Fig. 3 Factor analysis (Statistica 6.0) reveals the different trace element accumulations of the ochre/streak and the black mat samples

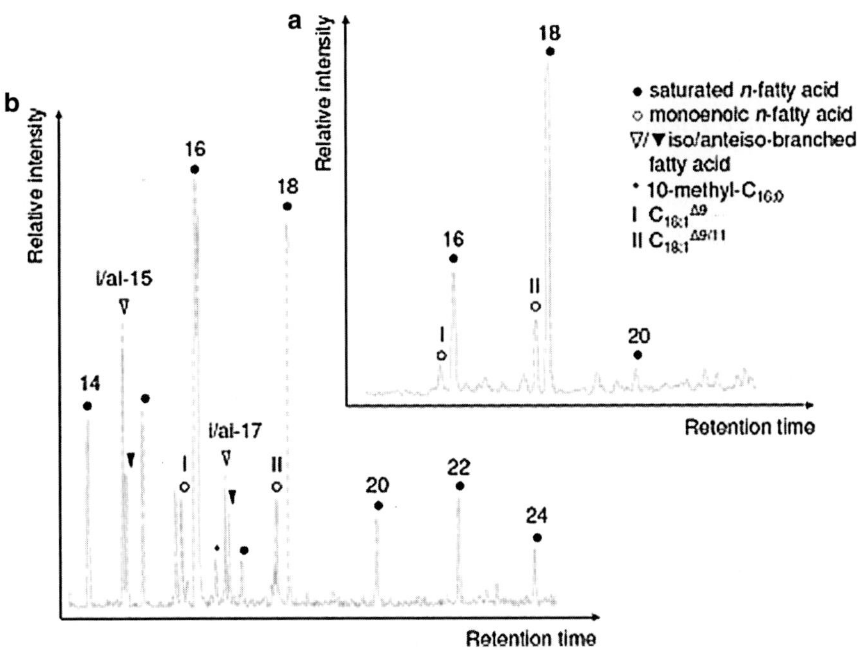

Fig. 4 Partial gas chromatograms (*m/z* 74) showing representative fatty acid (FAME) patterns from the ochre/streak mat (**a**) and the black mat (**b**). High abundances of methyl-branched C_{15}, C_{16} and C_{17} fatty acids are interpreted as biosignatures from SRB

Indeed, hematite was identified with Raman spectroscopy, together with magnetite (Fe_3O_4). The presence of magnetite plausibly explains the observed magnetism of the mat samples. Fe concentration has shown to be similar in both mats, whereas the Mn content was slightly higher in the streak mats compared to Fe. These high Mn contents suggest that the dark colour of the streak mats may result from Mn-oxide precipitates, namely MnO_2. The slightly different accumulation behaviour of the streak mats compared to the ochre mats may reflect minor differences in the bacterial consortia, e.g. the local occurrence of manganese-oxidizing bacteria like *Leptothrix sp.*, that have previously been observed in the Äspö tunnel (Hallbeck et al. 2006). The black mat contains much more diorite-derived minerals like quartz, albite and apatite than the ochre/streak *Gallionella* mats as revealed by SEM, XRD and Raman spectroscopy. Whereas the microbial cells seem to preferentially settle on μm-sized mineral particles in the further, the stalks of *Gallionella*, consisting of extracellular polymeric substances (EPS), may provide a stabilizing matrix for microbial settlement in the latter. Pyrite framboids are most likely linked to the metabolism of sulphate-reducing bacteria (SRB), and may therefore be used as biomarkers (Popa et al. 2004) for sulfate reduction. Some SRB are known to precipitate the magnetic mineral greigite (Fe_3S_4) inside of the cell as a magnetosome (Pósfai et al. 1998; Reitner et al. 2005). After cell death the magnetic minerals may cluster to form a sphere and later recrystallize to framboidal pyrite. Although not directly observed by the spectral methods used, we speculate that greigite in the black mat accounts for magnetism of the black mat sample, rather than magnetite.

4.2 Trace Elements

The question emerges how the different trace element patterns observed for the ochre/streak *Gallionella* mats and for the black mat, can be interpreted. There are two concepts explaining the accumulation of trace elements by bacteria, specifically (1) binding to their exopolymeric substances and (2) adsorption and absorption onto and into the biominerals they produce. Sorption effects of *Gallionella* EPS stalks likely account for the capturing of ferrous iron necessary for the microbial metabolism, i.e. dissimilatory iron oxidation (Hanert 2002). At the same time, the extracellular stalks may account for a most effective accumulation of trace elements. Anderson and Pedersen (2003) pointed out that heavy metals like REE and U compete with iron for the adsorption on the stalk material. Therefore, while steadily growing, *Gallionella* has to produce even more stalks to secure the iron supply for its metabolism.

Minerals produced or induced by SRB, namely iron-sulfides observed in the black mat are known to have excellent adsorption capacities e.g. for Co, Ni, Cu, Zn, Cd and Pb (Watson et al. 1995). Other studies showed Cu, As, Mn and Mo as preferably adsorbed (Neumann et al. 2006) and Mo to be preferred over Cr and Ni (Beech and Cheung 1995). These observations could be confirmed, except for Cd and As, which were below the detection limits in all samples studied. The black mat's trace element distribution pattern is also enriched in Ti, Rb, Pb and U. However, it should be considered that these elements are abundant in mica and clay minerals, both being present in the black mat, and may thus bias the pattern of the microbial system. A remarkable finding is the W-tetrad effect in the REE plot of the *Gallionella* mat. Though as yet poorly understood, the tetrad effect has been observed to be linked with selective REE accumulation by bacteria in waters and can possibly be used as a biosignature (Takahashi et al. 2002, 2007). It is concluded that the significant differences in the trace element distribution patterns are the result of a complex sorption system including the biominerals, the EPS, and the oxidizing or reducing conditions respectively. In addition, the presence of mineral particles from the surrounding environment like mica and clay most likely influences the inorganic chemistry of the microbial mats, as these minerals are naturally rich in trace elements.

4.3 Biomarkers

The *Gallionella* mats (ochre and streak) and the black mat can be clearly distinguished by their biomarkers. In *Gallionella* mats, the linear $C_{16:0}$, $C_{18:0}$, $C_{16:1}{}^{\Delta 9c}$ and $C_{18:1}{}^{\Delta 11c}$ *n*-fatty acids predominate. These widespread compounds however, exhibit a low specifity and are thus not unique for this system. Likewise, the presence of the hopanoids observed in the ochre/streak *Gallionella* mats reveals no particular feature as these compounds are widely observed in any kind of microbial environment. A tangible lipid biomarker for *Gallionella* and/or iron-oxidizing microbial systems thus still remains to be identified. In the black mat, on the other hand, methyl-branched *i*/*ai*-$C_{15:0}$ and *i*/*ai*-$C_{17:0}$ fatty acids occur, which are often, though not exclusively found in SRB, and have, for instance been observed in the genera *Desulfovibrio* (Edlund et al. 1985) and *Desulfomicrobium* (Vainshtein et al. 1992). 10-methyl $C_{16:0}$ was found in *Desulfobacterium* (Vainshtein et al. 1992) and *Desulfobacter* (Dowling et al. 1986). All of these genera have been identified in the Äspö HRL during earlier microbial studies (Pedersen et al. 1996; Anderson et al. 2006). The results are also in full agreement with the occurrence of DAGE, which are often found in observed in extreme environments such as hot springs and cold seeps and have been tentatively associated with SRB (Pancost et al. 2001).

5 Conclusion

160 m below the Baltic Sea floor, a natural pond containing by iron mineral-precipitating microbial mats has established beneath a fluid source. Ochre microbial mats thriving within this pond are dominated by the iron-oxidizing bacterium *Gallionella ferruginea* and largely consist of strongly intertwined amorphous

iron-oxides. Changes in the colour and trace element composition are expected to result from changes in the bacterial consortia. In a second, black-coloured mat type, diorite particles and iron-sulfides (namely pyrite framboids) are predominant. The latter, together with methyl-branched fatty acids and ether lipids (DAGE) are consistent with the prominent occurrence of SRB in the black mat. Statistical methods revealed differences between the microbial mats in biomarker patterns and trace element accumulation behaviours, such as a marked W-type tetrad effect. The characterisation of recent or fossil iron-oxidizing microbial mats by their different element and lipid patterns is a new and interesting field in geobiology. However, the collection of more data is obligate to establish reliable biosignatures for these systems. More iron-metabolizing microbial mats from other locations should therefore be investigated and compared to the ones from Äspö HRL to transfer the data from a local to a global scale and possibly, to the fossil record.

Acknowledgements We are grateful to the Äspö HRL for excellent assistance and the admission to work in the tunnel of Äspö. The authors acknowledge Dr. B. Schmidt (Mineralogy, GZG Univ. Göttingen) for help with the Raman spectrometer and Dr. K. Techmer (Crystallography, GZG Univ. Göttingen) for help with the ESEM. This study was financed by the German Research Foundation (DFG) through the Research Unit FOR 571/1&2. This is publication no. 50 of the DFG Research Unit FOR 571 "Geobiology of Organo- and Biofilms"

References

Anderson CR, Pedersen K (2003) *In situ* growth of *Gallionella* biofilms and partitioning of lanthanides and actinides between biological material and ferric oxyhydroxides. Geobiology 1:169–178

Anderson CR, James RE, Fru EC, Kennedy CB, Pedersen K (2006) *In situ* ecological development of a bacteriogenic iron oxide-producing microbial community from a subsurface granitic rock environment. Geobiology 4:29–42

Beech IB, Cheung CWS (1995) Interactions of exopolymers produced by sulphate-reducing bacteria with metal ions. International Biodeterioration & Biodegradation 35: 59–72

Dowling NJE, Widdel F, White DC (1986) Phospholipid ester-linked fatty acid biomarkers of acetate-oxidizing sulphate-reducers and other sulphide-forming bacteria. Journal of General Microbiology 132:1815–1825

Edlund A, Nichols PD, Roffey R, White DC (1985) Extractable and lipopolysaccaride fatty acid and hydroxyl acid profiles from *Desulfovibrio* species. Journal of Lipid Research 26:982–988

Ferris FG, Konhauser KO, Lyvén B, Pedersen K (1999) Accumulation of metals by bacteriogenic iron oxides in a subterranean environment. Geomicrobiology Journal 16:181–192

Hallbeck L, Grivé M, Gaona X, Duro L, Bruno J (2006) Main organic materials in a repository for high level radioactive waste. SKB report R-06-104

Hallberg R, Ferris FG (2004) Biomineralization by *Gallionella*. Geomicrobiology Journal 21:325–330

Hanert HH (2002) Bacterial and chemical iron oxide deposition in a shallow bay on Palaea Kameni, Santorini, Greece: microscopy, electron probe microanalysis, and photometry of *in situ* experiments. Geomicrobiology Journal 19:317–342

Neumann T, Ostermaier M, Kramar U, Simon R (2006) Formation of framboidal pyrite in estuarine sediments of the Achterwasser lagoon, SW Baltic Sea, and implications on trace metal mobility. Geophysical Research Abstracts: 8/01163

Pancost RD, Bouloubassi I, Aloisi G, Sinninghe Damsté JS (2001) Three series of non-isoprenoidal dialkyl glycerol diethers in cold-seep carbonate crusts. Organic Geochemistry 32:695–707

Pedersen K (1997) Microbial life in deep granitic rock. FEMS Microbiology Reviews 20: 399–414

Pedersen K, Arlinger J, Ekendahl S, Hallbeck L (1996) 16S rRNA gene diversity of attached and unattached bacteria in boreholes along the access tunnel to the Äspö hard rock laboratory, Sweden. FEMS Microbiology Ecology 19:249–262

Popa R, Kinkle BK, Badescu A (2004) Pyrite framboids as biomarkers for iron-sulfur systems. Geomicrobiology Journal 21:193–206

Pósfai M, Buseck, PR, Bazylinsky DA, Franckel RB (1998) Iron sulfides from magnetotactic bacteria: structure, composition and phase transitions. American Mineralogist 83:1469–1481

Reitner J, Peckmann J, Blumenberg M, Michaelis W, Reimer A, Thiel V (2005) Anatomy of methane-derived carbonates and associated microbial communities in Black Sea sediments. Palaeogeography, Palaeoclimatology, Palaeoecology 227:18–30

Rönnback P (2005) Äspö Hard Rock Laboratory: an overview of the geology, hydrology and hydrochemistry. ESS Bulletin 3:32–47

Takahashi Y, Yoshida H, Sato N, Hama K, Yusa Y, Shimizu H (2002) W- and M-type tetrad effects in REE patterns for water-rock systems in the Toro uranium deposit, central Japan. Chemical Geology 184:311–335

Takahashi Y, Hirata T, Shimizu H, Ozaki T, Fortin D (2007) A rare earth element signature of bacteria in natural waters? Chemical Geology 244:569–583

Vainshtein M, Hippe H, Kroppenstedt RM (1992) Cellular fatty acid composition of *Desulfovibrio* species and its use in classification of sulphate-reducing bacteria. Systematic and Applied Microbiology 15:554–566

Watson JHP, Ellwood DC, Deng Q, Mikhalovsky S, Hayter CE, Evans J (1995) Heavy metal adsorption on bacterially produced FeS. Minerals Engineering 8/10:1097–1108

Wedepohl KH (1995) The composition of the continental crust. Geochimica et Cosmochimica Acta 59/7:1217–1232

Microbial Euendolithic Assemblages and Microborings in Intertidal and Shallow Marine Habitats: Insight in Cyanobacterial Speciation

Gudrun Radtke and Stjepko Golubic

1 Introduction

The term "stromatolite" was coined by Kalkowsky (1908) while describing coastal paleoenvironments of a Lower Triassic (Buntsandstein) epicontinental sea. Stable, finely laminated structures appeared together with massive deposits of shoaling ooid sands, and later consolidated into rocks ("Stromatolith und Oolith"). Ooid sands and coated grains are often found associated as well as incorporated in stromatolitic structures (e.g. Riding et al. 1991). Modern examples include subtidal (0–6 m depth; Dravis 1983; Dill et al. 1986) and intertidal stromatolites (microbialites) of the Bahama Bank (Reid and Browne 1991) and Shark Bay, Australia (Reid et al. 2003). They are formed by entrapment and rapid stabilization of ooids (Reid and Browne 1991; Reid et al. 1995; Golubic and Browne 1996; Paterson et al. 2008).

Stromatolites and ooids are hard calcareous substrates, which were the earliest to be settled and bioeroded by microboring or euendolithic microorganisms (see Golubic et al. 1981 for definition). Fossil evidence of microbial euendoliths, preserved as trace fossils as well as body fossils in ooid sand grains, goes back to the Neoproterozoic (Campbell 1982; Knoll et al. 1986; Al-Thukair and Green 1988), and in stromatolites even further back to the Mesoproterozoic times (Zhang and Golubic 1987; Knoll and Golubic 1992; Golubic and Seong-Joo 1999). With the evolution of skeleton-forming metazoans and calcareous algae, a multiplicity of new substrates became available and subjected to microbial bioerosion. Metazoan evolution generated also a variety of grazers that challenged the very existence of epiliths and endoliths alike. Euendoliths and grazers are jointly

G. Radtke (✉)
Hessisches Landesamt für Umwelt und Geologie, Rheingaustr. 186, 65203 Wiesbaden, Germany
e-mail: gudrun.radtke@hlug.hessen.de

S. Golubic
Biological Science Center, Boston University, 05 Cummington Street, Boston, MA 02215, USA

J. Reitner et al., *Advances in Stromatolite Geobiology*, Lecture Notes in Earth Sciences 131, DOI 10.1007/978-3-642-10415-2_16, responsible for the current high rates of removal and recycling of carbonates (Tribollet and Golubic 2005) as well as for bioerosive shaping of coastal limestone in the formation of biokarst (Schneider 1976, 1977). Organically preserved euendoliths with discernable ultrastructure were found in crinoid ossicles of Silurian age (Kazmierczak and Golubic 1976; Campbell et al. 1979; Campbell 1980) and microboring traces were since found abundantly in shells and shell fragments throughout the Phanerozoic (Vogel et al. 1987; Glaub et al. 2007; Wisshak et al. 2008) (Table 1).

The interest for microbial euendoliths has been generated by the potential to distinguish between phototrophic and heterotrophic taxa based on the shape of their traces. Such distinction may provide reliable indication of the water depth of ancient deposits. Microborings conform closely with the outlines of the microorganisms that make them, so that they reveal the size, growth orientation and strategy of the microborer. Many modern and ancient microborings remain empty and can be faithfully replicated using polymerizing resins, which permits comparisons between modern microorganisms and their boreholes and, by extension, an interpretation of fossil boring traces. In modern environments, the interpretation was aided by studying euendolith assemblages in naturally occurring limestone and skeletal carbonate substrates (Golubic et al. 1975; Le Campion-Alsumard et al. 1979; Günther 1990; Radtke 1993; Le Campion-Alsumard et al. 1995; Radtke and Golubic 2005; Tribollet et al. 2006), as well as in substrates experimentally exposed to euendolith colonization (Golubic 1969; Perkins and Tsentas 1976; May et al. 1982; Vogel et al. 2000; Wisshak 2006; Gektidis et al. 2007). The role of microbial endoliths in consolidation of microbialites has been introduced in more recent papers (Reid et al. 1995; Feldmann and McKenzie 1998; Reid and Macintyre 2000; Macintyre et al. 2000; Reid et al. 2000; Decho et al. 2005).

Ecological studies of phototrophic microboring organisms have first focused on coastal limestones of the Mediterranean area (Ercegović 1932; Le Campion-Alsumard 1966; Schneider 1976, 1977), and then extended to shallow habitats and to tropical seas (Golubic 1969; Perkins and Tsentas 1976), including euendoliths in bivalve shells and shell fragments (Golubic 1990; Radtke 1993; Radtke and Golubic 2005). Studies of modern euendoliths in shoaling ooid sands of the Bahamas and the Arabian/Persian Gulf identified particular assemblages of cyanobacteria, some of which were described as new genera and species (Lukas and Golubic 1981, 1983; Lukas and Hoffman 1984; Al-Thukair and Golubic 1991a, b, 1996; Al-Thukair et al. 1994; Gektidis and Golubic 1996). These microorganisms leave characteristic boring patterns in the carbonate substrates they penetrate, which were recognized and studied as resin replicas using scanning electron microscopy (Golubic et al. 1970, 1983) and described according to the rules of nomenclature for trace fossils (Vogel et al. 1987; Bertling 2007).

The present contribution analyses the morphotype diversity of euendoliths of coastal marine environments with special reference to cyanobacterial genera *Solentia* and *Hyella*, in comparison with boring traces identified as ichnogenera *Scolecia* and *Fascichnus*.

Table 1 Time distribution of the fossil traces *Fascichnus* and *Scolecia*

Time	Ichnotaxa							References
	Fascichnus parvus	*Fascichnus dactylus*	*Fascichnus frutex*	*Fascichnus acinosus*	*Fascichnus rogos*	*Fascichnus* sp.	*Scolecia maeandria*	
Tertiary	x	x	x				x	Radtke (1991, 2007)
		x						Vogel and Marincovich (2004)
Cretaceous		x[a]						Hofmann (1996)
Jurassic		x	x	x				Glaub (1994)
Triassic	x	x		x		x		Schmidt (1992)
Permian		x	x					Balog (1996, 1997)
Carboniferous		x	x					Wisshak et al. (2008)
Devonian		x[b]						Vogel et al. 1987
Silurian			x		x	x		Bundschuh (2000)
Ordovician		x				x		Vogel and Brett (2009)

[a]As a small bush-form (“kleine Büschelform”)
[b]As “Hyellomorpha”

2 Materials and Methods

Samples of carbonate substrates, including coastal limestone, ooid sand grains and bivalve shells and shell fragments were collected from the supratidal ranges to shallow marine environments down to a depth of 10 m. These carbonate substrates were examined for the presence of euendolithic microorganisms and their traces. The materials from modern environments include: Korčula Island, Croatia, Safaga Bay (Egypt, Red Sea), Lee Stocking Island, Bahamas, Cozumel (Mexico, Caribbean Sea), One Tree Island (Australia, Great Barrier Reef); they were compared with fossil microborings from the Tertiary of the Paris Basin.

The samples were fixed in 3% formaldehyde solution and the microorganisms extracted by dissolving the carbonate matrix in dilute (3%) HCl, mounted on microscope slides and analysed by transmission light microscopy (LM) using Zeiss Universal Microscope (Zeiss, Oberkochen) equipped with digital camera.

Microborings were replicated in polymerizing resin (Araldite Durcupan, and SPURRS low viscosity medium). The procedure for embedding and casting (Golubic et al. 1970, 1983) includes gradual dehydration by acetone series followed by an equally gradual replacement by a complete (4 component) mixture of the resin under vacuum. The resin is hardened at 60°C for 3 days, the hardened blocks including the embedded sample is cut open and the carbonate matrix dissolved by dilute HCl. Microborings were examined using scanning electron microscopy (SEM). Organic component of the substrate, if present, were removed by potassium hypochlorite (Clorox bleach).

Microbial cells and microborings, observed by LM and SEM were imaged and measured for each population or cluster of borings, expressed as mean, standard deviation, standard error, confidence level of 90 and 99%, maximum and minimum values and sample size using measuring software of the digital Motic (Global Motic Group, Xiamen, China) camera and Sigma Scan (Sausalito, CA). Intra- and inter-population distribution of cellular dimensions and tunnel diameters respectively have been recorded, expressed in terms of mean ± standard deviation and presented graphically as two-dimensional cross diagrams, comparing two parameters at a time with population means in the intersection of the cross with one standard deviation on each side of the mean. Taxonomic determinations were carried out by consulting phycological and microbiological determination manuals (Geitler 1932; Komárek and Anagnostidis 1999; Rippka et al. 2001).

3 Morphotypic and Ecotypic Diversity and Variability of Coastal Euendolithic Cyanobacteria

Comprehensive studies of epilithic and endolithic assemblages of coastal limestones have been carried out in 1930s by Ante Ercegović on the Adriatic coast of Croatia near Split. Ercegović described a number of new genera and species of cyanobacteria that participate in bioerosion and explained their distribution in the context of the ecology of wave-exposure (Ercegović 1932). Later studies of microbial euendoliths on the coast of Istria in northern Adriatic included grazers as an integral component of bioerosion (Schneider 1976, 1977) and extended more recently to study of coral reefs (Radtke et al. 1996, 1997; Tribollet and Golubic 2005; Tribollet et al. 2006; Tribollet 2008).

The wave-exposed limestone coast of the Adriatic Sea at Vela Luka, Korčula, Croatia (Fig. 1a), the object of the present study (not far from the classic location of Ercegović), illustrates the zone of most intensive activity of coastal euendoliths. The natural color of the upper Cretaceous limestone is light beige as seen in the upper right corner of the picture; dark color zones to the left, ranging from dark brown to steel-blue and olive tones originate from pigments produced by dense populations of euendolithic cyanobacteria. The combination of extreme and changing conditions characterize the microenvironments across the intertidal and supratidal ranges, expressed primarily by alternation of exposure to the atmosphere, water supply by waves and tides, and by drainage. Water retention and subsequent evaporation alter the salinity and indirectly the solubility of nutrients and gases. These microenvironmental conditions support specialized niches dominated by those euendolithic taxa that are most competitive under such local conditions. Bioerosive activities modify the habitat physically and chemically, and by altering the local conditions they lead to replacements of euendolithic microflora. The result is a "color-coded" mosaic-like patchwork of microenvironments and their inhabitants.

The dark brown (almost black) colored rock surfaces in the upper reaches of the wave spray (Fig. 1a–1) are occupied by clusters of small cells with partial penetration into the rock, which have been described as a separate genus *Hormathonema* (Ercegović 1929). Below this range, the exposed, well drained brown-colored surfaces (Fig. 1a–2) are dominated by *Hyella* morphotypes (mostly *H. balani*). The pitted blue-colored areas (Fig. 1a–3) are dominated by morphotypes of *Solentia*.

Coccoid cyanobacteria that penetrate carbonate substrates are classified in the order Pleurocapsales (Geitler 1932; Rippka et al. 2001, Subsection 2 in Bergey's Manual). These organisms reproduce by binary as well as multiple fission (Waterbury and Stanier 1978; Le Campion-Alsumard et al. 1996). The growth is coordinated so that binary fission accompanies penetration of carbonate substrate and formation of colonies with predominant apical growth, whereas the reproductive phase consists of formation and release of large numbers of small cells (baeocytes), which are a product of multiple fission. Linear (uni-dimensional) binary fission in *Solentia* and *Hyella* results in formation of cell series, or pseudofilaments, which produce cylindrical tunnels in the substrate. In addition, the cell division is often accompanied by directional production of gelatinous envelopes (EPS), which separates the cells and accentuates the directionality of substrate penetration. This is in contrast to truly filamentous cyanobacteria, which form multicellular trichomes. However, euendoliths of both types produce branched and unbranched tunnels in carbonate substrates.

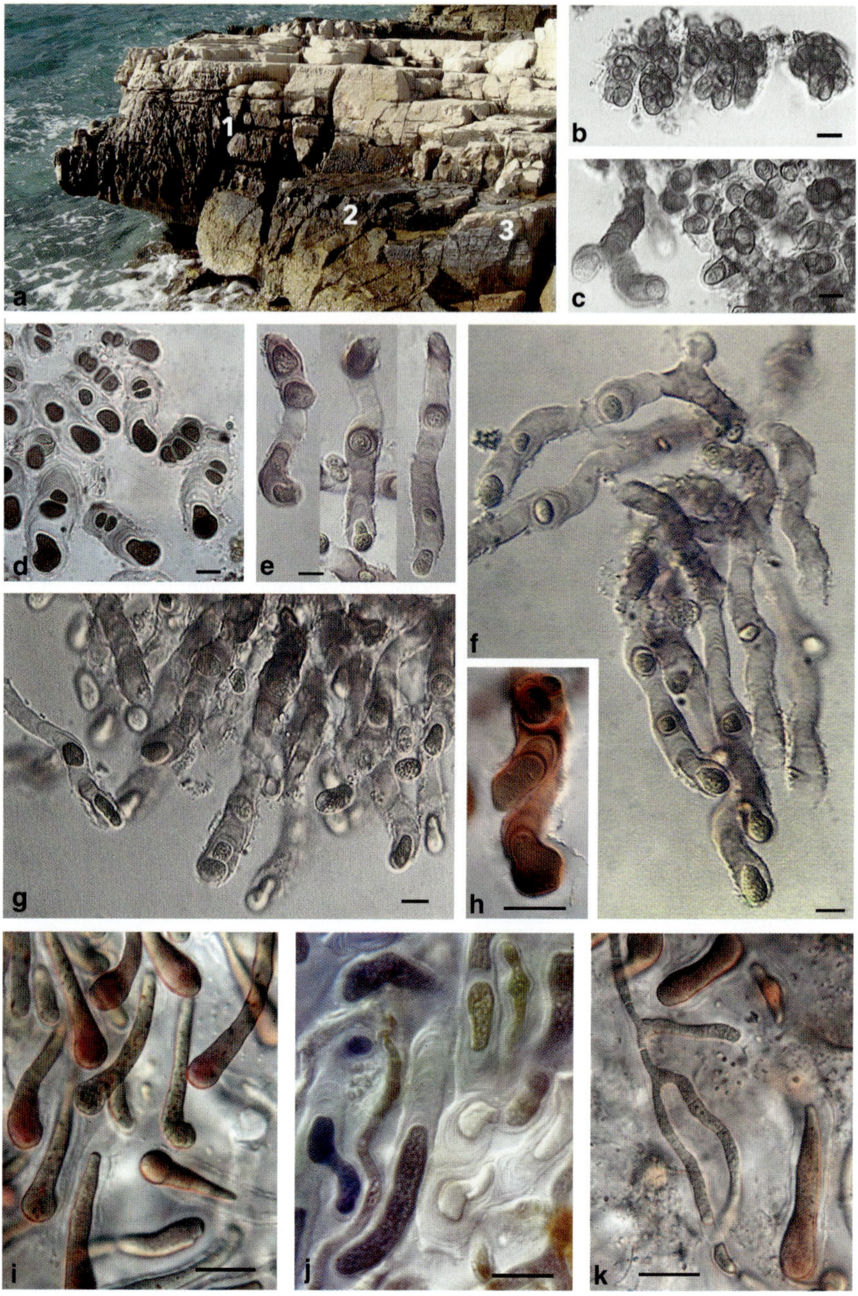

Fig. 1 Euendoliths in coastal limestone (Upper Cretaceous). (**a**) About 1 m high intertidal coastal profile, Adriatic Sea at Vela Luka, Korčula, penetrated by euendolithic cyanobacteria. Different cyanobacterial assemblages appear color-coded, with areas dominated by *Hormathonema* (1), *Hyella* (2) and *Solentia* (3) in dependence on water supply and wave exposure. (**b**) Uppermost

Hormathonema Ercegović with the species *H. luteo-brunneum* protected by the brown "sunscreen" pigment scytonemin and *H. violaceo-nigrum* (Fig. 1b, c, right) by the blue pigment gloeocapsin occur in the uppermost ranges of the colored zone. Both pigments are extracellular, incorporated in the gelatinous EPS-envelopes to absorb UV radiation (Ehling-Schulz and Scherer 1999; Castenholz and Garcia-Pichel 2000). The colonies are for the most part epilithic, extending into the substrate with short series of cells separated by EPS. In habitats with more frequent wetting by waves, *Hormathonema* spp. are replaced by *Solentia* (Fig. 1c, left), an entirely endolithic organism, which also prevails in rock pools and small pits in the lower intertidal ranges. The morphotypes in transition between these two genera are represented by *Solentia paulocellulare* (Le Campion-Alsumard and Golubic 1985a) which was originally described within the genus *Hormathonema* (Ercegović 1929).

Solentia Ercegović carries one of the ancient Roman names for the island Sulet or Šolta in central Dalmatia (Ercegović 1927). The pseudofilaments of *Solentia* are produced mainly by uni-directional production of EPS in the wake of cells, forming a stalk that pushes the cells forward (Fig. 1d–g). The tunnels branch when intercalary cells slip out of the alignment and become secondary apical cells of a branch (Fig. 1h). The extracellular pigment in *Solentia* is gloeocapsin, which changes the coloration from blue to red during acid-extraction (Fig. 1h). No extracellular pigment is produced by deep penetrating *Solentia achromatica*, which is characterized by large droplet-shaped phycoerythrin-rich apical cells (Fig. 1i–k). *Solentia* was very abundant across shallow ranges of the Bahama carbonate platform at Lee Stocking Island, Exuma islands (Kiene et al. 1995); it has been isolated in culture from moving ooid grains (Golubic et al. 1996).

Hyella Bornet and Flahault (1889) is the most diverse and most abundant euendolithic cyanobacterium. The pseudofilaments in *Hyella* are comprised of series of cells separated by moderate bidirectional production of EPS that separates them (Fig. 2), but the degree of cell separation varies from taxon to taxon. For example, *Hyella* sp. found in shoaling ooids and *H. caespitosa*, most common in subtidal ranges, both have cells separated by stretches of EPS, which often exceed the lengths of individual cells, whereas the cells of *H. pyxis* and *H. racemus* are invariably packed in tight EPS envelopes (Lukas and Hoffman 1984; Al-Thukair et al. 1994). The genus is highly diversified (over 25 species formally described) with numerous morphotypes that differ in cell and colony size, growth orientation, branching pattern, pigmentation, and mode of reproduction (Table 2).

Fig. 1 (continued) wave-spray zone is occupied by the coccoid *Hormathonema violaceo-nigrum*. (**c**) *H. violaceo-nigrum* and *Solentia paulocellulare* (*left*) in protected depressions. (**d–g**) *Solentia* morphotypes across the intertidal zone range from short borings of *S. paulocellulare* to extended tunnels of *S. foveolarum*, both with EPS stained by extracellular gloeocapsin pigment. (**h**) *S. paulocellulare* with gloeocapsin sheath pigment – red when extracted by acid showing false branch by cell slippage. (**i–k**) Long, deeply penetrating terminal portions of *S. achromatica*. Cells are rich in phycoerythrin. With filaments of *Mastigocoleus testarum* in **k**. Scale bars in all pictures are 10 μm long

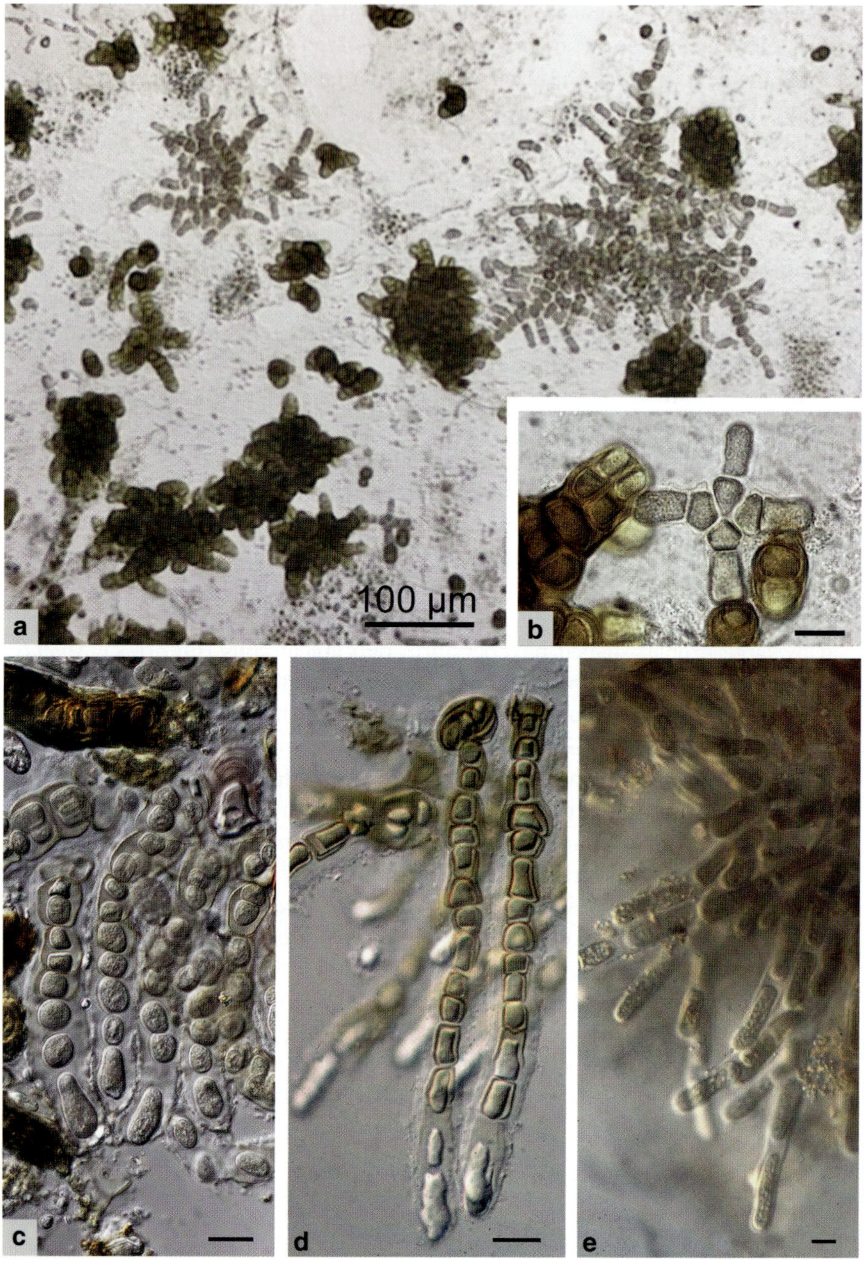

Fig. 2 Morphological diversity of *Hyella*. (**a**) Colonies of two *Hyella* species extracted from a mussel shell. Dark extracellular scytonemin pigment is produced only in one of the two taxa: *H. balani*. (**b**) Detail from **a**. (**c–e**) Three populations of *H. caespitosa* in side view with downward direction of boring. The colonies are at different stages of cell differentiation. Scytonemin is produced only close to the irradiated substrate surface. Scale bar in all pictures is 10 µm, unless stated otherwise

Table 2 Cell dimensions of coccoid cynaobacteria and tunnel diameters of boring traces

Species/Ichnospecies	Cell width (μm)	References
Euendoliths		
***Hyella*, Bornet et Flahault 1889**		Bornet and Flahault (1889)
H. caespitosa, Bornet et Flahault 1889	4.0–10.0 (4.2 ± 0.8)[a]	Le Campion-Alsumard and Golubic (1985b)
H. caespitosa var. *arbuscula*[b]	2.5–5.7 (4.1 ± 0.8)	Al-Thukair and Golubic (1996)
H. balani, Lehmann 1903	4.0–10.0 (5.1 ± 1.3)	Le Campion-Alsumard and Golubic (1985b)
H. tenuior, Ercegović 1927	2.0–5.0	Ercegović (1932)
H. gigas, Florida-Bahamas	4.5–28.5 (9.8 ± 3.2)	Lukas and Golubic (1983)
H. gigas, Mediterran	7.0–19.0 (9.9 ± 2.0)	Lukas and Golubic (1983)
H. pyxis, Florida	4.2–9.4 (6.8 ± 1.3)	Lukas and Hoffman (1984)
H. pyxis, Bermuda	3.9–13.1 (8.5 ± 2.3)	Lukas and Hoffman (1984)
H. immanis	7.5–22.3 (14.9 ± 3.7)	Al-Thukair and Golubic (1991a)
H. inconstans	3.4–8.2 (5.8 ± 1.2)	Al-Thukair and Golubic (1991b)
H. salutans	6.5–17.7 (12.1 ± 2.8)	Al-Thukair and Golubic (1991b)
H. reptans	2.7–8.2 (5.5 ± 1.4)	Al-Thukair and Golubic (1991b)
H. conferta	4.1–10.9 (7.5 ± 1.7)	Al-Thukair and Golubic (1991b)
H. stella	6.3–16.3 (11.3 ± 2.5)	Al-Thukair and Golubic (1991b)
H. vacans	9.2–14.5 (11.8 ± 1.5)	Gektidis and Golubic (1996)
H. racemus	5.9–21.5 (13.9 ± 3.9)	Al-Thukair et al. (1994)
Hyella sp. 1, GBR, Australia	5.0–8.5 (6.8 ± 1.1)	Le Campion-Alsumard (1991)
Hyella sp. 2, GBR, Australia	8.5–17 (12.0 ± 1.2)	Le Campion-Alsumard (1991)
Hyella sp. 3, GBR, Australia	3.5–8.0 (6.31 ± 0.63)	Le Campion-Alsumard (1991)
***Solentia*, Ercegović 1927**		Ercegović (1927)
S. intricata	(2–) 3.0–4.0 (–6)	Ercegović (1927)
S. stratosa, Ercegović 1927	(3–) 6.0–9.0 (–14)	Ercegović (1932)
S. foveolarum	5.0–20.0	Ercegović (1932)
S. achromatica	7.0–16.0	Ercegović (1932)
S. paulocellulare, (Ercegović 1929)	4.0–9.0	Le Campion-Alsumard and Golubic (1985a)
S. sanguinea	11.0–20.0 (14.8 ± 2.5)	Golubic et al. (1996)
Boring traces		
***Fascichnus (Fasciculus)*, Radtke 1991**		Radtke and Golubic (2005)
F. parvus	1.5–4.0	Radtke (1991)
F. dactylus	4.0–9.0	Radtke (1991), Radtke and Golubic (2005)
F. acinosus	4.0–10.0	Glaub (1994)
F. frutex	11.0–25.0	Radtke (1991)
F. rogus	15.2 ± 1.53	Bundschuh and Balog (2000)
F. grandis	40–160	Radtke (1991), Radtke et al. (1997)

[a]Range (mean ± SD)
[b]The authors are omitted if the reference contains formal description

We observed consistent differences among populations occupying the same habitat, i.e. penetrating and living in the same bivalve shell (Fig. 2a) indicating a fine scale differentiation of ecological niches. In this habitat, under the same illumination, *H. balani* formed colonies pigmented by the dark brown extracellular UV and light-screening pigment (scytonemin) in contrast to *Hyella* sp. (undescribed taxon), which remained consistently without extracellular pigmentation (Fig. 2b). These taxa differed also in branching patterns resulting in colonies of different density (compact vs. loosely branched).

The ability to produce scytonemin in response to solar irradiation is a valid taxonomic marker distinguishing otherwise similar taxa although this property is not expressed in shaded habitats or deeper inside the rock. In *H. caespitosa* (Fig. 2c–e), which has a wider water depth distribution and penetrates deeper into the rock, the pigmentation is restricted only to the parts of the colony close to the substrate surface. In the course of the development and substrate penetration, the cells of this organism become progressively elongated – another property that is not shared with other species of *Hyella* (Le Campion-Alsumard and Golubic 1985b). These properties are maintained even when the organisms are grown together and exposed to similar conditions, indicating an involvement of genetic control of their expression.

Species of *Hyella* morphologically distinct from those known from the Mediterranean region were found in the Great Barrier Reef of Australia (Le Campion-Alsumard 1991), suggesting that some of these taxa may be regional rather than cosmopolitan or pantropical in distribution. Large cell dimensions characterize *H. immanis*, *H. salutans* (Al-Thukair and Golubic 1991a, b) and *H. gigas* (Fig. 3a) with several geographically different variants (Lukas and Golubic 1981). A newly discovered *Hyella* sp. extracted from the intertidal limestone on Adriatic islands produces the blue sheath pigment gloeocapsin rather than the more common scytonemin (Fig. 3b). Cell division in alternate planes with limited production of EPS, produce dense grape-like clusters in *H. racemus* (Fig. 3c) (Al-Thukair et al. 1994). Dichotomous branching prevails in *H. stella* (Fig. 3d) and lateral branching in *H. reptans* (Fig. 3e). While in most species multiple fission and baeocyte release occurs only in cells close to the substrate surface, in *H. vacans*, multiple fission is triggered sequentially from the substrate surface down, so that ultimately all cells are converted into baeocytes (Gektidis and Golubic 1996).

Like many coccoid cyanobacteria that produce EPS envelopes, *Hyella* forms clonal colonies, which originate from single baeocytes, the progeny of which is held together by the EPS intermittently produced. *Hyella* sp. and *H. balani* compared in Fig. 2a, b, for example, have similar cell dimensions, but differ in branching pattern and pigmentation (Fig. 3f vs. g). Stable distinctions observed in natural populations are also maintained in culture (Al-Thukair and Golubic 1991a). However, each of these taxa is also subject to ecologically induced morphological variation in the course of its development, but within the range of genetically determined tolerance. Ecologically induced differences in cell arrangement, cell size distribution and pigment production within colonies along a gradient of changing water supply on the rocky intertidal habitats have been documented earlier (Golubic and Marčenko 1965; Le Campion-Alsumard and Golubic 1985b).

100 µm
a
b
c
d
e
f
g

4 The Identity of Microboring Traces

Microbial euendoliths are an ecologically important group of benthic organisms, which left their overprint mark on limestone rocks and skeletal carbonates in modern marine environments as well as in the fossil record (Golubic et al. 1984). These rather precise and instantly preservable microboring traces may contain paleoenvironmental information about the depositional depth. Both phototrophic and organotrophic members of euendolithic community leave their traces as fossil borings, but only the light-dependent phototrophs have paleobathymetric indicator value (Vogel et al. 1995), i.e. the organotrophic euendoliths could occur at any depth, whereas the obligatory phototrophic ones do not occur in aphotic environments. Thus, it is important to establish the identity between the microborers and their traces. Convergent evolution among euendoliths has resulted in similar boring patterns, and such convergent morphologies need to be distinguished in characterization of facies fossils and their traces.

The euendoliths *Hyella* and *Solentia* in the present study are coccoid cyanobacteria, which grow as EPS-supported cell series, but they produce tubular borings similar to those made by truly filamentous organisms. Both show cell differentiation in form and function following the first division of the settled baeocyte: One of the daughter cells remains at the substrate surface and later undergoes multiple fission, while the other grows larger and continues to divide by binary fission.

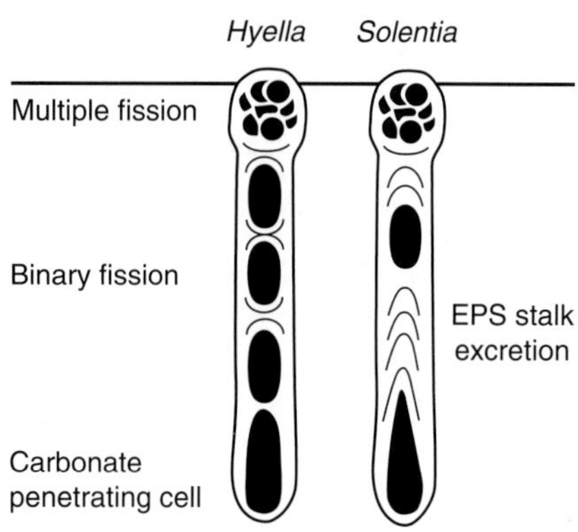

Fig. 4 Schematic presentation of cellular organization and EPS excretion in *Hyella* and *Solentia* during carbonate penetration. Note that both organisms produce borings (*Fascichnus*) with similar outlines

Fig. 3 Morphological diversity of *Hyella* (cont.): (**a**) *Hyella gigas*; (**b**) *Hyella* sp. with gloeocapsin as a UV-protecting pigment in the sheaths. (**c**) *H. racemus*, densely branched clusters with baeocyte forming cells close to the substrate surface. (**d**) *H. stella*, with dichotomous branching of pseudofilaments. (**e**) *H. reptans*, differentiates horizontal pseudofilaments parallel to the surface and lateral, deeper boring branches. (**f**) *Hyella* sp. showing formation of lateral branch by cell slipping out of the alignment. (**g**) *H. balani* forms branches by longitudinal fission of the apical cells. Scale bar in all pictures is 10 μm, unless stated otherwise

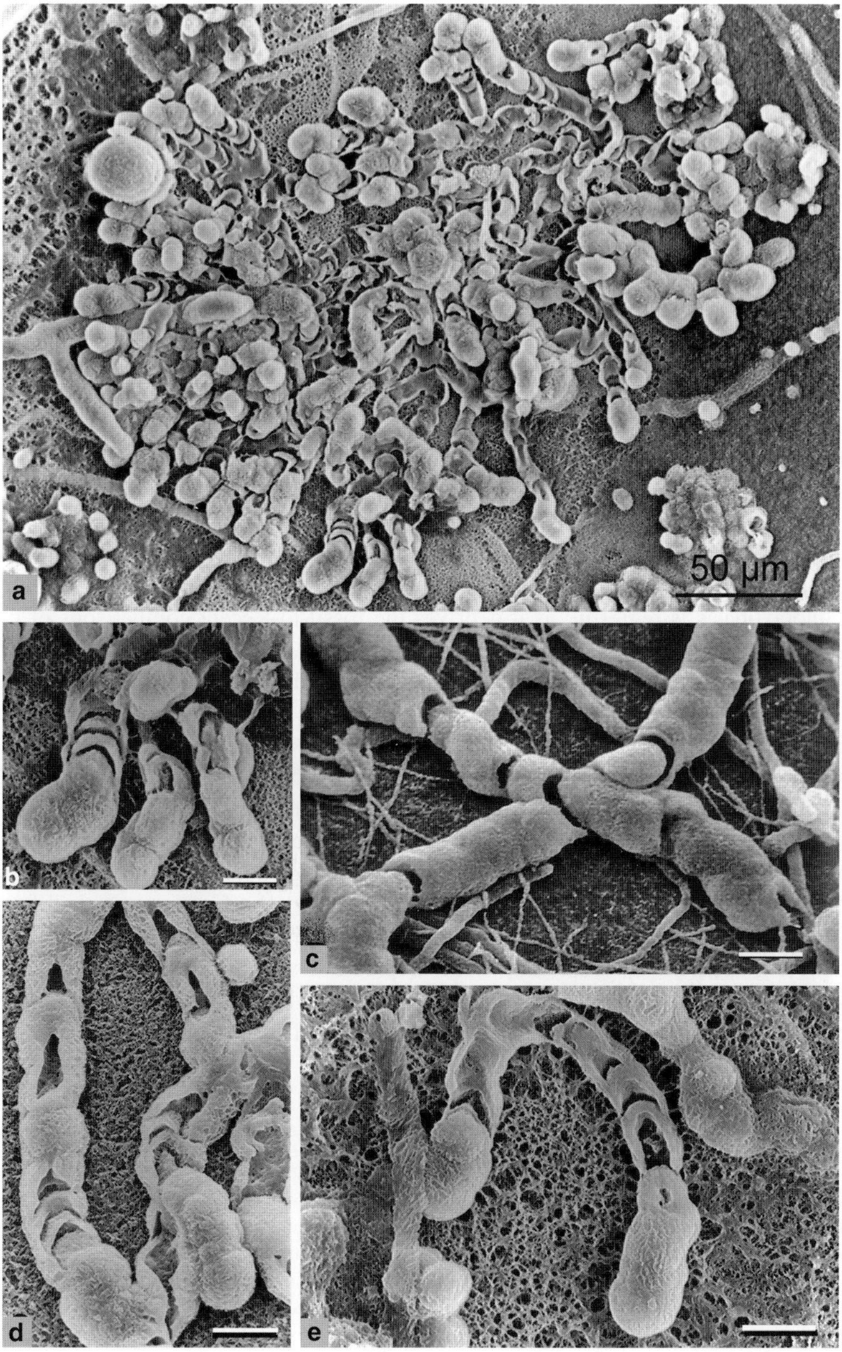
a
50 µm
b
c
d
e

Hyella and *Solentia* differ from each other in cell arrangements (Fig. 4), but the external outlines of their bodies, which are in contact with the walls of the tunnel they produced, are similar. Such subtle differences still need to be resolved. The pseudofilaments in *Hyella* consist of cell series (Fig. 4, left), whereas in *Solentia* they consist mostly of layered EPS (Fig. 4, right). These distinctions, however, are not necessarily revealed by the outlines of the tunnels, so that they are usually not replicated by natural or artificial casts.

Most traces in modern settings classified under ichnogenus *Fascichnus* could be compared with borings of *Hyella* spp. (Radtke and Golubic 2005: fig. 5, Table 1), although a few were found to have been produced by species of *Solentia.* Such uncertainties in identifying the trace with its maker justify the naming and classifying the traces by a separate classification system, the ichnotaxonomy (Bromley 1990). However, we view microbial boring traces as sources of biological information, revealing not only the behavior of the organism, but often its taxonomic identity, including variability and changes in the course of its development and life cycle. In addition, borings differ from burrows in soft sediments by their structural stability. The activity of euendoliths results often in generating instant fossils, which may persist unchanged over geological time spans. For these reasons, the ruling of ichnotaxonomy that prohibits the use of modern boring traces as ichnotaxobasis is regretful; as it deprives the fossil-to-Recent comparisons of the most informative reference.

Microborings are studied mostly by Scanning Electron Microscopy of tunnel replicas cast in polymerizing resins (Figs. 5–10) which reveal their three-dimensional display within the bored substrate. The resident organisms and secondary precipitates within the tunnels are not revealed by the casts, and require application of complementary preparation techniques, such as double embeddings (Tribollet et al. 2010: fig. 1a) or etched petrographic thin sections (e.g. Macintyre et al. 2000). Resin-cast borings that clearly identify *Solentia* as trace maker are rarely preserved (Fig. 5). A large radially sprawling colony, ca 0.5 mm in diameter (Fig. 5a) in a shell from Great Barrier Reef, Australia shows several tunnels that appear flattened and interrupted by crescent-shaped breaks, which separate U-shaped sections of the tunnel. The tunnels terminate with a solid oval, slightly wider swelling, suggesting the position of the apical cell of a *Solentia* at the end of its EPS-stalk (Fig. 5b, d, e).

Fig. 5 Comparison of resin-cast modern and fossil boring traces that reveal morphological properties of *Solentia.* (**a**) Radially expanding colony of filaments with terminal swellings, indicating the positions of the terminal cells of *Solentia* at the ends of stalks, which are comprised of U-shaped sections. One Tree Island, Great Barrier Reef, Australia, in a shell collected on a patch reef at 2 m depth. (**b**) Detail of the margin of the colony in **a**. Note that the stalk sections are interrupted by crescent-shaped gaps. (**c**) Similar crescent-shaped gaps, alternating with stretches of intact tunnels were observed in fossil trace, *Scolecia maeandria*, Paris Basin, Eocene, Middle Lutetian. (**d–e**) Variation in terminal portions of tunnels produced by modern *Solentia* from Great Barrier Reef with short stretches of intact tunnels. Scale bars in all pictures are 10 μm long, unless stated otherwise

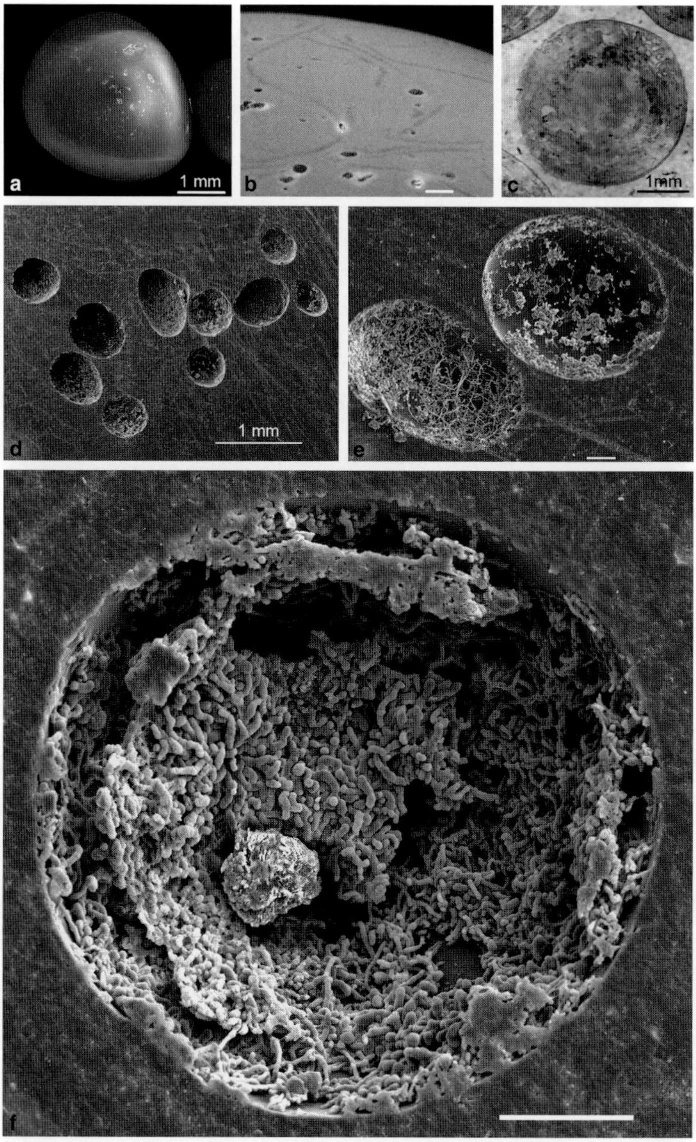

Fig. 6 Microborings in ooid sand grains. (**a**) Typical shoaling ooid grain, 0–0.75 m deep, Cat Cay, Andros Island, Bahamas, with smooth polished surface. (**b**) Scattered small boreholes that extend under the surface in the interior of the grain. (**c**) Petrographic thin section showing two generations of microborings as concentric rings. (**d**) A random selection of ooids with resin-cast microborings. (**e**) Detail view of two ooids with different assemblages of microborings. (**f**) Ooid with preserved two generations of borings (compare with **c**). Scale bar in all pictures is 100 μm, unless stated otherwise

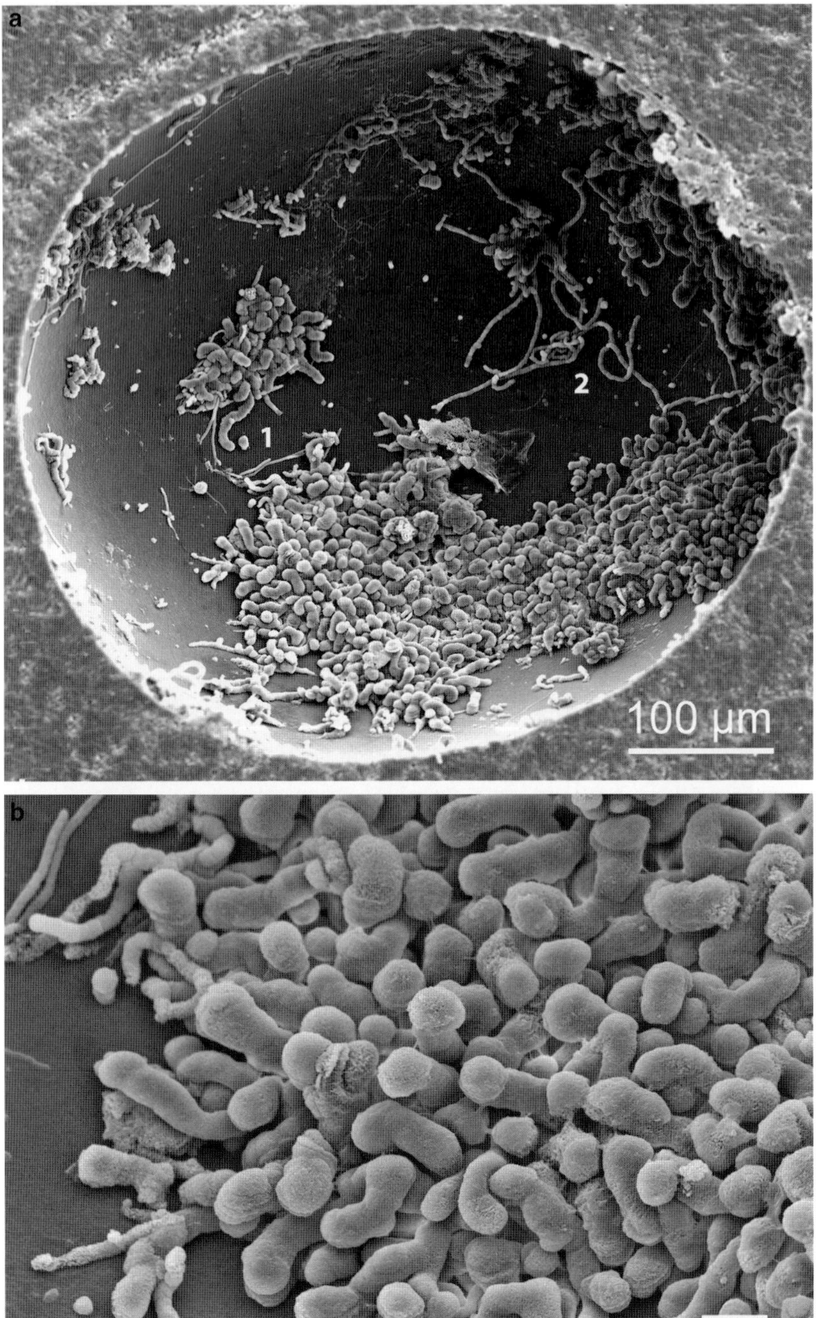
a
1
2
100 µm
b

The structures compare closely with the mode of growth and penetration of *Solentia* (comp. with Figs. 1d–f and 4, right).

The presence of semilunar breaks along the tunnel replicas in Fig. 5 suggests that the acid removed also the carbonate that precipitated as a part of the *Solentia's* activity within its EPS stalk. This interpretation is consistent with the observations that *Solentia* re-deposits the carbonate that was removed in the process of its boring activity, a mechanism proposed to contribute to fortification of stromatolitic structures following entrapment of ooids in the Bahamas (Reid et al. 2000; Macintyre et al. 2000). We assume that the precipitate in the sample shown in our Fig. 5 has been dissolved in places where it adhered to the tunnel walls. This interpretation suggests that a substantial amount of carbonate was deposited between the layers of the illustrated *Solentia* stalk. Similar semi-lunar breaks along replicated tunnels, identified as *Scolecia maeandria*, have been observed in fossil shells from the Paris Basin (Tertiary) (Fig. 5c). These are less regular, including stretches of completely replicated tunnels. Uninterrupted sections of tunnel casts are also seen in Fig. 5a. They may result if no carbonate precipitation took place in the tunnel, or if the precipitation did take place, but the deposit became completely enclosed in resin and thus protected from the acid. In either case, the trace resembles *Fascichnus*, which are produced by *Hyella* spp.

5 Microboring Traces in Ooid Sand Grains

Carbonate sediment particles are exposed to settlement and penetration by microbial euendoliths in all depths above the lysocline, but phototrophic microbial euendoliths are limited to the euphotic zone, the depth of which varies with the clarity of the water column (Perry and Macdonald 2002; Wisshak 2006; Gektidis et al. 2007). Shoaling ooids including those incorporated into stromatolitic structures form invariably in shallow agitated subtidal ranges, where they are bored by a particular subset of phototrophic euendoliths. Together with their substrate, this assemblage defines rather precisely a shallow environment common in tropical seas. The dynamic nature of this environment affects the settlement of organisms thus providing a unique selective pressure on the endolith assemblage, which does not exist on solid hardgrounds or in settings below the wave base. We are addressing here the diversity of borings in loose grains prior to their consolidation in stromatolite or beach-rock formations.

Fig. 7 Microborings in ooids (cont.) (**a**) Interior surface of an ooid grain with borings of at least 2 euendolithic morphotypes, dominated by colonies of *Fascichnus dactylus* (1) with a filamentous boring *Eurygonum nodosum* (2), formed by the heterocystous cyanobacterium *Mastigocoleus testarum*. (**b**) Detail at the lower part of **a**, showing uniform distribution of densely packed borings producing a lens-shaped colony. Scale bar in all pictures is 10 μm, unless stated otherwise

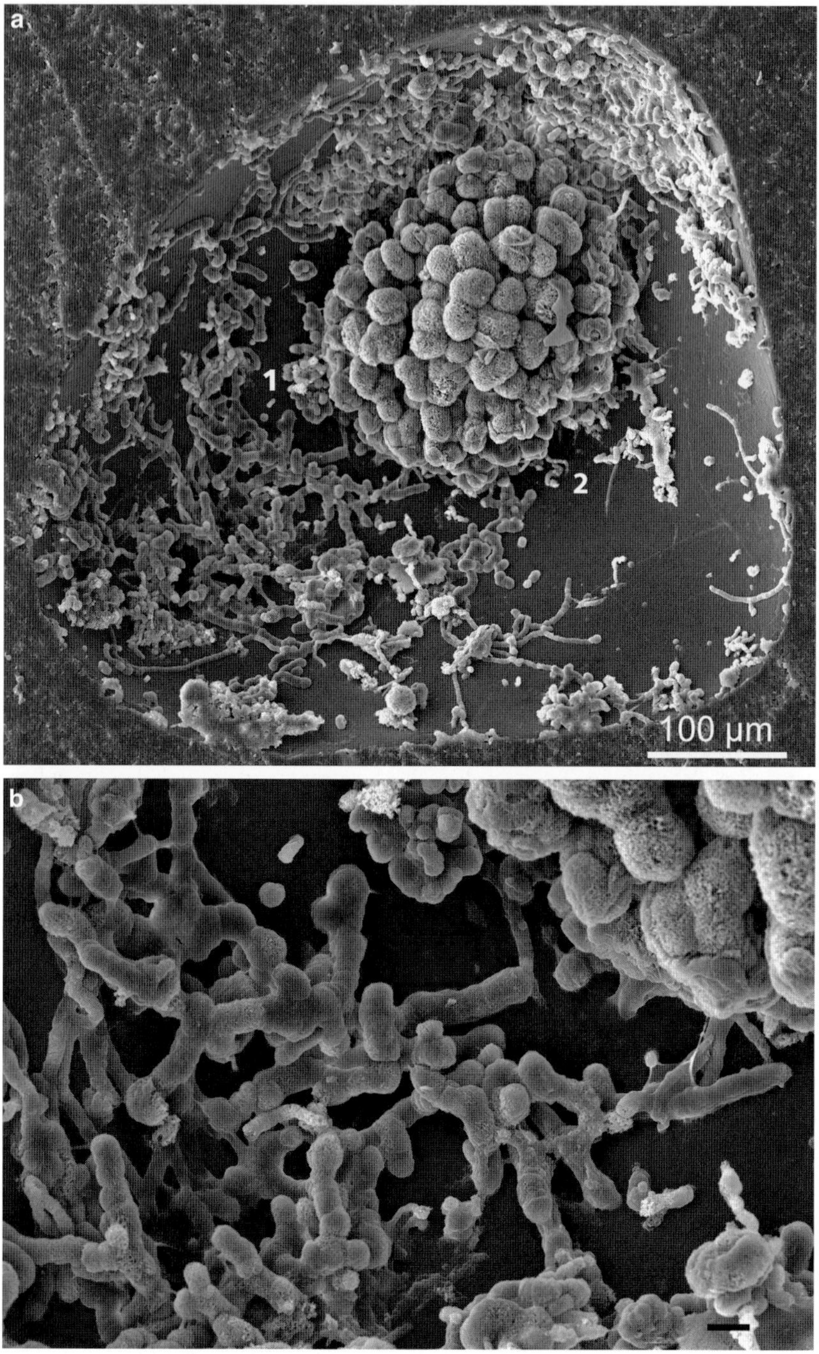
a
1
2
100 µm
b

The surfaces of loose ooid sand grains are smooth and free of epilithic growth (Fig. 6a), an advantage to microborers able to penetrate the grain surface rapidly and then proceed to grow in the interior of the grain (Fig. 6b). However, these microborers need to be able to persist under the conditions of transportation by currents, including repeated burial and the associated light limitation. Ooids represent a dynamic system subject to periods of carbonate accretion, interrupted by longer periods of stases during which they are exposed to microbial boring. Horizons with higher density of borings observed in petrographic sections and resin casts (Fig. 6c, f) document such periods of stasis, which may relate to changes in ocean chemistry.

In modern ooid shoals of the Bahamas almost every ooid grain is bored (Fig. 6d), often by a different microbial assemblage (Fig. 6e), but only a small proportion of euendoliths is found active at the time of collection. Thus, microboring assemblages in ooid grains are a cumulative record of euendolith activity. A diverse assemblage of euendoliths has been described from the Arabian/Persian Gulf (Al-Thukair and Golubic 1991a, b; Al-Thukair et al. 1994) with a number of active ones declining following oil spills (Al-Thukair 2002).

Microbial euendoliths extracted from ooids are mostly different species of *Hyella*. The casts of their borings are classified within the ichnogenus *Fascichnus* (Radtke 1991; Radtke and Golubic 2005). They produce distinct loose shrub-like clusters or densely branched lens-shaped colonies (Fig. 7a–1). Other borings may be solitary or filamentous (Fig. 7a–2). The traces compared in detail view (Fig. 7b) show that the diameters of tunnels within a cluster are fairly uniform, but may differ significantly from cluster to cluster (Fig. 8). The loosely branched cluster of ca 10 μm wide borings (Fig. 8a–1) corresponds to *F. dactylus,* whereas the large compact grape-like cluster of borings (Fig. 8a–2) was described as *Fascichnus rogus* (Bundschuh and Balog 2000). The latter is produced by the cyanobacterium *Hyella racemus* (Fig. 3c). The detail (Fig. 8b) compares the above two borings.

6 Diversity of *Fascichnus* Traces of Coccoid Cyanobacteria in Shallow Marine Environments

Fascichnus traces constitute a broad spectrum of forms that share basic features such as inward radiating shrub-like colony shape, with the origin from a single point of entry into the substrate, but vary widely in dimensions. Other distinctions include branching pattern and density, and change in orientation of boring in the course of

Fig. 8 Microborings in ooids (cont.). (**a**) Mixed assemblage of various coccoid (1) and filamentous microborings including a large colony of *Fascichnus rogus* (2) produced by the cyanobacterium *Hyella racemus* (Fig. 3c). (**b**) Detail of **a**-1, showing loosely branched *Fascichnus* isp. Scale bar in all pictures is 10 μm, unless stated otherwise

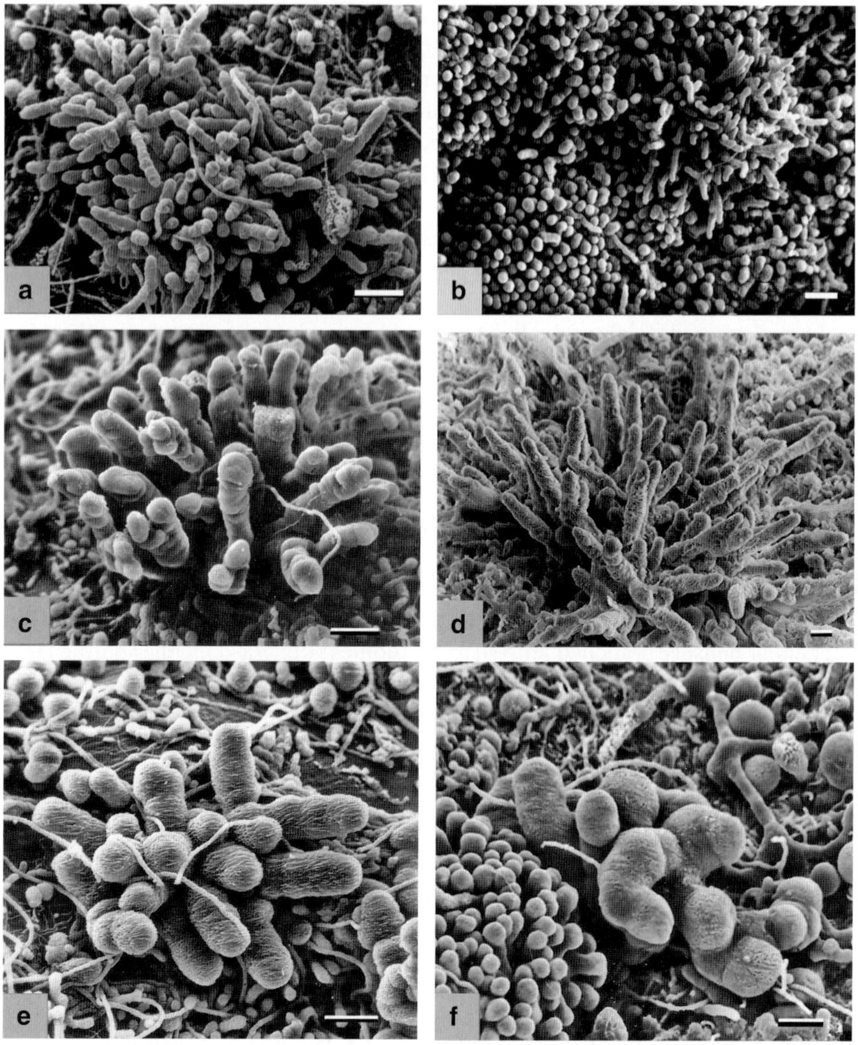

Fig. 9 *Fascichnus* clusters from modern shallow marine environments, assorted in the sequence of increasing tunnel diameters. (**a**) *Fascichnus parvus*, Lee Stocking Island, Bahamas, windward reef with *Acropora palmata*, 2–3 m water depth. (**b**) Dense clusters of *Fascichnus dactylus*, with deeper penetration in upper right; same site as **a**, 2–3 m deep. (**c**) Diverging colony of *Fascichnus frutex,* same site as **a**, 2–3 m deep. (**d**) *Fascichnus frutex*, One Tree Island, Great Barrier Reef, Australia, patch reef, 2 m deep. (**e**) *Fascichnus frutex*, Lee Stocking Island, Bahamas, windward *Acropora palmata* reef, 2–3 m deep. (**f**) *Fascichnus frutex*, next to a much smaller *F. dactylus*, Cozumel, Mexico, Bahia Columbia, 10 m deep (after Günther 1990). Clusters in **a**–**c** and the smaller cluster in **f** show repeated branching and formation of dense clusters, whereas those in **d** and **e** branch early, but then continue radiating into the substrate without branching. Scale bar in all pictures is 20 μm long

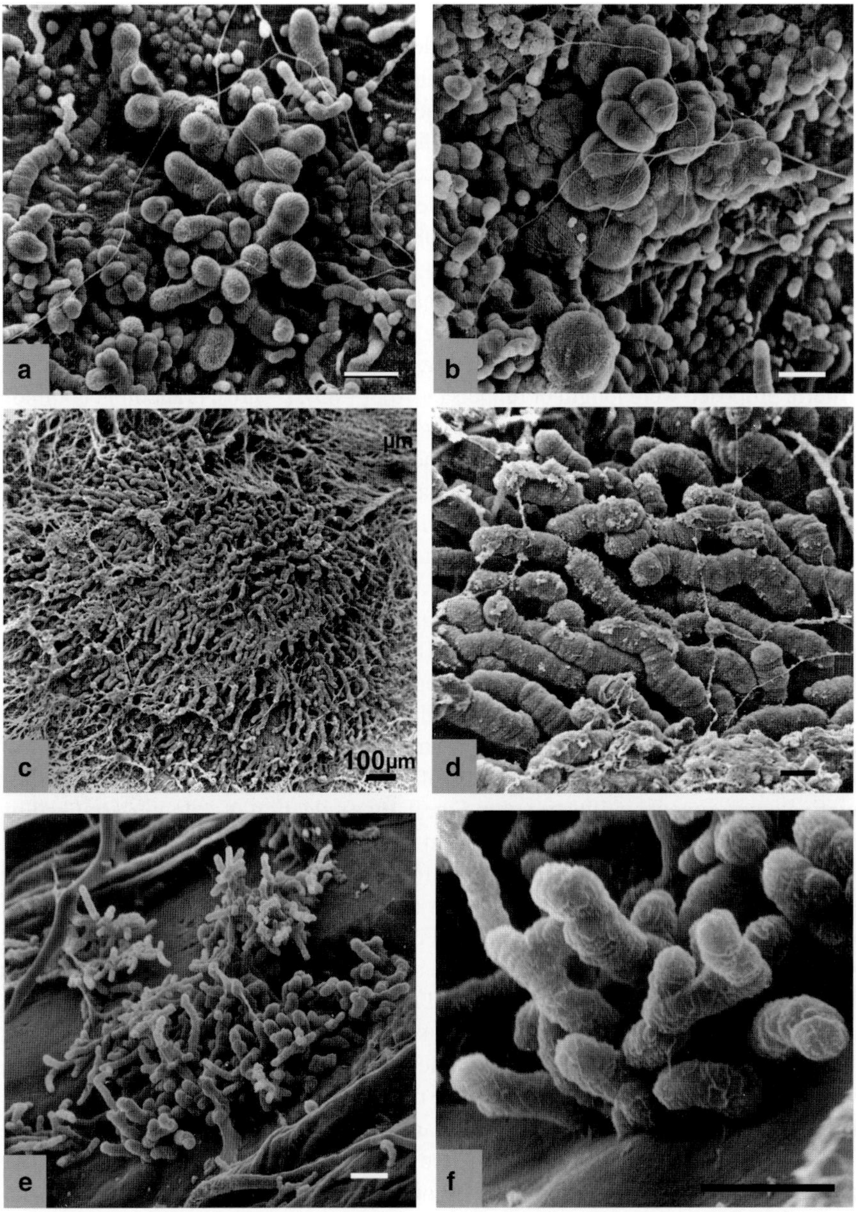

Fig. 10 *Fascichnus* clusters from fossil Tertiary sites. (**a**) A large cluster of *F. frutex*, surrounded by smaller traces of *F. dactylus* (*lower left*) and *F. parvus* (*above*) in a fossil *Cardium* shell. Villier-St.-Frédéric, Paris Basin, Middle Eocene (Lutetian), sea grass meadow, ca. 10–20 m. (**b**) A cluster of *F. rogus* with typical branches of mutually compressed cells. Same shell as in **a**. Microborings in a fossil. (**c**) Large *Fascichnus frutex* cluster surrounded by a dense network of fine filamentous borings in a *Callista* shell, Paris Basin, Auvers-St.-Georges, Oligocene, Stampian (shallow marine, sea grass meadow). (**d**) Detail of **c**, fine filamentous borings are *Scolecia filosa*, possibly

the colony development. Borings in some *Fascichnus* ispp. expand initially parallel to the substrate surface and later explore deeper layers of the substrate, while others bore immediately perpendicular to the surface or radiate inward forming clusters different in size and density. Changes that occur in the course of the lifetime of an organism (ontogeny) are known to be difficult to ascertain in fossils and especially in fossil traces. However, the abundance of well preserved microboring traces provides for reliable reconstructions of even complex life histories of the organisms that made them (Golubic and Radtke 2008).

Size is the most conspicuous property used to distinguish between *Fascichnus* traces. The tunnel diameters were originally used among other characters to separate ichnospecies with the following ranges proposed for *F. parvus*: 1.5–3.0 μm; for *F. dactylus*: 4.0–9.0 μm, and for *F. frutex*: 11–25 μm (Radtke 1991). As additional criteria proposed were length or tunnels (−40); (−85); (−110); branching: common in *F. parvus* and *F. frutex* but rare in *F. dactylus* and rounded to dome-shape ending of tunnels in all three ichnotaxa, sometimes club-shaped in *F. dactylus*.

The choice of this criterion was pragmatic and intuitive. Boring traces of similar shapes and sizes collected in modern environments could be measured and related to the microorganisms that produced them. In the majority of cases such traces were produced by euendolithic cyanobacteria and corresponded to different species of *Hyella* (e.g. Wisshak 2006), with the exception of the largest trace described as *F. grandis* (diameter ranging from 40 to 160 μm), which was found to be produced by the euendolithic rhizoids of the chlorophyte *Acetabularia* (Radtke et al. 1997).

The predominance of transverse cell division in *Hyella* and *Solentia* produced pseudofilaments and tunnels in the substrate with a specifically narrow range of diameter variation. The wide size range of *Fascichnus* traces from modern shallow marine environments is illustrated in Fig. 9, by a selection of colonies arranged by size in an increasing order, with a 20 μm long scale bar in all compared pictures. These pictures also show a relative uniformity in size within each colony. The smallest cluster corresponds to *F. parvus* (Fig. 9a), the next higher group is in the range of the ichnospecies *F. dactylus* (Fig. 9b, f left), followed by the larger *F. frutex* (Fig. 9c–f center). The contrast in size is most conspicuous in two clusters formed next to each other (Fig. 9f). Fossil *Fascichnus* traces from Tertiary deposits of Paris Basin with spectrum of forms is shown in Fig. 10.

In order to test the proposed taxonomic boundaries of *Fascichnus* ichnospecies, we have engaged first in determining size distribution of tunnel diameters in a large number of *Fascichnus* clusters, in comparison with size values published for cell dimensions of acid-extracted and cultured *Hyella* spp. (Table 2). In representative populations, selected on the basis of described ichnotaxa, the minimum vs. maximum diameter of each tunnel was plotted in a two-dimensional display and then summarized cluster by cluster by plotting average ± standard deviation (Fig. 11).

Fig. 10 (continued) produced by the filamentous cyanobacterium *Plectonema terebrans*. (**e**) Several branched clusters of *Fascichnus dactylus*, Paris Basin, Ferme de l'Orme, Middle Eocene (Lutetian). (**f**) Detail of **e**. Scale bar in all pictures is 20 μm, unless stated otherwise

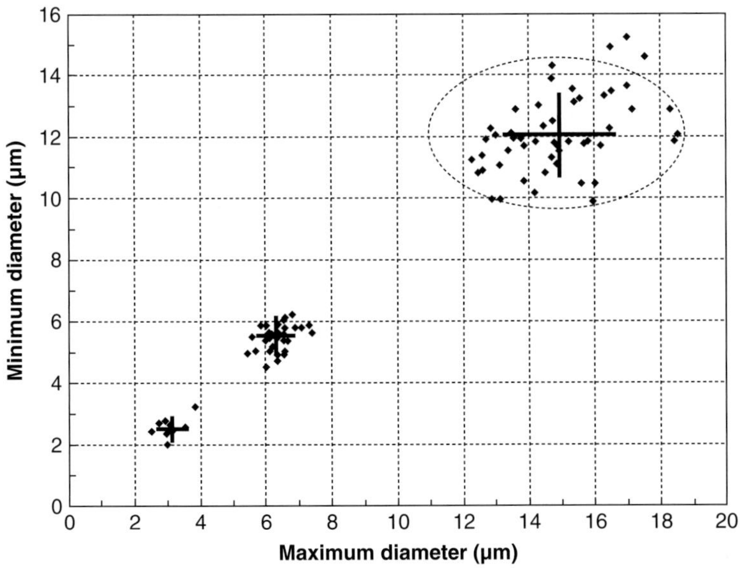

Fig. 11 Plots of maximum vs. minimum tunnel diameters of three randomly selected *Fascichnus* colonies. Note the narrow variation of data (*dots*) around mean ± standard deviation (cross diagrams) within each colony and clear separation between the colonies

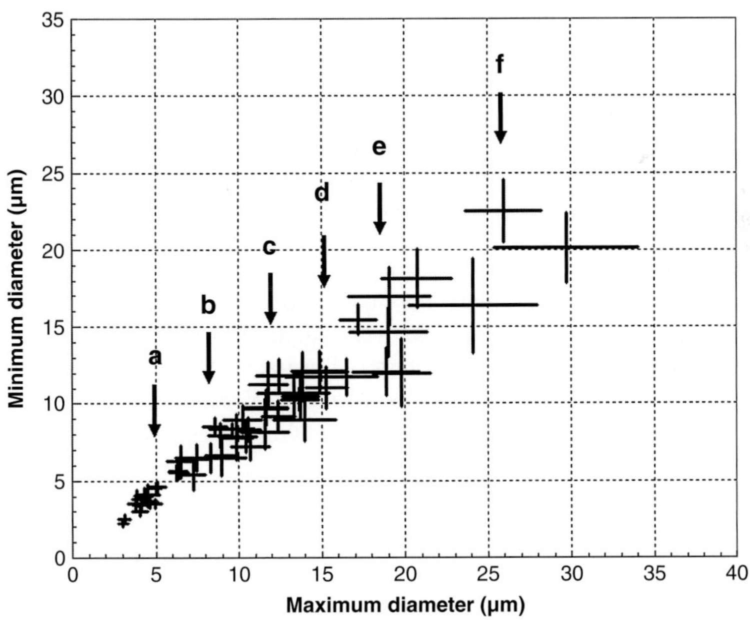

Fig. 12 Cross diagram plots (mean ± standard deviation) of a large sample of 54 *Fascichnus* clusters, showing overlapping data and lack of gaps between clusters. The positions of clusters illustrated in Fig. 9 are marked (**a–f**)

The plot of individual measurements produced a cloud of data points for each population, scattered along a linear function inclined 45° that would describe perfectly cylindrical borings. The distribution of measured parameters is close to normal, so that basic statistics is applied with the average values of selected populations close to the center of the respective data cloud, and the variability within each population expressed by standard deviation (arms of the cross). Area outlined by two standard deviations from the mean (dotted-lined oval) is expected to include about 95% of the sample, thus approximating the range of sizes for that population. The three selected populations had morphological features consistent with the description of the ichnogenus *Fascichnus*; they showed similar pattern in data distribution and were clearly distinct in size without overlap of measurements. They were well within size ranges of the three described *Fascichnus* ichnospecies: *F. parvus*, *F. dactylus* and *F. frutex*. However, such clear distinction could not be maintained when a large number of *Fascichnus* populations were measured. The added populations plotted in between leaving no gaps to separate the data clusters (Fig. 12).

By including a total of 54 *Fascichnus* populations in the sample, a pattern emerged in which tunnel diameter values of populations were arranged along a continuum, covering a wide total range of dimensions from 2 to 40 μm. Each cluster covered a part of that range, each showed close to normal distribution of values and relatively narrow intra-population variability (11.2 ± 3.7% of the mean) but also significant overlaps in tunnel diameter between clusters. The plot did not indicate any grouping or gaps in distribution of means that would support a separation of the measured populations by size. A similar pattern in size distribution within and between clonal populations has been reported earlier for *Gloeocapsa* among coccoid cyanobacteria inhabiting wet cliffs (Golubic 1965, 1966). The clonal nature of these populations was assured by enclosures in EPS envelopes.

Yet, the illustrated *Fascichnus* clusters suggest a higher diversity of forms, which differ regarding proximal to distal change in tunnel diameter, presence or absence of repeated constrictions along tunnels, or tunnel density, all properties not necessarily correlated with size or not in the same combination as in the formally described ichnotaxa. For example, the two clusters illustrated in Fig. 9b are both within the size range of *Fascichnus dactylus*, but differ in density and organization of tunnels in the colony. Similarly clusters illustrated in Fig. 9c–f each covers a different part of the size range originally ascribed to *F. frutex* but differ in direction of diameter change (compare Fig. 9c and 9d) and form (Fig. 9e and 9f).

7 Microboring Cyanobacteria and Prokaryote Speciation

Cell dimensions are routinely employed in characterizing taxa of cyanobacteria (and eukaryotic algae) and determination keys are constructed so as to separate taxa based on size differences with a few additional diagnostic characters employed (e.g. Geitler 1932; Komárek and Anagnostidis 1999, 2005). In our study of an ecologically defined group of prokaryotes (coccoid euendolithic cyanobacteria), a distinction based on size alone was lacking. Measurements were carried out on clonal populations, held together by EPS when extracted from the matrix, each characterized by a narrow normal distribution of sizes, supporting the existence of genetically stable units, with neighboring population overlapping in size. This pattern was faithfully reflected in tunnel diameter distribution of microboring clusters. The total range of cell sizes within forms of the genus *Hyella* and, similarly of the diameters of their boreholes between 2 and 40 µm and clear distinction between the end members of the range argued against inclusion of all these populations into a single taxon. Thus, the distinction between taxonomic units above clones but within established genera needs to be sought in other properties but size, which may or may not show similarly smooth transitions between populations. Correlation of qualitative and quantitative properties may then indicate the centers of diversification rather than gaps between them.

We recognize here a pattern of a continuous, fan-like diversification of populations regarding properties (cell size and tunnel diameters in this case), which are apparently neutral with respect to natural selection. Accordingly, these properties are inherited and distributed uniformly among successful descendent populations, forming continuous spectra of forms. The pattern is consistent with the asexual reproduction of cyanobacteria, which provides instant reproductive isolation of these prokaryotic organisms and a gradual divergence of properties over time. It also indicates the absence or ineffectiveness of lateral exchange of genetic information at the level of metapopulations. This is in contrast to regular lateral transfer of genetic information realized in sexually reproducing eukaryotes that results in formation of a common gene pool of sexually compatible population members, marked by sharper boundaries toward sexually incompatible or reproductively isolated ones (De Queiros 2005).

The observed pattern is consistent with the proposed clonal dynamics and population replacement mechanism in the speciation process of prokaryotic organisms (Cohan 2002). Similar pattern of speciation among sexually reproducing eukaryotes exists in the so called ring species where the gene flow lags behind the geographic spreading rate of populations (Moritz et al. 1992). In the case of ring species, the populations show distinct characters, but these change gradually along continuous geographic gradients. These groups are usually categorized as subspecies, although the end-members may become reproductively incompatible, thus qualifying for the species status in terms of the Biological Species Concept (Mayr 1982; De Queiros 2005). Clear separations by gaps (hiatus) between taxa are a convenience in estimation of diversity introduced by extinctions but also by uneven distribution of taxa (demographic replacements, *sensu* Templeton 1989).

The present study explored the phenotypic expression side of a polyphasic approach to cyanobacterial diversity, the molecular aspects yet to be employed in the study of microbial euendoliths. The experiences thus far project a significantly higher diversity of taxa then are formally described as suggested by the observations of cell organization pigmentation and boring behavior of coccoid cyanobacteria. It is expected that the observed phenotypes are supported with corresponding high number of molecular genotype signatures as has been obtained in recent studies of endoliths in travertine systems of the Yellowstone National Park (Norris and Castenholz 2006), microbial mats of the Antarctica (Jungblut et al. 2005) and marine stromatolites (Foster et al. 2009).

Acknowledgments The international and interdisciplinary cooperation has been supported by the Alexander-von-Humboldt Foundation and by the Hanse Institute for Advanced Research. We thank Prof. Dr. J. Reitner (Univ. Göttingen) for support during the International Kalkowsky Symposium in 2008 and Dr. R.P. Reid (Univ. Miami) for stimulating discussions regarding endoliths and stromatolites.

References

Al-Thukair AA (2002) Effect of oil pollution on euendolithic cyanobacteria of the Arabian Gulf. Environmental Microbiology 4:125–129

Al-Thukair AA, Golubic S (1991a) Five new *Hyella* species from the Arabian Gulf. Algological Studies 64:167–197

Al-Thukair AA, Golubic S (1991b) New endolithic cyanobacteria from the Arabian Gulf. I. *Hyella immanis* sp. nov. Journal of Phycology 27:766–780

Al-Thukair AA, Golubic S (1996) Characterization of *Hyella caespitosa* var. *arbuscula* var. nov. (Cyanophyta, Cyanobacteria) from shoaling ooid sand grains, Arabian Gulf. Nova Hedwigia Beihefte 112:83–91

Al-Thukair AA, Green J (1988) New endolithic taxa in modern and ancient (late Proterozoic) shallow water marine environments. Journal of Phycology 24:12

Al-Thukair AA, Golubic S, Rosen G (1994) New euendolithic cyanobacteria from the Bahama Bank and the Arabian Gulf: *Hyella racemus* sp. nov. Journal of Phycology 30:764–769

Balog S-J (1996) Boring thallophytes in some Permian and Triassic reefs: bathymetry and bioerosion. Gättinger Arbeiten für Geologie und Paläontologie Sb 2:305–308

Balog S-J (1997) Mikroendolithen im Capitan Riff Komplex (New Mexico, USA). Courier, Forschungsinstitut Senckenberg 201:47–55

Bertling M (2007) What's in a Name? Nomenclature, Systematics, Ichnotaxonomy. In: Miller W (ed) Trace Fossils: Concepts, Problems, Prospects. Elsevier, Amsterdam, pp 81–91

Bornet ME, Flahault C (1889) Sur quelques plantes vivant dans le test calcaire des mollusques. Bulletin de la Société Botanique de France 36:147–176

Bromley RG (1990) Trace Fossils: Biology and Taphonomy. Unwin Hyman, London, pp 280

Bundschuh M (2000) Silurische Mikrobohrspuren – ihre Beschreibung und Verteilung in verschiedenen Faziesräumen (Schweden, Litauen, Großbritannien und U.S.A.). Ph.D. Thesis, FB Geowissenschaften, Johann Wolfgang Goethe Universität, Frankfurt am Main, Germany, pp 1–129

Bundschuh M, Balog S (2000) *Fasciculus rogus* nov. isp., an endolithic trace fossil. Ichnos 7:149–152

Campbell SE (1980) *Palaeoconchocelis starmachii*, a carbonate boring microfossil from the Upper Silurian of Poland (425 million years old): implications for the evolution of the Bangiaceae (Rhodophyta). Phycologia 19:25–36

Campbell SE (1982) Precambrian endoliths discovered. Nature 299:429–431

Campbell SE, Kazmierczak J, Golubic S (1979) *Palaeoconchocelis starmachii* n. gen., n. sp., a Silurian endolithic rhodophyte (Bangiaceae). Acta Paleontologica Polonica 24:405–408

Castenholz RW, Garcia-Pichel F (2000) Cyanobacterial responses to UV-radiation. In: Whitton BA, Potts M (eds) Ecology of Cyanobacteria: Their Diversity in Time and Space. Kluwer Acad Publ, Dordrecht, pp 591–611

Cohan FM (2002) What are bacterial species? Annual Review of Microbiology 56:457–487

De Queiros K (2005) Ernst Mayr and the modern concept of species. Proceedings of the National Academy of Sciences of the United States of America 102:6600–6607

Decho AW, Visscher PT, Reid RP (2005) Production and cycling of natural microbial exopolymers (EPS) within a marine stromatolite. Palaeogeography, Palaeoclimatology 219:71–86

Dill RF, Shinn EA, Jones AT, Kelly K, Steinen RP (1986) Giant subtidal stromatolites forming in normal salinity waters. Nature 324:55–58

Dravis JJ (1983) Hardened subtidal stromatolites, Bahamas. Science 219:385–386

Ehling-Schulz M, Scherer S (1999) UV protection in cyanobacteria. European Journal of Phycology 34:329–338

Ercegović A (1927) Trois noveaux genres des Cyanophycées lithophytes de la côte adriatique. Acta Botanica 2:78–84

Ercegović A (1929) Sur quelques nouveaux types de Cyanophycées lithophytes de la côte adriatique. Archiv für Protistenkunde 66:164–174

Ercegović A (1932) Études écologique et sociologique des Cyanophycées lithophytes de la côte Yugoslave de l'Adriatique. Bulletin International de l'Académie Yougoslave de la Sciences des Arts, Classe Mathematic et Naturelles 26:33–56

Feldmann M, McKenzie JA (1998) Stromatolite-thrombolite associations in a modern environment, Lee Stocking Island, Bahamas. Palaios 13:201–212

Foster JS, Green SJ, Ahrendt SR, Golubic S, Reid RP, Hetherington KL, Bebout L (2009) Molecular and morphological characterization of cyanobacterial diversity in the stromatolites of Highborne Cay, Bahamas. The ISME Journal 3(5):573–587

Geitler L (1932) Cyanophyceae. In: Rabenhorst L (ed) Kryptogamen-Flora von Deutschland, Österreich und der Schweiz. Akademische Verlagsgesellschaft, Leipzig, pp 1196

Gektidis M, Golubic S (1996) A new endolithic cyanophyte/cyanobacterium: *Hyella vacans* sp. nov. from Lee Stocking Island, Bahamas. Nova Hedwigia Beihefte 112:91–98

Gektidis M, Dubinsky ZNE, Goffredo S (2007) Microendoliths of the Shallow Euphotic Zone in open and shaded habitats at 30°N – Eilat, Israel – paleoecological implications. Facies 53:43–55

Glaub I (1994) Mikrobohrspuren in ausgewählten Ablagerungsräumen des europäischen Jura und der Unterkreide (Klassifikation und Palökologie). Courier Forschungsinstitut Senckenberg 174:1–324

Glaub I, Golubic S, Gektidis M, Radtke G, Vogel K (2007) Microborings and microbial endoliths: Geological implications. In: Miller W (ed) Trace Fossils: Concepts, Problems, Prospects. Elsevier, Amsterdam, pp 368–381

Golubic S (1965) Zur Revision der Gattung *Gloeocapsa* Kützing (Cyanophyta). Schweizerische Zeitschrift für Hydrologie 27:218–232

Golubic S (1966) Über den Aufbau des Taxons bei den Cyanophyten anhand der Beispiele *Tolypothrix, Oscillatoria* und *Gloeocapsa*. Verhandlungen, Internationaler Verein für Limnologie 16:1577–1581

Golubic S (1969) Distribution, taxonomy, and boring patterns of marine endolithic algae. American Zoologist 9:747–751

Golubic S (1990) Shell boring microorganisms. In: Boucot A (ed) The Evolutionary Paleobiology of Behavior and Coevolution. Elsevier, Amsterdam, pp 347–352

Golubic S, Browne KM (1996) *Schizothrix gebeleinii* sp. nov. builds subtidal stromatolites, Lee Stocking Island, Bahamas. Algological Studies 83:273–290

Golubic S, Marčenko E (1965) Über Konvergenzerscheinungen bei Standortsformen der Blaualgen unter extremen Lebensbedingungen. Schweizerische Zeitschrift für Hydrologie 27: 207–217

Golubic S, Radtke G (2008) The trace *Rhopalia clavigera* isp. n. reflects the development of its maker *Eugomontia sacculata* Kornmann, 1960. In: Wisshak M, Tapanila L (eds) Current Developments in Bioerosion. Springer, Berlin, pp 95–108

Golubic S, Seong-Joo Lee (1999) Early cyanobacterial fossil record: preservation, palaeoenvironments and identification. European Journal of Phycology 34:339–348

Golubic S, Brent G, Le Campion-Alsumard T (1970) Scanning electron microscopy of endolithic algae and fungi using a multipurpose casting-embedding technique. Lethaia 3:203–209

Golubic S, Perkins RD, Lukas KJ (1975) Boring microorganisms and microborings in carbonate substrates. In: Frey RW (ed) The Study of Trace Fossils. Springer, Heidelberg, pp 229–259

Golubic S, Friedmann I, Schneider J (1981) The lithobiontic ecological niche, with special reference to microorganisms. Journal of Sedimentary Petrology 51:475–478

Golubic S, Campbell SE, Spaeth C (1983) Kunstharzausgüsse fossiler Mikroben-Bohrgänge. Der Präparator 29:197–200

Golubic S, Campbell SE, Drobne K, Cameron B, Balsam WL, Cimerman F, Dubois L (1984) Microbial endoliths: a benthic overprint in the sedimentary record, and a paleobathymetric cross-reference with foraminifera. Journal of Paleontology 58:351–361

Golubic S, Al-Thukair AA, Gektidis M (1996) New endolithic cyanobacteria from the Arabian Gulf and the Bahama Bank: *Solentia sanguinea* sp. nov. Algological Studies 83:291–301

Günther A (1990) Distribution and bathymetric zonation of shell-boring endoliths in Recent reef and shelf environments: Cozumel, Yucatan (Mexico). Facies 22:233–262

Hofmann K (1996) Die mikroendolithischen Spurenfossilien der borealen Oberkreide Nordwest-Europas und ihre Faziesbeziehungen. Geologisches Jahrbuch A 136:3–153

Jungblut A-D, Hawes I, Mountfort D, Dietrich D, Burns BP, Neilan BA (2005) Diversity within cyanobacterial mat communities in variable salinity meltwater ponds of McMurdo Ice Shelf, Antarctica. Environmental Microbiology 7:519–529

Kalkowsky E (1908) Oolith und Stromatolith im norddeutschen Buntsandstein. Zeitschrift der Deutschen geologischen Gesellschaft 60:68–125

Kazmierczak J, Golubic S (1976) Oldest organic remains of boring algae from Polish Upper Silurian. Nature 261:404–406

Kiene W, Radtke G, Gektidis M, Golubic S, Vogel K (1995) Factors controlling the distribution of microborers in Bahamian reef environments. In: Schuhmacher H, Kiene W, Dullo W-C (eds) Factors Controlling Holocene Reef Growth: An Interdisciplinary Approach. Facies 32: 174–188

Knoll AH, Golubic S (1992) Proterozoic and living cyanobacteria. In: Schidlowski M, Golubic S, Kimberley MM, McKirdy DM, Trudinger PA (eds) Early Organic Evolution, Implications for Mineral and Energy Resources. Springer, Berlin, pp 450–462

Knoll AH, Golubic S, Green J, Swett K (1986) Organically preserved microbial endoliths from the Late Proterozoic of East Greenland. Nature 321:856–857

Komárek J, Anagnostidis K (1999) Cyanoprokaryota – 1: Chroococcales. In: Ettl H, Gärtner G, Heynig H, Mollenhauer D (eds) Süβwasserflora von Mitteleuropa. 19/1, G Fischer, Jena, pp 548

Komárek J, Anagnostidis K (2005) Cyanoprokaryota – 2: Oscillatoriales. In: Ettl H, Gärtner G, Krienitz L, Schagerl M (eds) Süβwasserflora von Mitteleuropa. 19/2, Spectrum, Akademischer Verlag, Heidelberg, pp 759

Le Campion-Alsumard T (1966) Contribution à l'étude des Cyanophycées lithophytes des étages supralittoral et medio-littoral (Région de Marseille). Téthys 1:119–172

Le Campion-Alsumard T (1979) Les cyanophycées endolithes marines: Systématique, ultrastructure, écologie et biodestruction. Oceanologica Acta 2:143–156

Le Campion-Alsumard T (1991) Three *Hyella* taxa (endolithic cyanophytes) from tropical enviroments (Lizard Island, Great Barrier Reef). Algological Studies 64:159–166

Le Campion-Alsumard T, Golubic S (1985a) Ecological and taxonomic relationships between euendolithic cyanophytes *Hormathonema* and *Solentia*. In: Golubic S, Komárek J, Lhotsky O (eds) Cyanophyta (Cyanobacteria) Morphology, Taxonomy, Ecology. Algological Studies 38/39:115–118

Le Campion-Alsumard T, Golubic S (1985b) *Hyella caespitosa* Bornet et Flahault and *Hyella balani* Lehman (Pleurocapsales, Cyanophyta): a comparative study. In: Golubic S, Komárek J, Lhotsky O (eds) Cyanophyta (Cyanobacteria) Morphology, Taxonomy, Ecology. Algological Studies 38/39:119–148

Le Campion-Alsumard T, Golubic S, Hutchings P (1995) Microbial endoliths in skeleton of live and dead corals: *Porites lobata* (Moorea, French Polynesia). Marine Ecology Progress Series 117:149–157

Le Campion-Alsumard T, Golubic S, Pantazidou A (1996) On the euendolithic genus *Solentia* Ercegovic (Cyanophyta/Cyanobacteria). Algological Studies 83:108–127

Lukas KJ, Golubic S (1981) New endolithic cyanophytes from the North Atlantic Ocean: I. *Cyanosaccus piriformis* gen. et sp. nov. Journal of Phycology 17:224–229

Lukas KJ, Golubic S (1983) New endolithic cyanophytes from the North Atlantic Ocean: II. *Hyella gigas* sp. nov. from the Florida continental margin. Journal of Phycology 19:129–136

Lukas KJ, Hoffman EJ (1984) New endolithic cyanophytes from the North Atlantic Ocean. III: *Hyella pyxis* Lukas & Hoffman sp. nov. Journal of Phycology 20:515–520

Macintyre IG, Prufert-Bebout L, Reid RP (2000) The role of endolithic cyanobacteria in the formation of lithified laminae in Bahamian stromatolites. Sedimentology 47:915–921

May JA, MacIntyre IG, Perkins RD (1982) Distribution of microborings within planted substrates along a barrier reef transect, Carrie Bow Cay, Belize. In: Rützler K, MacIntyre IG (eds) The Atlantic Barrier Reef Ecosystem at Carrie Bow Cay, Belize, I. Structure and Communities. Smithsonian Contribution of the Marine Sciences 12:93–107

Mayr E (1982) The Growth of Biological Thought: Diversity, Evolution, and Inheritance. Harvard University Press, Cambridge MA, pp 974

Moritz C, Schneider CJ, Wake DB (1992) Evolutionary relationships within the *Ensatina eschscholtzii* complex confirm the ring species interpretation. Systematic Biology 41: 273–291

Norris TB, Castenholz RW (2006) Endolithic photosynthetic communities within ancient and recent travertine deposits in Yellowstone National Park. FEMS Microbiological Ecology 57:470–483

Paterson DM, Aspden RJ, Visscher PT, Consalvey M, Andres MS, Decho AW, Stolz J, Reid P (2008) Light-dependant biostabilisation of sediments by stromatolite assemblages. PLoS ONE 3(9):3176

Perkins RD, Tsentas CI (1976) Microbial Infestation of Carbonate Substrates Planted on the St. Croix Shelf, West Indies. Geological Society of America, Bulletin 87:1615–1628

Perry CT, Macdonald IA (2002) Impacts of light penetration on the bathymetry of reef microboring communities: implications for the development of microendolithic trace assemblages. Palaeogeography, Palaeoclimatology, Palaeoecology 186:101–113

Radtke G (1991) Die mikroendolithischen Spurenfossilien im Alt-Tertiär West-Europas und ihre palökologische Bedeutung. Courier Forschungsinstitut Senckenberg 138:1–185

Radtke G (1993) The distribution of microborings in molluscan shells from Recent reef environments at Lee Stocking Island, Bahamas. Facies 29:81–92

Radtke G (2007) Mikroendolithen in Molluskenschalen aus den Mittleren Pechelbronn-Schichten (Mainzer Becken). In: Radtke G, Martini E (eds) Die Bohrungen Wallau im nordöstlichen Mainzer Becken (Rotliegend, Pechelbronn-Gruppe, Bodenheim-Formation). Geologische Abhandlungen Hessen 116:141–155

Radtke G, Golubic S (2005) Microborings in mollusk shells, Bay of Safaga, Egypt: Morphometry and ichnology. Facies 51:125–141

Radtke G, Le Campion-Alsumard T, Golubic S (1996) Microbial assemblages of the bioerosional "notch" along tropical limestone coasts. Algological Studies 83:469–482

Radtke G, Gektidis M, Golubic S, Hofmann K, Kiene WE, Le Campion-Alsumard T (1997) The identity of an endolithic alga: *Ostreobium brabantium* Weber-van Bosse is recognized as

carbonate-penetrating rhizoids of *Acetabularia* (Chlorophyta, Dasycladales). Courier Forschungsinstitut Senckenberg 201:341–347
Reid RP, Browne KM (1991) Intertidal stromatolites in a fringing Holocene reef complex, Bahamas. Geology 19:15–18
Reid RP, Macintyre IG (2000) Microboring versus recrystallization: Further insight into the micritization process. Journal of Sedimentary Research 70:24–28
Reid RP, Macintyre IG, Browne KM, Steneck RS, Miller T (1995) Stromatolites in the Exuma Cays, Bahamas: Uncommonly common. Facies 33:1–18
Reid RP, Visscher PT, Decho AW, Stolz JF, Bebout BM, Dupraz CP, Macintyre IG, Paerl HW, Pinckney JL, Prufert-Bebout L, Steppe TF, DesMarais DJ (2000) The role of microbes in accretion, lamination and early lithification of modern marine stromatolites. Nature 406:989–992
Reid RP, James NP, Macintyre IG, Dupraz CP, Burne RV (2003) Shark Bay stromatolites: Microfabrics and reinterpretation of origins. Facies 49:299–324
Riding R, Braga JC, Martin JM (1991) Oolite stromatolites and thrombolites; Miocene, Spain: analogues of recent giant Bahamian examples. Sedimentary Geology 71:121–127
Rippka R, Waterbury JB, Herdman M, Castenholz RW (2001) Subsection II (Formerly Pleurocapsales Geitler 1925, emend Waterbury Stanier 1978). In: Boone DR, Castenholz RW, Garrity GM (eds) Bergey's Manual of Systematic Bacteriology, 2nd ed, vol. 1, Springer, Berlin, pp 514–539
Schmidt H (1992) Mikrobohrspuren ausgewählter Faziesbereiche der tethyalen und germanischen Trias (Beschreibung, Vergleich, bathymetrische Interpretation). Frankfurter geowissenschaftliche Arbeiten A12:1–228
Schneider J (1976) Biological and inorganic factors in the destruction of limestone coasts. Contribution Sedimentology 6:1–112
Schneider J (1977) Carbonate construction and decomposition by epilithic and endolithic microorganisms in salt and freshwater. In: Flügel E (ed) Fossil Algae. Springer, Berlin, pp 248–260
Templeton AR (1989) The meaning of species and speciation: A genetic perspective. In: Otte D, Endler JA (eds) Speciation and its Consequences. Sinauer, Sunderland, MA, pp 3–27
Tribollet A (2008) The boring microflora in modern coral reef ecosystems: a review of its roles. In: Wisshak M, Tapanila L (eds) Current Developments in Bioerosion. Springer, Berlin, pp 67–94
Tribollet A, Golubic S (2005) Cross-shelf differences in the pattern and pace of bioerosion of experimental carbonate substrates exposed for 3 years on the northern Great Barrier Reef, Australia. Coral Reefs 24:422–434
Tribollet A, Langdon C, Golubic S, Atkinson M (2006) Endolithic microflora are major primary producers in dead carbonate substrates of Hawaiian coral reefs. Journal of Phycology 42:292–303
Tribollet A, Golubic S, Radtke G, Reitner J (2010) On Microbiocorrosion. In: Reitner J, Queric N-V, Arp G (eds) Advances in Stromatolite Geobiology. Lecture Notes in Earth Sciences 131, Springer, Berlin, pp 243–253
Vogel K, Brett CE (2009) Record of microendoliths in different facies of the Upper Ordovician in the Cincinnati Arch region USA: The early history of light-related microendolithic zonation. Palaeogeography, Palaeoclimatology, Palaeoecology 281:1–24
Vogel K, Marincovich L Jr (2004) Paleobathymetric implications of microborings in Tertiary strata of Alaska, USA. Palaeogeography, Palaeoclimatology, Palaeoecology 206:1–20
Vogel K, Golubic S, Brett CE (1987) Endolith associations and their relation to facies distribution in the Middle Devonian of New York State, USA. Lethaia 20:263–290
Vogel K, Bundschuh M, Glaub I, Hofmann K, Radtke G, Schmidt H (1995) Hard substrate ichnocoenoses and their relation to light and marine bathymetry. Neues Jahrbuch für Geologie und Paläontologie, Abhandlungen 195:49–61

Vogel K, Gektidis M, Golubic S, Kiene WE, Radtke G (2000) Experimental studies on microbial bioerosion at Lee Stocking Island, Bahamas and One Tree Island, Great Barrier Reef, Australia: implications for paleoecological reconstructions. Lethaia 33:190–204

Waterbury JB, Stanier RW (1978) Patterns of growth and development in Pleurocapsalean cyanobacteria. Microbiological Reviews 42:2–44

Wisshak M (2006) High-latitude bioerosion: the Kosterfjord experiment. Lecture Notes in Earth Sciences 109:1–202

Wisshak M, Seuβ B, Nützel A (2008) Eolutionary implications of an exceptionally preserved Carboniferous microboring assemblage in the Buckhorn Asphalt lagerstätte (Oklahoma, USA). In: Wisshak M, Tapanila L (eds) Current Developments in Bioerosion. Springer, Berlin, pp 21–54

Zhang Y, Golubic S (1987) Endolithic microfossils (cyanophyta) from early Proterozoic stromatolites, Hebei, China. Acta Micropaleontologica Sinica 4:1–12

On Microbiocorrosion

Aline Tribollet, Stjepko Golubic, Gudrun Radtke, and Joachim Reitner

1 Introduction

Destruction of rocks and minerals by biological activities has been termed bioerosion (Neumann 1966). It includes mechanical as well as chemical effects, i.e. bioabrasion and biocorrosion (Schneider 1976; Golubic and Schneider 1979). However both processes often co-occur; they are functionally interconnected and mutually supportive. Biocorrosion can result from the activity of macro- or micro- organisms and thus is called macrobiocorrosion and microbiocorrosion. Microbiocorrosion can also be closely associated with microbial rock formation and consolidation in stromatolitic structures (Reid et al. 2000; Macintyre et al. 2000; Garcia-Pichel et al. 2004; Dupraz and Visscher 2005). In fact, the oldest known fossils of microboring organisms were located in lithified horizons of silicified stromatolites (Zhang and Golubic 1987).

2 Niches

Microbiocorrosion is due to various phototrophic and organotrophic organisms such as bacteria, cyanobacteria, red and green eukaryotic algae, lichens and fungi. Depending on their relation to the substrate, the following niches are

A. Tribollet (✉)
LOPB, Institut de Recherche pour le Développement, BP A5, 98848 Nouméa Cedex, New Caledonia
e-mail: aline.tribollet@ird.fr

S. Golubic
Biological Science Center, Boston University, 05 Cummington Str, Boston, MA 02215, USA

G. Radtke
Hessisches Landesamt für Umwelt und Geologie, Rheingaustr, 186, 65203 Wiesbaden, Germany

J. Reitner
Department of Geobiology and Center for Geosciences, University of Göttingen, Goldschmidtstr. 3, 37077 Göttingen, Germany

J. Reitner et al., *Advances in Stromatolite Geobiology*,
Lecture Notes in Earth Sciences 131, DOI 10.1007/978-3-642-10415-2_17,

distinguished: organisms, which occupy the surface of rocky substrates are called epiliths, those located within rocks are endoliths. Among the latter, chasmo- and crypto-endoliths occupy fissures and pore spaces in the rock respectively; while those which actively penetrate rocks and other carbonate substrates and leave microboring traces are called euendoliths (Golubic et al. 1981). Assessment of diversity of microbial assemblages using molecular methods has to date been carried out in a few terrestrial cryptoendolithic environments (De la Torre et al. 2003; Walker et al. 2005; Walker and Pace 2007a, b), but very few studies dealt with marine and freshwater euendolithic communities (e.g. Chacón et al. 2006). We support the polyphasic approach, as applied to marine benthic cyanobacterial populations (Abed et al. 2003, 2006, 2008), whereby phenotypic and genotypic properties are compared at the population level. Several methods in the study of natural populations of euendoliths are available for qualitative and quantitative assessments.

Euendoliths, similar to biofilms, and crustose coralline algae, occupy interface habitats on the contact between rocky substrate and water. The most conspicuous distinction among euendolithic microorganisms is between phototrophic and organotrophic (heterotrophic) euendoliths with respect to their behavior and distribution which, in turn, reflects their dependence of energy sources. Phototrophs are restricted to illuminated environments, although they show a wide range of tolerance both to excessive and deficient light as well as to its spectral quality. They do not occur in completely aphotic environments of the deep sea, nor in the dark interior of the substrate. Densities of phototrophic populations usually diminish sharply when observed in rock sections with different species being often zonally arranged according to their light requirements. In contrast, selectivity and distribution of organotrophs depends on the presence of organic matter that was either originally incorporated into the substrate (e.g. in skeletal carbonates), or became secondarily available by the growth of phototrophic euendoliths.

Distribution of microborings in the rock can be observed in petrographic thin sections (Fig. 1a), including the assessment of their abundance and distribution as well as the amounts of carbonate removed. Microborings can be detected in transparent substrates (Fig. 1b, d), but their spatial display of borings within the substrate is better resolved when studied by Scanning Electron Microscopy (SEM) of resin-cast borings (Fig. 1c, e). The resident euendoliths are identified after they have been acid-extracted from the rock. However, they could be best related to the shape of their borings in double embedded preparations (Fig. 1f, g), which preserved the organisms *in situ* using techniques developed for transmission electron microscopy.

3 Substrates

Various types of substrates are affected by biocorrosion. Limestone, carbonate skeletons (of corals, foraminifera, bryozoa, echinodermata, brachiopoda, mollusca as well as the thalli of encrusting algae) and sand grains are all subject to

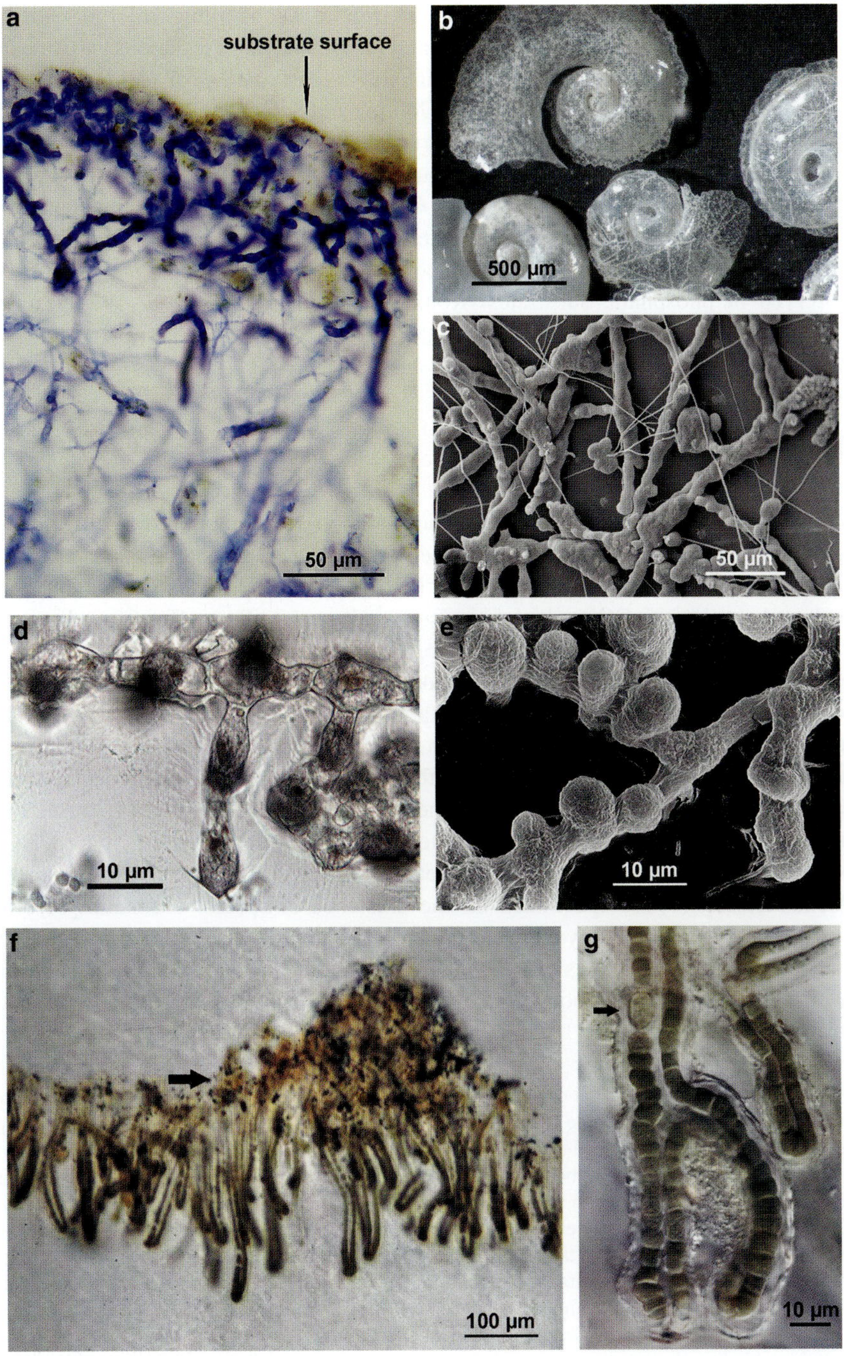
a
substrate surface
50 µm
b
500 µm
c
50 µm
d
10 µm
e
10 µm
f
100 µm
g
10 µm

biocorrosion by phototrophic and organotrophic euendolithic microorganisms (Perry 1998; Vogel et al. 2000; Tribollet and Payri 2001). For instance, a significant abundance and diversity of euendoliths inhabit extensive areas of seafloor covered by sand (Al-Thukair and Golubic 1991; Golubic and Schneider 2003; Rasheed et al. 2004; Radtke and Golubic 2010). Wild et al. (2006) quantified the microbial cells in reef before and after carbonate dissolution and found that the number doubled when euendoliths were added to the census. Abundance of euendolithic filaments can reach half a million per square centimeter in coastal limestone (Schneider and Le Campion-Alsumeard 1999). Microbiocorrosion by euendoliths has also been observed in phosphatic substrates (Roux 1887; Underwood et al. 1999; Zhang and Pratt 2008).

4 Mechanisms

The exact mechanisms involved in biocorrosion remain unknown although a number of plausible hypotheses have been proposed. It is probable that more than one of several possible mechanisms is involved. A simple mechanism involves hydrated respiratory CO_2, which acts as a weak acid, still an effective solvent of calcium carbonate with significant geological impact (Garcia-Pichel 2006). Other mechanisms may include local application of organic acids (Alexandersson 1975), which are regular products of fermentations and, like respiration, involved in microbial energy-generating metabolism. Microorganisms modify the pH of their immediate microenvironment changing the mineral solubility in the process. More complex is the action of phototrophic euendoliths, because their activity results in pH elevation which stimulates carbonate deposition rather than dissolution. It is possible that the two processes are separated in time (day vs. night) or space. Alexandersson (1975) speculated on the existence of special organelles of biocorrosion at the tips of cyanobacterial boring filaments. Ultrastructures of EPS surrounding the boring cells of *Solentia* showed textures localized at the dissolution front which suggest transport of some solvent or enzyme through the sheath and

---

Fig. 1 Microboring organisms or euendoliths in various carbonate substrates. (**a**) Phototrophic euendoliths penetrating from the surface of dead coral skeleton into its interior. Petrographic thin section stained with toluidine blue. (**b**) Shells of pteropods from a sediment sample Red Sea, Port Sudan where 1,558 m deep, bored by organotrophic microbial euendoliths. (**c**) Resin casts replicate the microborings from the sample in B. (**d**) Septate chlorophyte *Eugomontia sacculata* Kornmann boring into the calcite spar. Note the shadows of lateral bags descending deeper into the substrate and the conforming of the boring front with the calcite mineralogy. (**e**) Display of borings of *Eugomontia sacculata* and its lateral bags in the view from the interior of the substrate. (**f**) Euendolithic heterocystous cyanobacterium *Kyrtuthrix dalmatica* Ercegović penetrating coastal limestone: In situ position of euendolith filaments penetrating the rock with peculiar loop-shaped branches. Arrow indicates the position of the rock surface between the water column (above) and the rock (below). (**g**) Detail showing C-fixing trichome cells as distinct from N-fixing heterocyst (*arrow*). Double embedded preparation, courtesy of Therese Le Campion-Alsumard (Marseille)

Fig. 2 Impacts of microbiocorrosion and microbioabrasion. (**a**) Lithified iron-manganese crusts lining internal cavities in fossil reef: Lower Jurassic of the Northern Calcareous Alps (Pliensbachian, Adnet, Austria). Note the irregular outlines of coated grains with enclosed fossils. (**b**) Iron-

a network-pattern of carbonate dissolution at a sub-micrometer scale (Le Campion-Alsumard et al. 1996). Dissolution processes may also be enzymatically activated and/or regulated, e.g. by carbonic anhydrase (Hatch 1980) or involving calcium pumps (Garcia-Pichel 2006). The interest for microbial biocorrosion was for many years subject of the material science research, mainly in conjunction with biocorrosion of metals (e.g. Heitz et al. 1996; Videla and Herrera 2005). Microbial biofilms were found to be associated with promotion as well as prevention of corrosion of metal surfaces, some associated with the marine bacterium *Vibrio* (Videla 1996). These aspects of biocorrosion are beyond the interest of the present review, although the same principles of electrochemical dissolution may apply to carbonates associated with deposits of hydrated Fe and Mn oxides as observed on internal coatings and biofilms in cavities within coral reefs (Fig. 2a). The metal biofilms, the underlying calcium carbonate rocks, and the sea water act as a natural electrochemical cell with the calcium carbonate rocks and the biofilms having anode and cathode functions, respectively. These coatings are layered (pseudo-stromatolitic) and often enriched in metals and clay minerals; they appear to be microbially formed hard grounds which exhibit an irregularly shaped baseline strongly reminiscent of corrosion fabrics (Fig. 2b). Reitner (1993) and Reitner et al. (2000) described this process in modern and Holocene cryptic cave carbonates from the Great Barrier Reef (Australia) and the Salt River Canyon of St. Croix (Caribbean). They calculated that within ca. 3000 y, roughly 50% of the microbially formed carbonates may have been dissolved due to electrochemical corrosion.

5 Rates

Biocorrosion rates have essentially been measured by exposing blocks of carbonate (mostly carved out of skeleton of hermatypic corals such as *Porites*) to settlement and penetration by carbonate-boring organisms. The removal rates of carbonate has been determined by the weight loss over extended periods of time (Kiene and Hutchings 1994) or by quantifying the volume of boreholes along

---

Fig. 2 (continued) manganese biofilm in a modern microbialite (Lizard Island reef caves) with characteristic concave outlines of corrosional pattern (Reitner 1993; Reitner et al. 2000). (**c**) Grazing gastropod *Patella* sp. ca. 3 cm in diameter, leaving white denuded trace. (**d**) Closer look at gastropod grazing tracks. The toe-nail is about 1 cm wide. (**e**) Sea urchin *Diadema* sp. heavily impacting the reef. (**f**) Coastal bioerosional notch on Farmer's Cay, Exuma Archipelago, Bahamas illustrates almost diagrammatically the impact of combined microbioerosion and bioabrasion by grazing echinoderms, polyplacophores and gastropods, with maximum within the intertidal range. (**g**) Microbiocorrosion starts in the supratidal wave spray zone, and expands along water-retaining fissures dominated by coccoid cyanobacterium *Hormathonema*. (**h**) Microbiocorrosion by euendolithic cyanobacteria combined with gastropod grazing pressure modifies the water retention on the carbonate surfaces positively reinforcing the bioerosion effect and resulting in biokarstic landscape profiles cut perpendicular to substrate surface, stained and imaged by light microscopy (Fig. 1a) and subsequently by scanning electron microscopy (SEM) (Sammarco and Risk 1990; Chazottes et al. 1995, 2009; Tribollet et al. 2002; Tribollet 2008a).

Values obtained in measuring rates of microbial destruction of rocks refer to the net or residual biocorrosion (Chazottes et al. 1995), because it does not include the dissolution in layers of carbonate that were already consumed by grazing gastropods, polyplacophors, echinoderms and scarid fishes. The measured biocorrosion expresses the lag between sequentially occurring processes: the microbial settlement, substrate penetration and dissolution, followed by the removal of both substrate and microbial communities by grazing animals (Fig. 2c, d). The processes of microbiocorrosion and bioabrasion by grazers are mutually regulated. Microbial corrosion in the absence of grazers is a light-limited process confined to the substrate surface; it becomes a progressive destructive force only in combination with grazing activity, which transfers euendolithic photosynthesis deeper into the rock. Similarly regulating is the effect of overgrazing, with grazing resuming only after the substrate was re-colonized by epi- and endoliths (Schneider 1976). Thus microbiocorrosion is an integral part of biological processes that modify the carbonate structure and surrounding sedimentary environment and are viewed as a part of the taphonomy of the coral reefs (Scoffin 1992; Scoffin and Brandshaw 2000).

Microbiocorrosion varies tremendously with abiotic and biotic factors such as water quality (turbidity, light, eutrophication, pollution, temperature, pH, etc.), grazers' activity and colonization of substrate surfaces by epilithic organisms (see review by Tribollet 2008b). For instance, in shallow waters on the Great Barrier Reef (Australia), microbiocorrosion rates in conjunction with bioabrasion by grazers (mostly gastropod and echinoderms in the Mediterranean, and echinoderms and fish in tropical regions), increased from near-shore to offshore environments (Tribollet and Golubic 2005). The lowest rates of microbiocorrosion were found at the inshore sites because substrates were covered by a thin layer of mud due to the high turbidity of waters (*0.13 $\pm$ 0.31 kg $CaCO_3$ m^{-2} of exposed substrate surface $year^{-1}$), while the highest rates were found at the outer barrier sites (1.35 $\pm$ 0.48 kg $CaCO_3$ m^{-2} $year^{-1}$) in spite of a high degree of ground cover by crustose coralline algae (Tribollet 2008a). The opposite trend was found for macroborers which are largely filter-feeding organisms such as sponges and bivalves, and depend on and respond to production increases in the water column, therefore increasingly abundant in eutrophic near-shore environments, while their abundance declines offshore (Sammarco and Risk 1990; Risk et al. 1995; Tribollet and Golubic 2005). Bioerosion as a whole on the Great Barrier Reef is driven by the synergistic activity of euendoliths and grazers (Tribollet and Golubic 2005). In other shallow tropical regions (Indian, Pacific and Atlantic Oceans), rates of microbiocorrosion (also called microbioerosion) vary between 0.03 kg $CaCO_3$ m^{-2} of exposed substrate surface $year^{-1}$ and 0.65 kg m^{-2} y^{-1} depending on the type of substrate (shell, coral skeleton, limestone), environmental conditions and period of exposure of substrates to colonization by euendoliths (Chazottes et al. 1995, 2002; Vogel et al. 2000; Carreiro-Silva et al. 2005; Tribollet 2008a; Tribollet et al. 2009). Devastating rates of external bioerosion consisting of microbial biocorrosion and bioabrasion by grazers such as sea urchins (Fig. 2e) can also be recorded in these other regions, like in the Galapagos Islands, amounting on the average 25.4 kg m^{-2} $year^{-1}$ (Reaka-Kudla et al. 1996).

6 Global Impact

In addition to their role in carbonate dissolution, marine euendoliths can be important primary producers like in coral reefs (Tribollet et al. 2006) and as such serve as food source to a number of grazing animals, which remove mechanically (bioabrasion) layers of rocks loosened by microbiocorrosion (Schneider and Torunski 1983; Tribollet and Golubic 2005).

Microbiocorrosion is ubiquitous in terrestrial and aquatic environments. In soils, bacterial respiratory activity is probably the most effective biocorrosional power, responsible for karstification of large volumes of limestone bedrocks (Golubic and Schneider 1979) thereby contributing to the enrichment of groundwater with calcium carbonate. This calcium carbonate is subsequently deposited in caves as dripstone as well as in hot and cold springs in form of calcareous tufa or travertines (e.g. Golubic et al. 2008). The interconnection between carbonate deposition in rivers and biocorrosion is best documented by the river Volturno in the Apennines, where large terraces of calcareous tufa were deposited during the interglacial and early Holocene times, but the carbonate deposition ceased completely following deforestation and soil erosion in the river's watershed (Golubic et al. 1993, 1995). In the hypolimnion of lakes, the microbiocorrosion is expressed as aggressive CO_2, which originated from bacterial degradation of organic matter in the water column and sediments (Ruttner 1962). At a much finer scale, microbiocorrosion takes place underneath microbial mats and biofilms (e.g. Verrecchia et al. 2003; Ortega-Morales et al. 2004). It also affects and modifies the interstitial spaces in porous rocks, which are inhabited by chasmo- and cryptoendolithic microbes, and deeper inside the rock coated by metal-enriched biofilms as described above. However, the most direct biocorrosion of carbonate that shapes the landscape of limestone coastal rocks is the combined work of microbial phototrophic euendoliths and their animal grazers, resulting in bioerosional notch formation (Fig. 2f) with zonal arrangement of ecological niches (Radtke et al. 1996; Radtke and Golubic 2010), and in formation of the bizarre sharp-edged coastal biokarst (Fig. 2g, h).

7 Current Concerns

The current concern about the operation of the carbon cycle revolves around the problem of global change including global warming and ocean acidification, and impacts of anthropogenic activities (Falkowski et al. 2000). Biocorrosion is an integral part of carbon cycling, which is affected by various factors such as an increase in nutrient concentrations (Carreiro-Silva et al. 2005, 2009) and partial pressure of atmospheric CO_2 (Tribollet et al. 2009). Recently, it has been shown that rates of carbonate dissolution by euendolithic communities measured under elevated pCO_2 were 48% higher than under ambient pCO_2 (0.46 kg $CaCO_3$ dissolved $m^{-2}\,y^{-1}$ vs. 0.31 kg $m^{-2}\,y^{-1}$). Such dissolution in coral reefs, also called biogenic dissolution (Tribollet et al. 2009), may become a major mechanism of carbonate dissolution in a more acidic ocean.

References

Abed RMM, Golubic S, Garcia-Pichel F, Camoin GF, Sprachta S (2003) Characterization of microbialite-forming Cyanobacteria in a tropical lagoon: Tikehau Atoll, Tuamotu, French Polynesia. Journal of Phycology 39:862–873

Abed RMM, Palinska KA, Camoin G, Golubic S (2006) Common evolutionary origin of planktonic and benthic nitrogen-fixing oscillatoriacean cyanobacteria from tropical oceans. FEMS Microbiology Letters 260:171–177

Abed RMM, Kohls K, Schoon R, Scherf A-K, Schacht M, Palinska KA, Rullköter J, Golubic S (2008) Lipid biomarkers, pigments and cyanobacterial diversity of microbial mats across intertidal flats of the arid coast of the Arabian Gulf (Abu Dhabi, UAE). FEMS Microbiological Ecology 65:449–462

Alexandersson ET (1975) Marks of unknown carbonate-decomposing organelles in cyanophyte borings. Nature 254:212–238

Al-Thukair AA, Golubic S (1991) New endolithic cyanobacteria from the Arabian Gulf. I. *Hyella immanis* sp. nov. Journal of Phycology 27:766–780

Carreiro-Silva M, McClanahan TR, Kiene WE (2005) The role of inorganic nutrients and herbivory in controlling microbioerosion of carbonate substratum. Coral Reefs 24: 214–221

Carreiro-Silva M, McClanahan TR, Kiene WE (2009) Effect of inorganic nutrients and organic matter on microbial euendolithic community composition and microbioerosion rates. Marine Ecology Progress Series 392:1–15

Chacón E, Berrendero E, Garcia-Pichel F (2006) Biogeological signatures of microboring cyanobacterial communities in marine carbonates from Cabo Rojo, Puerto Rico. Sedimentary Geology 185:215–228

Chazottes V, Le Campion-Alsumard T, Peyrot-Clausade M (1995) Bioerosion rates on coral reefs: interaction between macroborers, microborers and grazers (Moorea, French Polynesia). Palaeogeography, Palaeoclimatology, Palaeoecology 113:189–198

Chazottes V, Le Campion-Alsumard T, Peyrot-Clausade M, Cuet P (2002) The effects of eutrophication-related alterations to coral reef communities on agents and rates of bioerosion (Reunion Island, Indian Ocean). Coral Reefs 21:375–390

Chazottes V, Cabioch G, Golubic S, Radtke G (2009) Bathymetric zonation of modern microborers in dead coral substrates from New Caledonia – Implications for paleodepth reconstructions in Holocene corals. Palaeogeography, Palaeoclimatology, Palaeoecology 280: 456–468

De la Torre JR, Goebel BM, Friedmann EI, Pace NR (2003) Molecular characterization of cryptoendolithic communities from the McMurdo Dry Valleys, Antarctica. Applied Environmental Microbiology 69:3858–3867

Dupraz C, Visscher PT (2005) Microbial lithification in marine stromatolites and hypersaline mats. Trends in Microbiology 13:429–338

Falkowski P, Scholes RJ, Boyle E, Canadell J, Canfield D, Elser J, Gruber N, Hibbard K, Högberg P, Linder S (2000) The global carbon cycle: a test of our knowledge of Earth as a system. Science 290:291–296

Garcia-Pichel F (2006) Plausible mechanisms for the boring on carbonates by microbial autotrophs. Sedimentary Geology 185:205–213

Garcia-Pichel F, Al-Horani F, Ludwig R, Farmer J, Wade B (2004) Balance between calcification and bioerosion in modern stromatolites. Geobiology 2:49–57

Golubic S (1969) Distribution, taxonomy and boring patterns of marine endolithic algae. American Zoologist 9:747–751

Golubic S, Schneider J (1979) Carbonate dissolution. In: Trudinger PA, Swaine DJ (eds) Biogeochemical Cycling of Mineral-forming Elements. Elsevier, Amsterdam, pp 107–129

Golubic S, Schneider J (2003) Microbial endoliths as internal biofilms. In: Krumbein WE, Dornieden T, Volkmann M (eds) Fossil and Recent biofilms. Kluwer Academic, Dordrecht, pp 249–263

Golubic S, Friedmann I, Schneider J (1981) The lithobiontic ecological niche, with special reference to microorganisms. Journal of Sedimentary Petrology 51:475–478

Golubic S, Violante C, Ferreri V, D'Argenio B (1993) ***Algal control and early diagenesis in Quaternary travertine formation (Rocchetta a Volturno, central Apennines). In: Baratolo F, De Castro P, Parente M (eds) Studies on Fossil Benthic Algae. Bolletino, Societa Paleontologica Italiana, Spec 1:231–247

Golubic S, Violante C, D'Argenio B (1995) Travertine biota: a record of seasonal changes. In: Ubertini L, Castelli F, Bras RL (eds) Climate Change and Hydrogeological Hazards in the Mediterranean Area. Natural Research Council, Italy, pp 25–30

Golubic S, Violante C, Plenkovic A, Grgasovic T (2008) Travertines and calcareous tufa deposits: an insight into diagenesis. Geologica Croatica 61:363–378

Hatch WI (1980) The implication of carbonic anhydrase in the physiological mechanism of penetration of carbonate substrata by the marine burrowing sponge *Cliona celata* (Demospongiae). Biological Bulletin 159:135–147

Heitz E, Flemming H-C, Sand W (eds) (1996) Microbially Influenced Corrosion of Materials. Springer, Berlin, pp 475

Kiene WE, Hutchings PA (1994) Bioerosion experiments at Lizard Island, Great Barrier Reef. Coral Reefs 13:91–98

Le Campion-Alsumard T, Golubic S, Pantazidou A (1996). On the euendolithic genus *Solentia* Ercegovic (Cyanophyta/Cyanobacteria). Algological Studies 83:108–127

Macintyre IG, Prufert-Bebout L, Reid RP (2000) The role of endolithic cyanobacteria in the formation of lithified laminae in Bahamian stromatolites. Sedimentology 47:915–921

Neumann AC (1966) Observations on coastal erosion in Bermuda and measurements of the boring rate of the sponge *Cliona lampa*. Limnology and Oceanography 11:92–108

Ortega-Morales BO, Narváez-Zapata JA, Schmalenberger A, Sosa-Loó A, Tebbe CC (2004) Biofilms fouling ancient limestone Mayan monuments in Uxmal, Mexico: a cultivation-independent analysis. Biofilms 1:79–90

Perry CT (1998) Grain susceptibility to the effects of microboring, implications for the preservation of skeletal carbonates. Sedimentology 45:39–51

Radtke G, Golubic S (2010) Microbial euendolithic assemblages and microborings in intertidal and shallow marine habitats: insight in cyanobacterial speciation. In: Reitner J, Queric, N-V, Arp G (eds) Advances in Stromatolite Geobiology. Lecture Notes in Earth Sciences 131, Springer, Berlin, pp 213–241

Radtke G, Le Campion-Alsumard T, Golubic S (1996) Microbial assemblages of the bioerosional "notch" along tropical limestone coasts. Algological Studies 83:469–482

Rasheed M, Wild C, Franke U, Huettel M. (2004) Benthic photosynthesis and oxygen consumption in permeable carbonate sediments at Heron Island, Great Barrier Reef, Australia. Estuarine, Coastal and Shelf Science 59:139–150

Reaka-Kudla ML, Feingold JS, Glynn W (1996) Experimental studies of rapid bioerosion of coral reefs in the Galapagos Islands. Coral Reefs 15:101–107

Reid RP, Visscher PT, Decho AW, Stolz JF, Bebout BM, Dupraz CP, Macintyre IG, Paerl HW, Pinckney JL, Prufert-Bebout L, Steppe TF, DesMarais DJ (2000) The role of microbes in accretion, lamination and early lithification of modern marine stromatolites. Nature 406:989–992

Reitner J (1993) Modern cryptic microbialite/metazoan facies from Lizard Island (Great Barrier Reef, Australia) – Formation and Concepts. Facies 29:3–40

Reitner J, Thiel V, Zankl H, Michaelis W, Wörheide G, Gautret P (2000) Organic and biochemical pattern in cryptic microbialites. In: Riding RE, Awramik SM (eds) Microbial Sediments. Springer, Berlin, 149–160

Risk MJ, Sammarco PW, Edinger EN (1995) Bioerosion in *Acropora* across the continental shelf of the Great Barrier Reef. Coral Reefs 14:79–86

Roux W (1887) Über eine in Knochen lebende Gruppe von Fadenpilzen (*Mycelites ossifragus*). Zeitschrift für wissenschaftliche Zoologie 45:227–254

Ruttner F (1962) Grundriss der Limnologie, 3rd ed. Walter de Gruyter, Berlin, pp 332

Sammarco PW, Risk MJ (1990) Large-scale patterns in internal bioerosion of *Porites*: cross continental shelf trends on the Great Barrier Reef. Marine Ecology Progress Series 59:145–156

Schneider J (1976) Biological and inorganic factors in the destruction of limestone coasts. Contributions Sedimentology 6:1–112

Schneider J, Le Campion-Alsumeard T (1999) Construction and destruction of carbonates by marine and freshwater cyanobacteria. European Journal Phycology 34:417–426

Schneider J, Torunski H (1983) Biokarst on limestone coasts, morphogenesis and sediment production. Marine Ecology 4:45–63

Scoffin TP (1992) Taphonomy of coral reefs: a review. Coral Reefs 11:57–77

Scoffin TP, Brandshaw C (2000) The taphonomic significance of endoliths in dead versus live coral skeletons. Palaios 15:248–254

Tribollet A (2008a) Dissolution of dead corals by euendolithic microorganisms across the northern Great Barrier Reef (Australia). Microbial Ecology 55:569–580

Tribollet A (2008b) The boring microflora in modern coral reef ecosystems: a review of its roles. In: Wisshak M, Tapanila L (eds) Current Developments in Bioerosion. Springer, Berlin, 67–94

Tribollet A, Golubic S (2005) Cross-shelf differences in the pattern and pace of bioerosion of experimental carbonate substrates exposed for 3 years on the northern Great Barrier Reef, Australia. Coral Reefs 24:422–434

Tribollet A, Payri C (2001) Bioerosion of coralline alga *Hydrolithon onkodes* by microborers in the coral reefs of Moorea, French Polynesia. Oceanologica Acta 24:329–342

Tribollet A, Decherf G, Hutchings PA, Peyrot-Clausade M (2002) Large-scale spatial variability in bioerosion of experimental coral substrates on the Great Barrier Reef (Australia): importance of microborers. Coral Reefs 21:424–432

Tribollet A, Langdon C, Golubic S, Atkinson M (2006) Endolithic microflora are major primary producers in dead carbonate substrates of Hawaiian coral reefs. Journal of Phycology 42: 292–303

Tribollet A, Godinot C, Atkinson M, Langdon C (2009) Effects of elevated pCO_2 on dissolution of coral carbonates by microbial euendoliths. Global Biogeochemical Cycles 23:1–7

Underwood CJ, Mitchell SF, Veltkamp CJ (1999) Microborings in mid-Cretaceous fish teeth. Proceedings Yorkshire Geological Society 52:269–274

Verrecchia EP, Loisy C, Braissant O, Gorbushina AA (2003) The role of fungal biofilm and networks in the terrestrial calcium carbonate cycle. In: Krumbein WE, Dornieden T, Volkmann M (eds) Fossil and Recent biofilms. Kluwer Academic, Dordtrecht, 363–369

Videla HA (1996) Manual of Biocorrosion. CRC Lewis Publishers, Boca Raton, pp 273

Videla HA, Herrera LK (2005) Microbiologically influenced corrosion: looking to the future. International Microbiology 8:169–180

Vogel K, Gektidis M, Golubic S, Kiene WE, Radtke G (2000) Experimental studies on microbial bioerosion at Lee Stocking Island, Bahamas and One Tree Island, Great Barrier Reef, Australia: implications for paleoecological reconstructions. Lethaia 33:190–204

Walker JJ, Pace, NR (2007a) Phylogenetic composition of Rocky Mountain endolithic microbial ecosystems. Applied Environmental Microbiology 73:3497–3504

Walker JJ, Pace NR (2007b) Endolithic microbial ecosystems. Annual Reviews Microbiology 61:331–347

Walker JJ, Spear JR, Pace NR (2005) Geobiology of a microbial endolithic community in the Yellowstone geothermal environment. Nature 434:1011–1014

Wild C, Laforsch C, Huettel M. (2006) Detection and enumeration of microbial cells within highly porous calcareous reef sands. Marine and Freshwater Research 57:415–420

Zhang Y, Golubic S (1987) Endolithic microfossils (cyanophyta) from early Proterozoic stromatolites, Hebei, China. Acta Micropaleontologica Sinica 4:1–12

Zhang X-G, Pratt BR (2008) Microborings in Early Cambrian phosphatic and phosphatized fossils. Palaeogeography, Palaeoclimatology, Palaeoecology 267:155–195

The Deep-Sea Chemoautotroph Microbial World as Experienced by the Mediterranean Metazoans Through Time

Marco Taviani

1 Preface

This paper is focused upon the Miocene-to-Recent deep-sea chemosynthetic ecosystems of the Mediterranean basin with emphasis on their metazoan associates. The life ingredients of this story are basically the "microbes" and the "metazoans", in particular molluscs, whose mutual interplay resulted in a superb variety of situations in response to geologic and oceanographic factors directing the production and availability of geofluids usable by microbial consortia. The astounding complexity and diversity of modern metazoan-bearing deep-sea ecosystems inhabiting the Mediterranean is a step of a long journey whose beginning dates back to more than 3 billion years ago and whose major events are resumed below.

2 The March of Life

The backbone of Life on Earth is centred on microbes, unicellular prokaryotes (mostly bacteria and archaea) whose first documentation is thought to date back to Archean times (e.g. Knoll 2003; Allwood et al. 2006; Noffke et al. 2006; Schopf and Bottjer 2009; Sugitani et al. 2009 with references) and since then evolved through the Eons in exploiting successfully a vast array of habitats from extremely hot to extremely cold, from oxygen-deficient to oxygen-plenty (e.g. Eigenbrode and Freeman 2006; Canfield et al. 2007; Blank 2009). Early prokaryotic microorganisms have been reported from rocks as old as 3.7 Ga (e.g. Schopf 1993, 1999; Pflug 2001; Van Kranendonk 2006; Payne et al. 2009 with references), although such putative early occurrences are still contrasted (e.g., Brasier et al. 2002; Appel et al. 2003; Westall and Folk 2003; Pinti et al. 2009). It is suggested that early

M. Taviani
ISMAR-CNR, via Gobetti 101, 40129 Bologna, Italy
e-mail: marco.taviani@bo.ismar.cnr.it

J. Reitner et al., *Advances in Stromatolite Geobiology*,
Lecture Notes in Earth Sciences 131, DOI 10.1007/978-3-642-10415-2_18,

prokaryotes, inhabiting an Earth marked by "prohibitive" conditions when compared to present ones, were chemoautrophs satisfying their energetic needs through chemosynthetic processes at the expenses of hydrogen sulphide, methane and other compounds in absence of sunlight. In chemoautotrophy the more common energetic pathway is the microbial transformation of CO_2 into organic matter through the oxidation of reduced sulphur compounds. At methane-seeps the major process is the anaerobic oxidation of methane accomplished by microbial consortia of sulphate-reducing bacteria and methanotrophic archaea. The current view is that at the beginning and for a while atmosphere and oceans alike might have been reducing, and related microbial ecosystems necessarily anaerobic (see Poulton et al. 2004; Xiaoying et al. 2008, for an account) with aerobic microbial ecosystems appearing in the late Archean (Eigenbrode and Freeman 2006). Photosynthesis is thought to emerge later, although arguments for (at least temporarily) oxygenated seas as old as ca. 3.5 Ga ago have been presented (Hoashi et al. 2009).

Unicellular eukaryotes are assumed to appear also in the Archean, achieving a discrete level of complexity and diversification by Paleoproterozoic times, based on body fossil and molecular evidence (e.g., Javaux et al. 2003; Javaux 2007; Knoll et al. 2006; Lamb et al. 2009 with references), although the fidelity of some such fossil records is controversial (e.g. Lamb et al. 2007; Rasmussen et al. 2008). One current theory considers the possibility that eukaryotic cells derived from the association of free-living bacteria, a first case of bacterial symbiosis (see Esteve and Gaju 1999). Both chemosynthetic and photosynthetic microbial organisms set the pace for Life on Earth for a prolonged period of time. How widespread and relevant was the chemosynthetic life in pre-Phanerozoic oceans is not known. However, abundant body fossils of giant sulfur bacteria similar to *Thiomargarita* have been reported from Neoproterozoic phosphorites (Bailey et al. 2007).

The evolution of basal metazoans (animals), multicelled heterotrophic eukaryotes, is suggested to have taken place in the Neoproterozoic, between ca 1,000–660 Ma (Xiao et al. 2000; Peterson and Butterfield 2005; Arthur 2006; Qun et al. 2007; Canfield et al. 2007). Although the earliest animals could have been cnidaria-like organisms (Dunn et al. 2008; Brocks and Butterfield 2009), first documented metazoans (only based on 24 isopropyle colestane) are marine Porifera (Demospongiae) inhabiting the Cryogenian seas (Love et al. 2009).

The hypothesized establishment of permanent oxic conditions in the deep sea after the Neoproterozoic Gaskiers glaciation (<580 Ma) may have been the veritable turning point for metazoan evolution (Canfield et al. 2007) and propitious to the evolution of metazoans associated with pre-existing chemotrophic microbial communities.

Metazoans have a reliable paleontological documentation since the Precambrian (Vendian) with many phyla appearing at the dawn of Paleozoic (Cambrian), also the time of the booming of organisms equipped with a mineralized shell, after some Ediacaran precursors (Wood et al. 2002; Porter 2010). Depending upon the interpretation of body fossils in the Ediacara fauna, first molluscs might be present as early as the Vendian (*Kimberella*: Fedonkin and Waggoner 1997). At any event, shelled molluscs are positively confirmed since the early Cambrian (Peel 1991;

Gubanov et al. 2004; Vinther and Nielsen 2005; Skovsted et al. 2007). First bivalves are also of this age but the major radiation of this class starts with the Ordovician (Runnegar and Pojeta 1992; Cope and Babin 1999; Fang 2006). Brachiopods and "worms" are equally well documented since Paleozoic times (e.g., Little 2002; Campbell 2006).

3 The Phanerozoic Record of Metazoans in Deep-Sea Chemoautotrophic Communities

Beginning with the Paleozoic the major prokaryotic and eukaryotic actors eventually involved in the establishment of marine chemosynthetic habitats as we know today are on the scene, adapting, radiating, getting extinct, evolving. Thus, since more than 400 Ma, from the Silurian period onwards, microbial chemoautotrophs continue the successful, rewarding and long-lasting symbiosis with marine metazoans (Felbeck et al. 1981; Cavanaugh 1983; Childress et al. 1986; Stein et al. 1988; Fisher 1990; Cavanaugh et al. 2006). As we can experience today, the net result is the immense number of complex chemosynthetic ecosystems from hot hydrothermal vents to hydrocarbon cold seeps that, still in perpetual evolution, inhabit the world's marine realm from shallow to hadal depths (Paull et al. 1984; Suess et al. 1985; Turner 1985; Brooks et al. 1987; Aharon 1994; Fujikura et al. 1999; Van Dover 1990, 2000; Sibuet and Olu-Le Roy 1998, 2002; Tunnicliffe et al. 2003; Bernard et al. 2004; Campbell 2006; Judd and Hovland 2007; MacAvoy et al. 2008 and many others).

Various metazoans are known to be exclusively associated with chemosynthetic marine habitats. Those include mostly molluscs, brachiopods and "worms", all with a relatively good paleontological documentation since the Silurian (Stein et al. 1988; Reid 1990; Campbell and Bottjer 1995; Distel 1998; Little et al. 1998; Callender and Powell 1999; Goedert et al. 2000; Peckmann et al. 2001, 2007; Little 2002; Van Dover et al. 2002; Barbieri et al. 2004; Peckman and Goedert 2005; Kiel and Little 2006; Campbell 2006; Jenkins et al. 2007; Kaim et al. 2008; Kiel and Dando 2009 etc.). The chemoautrophic bivalves are among the oldest around and apparently some did not undergo major morphologic changes since the early Paleozoic. A remarkable case in point are the primitive clams of family Solemyidae (Reid and Bernard 1980; Kuznestov et al. 1990; Beninger and Le Pennec 1997; Krueger and Cavanaugh 1997; Gustafson and Reid 2004; Stewart and Cavanaugh 2006) known from cold seep sites as old as the Paleozoic (Peckmann et al. 2001; Taylor et al. 2008) and well established since the Mesozoic (Campbell 2006; Kiel and Little 2007). Most information on deep-sea chemosynthetic metazoans concern Cenozoic occurrences, however, whose communities also share basic taxonomic traits observed in recent counterparts (Goedert and Squires 1990; Taviani 1994; Goedert and Campbell 1995; Peckmann et al. 1999, 2002, 2004; Goedert et al. 2003; Neasbitt 2005; Majima et al. 2005; Gill et al. 2005; Kiel and Little 2006, Campbell 2006; Kiel and Peckmann 2007 and many others).

Sulphide-based community symbiosis between sulphur-oxidizing bacteria and bivalves is proven in seven extant families (Solemyidae, Manzanellidae, Mytilidae,

Thyasiridae, Lucinidae, Vesicomyidae, Teredinae: Pennec et al. 1995; Taylor and Glover 2009). These clams harbour in their tissues chemolitotrophic or methanotrophic bacteria capable to oxidize reduced compounds providing energy to their metazoan hosts. This strategy is so successful to have permitted through ages an almost cosmopolitan distribution of such communities and the production of substantial biomass in the otherwise rather oligotrophic deep-sea realm. Remarkably on many occasions such strategy drives also the achievement of large body size (Cosel and Olu 1998; Kelly et al. 2000; Taylor and Glover 2009). More commonly chemoautotrophic symbionts occur intracellularly within gill filaments, but extracellular association is also documented (Gros and Gaill 2007).

Mytilidae (Paleozoic to Recent: Little and Vrijenhoek 2003) present many representatives with sulphur-based chemosymbionts, and to a lesser degree methane oxidizing bacteria (alone or dual: Pruski et al. 2002; Salerno et al. 2005; Duperron et al. 2005, 2008, 2009) at times coupled with heterotroph feeding. Lucinidae, appeared possibly in the late Silurian and confirmed from the Mesozoic (Hautmann and Nützel 2005; Kiel and Little 2006) are obligatory symbiotic with sulphur-oxidizing bacteria (Taylor and Glover 2000, 2006). Vesicomyidae (Paleogene to Recent: Kiel and Little 2006; Kiel and Peckmann 2007; Kiel and Dando 2009), are also uniquely linked to thiotrophy as are many of the Mesozoic to Recent Thyasiridae (Zuschin et al. 2001; Kiel and Little 2006; Taylor et al. 2007).

An interesting issue regards the antiquity of microbes involved in chemosynthesis (Imhoff et al. 2003) and their metazoan hosts. Cospeciation seems to be case (Peek et al. 1998) but recent work by Won et al. (2008) on thiotrophic bathymodiolinids and endosymbiontic associates document that this is not necessarily the general rule and that geographic dispersal of chemoautotrophic bacteria took place long before the evolution of their present host metazoans.

4 The Deep Mediterranean Basin is Microbes' Paradise

A concentration of most deep sea chemosynthetic scenarios is documented in the Mediterranean basin and its precursors. The inventory includes hydrothermal vents, cold seeps, pockmarks, mud volcanoes, anoxic bottoms, sunken wood, fallen whales and shipwrecks. Many such occurrences present some metazoan associates. Thus far deep-sea chemosymbiontic animals are not (yet?) detected at active submarine volcanic apparatuses the Mediterranean in spite of various propitious situations where microbial life teems such as seamounts in the Tyrrhenian basin.

In the distant past (Late Cretaceous) the situation was different as documented by Little et al. (1999) that recorded hydrothermal chemosynthetic tubeworms and molluscs from the Late Cretaceous Troodos ophiolite in Cyprus.

By large, the deep-sea cold temperature chemosynthetic situations are best represented in the geologic archives. In fact, deep-sea defluidisation (hydrocarbon-rich and H_2S fluids) is documented in the Mediterranean basin *sensu lato* since the Eocene at least (see Taviani 2001 for a review).

The best starting point is to observe the present situation of deep sea chemosynthetic records in the semi-enclosed Mediterranean Sea. All major areas known to be site of deep-sea metazoan bearing chemosynthetic-based ecosystems are summarized in Fig. 1.

Since the 1990s mud volcanoes, anoxic pools, hydrates, hydrocarbon and hydrogen sulphide emissions and microbial mats were discovered in the deep Eastern Mediterranean (e.g., De Lange and Brumsack 1998; Charlou et al. 2003). The occurrence of chemosynthetic metazoans was detected by Corselli and Basso (1996), who reported chemosynthetic bivalves at the Napoli mud volcano (including *Vesicomya* sp., later shown to be *Isorropodon perplexum* (Sturany 1896), in the family Vesicomyidae: Cosel and Salas 2001). Coleman and Ballard (2001) first observed with a ROV in the Southeastern Mediterranean, down to 800 m off the Egyptian coasts, mussels, clams and polychaetes around hydrocarbon seeps characterized also by bacterial mats. The continuation of research in the eastern Mediterranean led to the formal definition of another important component of the bathyal cold seep macrofauna, namely the lucinid bivalve *Lucinoma kazani* Salas and Woodside (2002) from the Kazan mud volcano south of Turkey. The last years witnessed a strong acceleration of research devoted to deep-sea chemosynthesis in the Eastern Mediterranean and its products aided also by the implementation of more refined technology for exploration and collaborative international

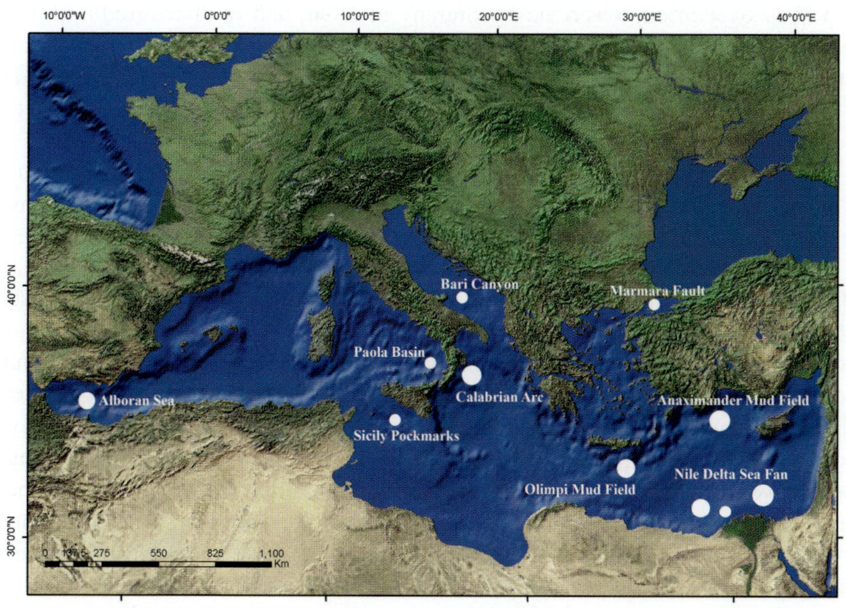

Fig. 1 Map showing major areas of intense chemosynthetic activity in the Mediterranean deep-sea floor (*cold seeps*) proven or suspected of sustaining specialized metazoan communities as known at present (*see text*)

programmes (Werne et al. 2004; Foucher et al. 2009). Regarding metazoans the current picture is the following.

Olu-Le Roy et al. (2004) record various chemosynthetic bivalves and polychaetes at depths of ca 2,000 m from the Olimpi Mud Field (Napoli, Milano, Maidstone, Moscow mud volcanoes), south of Crete along the Mediterranean Ridge, and from the Anaximander Mud Field (Kula, Kazan, Amsterdam and Faulted Ridge), south of Turkey between 1,700 and 2,000 m. Their list includes siboglinid polychaetes (*Lamellibrachia*, *Siboglinum*), *Isorropodon perplexum*, *Myrtea amorpha* (Sturany 1896), *Idas modiolaeformis* (Sturany 1896), *Lucinoma kazani* Cosel and Salas 2001, *Thyasira striata* (Sturany 1896) (only at AMV), the gastropods *Lurifax vitreus* Warén and Bouchet 2001, *Clelandella* n.sp., *Lepetella* sp., the decapods *Chaceon mediterraneus* Manning and Holthuis 1989, *Munidopsis* sp., and few others. *Lucinoma kazani* shells, other bivalve and worm fragments are recorded by Lykousis et al. (2004) in a sediment core at 1,783 m from the Athina mudvolcano in the Anaximander mountains. Further information on the megafauna from this area, including sponges, is reported by Vanreusel et al. (2009).

The on-going exploration of the astounding chemosynthetic grounds of the Nile Deep Sea Fan is also revealing a very similar metazoan fauna: Dupré et al. (2007) report *Isorropodon perplexum*, *Myrtea amorpha*, *Idas modiolaeformis*, *Lucinoma kazani*, tubicolous polychaetes and *Chaceon* crabs from Central Nile Deep Sea Fan. Quite interestingly some of such deep-sea chemosynthetic bivalves were known to science since more than a century based upon collections made in this region by the ship *Pola* expedition, described by Sturany in 1896, and re-discovered with the beginning of modern deep-sea exploration of the Nile Delta Fan. There is an interesting parallelism here with the rather similar first discovery of deep-sea corals at the Santa Maria di Leuca hotspot, off the Apulian Ionian equally to be credited to the *Pola* expedition (Taviani et al. 2005).

Multidisciplinary exploration of the Marmara Fault in the Sea of Marmara (Turkey) stepped over gas and water seep discharge, including hydrocarbons, between ca 1,000 and 1,200 m (Armijo et al. 2005). To such setting is associated a chemosynthetic fauna (siboglinids and bivalves, including *Lucinoma kazani*, *Myrtea amorpha*, and *Isorropodon perplexum*: Zitter et al. 2008; Ritt et al. 2010) whose detailed study is in progress.

Promising sites with active fluid emissions are also scattered in the central and western Mediterranean (e.g., Holland et al. 2003) but no biological information on associated macrofauna is available at present.

The recent most discovery of deep-water chemosynthetic communities, including tubeworms and lucinid bivalves, is the Alboran basin in the western Mediterranean (Ivanov et al. 2009; Blinova, Ivanov and coworkers, in progress) This finding is of utmost importance since practically connecting the Mediterranean deep-sea chemosynthetic cold-seep world, before documented for its eastern reaches only, to the Gulf of Cadiz and therefore the open ocean, punctuated by many such occurrences (e.g., Cosel 2006; Cosel and Olu 2009; Cosel and Salas 2001; Olu-Le Roy et al. 2007; Génio et al. 2008; Rodrigues et al. 2008, 2010; Foucher et al. 2009; Vanreusel et al. 2009; Sibuet and Vangriesheim 2009).

The overview would not be complete without mentioning the finding of the chemosynthetic *Lamellibrachia* (Polychaeta: Siboglinidae) from the deep-seated shipwreck of the SS *Persia* at 2,800 m depth south of Crete (Hughes and Crawford 2008). To this, a number of citations variously reporting molluscs of chemosynthetic significance (*Idas*, *Lucinoma*, *Lurifax*), some of still unknown cold-seep sites in the Tyrrhenian Sea, can be found in grey literature.

The Black Sea, easternmost marine basin, also houses an important deep-water chemosynthetic world of its own. Due to its anoxic conditions beneath ca. 200 m, the microbes dominate the habitats resulting in spectacular prokaryotic domes (Michaelis et al. 2002; Reitner et al. 2005). The occurrence of living metazoans (Nematoda, Polychaeta, Harpacticoida, Acarina, Bivalvia and Gastropoda) associated with seeps between 220 and 250 m (Sergeeva and Gulin 2009 with references) seems not yet convincingly substantiated (Yanko-Hombach et al. 2009) and, at least for some groups, unlikely because of the absence of oxygen, crucial to respiration.

In synthesis, the present-day Mediterranean metazoan macrofauna associated with deep-water cold seeps taxonomically comprehends most of the "classic" groups: Mytilidae, Lucinidae, Thyasiridae, Vesicomyidae, tube-worms. Others are absent (or not yet found), the most remarkable case being members of the Cretaceous to Recent gastropod family Provannidae (Kaim et al. 2008) and large Solemyidae (*Acharax*), both with modern representatives in the eastern Atlantic (Ravara et al. 2007; Warén and Bouchet 2009). Compared to ocean standards, the diversity of Mediterranean chemosynthetic metazoans is quite low and taxa are often lilliputians in their respective families, with some important ingredients missing, e.g., large bathymodiolinid mussels and *Calyptogena*. Maybe the most significant trait of the present Mediterranean chemosynthetic deep-sea metazoans concerns their nutritional strategy that is thiotrophy, at times coupled with functional heterotroph feeding. Thus far metazoans adapted to methanotrophic-based or dual metabolism as such observed in the adjacent Atlantic ocean (Duperron et al. 2005, 2008, 2009) have not been identified in the deep-sea chemosynthetic communities of the present Mediterranean in spite of the pervasiveness of methane degassing here (Charlou et al. 2003; Foucher et al. 2009). However, in earlier times this should not have been the case.

5 Deep Sea Chemosynthetic Environments in the Mediterranean Neogene

The fossil legacy of deep-sea chemosynthetic habitats tells us a different story by documenting situations that are remarkably different from the present Mediterranean Sea.

Tertiary rocks and sediments in the Apennine chain (Fig. 2) and Sicily show remarkable and impressive geobiological imprinting of various cold seep emissions onto the deep sea bottom, the most known being the "Calcari a Lucina" (Clari et al. 1988; Aharon 1994; Taviani 1994, 2001; Berti et al. 1994; Conti and Fontana 2005;

Fig. 2 Examples of Miocene chemosynthetic sites in the Italian Apennines: (**a**) The impressive methanogenetic chemoherm (*sensu* Aharon 1994) of Montepetra (Forlì), Tortonian; (**b**) the ancient reducing seabottom inhabited by dense and diverse populations of thiotrophic bivalves (solemyids, lucinids, thyasirids, vesicomyids), Fosso creek (Rocca Santa Maria, Modena), Serravallian

Clari et al. 2004; Conti et al. 2004; Lucente and Taviani 2005 with references). Most of the information about type and function of microbes at such ancient chemosynthetic sites necessarily derive from petrographic, geochemical and, mostly, biomarker arguments since only in exceptional cases microbes are fossilized as body fossils (Peckmann et al. 2004) or biofilms and fabrics (e.g. Cavagna et al. 1999; Peckmann et al. 1999). The only body fossil evidence is a sulphur bacterium (similar to *Beggiatoa*: Peckmann et al. 2004) identified in a Miocene methanogenetic limestone in the northern Apennines. Chemosynthetic tubeworms ("vestimentiferans") are equally rare in the Miocene cold seeps because of their lower preservation potential (best example is the Marmorito limestone in Piedmont: Peckmann et al. 1999), as are other polychaetes and decapods. By large the overwhelming paleontological documentation of Mediterranean Tertiary-age deep-sea cold seep metazoans is heralded by molluscs, especially bivalves (Taviani 1994, 2001; Venturini et al. 1998; Clari et al. 2004; Lucente and Taviani 2005). As for the Mediterranean at present, most of the important families (Solemyidae, Mytilidae, Lucinidae, Thyasiridae, Vesicomyidae) are well represented in the Miocene but, differently from today, they are more diverse and often their representatives are of larger size. A case in point is Lucinidae that include very large-size taxa and various genera [when compared with recent work by Cosel (2006)]. But the most astounding trait is the occurrence of various Vesicomyidae attributable to *Calyptogena* and Mytilidae, including putative bathymodiolinids (Taviani 1994, 2001, and work in progress). As said before, known solemyids, lucinids and vesicomyids are thiotrophic and their presence marks the availability of hydrogen sulphide habitats. The occurrence of bathymodiolinids is remarkable in the sense that they seem morphologically close to the recent amphiatlantic *Bathymodiolus childressi* complex (Olu-Le Roy et al. 2007) known to colonize hydrocarbon-seep habitats. This species is known to rely upon dual symbiosis at cold vent sites (Duperron et al. 2005, 2009). This was likely the strategy of its Miocene counterpart too highlighting the importance of intense hydrocarbon defluidization during the

Late Miocene, capable to sustain biota also through the anaerobic oxidation of methane. The stable carbon geochemical signatures, biomarkers and the concurring presence of the cold-seep gastropod *Thalassonerita* [considered a younger synonym of recent *Bathynerita* by Taviani (2001)] at Miocene-age bathymodiolinid sites lends further support to this hypothesis. Recent studies suggest in fact that *Bathymerita* occurring at hydrocarbon seeps in the Gulf of Mexico, while not methanotrophic itself, is uniquely linked to chemosynthetic hyperhaline microhabitats in methanotrophic *Bathymodiolus childressi* beds, grazing bacteria and decomposing mussel's periostracum (Dattagupta et al. 2007; Van Gaest et al. 2007). Therefore, a rather similar situation might be deducted from the Apennine Late Miocene occurrences showing co-occurring *Thalassonerita* and bathymodiolinids encased in methanogenetic limestones (Taviani 1994).

It has been argued that the final Miocene Messinian salinity crisis might have sealed the fate of the Mediterranean deep-sea stenoecious biota through their complete annihilation, including the chemosynthetic metazoan communities mentioned above (Taviani 1994, 2001, 2004).

The Pliocene renovation can only be evaluated at a few fossil sites (Taviani et al. 1997; Taviani 2001; Monegatti et al. 2001; Dominici et al. 2009) and included solemyids (*Acharax*: Taviani et al. 2008), mytilids (but not bathymodiolinids), lucinids (but not giant ones), vesicomyids (but not *Calyptogena*); *Thalassonerita* is equally absent.

With end of the Miocene the Mediteranean deep sea chemosynthetic metazoan communities largely lost their previous ocean affinity, assuming the basic aspect of the habitats we know at present, as best documented by the taxonomy of its mollusc fauna (Fig. 3).

6 Postface

This article has been assembled by a metazoan heterotroph whose nutritional strategy is largely photosynthetic-based. Although not yet definitely assessed, the geologic antiquity of this taxon is rather modest, perhaps a bare 200 kyrs or so. Apparently this is the only metazoan known to run conscious scientific research and also to be able to write and evolve abstract theories. These characteristics may conspire in biasing beyond tolerable levels its perception of the natural world with the distinct possibility of overemphasising its own ecological niche to the extent of considering its extreme ecological tolerance as the "normal" one. Deviations from its standard contour conditions are aberrant and therefore meriting to be labelled as abnormal, unusual or even extreme. Although somehow frustrating from the standpoint of a Phanerozoic metazoan breathing oxygen and enjoying a nutritional strategy based upon photosynthesis, it could be fair to recognize that latecomers like us are the real extremophiles of our planet. From a metazoan-centric point of view we favour the dignified idea that we metazoans choose the microbes for the symbiosis. From a microbial-centric point of view it

Fig. 3 Representative Mediterranean deep-sea chemosynthetic molluscs, Miocene to Present MIOCENE: (**a**) "*Lucina*" sp., Case Rovereti (Forlì, Romagna Apennine), "Calcari al Lucina" (Tortonian); (**b**) *Thalassonerita megastoma* Moroni 1966 (= *Bathynerita*), (Case Rovereti, Forlì, Romagna Apennine, Italy), "Calcari a Lucina" (Tortonian); (**c**) "*Bathymodiolus*" cf *exbrocchii* (Sacco 1898): Ca' Piantè (Faenza, Ravenna, Romagna Apennine), "Calcari a Lucina" (Tortonian); (**e**) *Calyptogena* n. sp. Ca' Piantè, Ravenna, Romagna Apennine, "Calcari a Lucina" (Tortonian); PLIOCENE: (**d**) "*Acharax*" cf. *doderleini* (Mayer 1861): Castelnuovo Berardenga Scalo (Siena, Tuscany, Italy), Middle Pliocene; (**h**) *Lucinoma* sp., Middle Pliocene, Stirone river (Scipione, Parma, Italy); RECENT: (**f**) *Isorropodon perplexum* (Sturany 1896), Eastern Mediterranean; (**g**) *Idas modiolaeformis* (Sturany 1896), Eastern Mediterranean; (**i**) *Lucinoma kazani* Salas and Woodside (2002), Eastern Mediterranean. Bar = 1 cm

might be that the microbes as dominant beings of Life on Earth substantially direct metazoan adaptation. That if all this nonsense is only partly true (B. Purser, unpublished quotation, 1991).

Acknowledgements This article is an overgrowth of an invited presentation to the International Kalkowsky-Symposium, held in Goettingen, October 4–11, 2008. My warmest thanks go to Joachim Reitner for inviting me to the symposium and for his patience while waiting for this article. Lorenzo Angeletti helped with photography and preparation of figures and Alessandro Ceregato with bibliography. Special thanks to Paul Aharon, Stefano Conti, Joern Peckmann, Marco Sami, Gian Battista Vai and all microbes and metazoans associated with cold seeps. This paper is a contribution to the EU Hermione programme (contract number 226354) and is ISMAR-Bologna scientific contribution n. 1670.

References

Aharon P (1994) Geology and biology of modern and ancient submarine hydrocarbon seeps and vents. Geo-Marine Letters 14:69–230

Appel PWU, Moorbath S, Myers JS (2003) *Isuasphaera isua* (Pflug) revisited. Precambrian Research 126:309–312

Allwood AC, Walter MR, Kamber BS, Marshall CP, Burch IW (2006) Stromatolite reef from the Early Archean era of Australia. Nature 441:714–715

Armijo R, Pondard N, Meyer B, Uçarkus G, Mercier de Lépinay B, Malavieille J, Dominguez S, Gustcher M-A, Schmidt S, Beck C, Çagatay N, Çakir Z, Imren C, Eris K, Natalin B, Özalaybey S, Tolun L, Lèfevre I, Seeber L, Gasperini L, Rangin C, Emre O, Sarikavak K (2005) Submarine fault scarps in the Sea of Marmara pull-apart (North Anatolian Fault): implications for seismic hazard in Istanbul. Geochemistry, Geophysics, Geosystems 6: Q06009, doi:10.1029/2004GC000896

Arthur W (2006) Creatures of Accident: the Rise of the Animal Kingdom. Hall & Wang, New York: 243 pp

Bailey JV, Joye SB, Kalanetra KM, Flood BE, Corsetti FA (2007) Evidence of giant sulphur bacteria in Neoproterozoic phosporites. Nature 445:198–201

Barbieri R, Ori GG, Cavalazzi B (2004) A Silurian cold-seep ecosystem from the Middle Atlas, Morocco. Palaios 19:527–542

Beninger PG, Le Pennec M (1997) Reproductive characteristics of a primitive bivalve from a deep-sea reducing environment: giant gametes and their significance in *Acharax alinae* (Cryptodonta: Solemyidae). Marine Ecology Progress Series 157:195–206

Bernard B, Brooks J, de Faragó M (2004) Asphalt volcanism and chemosynthetic life in the Campeche Knolls, Gulf of Mexico. Science 304:999–1002

Berti M, Cuzzani MG, Landuzzi A, Taviani M, Aharon P, Vai GB (1994) Hydrocarbon-derived imprints in olistostromes of the Early Serravallian Marnoso-arenacea Formation, Romagna Apennines (Northern Italy). Geo-Marine Letters 14:192–200

Blank C (2009) Not so old Archaea-the antiquity of biogeochemical processes in the archaeal domain of life. Geobiology 7:495–514

Brasier MD, Green OR, Jephcoat AP, Kleppe AK, Van Kranendonk MJ, Lindsay JF, Steele A, Grassineau NV (2002) Questioning the evidence for Earth's oldest fossils. Nature 416:76–81

Brocks JJ, Butterfield NJ (2009) Early animals out in the cold. Nature 457:672–673

Brooks JM, Kennicutt II MC, Fisher CR, Macko SA, Cole K, Childress JJ, Bidigare RR, Vetter RD (1987) Deep-sea hydrocarbon seep communities: evidence for energy and nutritional carbon sources. Science 238:1138–1142

Callender RW, Powell EN (1999) Why did ancient chemosynthetic seep and vent assemblages occur in shallower water than they do today? International Journal of Earth Sciences 88:377–391

Campbell KA (2006) Hydrocarbon seep and hydrothermal vent paleoenvironments and paleontology: Past developments and future research directions. Palaeogeography, Palaeoclimatology, Palaeoecology 232:362–407

Campbell KA, Bottjer DJ (1995) Brachiopods and chemosymbiontic bivalves in Phanerozoic hydrothermal vent and cold seep environments. Geology 23:321–324

Canfield DE, Poulton SW, Narbonne GM (2007) Late Neoproterozoic deep-ocean oxygenation and the rise of animal life. Science 315:92–95

Cavagna S, Clari P, Martire L (1999) The role of bacteria in the formation of cold seep carbonates: geological evidence from Monferrato (Tertiary, NW Italy). Sedimentary Geology 126: 253–270

Cavanaugh CM (1983) Symbiotic chemoautotrophic bacteria in marine invertebrates from sulphide-rich habitats. Nature 302:58–61

Cavanaugh CM, McKiness ZP, Newton ILG (2006) Marine Chemosynthetic Symbioses. In: Dworkin M, Falkow S, Rosenberg E, Schleifer KH, Stackebrandt E (eds) The Prokaryotes. Third Edition. A Handbook on the biology of Bacteria: Symbiotic associations, Biotechnology, Applied Microbiology, Springer New York: 475–507

Charlou JL, Donval JP, Zitter T, Roy N, Foucher JP, Woodside J (2003) Evidence of methane venting and geochemistry of brines in Mud Volcanoes of the Eastern Mediterranean Sea. Deep Sea Research Part I: Oceanographic Research Papers 50:941–958

Childress JJ, Fisher CR, Brooks JM, Kennicutt II MC, Bidigare R, Anderson AE (1986) A methanotrophic marine molluscan (Bivalvia, Mytilidae) symbiosis: mussels fuelled by gas. Science 233:1306–1308

Clari P, Conti S, Fontana D, Taviani M (2004) Fluid expulsion and authigenic products in the Miocene foredeep and satellite basins of the Northern Apennines, Italy. Field Guide Book-Post-Congress P07, 32nd IGC, August 20–28, Florence-Italy 2:B31–B60

Clari P, Gagliardi C, Governa ME, Ricci B, Zuppi GM (1988) I Calcari di Marmorito: una testimonianza di processi diagenetici in presenza di metano. Bollettino del Museo Regionale di Scienze Naturali, Torino 6:197–216

Coleman DF, Ballard RD (2001) A highly concentrated region of cold hydrocarbon seeps in the southeastern Mediterranean Sea. Geo-Marine Letters 21:162–167

Conti S, Fontana D (2005) Anatomy of seep-carbonates: ancient examples from the Miocene of the northern Apennines (Italy). Palaeogeography, Palaeoclimatology, Palaeoecology 227:156–175

Conti S, Fontana D, Gubertini A, Sighinolfi G, Tateo F, Fioroni C, Fregni P (2004). A multidisciplinary study of middle Miocene seep-carbonates from the northern Apennine foredeep (Italy). Sedimentary Geology 169:1–19

Cope JCW, Babin C (1999) Diversification of bivalves in the Ordovician. Geobios 32:175–185

Corselli C, Basso D (1996) First evidence of benthic communities based on chemosynthesis on the Napoli mud volcano (Eastern Mediterranean). Marine Geology 132:227–239

Cosel R (2006) Taxonomy of tropical West African bivalves. VI. Remarks on Lucinidae (Mollusca, Bivalvia), with description of six new genera and eight new species. Zoosystema 28:805–851

Cosel R, Olu K (1998) Gigantism in Mytilidae. A new *Bathymodiolus* from cold seep areas on the Barbados accretionary prism. Comptes Rendus de l'Académie des Sciences – Series III – Sciences de la Vie 321:655–663

Cosel R, Olu K (2009) Large Vesicomyidae (Mollusca: Bivalvia) from cold seeps in the Gulf of Guinea off the coasts of Gabon, Congo and northern Angola. Deep Sea Research Part II: Topical Studies in Oceanography 56:2350–2379

Cosel R, Salas C (2001) Vesicomyidae (Mollusca: Bivalvia) of the genera *Vesicomya*, *Waisiuconcha*, *Isorropodon* and *Callogonia* in the eastern Atlantic and the Mediterranean. Sarsia 86:333–366

Dattagupta S, Martin J, Liao S, Carney RS, Fisher CR (2007) Deep-sea hydrocarbon seep gastropod *Bathynerita naticoidea* responds to cues from the habitat-providing mussel *Bathymodiolus childressi*. Marine Ecology 28:193–198

De Lange G, Brumsack HJ (1998) The occurrence of gas hydrates in eastern Mediterranean mud dome structures as indicated by pore-water composition. In: Henriet JP, Mienert J (eds) Gas Hydrates: Relevance to World Maring Stability and Climate Change. Geological Society London, Special Publication 137:167–175

Distel DL (1998): Evolution of chemoautotrophic endosymbiosis in bivalves. Bioscience 48: 277–286

Dominici S, Cioppi E, Danise S, Betocchi U, Gallai G, Tangocci F, Valleri G, Monechi S (2009) Mediterranean fossil whale falls and the adaptation of molluscs to extreme habitats. Geology 37:815–818

Dunn CW, Hejnoll A, Matus DQ, Pang K, Browne WE, Smith SA, Seaver E, Rouse GW, Obst M, Edgecombe GD, Sørensen MV, Haddock SHD, Schmidt-Rhaesa A, Okusu A, Møbjerg Kristensen R, Wheeler WC, Martindale MQ, Giribet G (2008) Broad phylogenomic sampling improves resolution of the animal tree of life. Nature 452:745–749

Duperron D, Halary S, Lorion J, Sibuet M, Gaill F (2008) Unexpected co-occurrence of six bacterial symbionts in the gills of the cold seep mussel *Idas* sp. (Bivalvia: Mytilidae). Environmental Microbiology 10:433–445

Duperron S, Lorion J, Samadi S, Gros O, Gaill F (2009) Symbioses between deep-sea mussels (Mytilidae: Bathymodiolinae) and chemosynthetic bacteria: diversity, function and evolution. Comptes Rendus Biologies 332:298–310

Duperron S, Nadalig T, Caprais JC, Sibuet M, Fiala-Médioni A, Amann R, Dubilier N (2005) Dual symbiosis in a *Bathymodiolus* sp. Mussel from a methane seep on the Gabon continental margin (Southeast Atlantic): 16S rRNA phylogeny and distribution of the symbionts in gills. Applied and Environmental Microbiology 71(4):1694–1700

Dupré S, Woodside J, Foucher J-P, de Lange G, Mascle J, Boetius A, Mastalerz V, Stadnitskaia A, Ondréas H, Huguen C, Harmeniégnies F, Gontharet S, Loncke L, Deville E, Niemann H, Omoregie E, Olu-Le Roy K, Fiala-Médioni A, Dählmann A, Caprais J-C, Prinzhofer A, Sibuet M, Pierre C, Damsté JS, the NAUTINIL Scientific Party (2007) Seafloor geological studies above active gas chimneys off Egypt (Central Nile Deep Sea Fan). Deep Sea Research Part I: Oceanographic Research Papers 54:1146–1172

Eigenbrode JL, Freeman KH (2006) The late Archean rise of microbial aerobic ecosystems. Proceedings of the National Academy of Sciences of the United States of America 103:15759–15764

Esteve I, Gaju N, 1999 Bacterial symbioses. Predation and mutually beneficial associations. International Microbiology 2, 81–86

Fang Z-J (2006) An introduction to Ordovician bivalves of southern China, with a discussion of the early evolution of the Bivalvia. Geological Journal 41: 303–328

Fedonkin MA, Waggoner BM (1997) The Late Precambrian fossil *Kimberella* is a mollusc-like bilaterian organism. Nature 388:868–871

Felbeck H, Childress JJ, Somero GN (1981) Calvin-Benson cycle and sulphide oxidation enzymes in animals from sulphide-rich habitats. Nature 293:291–293

Fisher CR (1990) Chemoautotrophic and methanotrophic symbiosis in marine invertebrates. CRC Critical Reviews in Aquatic Sciences 2:399–436

Foucher J-P, Westbrook GK, Boetius A, Ceramicola S, Dupré S, Mascle J, Mienert J, Pfannkuche O, Pierre C, Praeg D (2009). Structures and drivers of cold seep ecosystems. Oceanography 22:58–74

Fujikura K, Kojima S, Tamaki K, Maki Y, Hunt J, Okutani T (1999) The deepest chemosynthesis-based community yet discovered from the hadal zone, 7326 m deep, in the Japan Trench. Marine Ecology Progress Series 190:17–26

Génio L, Johnson SB, Vrjenhoek RC, Cunha MR, Tyler PA, Kiel S, Little CTS (2008) New records of "*Bathymodiolus*" *mauritanicus* Cosel 2002 from the Gulf of Cadiz (NE Atlantic) mud volcanoes. Journal of Shellfish Research 27:53–61

Gill FL., Harding IC, Little CTS, Jonathan AT. (2005): Palaeogene and Neogene cold seep communities in Barbados, Trinidad and Venezuela: An overview. Palaeogeography, Palaeoclimatology, Palaeoecology 227:191–209

Goedert JL, Campbell KA (1995) An Early Oligocene chemosynthetic community from the Makah Formation, Northwestern Olympic Peninsula, Washington. The Veliger 38:22–29

Goedert JL, Peckmann J, Reitner J (2000) Worm tubes in an allochtonous cold-seep carbonate from lower Oligocene rocks of western Washington. Journal of Paleontology 74:992–999

Goedert JL, Squires RL (1990) Eocene deep-sea communities in localized limestones formed by subduction-related methane seeps, southwestern Washington. Geology 18:1182–1185

Goedert JL, Thiel V, Schmale O, Rau WW, Michaelis W, Peckmann J (2003) The Late Eocene 'Wiskey Creek' methane-seep deposit (Western Washington State). Part I: Geology, palaeontology and molecular geobiology. Facies 48:223–240

Gros O, Gaill F (2007) Extracellular bacterial association in gills of "wood mussels". Cahiers de Biologie Marine 48:103–109

Gubanov AP, Skosted CB, Peel JS (2004) Early Cambrian molluscs from Sierra de Cordoba (Spain). Geobios 37:199–215

Gustafson RG, Reid RGB (2004) Larval and post-larval morphogenesis in the gutless protobranch bivalve *Solemya reidi* (Cryptodonta: Solemyidae). Marine Biology 97:373–387

Hautmann M, Nützel A (2005) First record of a heterodont bivalve (Mollusca) from the Early Triassic: palaeoecological significance and implications for the 'Lazarus problem'. Palaeontology 48:1131–1138

Hoashi M, Bevacqua DC, Otake T, Watanabe Y, Hickman AH, Utsunomiya S, Ohmoto H (2009) Primary haematite formation in an oxygenated sea 3.46 billion years ago. Nature Geoscience 2:301–306

Holland CW, Etiope G, Milkov AV, Michelozzi E, Favali P (2003) Mud volcanoes discovered offshore Sicily. Marine Geology 199:1–6

Hughes DJ, Crawford M (2006) A new record of the vestimentiferan *Lamellibrachia* sp. (Polychaeta: Siboglinidae9 from a deep shipwrek in the eastern Mediterranean. Marine Biodiversity Records 1:1–3

Imhoff JF, Sahling H, Süling J, Kath T (2003) 16s rDNA-based phylogeny of sulphur-oxidising bacterial endosymbionts in marine bivalves from cold-seep habitats. Marine Ecology Progress Series 249:39–51

Ivanov M, Kenyon N, Comas M, Pinheiro L, Laberg J-S, and shipboard Scientific party (2009) Introduction to TTR-17 results. Geo-Marine Research on the Mediterranean and European-atlantic Margins International Conference and TTR-17th Post-Cruise Meeting of the Training-through-Research Programme, Workshop Rep 220:1–51

Javaux EJ (2007) Patterns of diversification in early eukaryotes. Carnet de Géologie/Notebooks in Geology, Brest, Mem 2007/01, Abstract 06 (CG2007_M01/06)

Javaux EJ, Knoll AH, Walter M (2003) Recognizing and interpreting the fossils of early Eukaryotes. Origins of Life and Evolution of Biospheres 33:75–94

Jenkins RG, Kaim A, Hikida Y, Tanabe K (2007) Methane-flux-dependent lateral fauna changes in a Late Cretaceous chemosymbiotic assemblage from the Nakagawa area of Hokkaido, Japan. Geobiology 5:127–139

Judd AG, Hovland M (2007) Seabed Fluid Flow. Impact on Geology, Biology, and the marine Environment. Cambridge University Press, Cambridge: 470 pp

Kaim A, Jenkins RG, Warén A (2008) Provannid and provannid-like gastropods from the Late Cretaceous cold seeps of Hokkaido (Japan) and the fossil record of the Provannidae (Gastropoda: Abyssochrysoidea). Zoological Journal of the Linnean Society 154:421–436

Kelly SRA, Blanc E, Price SP, Whitham AG (2000) Early Cretaceous giant bivalves from seep-related limestone mounds, Wollaston Forland, Northeast Greenland. Special Publications, Geological Society London 177:227–246

Kiel S, Dando PR (2009) Chaetopterid tubes from vent and seep sites: Implications for fossil record and evolutionary history of vent seep annelids. Acta palaeontologica Polonica 54:443–448

Kiel S, Little CST (2006) Cold-Seep molluscs are older than the general marine mollusc fauna. Science 313:1429–1431

Kiel S, Peckmann J (2007) Chemosymbiotic bivalves and stable carbon isotopes indicate hydrocarbon seepage at four unusual Cenozoic fossil localities. Lethaia 40:345–357

Knoll AH (2003) Life on a Young Planet. The First Three Billion Years of Evolution on Earth. Princeton University Press, Princeton, Chichester: 277 pp

Knoll AH, Javaux EJ, Hewitt D, Cohen P (2006) Eukaryotic organisms in Proterozoic Oceans. Philosophical Transactions of the Royal Society of London, Series B 361:1023–1038

Krueger DM, Cavanaugh CM (1997) Phylogenetic diversity of bacterial symbionts of *Solemya* hosts based on comparative sequence analysis of 16s rRNA genes. Applied and Environmental Microbiology 63:91–98

Kuznestov AP, Ohta S, Endow K (1990) Morphofunctional consequences of bacterial symbiotrophy in *Solemya* (*Petrasma*) *pusilla* (Protobranchia, Bivalvia) from Sagami Bay (Central Japan). Izvestiia Akademii nauk SSSR. Seriia biologicheskaia 6:895–903

Lamb DM, Awramik SM, Chapman DJ, Zhu S (2007) Paleoproterozoic compression-like structures from the Changzhougou Formation, China: Eukaryotes or clasts? Precambrian Research 154:236–247

Lamb DM, Awramik SM, Chapman DJ, Zhu S (2009) Evidence for eukaryotic diversification in the – 1800 million-year-old Changzhougou Formation, North China. Precambrian Research 173:93–104

Little CTS (2002) The fossil record of hydrothermal vent communities. Cahiers de Biologie Marine 43:313–316

Little CTS, Vrijenhoek RC (2003) Are hydrothermal vent animals living fossils? TRENDS in Ecology and Evolution 18:582–588

Little CTS, Herrington RJ, Maslennikov VV, Zaykov VV (1998) The fossil record of hydrothermal vent communities. In: Mills RA, Harrison K. (Eds) Modern Ocean Floor Processes and the Geological Record. Special Publications, Geological Society London 148:259–270

Little CTS, Cann JR, Herrington RJ & Morisseau M (1999) Late Cretaceous hydrothermal vent communities from the Troodos ophiolite, Cyprus. Geology 27:1027–1030

Love GD, Grosjean E, Stalvies C et al (2009) Fossil steroids record the appearance of Demospongiae during the Cryogenian period. Nature 457:718–721

Lucente C, Taviani M (2005) Chemosynthetic communities as fingerprints of submarine sliding-linked fluid seepage, Miocene deep-sea strata of the Romagna Apennines, Italy. Palaeogeography, Palaeoclimology, Palaeoecology 227:176–190

Lykousis V, Alexandri S, Woodside J, Nomikou P, Perissoratis C, Sakellariou D, de Lange G, Dahlmann A, Casas D, Rousakis G, Ballas D, Ioakim C (2004) New evidence of extensive active mud volcanism in the Anaximander mountains (Eastern Mediterranean): The "*ATHINA*" mud volcano. Environmental Geology 46:1030–1037

MacAvoy SE, Morgan E, Carney RS, Macko SA (2008) Chemoautotrophic production incorporated by heterotrophs in Gulf of Mexico hydrocarbon seeps: an examination of mobile benthic predators and seep residents. Journal of Shellfish Research 27:153–161

Majima R, Nobuhara, T, Kitazaki T (2005) Review of fossil chemosynthetic assemblages in Japan. Palaeogeography, Palaeoclimatology, Palaeoecology 227:86–123

Michaelis W, Seifert R, Nauhaus K, Treude T, Thiel V, Blumenberg M, Knittel K, Gieske A, Peternecht K, Pape T, Boetius A, Amann R, Jorgensen BB, Widdel F, Peckmann JR, Pimenov NV, Gulin MB (2002) Microbial reefs in the Black Sea fuelled by anaerobic oxidation of methane. Science 297:1013–1015

Monegatti P, Raffi S, Roveri M, Taviani M (2001) One day trip in the outcrops of Castell'Arquato Plio-Pleistocene Basin: from the Badland of Monte Giogo to the Stirone River. Paleobiogeography & Paleoecology 2001, International Conference May 31-June 2, 2001 Piacenza & Castell'Arquato Italy, Bologna: 22 pp

Neasbitt EA (2005) A novel trophic relationship between cassid gastropods and mysticete whale carcasses. Lethaia 38:17–25

Noffke N, Eriksson KA, Hazen RM et al (2006) new window into Early Archean life: microbial mats in Earth's oldest siliclastic tidal deposits (3.2 Ga Moodies group, South Africa). Geology 34:235–256

Olu-Le Roy K, Sibuet M, Fiala-Médioni A, Gofas S, Salas C, Mariotti A, Foucher JP, Woodside J (2004) Cold seep communities in the deep eastern Mediterranean Sea: composition, symbiosis and spatial distribution on mud volcanoes. Deep Sea Research Part I: Oceanographic Research Papers 51:1915–1936

Olu-Le Roy K, Cosel R, Hourdez S, Carney SL, Jollivet D (2007) Amphi-Atlantic cold-seep *Bathymodiolus* species complexes across the equatorial belt. Deep Sea Research Part I: Oceanographic Research Papers 54:1890–1911

Paull CK, Hecker B, Commeau R, Freemanlynde RP, Neumann C, Corso WP, Golubic S, Hook JE, Sikes E, Curray J (1984) Biological communities at the Florida Escarpment resemble hydrothermal vent taxa. Science 226:965–967

Payne JL, Boyer AG, Brown JH, Finnegan S, Kowalewski M, Krause RA Jr, Lyons KS, McLain CR, McShea DW, Novack-Gottshall PM, Smith FA, Stempien JA, Wang SC (2009) Two-phase increase in the maximum size of life over 3.5 billion years reflects biological innovation and environmental opportunity. Proceedings of the National Academy of Sciences of the United States of America 106:24–27

Peckmann J, Thiel V, Michaelis W, Clari P, Gaillard C, Martire L, Reitner J (1999) Cold seep deposits of Beauvoisin (Oxfordian; southeastern France) and Marmorito (Miocene; northern Italy): microbially induced authigenic carbonates. International Journal of Earth Sciences 88:60–75

Peckmann J, Girschler E, Oschmann W, Reitner J (2001) An Early Carboniferous seep community and hydrocarbon-derived carbonates from the Harz Mountains, Germany. Geology 29: 271–274

Peckmann J, Goedert JL, Thiel V, Michaelis W, Reitner J (2002) A comprehensive approach to the study of methane-seep deposits from the Lincoln Creek Formation, western Washington State, USA. Sedimentology 49:855–873

Peckmann J, Thiel V, Reitner J, Taviani M, Aharon P, Michaelis W (2004) A microbial mat of a large sulphur bacterium preserved in a Miocene methane-seep. Geomicrobiology Journal 21: 247–255

Peckmann J, Goedert JL (2005) Geobiology of ancient and modern methane-seeps. Palaeogeography, Palaeoclimatology, Palaeoecology 227:1–5

Peckmann J, Campbell KA, Walliser OH, Reitner J (2007) A late Devonian hydrocarbon-seep dominated by dimerelloid brachiopods, Morocco. Palaios 19:114–122

Peek AS, Feldman RA, Lutz RA, Vrijenhock RC (1998) Cospeciation of chemoautotrophic bacteria and deep sea clams. Proceedings of the National Academy of Sciences of the United States of America 95:9962–9966

Peel JS (1991) Functional morphology of the Class Helcionelloida nov, and the early evolution of Mollusca. In: Simonetta AM, Conway S (eds) The early evolution of Metazoa and the significance of problematic taxa. Cambridge University Press, Cambridge: 157–177

Pennec M, Beninger PG, Herry A (1995) Feeding and digestive adaptations of bivalve molluscs to sulphide-rich habitats. Comparative Biochemistry and Physiology 111A:183–189

Peterson KJ, Butterfield NJ (2005) Origin of the Eumetazoa: testing ecological predictions of molecular clocks against the Proterozoic fossil record. Proceedings of the National Academy of Sciences of the United States of America 102:9547–9552

Pinti DL, Mineau R, Clement V (2009) Hydrothermal alteration and microfossil artefacts of the 3,465-million-year-old Apex chert. Nature Geoscience 2:640–643

Porter SM (2010) Calcite and aragonite seas and the *de novo* acquisition of carbonate skeletons. Geobiology 8:256–277

Pflug HD (2001) Earliest organic evolution. Essay to the memory of Batholomew Nagy. Precambrian Research 106:79–91

Poulton SW, Frolick PW, Canfield DE (2004) The transition to a sulphidic ocean approximately 1.84 billion years ago. Nature 431:173–177

Pruski AM, Rousse N, Fiala-Médioni A, Boulègue J (2002) Sulphur signature in the hydrothermal vent mussel *Bathymodiolus azoricus* from the Mid-Atlantic Ridge. Journal of the Marine Biological Association of the United Kingdom 82:463–468

Qun Y, Junye M, Xiaoyan S, Peiyun S (2007) Phylochronology of early metazoans: combined evidence from molecular and fossil data. Geological Journal 42:281–295

Rasmussen B, Fletcher IR, Brocks JJ, Kilburn MR (2008) Reassessing the first appearance of eukaryotes and cyanobacteria. Nature 455:1101–1104

Ravara A, Cunha MR, Rodrigues CL (2007) The occurrence of *Natsushima bifurcata* (Polychaeta: Nautiliniellidae) in *Acharax* hosts from mud volcanoes in the Gulf of Cadiz (south Iberian and north Moroccan Margins). Scientia Marina 7:95–100

Reid RGB (1990) Evolutionary implications of sulphide-oxidizing symbionts in bivalves. In: Morton B (ed) The Bivalvia – Proceedings of a Memorial Symposium in Honour of Sir Charles Maurice Yonge, Edinborough, 1986: 127–140, Hong Kong University Press, Hong Kong

Reid RGB, Bernard FR (1980) Gutless bivalves. Science 208:609–610

Reitner J, Peckmann J, Reimer A, Schumann G,Thiel V (2005). Methane-derived carbonate build-ups and associated microbial communities at cold seeps on the lower Crimean shelf (Black Sea). Facies 51:66–79

Ritt B, Sarrazin J, Caprais J-C, Noël P, Gauthier O, Pierre C, Henry P, Desbruyères D (2010) First insights into the structure and environmental setting of cold-seep communities in the Marmara Sea. Deep-Sea Research Part I, Oceanographic Research Papers doi.10.1016/j.dsr.2010.05.011

Rodrigues CF, Webster G, Cunha MR, Duperron S, Weightman AJ (2010) Chemosynthetic bacteria found in bivalve species from mud volcanoes of the Gulf of Cadiz. Microbiology Ecology 73:486–499
Rodrigues CL, Oliver PG, Cunha MR (2008) Thyasiroidea (Mollusca: Bivalvia) from the mud volcanoes of the Gulf of Cadiz (North-east Atlantic). Zootaxa 1572:41–56
Runnegar B, Pojeta J Jr (1992) The earliest bivalves and their Ordovician descendants. American Malacological Bulletin 9:117–122
Salas C, Woodside J (2002) *Lucinoma kazani* n.sp. (Mollusca: Bivalvia): evidence of a living benthic community associated with a cold seep in the eastern Mediterranean Sea. Deep Sea Research Part I: Oceanographic Research Papers 50:1391–1409
Salerno JL, Macko SA, Hallam SJ, Bright M, Won Y-J, McKiness Z, Van Dover CL (2005) Characterization of symbiont populations in life-history stages of mussels from chemosynthetic environments. Biological Bulletin 208:145–155
Schopf JW (1993) Microfossils of the early Archean Apex chert: new evidence of the antiquity of life. Science 260:640–646
Schopf JW (1999) Cradle of Life: the Discovery of Earth's Earliest Fossils. Princeton University Press, Princeton: 367 pp
Schopf JW, Bottjer DJ (2009) World summit on ancient microscopic fossils. Precambrian Research 173:1–3
Sergeeva N, Gulin S (2009) Benthic fauna of methane seeps in the Dniester Palaeo-Delta: comparative analysis. IGCP 521-INQA 0501 Fifth Plenary Meeting and Field Trip, Istanbul-Izmir-Çanakkale (Turkey) 22–31 August 2009, Extended Abstracts:158–159
Sibuet M, Olu-Le Roy K (1998) Biogeography, biodiversity and fluid dependence of deep-sea cold-seep communities at active and passive margins. Deep Sea Research Part II: Topical Studies in Oceanography 45:517–567
Sibuet M, Olu-Le Roy K (2002) Cold Seep Communities on continental margins: structure and qauantitative distribution relative to geological and fluid venting patterns. In: Wefer G, Billett D, Hebbeln D, Jørgensen BB, Schlüter M, Van Weering T (eds) Ocean Margin Systems. Springer-Verlag, Berlin, Heidelberg:235–251
Sibuet M, Vangriesheim A (2009) Deep-sea environment and biodiversity of the West African Equatorial margin. Deep Sea Research Part II: Topical Studies in Oceanography 56:2156–2168
Skovsted CB, Brock GA, Lindstroem A, Peel JS, Paterson JR, Fuller MK (2007) Early Cambrian record of failed durophagy and shell repair in an epibenthic mollusc. Biology Letters 3:314–317
Stein JL, Cary CS, Hessler RR, Ohta S, Vetter RD, Childress JJ, Felbeck H (1988) Chemoautotrophic symbiosis in a hydrothermal vent gastropod. Biological Bulletin 174 373–378
Stewart FJ, Cavanaugh CM (2006): Bacterial endosymbioses in *Solemya* (Mollusca: Bivalvia) – Model systems for studies of symbiont-host adaptation. Antonie van Leeuwenhoek 90:343–360
Sturany R (1896) Zoologische Ergebnisse VII Mollusken I (Prosobranchier und Opistobranchier: Scphopoden; Lamellibranchier) gesammelt von SM Schiff "Pola" 1890–1894. Denkschriften Akademie der Wissenschaften in Wien, Mathematisch-Naturwissenschaftliche Klasse 63:1–36
Suess E, Carson B, Ritger SD, Moore C, Jones ML, LD Kulm, Cochrane GR (1985) Biological communities at vent sits along the subduction zone off Oregon. Bulletin of the Biological Society of Washington 6:475–484
Sugitani K, Grey K, Nagaoka T, Mimura K, Walter MR (2009) Taxonomy and biogenicity of Archean spheroidal microfossils (ca. 3.0 Ga) from the Mount Goldsworthy-Mount Grant area in the northeastern Pilbara Craton, Western Australia. Precambrian Research 173:50–59
Taviani M (1994) The "calcari a *Lucina*" macrofauna reconsidered: Deep-sea faunal oases from Miocene-age cold vents in the Romagna Apennine, Italy. Geo-Marine Letters 14:185–191
Taviani M (2001) Fluid venting and associated processes. In: Vai GB, Martini PI (eds) Anatomy of an Orogen: the Apennines and Adjacent Mediterranean Basins. Kluwer Academic Publishers, Great Britain:351–366

Taviani M (2004) Shaping the biogeography of the Mediterranean basin: one geologist's perspective. In: Marine Biogeography of the Mediterranean Sea: patterns and dynamics of biodiversity. Biogeographia 24(2003):15–22

Taviani M, Roveri M, Aharon P, Zibrowius H (1997) A Pliocene deep-water cold seep (Stirone River, N. Italy). COLD-E-VENT International Works Hydrocarbon Seep Chemosynt Tethyan Relic Basins: Products, Processes and Causes Abstracts:20

Taviani M, Corselli C, Freiwald A, Malinverno E, Mastrototaro F, Remia A, Savini A, Tursi A (2005) First geo-marine survey of living cold-water *Lophelia* reefs in the Ionian Sea (Mediterranean basin). Facies 50:409–417

Taviani M, Angeletti L, Ceregato A (2008) Past and present chemosynthetic bivalves (family Solemyidae) inhabiting deep-sea cold-vent and reducing environments in the Neogene of the Mediterranean basin. 33rd International Geological Congress Oslo 6–10 August 2008 Abstracts

Taylor JD, Glover EA (2000) Functional anatomy, chemosymbiosis and evolution of the Lucinidae. In: Harper EM, Taylor JD, Crame JA (Eds) The Evolutionary Biology of the Bivalvia. Special Publications, Geological Society London 177:207–225

Taylor JD, Glover EA (2006) Lucinidae (Bivalvia) – the most diverse group of chemosymbiotic molluscs. Zoological Journal of the Linnean Society 148:421–438

Taylor JD, Glover EA (2009) A giant lucinid bivalve from the Eocene of Jamaica – Systematics, life habits and Chemosymbiosis (Mollusca: Bivalvia: Lucinidae). Palaeontology 52:95–109

Taylor JD, Glover EA, Williams ST (2008) Ancient chemosynthetic bivalves: systematics of Solemyidae from eastern and southern Australia (Mollusca: Bivalvia). Memoirs Queensland Museum, Nature:75–104

Taylor JD, Williams ST, Glover AE (2007) Evolutionary relationships of the bivalve family Thyasiridae (Mollusca: Bivalvia), monophyly and superfamily status. Journal of the Marine Biological Association of the United Kingdom 87:565–574

Tunnicliffe V, Juniper SK, Sibuet M (2003) Reducing environments of the deep-sea floor. In: Tyler PA (ed) Ecosystems of the Deep Oceans. Elsevier, Amsterdam:81–110

Turner RD (1985) Notes on molluscs of deep-sea vents and reducing sediments. American Malacological Bulletin Spec Edition 1:23–34

Van Dover CL (1990) Biogeography of hydrothermal vent communities along seafloor spreading centers. TREE 5:242–246

Van Dover CL (2000) The ecology of deep-sea hydrothermal vents. Princeton University Press, Princeton, New Jersey:1–424 pp

Van Dover CL, German CR, Speer KG, Parson LM,Vrijenhoek RC (2002) Evolution and biogeography of deep-sea venting and seep invertebrates. Science 295:1253–1257

Van Gaest AL, Young CM, Young JJ, Helms AR, Arellano SM (2007) Physiological and behavioral responses of *Bathynerita naticoidea* (Gastropoda: Neritidae) and *Methanoaricia dendrobranchiata* (Polychaeta: Orbiniidae) to hypersaline conditions at a brine pool cold seep. Marine Ecology 28:199–207

Van Kranendonk MJ (2006) Volcanic degassing, hydrothermal circulation and the flourishing of early life on Earth: a review of the evidence from c. 3490–3240 Ma rocks of the Pilbara Supergroup, Pilbara Craton, Western Australia. Earth-Science Reviews 74:197–240

Vanreusel A, Andersen AC, Boetius A, Connelly D, Cunha MR, Decker C, Hilario A, Kormas KA, Maignien L, Olu K, Pachiadaki M, Ritt B, Rodrigues C, Sarrazin J, Tyler P, Van Gaever S, Vanneste H (2009) Biodiversity of cold seep ecosystems along the European margins. Oceanography 22:110–127

Venturini S, Selmo E, Tarlao A, Tunis G (1998) Fossiliferous methanogenetic limestones in the Eocene flysch of Istria (Croatia). Giornale di Geologia 60: 219–234

Vinther J, Nielsen C (2005) The Early Cambrian *Halkeria* is a mollusc. Zoologica Scripta 34: 81–89

Warén A, Bouchet P (2001) Gastropoda and Monoplacophora from hydrothermal vents and seeps, new taxa and records. The Veliger 44:116–231

Warén A, Bouchet P (2009) New gastropods from deep-sea hydrocarbon seeps off West Africa. Deep Sea Research Part II: Topical Studies in Oceanography 56:2326–2349

Werne JP, Haese RR, Zitter T, Aloisi G, Bouloubassi I, Heijfs S, Fiala-Médiconi A, Pancost RD, Sinninghe Damsté JS, de Lange G, Forney LJ, Gottschal JC, Foucher JP, Mascle J, Woodside J (2004) Life at cold seeps: a synthesis of biogeochemical and ecological data from Kazan mud volcano, eastern Mediterranean Sea. Chemical Geology 205:367–390

Westall F, Folk RL (2003) exogenous carbonaceous microstructures in Early Archaen cherts and BIFs from the Isua Greenstone Belt: implications for the search for life in ancient rocks. Precambrian Research 126:313–330

Won Y-J, Jones WJ, Vrijenhoek RC (2008) Absence of cospeciation between deep-sea Mytilids and their thiotrophic endosymbionts. Journal of Shellfish Research 27:129–138

Wood RA, Grotzinger JP, Dickson JAD (2002) Proterozoic modular biomineralized metazoan from the Nma Group, Namibia. Science 296:2383–2386

Xiao S, Yuan X, Knoll AH (2000) Eumetazoan fossils in terminal Proterozoic phosporites? Proceedings of the National Academy of Sciences of the United States of America 97: 13684–13689

Xiaoying S, Chuanheng Z, Ganqing J, Juan L, Li W, Dianbo L (2008) Microbial mats in the Mesoproterozoic carbonates of the North China platform and their potential for hydrocarbon generation. Journal of China University of Geosciences 19:549–566

Yanko-Hombach et al. (2009) Response of biota to methane emissions in the Black Sea: preliminary results from complex geological, geochemical, palaeontological, and biological study. IGCP 521-INQA 0501 Fifth Plenary Meeting and Field Trip, Istanbul-Izmir-Canakkale (Turkey) 22–31 August 2009, Extended Abstracts:181–184

Zitter TAC, Henry P, Aloisi G, Delaygue G, Çagatay MN, Mercier de Lepinay B, Al-Samir M, Fornacciari F, Tesmer M, Pekdeger A, Wallmann K, Lericolais G (2008) Cold seeps along the main Marmara Fault in the Sea of Marmara (Turkey). Deep Sea Research Part I: Oceanographic Research Papers 55:552–570

Zuschin M, Mandic O, Harzhauser M, Persvesler P (2001) Fossil Evidence for chemoautotrophic bacterial symbiosis in the thyasirid bivalve *Thyasira michelottii* from the middle Miocene (Badenium) of Austria. Historical Biology 15:223–234

Gypsum Microbialite Domes Shaped by Brine Currents from the Badenian Evaporites of Western Ukraine

Maciej Bąbel, Danuta Olszewska-Nejbert, and Andrii Bogucki

1 Introduction

Microbialite deposits are considered as a product of the interactions between activity of benthic microorganisms and physicochemical properties of environment. Activity of currents is one of the most important physical processes influencing the accretion of microbialite domes. The domes growing in current-swept areas usually show elongated shapes parallel to the dominating water flow direction. Current-oriented microbialite domes are well known both from modern carbonate environments, particularly from tidal zones (Hoffman 1967, 1976; Hofmann 1973; Gunatilaka 1975; Playford and Cockbain 1976; Park 1977; Andres and Reid 2006), as well as were found in many ancient settings (e.g. Trompette 1969; Grotzinger 1986; Feldmann and McKenzie 1997; Zamarreño et al. 1997). Non-carbonate microbialites showing current-related features are not so commonly recorded (e.g. Walter et al. 1976) and they include gypsum microbialites from the Badenian of the Carpathian Foredeep (Kwiatkowski 1970, 1972; Bąbel 2002).

These Badenian (and some other) evaporite deposits show also the other unique structures interpreted as influenced by flowing water. The bottom-grown gypsum crystals, (selenites), are concordantly oriented and inclined in some beds, with apices directed in the similar azimuth interpreted as the upstream direction. The crystals grew preferentially oriented upstream into the calcium sulphate over-saturated brine flowing over the bottom (Dronkert 1985; Bąbel and Becker 2006, with references). Both types of structure, those promoted by physical and these related to chemical activity of the flowing water, are important for paleocurrent analysis in evaporite deposits.

M. Bąbel (✉) and D. Olszewska-Nejbert
Institute of Geology, Warsaw University, Al. Żwirki i Wigury 93, PL-02-089 Warszawa, Poland
e-mail: m.babel@uw.edu.pl

A. Bogucki
Faculty of Geography, Ivan Franko National University of Lviv, P. Doroshenka 41, 79000 Lviv, Ukraine

J. Reitner et al., *Advances in Stromatolite Geobiology*,
Lecture Notes in Earth Sciences 131, DOI 10.1007/978-3-642-10415-2_19,

In this paper we describe the new and specific directional structures – the gypsum microbialite domes that are shaped by both chemical and physical action of the brine current. This unique structure was found only in a few outcrops in the Badenian basin and in this paper we present the locality where the structure occurs in association with the other current related features which all together allow consistent interpretation of the origin of the directional structures.

2 Terminology

In this study we use the definition of microbialites by Burne and Moore (1987) and accept that the microbialites are "organosedimentary deposits that have accreted as a result of a benthic microbial community trapping and binding detrital sediment and/or forming the locus of mineral precipitation". In particular we include into the microbialites the deposits formed by inorganic and passive precipitation of gypsum within the locus of benthic microbial communities (see Arp et al. 2004). Further the broadest meaning of the term microbialite is also used for the rocks, in which the "influence of algae on the rock is more commonly inferred than observed" (Burne and Moore 1987, p. 242).

The term selenite, in the paper, is applied to gypsum crystals over 2 mm in size (cf. Warren 1982b). These are spacifically primary gypsum ctystals grown from solution or brine directly on the bottom of the basin, or within the soft deposits, microbial mats, or organic sediments as displacive or incorporative crystals.

3 The Badenian Gypsum Deposits in Carpathian Foredeep

The Badenian evaporites of the Carpathian region originated during a major evaporite event (salinity crisis) which took place in the Central Paratethys Sea during its separation from the Mediterranean in the Badenian time, presumably about 13.6 Ma ago (Dudek et al. 2006; Peryt 2006). The Carpathian Foredeep basin is the largest Badenian evaporite basin, although its Carpathian margin has been destroyed by erosion related to post-Badenian tectonic shortening, overthrusting, and uplift of the Carpathian orogen (Fig. 1). In the northern Carpathian Foredeep the Badenian evaporites comprise mainly calcium sulphate sediments (exposed on the platformal margin of the basin) and sodium chloride deposits (buried in the foredeep) and are ca 10–70 m in thickness. The sulphate facies are represented by the primary gypsum deposits, mainly the selenites, gypsum microbialites, and clastic gypsum. These deposits show a nearly undisturbed layer-cake stratigraphy. The particular layers or groups of layers are laterally continuous and traced over distances of tens to hundreds kilometers. Several lithostratigraphical units, coinciding with typical gypsum facies, were distinguished within these deposits and several thin marker beds allowed more precise stratigraphical correlation (Kubica 1992; Bąbel 2005a; Peryt 2006). The section is bipartite and comprises

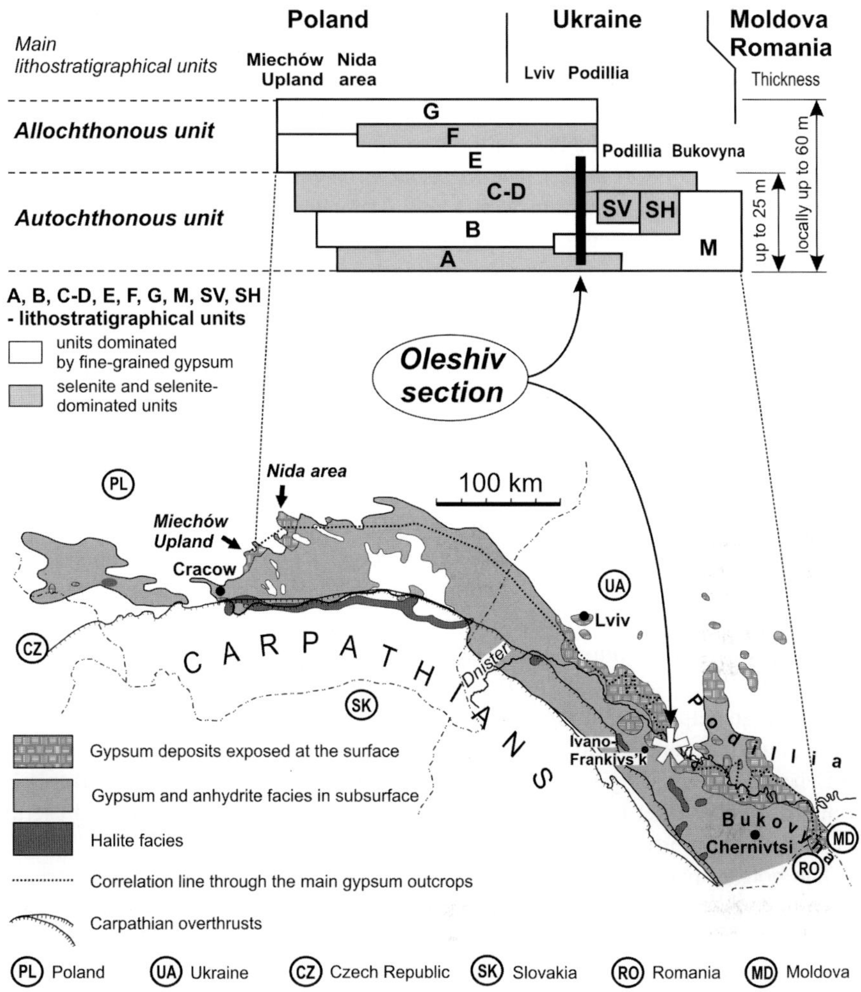

Fig. 1 Distribution of Badenian (Middle Miocene) evaporites in the northern Carpathian Foredeep [after many sources cited in Peryt (2006) and Bąbel (2005a), On-Line Appendix], lithostratigraphy of the marginal gypsum deposits [lettered units after Kubica (1992) and Bąbel (2005a)] and location of the studied section at Oleshiv

the lower autochthonous unit, containing mainly selenite and gypsum microbialite facies, and the upper allochthonous unit which is dominated by clastic gypsum.

4 Geology of the Study Area

The Badenian gypsum deposits described in this study are exposed between the Dnister river and Tlumach stream, the small right tributary of Dnister (E of Ivano-Frankivs'k; Fig. 2). The best outcrops occur along the high escarpment running meridionally along the Tlumach stream bank. The studied section at Oleshiv, situated in the northwestern part of the escarpment, is representative of that part of the basinal platform zone and formerly was described by Alth (1877), Łomnicki (1881), Peryt et al. (1994, figs. 4, 5), Peryt (1996, figs. 4, 7, 11a, c–e), Bąbel (2005a, fig. 5 and On-line Appendix) and Bąbel (2007, figs. 15f, 16, 17). The base of the escarpment is made up by horizontally lying Cretaceous limestones (Fig. 2a). The Mesozoic is capped by Miocene deposits, lying above an erosional gap, starting with a few decimetres of Badenian marine marly sandstones that contain fragmented *Pecten* shells, as observed at Palahychi (Łomnicki 1881; Siemiradzki 1909). In some places these sandstones are underlain with a very thin layer of clays of supposed freshwater origin (Malicki 1938). The sandstones are covered with thick layers of Badenian gypsum deposits, best exposed in abandoned quarries at the top part of the escarpment. Gypsum deposits are overlain by poorly exposed, barren, massive limestone (Alth 1877; Peryt 1996). Alth (1877), however, recognised borings of mollusk *Pholas* in this part of the section. The top of section is eroded and covered by Quaternary clastics and loess.

The gypsum section at Oleshiv shows a similar sequence of lithostratigraphical units and arrangement of facies as in the whole northern margin of the Badenian basin, particularly between Lviv and Odaiv (Figs. 1 and 2). The section (Fig. 3) starts with secondary alabaster replacing the giant gypsum intergrowths (unit A). It is covered by gypsum microbialites (unit M), grass-like, banded, gypsum (unit B), composed of rows of bottom-grown selenite crystals intercalated with fine-grained gypsum, sabre gypsum (unit C–D) that comprise the lower autochthonous part of the Badenian sulphates. Further up in the section, there is a clastic gypsum (unit E) belonging to the upper allochthonous unit. The section is not fully exposed, unit C–D being disturbed by some karst collapse or faults. The lower part of unit B contains set of marker beds traced across nearly the whole basin margin and useful for isochronous correlation (Bąbel 2005a).

The bottom part of the microbialite unit M at Oleshiv, below a characteristic horizontal discontinuity surface (in places covered with clay), contains sub-horizontal selenite crystals. The contact of these deposits with the underlying giant gypsum intergrowths is very clear. In the southernmost exposure (between Oleshiv S1 and S2, Fig. 2a, Bąbel 2005a, in the On-line Appendix) the giant gypsum intergrowths show numerous flat dissolution surfaces and their upper surface is scalloped by dissolution and/or erosional features (Fig. 4b). Northwards (Oleshiv S1; Figs. 2 and 4a), the layers of grass-like gypsum (rows of bottom-grown selenite crystals with top surface flatted by dissolution) appears in between the gypsum intergrowths and the microbialite layer. Further north (Oleshiv N; Fig. 2) the transition from the giant intergrowths to the microbialites is gradual and without traces of dissolution surfaces (Bąbel 2005a, On-line Appendix). Going up the section the number and sizes of selenite crystals decrease and fine-grained gypsum gradually becomes the main component. Transitions similar to this are observed throughout the basin, and were interpreted as the result of shallowing of the permanently submerged several metre deep saline pan, where the giant gypsum intergrowths were crystallizing, followed by the establishment of the semi-emerged

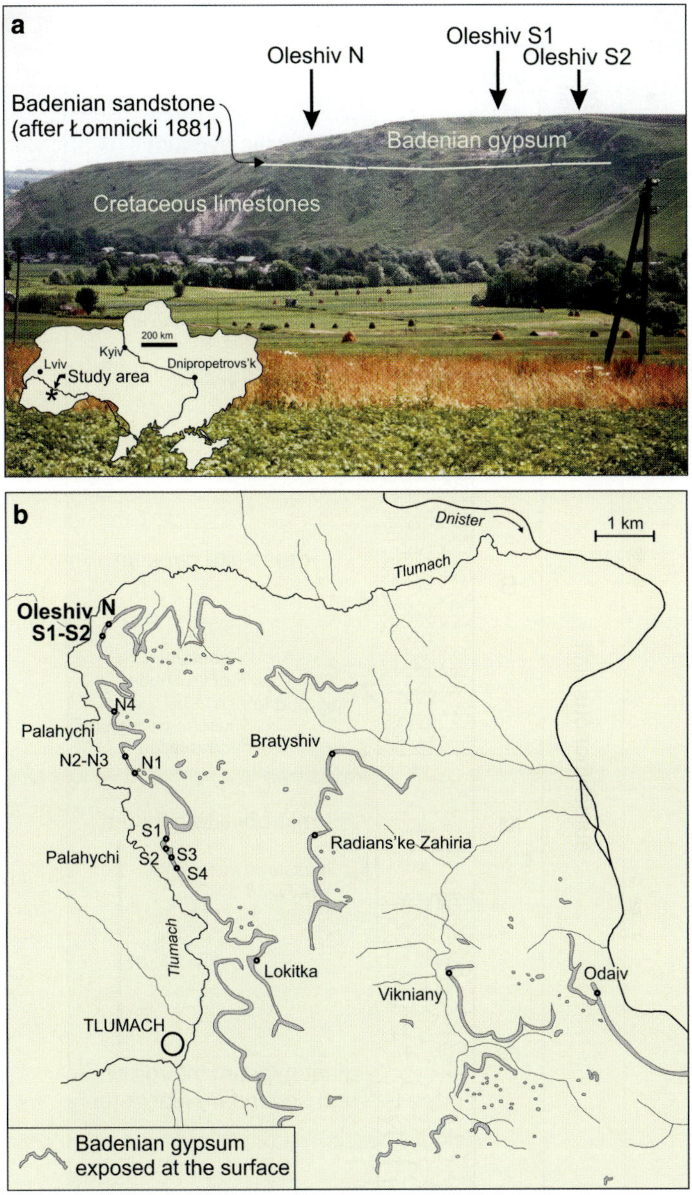

Fig. 2 (**a**) Panoramic view of the Oleshiv outcrop, Ukraine. (**b**) Map showing exposures of Badenian gypsum deposits (after Bieniasz 1885) and representative gypsum sections in vicinity of Oleshiv (see Bąbel 2005a, On-line Appendix for detailed description of the sections)

evaporite shoal on which microbialites were deposited (Peryt 1996; Bąbel 2005b, 2007). The laterally differentiated transitions at Oleshiv suggest that there had been some bottom relief.

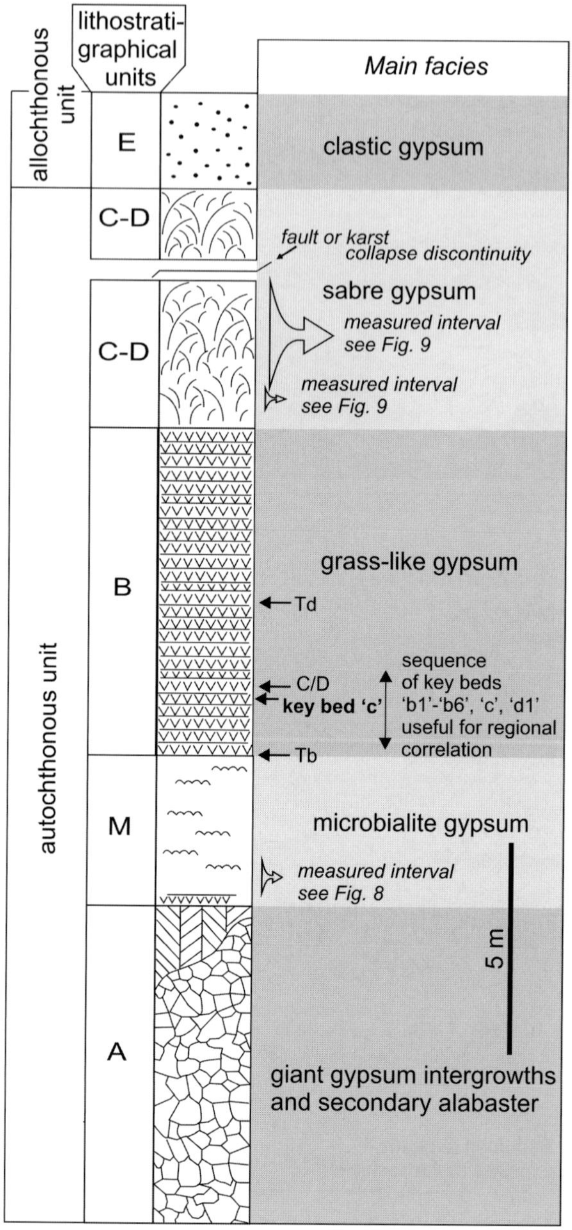

Fig. 3 Stratigraphy of the Badenian gypsum deposits at Oleshiv with distribution of the measured paleocurrent indicators; hachure reflects crystal arrangement; main lithostratigraphical units (lettered A, M, B, C–D, and E) after Kubica (1992) and Bąbel (2005a); key beds b1–b6, c, d1, and isochronous surfaces Tb, C/D, Td after Bąbel (2005a); complete documentation of the section and details are shown at www.geo.uw.edu.pl/agp/table/appendixes/55-1/

Fig. 4 (**a**) Panoramic view of the Badenian gypsum deposits at Oleshiv S1 locality, Ukraine (see Fig. 2a), showing distribution of the measured fields with gypsum microbialite domes; key bed "c" important for regional correlation is marked; photo taken in 1994. (**b**) The giant gypsum intergrowths (unit A) overlaid with gypsum microbialites with scattered selenite crystals (unit M) exposed in between S1 and S2 locality at Oleshiv (see Fig. 2a). (**c**) Details of Fig. 2b showing channel-like scours (outlined in *red*) which cut the top surface of the giant gypsum intergrowths and are infilled with gypsum microbialite and clastic deposits

5 Sedimentology of Badenian Gypsum Deposits and Microbialites

Sedimentological studies of the marginal sulphate deposits indicate that the lower autochthonous unit was deposited on the vast evaporite shoals (or platforms) and in shallow (<10 m) saline pans (Kasprzyk 1993, 1999; Peryt 1996, 2001, 2006;

Bąbel 2007). Widespread selenite facies represent permanently subaqueous environments whereas microbialites form mainly on semi-emerged shoals. The microbialite deposits and structures are very common in the Badenian gypsum deposits in the Czech Republic, Poland, Ukraine and Moldova (Bąbel 2005b), and are also documented in Bulgaria (Trashliev 1969). Gypsum stromatolite domes were recognized in the Nida river area in Poland (Kwiatkowski 1970, 1972) and many varieties of microbialites were recognized in drill cores (Kasprzyk 1993). The variable nearshore environments of the Badenian basin were subdivided into two main types; the "wet shore" and the "dry shore" environment, depending on the influx of continental water from land recorded by clay intercalations (Schreiber et al. 2007). The "wet shore" microbialite environments are mostly represented in Poland and "dry-shore" ones in Ukraine. The gypsum microbialites from both these zones apparently differ in morphology and structure [compare Kwiatkowski (1972), Kasprzyk (1993) and Peryt (1996)]. The microbialites are very abundant in Ukraine, particularly on the east part of the basin, and they were mostly interpreted as the result of gypsification of microbial mats (Peryt 1996). It seems that gypsification of microbial mats was easier there because of the warmer local climate and hence greater evaporation rate.

6 The "Dry Shore" Gypsum Microbialites (Unit M)

The gypsum microbialites in Oleshiv represent the same type of microbialites as those present in the other outcrops of unit M stretching from Lviv to Moldova and described by Peryt (1996). They are also similar to microbialites occurring in marker bed "c" (Kasprzyk 1993, 1999; Peryt 2001; Bąbel 2005a). These microbialites are composed of white to light brown, fine-grained gypsum (alabaster) showing more or less pronounced crenulated wavy lamination (with laminae from 1 to several mm thick, commonly obliterated and visible only as the ghosts). The crenulated lamination is generally horizontal. Crenulations reflect presence of very small (<5 cm in size) flattened domal structures. In some places the crenulated laminae form waves recording the occurrence of larger scale domal structures up to 0.6 m in diameter (Fig. 5b). The synoptic relief of these domal structures is usually low (commonly only several cm) and it rises in places of transition to the selenite facies (Fig. 5b). Locally, the beds of laminae form very complex fold structures which appear be the product of gypsification of the microbial mats of the irregular mamillate morphology possibly related to the lateral expansive growth of microbial benthic communities (Fig. 5a, see e.g. Grotzinger and Knoll 1999; Bąbel 2004, pl. 1, fig. 1; fig. 8a; Douglas et al. 2008). In some places, below some convex laminae, fenestral pore structures appear filled with coarse gypsum cement (such structures were however unnoticed at Oleshiv). A very characteristic feature is occurrence of shallow flat-bottomed channels filled with laminated clastic gypsum (Peryt 1996, figs. 12, 13; Bąbel 2005b, pl. 3, fig. 2, 2007, figs. 21, 22) which are however poorly represented at Oleshiv (Fig. 4c).

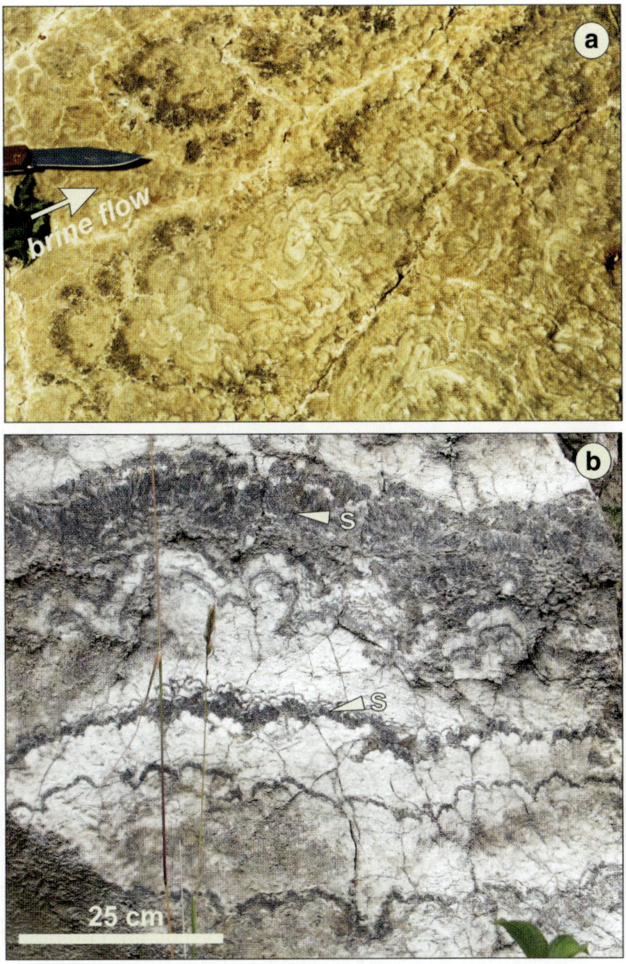

Fig. 5 (**a**) Elongated gypsum microbialite domes with asymmetric crusts of selenite crystals (*dark*) seen on the horizontal erosion surface. Folded laminae in the interior of the dome presumably represent gypsified wrinkled microbial mats. Brine flow direction is interpreted as parallel to the dome elongation and facing selenite crusts. Locality Oleshiv S1-c, Ukraine. (**b**) Exemplary gypsum microbialite domes covered with crusts of selenite crystals (marked "s") displaying variable morphology and many orders of crenulations. Note that synoptic relief of the particular domes is rising going up the section which is interpreted as deepening. Verenchanka quarry, Ukraine, layers directly below key bed c

Petrographical investigations do not detect any traces or ghosts of actual microbial structures (like "algal" filaments; Panieri et al. 2008; Oren et al. 2009) in the gypsum microbialites. SEM observations revealed however that dolomite is common within unit M and occurs as spheroidal or sub-spheroidal bodies (crystal aggregates) scattered among gypsum crystals or grouped into clusters or chains, in places associated with clay particles (Fig. 6). Very rarely imperfect rhombic

Fig. 6 (**a–d**) SEM photographs of the polished and water etched surfaces of gypsum microbialite samples from measured interval of unit M at Oleshiv, Ukraine [taken from "dark" microbialite significant for local correlation and marked in Bąbel (2005a, On-line Appendix, fig. 7, hachure no. 38, and fig. 8D)] showing various aspects of dolomite spheroids scattered among gypsum crystals. Note porous structure of spheroids, clustered spheroids, and organic filament (**a** and **b**) overgrown with dolomite (marked Or in **b**). *Dol* dolomite, *Gp* gypsum, *CM* clay minerals, *Or* organic matter. Depositional top is towards the *upper sides* of the photos

crystals are observed. The spheroids and larger crystals are composed of smaller dolomite plates or granules. Some spheroids show porous structure or hollow round centre suggesting that the dolomite could have nucleated on some rounded objects which were later disintegrated or dissolved. The sizes of spheroids are relatively uniform and range from 5 to 15 μm (most commonly 10 μm). We interpret these spheroids as primary precipitates, possibly related to presence of benthic microbial communities. Primary or synsedimentary (early diagenetic) dolomite spheroids similar in sizes and structure have been frequently reported from modern and ancient saline (or brackish) lakes or lagoon deposits associated with sulphate waters and microbial mats, and interpreted as products of microbially mediated precipitation related to bacterial sulphate reduction (e.g. Bellanca et al. 1993; Feldmann and McKenzie 1997; van Lith et al. 2002; Rao et al. 2003; Sanz-Montero et al. 2006; Ayllón-Quevedo et al. 2007).

The discussed microbialites were interpreted as "gypsified stromatolites" by Peryt (1996) and similar facies from Poland as products of gypsification of microbial mats (Kasprzyk 1999). Most common stratiform gypsum microbialites presumably were deposited on semi-emerged shoals flooded by brine-sheets (Logan 1987; Peryt 1996; Bąbel 2007) where accretion of gypsum (and supposed growth of microbial mats) was controlled by water level (e.g. Horodyski 1977). The microbialites with domes showing some relief and covered with tiny selenites (Fig. 5b) suggests permanent cover of gypsum-saturated brine over the bottom, i.e. deeper saline pan environment (cf. Southgate 1989).

7 Structures Indicating the Direction of Water Flow

Structures indicating the water flow are abundant at Oleshiv and include: (1) sabre gypsum crystals with horizontally oriented apices (Fig. 7a, Bąbel 2002), (2) sub-horizontal gypsum crystals and intergrowths within gypsum microbialite (Fig. 7b),

Fig. 7 (**a**) Sabre gypsum facies showing conformable orientation of crystal apices interpreted as a result of accelerated upstream growth of crystals. Holovchyntsi, Ukraine. (**b**) Gypsum microbialite deposits with sub-horizontally oriented selenite crystals overlying the selenite bed, lowermost part of the measured interval at locality Oleshiv S1, Ukraine (see Fig. 8)

(3) elongated gypsum microbialite domes, and (4) elongated gypsum microbialite domes with asymmetric selenite covers (Fig. 5a). Most structures indicate precise direction of the current, except of the elongated domes which show only streamlines of the current. The oriented sabre crystals occur in the unit C–D, the other structures in the lower part of unit M within the 1 m thick interval of the section (Figs. 3, 7 and 8). The sabre crystals were measured in outcrop S1. The domes were measured in three fields: a, b, c in outcrop S1 at approximately the same

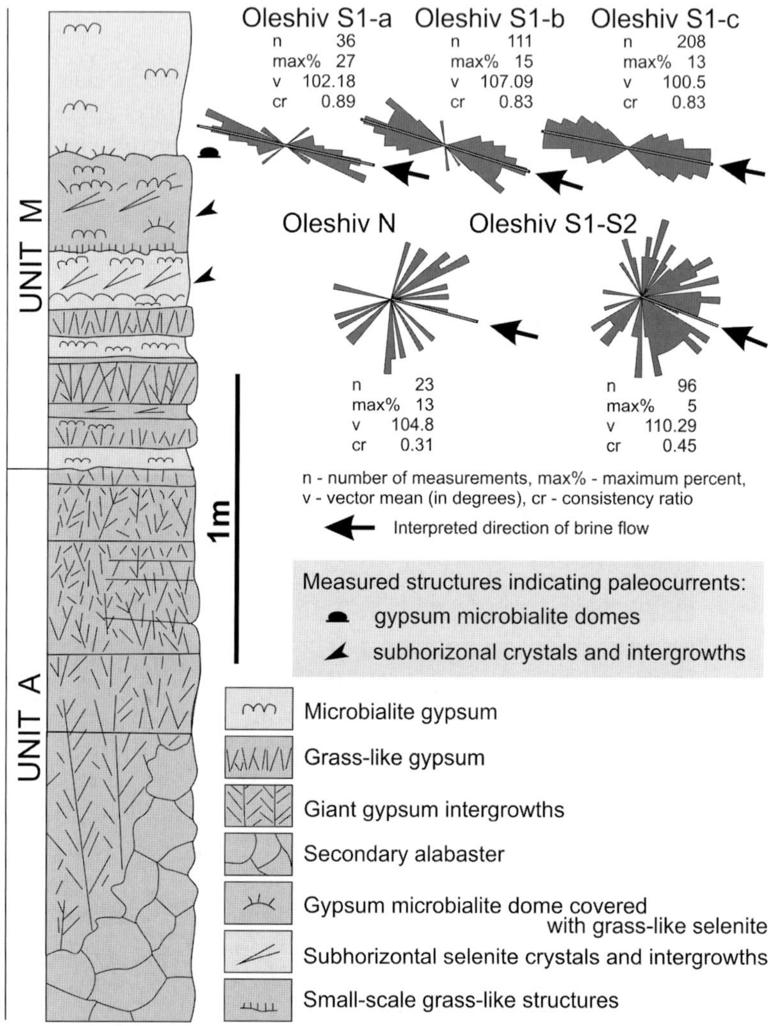

Fig. 8 Section of the giant gypsum intergrowths and gypsum microbialites at Oleshiv S1 locality, Ukraine, with location of the intervals and stratigraphical levels measured for paleocurrent analysis (see Figs. 2a and 4) and rose diagrams showing azimuth of gypsum microbialite domes and azimuths of sub-horizontal selenite crystals apices. Interpreted flow directions are parallel to mean vectors

stratigraphical level (Fig. 4a). Field S1-a contains poorly exposed, elongated domes without selenite crusts. Field S1-b (8 × 4 m) is situated ca 80 m south of S1-a and contains domes without selenites as well as domes with selenites (ca 16%). Nearly all domes and sub-domes in field S1-c (14 × 5 m) are with asymmetric selenite crusts. The latter two fields represent just top surfaces below the above mentioned correlating discontinuity (dissolution?) surface. The sub-horizontal crystals within microbialites were measured in the same interval of the section, below the discontinuity surface in outcrop N, S1 and S2 (Fig. 2a).

7.1 Sabre Gypsum Crystals with Concordant Orientation of Apices

The current related structures do occur more frequently within unit C–D, the so-called sabre gypsum, where the rows of bottom-grown curved crystals show sabre-like shapes. As in the whole basin the sabre gypsum crystals in unit C–D at Oleshiv show conformable orientation of apices turned in the same or similar horizontal direction as illustrated by Peryt (1996, fig. 7). This widespread directional structure was described and interpreted in detail in the previous papers (Bąbel and Becker 2006; Bąbel and Bogucki 2007) and at Oleshiv it was measured and interpreted in the same way (Fig. 9).

7.2 Sub-horizontal Selenite Crystals Within Microbialites

The investigated microbialites contain single and aggregated gypsum crystals and crystal intergrowths (<15 cm long) similar in the crystallographical structure to the giant gypsum intergrowths below (Figs. 4b, c and 7b). These crystals commonly grew in depressions between the microbialite domes, or on slopes of the domes, and are oriented obliquely to layering or sub-horizontally. They are honey-coloured presumably due to included organic matter (which is typical feature of such crystals; Cody 1979; Cody and Cody 1988). These crystals are interpreted as growing concurrently with deposition of surrounding fine-grained gypsum (microbialite). They presumably grew at least partly under the cover of, or within the microbial mats and hence could include organic matter during the growth (Bąbel 2005b). Their oblique orientation suggests that their growth could be also controlled by the flowing brine. The azimuths of such crystals apices were measured for paleocurrent analysis (Fig. 8).

7.3 Elongated Gypsum Microbialite Domes

The gypsum microbialite domes are exposed in the quarry on the flat shelves, on the eroded top surfaces of the gypsum beds, where the shape and internal structure of

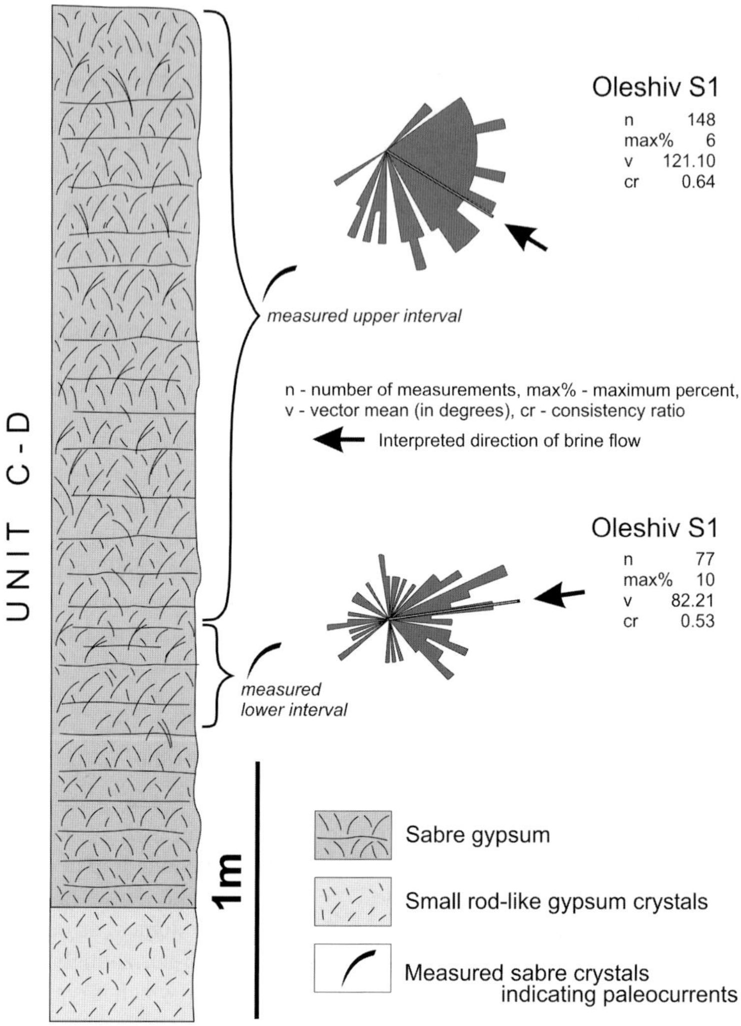

Fig. 9 Section of the selenite sabre gypsum deposits at Oleshiv S1, Ukraine (see Fig. 2a), with location of measured intervals and rose diagrams showing azimuths of sabre crystal apices. Interpreted brine flow directions are parallel to mean vectors

the domes is visible in the plan view (Figs. 4 and 10). The domes show crude internal wavy lamination or banding reflected by slight colour differences of the fine-grained gypsum. The largest domes are about 35 cm in length and they are flattened – the synoptic relief can be estimated as less than 10 cm (Fig. 11). The smaller domes, several cm wide, occur in between some of the larger ones. The large domes are commonly composed of several smaller sub-domes (compare Figs. 5b, 10 and 12). The domes seen in horizontal section show irregularly circular, oval or elliptic shapes, elongated in similar direction. The large circular domes are

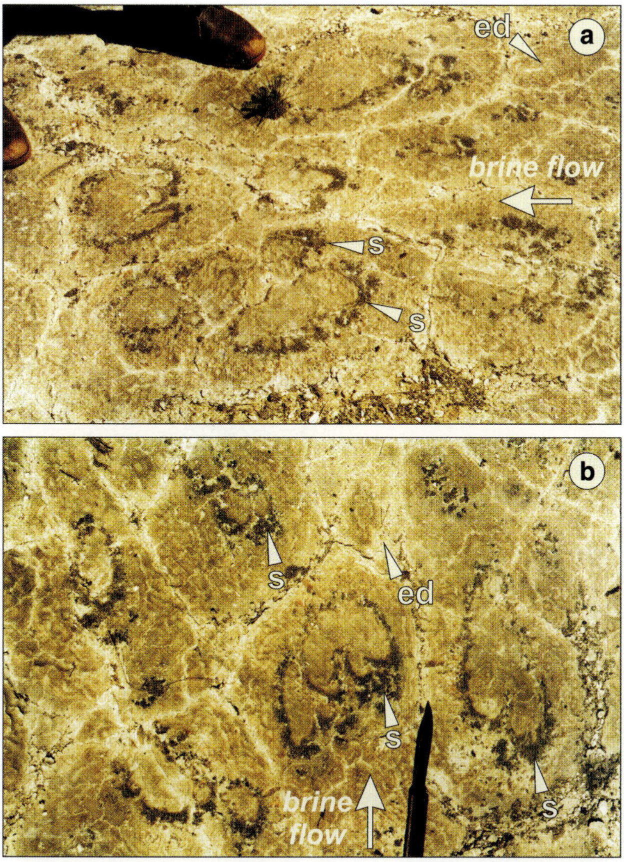

Fig. 10 (**a–b**) Pattern of elongated gypsum microbialite domes with asymmetric (upstream oriented) crusts of bottom-grown selenite crystals (marked "s"), resembling traces of hoofs, seen on the horizontal erosion surface; note elongated domes without selenite crusts (marked "ed"); locality Oleshiv S1-c, Ukraine (see Figs. 4 and 5)

composed of elliptical sub-domes also oriented parallel or sub-parallel. Many of the domes and sub-domes show apparent streamlined slightly asymmetric shapes. One elongated side is blunted or well rounded and thicker and the opposite side is narrower. The relation of long to short axis of the domes is less than 2:1.

7.4 Elongated Gypsum Microbialite Domes with Asymmetric Selenite Covers

These domes are similar to previously described except of the presence of a selenite crystal cover. The crystals coat the domes in an asymmetric way. They are

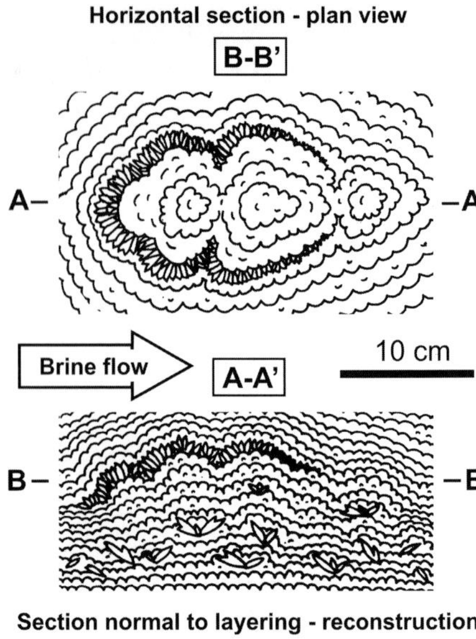

Fig. 11 Scheme showing horizontal and interpreted vertical section through the elongated gypsum microbialite dome with asymmetric cover of selenite crystals

apparently more abundant and larger on one side of the dome, just the side which is elongated and commonly blunted (Figs. 5a, 10, 11 and 12). The opposite side is usually without selenite crystals at all. The selenite always covers the same side of the domes in all these structures. Such domes seen on horizontal surface resembles imprints of hoofs indicating one direction of march (Fig. 10).

8 Interpretation of the Oriented Badenian Selenite Crystals

The structures having containing oriented selenite crystals are interpreted as a result of the influence of the gypsiferous brine, flowing over the bottom, on the crystals growing directly on the bottom, having namely an accelerated upstream crystal growth. The arguments for such an interpretation were summarized by Bąbel and Becker (2006, with references), see also: Boudreau (1988), Sizaret et al. (2006, 2009) and Uchymiak et al. (2008).

The measurements made in the lower and upper interval of the unit C–D showed a different orientation of mean growth vectors (Fig. 9), suggesting that the average flow direction changed with time, and shifted clockwise about 27°. Shapes of rose diagrams and consistency ratio values suggest that the flow directions were more uniform during deposition of the upper interval.

The measurements in unit M revealed that sub-horizontal crystals show also preferred orientation of apices although the statistical results are less univocal

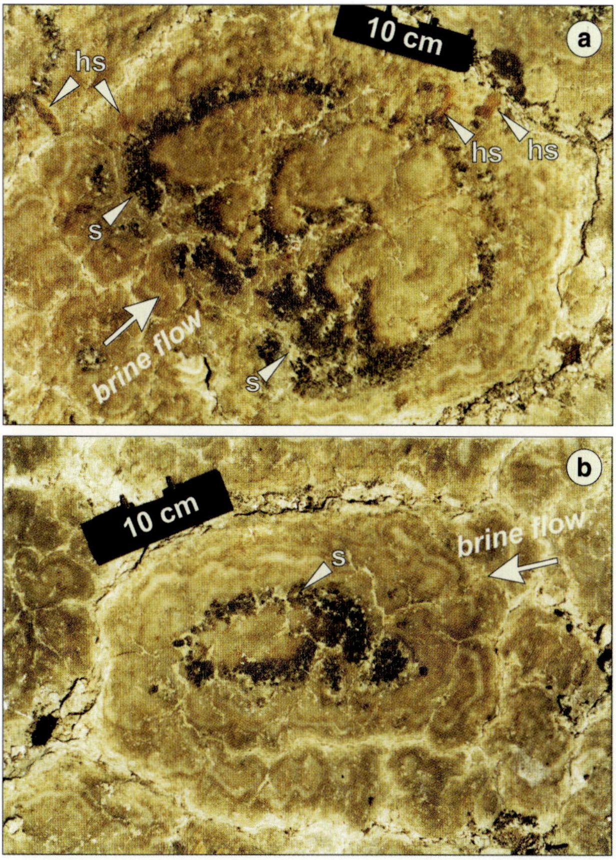

Fig. 12 (**a–b**) Elongated gypsum microbialite domes with asymmetric (upstream oriented) crusts of bottom-grown selenite crystals (marked "s") seen on the horizontal erosion surface; honey gypsum crystals are marked by "hs"; locality Oleshiv S1-c, Ukraine (see Figs. 4 and 5)

than for the sabre crystals as suggested by low number of measurements in Oleshiv N locality ($n = 23$) and relatively low consistency ratio in this and S1–S2 locality (0.31 and 0.45 respectively). Nevertheless, the mean vectors obtained are very similarly oriented and nearly parallel to the vectors from the other intervals of the Oleshiv section, including vectors representing elongated microbialite domes. All these vectors together suggest flow of brine permanently from ESE to WNW during deposition of the autochthonous gypsum in this locality and this direction perfectly coincide with the brine paleocurrents detected in adjacent localities (Fig. 2b, Bąbel et al. 1999; Bąbel and Bogucki 2007). The similar, constant direction of brine flow during deposition of B and C–D units, was recorded on the Nida area in Poland (Kwiatkowski 1970, 1972; Bąbel 2002). Such concurrence suggests that the currents shaping all these structures were similar in nature.

9 Origin and Environment of the Elongated Domes and Selenite Directional Structures

The elongated shape of gypsum microbialite domes apparently was produced by the unidirectional bottom currents acting during their growth. They are elongated parallel to the current similarly as many recent microbialite domes, particularly those growing in tidal zone (Hoffman 1967, 1976; Hofmann 1973; Gunatilaka 1975; Playford and Cockbain 1976; Walter et al. 1976; Park 1977; Andres and Reid 2006). The currents shaping gypsum domes from Oleshiv were not driven by tides. Tides were certainly negligible in the Badenian Foredeep evaporite basin which, most probably, was a closed basin separated by some barriers from the sea (Peryt 2001). Elongated gypsum stromatolite domes present in the Nida area in Poland, different than those from Oleshiv (containing many rows of bottom-grown gypsum crystals), were interpreted as parallel to the currents (Kwiatkowski 1970, 1972; Bąbel 2002).

It seems that at Oleshiv the same currents which mechanically shaped the domes were also able, in some periods, to promote the accelerated growth of selenite crystals on one side of the domes (Figs. 11 and 13). The coincidence of the elongation of the domes with the direction of the accelerated crystal growth seems crucial for the interpretation of all the studied directional structures. This feature strongly supports the view that it was really the bottom brine flow which produced the beds of selenite crystals with conformably oriented apices – adding the strong argument in favour of such an interpretation (see Bąbel and Becker 2006). Moreover, at Oleshiv the direction of flow can be deduced from the stream-lined shapes of some of the domes (see domes with downstream "tails" in Fig. 5a,

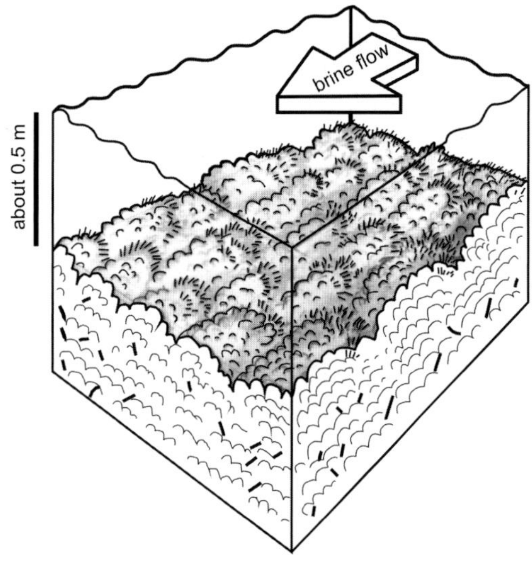

Fig. 13 Scheme showing interpreted origin of the elongated gypsum microbialite domes with asymmetric covers of selenite crystals

and right in Fig. 10b), or from position of blunted, more rounded side of the dome, resembling "crescent marks" (compare Figs. 5a, 10, 11 and 12 with Vidal 1972, figs. 5, 6). This further supports the view that the selenite crystals really were able to grow with an upstream orientation.

The brine currents that shaped the domes and influenced the growth of selenite crystals in unit M were probably similar in nature to those interpreted for the sabre gypsum (Bąbel and Bogucki 2007). They represent oriented, continuous water flow reflecting longshore (cyclonic) brine circulation or mean drift in evaporite basin (Bąbel and Becker 2006).

The elongated domes appear to be rare in unit M. The lack of such domes can be puzzling. They can be only weakly recorded because the horizontally exposed gypsum surfaces are rare; it is very difficult to recognize them in vertical sections, on walls of the outcrops. Alternatively, it is equally possible that the water or brine normally was flowing relatively slowly along the flat basin shoals and its mechanical influence on the morphology of the substrate, possibly covered with cohesive microbial mats, was negligible. The currents could, however, increase their speed and erosive power passing some narrows between more elevated areas. In such narrows they could influence the shape of the growing domes and accelerate the growth of selenite crystals. The Oleshiv outcrop can represent such a place, namely the gentle slope of the channel-like restriction or some broad depression between slightly more elevated shoals. The directional structures become more and more pronounced going to the southernmost part of this exposure, probably located closer to the axis of the presumed narrow channel.

10 Modern Analogues of the Badenian Gypsum Domes

In the primary gypsum deposits the bottom-grown crystals commonly create domal structures. Small, from centimetres to a few decimetres large selenite domes are typical forms crystallizing in the shallow (<1 m deep) gypsum pans in the marine saltworks, usually in the salinity range 200–300‰ (Ortí Cabo et al. 1984). Much larger and more complex selenite domes crystallize at depths of 1–3 m, on the floor of the Lake Inneston, York Peninsula, Australia (Warren 1982b). They are up to several meters in diameter and tens of centimeters in synoptic relief. They were described as "gypsum-cyanobacterial mounds" (De Deckker 1988; Bąbel 2005b and references therein), or "gypsiferous domal stromatolites" (Southgate 1989). The selenite domes are coated by microbial slime (cyanobacteria and diatoms) and "can be considered to be a type of coarsely-layered gypsum stromatolite" (Warren 2006, p. 229). The ancient Messinian gypsum deposits show even greater and more complex domal selenite structures (e.g. Dronkert 1985). In the Badenian evaporite basin such large selenite domes commonly are intercalated with fine-grained gypsum, interpreted as gypsum microbialite, and create complex microbialite-selenite structures (Bąbel 2005b). Well described ancient gypsum microbialite deposits creating small-scale domal structures are known from the Messinian of

Cyprus (Rouchy and Monty 1981). Gypsum microbialites related to gypsification (gypsum encrustation) of microbial mats are recorded from many modern shallow and semi-emerged evaporite environments (Eckstein 1970, figs. 5, 6; Krumbein and Cohen 1977; Ortí Cabo et al. 1984; Reineck et al. 1990; Taher et al. 1995; Gerdes et al. 2000). The origin of these and associated deposits is much less recognized than the carbonate microbialites and most authors consider them as products of passive precipitation of gypsum on, or within the mat (Rouchy and Monty 2000).

Modern gypsum domes very similar to the described Badenian structures from Oleshiv were noted by West et al. (1985) in saline pans ("lagoons") near El Alamein, on the Mediterranean coast of Egypt. The marine water in these pans, about 1 m deep, attained salinity 168 g/l during the study, which is the salinity at the beginning of the gypsum saturation field for evaporating marine brine and also salinity yet tolerable for many microorganisms, and cyanobacteria in particular (see references in Bąbel 2004). The gypsum domes were composed of aggregates of millimetre-size crystals and some of them were covered with crusts of larger selenite crystals (<3 cm). Like in many similar environments (Oren et al. 2009), microbial communities were developed within these crystalline gypsum deposits with a marked green colour, presumably cyanobacterial communities present at the bottom side of the selenite crust, 2–3 cm down from its surface (West et al. 1985, fig. 6c). Similar to the studied domes from Oleshiv, the selenite crystals were light-brown in colour. The domes growing in these pans were about 10 cm high and 15–50 cm wide and they were "particularly well developed where small waves break in a few centimeters of water" (West et al. 1985, p. 176, fig. 5). "In plan-view many of the mounds have roughly lunate forms which are convex toward the lagoon" (West et al. 1985, p. 176). The description suggests that the lunate domes accreted towards the propagating incoming waves. Their shape was apparently the product of the increased accretion of gypsum deposits and growth of selenite crystals on the upstream sides of the domes in the very shallow water. The lunate domes are similar to the hoof-like selenite crusts on the domes from Oleshiv and in both cases dominating current direction is indicated by the streamline shape of these structures, although in the domes from Egypt this current is oriented normally to the shoreline.

The other partial analogue for the elongated Badenian microbialite domes is Deep Lake salina in Australia (Warren 1982a, b). Gypsum is deposited in the centre of this 1 m deep lake and microbial mats are growing in "algal channels" on the lake margin forming low-relief ridges (2 m wide, 10 m long) oriented parallel to axis of the channel.

11 Final Remarks

The growth of large selenite crystals on the surface of the microbialite domes was evidently accelerated by influx of unidirectional calcium sulphate oversaturated currents carrying "nourishing" ions to the depositional surface. It seems very likely that the periodic action of similar currents also influenced (promoted or accelerated) the growth of smaller micro- and fine-crystalline gypsum precipitating at the water-sediment interface, on or within benthic microbial community, biofilm, or microbial mat, contributing to its gypsification. The Badenian gypsum microbialite domes with selenite covers can be considered as some analogue of the carbonate stromatolitic structures containing large crystal fans (calcite pseudomorphs after aragonite) described as "chemical" (Demicco and Hardie 1994) or "crystal stromatolites" (James et al. 2001), common in some Precambrian deposits. Similarly, they represent complex hybrid structures which are both organic (microbial) and inorganic (chemical) in origin (Riding 2008). The Precambrian carbonate crystal stromatolites also created elongated domal structures the origin of which was connected with action of subtidal currents. It is interesting that just within this elongated stromatolite facies the crystal fans "are best developed" (Sumner and Grotzinger 2004, p. 1293).

Acknowledgements The authors are greatly indebted to Hala and Vasyl Tycholiz, and Andrii Yatsyshyn for assistance in the field. We thank Charlotte Schreiber, University of Washington, and an anonymous reviewer for corrections which improved the final version of the text. This research was supported by funds of the Dean of the Faculty of Geology, University of Warsaw, grant BST 1241/6, and by the Polish Scientific Committee (KBN), grant 6 P04D 038 09.

References

Alth A (1877) Sprawozdanie Dra Alojzego Altha, z podróży odbytej w r. 1875 w niektórych częściach Podola Galicyjskiego. Akademija Umiejętności w Krakowie, Sprawozdanie Komisyi Fizjograficznéj 11:198–218 (In Polish)

Andres SM, Reid P (2006) Growth morphologies of modern marine stromatolites: A case study from Highborne Cay, Bahamas. Sedimentary Geology 185:319–328

Arp G, Reimer A, Reitner J (2004) Microbialite formation in seawater of increased alkalinity, Satonga Crater Lake, Indonesia – reply. Journal of Sedimentary Research 74:318–325

Ayllón-Quevedo F, Souza-Egipsy V, Sanz-Montero ME, Rodríguez-Aranda JP (2007) Fluid inclusion analysis of twinned selenite gypsum beds from the Miocene of the Madrid basin (Spain). Implication on dolomite bioformation. Sedimentary Geology 201:212–230

Bąbel M (2002) Brine palaeocurrent analysis based on oriented selenite crystals in the Nida Gypsum deposits (Badenian, southern Poland). Geological Quarterly 46:49–62

Bąbel M (2004) Models for evaporite, selenite and gypsum microbialite deposition in ancient saline basins. Acta Geologica Polonica 54:219–249

Bąbel M (2005a) Event stratigraphy of the Badenian selenite evaporites (Middle Miocene) of the northern Carpathian Foredeep. Acta Geologica Polonica 55:9–29. On-Line Appendix: http://www.geo.uw.edu.pl/agp/table/appendixes/55-1/

Bąbel M (2005b) Selenite-gypsum microbialite facies and sedimentary evolution of the Badenian evaporite basin of the northern Carpathian Foredeep. Acta Geologica Polonica 55:187–210

Bąbel M (2007) Depositional environments of a salina-type evaporite basin recorded in the Badenian gypsum facies in northern Carpathian Foredeep. In: Schreiber BC, Lugli S, Bąbel M (eds) Evaporites Through Space and Time. Journal of Geological Society, Special Publication 285, pp 107–142

Bąbel M, Becker A (2006) Cyclonic brine-flow pattern recorded by oriented gypsum crystals in the Badenian evaporite basin of the northern Carpathian Foredeep. Journal of Sedimentary Research 76:996–1011

Bąbel M, Bogucki A (2007) The Badenian evaporite basin of the northern Carpathian Foredeep as a model of a meromictic selenite basin. In: Schreiber BC, Lugli S, Bąbel M (eds) Evaporites Through Space and Time. Journal of Geological Society, Special Publication 285, pp 219–246
Bąbel M, Boguckiy A, Vizna S, Yatsyshyn A (1999) Reconstruction of brine paleocurrents in the Middle Miocene evaporitic basin of the Carpathian Foredeep. Biuletyn Państwowego Instytutu Geologicznego 387:12–13
Bellanca A, Karakas Z, Neri R, Varol B (1993) Sedimentology and isotope geochemistry of lacustrine dolomite-evaporite deposits and associated clays (Neogene, Turkey): environmental implications. Mineralogica et Petrographica Acta 36:245–264
Bieniasz F (1885) Tyśmienica – Tłumacz, 1:75000. Atlas Geologiczny Galicyi. Zakład wojskowy geograficzny w Wiedniu
Boudreau BP (1988) Mass-transport constraints on the growth of discoidal ferromanganese nodules. American Journal of Science 288:777–797
Burne RV, Moore LS (1987) Microbialites: organosedimentary deposits of benthic microbial communities. Palaios 2:241–254
Cody RD (1979) Lenticular gypsum: occurrences in nature, and experimental determination of soluble green plant material on its formation. Journal of Sedimentary Petrology 49:1015–1028
Cody RD, Cody AM (1988) Gypsum nucleation and crystal morphology in analog saline terrestrial environments. Journal of Sedimentary Petrology 58:247–255
De Deckker P (1988) Biological and sedimentary facies of Australian salt lakes. Palaeogeography, Palaeoclimatology, Palaeoecology 62:237–270
Demicco RV, Hardie LA (1994) Sedimentary structures and early diagenetic features of shallow marine carbonate deposits. SEPM Atlas Series 1:1–255
Douglas S, Abbey W, Mielke R, Conrad P, Kanik I (2008) Textural and mineralogical biosignatures in an unusual microbialite from Death Valley, California. Icarus 193:620–636
Dronkert H (1985) Evaporite models and sedimentology of Messinian and Recent evaporites. GUA Papers of Geology, Ser. 1/24, Utrecht, pp 1–283
Dudek K, Bukowski K, Wiewiórka J (2006) Radiometric dating of Badenian pyroclastic sediments from the Wieliczka-Bochnia area. In: 8 Ogólnopolska Sesja Naukowa. Datowanie Minerałów i Skał. Kraków 18–19.11.2006, pp 19–26 (in Polish with English abstract)
Eckstein Y (1970) Physicochemical limnology and geology of a meromictic pond on the Red Sea shore. Limnology and Oceanography 15:363–372
Feldmann M, McKenzie JA (1997) Messinian stromatolite-thrombolite associations, Santa Pola, SE Spain: an analogue for the Palaeozoic? Sedimentology 44:893–914
Gerdes G, Krumbein WE, Noffke N (2000) Evaporite microbial sediments. In: Riding RE, Awramik SM (eds) Microbial sediments. Springer, Berlin, pp 196–208
Grotzinger JP (1986) Evolution of an early Proterozoic passive margin carbonate platform, Rocknest Formation, Wopmay Orogen, Northwest Territories, Canada. Journal of Sedimentary Petrology 56:831–847
Grotzinger JP, Knoll AH (1999) Stromatolites in Precambrian carbonates: evolutionary mileposts or evolutionary dipsticks? Annual Review of Earth and Planetary Sciences 27:313–358
Gunatilaka A (1975) Some aspects of the biology and sedimentology of laminated algal mats from Mannar Lagoon, north-west Ceylon. Sedimentary Geology 14:275–300
Hoffman P (1967) Algal stromatolites: use in stratigraphic correlation and paleocurrent determination. Science 157:1043–1045
Hoffman P (1976) Stromatolite morphogenesis in Shark Bay, western Australia. In: Walter MR (ed) Stromatolites. Developments in Sedimentology 20. Elsevier, Amsterdam, pp 261–271
Hofmann HJ (1973) Stromatolites: characteristics and utility. Earth-Science Reviews 9:339–373
Horodyski RJ (1977) *Lyngbya* mats at Laguna Mormona, Baja California, Mexico: comparison with Proterozoic stromatolites. Journal of Sedimentary Petrology 47:1305–1320
James NP, Narbonne GM, Kyser TK (2001) Late Neoproterozoic cap carbonates: Mackenzie Mountains, northwestern Canada: precipitation and global glacial meltdown. Canadian Journal of Earth Sciences 38:1229–1262

Kasprzyk A (1993) Stromatolitic facies in the Badenian (middle Miocene) gypsum deposits of southern Poland. Neues Jahrbuch für Geologie und Paläontologie, Abhandlungen 187: 375–395

Kasprzyk A (1999) Sedimentary evolution of Badenian (Middle Miocene) gypsum deposits in the northern Carpathian Foredeep. Geological Quarterly 43:449–465

Krumbein WE, Cohen Y (1977) Primary production, mat formation and lithification: contribution of oxygenic and facultative anoxygenic cyanobacteria. In: Flügel E (ed) Fossil algae. Springer, Berlin, pp 37–56

Kubica B (1992) Lithofacial development of the Badenian chemical sediments in the northern part of the Carpathian Foredeep. Prace Państwowego Instytutu Geologicznego 133:1–64 (in Polish with English summary)

Kwiatkowski S (1970) Origin of alabasters, intraformational breccias, folds and stromatolites in Miocene gypsum of Southern Poland. Bulletin de l'Académie Polonaise des Sciences, Série des Sciences Géologiques et Géographiques 18:37–42

Kwiatkowski S (1972) Sedimentation of gypsum in the Miocene of southern Poland. Prace Muzeum Ziemi 19:3–94. Warszawa (in Polish with English summary)

Logan BW (1987) The MacLeod evaporite basin, western Australia. Holocene environments, sediments and geological evolution. AAPG Memoir 44:1–140

Łomnicki M (1881) Formacyja gipsu na zachodnio-południowéj krawędzi płaskowzgórza Podolskiego. Kosmos 6:174–201 (in Polish)

Malicki A (1938) Karst phenomena in the gypsum beds of Pokutian Podolia. Prace Geograficzne Wydawane Przez Prof. E. Romera 18:1–48. Lwów (in Polish with English summary)

Oren A, Sørensen KB, Canfield DE, Teske AP, Ionesco D, Lipski A, Altendorf K (2009) Microbial communities and processes within a hypersaline gypsum crust in a saltern evaporation pond (Eilat, Israel). Hydrobiologia 626:15–26

Ortí Cabo F, Pueyo Mur JJ, Geisler-Cussey D, Dulau N (1984) Evaporitic sedimentation in the coastal salinas of Santa Pola (Alicante, Spain). Revista d'Investigacions Geologiques 38–39:169–220

Panieri G, Lugli S, Manzi V, Palinska KA, Roveri M (2008) Microbial communities in Messinian evaporite deposits of the Vena del Gesso (northern Apennines, Italy). Stratigraphy 5:343–352

Park RK (1977) The preservation potential of some recent stromatolites. Sedimentology 24:485–506

Peryt TM (1996) Sedimentology of Badenian (middle Miocene) gypsum in eastern Galicia, Podolia and Bukovina (West Ukraine). Sedimentology 43:571–588

Peryt TM (2001) Gypsum facies transitions in basinal-marginal evaporites: middle Miocene (Badenian) of west Ukraine. Sedimentology 48:1103–1119

Peryt TM (2006) The beginning, development and termination of the Middle Miocene Badenian salinity crisis in Central Paratethys. Sedimentary Geology 188-189:379–396

Peryt TM, Poberezski AW, Jasionowski M, Petryczenko OI, Peryt D, Ryka W (1994) Badenian gypsum facies in the Nida and Dnister river-basins (southern Poland and SW Ukraine). Przegląd Geologiczny 42:771–776 (in Polish)

Playford PE, Cockbain AE (1976) Modern algal stromatolites at Hamelin Pool, a hypersaline barred basin in Shark Bay, Western Australia. In: Walter MR (ed) Stromatolites. Developments in Sedimentology 20. Elsevier, Amsterdam, pp 389–411

Rao VP, Kessarkar PM, Krumbein WE, Krajewski KP, Schneider RJ (2003) Microbial dolomite crusts from the carbonate platform off western India. Sedimentology 50:819–830

Reineck HE, Gerdes G, Claes M, Dunajtschik K, Riege H, Krumbein WE (1990) Microbial modification of sedimentary surface structures. In: Heling D, Rothe P, Förstner U, Stoffers P (eds) Sediments and environmental geochemistry. Selected aspects and case histories. Springer, Berlin, pp 254–276

Riding R (2008) Abiogenic, microbial and hybrid authigenic carbonate crusts: components of Precambrian stromatolites. Geologia Croatica 61:73–103

Rouchy JM, Monty C (1981) Stromatolites and cryptalgal laminites associated with Messinian gypsum of Cyprus. In: Monty C (ed) Phanerozoic stromatolites. Springer, Berlin, pp 155–180
Rouchy JM, Monty C (2000) Gypsum microbial sediments: Neogene and modern examples. In: Riding RE, Awramik SM (eds) Microbial sediments. Springer, Berlin, pp 209–216
Sanz-Montero ME, Rodríguez-Aranda JP, Calvo JP (2006) Mediation of endoevaporitic microbial communities in early replacement of gypsum by dolomite: a case study from Miocene lake deposits of the Madrid Basin, Spain. Journal of Sedimentary Research 76:1257–1266
Schreiber BC, Bąbel M, Lugli S (2007) Introduction and overview. In: Schreiber BC, Lugli S, Bąbel M (eds) Evaporites Through Space and Time. Journal of Geological Society, Special Publication 285:1–13
Siemiradzki J (1909) Geologia Ziem Polskich, v. 2. Formacye młodsze (Kreda-Dyluwium). Nakładem Muzeum im. Dzieduszyckich, Lwów (in Polish)
Sizaret S, Fedioun I, Barbanson L, Chen Y (2006) Crystallization in flow – II. Modelling crystal growth kinetics controlled by boundary layer thickness. Geophysical Journal International 167:1027–1034
Sizaret S, Branquet Y, Gloaguen E, Chauvet A, Barbanson L, Arbaret L, Chen Y (2009) Estimating the local paleo-fluid flow velocity: New textural method and application to metasomatism. Earth and Planetary Science Letters 280:71–82
Southgate PN (1989) Relationships between cyclicity and stromatolite form in the Late Proterozoic Bitter Spring Formation, Australia. Sedimentology 36:323–339
Sumner DY, Grotzinger JP (2004) Implications for Neoarchaean ocean chemistry from primary carbonate mineralogy of the Campbellrand-Malmani Platform, South Africa. Sedimentology 51:1273–1299
Taher AG, Abd El Wahab S, Philip G, Krumbein W, Wali AM (1995) Evaporitic sedimentation and microbial mats in a salina system (Port Fouad, Egypt). International Journal of Salt Lake Research 4:95–116
Trashliev S (1969) Structures of the Tortonian gypsum rocks in northwestern Bulgaria. Sedimentary Geology 3:291–316
Trompette R (1969) Les stromatolites du "Précambrien Supérieur" de l'Adrar Mauritanie (Sahara Occidental). Sedimentology 13:123–154
Uchymiak M, Lyster E, Glater J, Cohen Y (2008) Kinetics of gypsum crystal growth on a reverse osmosis membrane. Journal of Membrane Science 314:163–172
van Lith Y, Vasconcelos C, Warthmann R, Martins JCF, McKenzie JA (2002) Bacterial sulfate reduction and salinity: two controls on dolomite precipitation in Lagoa Vermelha and Brejo do Espinho (Brazil). Hydrobiologia 485:35–49
Vidal G (1972) Algal stromatolites from the Late Precambrian of Sweden. Lethaia 5:353–368
Walter MR, Bauld J, Brock TD (1976) Microbiology and morphogenesis of columnar stromatolites (*Conophyton*, *Vacerrilla*) from hot springs in Yellowstone National Park. In: Walter MR (ed) Stromatolites. Developments in Sedimentology 20. Elsevier, Amsterdam, pp 273–310
Warren JK (1982a) The hydrological significance of Holocene tepees, stromatolites, and boxwork limestones in coastal Salinas in South Australia. Journal of Sedimentary Petrology 52:1171–1201
Warren JK (1982b) The hydrological setting, occurrence and significance of gypsum in late Quaternary salt lakes in South Australia. Sedimentology 29:609–637
Warren JK (2006) Evaporites: sediments, resources and hydrocarbons. Springer, Berlin, Heidelberg, 1036 p
West IM, Ali YA, Hilmy ME (1985) Facies associated with primary gypsum nodules of northern Egyptian sabkhas. In: Schreiber BC, Harner HR (eds) 6th International Symposium on Salt, Toronto, Ontario 1983/1:171–183. The Salt Institute, Alexandria, Virginia
Zamarreño I, Anadón P, Utrilla R (1997) Sedimentology and isotopic composition of Upper Palaeocene to Eocene non-marine stromatolites, eastern Ebro Basin, NE Spain. Sedimentology 44:159–176

The Microbialite-Vermetid Community of the Salento Peninsula in Southern Italy: A Late Miocene Example of Automicrite Deposition

Alessandro Vescogni, Adriano Guido, Adelaide Mastandrea, and Franco Russo

1 Introduction

A carbonate facies made up mainly of micrite and vermetid shells has been identified in the early Messinian coral reef complex of the Novaglie Formation on the Salento Peninsula of southern Italy. Vermetids (Mollusca, Prosobranchia, and Caenogastropoda) can form large colonies that are usually attached to hard substrates (Vescogni et al. 2008). The micrite-dominated facies with a large number of vermetid shells (most of them horizontal) is unusual. Moreover, this facies shows completely different sediment textures and organism compositions than the associated reefal carbonates. To explain the origin of this facies and identify the source of the micrites, field, microfacies and epifluorescence observations were carried out along with scanning electron microscopy (SEM), EDS and FT-IR analyses.

2 Geological Setting

The Salento Peninsula is the subaerial portion of the Apulia Platform, a very large carbonate bank that corresponds to the foreland of the Apennine thrust chain (Fig. 1). This area contains outcrops of a spectacular succession of pure carbonate sequences. These carbonate sequences are separated by major unconformities and span from the Cretaceous to the Pleistocene (Bosellini et al. 1999). The Late Miocene reefal unit is represented by the early Messinian Novaglie Formation,

A. Vescogni (✉)
Dipartimento di Scienze della Terra, Università di Modena e Reggio Emilia, Largo S. Eufemia 19, 41100 Modena, Italy
e-mail: alessandro.vescogni@unimore.it

A. Guido, A. Mastandrea, and F. Russo
Dipartimento di Scienze della Terra, Università della Calabria, Via Bucci Cubo 15b, 87036 Rende (CS), Italy

J. Reitner et al., *Advances in Stromatolite Geobiology*,
Lecture Notes in Earth Sciences 131, DOI 10.1007/978-3-642-10415-2_20,

Fig. 1 Location map

which is exposed along the eastern coast of the Salento from Porto Tricase to S. Maria di Leuca (Fig. 1). This Late Miocene reefal unit lies unconformably over the pre-Miocene formations, and its age (early Messinian) has been determined by the occurrence of benthic foraminifera (*Bulimina echinata* zone) and ostracod associations (Bosellini et al. 1999).

The Novaglie Formation corresponds to the superimposition of three depositional sequences, which are separated by erosional surfaces and interpreted as the result of fourth order sea level fluctuations (Fig. 2a). Each sequence consists of a fringing reef unit, which is mainly bioconstructed by corals, and a variable number of associated organisms (mainly vermetids, coralline algae and *Halimeda*). A detailed description of the stratigraphic setting and reef facies analysis has been carried out by Bosellini et al. (2001, 2002) and Vescogni et al. (2008).

A microbialite-vermetid community occurs at the base of the Novaglie Formation on the pre-Miocene substrate (Fig. 2b). This microbialite-vermetid community ranges in thickness from 30 cm up to 2.5 m and grades upward into biocalcarenite deposits that are related to the middle to distal part of the early Messinian reef slope.

3 Methodology

Carbonate characterisation was performed by optical and scanning electron microscope (SEM) microfacies observations and energy-dispersive X-ray spectrometer (EDS) microanalyses. Chemical compositions were determined using a

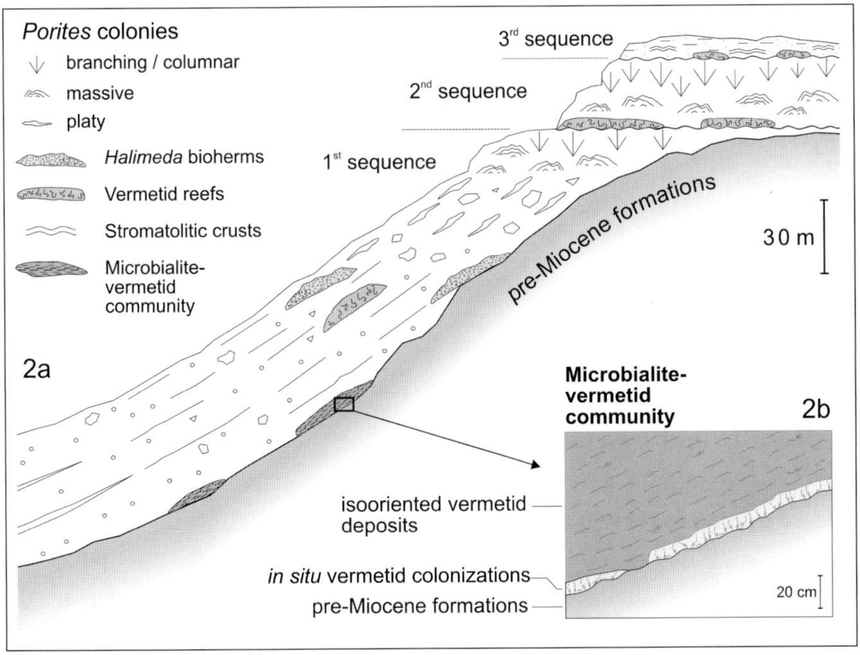

Fig. 2 (**a**) Synthetic profile of the Messinian reef complex [modified from Bosellini et al. (2001)] and (**b**) stratigraphic setting of the microbialite-vermetid community

FEI-Philips ESEM-FEG Quanta 200F scanning electron microscope linked to an EDAX Genesis 4000 EDS. The analysed samples were polished with 0.25-μm diamond-impregnated surfaces. Then, the samples were etched and carbon coated (about 250 Å). The voltage was 15 kV. The tilt angle was 0°, and the take-off angle was 36.01°.

The fluorescence of the uncovered thin-sections was checked for to reveal the distribution of the organic matter (OM) (Dravis and Yurewicz 1985; Machel et al. 1991; Neuweiler et al. 1997; Russo et al. 1997). Fluorescence was induced by a Hg vapour lamp linked to an Axioplan II imaging microscope (Zeiss) equipped with high performance wide bandpass filters (BP 436/10 nm/LP 470 nm for the green light and BP 450–490 nm/LP 520 nm for the yellow light).

Fourier transform infrared analytical techniques have been performed on bulk organic matter extracted from the mineral matrix with a mixture of dichloromethane/methanol (1:1). FT-IR spectroscopic analyses were performed using a Spot Light 200 Perkin Elmer spectrometer. Interferograms were acquired in the transmission mode within the range of 400–4,000 cm^{-1} by accumulating 256 scans at a spectral resolution of 4 cm^{-1}. Spectral bands were assigned with reference to the literature (Painter et al. 1981; Solomon and Carangelo 1988; Wang and Griffiths 1985; Sobkowiak and Painter 1992).

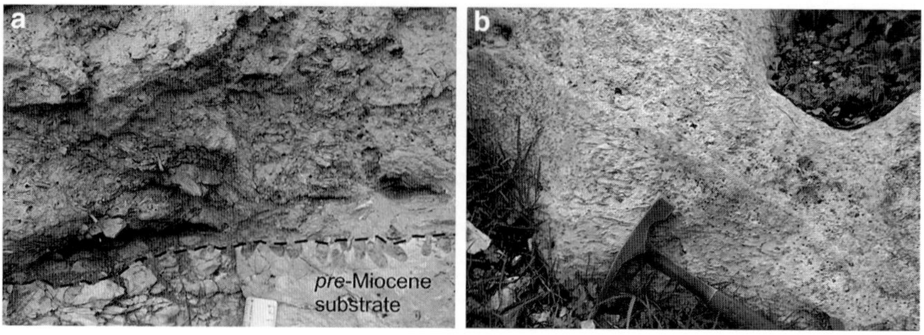

Fig. 3 (**a**) *In situ* vermetid colonisation on the pre-Miocene substrate and (**b**) iso-oriented vermetid facies

4 Results and Discussion

4.1 Facies Observations

Within the microbialite-vermetid association, vermetids are represented by the *Petaloconchus* monogeneric association. Two different facies can be distinguished. The first facies is characterised by vermetids in life position (in situ), and the other is characterised by "iso-oriented" vermetid shells organised in clinostratified deposits.

The in situ facies is arranged in discontinuous layers up to 10 cm thick, and vermetids and barnacles encrust the pre-Miocene substrate (Fig. 3a). The vermetids are about 1.5 cm in length and sparsely distributed. They sometimes form small aggregates that grow upright. Most of the shells have unwound "feeding tubes", which can be destroyed and rebuilt to cope with environmental changes. The upper part of this facies contains deposits of "iso-oriented" vermetids (Fig. 3b), which constitute the largest part of the microbialite-vermetid community. The most peculiar feature of these deposits is the large number of broken vermetid feeding tubes, which form iso-oriented accumulations parallel to the stratification. The *Petaloconchus* feeding tubes have an average diameter of 2 mm and lengths of up to 8 cm. Thin bryozoan crusts, serpulids, and young vermetids colonise the surface of the *Petaloconchus* shells.

Both facies are characterised by a matrix of fine-grained wackestone/packstone. Planktonic foraminifera, peloids, fragments of barnacles, bryozoans, serpulids, bivalves and coralline algae are also present.

4.2 Microfacies Observations

The following three types of micrite-rich microfacies have been recognised within the microbialite-vermetid community: (a) non- or low-fluorescent detritic micrite

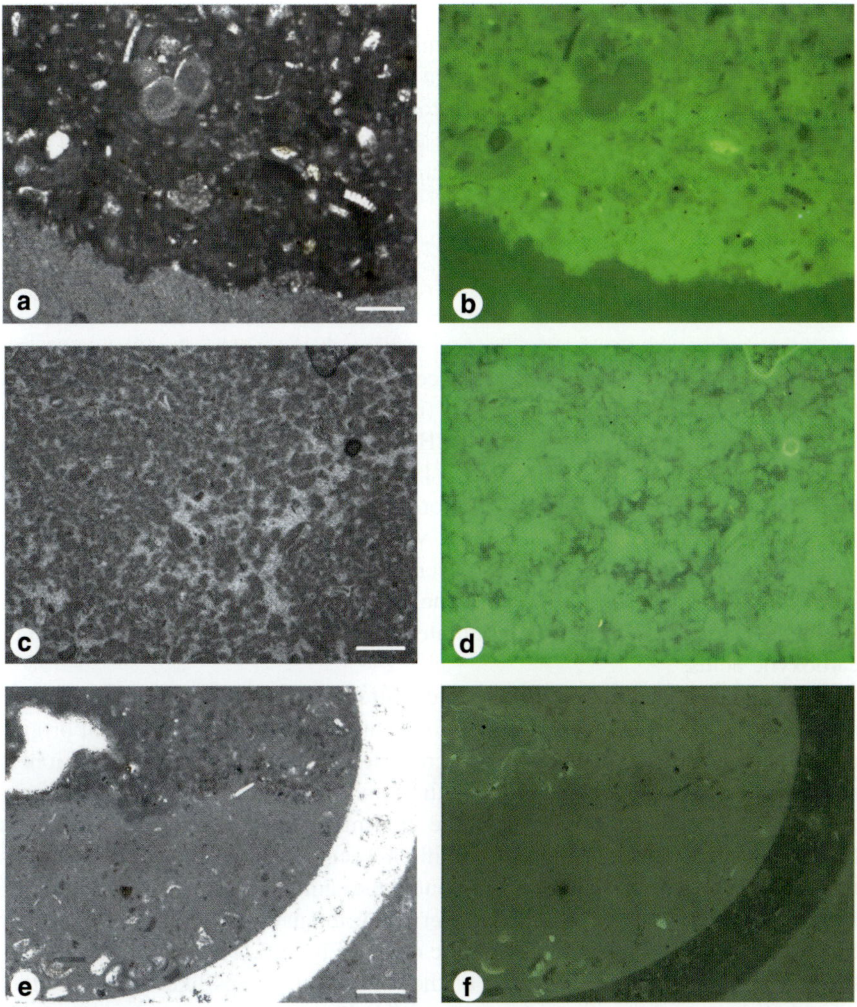

Fig. 4 On the left are transmitted light microphotographs, and on the right are UV epifluorescence of the main types of micrites. (**a**, **b**) Fluorescent mudstones/wackestones rich in bioclastic grains; (**c**, **d**) fluorescent thrombolitic peloidal organomicrite; and (**e**, **f**) low fluorescent detritic micrite with gravitative fabric. This organomicrite is filling a vermetid shell (scale bar 200 μm)

with a few fine bioclastic grains that sometime show gravitative fabric; (b) fluorescent mudstones/wackestones that are rich in fine bioclastic grains; and (c) fluorescent thrombolitic peloidal organomicrite (Fig. 4).

Basal layers with vermetids in life position show the dominance of the first type of micrite. Its detritic nature is supported by the occurrence of gradated geopetal fillings of several vermetid shells, indicating a gravitative deposition. This interpretation is corroborated by the low autofluorescence, which implies a low organic matter content. Iso-oriented vermetid deposits are characterised by the occurrence of types (b) and (c) of the micrite-rich microfacies. The autofluorescence can be explained by two different types of organic matter. The first type probably derives from decaying metazoan and planktonic organisms, and the second type is linked to the metabolic activity of microorganisms which have induced the biomineralisation of the thrombolites.

4.3 FT-IR Analyses

The infrared spectra show bands between 600 and 3,000 cm^{-1}. These spectra bands include stretching aliphatic bands (υCHali) at 2,949, 2,918 and 2,848 cm^{-1}, the deformation bands of methyl (δCH_3; 1,365 cm^{-1}) and both methyl and methylene [$\delta(CH_2 + CH_3)$; 1,458 cm^{-1}] groups (Fig. 5). The spectra also show the band associated with carbonyl and/or carboxyl groups ($\nu C{=}O$; 1,740 cm^{-1}). The band associated with the skeletal vibration of more than four methylene groups [$\delta(CH_2)_4$; 720 cm^{-1}] has also been observed. The νC–O vibration appears between 1,300 and 1,100 cm^{-1}. In some samples we observed the band $\nu C{=}C$, which is probably related to unsaturated compounds such as alkenes and/or carboxylic acids.

FT-IR was carried out on the organic matter that was extracted from the iso-oriented vermetid facies (Fig. 5). The different spectra show the different origins of the three types of micrite-rich microfacies. The intensity of absorption is related to the abundance of organic matter, and the fluorescent mudstone/wackestone presents higher peaks that the thrombolitic and detritic microfacies (Fig. 5a). This result confirms the epifluorescence observations and indicates that the organic matter content is distributed the same way like in all three microfacies. The detritic micrite shows a very low intensity for the organic matter functional groups, confirming its abiotic origin (Fig. 5d). The organic matter has the same intensity and distribution of the functional groups like in the peloidal thrombolitic micrite, external vermetid shells and filling vermetid shells (Fig. 5b, c). The FT-IR spectra of this type of micrite are different from those of the mudstone/wackestone and detritic micrite and are characterised by the presence of stretching C=C vibrations. The absence of bands in the 700–900 cm^{-1} region indicates that the $\nu C{=}C$ band does not belong to an aromatic compound. We attribute this band to alkenes and/or unsaturated carboxylic acids that may have been synthesised by microbial communities that precipitated carbonates through their metabolic activities.

These data, together with microfacies observations, indicate the microbial origin of most of the micrite within the iso-oriented vermetid facies. Organomicrite deposition is correlated with the occurrence of a large amount of decaying organic matter (Reitner and Neuweiler 1995). The occurrence of this organic matter can be linked to the death of a large number of *Petaloconchus* individuals.

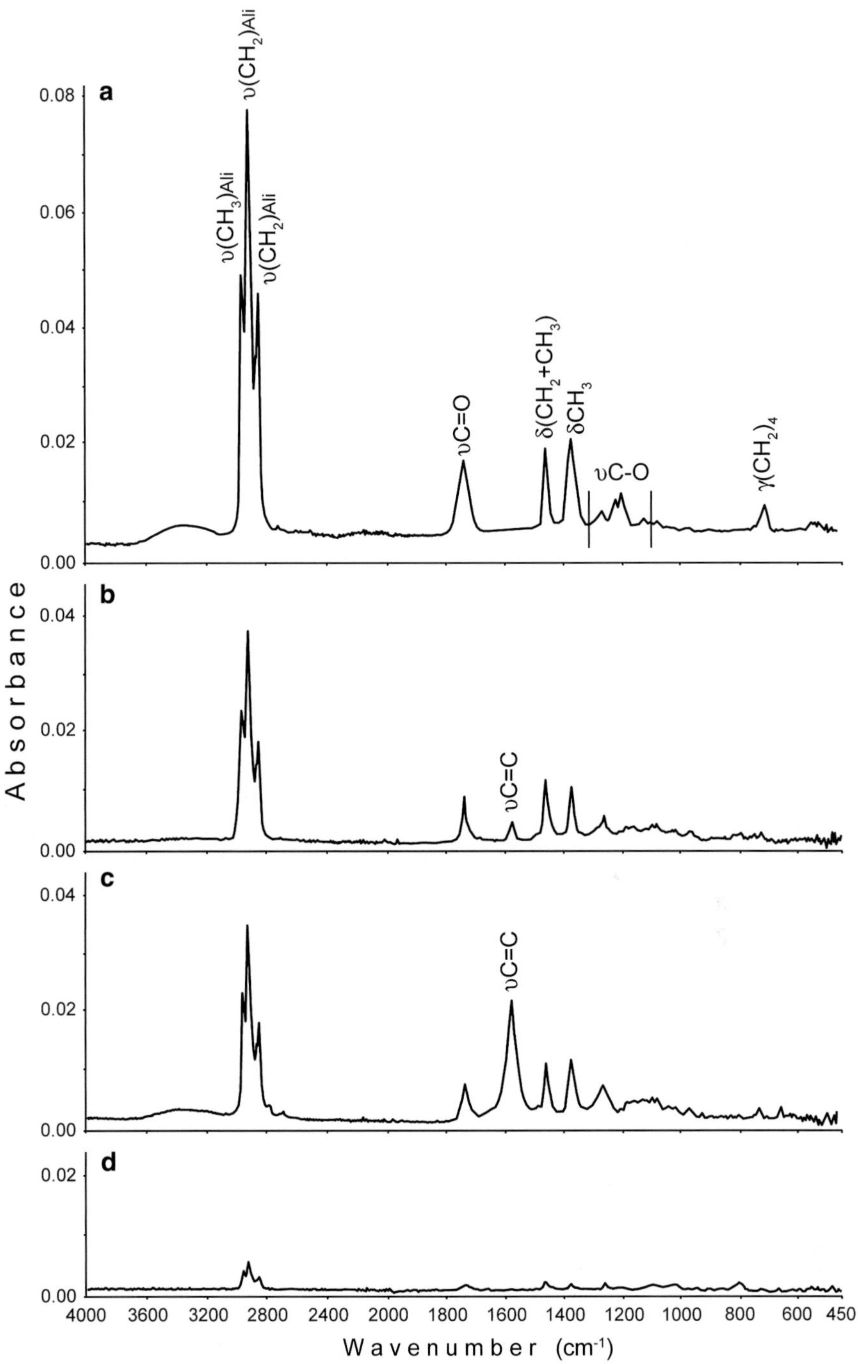

Fig. 5 Infrared spectra of extracted organic matter: (**a**) fluorescent mudstones/wackestones rich in fine bioclastic grains; (**b**) micrite filling vermetid shells; (**c**) fluorescent thrombolitic peloidal micrite; and (**d**) non-fluorescent detritic micrite

5 Conclusions

Microfacies and vermetid assemblages suggest that the deposition of the microbialite-vermetid community happened in the following two distinct phases: (1) vermetid colonisation and detritic micrite deposition and (2) redeposition of the previous assemblage as iso-oriented vermetid facies rich in mudstones/wackestone and thrombolitic organomicrite.

The first phase starts with the colonisation of pre-Miocene substrate by vermetids. Penecontemporaneous fast inputs of detritic micrite occurred, forcing the vermetids to produce remarkably long, upward oriented, feeding tubes to avoid being buried by sedimentation.

The second phase is the redeposition of the previous assemblage as iso-oriented vermetid facies rich in mudstones/wackestone and thrombolitic organomicrite. The iso-oriented setting of vermetid conchs is interpreted as the result of a slight redeposition of the in situ facies because of overloading or another destabilising process. Vermetid ecological demands do not support life within a muddy substrate. Therefore, the organomicrite deposition, which must have happened after the death of the vermetid community, may be correlated to the occurrence of a large amount of decaying organic matter (Reitner and Neuweiler 1995). The source of this organic matter can be linked to the death of a large number of *Petaloconchus* individuals.

The nature and distribution of organic matter confirm this interpretation. FT-IR analyses permit us to distinguish the following two main types of organic source: (1) a source that derives from decaying metazoan and planktonic organisms, which characterise the mudstones/wackestones microfacies and (2) a source linked to the metabolic activities of microorganism that induced thrombolite biomineralisation.

Acknowledgements We thank Prof. Dr. Joachim Reitner for the excellent organisation of International Kalkowsky Symposium on the Geobiology of Stromatolites. We are indebted to an anonymous reviewer for suggestions that greatly improved the manuscript. This research was funded by the Università della Calabria, PRIN 2007.

References

Bosellini A, Bosellini FR, Colalongo ML, Parente M, Russo A, Vescogni A (1999) Statigraphic architecture of the Salento coast from Capo d'Otranto to S. Maria di Leuca (Apulia, southern Italy). Rivista Italiana di Paleontologia e Stratigrafia 105:397–416

Bosellini FR, Russo A, Vescogni A (2001) Messinian reef-building assemblages of the Salento Peninsula (southern Italy): palaeobathymetric and palaeoclimatic significance. Palaeogeography, Palaeoclimatology, Palaeoecology 175:7–26

Bosellini FR, Russo A, Vescogni A (2002) The messinian reef complex of the Salento Peninsula (southern Italy). Stratigraphy, facies and paleoenvironmental interpretation. Facies 47:91–112

Dravis JJ, Yurewicz DA (1985) Enhanced carbonate petrography using fluorescence microscopy. Journal of Sedimentary Petrology 55:795–804

Machel HG, Mason RA, Mariano AN, Mucci A (1991) Causes and emission of luminescence in calcite and dolomite. In: Barker CE, Koop OC (Eds.) Luminescence Microscopy and Spectroscopy: Qualitative and Quantitative Applications. SEMP Short Course 25:9–25

Neuweiler F, Reitner J, Monty C (Eds.) (1997) Biosedimentology of Microbial Buildups – IGCP Project No. 380 – Proceedings of 2nd Meeting, Göttingen/Germany, 1996. Facies 36:195–284

Painter PC, Snyder RW, Starsinic M, Coleman MM, Kuehn DW, Davis A (1981) Concerning the application of FT-IR to the study of coal: A critical assessment of band assignments and the application of spectral analysis programs. Applied Spectroscopy 35:475–485

Reitner J, Neuweiler F (1995) Mud Mounds: a polygenetic spectrum of fine-grained carbonate buildups. Facies 32:1–69

Russo F, Neri C, Mastandrea A, Baracca A (1997) The mud-mound nature of the Cassian platform margins of the Dolomites. A case history: the Cipit Boulders from Punta Grohmann (Sasso Piatto Massif, Northern Italy). Facies 36:25–36

Sobkowiak M, Painter P (1992) Determination of the aliphatic and aromatic CH contents of coals by FT-i.r.: studies of coal extracts. Fuel 71:1105–1125

Solomon PR, Carangelo RM (1988) FT-i.r. analysis of coal. 2. Aliphatic and aromatic hydrogen concentration. Fuel 67:949–959

Vescogni A, Bosellini FR, Reuter M, Brachert TC (2008) Vermetid reefs and their use as palaeobathymetric markers: new insights from the Late Miocene of the Mediterranean (Southern Italy, Crete). Palaeogeography, Palaeoclimatology, Palaeoecology 267:89–101

Wang S-H, Griffiths PR (1985) Resolution enhancement of diffuse reflectance i.r. spectra of coals by Fourier-self deconvolution. 1. C–H stretching and bending modes. Fuel 64:229–236

The Characterisation of Sedimentary Organic Matter in Carbonates with Fourier-Transform Infrared (FTIR) Spectroscopy

Adelaide Mastandrea, Adriano Guido, Fabio Demasi, Silvestro Antonio Ruffolo, and Franco Russo

1 Introduction

Fourier-transform infrared spectroscopy (FTIR), an analytic technique based on the interaction between infrared electromagnetic radiation and matter, is of great importance in the characterisation of kerogens.

The chemical complexity of kerogens derives from the numerous sources of original organic matter (OM), the selective biodegradation or preservation of organic moieties, and physical and chemical processes leading to diagenesis (Lis et al. 2005a).

FTIR studies have revealed bands specific to the chemical structures of complex kerogen molecules (Painter et al. 1981, 1983; Baruah 1986; Černý 1955), and proven their diagnostic value for determining the maturity, type, and oil/gas generation potential of kerogens (Ganz and Kalkreuth 1987; Christy et al. 1989; Kister et al. 1990; Lin and Ritz 1993b; Chen et al. 1998).

FTIR is utilised for the identification of organic compounds, both in crystalline and amorphous forms, through the measurement of the energy absorbed by the molecules in transition to different vibrational states. Because each functional group tends to adsorb infrared radiation in a specific wavelength range, it is possible to detect and discriminate different chemical compounds. FTIR can be used to identify a broad range of chemical functional groups in a variety of physical states and it is complementary to other methods such as Raman spectroscopy and gas chromatography–mass spectroscopy (GC–MS). FTIR generally furnishes a better sensitivity for minor components than Raman spectroscopy and provides a comparable ability to classify and identify minerals.

A. Mastandrea (✉), A. Guido, F. Demasi, S.A. Ruffolo, and F. Russo
Dipartimento di Scienze della Terra, Università della Calabria, Via P. Bucci Cubo 15b, I-87036 Rende (CS), Italy
e-mail: a.mast@unical.it

J. Reitner et al., *Advances in Stromatolite Geobiology*,
Lecture Notes in Earth Sciences 131, DOI 10.1007/978-3-642-10415-2_21,

Infrared analysis, however, does not target particular molecules but rather detects classes of molecules. Each organic group (i.e., aliphatic C–H bonds) is present in many compounds with a particular wavelength and this provides a high sensitivity for mixtures (Anderson et al. 2005; Borrego et al. 1995). The major classes of biomolecules readily distinguished by FTIR include (a) proteins, peptides and enzymes, (b) lipids and fatty acids, including cell membranes, (c) carbohydrates and sugars, and (d) nucleic acids. These classes of functional groups have distinct infrared absorption-wavelength ranges (Anderson et al. 2005).

The sensitivity of IR analysis for organic molecules, however, depends on the mineral phase and the granular state, and it is increased if the organics are extracted or thermally desorbed and concentrated. With extraction, FTIR can detect organic compounds at parts per billion.

Organic matter type and abundance can be used as a measure of benthic secondary productivity, terrestrial/aquatic productivity and organic-matter flux in depositional systems (Palacios-Fest et al. 2005). Considering the close connection with the biological world, the reconstruction of depositional conditions and carbonate autogenesis processes based on organic-matter data is presently one of the main topics in carbonates research.

A broad variety of microorganisms, whether phototrophic or not, are implicated in carbonate autogenesis; however, their direct characterisation in ancient sedimentary systems is limited by the extremely low fossilisation potential of most microorganisms, especially bacteria. Therefore, the composition of the original community (primary producers, zooplankton, aerobic and anaerobic bacteria, benthic microorganisms, etc.) is difficult to evaluate by traditional morphological methods. Many studies have confirmed that organic geochemical techniques can be used to confirm the original presence of microorganisms.

The main goal of this study was to verify the applicability of the FTIR and μFTIR analyses to the geochemical characterisation of carbonates with respect to organic-matter content and thermal evolution.

2 Background Data

2.1 *Samples and Microfacies*

The analyses were performed on samples of the Calcare di Base (CdB), an outcropping in the Rossano Basin (Northern Calabria, Italy), which was previously studied and characterised through molecular fossils and other organic markers (Guido et al. 2007). The study samples were collected from an outcrop located near the Cropalati village. In this area the CdB succession consisted of two massive fine-grained calcareous beds interlayered with decimetric laminated marls. These calcareous beds overlay the diatomitic and marly layers of the Tripoli formation, which represent the first stage of basin restriction after the open marine conditions recorded by the Upper Tortonian/Early Messinian clays.

Optical and scanning electron microscope (SEM) observations and energy-dispersive X-ray spectrometer (EDS) microanalyses provided evidence of the homogenous dominant micrite microfacies and the preservation of the original aragonitic mineralogical phase. The microfacies were characterised by variably sized irregular peloids forming clots with an antigravitative fabric, in which faecal pellets are frequently dispersed. The peloidal micrite showed a typical thrombolitic microfacies, very rich in organic matter content as indicated by its bright epifluorescence. The organic matter follows the morphological pattern of peloids and appeared to be absent in the microsparite areas (Guido et al. 2007).

2.2 *Origin and Maturity of Organic Matter*

Organic-matter analyses were performed by Guido et al. (2007) using Rock-Eval pyrolysis, palynofacies observations and molecular-fossil recognition via gas chromatography–mass spectrometry.

The Rock-Eval pyrolysis data should be interpreted with caution, due to the low TOC values, but the Tmax value, at around 400°C, indicated the presence of unsaturated fatty acids and ββ-hopanoids, revealing that the sediment suffered very low thermal stress (Fig. 1).

The organic matter study revealed a transitional composition between Type II and Type III kerogen, suggesting a mixture of marine and terrigenous OM (Fig. 1). These different contributions are supported by biomarker analyses. The n-alkane distribution indicated the presence of three main biological signatures: algal (*n*C18–*n*C20 compounds), terrestrial (*n*C27–*n*C29), and bacterial (nC_{26}–nC_{28}, with no odd-even carbon-number predominance); see Fig. 2a. The autochthonous acidic fraction ($<C_{22}$) and relative level of unsaturated compounds validated the bacterial signature (Fig. 2c), further evidenced by the presence of hopanes (Fig. 2b) and the branched n-alkane distributions. These data point to a marine depositional scenario, interrupted by episodic freshwater inputs, in which carbonate precipitation was induced by bacterial ammonification of amino acids in aerobic conditions (Guido et al. 2007).

3 Analytical Techniques

Fourier-transform infrared analytical techniques were applied to thin sections, powdered sediment and organic matter extracted from the mineral matrix. FT-IR spectroscopic analyses were performed using a Perkin Elmer 100 interfaced with a Spot Light 200 spectrometer.

Interferograms were acquired in transmission mode within the range of 400–4,000 cm^{-1} by accumulating 256 scans at a spectral resolution of 4 cm^{-1}.

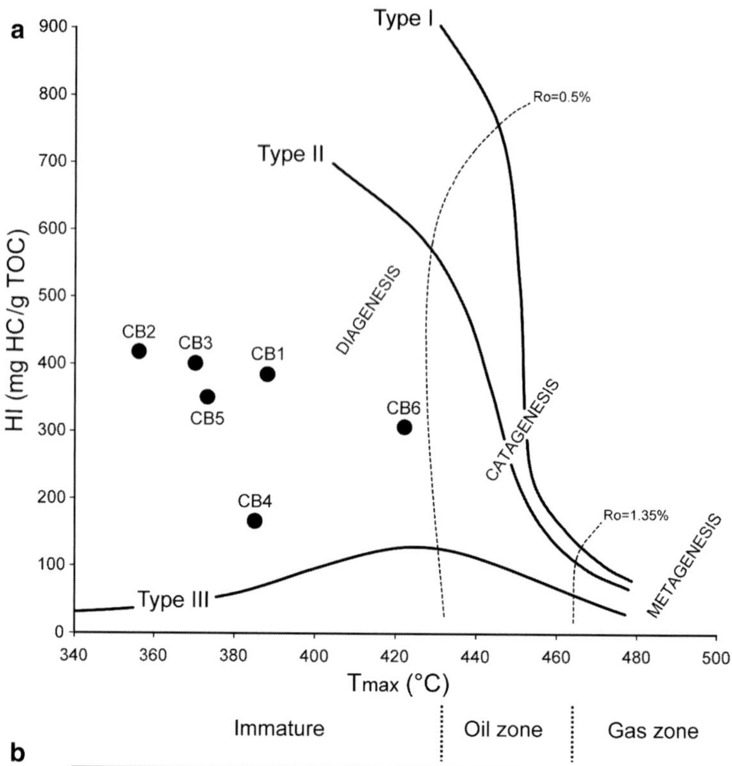

b

Samples	T_{max} (°C)	TOC (%)	HI (mg HC/g TOC)	IO (mg O_2/g TOC)	S1 (mg HC/g rock)	S2 (mg HC/g rock)
CB1	388	0,19	384	105	0,13	0,73
CB2	356	0,06	417	167	0,05	0,25
CB3	370	0,10	400	300	0,08	0,40
CB4	385	0,03	166	833	0,04	0,05
CB5	373	0,08	350	250	0,05	0,28
CB6	424	0,16	306	312	0,06	0,49

Fig. 1 (**a**) Classification of kerogens on a pseudo-van Krevelen diagram (HI-Tmax), the diagram shows the position in the immature region of the Calcare di Base organic matter [modified from Guido et al. (2007)]. (**b**) Rock-Eval pyrolysis data. *Tmax* temperature of maximum hydrocarbon generation, *TOC* total organic carbon, *HI* hydrogen index (mg HC/g TOC), *OI* oxygen index (mg O_2/g TOC), *S1* free oil (mg HC/g rock), *S2* hydrocarbons generated by thermal cracking (mg HC/g rock)

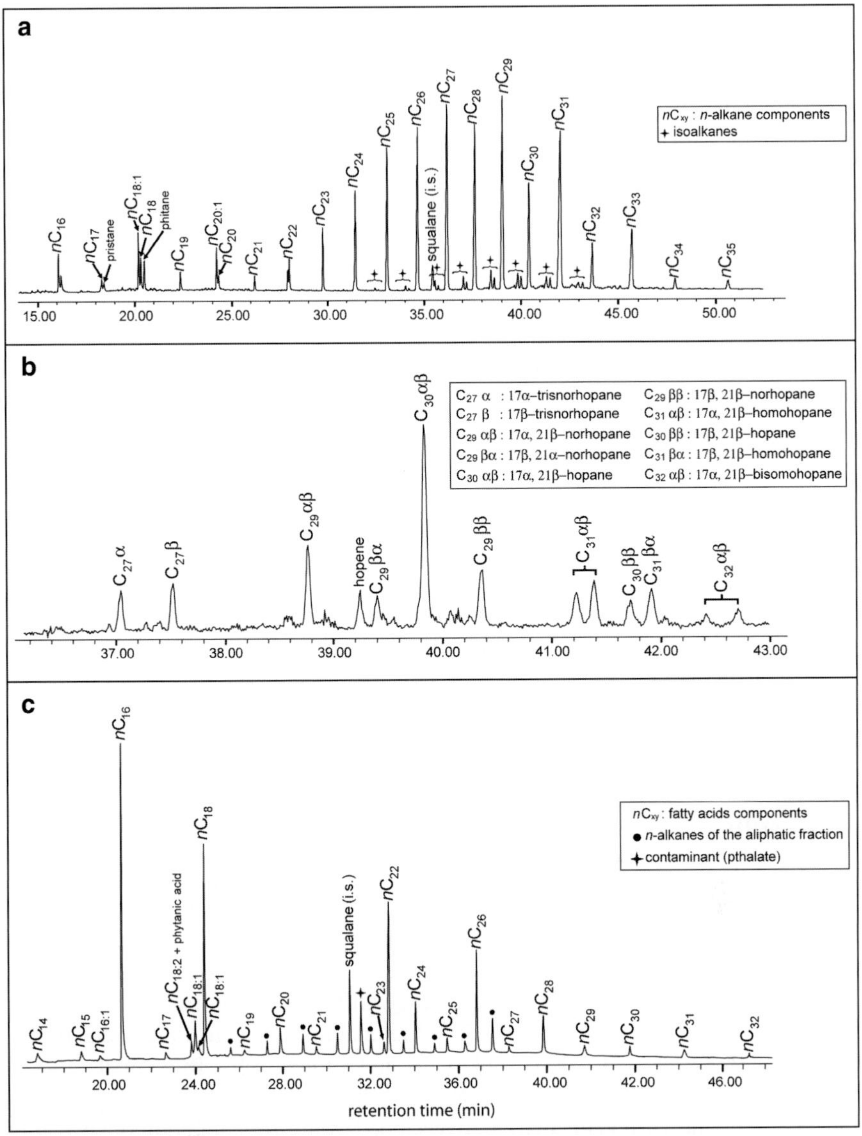

Fig. 2 Example of biomarker distributions in the hydrocarbon and fatty acid fractions [modified from Guido et al. (2007)]: (**a**) partial gas chromatogram of the hydrocarbon fraction showing the distribution of *n*-alkanes, iso-alkanes and isoprenoid hydrocarbons; (**b**) partial *m*/*z* 191 mass fragmentogram of the hydrocarbon fraction illustrating the distribution of hopanoids; (**c**) partial gas chromatogram of the acidic fraction showing the distribution of fatty acids

Spectral bands were assigned by reference to the literature (Painter et al. 1981; Solomon and Carangelo 1988; Wang and Griffiths 1985; Sobkowiak and Painter 1992).

The following absorption bands were analysed: aromatic C–H stretching region (3,100–3,000 cm^{-1}), aliphatic C–H stretching region (3,000–2,800 cm^{-1}), oxygenated groups and aromatic/olefinic region (1,550–1,750 cm^{-1}), CH_2 and CH_3 bending modes (peak at 1,450 cm^{-1}), CH_3 absorption bands (peak at 1,375 cm^{-1}), and aromatic out-of-plane C–H deformation region (700–900 cm^{-1}).

Carbonate ions show strong bands in the region near 1,400 cm^{-1} due the asymmetric C–O stretching modes and weak bending mode frequencies in the regions near 785 cm^{-1} and 700 cm^{-1} (Anderson et al. 2005). These bands do not allow for a clear assignment of the organic macromolecules that vibrate in the same range as the mineral matrix. To avoid mineral disturbance we isolated organic compounds from the carbonate and silicate phases via acidic separation.

Small fragments of carbonate samples, visibly free of alteration, were hand ground in an agate mortar and three grams of powder were ultrasonically treated three times with a mixture of dichloromethane/methanol (1:1). Samples were centrifuged after each treatment and the supernatant was collected. To concentrate the organic compounds, the solution was evaporated and the dry residue was resuspended in a few millilitres of dichloromethane/methanol. Several drops of this extract was placed onto the crystal of the ATR (attenued total reflectance) apparatus and dried under a nitrogen flow.

Precautions were taken to avoid contamination either from recent organic matter or from industrial contaminants during laboratory treatment. To verify the reliability of the data they were compared with those reported in the literature for similar sedimentary conditions and/or biological sources.

4 Results

Infrared analyses (FT-IR) were carried out to study the distribution of functional groups in the kerogens. Considering that the absorption bands of kerogen overlap in some regions of the spectrum, we used Fourier self-deconvolution (Kauppinen et al. 1981a, b; Wang and Griffiths 1985) to semiquantitatively determine the content of functional groups (Starsinic et al. 1984; Solomon and Carangelo 1988; Landais and Rochdi 1990; Kister et al. 1990; Sobkowiak and Painter 1992; Ruau et al. 1997).

As previously mentioned, the strong bands, due to the asymmetric C–O stretching modes of the carbonate matrix, overlapped some organic-compound bands. FTIR and μFTIR analyses performed on powdered sediments and thin sections, respectively, were not able to detect organic compounds having strong bands at 1,445, 1,081, 853, 713 or 697 cm^{-1}, all attributable to the functional groups of aragonite (Fig. 3a). IR spectra on bulk sediment also evidenced bands at 640 and 607 cm^{-1}. These are characteristic of the skeletal vibrations of sulphate minerals probably related to celestine ($SrSO_4$).

Interference between the bands of the carbonate matrix and the organic matter is quite critical, especially when the OM content is very low as in our samples.

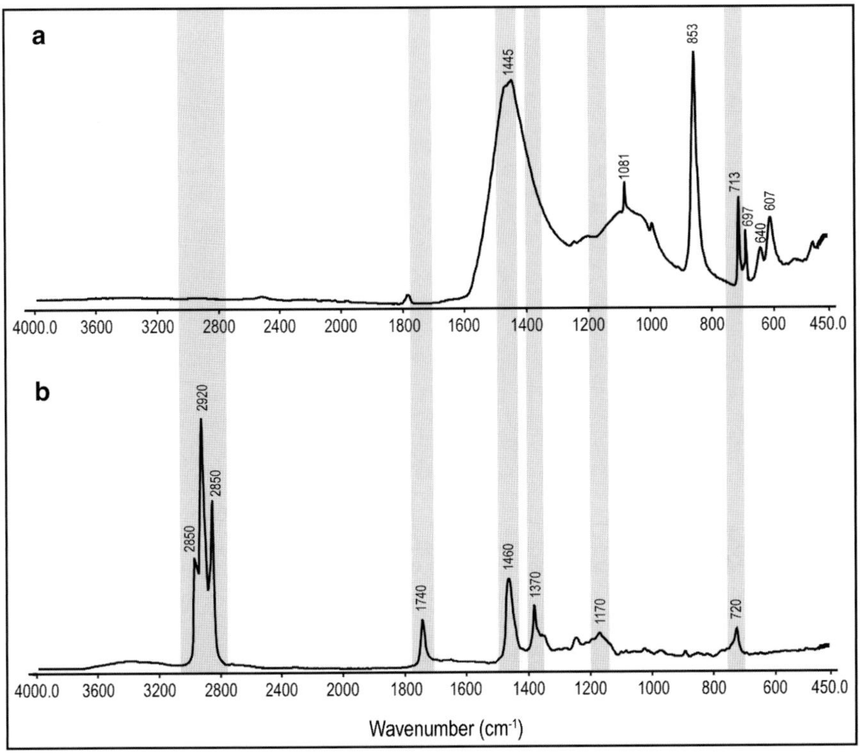

Fig. 3 Comparison between IR spectra on (**a**) powdered dry sediment and (**b**) extracted organic fraction (sample CB1)

Likewise, it is very difficult to detect organic compounds with absorption bands without interference but with intensities below the detection limit. To overcome this problem and to increase the sensitivity of the IR analysis it was necessary to extract and concentrate the organic compounds.

Fourier self-deconvolution resolved the region around 700 cm^{-1}, discriminating a peak at 720 cm^{-1} due to the skeletal vibration of more than four methylene groups [$\gamma(CH_2)_4$]. The aliphatic C–H stretching region (3,000–2,800 cm^{-1}) was resolved into three spectral bands at 2,950 (asymmetrical CH_3 stretching), 2,920 (asymmetrical CH_2 stretching), and 2,850 cm^{-1} (symmetrical CH_2 stretching); see Fig. 4. The CH_2/CH_3 ratio was calculated using the asymmetrical stretching of these bands (2,920/2,950 cm^{-1}) following Lin and Ritz (1993a). Fourier deconvolution revealed methyl (δCH_3; 1,370 cm^{-1}) and methyl-methylene [$\delta(CH_2 + CH_3)$; 1,460 cm^{-1}] groups (Fig. 4).

The spectra displayed the band assigned to carbonyl and/or carboxyl groups ($\nu C{=}O$; 1,740 cm^{-1}). The $\nu C{-}O$ vibration appeared between 1,300 and 1,100 cm^{-1}.

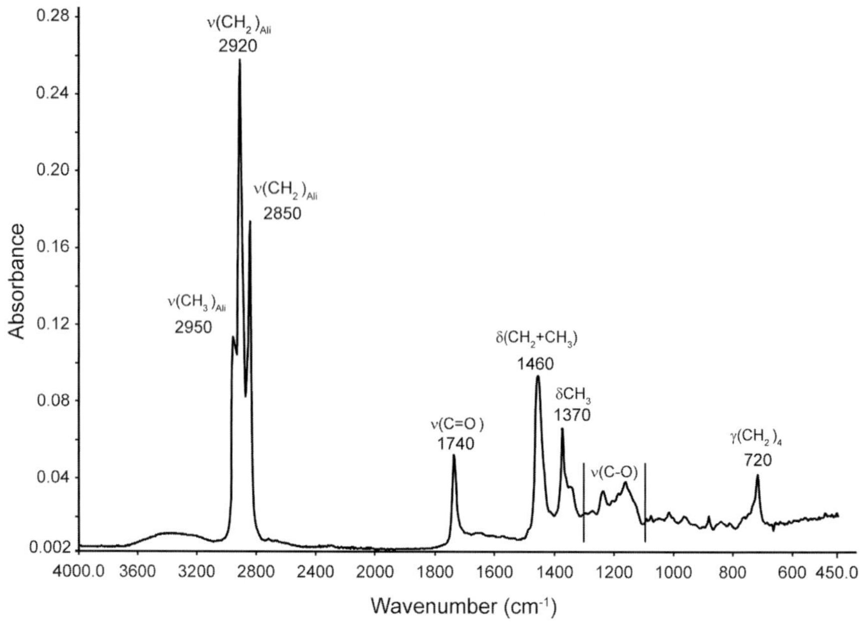

Fig. 4 Example of an infrared spectrum of extracted organic matter (sample CB1)

5 Interpretation and Discussion

The ratios of several peak heights of the FTIR bands allowed us to characterise the CdB kerogen. The ratio of carbonyl to aliphatic groups ($\nu C{=}O/\nu CH_{ali}$) permitted us to distinguish the marine vs. continental inputs, whereas the ratios for chain length ($\delta CH_2/\delta CH_3$) and degree of branching [$\nu(CH_2)_4/\nu CH_{ali}$] allowed us to evaluate the thermal maturity of the organic matter (Table 1).

The macromolecular structure of kerogen changes during thermal maturation, and this evolution is evidenced by the FTIR spectroscopic shifts of several bands (Brown 1955; Kuehn et al. 1982). Ganz and Kalkreuth (1987) proposed the "A Factor" (2,930 + 2,860 cm^{-1})/(2,930 + 2,860 + 1,630 cm^{-1}) and the "C Factor" (1,710 cm^{-1})/(1,710 + 1,630 cm^{-1}) to quantify the changes in the abundances of aliphatic and carbonyl/carboxyl groups, respectively, and to describe kerogen type and maturity.

Painter et al. (1981) separated the C–H aliphatic stretching region into five individual bands: 2,956 and 2,864 cm^{-1} for CH_3, 2,923 and 2,849 cm^{-1} for CH_2, and 2,891 cm^{-1} for CH. They also proposed that the bands at 2,920 and 2,955 cm^{-1} are diagnostic for the CH_2 and CH_3 groups, respectively. The CH_2/CH_3 ratio (2,920 cm^{-1}/2,955 cm^{-1}) is sensitive to the degree of carbon-skeleton branching and to the length of the aliphatic chains.

Table 1 Ratios calculated from peak heights of the infrared bands of the Calcare di Base kerogens

Samples	Carbonil to aliphatic $\frac{\upsilon C=O}{\upsilon CH_{Ali}}$	Degree of branching $\frac{\upsilon(CH_2)_4}{\upsilon CH_{Ali}}$	Chain length $\frac{\upsilon(CH_2)_{Ali}}{\upsilon(CH_3)_{Ali}}$
CB1	0.1893	0.1134	2.4257
CB2	0.1943	0.1094	2.4458
CB3	0.3328	0.1002	2.4479
CB4	0.2206	0.1162	2.4427
CB5	0.2041	0.1000	2.4068
CB6	0.2261	0.0867	2.5126

The ratio of aromatic to aliphatic absorption (i.e., aromaticity) increases with maturity (Fuller et al. 1982; Mastalerz and Bustin 1993; Iglesias et al. 1995) and with increasing carbon content in coals (Riesser et al. 1984), with fully aromatic pure graphite being the ultimate end product. Lis et al. (2005b) calculated nine FTIR parameters for aromaticity and evaluated their utility as proxies for maturity.

In the Calcare di Base samples, the CH_2/CH_3 ratio was determined using Fourier deconvolution of the 2,800–3,000 cm^{-1} aliphatic-stretching region (Painter et al. 1981; Lin and Ritz 1993a). The value of this ratio ranged from 2 to 2.5, confirming the immaturity of the organic compounds, which was already deduced from the biomarker data and maximum hydrocarbon-generation temperature (Tmax).

Considering that the kerogens (lacustrine, marine and continental) were distinguishable by their different H/C and O/C ratios, we propose that it is possible to obtain a preliminary indication of the origin of organic matter in kerogen through the analysis of the infrared spectra of the functional groups therein.

Type I kerogen (lacustrine), which includes few aromatic and heteroaromatic compounds and is characterised by a high H/C ratio (~1.5), a high Rock-Eval HI (>600 mg HC/g TOC) and a low O/C (<0.1) ratio, would contain more intense aliphatic bands in comparison to Types II and III.

Type II kerogen, which originates from mixed phytoplankton, zooplankton, and bacterial debris, has a high H/C ratio (1.2–1.5), a high Rock-Eval HI (300–600 mg HC/g TOC) and a low O/C ratio compared with types III and IV. This kerogen should exhibit FTIR spectra similar to Type I but with more intense carbonyl and carboxylic bands.

Type III kerogen (continental) is characterised by a low atomic H/C ratio (0.7–1.0), a low Rock-Eval HI (50–200 mg HC/g TOC) and a high O/C ratio (up to ~0.3). This kerogen should contain more intense carbonyl and carboxylic bands than Types I and II.

To test the reliability of FTIR analyses for the determination of kerogen origin in our carbonate samples, we compared the carbonyl/aliphatic ($\nu C{=}O/\nu CH_{ali}$) ratio with the Rock-Eval data (hydrogen index and oxygen index). The carbonyl groups, being formed mainly of C=O bonds, were correlated with the oxygen index, while the aliphatic groups, being formed mainly of C-H bonds, were correlated with the hydrogen index.

In the CdB sections, the $\nu C{=}O/\nu CH_{ali}$ ratio tended to show the same trends in hydrogen and oxygen indices that Guido et al. (2007) attributed to periodic increases in the deposition of continental organic matter.

6 Conclusions

The capability of FTIR in the determination of organic molecules has been previously reported in the literature. In this paper, we demonstrated the potential of FTIR in the characterisation of fossil organic matter preserved in carbonates.

- IR spectrometry allowed for fast analyses, useful for correlating samples from different basins.
- The FTIR methodology allowed for the detection of minor mineralogical phases in a carbonate matrix, such as celestine, not easily identifiable by optical microscopy.
- The spectra obtained from powder sediments and/or thin sections, because of interference with the carbonate functional groups, did not yield useful information on the content of organic compounds.
- The soundness of the FTIR methodology was appreciable when organic matter was extracted from the carbonate mineral phase via acidic separation.
- FTIR analysis made it possible to recognise the origin and thermal maturity of the organic matter preserved in carbonates sediments.
- FTIR can be considered as a complementary technique to the Rock-Eval pyrolysis and gas chromatography–mass spectrometry in the geochemical characterisation of the main kerogen types.

Acknowledgments Our thanks to Prof. Dr. Joachim Reitner for the excellent organisation of the International Kalkowsky Symposium: Geobiology of Stromatolites.

We are indebted to the anonymous reviewer for his suggestions, which greatly improved the manuscript.

This research was founded by Università della Calabria, PRIN 2007.

References

Anderson MS, Andringa JM, Carlson RW, Conrad P, Hartford W, Shafer M, Soto A, Tsapin AI (2005) Fourier transform infrared spectroscopy for Mars science. Review of Scientific Instruments 76:1–9

Baruah MK (1986) Assignment of the i.r. absorption band at 1050 cm^{-1} in lignite humic acid. Fuel 65:1756–1759

Borrego AG, Blanco GC, Prado JG, Díaz C, Guillén MD (1995) Study of the bitumenes and pirolizates from some Spanish oil shale by ^{1}H NMR and FTIR spectroscopy. In: Grimalt JO, Dorronsoro C, Editors, Organic Geochemistry: Developments and Applications to Energy, Climate, Environmental and Human History, A.I.G.O.A., San Sebastian, 798–800

Brown JK (1955) The infrared spectra of coals. Journal of the Chemical Society: 744–752

Černý J (1955) Aliphatic C–H bond responses in the 900–700 cm^{-1} region of the FTIR spectra of coal tars. Fuel Science and Technology International 13:807–818
Chen J, Ping L, Jinchao L (1998) Using kerogen FTIR parameters for determination of organic facies. Chinese Science Bulletin 43:681–684
Christy AA, Hopland AL, Barth T, Kvalheim OM (1989) Quantitative determination of thermal maturity in sedimentary organic matter by diffuse reflectance infrared spectroscopy of asphaltenes. Organic Geochemistry 14:77–81
Fuller MP, Hamadeh IM, Griffiths PR, Lowenhaupt DE (1982) Diffuse reflectance infrared spectroscopy of powdered coals. Fuel 61:529–536
Ganz H, Kalkreuth W (1987) Application of infrared spectroscopy to the classification of kerogen-types and the evolution of source rock and oil shale potentials. Fuel 66:708–711
Guido A, Jacob J, Gautret P, Laggoun-Défarge F, Mastandrea A, Russo F (2007) Molecular fossils and other organic markers as palaeoenvironmental indicators of the Messinian CdB formation: normal versus stressed marine deposition (Rossano Basin, Northern Calabria, Italy). Palaeogeography, Palaeoclimatology, Palaeoecology 255:265–283
Iglesias MJ, Jiménez A, Laggoun-Défarge F, Suárez-Ruiz I (1995) FTIR study of pure vitrains and associated coals. Energy & Fuels 9:458–466
Kauppinen JK, Moffatt DJ, Mantsch HH, Cameron DG (1981a) Fourier transforms in the computation of selfdeconvoluted and first-order derivative spectra of overlapped band contours. Analytical Chemistry 53:1454–1457
Kauppinen, JK, Moffatt DJ, Mantsch HH, Cameron DG (1981b) Fourier self-deconvolution: a method for resolving intrinsically overlapped bands. Applied Spectroscopy 35:271–276
Kister J, Guiliano M, Largeau C, Derenne S, Casadevall E (1990) Characterization of chemical structure, degree of maturation and oil potential of Torbanites (type I kerogens) by quantitative FT-i.r. spectroscopy. Fuel 69:1356–1361
Kuehn DW, Snyder RW, Davis A, Painter PC (1982) Characterization of vitrinite concentrates. 1. Fourier Transform infrared studies. Fuel 61:682–694
Landais P, Rochdi A (1990) Reliability of semiquantitative data extracted from transmission microscopy-Fourier transform infrared spectra of coal. Energy & Fuels 4:290–295
Lin R, Ritz GP (1993a) Reflectance FT-IR microspectroscopy of fossil algae contained in organic-rich shales. Applied Spectroscopy 47:265–271
Lin R, Ritz GP (1993b) Studying individual macerals using i.r. microspectroscopy, and implications on oil versus gas/condensate proneness and "low-rank" generation. Organic Geochemistry 20:695–706
Lis GP, Schummelmann A, Mastalerz M (2005) D/H ratios and hydrogen exchangeability of type-II kerogens with increasing thermal maturity. Organic Geochemistry 27:342–353
Lis GP, Mastalerz M, Schimmelmann A, Lewan MD, Stankiewicz BA (2005) FTIR absorption indices for thermal maturity in comparison with vitrinite reflectance R_0 in type-II kerogens from Devonian black shales. Organic Geochemistry 36:1533–1552
Mastalerz M, Bustin RM (1993) Variation in maceral chemistry within and between coals of varying rank: an electron microprobe and micro-Fourier transform infra-red investigation. Journal of Microscopy 171:153–166
Painter PC, Snyder RW, Starsinic M, Coleman MM, Kuehn DW, Davis A (1981) Concerning the application of FT-IR to the study of coal: a critical assessment of band assignments and the application of spectral analysis programs. Applied Spectroscopy 35:475–485
Painter PC, Starsinic M, Squires E, Davis AA (1983) Concerning the 1600 cm^{-1} region in the i.r. spectrum of coal. Fuel 62:742–744
Palacios-Fest MR, Cohen AS, Lezzar KE, Nahimana L, Tanner BM (2005) Paleolimnological investigations of anthropogenic environmental change in Lake Tanganyika: III. Physical stratigraphy and charcoal analysis. Journal of Paleolimnology 34:31–49
Riesser B, Starsinic M, Squires E, Davis A, Painter PC (1984) Determination of aromatic and aliphatic CH groups in coal by FT-i.r. 2. Studies of coals and vitrinite concentrates. Fuel 63:1253–1261

Ruau O, Landais P, Gardette JL (1997) Quantitative analysis of powdered organic matter by transmission infrared microspectroscopy using a diamond-window compression cell. Fuel 76:645–653

Sobkowiak M, Painter P (1992) Determination of the aliphatic and aromatic CH contents of coals by FT-i.r.: studies of coal extracts. Fuel 71:1105–1125

Solomon PR, Carangelo RM (1988) FT-i.r. analysis of coal. 2. Aliphatic and aromatic hydrogen concentration. Fuel 67:949–959

Starsinic M, Otake Y, Walker Jr, PL, Painter PC (1984) Application of FT-i.r. spectroscopy to the determination of COOH groups in coal. Fuel 63:1002–1007

Wang S-H, Griffiths PR (1985) Resolution enhancement of diffuse reflectance i.r. spectra of coals by Fourier-self deconvolution. 1. C–H stretching and bending modes. Fuel 64:229–236

Interactions Between Microbes and Siliceous Sponges from Upper Jurassic Buildups of External Prebetic (SE Spain)

Matías Reolid

1 Introduction

The Upper Jurasic buildups dominated by siliceous sponges and microbialites are widely known from the Tethyan realm and North Atlantic (Trammer 1989; Gaillard 1983; Matyszkiewicz 1989; Pomoni-Papaioannou et al. 1989; Brachert 1991; Lang 1991; Leinfelder et al. 1993, 1994; Reitner and Neuweiler 1995; Olóriz et al. 2003; Olivier et al. 2004). They were located and particularly widespread in deep-shelf settings. Such buildups are unknown in Modern Ocean and only recently, some sponge buildups have been discovered, but microbialites have not an important role as builders (Conway et al. 2001; Whitney et al. 2005; Krautter et al. 2006). In the Jurassic, sponges usually were associated to important amount of microbialites which favouring the development of buildups as well as the preservation of the sponge skeletons (Reolid 2007). Upper Jurassic sponge–microbialite buildups are thus a very good opportunity to study (a) the relationships between sponges and microbes and (b) related palaeoenvironmental conditions. The objectives of the paper are precise the architecture of sponge–microbialite buildups and discuss the equilibrium between microbes and sponges in terms of controlling factors. The characterization of the interactions between sponges and microbes in this research are based on Jurassic examples from the Betic Cordillera (SE Spain).

2 Geological Framework

The samples studied are located in the External Prebetic (SE Spain). The Prebetic Zone comprises the northernmost part of the Betic Cordillera (Fig. 1). The Oxfordian deposits of the External Prebetic represent the record of the mid shelf environments

M. Reolid
Departamento de Geología, Universidad de Jaén, 23071 Jaén, Spain
e-mail: mreolid@ujaen.es

J. Reitner et al., *Advances in Stromatolite Geobiology*, 343
Lecture Notes in Earth Sciences 131, DOI 10.1007/978-3-642-10415-2_22,

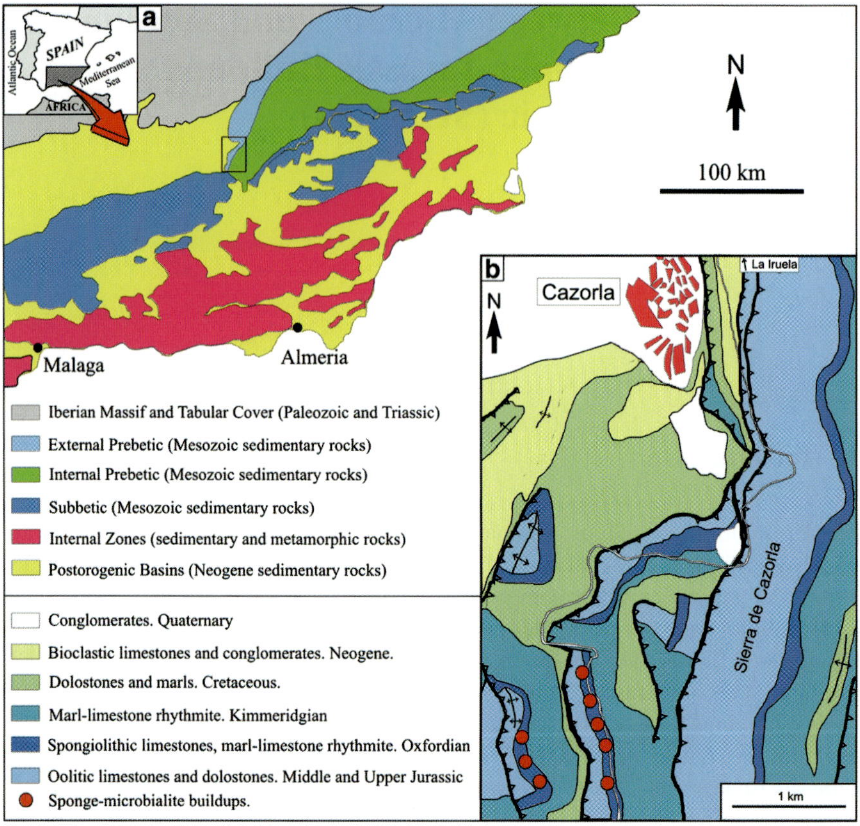

Fig. 1 Geological sketch of the Betic Cordillera (**a**) and detailed geological setting of the Sierra de Cazorla sector from the External Prebetic (**b**)

developed in the southwestern margin of the Tethys Domain. The buildups analysed occur in Sierra de Cazorla (Fig. 1).

The buildups of External Prebetic are registered in spongiolithic limestones and marl–limestone rhythmites, being better developed in the last one. The spongiolithic limestones are well stratified in beds 12–50-cm thick and are characterized by high abundance of sponges and other macroinvertebrates (Olóriz et al. 2002). The texture is a wackestone-packstone with lumps, tuberoids, oncoids and bioclasts. In the Sierra de Cazorla this lithofacies presents a thickness <2.25 m and correspond to the Transversarium Zone and the lower part of the Bifurcatus Zone.

The marl–limestone rhythmite consists of alternating beds (15–45 cm thick) of well-stratified limestones and marls with low abundance of fossil macroinvertebrates. Wackestones consist of very small (<250 μm) homometric grains (peloids, lumps and bioclasts). This lithofacies has a stratigraphic range that includes from Bifurcatus to middle Kimmeridgian but the buildups are only registered in the lower part of Bifurcatus Zone.

3 Sponge–Microbialite Buildups

3.1 Geometric Description

Buildups appear like sedimentary anomalies disturbing the regular and monotonous order of beds. Buildups are more or less massive, plano-convex, showing a low relief, and lenticular morphology. Buildups are larger in the marl–limestone rhythmites, with maximum height of 3.5 m and maximum diameter of 10 m. Those observed in the spongiolithic limestone are less than 1.5 m high and 3 m diameter. Strata adjacent to the buildups can show gradual transition and progressive overlap (bed truncation and pinch-out). The microfacies analysis confirms a gradual transition among buildups and stratified sediments, marked by a diminution on sponge spicules, microbial encrustations, tuberoids and crinoid remains. A variable degree of internal complexity exits in these buildups, being the simplest buildups those from spongiolithic limestone, while the more complex are from marl–limestone rhythmite. Therefore, in the Prebetic Zone, Reolid (2005) distinguished simple and complex buildups (Fig. 2). The simple buildups present less size and they correspond to lateral thickening of one or two levels of spongiolithic limestone. The more complex buildups are the biggest ones, and they are constituted of various smaller bodies that are more or less lenticular and separated by distinct planar or slightly convex surfaces. The complex buildups are a superposition of simple buildups.

3.2 Main Components

Three main components have been identified: siliceous sponges, microbialite and micritic mud. The last one represents the environmental sedimentation.

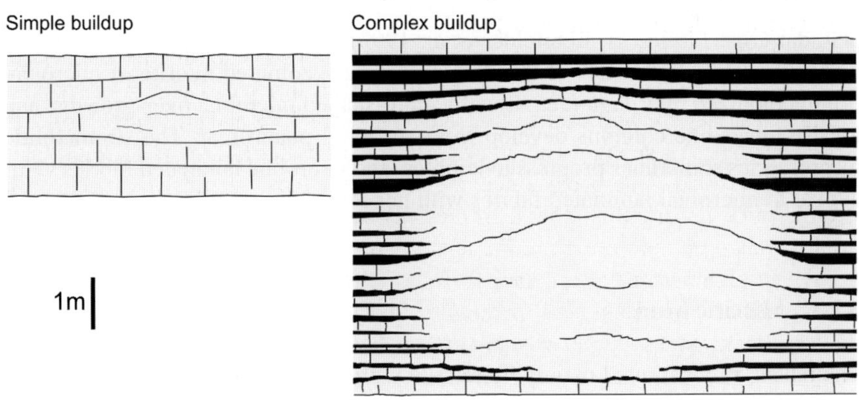

Fig. 2 Relationships of the simple and complex buildups with the surrounding stratification

3.2.1 Sponges

The studied sponges were siliceous (Hexactinellida and Demospongia). The sponge assemblage is dominated by Hexactinellida (77% Hexactinosida and 17% Lychniskosida). Lithistida makes up 6% of the recorded sponges. The mean size of the most frequent morphotype, the dish-shaped sponge, is 13 cm at the base and 16 cm at the top of the buildups. The largest specimen is 53 cm. The reticular morphology and the growth form of the sponge are usually preserved in calcite. The original siliceous spicules were dissolved and totally replaced by sparitic calcite (Reolid 2007).

3.2.2 Microbialite

The microbialite is the main component in the studied buildups. These are microbial encrustations developed on dish-shaped sponges. Microbialites make up most of the organo-sedimentary framework, and at a macroscopic scale correspond to leiolites and stromatolites. The leiolites are more frequents in complex buildups while the stromatolites are registered mainly in simple buildups. The microbialite can reach 75 mm thick and commonly corresponds to laminated fabric in thin section on both the stromatolites and the leiolites. The benthic microbial communities (BMC) constitute microbialites through the trapping and/or binding of fine detritic sediment or carbonate precipitation. Microbial encrustations with laminated fabrics can be constituted of BMC or of consortia made up of BMC and nubeculariids (sessile foraminifera). Thus, differentiation of two laminated fabrics is possible according to the dominant organisms (a) microbial laminated fabrics s. str. and (b) microbial laminated fabric with nubeculariids (Reolid et al. 2005). The microbial laminated fabrics s. str. (Reolid et al. 2005) consists of alternating light and dark laminae (result from grain size alternation). The light laminae comprise very small micritic particles with maximum size of 20–30 μm ("clotty structures" Monty 1967; "clotted, peloidal microfabrics" s. Dromart et al. 1994; "minipeloidal fabrics". Reitner and Schumann-Kindel 1997) surrounded by microsparitic matrix ("sparitic peloidal crust fabrics" s. Leinfelder et al. 1993). The microbial laminated fabric with nubeculariids consists of thin laminae of dark-coloured micrite of microbial origin colonized by *Vinelloidea*. Locally, club-shaped microbial over-growths, and locally finger-like columns develop from an initial planar mat. The stromatolitic macrofabrics sometimes progress to leiolites, and microbial laminated fabrics s. str. evolve to microbial laminated fabrics with nubeculariids.

3.2.3 Micritic Mud

The matrix is constituted mainly by micrite with mudstone to wackestone textures that includes small lumps, tuberoids, crinoids and peloids. Nevertheless, there are differences in grain size and sorting, depending of the area inside the buildups.

These three main components are placed according to an elementary sequence composed upward by (a) siliceous sponge, (b) microbialite, and (c) micritic sediment. This elementary sequence has never surpassed 9 cm. We can differentiate condensed and expanded elementary sequences (Olóriz et al. 2003) depending on thickness of microbialite and micritic mud.

The material that fills the interpores within the sponge meshwork consists of microbialites characterised by dense micrite and micropeloidal fabrics with microsparite, and it is known as automicrite (Leinfelder and Keupp 1995; Reitner and Neuweiler 1995; Reitner et al. 1995; Neuweiler et al. 2003; Reolid 2007). This micrite usually is darker than the micrite of the adjacent bedded wackestones, suggesting a higher proportion of organic matter and/or finely distributed pyrite (Reitner 1993; Warnkle 1995; Delecat et al. 2001; Delecat and Reitner 2005).

Epibionts from upward and downward surfaces of the sponges differ. Lower surfaces are colonized by annelids, bryozoans and some attached foraminifera. The surfaces of upward orientation are characterized by the microbial encrustations described above. The distribution of epibionts and encrustations is independent of the position of the sponge (life position or overturned). That distribution has been recognised in other buildups and sponge facies from other regions (Gaillard 1983; Matyszkiewicz 1989; Pomoni-Papaioannou et al. 1989).

3.3 Buildup Architecture

The simple buildups and the distinct rock bodies of the complex buildups of the Prebetic comprise two fundamental stages that represent a growth sequence (Figs. 3 and 4):

I.- First stage (<30 cm thick) is characterized by the dominance of the siliceous dish-shaped sponges (65%) where the microbial encrustations are scarce (6.5%). The sponges are directly growing on top of each other in condensed elementary sequences (Fig. 3a, b). In this stage the sponges are the main builder. This stage was called sponge bioherm by Olóriz et al. (2003).

II.- The second stage (the thickest one) is constituted by siliceous dish-shaped sponges with thick microbial encrustations and high abundance of micritic matrix. There is a lower proportion of sponges per volume of rock (20%) resulting in an expanded elementary sequence with BMC as the main builders (42% in complex buildups and 67% in simple ones). This stage was called microbial lithoherm by Olóriz et al. (2003). In the simplest, smallest buildups from spongiolithic limestone the macrofabric of the microbialite is mainly stromatolite (Fig. 3c, d). Larger buildups from marl–limestone rhythmite are characterized by a greater importance of leiolites.

The alternance of sponge dominate phases and microbialite dominate phases in the buildups is related to cycles of growth interruption and resumption. Sometimes, the top of such growth cycle is characterized by diminishing of microbial encrustations, jointly to increasing fragmentation degree of sponges and proportions of overturned sponges.

Fig. 3 Details of the different phases of the sponge–microbialite buildups: (**a**) and (**b**) correspond to the sponge bioherm phase; (**c**) and (**d**) correspond to stromatolites growing over siliceous sponges from the microbial lithoherm phase

4 Sponges vs. Microbes

4.1 Main Controlling Factors

4.1.1 Sedimentation Rate

A low sedimentation rate enabled the sponges to colonise the seabed and favours the dominance of large dish-shaped forms, as well as the presence of microencrusters and microbialites (Leinfelder et al. 1996; Krautter 1998; Dupraz and Strasser

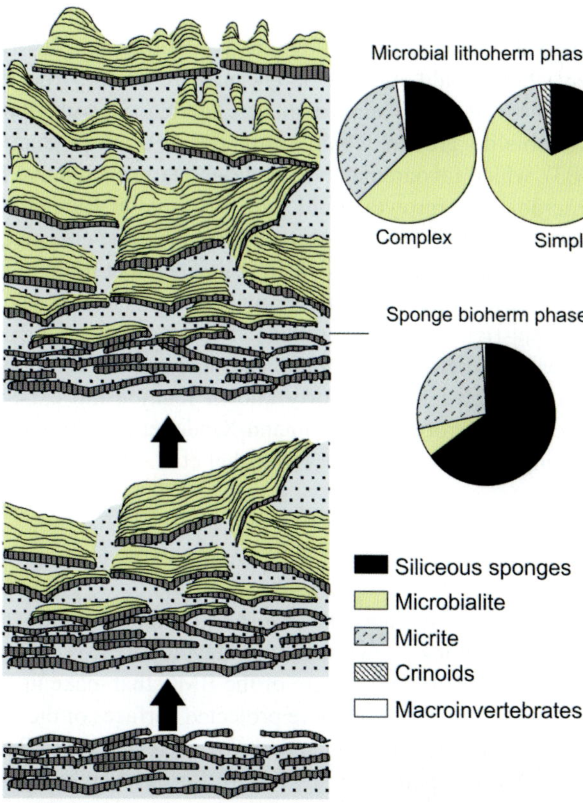

Fig. 4 Growth model of the Prebetic buildups, showing the proportions of the main components in the sponge bioherm phase and microbial lithoherm phase (in complex and simple buildups)

1999; Olóriz et al. 2003; Reolid and Gaillard 2007). The sedimentation rate and mainly, the siliciclastic influx from emerged areas, varied considerably in the different lithofacies studied (Olóriz et al. 2002). Spongiolithic limestones could be related to low sedimentation rates, which enabled prolonged colonization of the seabed by epibenthos and, especially, by sponges (Olóriz et al. 2003). In marl–limestone rhythmites the sedimentation rate, and particularly that of siliciclastic influx, was higher than in the spongiolithic limestones, which prevented the colonization of the seabed by epibenthic, sessile organisms. Thus, sponge colonization was limited to reduced areas of lower sedimentation and turbidity at the seabed. Here, small benthic islands or hillocks formed, suitable for colonization by sessile epibenthos and for the development of BMC.

4.1.2 Nutrients Availability

The proliferation of siliceous sponges is indicative of the type of organic matter available; thus, particulate organic matter favours to Lithistida whereas dissolved one favours to Hexactinellida (Krautter 1998). Osmotrophic dictyid sponges and typically dish-shaped sponges, depending on availability of dissolved organic matter (Krautter 1998), could be especially sensitive to the Reiswig's "biotic jungle" (1990). Thus, when the environment experienced wide fluctuations in productivity, microbialite growth would result during high-productivity phases (Olóriz et al. 2003), which favoured active bacterial overgrowth lethal for sponges. Alternatively, microbial overgrowth could start shortly after sponge death, but, in this case, some control on microbialite growth rate would be envisaged for living sponges. The distribution of encrusters, which is independent of the position of the sponge, indicates a post-mortem growth. Probably cyanobacteria and fungi took advantage of nutrient-rich sponge excretions. However both confrontation and commensal relationships between microbes and sponges are difficult to demonstrate from fossil assemblages [i.e. poorly known relationship existing between sponges and associated microbiotas (Schumann-Kindel et al. 1997), but increasing research is promising (Manz et al. 2000; Hentschel et al. 2002)].

4.1.3 Light

Microbial over-growths on sponges are registered exclusively on the exposed surface, and their morphology shows that such development is always upwards. This could be due to the photophilic nature of the BMC that make up the biogenic encrustation. The less-illuminated and more protected surfaces of the sponges were mainly occupied by serpulids, *Terebella* and bryozoa. According to Reolid (2007), the growth of BMC was comparatively aggressive for other epibionts.

4.2 Growth Model of Sponge–Microbialite Buildups

When the environment was suitable, and especially when the sedimentation rate was low enough for sponge growth, large areas of the seabed were colonized, forming sponge meadows, which later become spongiolithic limestones (Olóriz et al. 2003; Reolid and Gaillard 2007). Development of buildups was reduced to local proliferation of BMC.

A sedimentary environment such as that corresponding to the marl–limestone rhythmites presented adverse characteristics for sponge colonization, because of the high input of fine-grained detritic particles that caused turbidity at the seabed and the low consolidation degree (softground).

The occurrence of sponge–microbialite buildups was due to the sporadic colonization of the seabed by pioneer sponges as envisaged by Gaillard (1983). The growth of one or more very large pioneer sponges yielded an advantageous surface for subsequent colonization by other sponges and sessile organisms (serpulids, crinoids, bryozoa and brachiopods). At the same time, this yielded an optimum substrate for active BMC, thus favouring the development of thick microbial encrustation (Fig. 4). This favorable situation resulted in a greater growth of BMC with respect to the possibilities available in the surrounding muddy bottom (but see Noffke et al. 2001), and so a sponge–microbialite buildups occurred. Ostrowski (2002) indicates a context of competition for available surfaces between sponges and microbes which triggered the buildup growth. The lower part of these buildups, then, consisting of a large number of dish-shaped sponges (sponge bioherms s. Olóriz et al. 2003), represents the record of seabed colonization. The thicker upper interval (microbial lithoherms s. Olóriz et al. 2003), composed almost exclusively of microbialites (leiolites and stromatolites), corresponds to a phase of proliferating BMC, where nutrients availability could be a factor determinant triggering the growth of microbes (Olivier et al. 2004, 2007). The interruptions of growth of the buildups, registered as discontinuities that delimit growth cycles, imply the alternation of favorable and unfavorable phases for buildup growth.

Sedimentation rate, mainly of siliciclastic influx, controlled colonization and development of buildups. That also explains (1) the higher size of the sponge–microbialite buildups in marl–limestone rhythmites than in spongiolithic limestone, and (2) the higher proportion of micrite in buildups from rhythmite as a result of competition between microbial growth and background sedimentation. Pioneer sponges appeared in the rhythmites, providing a much more favorable substrate for the concentration of epibenthos and for biogenic encrustation.

4.3 Sponge Preservation and Benthic Microbial Communities

It is known from modern species, and reported in fossil counterparts, that microbially induced carbonate precipitation occurs in decaying sponge tissues. The precipitation of carbonate can occur both within and around the sponge, producing 'sponge mummies' (Schwamm-Mumien; Fritz 1958; Gwinner 1976; Flügel and Steiger 1981), while at the same time preserving the original arrangement of the siliceous spicules. Such rapid lithifying serves to protect fragile skeletons.

Calcite precipitation was induced by the consortium of heterotrophic, symbiotic bacteria that inhabited the mesohyl of the sponges (Delecat et al. 2001). Recent sponges are known to be occupied by bacteria (Simpson 1984). They most likely play a role in the supply of nutrients to sponges. Modern demosponge taxa present high amounts of bacterial biomass, as much as 50% of the weight of the sponge biomass (Reitner 1993; Reitner and Schumann-Kindel 1997). Otherwise, according to Reiswig (1990) the surface of siliceous sponges is covered by a dense veil of prostal spicules hosting a largely organic microbial layer. Sponges could take advantage of metabolic products of these microbes (Krautter 1998). High amounts of organic macromolecules, derived from both microbial communities (from mesohyl and surface prostal spicules) are supposed to be responsible for early organomineralization (Reitner et al. 1995). These automicrites are frequently darker than the surrounding wackestone, due to a higher content of organic matter and pyrite that developed during the calcification of decaying soft tissue of sponge under the influence of sulphate-reducing bacteria (Reitner 1993; Reitner and Schumann-Kindel 1997). However, Neuweiler et al. (2007) explain the calcification of modern siliceous sponges without microbial participation. Therefore, more studies are necessaries from fossil and recent siliceous sponges.

5 Conclusions

The relationships between siliceous sponges and microbes in the epicontinental shelf of the Prebetic were very important because they resulted in the development of sponge–microbialite buildups which constituted favourable settings for sessile organisms. Favourable conditions were low sedimentation rate, fluctuations in the trophic conditions and probably light availability. These paleoenvironmental oscillations would produce alternance of sponge bioherm-microbial lithoherm phases.

The interactions between microbes and siliceous sponges were not restricted to the development of microbial encrustations, but microbes inhabited the mesohyl and surface prostal spicules with relationships of commensalism and symbiosis. The microbes could be related to the precipitation of the automicrite among the sponge meshwork and then, the preservation of the remains when the siliceous skeleton is dissolved.

Acknowledgements This research was carried out with the financial support of Research Project CGL2005-06636-C0201/BTE (Spanish Ministry of Science and Technology) and Research group RNM-178 of the Junta de Andalucía. The author is indebted to Jean Louise Sanders (native English speaker) for grammar review.

References

Brachert TC (1991). Environmental control on fossilization of siliceous sponge assemblages: a proposal. In: Reitner J, Keupp H (eds) Fossil and recent sponges. Springer, Berlin, pp. 543–553

Conway KW, Krautter M, Barrie JV, Neuweiler M (2001) Hexactinellid sponge reefs on the Canadian continental shelf: A unique "living fossil". Geoscience Canada 28:71–78

Delecat S, Reitner J (2005) Sponge communities from the Lower Liassic of Adnet (Northern Calcareous Alps, Austria). Facies 51:385–404

Delecat S, Peckmann J, Reitner J (2001). Non-rigid cryptic sponges in oyster patch reefs (Lower Kimmeridgian, Langenberg/Oker, Germany). Facies 45:231–254

Dromart G, Gaillard C, Jansa LF (1994) Deep-marine microbial structures in the Upper Jurassic of western Tethys. In: Bertrand-Sarfati J, Monty C (eds) Phanerozoic stromatolites II. Kluwer Academic Publishers, The Netherlands, pp. 295–318

Dupraz C, Strasser A (1999) Microbialites and micro-encrusters in shallow coral bioherms (Middle to Late Oxfordian, Swiss Jura Mountains). Facies 40:101–130

Flügel E, Steiger T (1981) An Upper Jurassic sponge-algal buildup from the northern Frankenalb, West Germany. SEPM Special Publication 30:371–397

Fritz G (1958) Schammstotzen, Tuberolithe und Schuttbreccien im Weissen Jura der Schwabischen Alb. Arbeiten aus dem Geologisch-Paläontologischen Institut der Technischen Hochschule Stuttgart, Neue Folge 13:1–118

Gaillard C (1983) Les biohermes à spongiaires et leur environment dans l´Oxfordian du Jura meridional. Documents des Laboratoires de Géologie de Lyon 90, p. 515

Gwinner MP (1976) Origin of Upper Jurassic limestones of the Swabian Alb (Southwest Germany). Contributions to Sedimentology 5:1–75

Hentschel U, Hopke J, Horn M, Friedrich AB, Wagner M, Hacker J, Moore BS (2002) Molecular evidence for a uniform microbial community in sponges from different oceans. Applied Environmental Microbiology 68:4431–4440

Krautter M (1998) Ecology of siliceous sponges – Application to the environmental interpretation of the Upper Jurassic sponge facies (Oxfordian) from Spain. Cuadernos de Geología Ibérica 24:223–239

Krautter M, Conway KW, Barrie JV (2006) Recent hexactinosidan sponge reefs (silicate mounds) off British Columbia, Canada: Frame-building processes. Journal of Paleontology 80:38–48

Lang B (1991) Baffling, binding, or debris accumulation? Ecology of Upper Jurassic sponge-bacterial buidups (Oxfordian, Franconian Alb, Southern Germany). In: Reitner J, Keupp H (eds) Fossil and recent sponges. Springer, Berlin, pp. 516–521

Leinfelder RR, Keupp H (1995). Upper Jurassic mudmounds: Allochthonous sedimentation versus autochthonous carbonate production. In: Reitner J, Neuweiler F (coord.) Mud mounds: a polygenetic spectrum of fine-grained carbonate buildups. Facies 32:17–26

Leinfelder RR, Nose M, Schmid DU, Werner W (1993) Microbial crust of the late Jurassic: composition, palaeoecological significance and importance in reef constructions. Facies 29:195–230

Leinfelder RR, Krautter M, Laternser R, Nose M, Schmid DU, Schweigert G, Werner W, Keupp H, Brugger H, Herrmann R, Rehfeld-Kiefer U, Koch R, Zeiss A, Schweizer V, Christmann H, Menges G, Luterbacher H (1994) The origin of the Jurassic reefs: current research developments and results. Facies 31:1–56

Leinfelder RR, Werner W, Nose M, Schmid DU, Krautter M, Laternser R, Takacs M, Hartmann D (1996) Paleoecology, growth parameters and dynamics of coral, sponge and microbolite reefs from the Late Jurassic. Göttinger Arbeiten zur Geologie und Paläontologie Sonderband 2:227–248

Manz W, Arp G, Schumann-Kindel G, Szewzyk U, Reitner J (2000) Widefield deconvolution epifluorescence microscopy combined with fluorescence in situ hybridazion reveals the spatial arrangement of bacteria in sponge tissue. Journal of Microbiological Methods 40:125–134

Matyszkiewicz J (1989) Sedimentation and diagenesis of the Upper Oxfordian cyanobacterial-sponge limestones in Piekary near Kraków. Annales Societatis Geologorum Poloniae 59:201–232

Monty LV (1967) Distribution and structure of recent stromatolitic algal mats, eastern Andros Island, Bahamas. Annales de la Societe Geologique de Belgique 90:55–100

Neuweiler F, D'Orazio V, Immenhauser A, Geipel G, Heise K-H, Cocozza C, Miano TM (2003) Fulvic acid-like organic compounds control nucleation of marine calcite under suboxic conditions. Geology 31:681–684

Neuweiler F, Daoust I, Bourque PA, Burdige DJ (2007) Degradative calcification of a modern siliceous sponge from the Great Bahama Bank, the Bahamas: A guide for interpretation of ancient sponge-bearing limestones. Journal of Sedimentary Research 77:552–563

Noffke N, Gerdes G, Klenke T, Krumbein WE (2001) Microbially induced sedimentary structures – a new category within the classification of primary sedimentary structures. Journal of Sedimentary Research 71:649–656

Olivier N, Pittet B, Mattioli E (2004) Palaeoenvironmental control on sponge–microbialite reefs and contemporaneous deep-shelf marl–limestone deposition (Late Oxfordian, southern Germany). Palaeogeography Palaeoclimatology Palaeoecology 212:233–263

Olivier N, Pittet B, Gaillard C, Hantzpergue P (2007) High-frequency palaeoenvironmental fluctuations recorded in Jurassic coral- and sponge–microbialite bioconstructions. Comptes Rendus Palevol 6:21–36

Olóriz F, Reolid M, Rodríguez-Tovar FJ (2002) Fossil assemblages, lithofacies and taphofacies for interpreting depositional dynamics in epicontinental Oxfordian (Prebetic Zone, Betic Cordillera, southern Spain). Palaeogeography Palaeoclimatology Palaeoecology 185:53–75

Olóriz F, Reolid M, Rodríguez-Tovar FJ (2003) A Late Jurassic carbonate ramp colonized by sponges and benthic microbial communities (External Prebetic, Southern Spain). Palaios 18:528–545

Ostrowski S (2002) Development of sponge-microbial bioherms from Oxfordian of central and southern Poland. Abstracts VI Inter. Symp. Jurassic System, p. 140

Pomoni-Papaioannou F, Flügel E, Koch R (1989) Depositional environments and diagenesis of Upper Jurassic subsurface sponge and Tubiphytes reef limestones: Altensteig 1 Well, Western Molasse Basin, Southern Germany. Facies 21:263–284

Reiswig HM (1990) In situ feeding in two shallow-water hexactinellids sponges. In: Rützler K, MacIntyre VV, Smith KP (eds) New perspectives in sponge biology. Smithonian Institution Press, Washington, DC, pp. 504–510

Reitner J (1993) Modern cryptic microbialite/metazoan facies from Lizard Island (Great Barrier Reef, Australia). Formation and concepts. Facies 29:3–40

Reitner J, Neuweiler F (1995) Mud mounds: a polygenetic spectrum of fine-grained carbonate buildups. Facies 32:1–70

Reitner J, Schumann-Kindel G (1997) Pyrite in mineralized sponge tissue: product of sulfate reducing sponge related bacteria? Facies 36:272–276

Reitner J, Neuweiler F, Gautret P (1995) Modern and fossils automicrites: Implications for mud mounds genesis. In: Reitner J, Neuweiler F (coord) Mud mounds: a polygenetic spectrum of fine-grained carbonate buildups. Facies 32:4–17

Reolid M (2005) Dinámica eco-sedimentaria durante el Oxfordiense medio-Kimmeridgiense temprano en la Zona Prebética: Interpretación ecoestratigráfica y secuencial. PhD Thesis, Universidad de Granada 2003, Granada, p. 254

Reolid M (2007) Taphonomy of the Oxfordian-lowermost Kimmeridgian siliceous sponges of the prebetic zone (Southern Iberia). Journal of Taphonomy 5:71–90

Reolid M, Gaillard C (2007) Microtaphonomy of bioclasts and paleoecology of microencrusters from the Upper Jurassic spongiolithic limestones (External Prebetic, southern Spain). Facies 53:97–112

Reolid M, Gaillard C, Olóriz F, Rodríguez-Tovar FJ (2005) Microbial encrustations from the Middle Oxfordian-earliest Kimmeridgian lithofacies in the Prebetic Zone (Betic Cordillera, southern Spain): characterization, distribution and controlling factors. Facies 50:529–543

Schumann-Kindel G, Bergbauer M, Manz W, Szewzyk U, Reitner J (1997) Aerobic and anaerobic microorganims in modern sponges: a possible relationship to fossilization-processes. Facies 36:268–272

Simpson TL (1984) The cell biology of sponges. Springer, New York, p. 662

Trammer J (1989) Middle to upper Oxfordian sponges of the Polish Jura. Acta Geologica Polonica 39:49–91

Warnkle K (1995) Calcification processes of siliceous sponges in Visean Limestones (Counties Sligo and Leitrim, northwestern Ireland). Facies 33:215–228

Whitney F, Conway K, Thomson R, Barrie V, Krautter M, Mungov G (2005) Oceanographic habitat of sponge reefs on the western Canadian Continental Shelf. Continental Shelf Research 25:211–226

Aftermath of the Triassic–Jurassic Boundary Crisis: Spiculite Formation on Drowned Triassic Steinplatte Reef-Slope by Communities of Hexactinellid Sponges (Northern Calcareous Alps, Austria)

Stefan Delecat, Gernot Arp, and Joachim Reitner

1 Introduction

In Upper Rhaetian time, reef evolution culminated in highly diverse ecosystems that notably flourished along the northern shelf of the former Tethys ocean. Several of these buildups constructed by branching corals and coralline sponges (Demospongiae with secondary basal skeletons) are nowadays exposed in the Northern Calcareous Alps, where they are found both at the northern and southern margins of the Upper Triassic Carbonate Platform. Further, the record of capping Lower Liassic sediments show that all of the reefs must have been abruptly disappeared at the Triassic–Jurassic boundary. The reasons for their sudden decease which took place simultaneously with a world wide mass extinction event are still under discussion and less is even known about its effect on subsequent Liassic benthic faunas. After the extinction of important reef-building organisms most likely a sea level rise and changes in sedimentation led to a radiation of sponges from deep water environments (Mostler 1989a, b, 1990a, b) onto some of the drowned buildups. Böhm (1992) already assumed that Liassic sponge diversity in the alpine region was strongly influenced by the extant Triassic palaeorelief. In order to discover the crucial factors that controlled the recolonization of the drowned areas, Liassic sponge communities were studied from localities of the former Triassic reef slopes and adjacent basinal settings.

In this paper a first description of autochthonous Hettangian sponge communities situated on the upper slope of Steinplatte "Reef" (N of Waidring, Austria) is provided. Although most of the sponge skeletons are highly collapsed, they are preserved in situ and arranged in spicular mats with sediments partially bound by microbialites, so that the spiculite facies bears analogy to the Schnöll facies that is

S. Delecat, G. Arp, and J. Reitner (✉)
Geoscience Centre, University of Göttingen, Geobiology Division, Goldschmidtstraße 3, 37077, Göttingen, Germany
e-mail: jreitne@gwdg.de

J. Reitner et al., *Advances in Stromatolite Geobiology*,
Lecture Notes in Earth Sciences 131, DOI 10.1007/978-3-642-10415-2_23,

known from the slope of the Adnet Reef, near Hallein, Austria (Delecat and Reitner 2005). Both localities are favorite sites to examine entire Lower Jurassic Porifera communities as well as sponge taphonomy in fossil environments. Additional to the documentation of these rarely preserved sponge species, the paper's aim is to evaluate probable interactions between taphonomic processes in collapsing sponges and abiotic factors like variances of the Liassic geochemical milieu or aspects of sedimentation and diagenesis. Hereby a focus lays on the study of early diagenetic stromatactis cavities that are a characteristic feature of the Steinplatte spiculite and whose origin seems to be closely related to the decay of sponge organic matter.

2 Regional Setting

The Upper Triassic Carbonate Platform in the Northern Calcareous Alps extended in W–E direction and separated the former Tethyian deep water realm in the south from the northern epicontinental Kössen Basin. The famous Steinplatte "reef", whose crest is cropping out at the Sonnenwände hill north of Waidring and Lofer (Austria) was situated on the northern slope of this platform (Fig. 1). As facies analysis by Stanton and Flügel (1989, 1995) have shown, only a minor part of the Steinplatte buildup is formed by a real reef framework, thus the whole complex is rather described as an "accretionary distally steepened ramp" the top of which was partly overgrown by separate bushlike corals (capping facies) in a late Upper

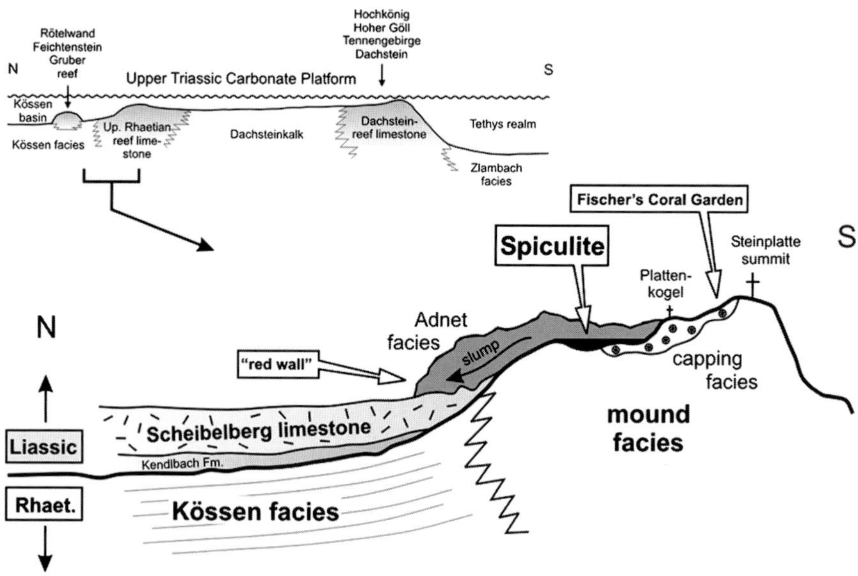

Fig. 1 Northern slope section of Steinplatte "Reef" (N of Waidring, Austria) and its paleogeographic position

Rhaetian stage. At the end of Triassic, the coral growth stopped concurrent to one of the five most severe mass extinction events in earths Phanerozoic history. The reason for that is difficult to evaluate. Although the palaeo-relief of the carbonate platform still existed until Middle Liassic time, just on top of the buildup the Triassic–Jurassic interval encompassing the decease of the coral fauna is concealed by a small sedimentary break. Only some traces of fresh-water diagenesis in the capping facies might indicate short-termed falls of the sea level at Steinplatte buildup (Stanton and Flügel 1989). In contrast to the elevated platform position, in adjacent basins sedimentation continuously passed into grey cherty limestones (Hettangian Kendlbach Formation and Hettangian-Sinemurian Scheibelberg Formation). Predominantly the latter Scheibelberg Limestones are characterized by varying, often high amounts of siliceous sponges and/or siliceous bulbs (Mostler 1986; Krainer and Mostler 1997). They are frequently exposed at the toe of Steinplatte buildup slope at Kammerköhr Alpe and also at several outcrops in the Unken syncline (e.g. at Scheibelberg locality and at Karnergraben) that extends north of the Steinplatte buildup.

First Hettangian to Sinemurian sedimentation at the northern slope of Steinplatte "reef" shows an onlap of red condensed limestones indicating a drowning of the platform during Lower Liassic time. A wedge of these sediments is cut by a trail surrounding the Plattenkogel hill, north of Steinplatte summit (Fig. 2). At the "red

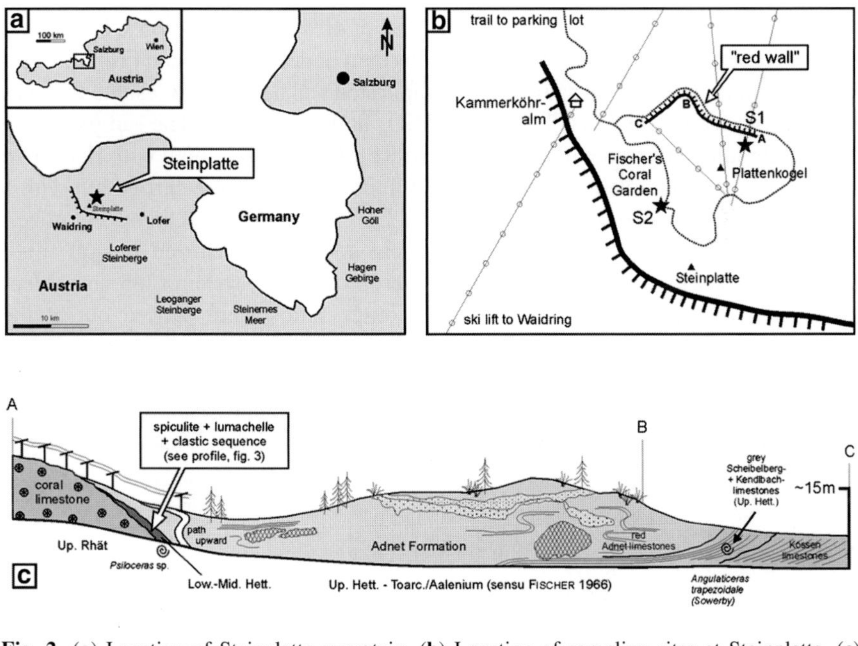

Fig. 2 (**a**) Location of Steinplatte mountain. (**b**) Location of sampling sites at Steinplatte. (**c**) Schematic sketch of the outcrop "red wall" at Kammerköhr-Alpe (tourist trail along northern slope of Steinplatte/Plattenkogel, view to the south). Jurassic onlap sedimentation comprises red Adnet limestones with several slump folds, slided megabreccias (*checkered*) and debrites (*dotted*)

wall" of this site a big massflow of Adnet limestones crops out comprising slump folds and megabreccias slided down onto grey Scheibelberg Formation (Garrison and Fischer 1969; Wächter 1987). At the south-eastern margin of this outcrop the Adnet limestone pinches out along Triassic coral "reef" limestone. The study on hand deals with some limestones representing the onset of Liassic sedimentation that is preserved in small crevices or interstices of the rough Triassic relief.

3 Methods

Facies analyses were carried out by numerous thin sections (up to 10 × 15 cm in size) that were stained partly by potassium ferricyanid and alizarin red, conventionally dissolved in 0.1% HCl solution. Automicrites were examined additionally by epifluorescence microscopy on a Zeiss Axioscope (lamp: HBO 50; filter: BP 450–490 FT 510) and by scanning electron microscopy (SEM) combined with an energy dispersive X-ray detection system (EDX). Mineral phases in the ferromanganese crusts were determined by X-ray diffraction (CuKα). Weight ratios of major and trace elements were analyzed by X-ray fluorescence analyses (XRF) and rare earth elements (REE) by inductively coupled plasma mass spectrometry (liquid and laser ICP-MS). The samples for liquid ICP-MS analyses were transfered first into a powder by a ball mill and than undergone a total chemical pulping. Hereby 100·mg of each powder sample were treated with 3 ml of 40% HF, 2 ml of 65% HNO_3 and 3 ml of $HClO_4$ first under standard conditions then under pressure and heat (150°C). The procedure was applied again on the evaporated and neutralized residue that was treated now with 1 ml of HCl, 2 ml of HNO_3 and 10 ml H_2O. The samples were afterwards supplemented with a standard (In/Re, 200 µl) and refilled with H_2O up to 100 ml. The measurements of the prepared liquid samples were carried out with a FISONS VG PQ STE instrument with a CETAC DIN-100 Direct Nebulizer. Comparative analyses by laser ICP-MS were additionally carried out on small polished slabs with a combined VG UV-Microprobe laser system (266 nm). Measurements of carbon and oxygen stable isotope ratios were carried out at the University of Erlangen with a Finnigan Mat 252 mass spectrometer (method with 100% phosphoric acid at 75°C and Carbo-Kiel carbonate preparation-technique). Reproducibility was checked by replicate analysis of laboratory standards. Average of standard deviation is 0.01–0.02 for $\delta^{13}C$ and 0.02–0.03 for $\delta^{18}O$ ($\pm 1\sigma$).

4 Localities

Concerning the main topic of this study, samples of the spiculite facies were taken at the northeastern slope of the Plattenkogel hill (locality S1), where Triassic "reef" limestone of the Steinplatte buildup is discordantly overlain by Lower Liassic sediments. The spiculite covers a small clastic sequence that was also sampled to evaluate the emergence of the sponge fauna. Furthermore the sequence most likely provides the oldest Jurassic limestones at Steinplatte buildup slope, thus giving a small insight into the sedimentary conditions closely after the Triassic–Jurassic boundary event. Another site at Steinplatte, showing the contact between Triassic and Jurassic sedimentation is located at the Steinplatte summit, called "Fischer's Coral Garden" (locality S2). It displays limestone of the capping facies in which gaps and interstices of the Triassic coral framework are partly filled with Liassic red sediments. In the alpine region, the T–J boundary interval is often characterized by sedimentary breaks and ferromanganese crusts. Since a small Fe/Mn crust was also found in association to the Liassic sequence at Steinplatte/Plattenkogel hill, it was of interest to compare its geochemistry with those of other Fe/Mn-crusts from famous localities of the Northern Calcareous Alps (localities S3 a, b, c; see supplement for Figure 9).

4.1 *Steinplatte Reef-Slope (Plattenkogel Hill)*

The Plattenkogel hill (locality S1) provides a small sequence of Lower Liassic sediments that is preserved in a shallow depression of the Upper Rhaetian coral limestone (Figs. 2–4a). A first profile from this site was published by Rakús and Lobitzer (1993). The Liassic section starts with a lumachelle layer and clastic sediments that fill up a small sink hole abutting a little fault. The lumachelle

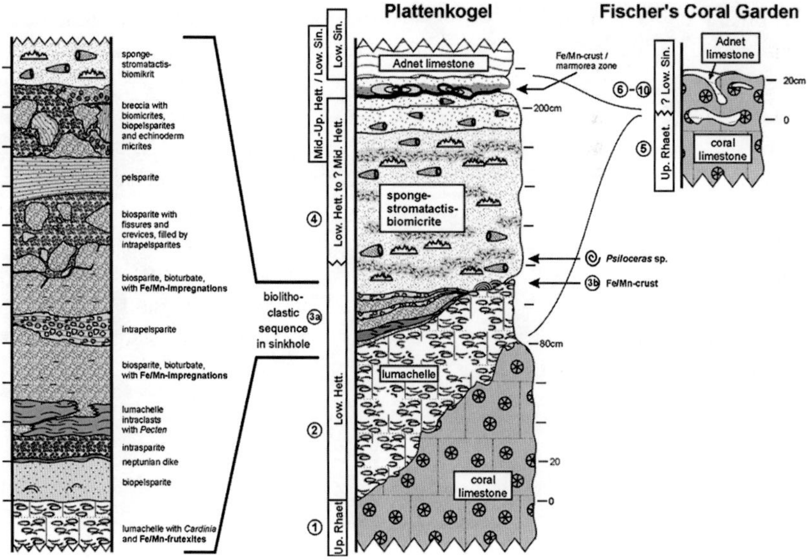

Fig. 3 Profiles of Triassic–Jurassic boundary sequences from Steinplatte localities: Plattenkogel hill (locality S1) and "Fischer's Coral Garden" (locality S2). Numbers of layers/samples correspond with Figs. 5 and 9

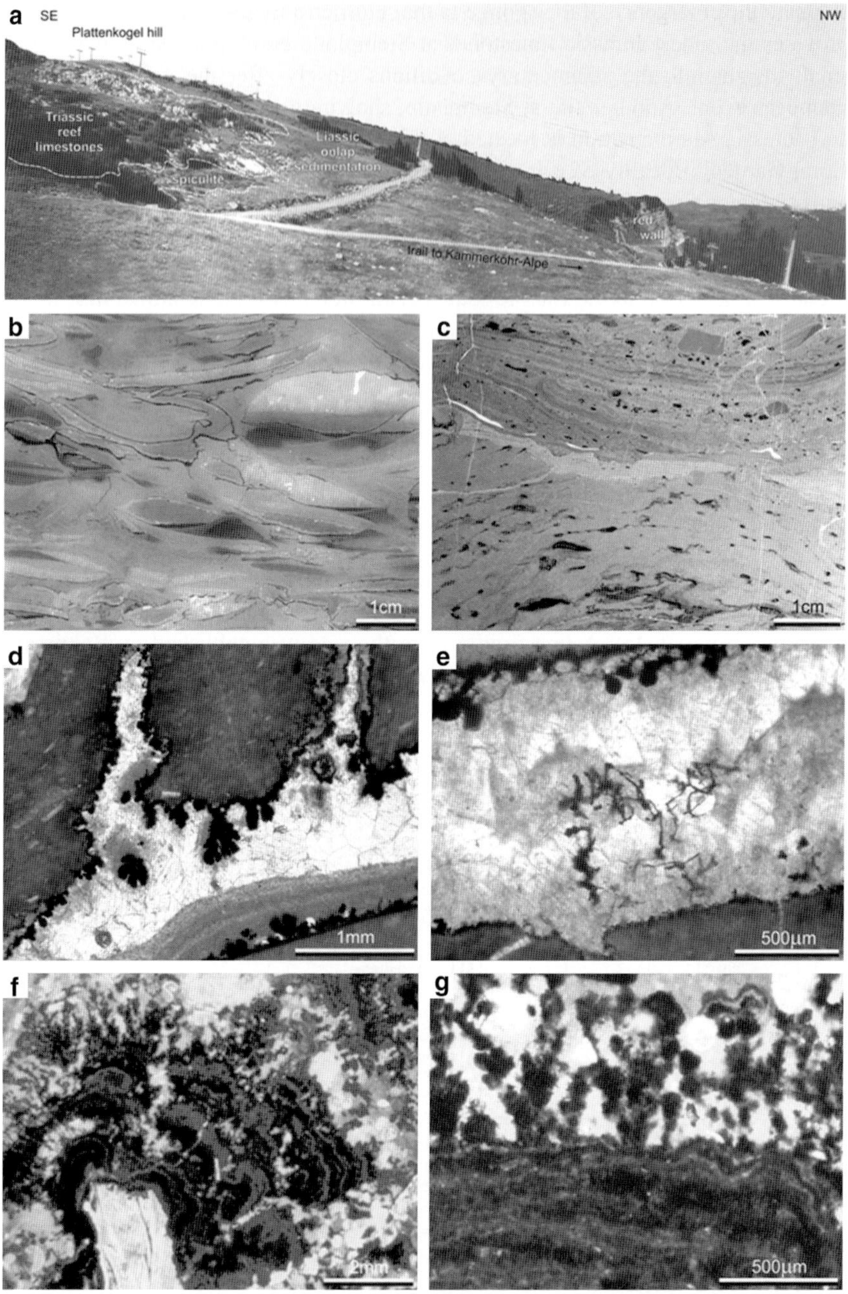

Fig. 4 (**a**) Panoramic view on the north-eastern slope of Steinplatte reef showing the Triassic–Jurassic boundary at Pattenkogel hill (locality S1). (**b**) Lumachelle layer from Steinplatte/Plattenkogel locality. Detrital matrix comprises articulated *Cardinia* shells that were later dissolved and replaced by infiltrated sediments and blocky spar. Some cavity walls are lined by dark

displays dissolved valves of transported, but often articulated *Cardinia* shells (3–4 cm in size, Fig. 4b). Sporadically iron/manganese "*Frutexites*" structures grew into the mould cavities, before remaining space was closed by fine detrital sediments and blocky sparite (Fig. 4d, e). Accessorial fauna comprises ostreoid, pterioid, and pectinoid bivalves and crinoidal fragments. Intercalated and more matrix-dominated layers show fine-grained biopelsparites/micrites and intrasparites. The upper part of the sink hole is filled by a small succession of different clastic sediments, partly fractured, reworked or brecciated, and showing several discontinuities (Fig. 3). Some of the layers display ferromangenese impregnations in small cavities of probably biogenous origin (Fig. 4c). Similar to the base of the clastic sequence, reworked and corroded clasts of a *Pecten*-lumachelle layer were found at the edge of the depression. The clasts are often covered by black to brown goethite crusts that consist of thin and curly laminae, alternating with 25–100 μm thick sedimentary layers. They grew in cauliflower-like to digitate structures of up to 5 mm thickness (Fig. 4f). From the crust surface black tree-like "*Frutexites*" structures protude non-geopetally into the surrounding sediment (Fig. 4g).

Finally the whole clastic sequence in the sink hole as well as parts of the Triassic reef limestone are discordantly overlain by a conspicuous spiculite layer up to 80 cm in thickness pinching out laterally after a few meters. It is red to pink colored and pervaded by a network of "white" spar-filled stromatactis cavities. Due to the very similar facies that is known from the Adnet reef slope near Hallein (Böhm et al. 1999; Delecat and Reitner 2005), the sponge-stromatactis-biomicrites should also be attributed to the Schnöll-Formation (Böhm et al. 1999) instead to the Hierlatz Facies, as it has been assumed by Mazzullo et al. (1990) who focussed their studies on the cements of the cavities. Stable isotopes of this horizon were measured also by Turnšek et al. (1999). As a new road-cut in summer 2003 has shown, the succession above the spiculite continues with some red crinoidal limestones, where a few isolated sponges appear but spicular mats are absent. They are followed by the *marmorea* crust (zone of *Schlotheimia marmorea*), an ammonite-rich and condensed marker horizon that is best known from several quarries in the Northern Calcareous Alps. The Sinemurian Adnet Formation above belongs to the Lienbacher Member (zone of *Arnioceras semicostatum*). The Liassic sequence ends in red nodular breccias.

Fig. 4 (Continued) ferromanganese precipitates. (**c**) Section of the clastic sequence that fills up a sinkhole at locality S1 (Fig. 3, 90 cm). Upper part shows intrapelsparites with angular clasts of red limestones, the lower half Fe/Mn-impregnated biosparites. (**d**) Close view on a cavity wall of the lumachelle layer (Fig. 4a) that is covered by "*Frutexites*" structures. Migration of manganese was probably induced by hydrothermal fluids. (**e**) Cement in a dissolution cavity of the lumachelle layer. Fungi-like meshwork suggests Fe/Mn-precipitation induced by filamentous microbes. (**f**) Ferromanganese crust at the contact of the lumachelle and the spiculite (Fig. 3-sample no. 3b). Rippled Fe/Mn laminae with cauliflower-like shapes alternate with calcitic layers. From the top of the crust, "*Frutexites*" structures protrude into the surrounding spiculite matrix. (**g**) Enlarged view of the rippled laminae and tree-like "*Frutexites*" structures on top of the ferromanganese crust

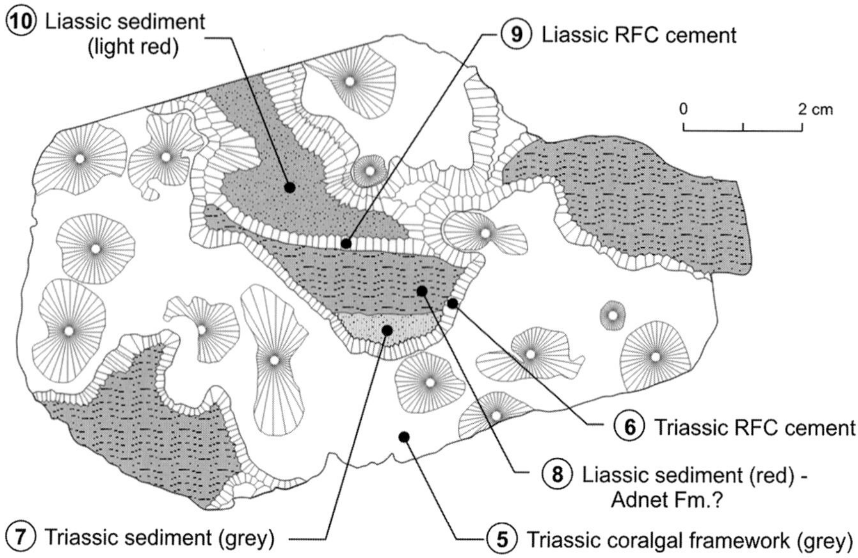

Fig. 5 Drawing of a thin section showing coralgal capping facies from "Fischer's Coral Garden" (locality S2). Remaining space between coral framework is closed by a succession of Jurassic sediments and radiaxial fibrous cements. Numbers refer to layers in Figs. 3 and 9

4.2 *Steinplatte Reef-Top ("Fischer's Coral Garden")*

Condensed red limestones that represent an advanced stage of Liassic onlap sedimentation were sampled from interstices of the Upper Rhaetian coral limestones (capping facies) at Steinplatte summit, called "Fischer's Coral Garden" (locality S2, Figs. 3 and 5). They are lacking sponge remains and presumably belong to the Adnet Limestone facies that is of Lower Sinemurian age.

4.3 *Triassic–Jurassic Boundary Sections (Salzburg Area, Osterhorn Block)*

For comparative geochemical analyses several Liassic ferromanganese crusts were sampled at four localities in the Osterhorn block (part of the Tyrolic Nappe, SE of Salzburg). The Upper Hettangian/Lower Sinemurian *marmorea* crust from Adnet reef slope was sampled at the Lienbacher Quarry (locality S3-a) and the Rot-Grau-Schnöll Quarry (locality S3-b) both situated NE of Adnet near Hallein. Another crust that formed in Lower Sinemurian time was also sampled at Adnet/Lienbacher Quarry (locality S3-a). Further samples encompasse a Hettangian crust from Luegwinkel (locality S3-c) situated SE of Golling, about 150 m NNE of the road to Paß

Lueg, and a Pliensbachian/Toarcian crust from Tannhauser Berg (locality S3-d, SSW of Golling, at the northern slope of the Hagengebirge).

5 The Spiculite at Steinplatte Reef-Slope

5.1 Facies

Situated at the northern slope of the Steinplatte, a wedge of Liassic limestones form the Plattenkogel hill. At its northeastern margin, about 15 m of the underlaying Triassic reef facies are covered by a 80–120 cm thick spiculite limestone. While its base and its top are more dominated by crinoidal debris, the whole sequence is characterized by a succession of irregularly layered sponge-biomicrites (wackestones) that are separated or intercalated by stromatactis cavities of different size (Fig. 6).

The sponge-biomicrites contain high amounts of solitary spicules from non-rigid siliceous sponges the material of which is dissolved and replaced by microspar. Sponge spicules are predominantly arranged in layers of globular to flat or irregular clusters representing skeletal remains of collapsed sponges in stages of advanced decay. Often their habit is nearly lost, but there are also samples still showing complete specimens or at least sponge fragments with spicules preserved in almost original configuration (Fig. 7) known from mesohyl, dermal, and gastral sponge "tissues". Although exact classifications are impossible due to the lack of microscleres, the interpolation of numerous cross sections allows to differentiate several species of globular to irregular bulbous shapes. Additionally one sample has shown circular transverse sections (Fig. 7a) indicating also the presence of sponges with tubular or cup-shaped growth forms.

Former areas of sponge tissues show clotted micropeloidal fabrics (Fig. 7b, c) that are typical features of in situ-formed organomicrites (Reitner 1993; Reitner et al. 1995; Reitner and Schumann-Kindel 1997; Schmid 1996; Vigener 1996; Delecat et al. 2001). Interspace between sponge-microbialites is closed by biodetrital matrix comprising fine shell debris, radiolarian, small gastropods, and a few foraminifers (mainly miliolids and lagenids, e.g. Nodosariidae, *Lingulina* sp., *Ammodiscus* sp., *Involutina liassica*, *Ophtalmidium* sp., a.o.). Bioturbation is rare. The sediment has also infiltrated the sponge remains, thus there is no distinct difference to the embedding matrix, hence corroded specimens are absent. The ratio between automicrites and allomicrites varies and often they are closely intertwined. Although the microbialites are well to distinguish by their microtexture from the allomicrites, the attempt to discriminate them by fluorescence or EDX analyses failed.

Most remains of the collapsed sponge skeletons merge into spicular mats of several millimeters up to a few centimeters in thickness (Fig. 6a, b). Stromatactis cavities particularly occur in between the sponge-microbialite-horizons, where two

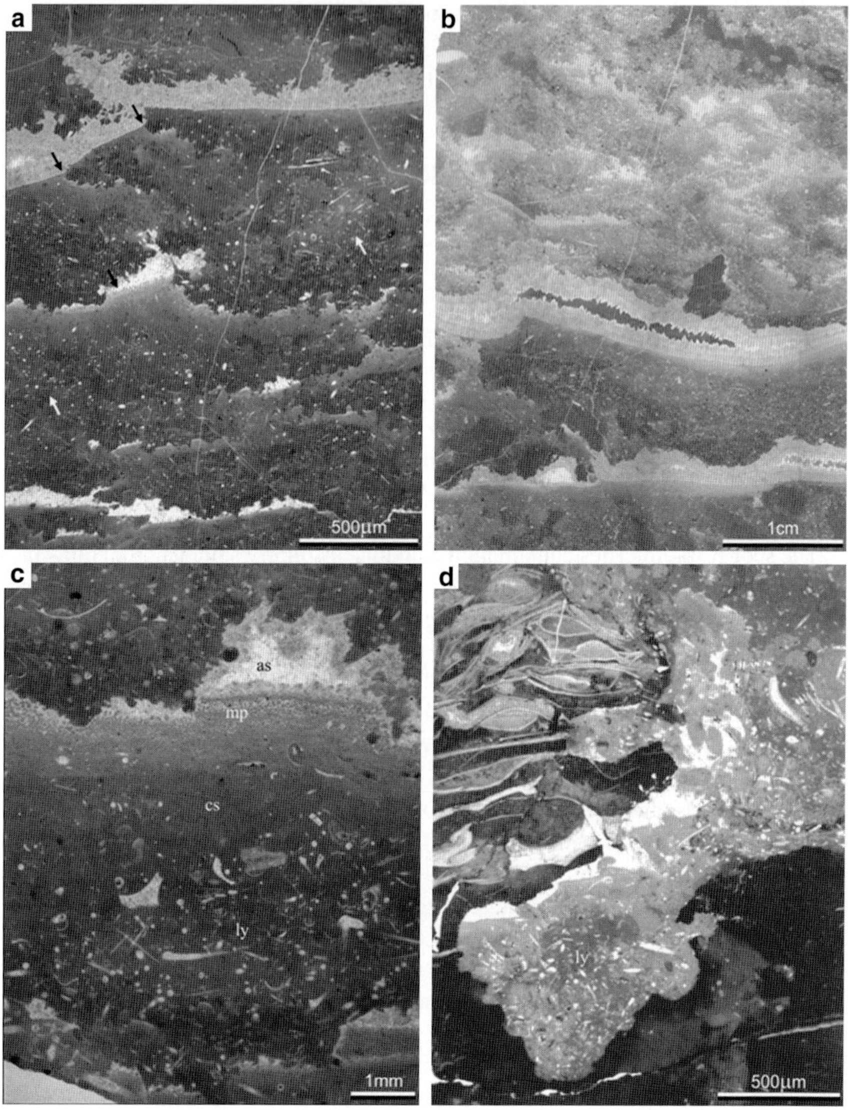

Fig. 6 (**a**) Typical cross section of the spiculite layer. Spicular mats with collapsed sponges (*white arrows*) are separated by different networks of stromatactoid cavities. Successive phases of compaction and sediment infiltration during syndiagenetic tilting are distinguishable (*black arrows*). (**b**) Cross section of three spicular mats. The intraclast in the center was hold in place by microbially fixed carbonates while below compaction of spongy horizons took place. Upper part of the spiculite is characterized by strongly collapsed sponges and densely associated micro-stromatactis cavities. (**c**): Detail of spicular mat showing the usual succession of a collapsed lyssacinosid sponge (ly), compacted sediment (cs), reworked micropeloidals (mp) and aborted stromatactis (as) (Neuweiler et al. 2001a) with radiaxial fibrous cements. (**d**) Intraclast of the *Pecten*-layer with ferromanganese incrustations. An eroded gap was colonized by a lyssacinosid sponge (ly)

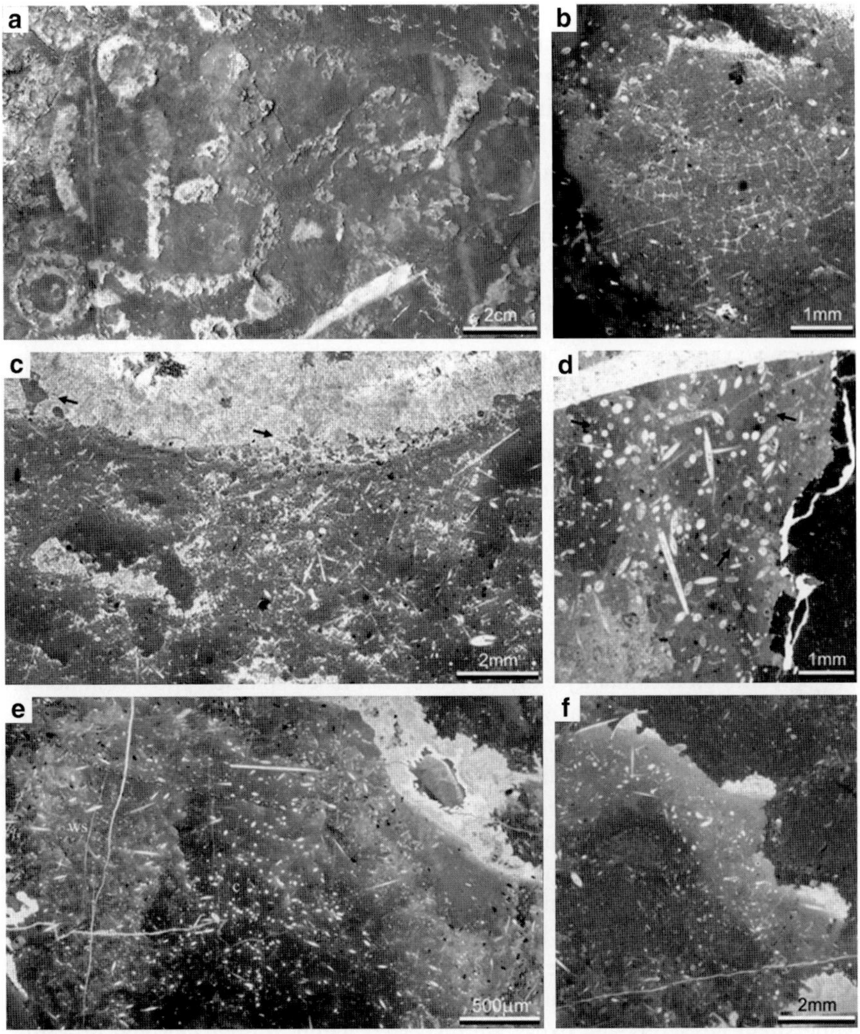

Fig. 7 (**a**) Eroded spiculite layer showing a dense population of in situ preserved sponges. Circular cross sections point to tubular or cup shaped species. (**b**) Dictyonal skeleton of an euretid sponge (Hexactinosa). Interspace shows the typical clotted fabric of micropeloidal automicrites. At the top stromatactis originated, because microbialites detached from the embedding matrix due to compaction and volume reduction in the decaying sponge tissue. (**c**) Part of a spicular mat displays a collapsed lyssacinosid sponge with numerous hexactine and monaxon spicules and several micro-stromatactis cavities. A bigger stromatactis cavity above shows small scaled clasts of reworked material (*arrows*). (**d**) Cross section of the center of a sponge body with bundles of thick diactin spicules, surrounding canals (*arrows*) of the former aquiferous system. (**e**) Cross section of a Lyssacinosida. Outer wall structure (ws) is characterized by mainly micro-hexactins embedded in light-colored microbial carbonates. The center (c) shows big diactins that were partly scattered during collapse of the sponge. (**f**) Dermal fragment of a lyssacinosid sponge. Different levels of micro-stromatactis prove that the overlaying sediment was already fixed, when compaction and microbialite formation continued in the sponge remain

kinds of different sizes are distinguishable. The bigger ones are some millimeters up to 1 cm in height and form horizontally arranged networks that do not cut any components of the sedimentary matrix (Fig. 6a, b). The network separates spicular layers displaying further cavities that are much smaller in size. Latter microstromatactis cavities often occur in several levels inside the sponge remains (Figs. 6b and 7c). Their distribution and irregular contour is determined by the sponge skeletons and the automicrites. Nearly all stromatactis cavities display irregular roofs that are defined by micropeloids as well as intraclasts, fossil debris, or sponge spicules. In contrast, bases of the bigger cavities are rather flat, due to fine detrital sediments or peloidal micrites that were either washed in or represent reworked material from cavity walls (Fig. 6a–c). Many cavities are nearly closed by these infiltrated sediments (aborted stromatactis, Neuweiler et al. 2001a, for further aspects of stromatactis formation see Neuweiler and Bernoulli 2005) that sometimes fade smoothly into subjacent micrites so that a definite original base of these cavities is not discernable. Remaining space of the cavities is closed first by radiaxial fibrous cements, secondary by internal sediments or blocky sparites.

5.2 *Sponge Fauna*

The spiculite predominantly exhibits loose, non-fused spicules. Most common are smooth hexactins and monaxon oxeas of different sizes. Less frequently pentactins and stauractins occur. The megascleres possess rays of 1–4 mm in length and 80–300 μm in diameter. Some hexactin spicules exhibit long rays in opposite position, while other rays are reduced to stubs. Smaller hexactin spicules are mostly found in micropeloidal automicrites and display thinner rays with diameters of 20–40 μm and either short equal rays or some strongly elongated and curved ones with lengths up to a few millimeters. Micro-oxeas are also slightly curved, 20–25 μm thick and 1–2 mm long.

Sponge fragments with uncollapsed spicule configurations often show monaxon and hexactin megascleres, surrounding voids or canals of the aquiferous system, while small hexactins and oxeas are dispersed within former mesohyl or dermal tissues. Some bundles of big diactins have been detected in the atrial center of sponge bodies, surrounded by wall structure with mainly micro-hexactins (Fig. 7e). Wall structures of the same dimension (0.5–0.8 cm) have been found in a sample in which a dense population of in situ preserved sponges show complete individuals of 2–3.5 cm in diameter (Fig. 7a).

Sparsely some small fused hexactins occur showing simple meshworks of euretoid kind (Fig. 7b). The spicules possess rays of 20–40 μm thickness and form grids of 200–300 μm in width. Furthermore a few isolated dichotriaen spicules from demosponges and some quite small triactins from probably Calcarea species have been found without any relation to other skeletal remains.

Investigated cross sections of sponge remains suggest the presence of species with flat and bulbous habits and less common tubular or cup-shaped specimens without stalk. The absence of any colonized omission surfaces or secondary hardgrounds, but the presence of monaxon megascleres scattered or arranged in clusters let assume a lophophytous substrate attachment on spicular mats or detrital soft- to firmgrounds.

The predominance of non-fused hexactinellid spicules and the absence of hexactinos or amphidiscosid spicules (mesamphidiscs, uncinates, pinules) let attribute most of the sponges to the order Lyssacinosida. In contrast to the Schnöll Formation from the Adnet reef slope (Delecat and Reitner 2005) where growth forms are better preserved, the taxonomy of Steinplatte sponges is more or less restricted to the classification of megascleres. Similar to Adnet, microscleres are lacking or were not preserved due to recrystallization processes, thus any specification is difficult. Although latest revision of recent sponge systematic divides lyssacinosid sponges into Euplectellidae, Leucopsacidae, and Rossellidae (Hooper and van Soest 2002) it is not possible to integrate the ones from Steinplatte/Plattenkogel buildup into that system: Predominantly recent Leucopsacidae usually show a basiphytous attachment. The species from Steinplatte show the same but the spicular inventory of the studied sponges corresponds better with that of some Euplectellidae and Rossellidae. The observed growth forms suggest that the detected species should rather be attributed to rossellid hexactinellids. On the other hand, most species of living rossellid hexactinellids are characterized by hypodermal pentactins, whereas in the present spiculite no prostalia pleuralia have been found.

Lacking microscleres and the high degree of decay also complicate exact estimations of the diversity and detailed correlations with sponge faunas from adjacent basinal settings. Unfortunately from deep water environments mainly dislocated spicules, but even microscleres in best condition of preservation were found (Mostler 1989a, 1990b), whereas slope settings often provide complete specimens, but do not show any microscleres due to diagenesis. In result, from basinal cherty limestones Mostler was able to identify spicules of poecilosclerid and astrophorid demosponges as well as several taxa of the Lyssacinosida, Hexactinosida, and Amphidiscophora. In contrast to these highly diverse faunas the biodiversity of Steinplatte slope facies apparently was low. Nevertheless it is known from recent studies that a lot of sponges can only be distinguished by their color respectively their sponge-specific bacteria or at least by molecular analysis (Dohrmann et al. 2008). Therefore the number of species in the spiculite should be estimated clearly higher than the fossil record suggests.

In comparison to the Schnöll-quarry facies at Adnet, at Steinplatte/Plattenkogel locality presumably the higher palaeoaltitude was responsible for a strongly condensed sequence. Low sedimentation rates caused high decomposition of the sponge bodies due to long-lasting decay in the open water column, whereas restricted water currents probably reduced dispersal of sponge larvae as a result of which diversity kept low. As analyses of the isotopic record and trace elements have shown (see below) the colonization by sponges as well as their disappearence shortly after were not linked directly to changes in the geochemical setting. Their migration onto the drowned buildups, that was prior to the extinction event suppressed by the competition of corals, was apparently controlled mainly by sedimentation rates and water currents. Thus, high sediment input as well as non-sedimentation during development of ferromanganese crusts obviously inhibited sponge settlement. The base of the spiculite layers shows that lyssacinosid species acted as pioneers (opportunists, sensu Harries et al. 1996) that were able to settle down on soft gravel-like substrates, whereas hexactinose species (Euretidae and tubular-shaped forms) first appeared in between, when the sediment was fixed by Lysscinosida and spicular mats already served as secondary firmgrounds.

Liassic faunas of non-rigid sponges were not restricted to the alpine region. Similar species are known from France (Termier et al. 1990) and studies in Morocco have shown that in the area of the proto-Atlantic, new territory was also occupied by lyssacinosid sponges (Du Dresnay et al. 1978; Evans and Kendall 1977; Neuweiler et al. 2001b).

5.3 *Diagenesis*

Sponge-dominated sediments are generally described as spongiolites, those with high amounts of isolated spicules are furthermore specified as spiculites (Geyer 1962). There are numerous examples known from all over the world and from Precambrian to Recent. If non-rigid sponges are involved, often a close relation with associated stromatactis-like cavities is documented, like at Steinplatte/Plattenkogel locality as well as in several Silurian and Devonian mud mounds from France, Canada, and Morocco (Burque and Gignac 1983; Flajs et al. 1996; Vigener 1996; Neuweiler et al. 2001b). Those cavities have previously and recently been discussed as being the result of volume reduction due to special interaction of sedimentary compaction and sponge taphonomy (Reitner 1986; Neuweiler et al. 2001a). Neuweiler and Bernoulli (2005) interpreted the formation of stromatacis as a maturation process of polymer gels which are forming fine grained carbonate particles within the sediment.

At Plattenkogel hill, a closer look on the spiculite also reveals that neither the distribution of sponge spicules is random nor did cavities develop primarily by dissolution activity. Clusters of spicules and skeletal remains with original spicule configurations prove the assumption of in situ-preservation: The highest density of spicules is related to horizons with lowest detrital input thus indicating a colonization by sponges in times of low sedimentation rate. Although predominantly loose, non-fused spicules are present, they never touch each other, though soft bodies of the non-rigid sponges must have been consolidated before complete collapse could happen. In the sediment spicules are often so frequent that skeletons of dead sponges formed dense spicular meshworks. It is known from recent counterparts that such layers of collapsed sponges (spicular mats) are excellent grounds for further larval attachment due to their protected micro-environments within the rough and spiny surface (Henrich et al. 1992). The inner pore space of such a spicule network gets either closed by fine sediments or partially keeps open to percolating waters. The shape of most stromatactis cavities at Steinplatte/Plattenkogel locality also implies an early cementation of the decaying sponges syngenetic to compaction and rework of intercalated detrital sediments. Horizons tore apart above shrinking sponge layers, whereas the roofs of arising cavities were protected against collapse mostly by microbially fixed carbonates or sometimes by embedded fossil shells, intraclasts, and at some spots by corroded ferromanganese crusts (Fig. 6d). Smooth transitions between infiltrated cavity material and subjacent sediments are interpreted as the result of successive and early diagenetic compaction and exclude the possibility of a development by diagenetic dissolution which would rather result in cavities with straight and sharp outlines. Smaller stromatactoid cavities inside the microbialite-cemented sponge skeletons often lack infiltrated sediments. Their bases are either irregular like their roofs or flattened by reworked peloidal microbialites. Presumably these little voids originated during the last steps of shrinkage coupled with the decay of sponge tissue and its fixation due to microbially induced carbonate precipitation (summerized in Fig. 8).

6 Geochemical Data

Porifera communities on drowned late Rhaetian alpine reefs represent the first attempt of Liassic metazoan faunas to recover free niches that arose by the Triassic–Jurassic extinction event. Since sequences at or close to the T–J boundary are scarce in the world, the here presented data provide additional information close to the critical interval and its aftermath. Like at Steinplatte buildup slope, most spiculites are characterized by low-diverse fauna and reduced sedimentary input consolidated by microbialites that formed during early diagenetic taphonomic processes. Thus, their composition displays ancient structures already known from ancestral communities, especially from Palaeozoic mud mounds some of which also show the onset of benthic associations following mass extinction events like it is the case at Devonian mud mounds of the Australian Canning Basin (Playford 1980; Stephens and Sumner 2003). Since the causes for the T–J event are not firmed and recovery strategies are badly understood so far, geochemical proxies are one important possibility to reconstruct the former paleoenvironmental conditions and to evaluate probable influences on sponge settlement e.g. by cold seeps or by hydrothermal fluids. The origin of some Liassic alpine ferromanganese crusts was already discussed as being either hydrologically or hydrothermally controlled (Böhm et al. 1999), so that the finding of Fe/Mn precipitates also at Steinplatte locality led to further studies of the crust's composition. Hereby measurements of stable isotopes and element analyses were additionally correlated with those of corresponding sediments, because all crusts are formed by alternating sedimentary and ferromanganese laminae that both are diagenetically overprinted (Germann 1971; Drittenbass 1979; Fels 1995).

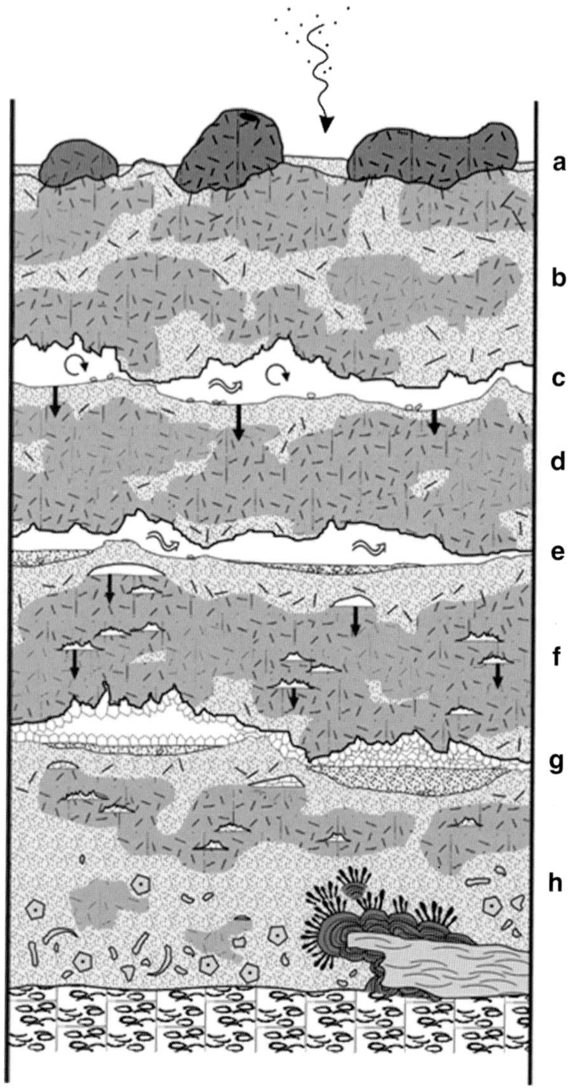

Fig. 8 (**a**) Settlement of lyssacinosid sponges takes place on spicular mats, emanated mainly from skeletal remains of collapsed predecessors. Low sedimentation rate. (**b**) Dead sponges collapse due to decomposition of their organic compound. Sediment infiltrates into the aquiferous systems and decayed sponge parts. Microbially induced carbonate precipitation slightly fixes skeletal remains and adjacent sediments. (**c**) Horizons tear apart between shrinking sponge layers whereas the roofs of developing cavities are protected against collapse by microbially fixed carbonates. Pore water reworks eroded material from cavity walls. (**d**) Decay of sponge tissues continues. (**e**) Fine detrital sediment trickles into some of the cavities. (**f**) Sediment compaction is completed, but decay and shrinkage of sponges successively continue resulting in several levels of micro-stromatactis cavities. (**g**) RFC cements cover cavity walls. Remaining space gets closed by infiltrated sediments and blocky sparites. (**h**) The base of the spiculite above the lumachelle layer is characterized by a higher content of crinoid fragments. *Pecten*-rich intraclasts are covered by Fe/Mn-crusts. On their top, manganese "*Frutexites*" structures protrude into the host rock

6.1 Stable Isotopes

Figure 9a combines the results of stable isotope measurements from the top ("Fischer's Coral Garden") and from the slope (Plattenkogel hill) of the former Steinplatte ramp. While the data of the Triassic limestones all rank almost identically at about $\delta^{13}C$ +3.0 and $\delta^{18}O$ −1, the following isotopic pathways of the Lower Jurassic sediments are split up, seen better in the cross-plots (Fig. 9b–d). At the Steinplatte/Plattenkogel locality a distinct trend to lower $\delta^{13}C$ values and slightly increasing $\delta^{18}O$ values is documented (Fig. 9b). Limestones at "Fischer's Coral Garden" first show increasing values of $\delta^{13}C$, then a backward trend of both kinds of isotopes (Fig. 9a). In addition to the crust at Plattenkogel hill stable isotopes of other Liassic ferromanganese crusts were measured and compared with adjacent micritic sediments, respectively (data in the supplement). In result, all crusts display displacements to lighter $\delta^{13}C$ and lighter $\delta^{18}O$ values (Fig. 9c) as a typical effect of late diagenetic overprint.

The Triassic coral limestones have apparently retained their original signals. They all concentrate in a small range (Fig. 9b) that is similar to the range of other Triassic limestones like that from Adnet (Böhm et al. 1999; Delecat and Reitner 2005), Sonntagkendlgraben (Morante and Hallam 1996) and from a T/J-section in Hungary (Pálfy et al. 2001). They all also match with mean values of T/J-marine

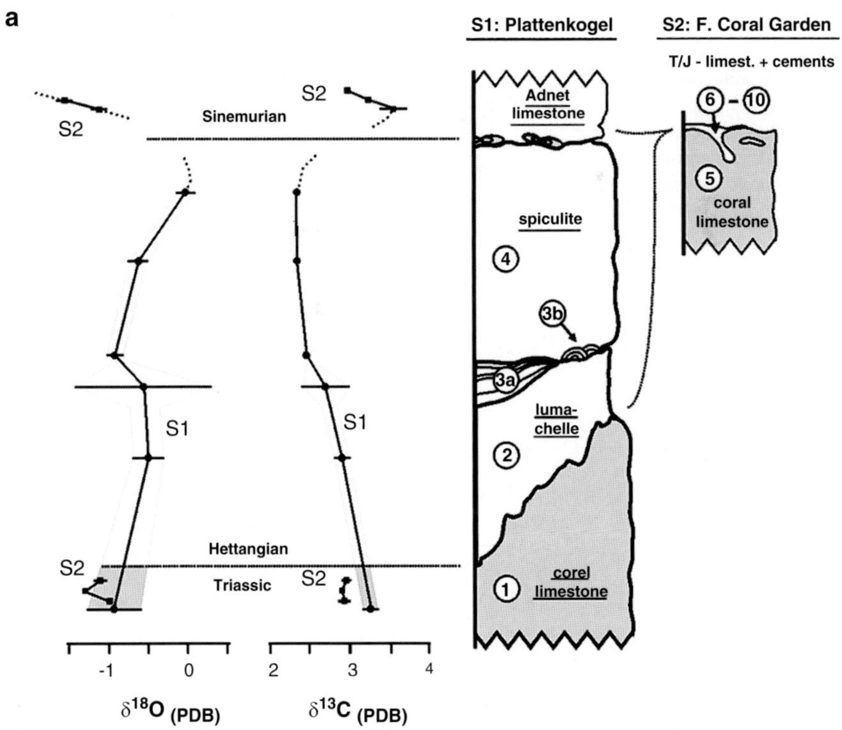

Fig. 9 Stable carbon and oxygen isotopes of the Triassic–Liassic boundary section of the Steinplatte (Rhaetian-Hettangian), Adnet (Hettangian-Sinemurian), and Luegwinkel (Hettangian). (**a**) Variations of carbon and oxygen isotope values across the Triassic-Liassic boundary at Plattenkogel and Fischer's Coral Garden, Steinplatte. (**b**) Cross-plots of $\delta^{13}C$ and $\delta^{18}O$ of carbonate sediments, demonstrating a shift to lower $\delta^{13}C$ from Rhaetian to Hettangian-Sinemurian carbonates. (**c**) Cross-plots of $\delta^{13}C$ and $\delta^{18}O$ of carbonates within Fe–Mn-crusts and micrite of adjacent sediments, showing a diagenetic alteration of the former as indicated by shift towards more negative $\delta^{18}O$ and less positive $\delta^{13}C$ values. (**d**) Cross-plots of $\delta^{13}C$ and $\delta^{18}O$ of Triassic–Liassic carbonate sediments, radiaxial-fibrous cements ("cement A") and late blocky calcite cement ("cement B"), demonstrating the diagenetic pathway towards more negative $\delta^{18}O$ values

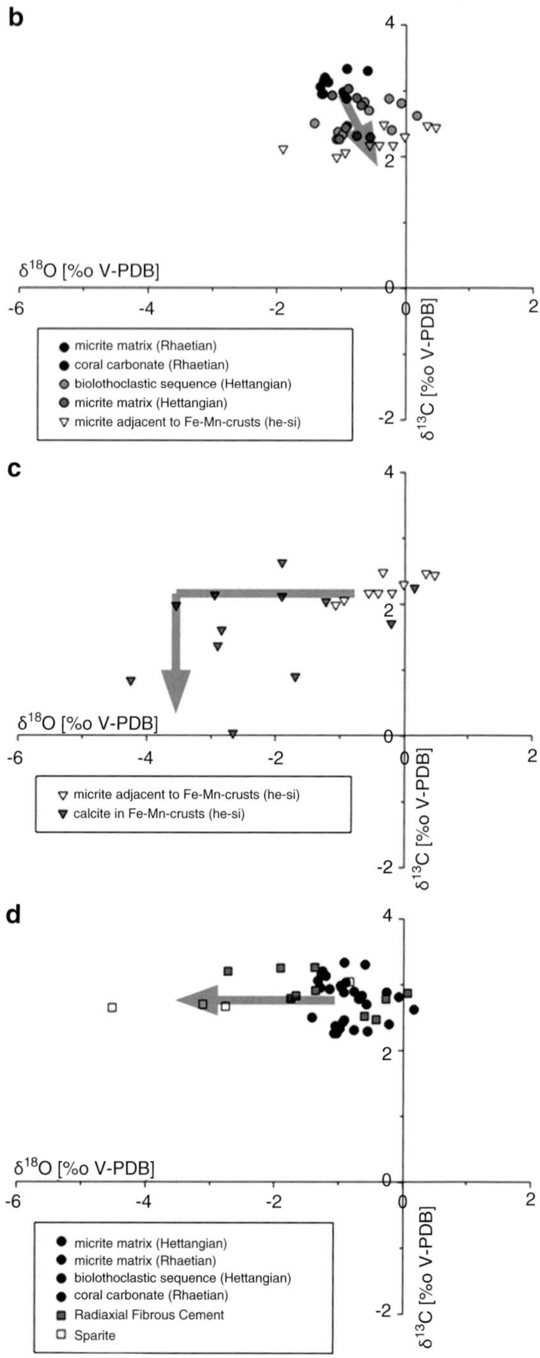

seawaters given by Veizer and Hoefs (1976). Regarding the values of the Liassic succession, it is necessary to consider that, when diagenesis happens, it usually causes first a strong ^{18}O depletion followed by a negativ $\delta^{13}C$ shift. The late diagenetic blocky sparites show a ^{18}O depletion only (Fig. 9d), while ferromanganese crusts, which are altered like all other Liassic crusts that were studied from the alpine region, shows a minor ^{13}C depletion, too (Fig. 9c). According to this, the isotopic pathway from "Fischer's Coral Garden" (Rhaetian) and Plattenkogel hill (Hettangian) both show a moderate early and late diagenetic overprint with $\delta^{13}C$ signals retained closely to original values. Although some Steinplatte/Plattenkogel data scatter broadly due to the high content of detrital components, a distinct negative $\delta^{13}C$ shift is discernible. Considering that microbialites, if not related to methanogenesis, methane oxidation, or photosynthesis, usually do not show any biologically isotope fractionation (e.g., Keupp et al. 1993), the automicrites of the spiculite in fact should provide the best results. The younger Adnet Limestones (Figs. 3 and 9-sediment no. 8) from "Fischer's Coral Garden" (interstices of the reef surface) again show values like the Triassic samples. It is postulated that there was a little $\delta^{13}C_{carb}$ excursion during Hettangian time that ends in the Lower Sinemurian. This hypothesis is supported by a more complete data set from Adnet localities, where the same trend is actually better documented (Delecat and Reitner 2005).

Negative excursions in $\delta^{18}O$, $\delta^{13}C_{carb}$ and $\delta^{13}C_{org}$ are known from several marine T–J boundary sections (Morante and Hallam 1996; Hallam and Goodfellow 1990; Pálfy et al. 2001; Guex et al. 2002; Ward et al. 2001). Possible reasons discussed for the isotopic shifts are e.g. sea level changes, catastrophic impacts but most convincingly the CO_2 release from either elevated volcanism in the central Atlantic magmatic province (CAMP) or by destabilization of methane hydrates due to increased temperatures (references in Pálfy et al. 2001; Delecat and Reitner 2005). In contrast, the negative carbon isotope shift that has been detected at Steinplatte/Plattenkogel locality is too small to explain global specific causes. Most likely it is related to local conditions that probably reflect processes during spiculite formation. As it has already been assumed for a similar isotope record at Adnet (Delecat and Reitner 2005), the high ratio of sponge automicrites in the spiculite could explain the small negative shift. These microbialites are formed by the degradation of sponge organic mater related with bacterial sulphate reduction that is lowering the $\delta^{13}C$ ratio significantly (Hoefs 1997; Londry and Des Marais 2003).

6.2 Element Analyses

Weight ratios of major and trace elements were measured on bulk samples of six ferromanganese crusts and four corresponding sediments. In result, the element analyses show that the distribution patterns are quite similar in both sediments and crusts, but the concentrations in the ferromanganese precipitates are much higher (Fig. 10). All samples show definite accumulations of predominantly V, Ni, Zn, Sr,

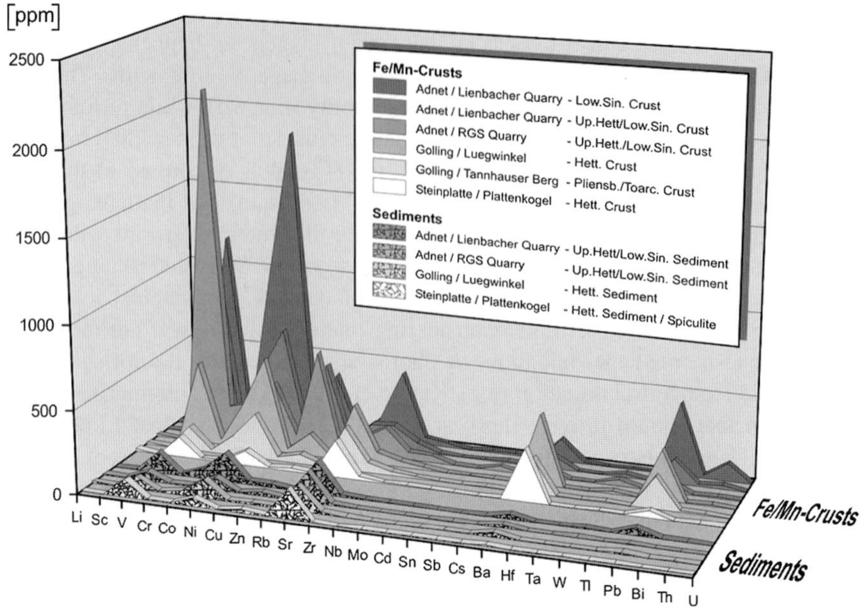

Fig. 10 Trace elements in Fe/Mn-crusts and corresponding sediments from Steinplatte/Plattenkogel (locality S1) and different localities of the Osterhorn block. Measured by liquid icp-mass spectrometry, except Zn by XRF-analyses. For all ICP-MS data see the supplement

Ba, and Pb. The different amounts are most likely caused by use of bulk samples that vary in their content due to the detrital character of the sediments and the laminated composition of the crusts. In order to verify the validity of bulk analyses, measurements of quantitative XRF- and liquid ICP-mass spectrometry were additionally compared with results of semi-quantitative point analyses by laser ICP-mass spectrometry. The comparison between bulk and point analyses revealed some strong differences in the measured amounts of trace elements, but not in the values of rare earth elements, thus the ratio of REE's seems to be homogeneously dispersed in all samples. In the sediments the REE's show quite uniform patterns (Fig. 11), whereas the crusts display very high values of Cerium (up to 1,500 ppm in the Adnet/Lienbacher Quarry – Low. Sin. Crust) (Moffett 1990). In contrast the crust from Plattenkogel hill exhibits a clearly negative cerium anomaly, also the "*Frutexites*" structures of the lumachelle layer (data in the supplement to Fig. 11b) and at a lower amount also the matrix of the overlaying spiculite. Furthermore the crust at Plattenkogel hill is characterized by high concentrations of barium and a Fe/Mn-ratio of about 3.3 (Fig. 12) that is lower than in most other Liassic crusts (values of Fe/Mn = 8–150, in accordance with Böhm et al. 1999).

Ferromanganese crusts and nodules are common features of Lower Liassic condensed sedimentation in the Northern Calcareous Alps (Wendt 1970; Germann 1971; Drittenbass 1979; Fels 1995). Their genesis has often been discussed and recently been interpreted as mainly hydrothermally in origin (Böhm et al.

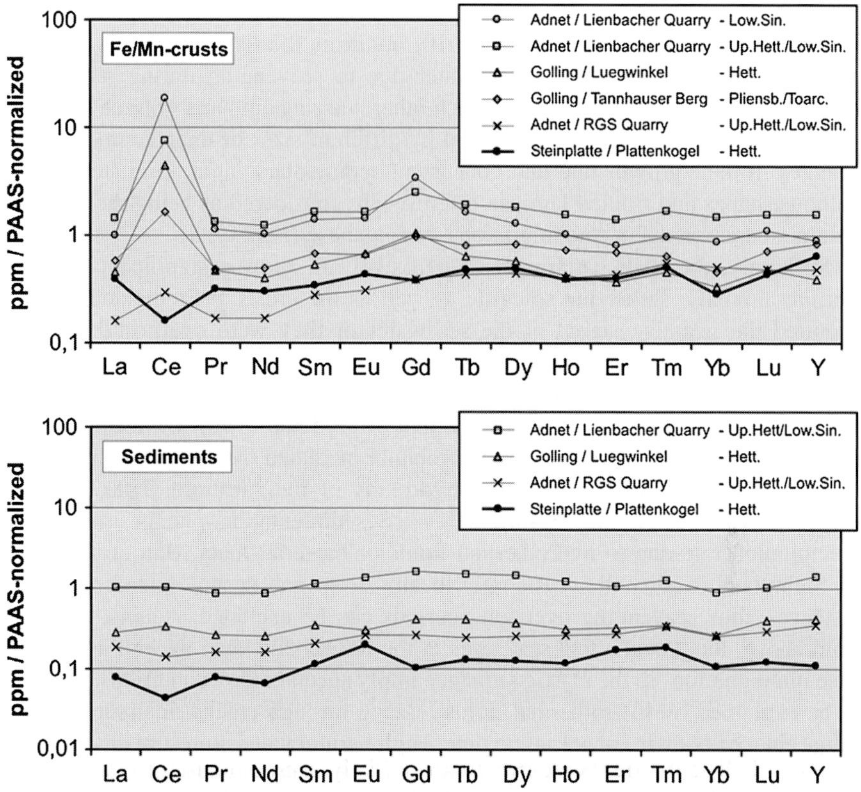

Fig. 11 PAAS-normalized (McLennan 1989) rare earth elements in Fe/Mn-crusts and corresponding sediments from Steinplatte/Plattenkogel (locality S1) and different localities of the Osterhorn block. Measured by liquid icp-mass spectrometry. For all ICP-MS data see the supplement

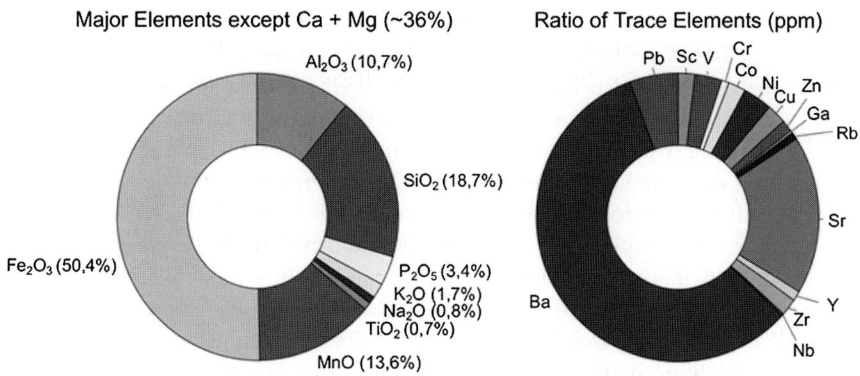

Fig. 12 Major and trace element weight ratios of the Fe/Mn-crust (Fig. 3/no. 3b) from Steinplatte/Plattenkogel (locality S1). Note the relatively low Fe/Mn ratio and the dominance of barium. Measured by X-ray fluorescence analysis. For all XRF data see the supplement

1999). X-ray diffraction shows that they mainly consist of the mineral goethite. The result of trace element analyses (Fig. 10) confirms the hypothesis of very slow precipitation of the crusts from seawater due to scavenging during reduced or starved sedimentation rates. Among each other, varying amounts of trace elements in the crusts could mainly be explained by different sizes of their laminae due to changes in their growth rate and concurrent sedimentary input. Digitate growth, bulbous shapes and rippled laminae of crusts are considered as being the result of volume reduction due to the migration of manganese (Fels 1995).

Strikingly, the REE patterns at Plattenkogel hill show a significant negative Cerium anomaly. Either the spiculite as well as the crusts at Steinplatte locality captured the negative signal of the sea water or they were overprinted by late diagenesis. At first glance, the occurrence of the anomaly in three different horizons suggests secondary changes, but its Y/Ho-ratio (~1.67) and Zr/Hf-ratio (~40) does not correspond with values of hydrothermal origin (Bau 1996). Oxidized cerium, the oxidation of which is obviously microbially mediated (Moffett 1990) preferentially gets trapped in the Fe/Mn-oxyhydroxids of the substrate. Thus negative values could also be the result of elevated sedimentation rate or accelerated precipitation rate due to hydrothermal fluids or bacterial mats (Bau et al. 1996). At Steinplatte locality the crusts only locally occur and do not extend over long distances, thus scavenging over long periods can be excluded. Against the facts from above, the "*Frutexites*" structures in the lumachelle layer as well as manganese impregnations in the clastic sequence imply special redox conditions, possibly to be explained by hydrothermal fluids leaking through cracks or fissures of the subjacent reef body in early Liassic time. High barium values (presumably fixed by biofilms), relatively low Ni + Cu + Co values, and the low Fe/Mn ratio would support this assumption (Bonatti et al. 1972).

7 Autochthonous Spiculites

Since the highest diversity of recent Hexactinellida is known from the bathyal zone (Tabachnick 1994), these taxa mostly elude direct observation. As a result rare recent autochthonous sponge communities from shallow environments as well as fossil assemblages present the best possibility for extensive investigations. The preservation of alpine autochthonous spiculites on drowned Triassic reefs reflects peculiar taphonomic processes that seem to be closely related to the synsedimentary environment. While spiculites formed on elevated position have previously been discussed as a model for mud mound genesis (Reitner and Neuweiler 1995) comparative studies with two other sponge communities are given here to point up the relevance of environmental conditions during spiculite formation. This is shown by spiculitic mats from the crest of Recent "Arctic Vesterisbanken Seamount" (Central Greenland Sea) (Henrich et al. 1992) and by Campanian sponge spicule-dominated sediments exposed on the top of the Albian "Murguia" Diapir (Province Alava, Basco-Cantabrian Basin, Spain). Throughout all localities sponge settlement was apparently controlled first by the water regime respectively by the nutrient supply via consistent currents. In the cold waters at Vesterisbanken these currents probably derived from a "streamlined down-welling vortex known as the Taylor column" (Henrich et al. 1992) whereas sites of temperate latitudes (Diapir of Murguia, Steinplatte/Plattenkogel) were more likely dominated by upwelling systems or contour currents as postulated for alpine spiculites at Adnet (Böhm et al. 1999; Delecat and Reitner 2005). Furthermore the faunal composition is different in all studied sites and depends on the given substrate. At Murguia e.g. (Reitner 1982, 1987), the nowadays sunken center of the diapir top exposes marls of Lower Campanian age showing the initial stage of recolonization by mainly non-lithistid Demosponges. Followed much later by a coral reef growth in the Upper Campanian Oro Member (von Stackelberg 1964; Engeser 1984) the pioneer sponge fauna comprises specimens with non-rigid skeletons formed by unfused spicules the remains of their collapsed skeletons formed spiculitic mats that served as secondary substrates for hexactinosid and demosponge lithistid taxa. The facies displays a high ratio of spicules and very low amount of embedding matrix thus implying a reduced sedimentation rate. Accordingly automicrites are scarce and only occur in the interior of some of the rigid sponge skeletons so that the sediment resembles the grey limestone facies from alpine basinal settings. In contrast to these sediments lacking diverse accessorial faunas, the ridge of the deep sea volcano "Vesterisbanken" is characterized by more complex biogenic mats with sponges, bryozoans, serpulids and hydrozoans. They formed on sands and coarse volcanic gravel that provided larger surfaces also for encrusting filter-feeding organisms. The sponge taxa encompasses several groups of Lyssacinosida (Rossellidae), non-rigid Demospongiae (Geodiidae) and a few poecilosclerid Demosponges. A closer look onto the spiculite mats show that fine-grained, microbial sediment is trapped within the meshwork of sponge spicules and branched bryozoan colonies. Furthermore the mats from Vesterisbanken are also characterized by spicules of dead predecessors, by microbially induced carbonate precipitation and by small interior cavities where larvae settlement of cryptic sponges occur. Thus the mats could be seen as modern counterparts to the Liassic alpine spiculite at Steinplatte with both displaying similarities in their formation and taphonomic structure.

8 Triassic–Jurassic Boundary

In the last decade research on the Triassic–Jurassic mass extinction was strongly enforced, currently by the IGCP-Project 458 – Triassic–Jurassic Boundary Events. The T–J event was characterized rather by several local changes than by one big turn-over, hence global effects have been discussed controversially (Hallam 1989, 2002; Marzoli et al. 1999; McElwain et al. 1999; Tanner et al. 2001; Hesselbo et al. 2002; Olsen et al. 2002; Ward et al. 2001; a.o.). Eustatic sea level changes caused high extinction rates especially in the shelf areas, but simultaneous lowstands were responsible for a lot of shallow marine sequences displaying a hiatus in sedimentation just where the event should have been documented. A relation between the Triassic–Jurassic boundary event and carbon isotope excursions in $\delta^{13}C_{carb}$ and $\delta^{13}C_{org}$ has also been mentioned from several localities (Germany: McRoberts et al. 1997; Turnšek et al. 1999; Hungary: Pálfy et al. 2001; Canada: Ward et al. 2001; England: Hesselbo et al. 2002). Disregarding those of apparently diagenetic origin (Morante and Hallam 1996; Ward et al. 2001), a hiatus in the fossil record often prevents exact stratigraphical correlations thus a lot of work still has to be done correlating the few present data from around the world in order to gain insight into the driving forces of the event (Guex et al. 2002; Pálfy et al. 2001).

At Steinplatte/Plattenkogel locality a new ammonite found of *Psiloceras* sp. from the spiculite base let allocate the stromatactis limestones to the Lower Hettangian. This also corresponds with the geochemical data, supposing that the isotopic pathway in sediments at Plattenkogel hill displays the beginning of the same amplitude that is shown in Adnet quarries (Delecat and Reitner 2005).

Nevertheless it keeps unknown so far whether the excursion at Steinplatte shows only one amplitude of probably a series of little fluctuations or just the end of a bigger carbon anomaly. Anyway, it should be considered that in contrast to some deep-water settings the isotopic records from shallow marine sediments of the Tethyian region could have been overprinted by several local paleoenvironmental changes. At Steinplatte for example an effect of meteoric diagenesis seems to decrease upward from the capping facies (Stanton and Flügel 1989) to the cements of the stromatactis cavities (Mazzullo et al. 1990). But the following succession is characterized by constantly increasing water depths so that an emergence later than the origin of the spiculite layer is unlikely. Probably the sediments of the Lower Liassic sequence were influenced afterwards by percolating waters that were trapped first in Upper Triassic time and then reactivated during Liassic burial.

9 Conclusions

- At Steinplatte/Plattenkogel locality (N of Waidring, Austria) the upper slope of the Triassic Carbonate Platform was colonized by Lower Hettangian fauna of mainly non-rigid lyssacinosid sponges.
- Settlement of sponges took place during reduced sedimentation on detrital soft- or firmgrounds in small depressions of the drowned former reef surface.
- There is no correlation between sponge settlement and geochemical data, thus colonization was exclusively controlled by sedimentation rate and water regime.
- A succession of several sponge generations emerged into a spiculite layer of about 1.5 m in thickness.
- Mats of dead sponges were fixed in situ by infiltrated sediments and microbially induced carbonate precipitation in decaying sponge tissues.
- Large stromatactis cavities originated due to compaction of non-cemented biomicrites in between already microbially-fixed parts of the spiculite. Smaller

cavities are the result of volume reduction due to the decay of organic matter and microbialite formation.

- At Steinplatte/Plattenkogel localitiy, a negative cerium anomaly is found in a ferromanganese crust and the overlaying Liassic spiculite.
- The Triassic–Jurassic boundary at Steinplatte is followed by sediments showing a minor but distinct negative $\delta^{13}C_{carb}$ – excursion that starts in Lower Hettangian and presumably ends in Lower Sinemurian time.
- The facies of sponge-stromatactis-biomicrites represents a pendant to the very similar Schnöll Quarry Formation of Adnet (E of Hallein, Austria).

Acknowledgements We thank Dr. M. Joachimski (Erlangen) for stable isotope analyses, Dr. Helmut Klein and Dr. A. Preusser (Göttingen) for XRD analyses, Dr. G. Hartmann (Göttingen) for XRF analyses, Dr. K. Simon (Göttingen) for ICP-MS measurements and Dr. T. Heinrichs (Göttingen) for EDX analyses. Ammonites were specified by Dr. G. Bloos (Stuttgart). The Deutsche Forschungsgemeinschaft is gratefully acknowledged for financial support (Re 665/17). This paper contributes to the IGCP- Project 458 – Triassic–Jurassic Boundary Events.

Appendix

Supplement to Fig. 9: Carbon and Oxygen Stable Isotope Datas

Thin section/ i.-sample	Locality/ no. of horizon	Stratigraphy	Micro facies	$\delta^{13}C_{carb}$ (PDB)	$\delta^{18}O_{carb}$ (PDB)
ST20/01	S2/5	Triassic reef limestone	Coral material	2.98	−0.97
ST20/01b	S2/5			2.96	−0.96
ST20/02	S2/5			2.95	−1.30
ST20/02b	S2/5			2.88	−0.92
ST20/03	S2/5		Microbial limestone	2.95	−1.28
ST20/03b	S2/5			3.13	−1.20
ST20/03c	S2/5			3.06	−1.32
ST20/04	S2/6	Triassic? Infill of reef cavity	RFC cement	2.79	−1.74
ST20/04b	S2/6			2.91	−1.36
ST20/04c	S2/6			2.83	−1.66
ST20/05	S2/7		Grey intern micrite	2.99	−1.01
ST20/05b	S2/7			2.95	−1.42
ST20/05c	S2/7			2.79	−1.29
ST20/06	S2/8	Liassic infill of reef cavity/ Adnet Fm.?	Red intern micrite	3.42	−1.05
ST20/06b	S2/8			3.56	−1.03
ST20/06c	S2/8			3.71	−1.18
ST20/07	S2/9		RFC cement	3.26	−1.37
ST20/07b	S2/9			3.20	−2.72
ST20/07c	S2/9			3.25	−1.90
ST20/08	S2/10		Red intern micrite	2.88	−3.27
ST20/08b	S2/10			2.90	−3.04
ST20/08c	S2/10			3.06	−1.65

(*continued*)

Thin section/ i.-sample	Locality/ no. of horizon	Stratigraphy	Micro facies	$\delta^{13}C_{carb}$ (PDB)	$\delta^{18}O_{carb}$ (PDB)
ST9/09	S1/1	Triassic reef	Coral material	3.15	−1.27
ST9/10	S1/1	limestone		3.20	−1.25
ST9/11	S1/1		Micrite	3.33	−0.91
ST9/12	S1/1			3.30	−0.59
ST9/13	S1/4?	Schnöll Fm./	Red biomicrite	2.93	−1.14
ST9/14	S1/4?	Low. Hettangian		3.03	−0.89
ST1/15	S1/2	Lumachelle	Bivalve – orig. mat.	1.21	−0.58
ST1/16	S1/2	layer/	Biomicritic matrix	2.78	−0.69
ST1/17	S1/2	*Cardinia*	Sediment infill	3.01	−0.29
ST1/18	S1/2	packstone	Sparite	2.65	−4.52
ST2/19	S1/3a	Clastic sequence	Bivalve biomicrite	2.89	−0.76
ST2/20	S1/3a		Micrite in nept. dike	2.99	0.33
ST2/21	S1/3a		Intrasparite	2.62	0.17
ST2/22	S1/3a		Clasts of *Pecten* biomicrite	2.50	−1.41
ST4/23	S1/3a		Bioturbate biosparite	2.83	−0.64
ST4/24	S1/3a		Intrapelsparite	2.88	−0.26
ST5/25	S1/3a		Bioturbate biosparite	2.70	−0.57
ST5/26	S1/3a		Intrapelsparite	2.81	−0.07
ST7-F3/27	S1/3a		Pelsparite	2.40	−0.22
ST7-F3/28	S1/3a		Biomicrite	2.37	−1.05
ST7-F3/29	S1/3a	Clasts of *Pecten*	Echinod. biomicrite	2.33	−0.98
ST-JR2/30	S1/3b	lumachelle	Micritic matrix	2.26	−1.07
ST-JR2/31	S1/3b	overlain by	Micritic matrix	2.26	−1.03
ST-JR2/32	S1/3b	Lower	*Pecten* shell	2.12	1.50
ST-JR2/33	S1/3b	Hettangian	Red ferromanganese micrite	2.43	−0.55
ST-JR2/34	S1/3b	spiculite	Brown-black ferromangan.	2.25	−2.49
ST-JR/157	S1/3b		crust	2.38	−1.61
ST-JR/158	S1/3b			2.39	−2.61
ST-JR2/35	S1/4	Lower	Micritic matrix	2.46	−0.91
ST-JR2/36	S1/4	Hettangian	Micritic matrix	2.43	−0.93
ST-JR2/37	S1/4	spiculite	Stromatactis sparite	2.26	−1.05
ST-JR2/38	S1/4		Sparite	2.70	−3.11
ST-JR2/39	S1/4			3.04	−0.83
ST-JR1/40	S1/4		Micritic matrix	2.31	−0.76
ST-JR1/41	S1/4		RFC cement	2.47	−0.42
ST-JR1/42	S1/4			2.87	0.07
ST-JR1/43	S1/4		Allomicrite in cavity	2.64	0.09
ST-JR1/44	S1/4		sparite	2.67	−2.76
ST-JR1/45	S1/4		RFC cement	2.52	−0.60
ST-JR1/46	S1/4			2.78	−0.27
ST-JR1/47	S1/4		Micritic matrix	2.29	−0.55
L4/70g	S3-a	Lower	Fe/Mn-Crust	2.25	0.15
L4/70h	S3-a	Sinemurian		1.71	−0.21
L4/70e	S3-a		Sediment adjacent to the	2.44	0.47
L4/70f	S3-a		crust	2.47	0.33
L6/66	S3-a		Fe/Mn-Crust	0.84	−4.25

(*continued*)

Thin section/ i.-sample	Locality/ no. of horizon	Stratigraphy	Micro facies	$\delta^{13}C_{carb}$ (PDB)	$\delta^{18}O_{carb}$ (PDB)
L11/70	S3-a	Up. Hett./Low.		0.04	−2.66
L11/70b	S3-a	Sin.		0.90	−1.69
L11/70c	S3-a			1.37	−2.90
L11/70d	S3-a			1.61	−2.84
L6/67	S3-a		Sediment adjacent to the	2.48	−0.34
L11/69	S3-a		crust	2.30	−0.02
L11/69b	S3-a			2.17	−0.20
SCH18/105	S3-b	Up. Hett./Low.	Fe/Mn-Crust	2.04	−1.22
SCH18/106	S3-b	Sin.		1.98	−3.55
SCH18/103	S3-b		Sediment adjacent to the	2.17	−0.56
SCH18/104	S3-b		crust	2.17	−0.41
LW12a/153	S3-c	? Hettangian	Fe/Mn-Crust	2.14	−2.95
LW12a/154	S3-c			2.63	−1.90
LW12a/161	S3-c			2.12	−1.90
LW12a/155	S3-c		Sediment adjacent to the	1.99	−1.07
LW12a/156	S3-c		crust	2.06	−0.94
LW12a/160	S3-c			2.12	−1.90

Carbon and oxygen isotope data of Upper Triassic/Lower Jurassic carbonates from Steinplatte localities

S1: Plattenkogel, S2: Fisher's Coral Garden, S3-a: Adnet/Lienbacher Quarry, S3-b: Adnet/Rot-Grau-Schnöll Quarry, S3-c: Luegwinkel

Supplement to Fig. 10: Liquid ICP-Mass Spectrometry Datas

Supplement to Fig. 11a: Liquid ICP-Mass Spectrometry Datas

Supplement to Fig. 11b: Laser ICP-Mass Spectrometry Datas

Supplement to Fig. 12: X-ray Fluorescence Analyses Datas

Elements (ppm)	Fe/Mn-crusts						Sediments adjacent to crusts			
	Adnet/ Lienbacher Quarry – Low. Sin. Crust	Adnet/ Lienbacher Quarry – Up. Hett./Low.Sin. Crust	Adnet/RGS Quarry – Up. Hett/Low. Sin. Crust	Golling/ Luegwinkel – Hett. Crust	Golling/ Tannhauser Berg – Pliensb./Toarc. Crust	Steinplatte/ Plattenkogel – Hett. Crust	Adnet/ Lienbacher Quarry – Up. Hett./ Low. Sin. Sediment	Adnet/RGS Quarry – Up. Hett/Low. Sin. Sediment	Golling/ Luegwinkel – Hett. Sediment	Steinplatte/ Plattenkogel – Hett. Sediment = Spiculite
Li	10.25	7.03	7.07	15.93	11.93	18.51	3.68	5.79	12.35	2.76
Sc	10.68	7.16	6.25	3.51	4.87	3.09	4.36	2.67	3.20	2.75
V	1,168.44	973.73	2,137.93	491.13	142.43	127.83	126.92	55.42	76.29	105.72
Cr	19.56	51.90	195.19	31.25	13.79	11.62	10.80	3.89	8.31	6.95
Co	967.34	401.58	237.54	264.66	122.57	19.47	61.46	12.62	27.39	4.91
Ni	1850.23	666.20	373.09	569.10	268.19	91.33	172.27	78.06	115.73	55.88
Cu	90.88	45.68	67.42	24.04	24.82	11.52	24.43	4.23	4.71	5.03
Rb	13.08	7.61	3.84	6.88	7.03	11.03	8.03	7.98	13.09	5.70
Sr	122.74	139.35	105.95	348.63	258.47	224.31	196.96	186.04	153.73	178.91
Zr	429.79	148.99	177.33	76.54	60.89	27.20	17.47	8.76	13.36	5.37
Nb	41.70	54.43	45.23	37.51	23.51	4.84	8.82	2.91	3.52	1.48
Mo	21.10	9.78	37.36	19.01	8.75	3.78	1.32	1.42	0.74	0.75
Cd	0.40	0.32	0.32	0.35	0.31	0.27	0.31	0.26	0.21	0.18
Sn	11.23	2.99	1.52	1.72	2.08	0.51	0.57	0.29	0.50	0.28
Sb	12.16	12.87	26.68	5.31	5.07	1.19	1.06	0.44	0.40	0.17
Cs	0.77	0.57	0.46	0.76	0.51	0.85	0.63	0.89	1.25	0.36
Ba	133.26	78.76	53.00	423.46	211.23	315.36	31.74	22.95	26.37	10.87
Hf	7.79	3.05	2.96	1.43	1.41	0.68	0.52	0.37	0.47	0.34
Ta	0.38	0.51	0.89	0.88	0.57	0.32	0.39	0.26	0.35	0.21
W	4.29	19.89	33.71	17.25	6.73	5.14	3.82	3.39	2.50	2.26
Ti	1.50	0.49	0.28	0.54	0.69	0.23	0.31	0.09	0.14	0.06
Pb	440.38	145.21	65.66	298.47	192.46	37.59	45.71	5.58	11.52	1.79
Bi	18.51	10.82	2.46	4.37	5.78	0.18	1.22	0.14	0.10	0.06
Th	109.84	23.30	5.26	11.79	6.23	1.51	1.77	0.93	1.32	0.52
U	1.32	0.98	2.28	2.17	0.65	0.63	0.32	0.23	0.31	0.12

Trace elements in Fe/Mn-crusts and corresponding sediments from Steinplatte/Plattenkogel (S1) and different localities of the Osterhorn block

Elements (ppm)	Fe/Mn-crusts						Sediments adjacent to crusts			
	Adnet/ Lienbacher Quarry – Low. Sin. Crust	Adnet/ Lienbacher Quarry – Up. Hett./ Low. Sin. Crust	Adnet/ RGS Quarry – Up.Hett/ Low. Sin. Crust	Golling/ Luegwinkel – Hett. Crust	Golling/ Tannhauser Berg – Pliensb./ Toarc. Crust	Steinplatte/ Plattenkogel – Hett. Crust	Adnet/ Lienbacher Quarry – Up. Hett./ Low. Sin. Sediment	Adnet/ RGS Quarry – Up. Hett/ Low. Sin. Sediment	Golling/ Luegwinkel – Hett. Sediment	Steinplatte/ Plattenkogel – Hett. Sediment = Spiculite
La	1.0044	1.4554	0.1622	0.4577	0.5769	0.3855	1.0257	0.1875	0.2838	0.0781
Ce	18.9089	7.4423	0.2926	4.4101	1.6618	0.1584	1.0266	0.1430	0.3355	0.0427
Pr	1.1433	1.3174	0.1686	0.4693	0.4847	0.3131	0.8610	0.1616	0.2665	0.0789
Nd	1.0157	1.2290	0.1679	0.3990	0.4971	0.3000	0.8598	0.1639	0.2562	0.0662
Sm	1.3935	1.6262	0.2748	0.5317	0.6752	0.3427	1.1538	0.2049	0.3477	0.1130
Eu	1.4516	1.6449	0.3078	0.6698	0.6617	0.4347	1.3455	0.2612	0.3009	0.1984
Gd	3.4145	2.4777	0.3874	1.0419	0.9774	0.3811	1.6319	0.2639	0.4200	0.1040
Tb	1.6360	1.9478	0.4324	0.6466	0.8009	0.4809	1.5337	0.2459	0.4113	0.1338
Dy	1.2904	1.8018	0.4422	0.5771	0.8368	0.4930	1.4522	0.2521	0.3726	0.1279
Ho	1.0346	1.5617	0.4014	0.4227	0.7234	0.3884	1.2199	0.2614	0.3185	0.1200
Er	0.8110	1.4141	0.4365	0.3706	0.7001	0.4041	1.0790	0.2721	0.3275	0.1758
Tm	0.9735	1.7089	0.5428	0.4535	0.6435	0.5021	1.2585	0.3371	0.3566	0.1894
Yb	0.8702	1.4884	0.5025	0.3374	0.4535	0.2832	0.9058	0.2583	0.2644	0.1073
Lu	1.1099	1.5408	0.4832	0.4804	0.7074	0.4336	1.0412	0.2924	0.3981	0.1219
Y	0.9048	1.5796	0.4780	0.3896	0.8289	0.6489	1.3995	0.3541	0.4198	0.1096

PAAS-normalizes (McLennan 1989) rare earth elements in Fe/Mn-crusts and corresponding sediments from Steinplatte/Plattenkogel (S1) and different localities of the Osterhorn block

Elements (ppm)	Fe/Mn-crusts								Sediments adjacent to crusts	
	Adnet/ Lienbacher Quarry – Low. Sin. Crust	Adnet/ Lienbacher Quarry – Up. Hett./Low. Sin. Crust	Adnet/RGS Quarry – Up. Hett/Low. Sin. Crust	Golling/ Luegwinkel – Hett. Crust	Golling/ Tannhauser Berg – Pliensb./ Toarc. Crust	Steinplatte/ Plattenkogel – Hett. Crust – 2 measurements on different samples		Steinplatte/ Plattenkogel – Frutexites on Hett. Crust	Adnet/RGS Quarry – Up. Hett/Low. Sin. Sediment	Golling/ Luegwinkel – Hett. Sediment
La	0.8020	0.2604	0.1534	0.2404	0.3335	0.3790	0.4051	1.0576	0.1796	0.3472
Ce	12.2423	0.5027	0.1934	1.2822	1.6692	0.1092	0.1588	0.2583	1.0123	0.3492
Pr	1.4119	0.3756	0.1425	0.2465	0.3053	0.2959	0.3641	1.1031	0.2545	0.3437
Nd	0.9619	0.2420	0.1317	0.1795	0.1815	0.2746	0.1891	0.5288	nd	0.2261
Sm	2.9574	0.8095	0.2952	0.5089	nd	0.6552	0.6448	3.0960	nd	0.4235
Eu	2.8128	0.7347	0.5114	0.4140	nd	0.5655	0.3987	2.5541	nd	0.5838
Gd	1.6884	0.3521	0.3623	0.3181	nd	0.6634	1.0047	6.6910	nd	0.2481
Tb	2.5453	0.2252	0.3249	0.3183	nd	0.5204	nd	1.1358	nd	0.1372
Dy	1.2141	0.0909	0.2926	0.1729	nd	0.3390	0.0170	0.4738	nd	0.0493
Ho	1.5122	0.0840	0.3325	0.4138	nd	0.3677	nd	0.5541	nd	nd
Er	1.0407	0.0779	0.3412	0.3142	nd	0.4651	nd	0.7084	nd	nd
Tm	1.8704	0.1032	0.6034	1.3521	nd	0.2033	nd	1.6080	nd	nd
Yb	1.2607	0.0984	0.5726	0.7023	nd	0.4550	nd	0.9589	nd	nd
Lu	1.6674	0.1156	0.3179	0.4269	nd	0.2288	nd	1.5926	nd	0.5183
Y	0.5979	0.1000	0.3134	0.2213	0.7482	0.5118	0.1419	0.2330	nd	0.2703

PAAS-normalizes (McLennan 1989) rare earth elements in Fe/Mn-crusts and corresponding sediments from Steinplatte/Plattenkogel (S1) and different localities of the Osterhorn block

Supposed mean values of Ca: 12% (Fe/Mn-crusts), and 37% (sediments)

Elements	Fe/Mn-crusts						Sediments adjacent to crusts			
	Adnet/ Lienbacher Quarry – Low. Sin. Crust	Adnet/ Lienbacher Quarry – Up. Hett./ Low. Sin. Crust	Adnet/ RGS Quarry – Up. Hett/ Low. Sin. Crust	Golling/ Luegwinkel – Hett. Crust	Golling/ Tannhauser Berg – Pliensb./ Toarc. Crust	Steinplatte/ Plattenkogel – Hett. Crust	Adnet/ Lienbacher Quarry – Up. Hett./ Low. Sin. Sediment	Adnet/ RGS Quarry – Up. Hett/ Low. Sin. Sediment	Golling/ Luegwinkel – Hett. Sediment	Steinplatte/ Plattenkogel – Hett. Sediment = Spiculite
Weight %										
Na_2O	0.1	0.11	0.15	0.04	0.07	0.06	0.05	0.04	0.08	Not measured
MgO	1.15	1.14	1.44	1.59	0.96	0.68	0.83	0.72	0.85	
Al_2O_3	4.0	3.6	4.30	4.2	1.2	0.80	0.9	1.00	1.3	
SiO_2	7.3	4.9	4.60	5.7	2.2	1.40	1.7	1.80	2.1	
P_2O_5	0.208	0.536	0.51	0.47	0.079	0.26	0.11	0.06	0.054	
K_2O	0.45	0.29	0.11	0.2	0.1	0.13	0.19	0.18	0.25	
CaO	14.49	19.47	6.48	17.08	47.16	50.40	52.17	52.30	51.76	
TiO_2	1.193	0.335	0.44	0.748	0.189	0.05	0.07	0.06	0.079	
MnO	5.323	3.87	0.43	0.309	0.649	1.02	0.17	0.06	0.095	
Fe_2O_3	45.9	42.21	64.12	51.29	7.7	3.78	0.85	0.75	0.76	
ppm										
Sc	8	17	8	4	29	31	38	29	30	Not measured
V	1,027	897	1,898	1,296	75	50	23	15	29	
Cr	53	176	523	116	22	12	19	11	11	
Co	1,013	969	296	255	122	31	68	11	17	
Ni	2,223	1,499	457	885	259	51	187	42	88	
Cu	157	76	132	71	37	35	3	0	1	
Zn	374	478	596	135	60	19	22	21	15	
Ga	14	19	21	10	4	4	3	3	1	
Rb	11	6	8	8	11	13	14	14	19	
Sr	107	124	80	156	266	265	177	172	161	
Y	23	54	12	25	20	19	25	8	11	

(continued)

Elements	Fe/Mn-crusts						Sediments adjacent to crusts			
	Adnet/ Lienbacher Quarry – Low. Sin. Crust	Adnet/ Lienbacher Quarry – Up. Hett./ Low. Sin. Crust	Adnet/ RGS Quarry – Up. Hett/ Low. Sin. Crust	Golling/ Luegwinkel – Hett. Crust	Golling/ Tannhauser Berg – Pliensb./ Toarc. Crust	Steinplatte/ Plattenkogel – Hett. Crust	Adnet/ Lienbacher Quarry – Up. Hett./ Low. Sin. Sediment	Adnet/ RGS Quarry – Up. Hett/ Low. Sin. Sediment	Golling/ Luegwinkel – Hett. Sediment	Steinplatte/ Plattenkogel – Hett. Sediment = Spiculite
Zr	414	169	194	243	66	30	22	23	27	
Nb	175	51	54	99	20	5	3	3	4	
Ba	164	793	37	106	198	912	33	14	19	
Pb	389	745	64	268	189	90	31	10	14	
Ratios										
Fe/Mn	7.79	9.84	135.88	149.46	10.78	3.34	4.54	13.00	7.57	Not measured
Fe/Al	15.14	15.54	10.43	16.16	8.42	6.29	1.23	0.98	0.77	
Ni/Ba	13.55	1.89	12.35	8.35	1.31	0.06	5.67	3.00	4.63	
Ni/Pb	5.71	2.01	7.14	3.3	1.37	0.57	6.03	4.20	6.29	
Cu/Pb	0.4	0,1	2,06	0.26	0.2	0.39	0,10	–	0.07	
Cu/Ba	0.96	0.1	3.57	0.67	0.19	0.04	0.09	–	0.05	

Major and trace element weight ratios of Fe/Mn-crusts and corresponding sediments from Steinplatte/Plattenkogel (S1) and different localities of the Osterhorn block

References

Bau M (1996) Controls on the fractionation of isovalent trace elements in magmatic and aqueous systems: evidence from Y/Ho, Zr/Hf, and lanthanide tetrad effect. Contributions to Mineralogy and Petrology 123: 323–333

Bau M, Koschinsky A, Dulski P, Hein JR (1996) Comparison of the partitioning behaviours of yttrium, rare earth elements, and titanium between hydrogenetic marine ferromanganese crusts and seawater. Geochimica et Cosmochimica Acta 60: 1709–1725

Böhm F (1992) Mikrofazies und Ablagerungsmilieu des Lias und Dogger der Nordöstlichen Kalkalpen. Erlanger Geologische Abhandlungen 121: 57–217

Böhm F, Ebli O, Krystyn L, Lobitzer H, Rakús M, Siblík M (1999) Fauna, stratigraphy and depositional environment of the Hettangian-Sinemurian (Early Jurassic) of Adnet (Salzburg, Austria). Abhandlungen der Geologischen Bundesanstalt Wien 56: 143–271

Bonatti E, Kraemer T, Rydell H (1972) Classification and genesis of submarine iron-manganese deposits. In: Horn D (ed) Ferromanganese Deposits on the Ocean Floor. National Science Foundation, Washington, 149–165

Burque P-A, Gignac H (1983) Sponge-constructed stromatactis mud mounds, Silurian of Gaspé, Québec. Journal of Sedimentary Petrology 53: 521–532

Delecat S, Reitner J (2005) Sponge communities from the Lower Liassic of Adnet (Northern Calcareous Alps, Austria). Facies 51: 385–404

Delecat S, Peckmann J, Reitner J (2001) Non-rigid cryptic sponges in Oyster patch reefs (Lower Kimmeridgian, Langenberg/Oker, Germany). Facies 45: 231–254

Dohrmann M, Janussen D, Reitner J, Collins AG, Wörheide G (2008) Phylogeny and evolution of glass sponges (Porifera, Hexactinellida). Systematic Biology 57: 388–405

Drittenbass W (1979) Sedimentologie und Geochemie von Eisen-Mangan führenden Knollen und Krusten im Jura der Trento-Zone (östliche Südalpen, Norditalien). Eclogae Geologica Helvetica 72: 313–345

Dresnay R du, Termier G, Termier H (1978) Les Hexactinellides (Lyssakides et Dictyonines) du Lias marocain. Géobios 11: 269–295

Engeser T (1984) Sedimentologische, fazielle und tektogenetische Untersuchungen in der Oberkreide des Basko-Kantabrischen Beckens (Nordspanien). Inaug Diss, Geowissenschaftliche Fakultät der Eberhard-Karls-Universität, Tübingen

Evans I, Kendall CGStC (1977) An interpretation of the depositional setting of some deep-water Jurassic carbonatesof the central High Atlas mountains, Morocco. In: Cook HE, Enos P (eds) Deep water carbonate environments. Society of Economic Paleontologists and Mineralogists, Special Publication 25: 249–261

Fels A (1995) Prozesse und Produkte geologischer Kondensation im Jura der westlichen Tethys. Profil 8: 363–472

Flajs G, Hüssner H, Vigener M (1996) Stromatactis Mud Mounds in the Upper Emsian of the Montagne Noire (France): Formation and Diagenesis of Stromatactis Structures. In: Reitner J, Neuweiler F, Gunkel F (eds) Global and regional controls on biogenic sedimentation. I. Reef evolution. Research reports. Göttinger Arbeiten zur Geologie und Paläontologie Sb2: 345–348

Garrison RE, Fischer AG (1969) Deep-water limestones and radiolarites of the Alpine Jurassic. In: Friedman GM (ed) Depositional environments in carbonate rocks. Society of Economic Paleontologists and Mineralogists, Special Publication 14: 20–56

Germann K (1971) Mangan-Eisen-führende Knollen und Krusten in jurassischen Rotkalken der Nördlichen Kalkalpen. Neues Jahrbuch für Geologie und Paläontologie, Monatshefte 1971: 133–156

Geyer OF (1962) Über Schwammgesteine (Spongiolith, Tuberolith, Spiculit und Gaizit): Hermann-Aldinger-Festschrift. Schweizerbart, Stuttgart, 51–59

Guex J, Bartolini A, Taylor D (2002) Discovery of *Neophyllites* (Ammonitina, Cephalopoda, Early Hettangian) in the New York Canyon sections (Gabbs Valley Range, Nevada) and discussion

of the $\delta^{13}C$ negative anomalies located around the Triassic–Jurassic boundary. Bulletin de la Société Vaudoise des Sciences Naturelles 88: 247–255
Hallam A (1989) The case for sea-level change as a dominant causal factor in mass extinction of marine invertebrates. Philosophical Transactions of the Royal Society of London B 325: 437–455
Hallam A (2002) How catastrophic was the end-Triassic mass extinction? Lethaia 35: 147–157
Hallam A, Goodfellow WD (1990) Facies and geochemical evidence bearing on the end-Triassic disappearance of the Alpine reef ecosystem. Historical Biology 4: 131–138
Harries PJ, Kauffman EG, Hansen TA (1996) Models for biotic survival following mass extinction. In: Hart MB (ed) Biotic recovery from mass extinction events. Geological Society, Special Publication 102: 41–60
Henrich R, Hartmann M, Reitner J, Schäfer P, Freiwald A, Steinmetz S, Dietrich P, Tiede J (1992) Facies belts and communities of the Arctic Vesterisbanken Seamount (Central Greenland Sea). Facies 27: 71–104
Hesselbo SP, Robinson SA, Surlyk F, Piasecki S (2002) Terrestrial and marine extinction at the Triassic–Jurassic boundary synchronized with major carbon-cycle perturbation: a link to initiation of massive volcanism? Geology 30: 251–254
Hoefs J (1997) Stable Isotope Geochemistry. Springer, Berlin
Hooper JNA, Van Soest RWM (eds) (2002) Systema Porifera: A Guide to the Classification of Sponges. Kluwer Academic, Plenum Publishers, New York
Keupp H, Jenisch A, Herrmann R, Neuweiler F, Reitner J (1993) Microbial carbonate crusts – a key to the environmental analysis of fossil spongiolites?. Facies 29: 41–54
Krainer K, Mostler H (1997) Die Lias-Beckenentwicklung der Unkener Synklinale (Nördliche Kalkalpen, Salzburg) unter besonderer Berücksichtigung der Scheibelberg Formation. Geologisch-Paläontologische Mitteilungen Innsbruck 22: 1–41
Londry KL, Des Marais DJ (2003) Stable carbon isotope fractionation by sulfate-reducing bacteria. Applied and Environmental Microbiology 69: 2942–2949
Marzoli A, Renne PR, Piccirillo EM, Ernesto M, Bellieni G, De Min A (1999) Extensive 200-million-year-old continental flood basalts of the central Atlantic Magmatic Province. Science 284: 616–618
Mazzullo SJ, Bischoff WD, Lobitzer H (1990) Diagenesis of radiaxial fibrous calcites in a subunconformity, shallow-burial setting: Upper Triassic and Liassic, Northern Calcareous Alps, Austria. Sedimentology 37: 407–425
McElwain JC, Beerling DJ, Woodward FI (1999) Fossil plants and global warming at the Triassic–Jurassic boundary. Science 285: 1386–1390
McLennan SM (1989) Rare earth elements in sedimentary rocks: influences of provenance and sedimentary processes. In: Lipin BR, McKay GA (eds) Geochemistry and mineralogy of rare earth elements. Reviews in Mineralogy 21: 169–200
McRoberts CA, Furrer H, Jones DS (1997) Palaeoenvironmental interpretation of a Triassic–Jurassic boundary section from western Austria based on palaeoecological and geochemical data. Palaeogeography, Palaeoclimatology, Palaeoecology 136: 79–95
Moffett JW (1990) Microbially mediated cerium oxidation in sea water. Nature 345: 421–423
Morante R, Hallam A (1996) Organic carbon isotopic record across the Triassic–Jurassic boundary in Austria and its bearing on the cause of the mass extinction. Geology 24: 391–394
Mostler H (1986) Neue Kieselschwämme aus den Zlambachschichten (Obertrias, Nördliche Kalkalpen). Geologisch-Paläontologische Mitteilungen Innsbruck 13: 331–361
Mostler H (1989a) Mikroskleren hexactinellider Schwämme aus dem Lias der Nördlichen Kalkalpen. Jahrbuch der Geologischen Bundesanstalt Wien 132/4: 687–700
Mostler H (1989b) Mit Zygomen ausgestattete Dermalia von Kieselschwämmen (Desmospongiae) aus pelagischen Sedimenten der Obertrias und des unteren Jura (Nördliche Kalkalpen). Jahrbuch der Geologischen Bundesanstalt Wien 132/4: 701–726
Mostler H (1990a) Mikroskleren von Demospongien (Porifera) aus dem basalen Jura der Nördlichen Kalkalpen. Geologisch-Paläontologische Mitteilungen Innsbruck 17: 119–142

Mostler H (1990b) Hexactinellide Poriferen aus pelagischen Kieselkalken (Unterer Lias, Nördliche Kalkalpen). Geologisch-Paläontologische Mitteilungen Innsbruck 17: 143–178

Neuweiler F, Bernoulli D (2005) Mesozoic (Lower Jurassic) red stromatactis limestones from the Southern Alps (Arzo, Switzerland): calcite mineral authigenesis and syneresis-type deformation. International Journal of Earth Sciences 94: 130–146

Neuweiler F, Bourque P-A, Boulvain F (2001a) Why is stromatactis so rare in Mesozoic carbonate mud mounds? Terra Nova 13: 338–346

Neuweiler F, Mehdi M, Wilmsen M (2001b) Facies of Liassic sponge mounds, central High Atlas, Morocco. Facies 44: 243–264

Olsen PE, Kent DV, Sues H-D, Koeberl C, Huber H, Montanari A, Rainforth EC, Fowell SJ, Szajna MJ, Hartline BW (2002) Ascent of dinosaurs linked to an iridium anomaly at the Triassic–Jurassic boundary. Science 296: 1305–1307

Pálfy J, Demény A, Haas J, Hetényi M, Orchard MJ, Vető I (2001) Carbon isotope anomaly and other geochemical changes at the Triassic–Jurassic boundary from a marine section in Hungary. Geology 29: 1047–1050

Playford PE (1980) Devonian 'Great Barrier Reef' of Canning Basin, Western Australia. American Association of Petroleum Geologists, Bulletin 62: 814–841

Rakús M, Lobitzer H (1993) Early Liassic ammonites from the Steinplatte-Kammerköhralm area (Northern Calcareous Alps/Salzburg). Jahrbuch der Geologischen Bundesanstalt Wien 136/4: 919–932

Reitner J (1982) Die Entwicklung von Inselplattformen und Diapir-Atollen im Alb des Basko-Kantabrikums (Nordspanien). Neues Jahrbuch für Geologie und Paläontologie 165/1: 87–101

Reitner J (1986) A comparative study of the diagenesis in diapir-influenced reef atolls and a fault block reef platform in the Late Albian of the Vasco-Cantabrian Basin (Northern Spain). In: Schroeder JH, Purser BH (eds), Reef Diagenesis. Springer, Berlin, 186–209

Reitner J (1987) Mikrofazielle, palökologische und paläogeographische Analyse ausgewählter Vorkommen flachmariner Karbonate im Basko-Kantabrischen Strike Slip Fault-Becken System (Nordspanien) an der Wende von der Unterkreide zur Oberkreide. Documenta naturae 40: 1–239

Reitner J (1993) Modern Cryptic Microbialite/Metazoan Facies from Lizard Island (Great Barrier Reef, Australia) – Formation and concepts. Facies 29: 3–40

Reitner J, Neuweiler F (1995) Mud mounds: a polygenetic spectrum of fine-grained carbonate buildups. Facies 32: 1–70

Reitner J, Schumann-Kindel G (1997) Pyrite in mineralized sponge tissue-product of sulfate reducing sponge related bacteria? In: Neuweiler F, Reitner J, Monty Cl (eds) Biosedimentology of Microbial Buildups IGCP Project No 380, Proceedings of 2nd Meeting, Göttingen/Germany 1996. Facies 36: 272–276

Reitner J, Gautret P, Marin F, Neuweiler F (1995) Automicrites in a modern microbialite – Formation model via organic matrices (Lizard Island, Great Barrier Reef, Australia). Bulletin de l'Institut oceanographique, Monaco, Special Number 14/2: 237–263

Schmid DU (1996) Marine Mikrobolithe und Mikroinkrustierer aus dem Oberjura. Profil 9: 101–251

Stanton RJ Jr, Flügel E (1989) Problems with Reef Models: The Late Triassic Steinplatte "Reef" (Northern Alps, Salzburg/Tyrol, Austria). Facies 20: 1–138

Stanton RJ, Flügel E (1995) An accretionary distally steepened ramp at an intrashelf basin margin: an alternative explanation for the Upper Triassic Steinplatte "reef" (Northern Calcareous Alps, Austria). Sedimentary Geology 95: 269–286

Stephens NP, Sumner DY (2003) Famennian microbial reef textures, Napier and Oscar Ranges, Canning Basin, Western Australia. Sedimentology 50: 1283–1302

Tabachnick KR (1994) Distribution of recent Hexactinellida. In: Soest RMW, Kempen van TMG, Braekman J-C (eds) Sponges in Time and Space. Balkema Press, Rotterdam, 225–232

Tanner LH, Hubert JF, Coffey BP, McInerney D (2001) Stability of atmospheric CO_2 levels across the Triassic/Jurassic boundary. Nature 411: 675–677

Termier G, Termier H, Thibieroz J (1990) *Hexactinella Lyssakida* Liasiques de la Bordure Sud-Est des Cevennes. Bulletin trimestriel de la Société Géologique de Normandie et des Amis du Muséum du Havre 77: 5–17

Turnšek D, Dolenec T, Siblík M, Ogorelec B, Ebli O, Lobitzer H (1999) Contributions to the fauna (corals, brachiopods) and stable isotopes of the Late Triassic Steinplatte reef/basin-complex, Northern Calcareous Alps, Austria. Abhandlungen der Geologischen Bundesanstalt Wien 56/2: 121–140

Veizer J, Hoefs J (1976) The nature of $^{18}O/^{16}O$ and $^{13}C/^{12}C$ secular trends in sedimentary carbonate rocks. Geochimica et Cosmochimica Acta 40: 1387–1395

Vigener M (1996) Mikrofazies, Zementation und Diagenese unterdevonischer Stromatactis-Mud Mounds in der Montagne Noire (Südfrankreich). Aachener Geowissenschaftliche Beiträge 14. Augustinus Buchhandlung, Aachen

von Stackelberg U (1964) Der Diapir von Murguía (Nordspanien). Beihefte zum Geologischen Jahrbuch 66: 63–94

Wächter J (1987) Jurassische Massflow- und Internbreccien und ihr sedimentär-tektonisches Umfeld im mittleren Abschnitt der Nördlichen Kalkalpen. Bochumer geologische und geotechnische Arbeiten 27. Institut für Geologie der Ruhr-Universität Bochum, Bochum

Ward PD, Haggart JW, Carter ES, Wilbur D, Tipper HW, Evans T (2001) Sudden productivity collapse associated with the Triassic–Jurassic boundary mass extinction. Science 292: 1148–1151

Wendt J (1970) Stratigraphische Kondensation in triadischen und jurassischen Cephalopodenkalken der Tethys. Neues Jahrbuch für Geologie und Paläontologie, Monatshefte 1970: 433–448

Microbes in Resinous Habitats: A Compilation from Modern and Fossil Resins

Christina Beimforde and Alexander R. Schmidt

1 Introduction

Amber can be considered as a window into the past allowing insights into the palaeoecology of Mesozoic and Cenozoic woodlands. Insects, spiders, fragments of higher plants and even small vertebrates were trapped by resins of conifers and angiosperms of the ancient "amber forests". Apart from these macroinclusions, amber may contain diverse and well-preserved fossil microorganisms (e.g., Waggoner 1994; Schönborn et al. 1999; Schmidt et al. 2006). Most often they are representatives of limnetic or humid microhabitats of the forest floor (Girard et al. 2009b). Although microorganisms associated with amber are known for a long time (e.g. Berkeley 1848; Caspary and Klebs 1907, Galippe 1920; Blunck 1929) extensive studies were not done until the end of the 1980s (e.g. Ting and Nissenbaum 1986; Waggoner 1993, 1994, 1996). Currently, continuous research on microinclusions is done at several institutions (e.g., Schmidt et al. 2006; Rikkinen and Poinar 2000; Girard et al. 2009b). Various fossil microorganisms such as bacteria, cyanobacteria, fungi, algae, protozoans and micrometazoans can be found associated in single amber pieces (Schmidt et al. 2006; Girard et al. 2008b). Amber preserved organisms from forest habitats where fossilization-processes rarely happened. Furthermore, besides of chert, amber is the only source of well-preserved fossil soft-bodied microorganisms. Because of the excellent preservation even of cell organelles, a morphological characterization and a direct comparison to determinable recent organisms is often possible (Dörfelt and Schäfer 2000; Dörfelt and Schmidt 2007, Schmidt et al. 2006). The reconstruction of microbial life cycles and interactions of plants and microorganisms of Mesozoic and Cenozoic ecosystems is of special interest in examining amber microinclusions. Furthermore, amber

C. Beimforde and A.R. Schmidt (✉)
Courant Research Centre Geobiology, Georg-August-Universität Göttingen, Goldschmidtstr 3, 37077 Göttingen, Germany
e-mail: alexander.schmidt@geo.uni-goettingen.de

J. Reitner et al., *Advances in Stromatolite Geobiology*, 391
Lecture Notes in Earth Sciences 131, DOI 10.1007/978-3-642-10415-2_24,

inclusions may help to trace the evolutionary history of groups of organisms that are otherwise rarely or never preserved.

Until now, only little research has been done about recent microorganisms that are associated with fresh resin and amber. Resinicolous fungi are found on resins and other exudates of vascular plants (Tibell and Titov 1995; Rikkinen 1999, 2003) and also bacteria can be observed growing into liquid resin (Schmidt and Dilcher 2007; Schmidt and Schäfer 2005). The first researcher who noticed modern microorganisms from fossil resin samples was Galippe (1920). He observed bacteria on Cretaceous and Cenozoic amber samples from various localities. Recently, a study by Girard et al. (2008a) emphasized the occurrence of recent bacteria and diatoms at surfaces and in fissures of mid-Cretaceous amber samples that were found at the sea shore and in freshwater lakes of southwestern France.

In the 1990s, several papers dealt with DNA enclosed in amber and from amber-preserved insect and plant inclusions (e.g., Cano et al. 1994; Poinar and Poinar 1994; Lambert et al. 1998; Greenblatt et al. 1999). However, critical reinvestigations suggest that the DNA obtained in these studies was modern contamination. Because even the amplification of DNA from copal failed, Austin et al. (1997) concluded that resin is not predestined for preservation of complex molecules such as DNA although morphological structures are recorded very well. Greenblatt et al. (2004) demonstrated that it is difficult or almost impossible to get the amber free of modern DNA and RNA since bacteria, fungi and other microbes remain in fissures even after surface sterilization by ultrasonic cleaning, H_2O_2 and ethanol as described by Lambert et al. (1998). Greenblatt et al. (1999, 2004) isolated putative fossil microorganisms from different ambers but characterisation of the supposed microfossils showed that they were actually much younger than the amber itself. Thereupon they suggested that the microorganisms exist as contaminants in amber cracks. The estimation of the exact date of these supposed microfossils seems to be impossible. Since non-degraded modern molecules are always preferred during the polymerase chain reaction, an amplification of fragments of ancient DNA is highly unlikely and today it is widely assumed that DNA does generally not persist more than thousands of years as amplifiable molecules (Hofreiter et al. 2001; Willerslev et al. 2004; Willerslev and Cooper 2005).

Investigating amber one has to consider that amber is an organic substance and thus potentially represents an attractive substrate by itself. Because of their habitat specificity, it has been suggested that resinicolous fungi use liquid resin as sole nutrient source. Thereby the fungal mycelium penetrates into upper regions of liquid resins (Tibell and Titov 1995; Rikkinen and Poinar 2000; Rikkinen et al. 2003). Greenblatt et al. (2004) postulate, that bacteria colonize small cracks of amber and possibly use the amber as a nutrient source.

These previous studies indicate that interactions of microorganisms with fresh and fossil resin may be very diverse and are sometimes poorly understood. We therefore examined microorganisms from inside and at the surface of various modern and fossil resins. In both, modern and fossil resins, we found microbes that (1) were enclosed by resin, (2) showed patterns of growth into the resin, and

(3) colonized the resin surfaces. It is the intent of this study to illustrate and to discuss these various associations of microbes and resins.

2 Materials and Methods

We investigated modern and fossil resins of various botanical sources that were colonized by microbial communities before and after solidification of the resin. The pieces of amber were ground and polished manually with a series of wet silicon carbide papers (grit from 600 to 4,000, firm Struers, Germany) to minimize light scattering for the investigation of the inclusions. The samples of modern and fossil resins and fragments thereof were examined under incident-light (Carl Zeiss Stemi 2000) and transmitted light microscopes (Carl Zeiss AxioScope A1 with long-distance objectives) equipped with Canon 450D digital cameras and under a scanning electron microscope (Carl Zeiss LEO 1530).

In laboratory experiments, fresh resin of *Cycas revoluta* and *Pinus strobus* was placed on a microscopic slide, coated with a cover slip and stored under humid conditions in a Petri dish for two weeks. Growth of microbes was observed daily.

For sequencing, electrophoresis was carried out on an ABI Prism 3100 Genetic Analyser. Sequences were assembled and edited in BioEdit version 7.0.8 (www.mbio.ncsu.edu/BioEdit/bioedit.html) and ARB version 4.0 (www.arb-home.de)

3 Modern Resin Investigated

3.1 Resin of Agathis lanceolata *Lindley ex Warburg and* Agathis ovata *(Moore) Warburg (Araucariaceae)*

The resin samples derive from southeastern New Caledonia and were collected by A.R.S. in 2005. These *Agathis* species have a natural massive resin production. Root-resin of *A. lanceolata* was found solidified at the forest floor. Solidified resin of *A. ovata* was collected at the base of the tree trunks. The samples are housed at the Geoscientific Museum of the University of Göttingen (GZG.BST 14975 to GZG.BST 14977).

3.2 Resin of Pinus elliottii *Engelmann (Pinaceae)*

The resin samples originate from "Dilcher's swamp forest", a swamp in a warm-temperate mixed forest east of the city of Gainesville in north central Florida. The resin samples were collected by A.R.S. and David L. Dilcher in 2006 in liquid and solid stage (for methods, see Schmidt and Dilcher 2007). The sample is housed at the Geoscientific Museum of the University of Göttingen (GZG.BST 14974).

3.3 Resin of **Prunus avium** *L. (Rosaceae)*

Resin was collected at a tree trunk near the city of Jena in central Germany. The sample is housed at the Geoscientific Museum of the University of Göttingen (GZG.BST 14973).

3.4 Resin of **Cycas revoluta** *Thunberg (Cycadaceae) and* **Pinus strobus** *L. (Pinaceae)*

Resin of potted plants was used for laboratory studies on the growth of bacteria, cyanobacteria and fungi in liquid resin.

4 Fossil Resin Investigated

4.1 Cretaceous Amber from Golling (Salzburg, Austria)

Amber from this famous locality is probably Hauterivian in age (for review, see Vávra 2005). Samples were kindly provided by Johann Peschl (Laufen) and are housed in the Museum für Naturkunde Berlin (MB. Pb 2009/344).

4.2 Cretaceous Amber from Myanmar (Burma)

Both, biostratigraphic data obtained from the amber-bearing sediment and the amber inclusions indicate a mid-Cretaceous age of the amber (Grimaldi et al. 2002; Cognato and Grimaldi 2009). Based on a recent re-investigation of the amber locality, Cruickshank and Ko (2003) suggest an Upper Albian age of the sediment. The samples are housed in the British Natural History Museum in London (BNHM In 20206).

4.3 Cretaceous Amber from Santander (Spain)

This Upper Albian amber was found north of the village of Ubiarco that is situated about 20 km west of Santander (Wilmsen 1997). The samples were kindly provided by Markus Wilmsen (Würzburg) and are housed in the Museum für Naturkunde

Berlin (MB. Pb 2009/345 and MB. Pb 2009/346) and in the Geoscientific Museum of the University of Göttingen (GZG.BST 14973).

4.4 *Cretaceous Amber from Kuji (Japan)*

This amber originates from the Late Cretaceous Tamagawa Formation near the city of Kuji and is probably Coniacian in age (see Krumbiegel and Krumbiegel 2005). The samples are housed in the British Natural History Museum in London (BNHM V.65206).

4.5 *Cretaceous Amber from the Cedar Lake (Canada)*

Cedar Lake amber is considered to be Upper Cretaceous (probably Campanian) in age (see McKellar et al. 2008). The samples were kindly provided by Ulf-Christian Bauer (Schliersee) and are housed in the Museum für Naturkunde Berlin (MB. Pb 2009/347).

4.6 *Eocene Baltic Amber from Russia*

The pieces of Baltic amber were found in the Kaliningrad (Königsberg) area on the eastern coast of the Baltic Sea (western Russia). The sediments containing the majority of Baltic amber in the Kaliningrad area are 47–38 million years old (Ritzkowski 1999, Standke 2008). One amber piece investigated was kindly loaned from Wolfgang Weitschat (Hamburg; collection number 1805 of the amber collection of the Institute of Geology and Palaeontology of the University of Hamburg, Germany). Eocene Baltic amber of the collection of the Geoscientific Museum of the University of Göttingen was also investigated (GZG.HST 19 and GZG.HST 21). These samples are part of the historic Blumenbach collection that was collected by Johann Friedrich Blumenbach in (1752–1840). These samples have been stored in the Göttingen collection for more than 200 years. During the whole storage period the amber was kept in the dark and exposed to low air humidity. Since 40 years the ambient conditions of the storage-room are kept at a constant level of 40–60% air humidity and 22°C.

5 Results and Discussion

Our studies revealed that microorganisms can be associated with plant resins in three different ways since (1) enclosure by resin, (2) growth into resins and (3) colonization of resin surfaces and fissures may take place.

5.1 Microorganisms Enclosed by Resin

Various limnetic and terrestrial microorganisms were found engulfed, i.e., entirely embedded in resins. Among others, representatives of conjugatophytes (Fig. 1a) and testate amoeba (testate lobose amoebae, order Arcellinida, Fig. 1b–c) were found inside resin flows of *Pinus elliottii* from a Florida swamp forest. Single celled

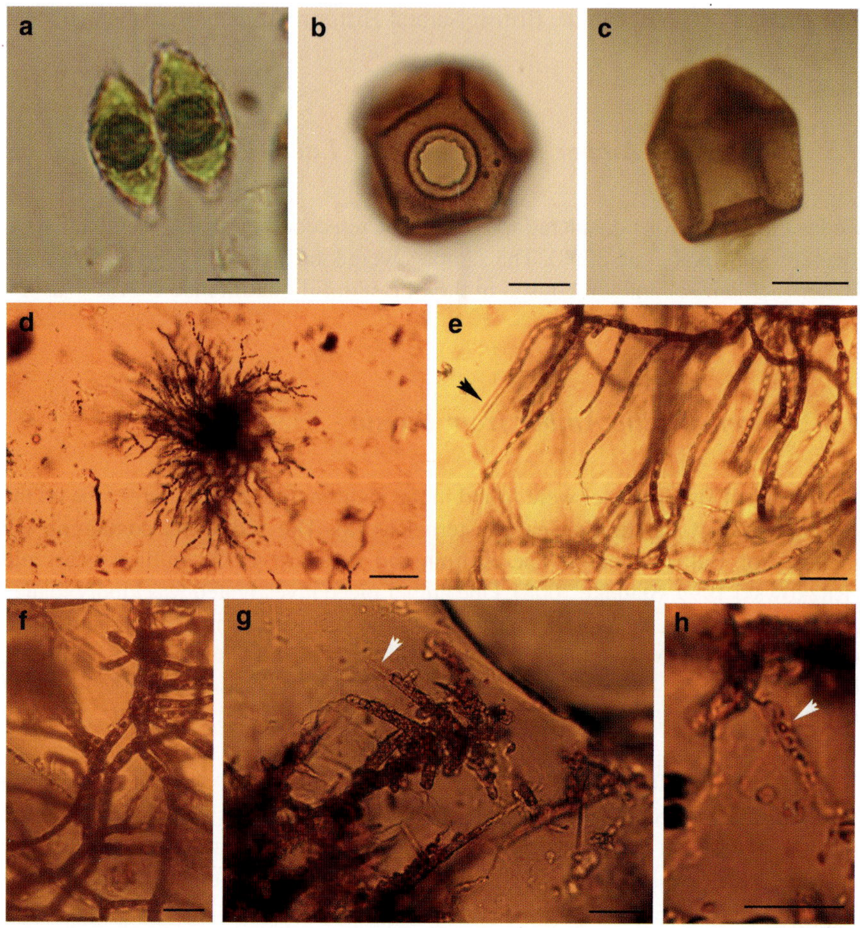

Fig. 1 Light-microscopical images of modern and fossil microorganisms that became engulfed by resin: (**a**) Conjugatophyte of the genus *Cosmarium* enclosed in modern resin from *Pinus elliottii*. (**b–c**) Testate amoeba of the genus *Arcella* in ventral (**b**) and lateral (**c**) view inside *P. elliottii* resin. (**d–f**) Epiphytic cyanobacteria enclosed in modern resin of *Prunus avium* (GZG.BST 14973). The filaments continued to grow for a short time after embedding. The *arrow* shows a narrow filament in which cross walls are absent. (**g–h**) Fossil cyanobacteria in Upper Cretaceous amber from Canada (MB.Pb.2009/347). The *arrow* in (**c**) indicates a narrow flament without cross walls; the *arrow* in (**h**) shows a regular filament. Scale bars represent 30 μm in (**a–c** and **e–h**), and 200 μm in (**d**)

limnetic organisms became enclosed, when resin flowed down the tree trunks and reached the humid forest floor. Some of these inclusions exhibit an excellent preservation of detailed structures such as chloroplasts (Fig. 1a). Schmidt and Dilcher (2007) found all major groups of limnic microorganisms of the swamp water embedded in resin of *Pinus elliottii*, a species that exhibits a massive natural resin production. Limnetic-terrestrial microorganisms have also repeatedly been reported from Cretaceous ambers. Aquatic bacteria, microalgae, conjugatophytes, ciliates and testate amoebae described from the Cenomanian Schliersee amber from the Bavarian Alps are examples hereof (see Schönborn et al. 1999; Dörfelt and Schäfer 2000; Schmidt et al. 2004).

Cyanobacteria are rarely enclosed in fossil resins. We found modern filamentous cyanobacteria inside resin at the trunk of a cherry tree (*Prunus avium*, Fig. 1d–f). After becoming entirely enclosed by the fresh resin, further growth of the filaments took place for a short time. During this limited growth, narrow filaments were produced in which cross walls were often absent. We discovered similar narrow filaments without cross walls in cyanobacterial mats inside Upper Cretaceous amber from the Cedar Lake in Canada (Fig. 1g–h). Therefore, we interpret these fossil cyanobacteria to be alive during embedding and a couple of hours to days afterwards in the liquid resin.

However, generally those entirely embedded organisms died off immediately after embedding and correspond to the classic term of inclusions since they do not touch the resin surface at all.

5.2 *Microorganisms Growing on and into the Liquid Resin*

We found many examples of modern and fossil fungal hyphae and filamentous bacteria that grew from the surface into the resin bodies until solidification occurred. Despite being preserved inside amber and modern resin, these inclusions do not correspond to the typical criteria of inclusions because they touch the surface at least at one point.

Sheathed bacteria occurred sometimes in fresh resin of *Pinus elliottii* that flowed into the ponds of a Florida swamp forest. Their sheaths are about 10 μm in diameter and exhibit a central lumen of about 1 μm diameter. Cell chains could sometimes not be identified inside the lumen. As in modern *Leptothrix* specimens, the observed sheaths are hyaline, slightly granulated and contain minute gas bubbles which might indicate metabolic activity after embedding in the resin (Fig. 2a–b). Spiral prokaryotic filaments and branched fungal hyphae were regularly found in *P. elliottii* resin that had contact to the swamp water (Fig. 2c–d).

Correspondingly, many Cretaceous amber pieces are interspersed with fossil bacterial filaments. We found filaments that are composed of rod-shaped cells (Fig. 3a–b), but mostly sheathed branched cell chains or empty sheaths that morphologically closely resemble modern representatives of the genus *Leptothrix* occurred in the ambers (Fig. 3c–i). The random orientation of the sheathed

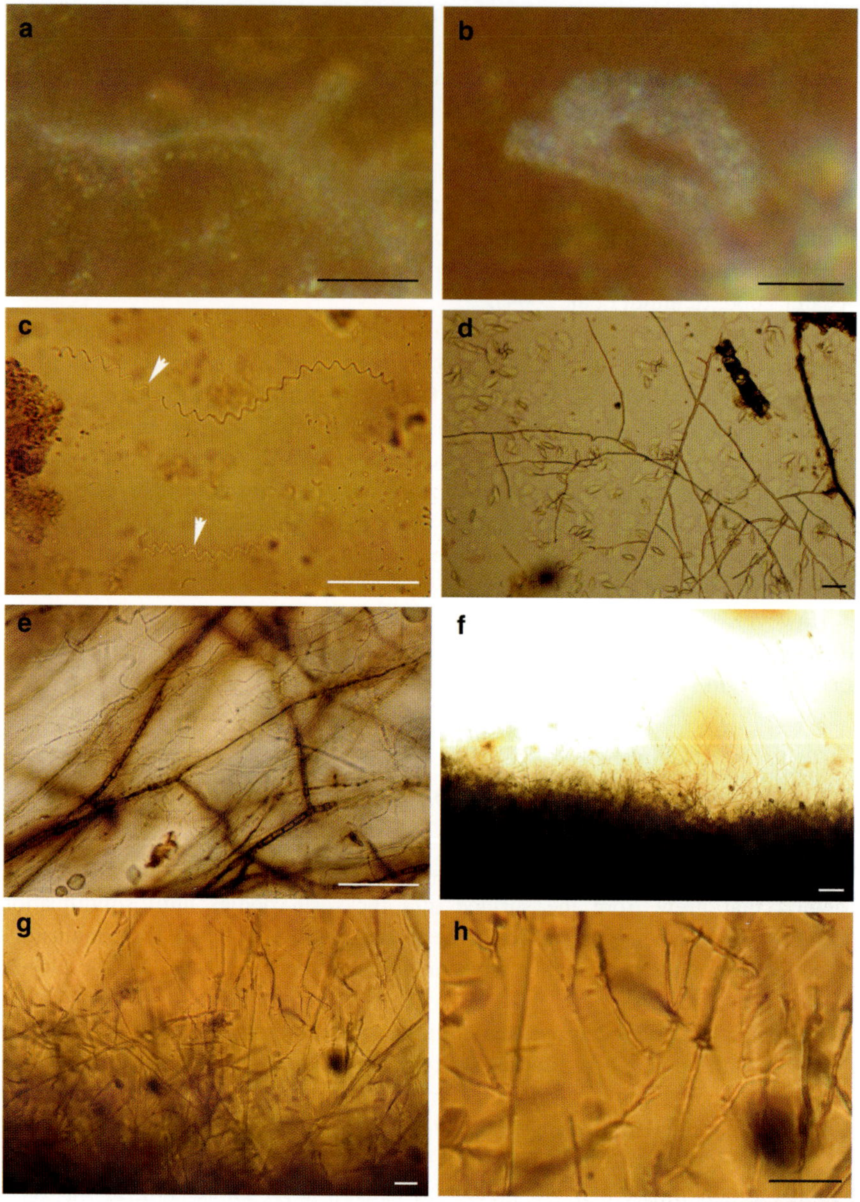

Fig. 2 Light-microscopical images of modern microorganisms that grew on and into liquid resin: (**a**–**b**) *Leptothrix*-like sheathed bacteria inside modern resin of *Pinus elliottii* (GZG.BST 14974). (**c**) Spiral prokaryotic filaments growing in resin of *Pinus elliottii*. (**d**) Fungal hyphae growing in resin of *Pinus elliottii*. (**e**) Fungal hyphae growing into liquid resin of *Cycas revoluta*. (**f**–**h**) Resinicolous fungi grown into modern resin of *Agathis lanceolata* in New Caledonia (GZG.BST 14975). Scale bars represent 10 μm in (**a**–**b**), 50 μm in (**c**–**e**), 100 μm in (**f**), and 20 μm in (**g**–**h**)

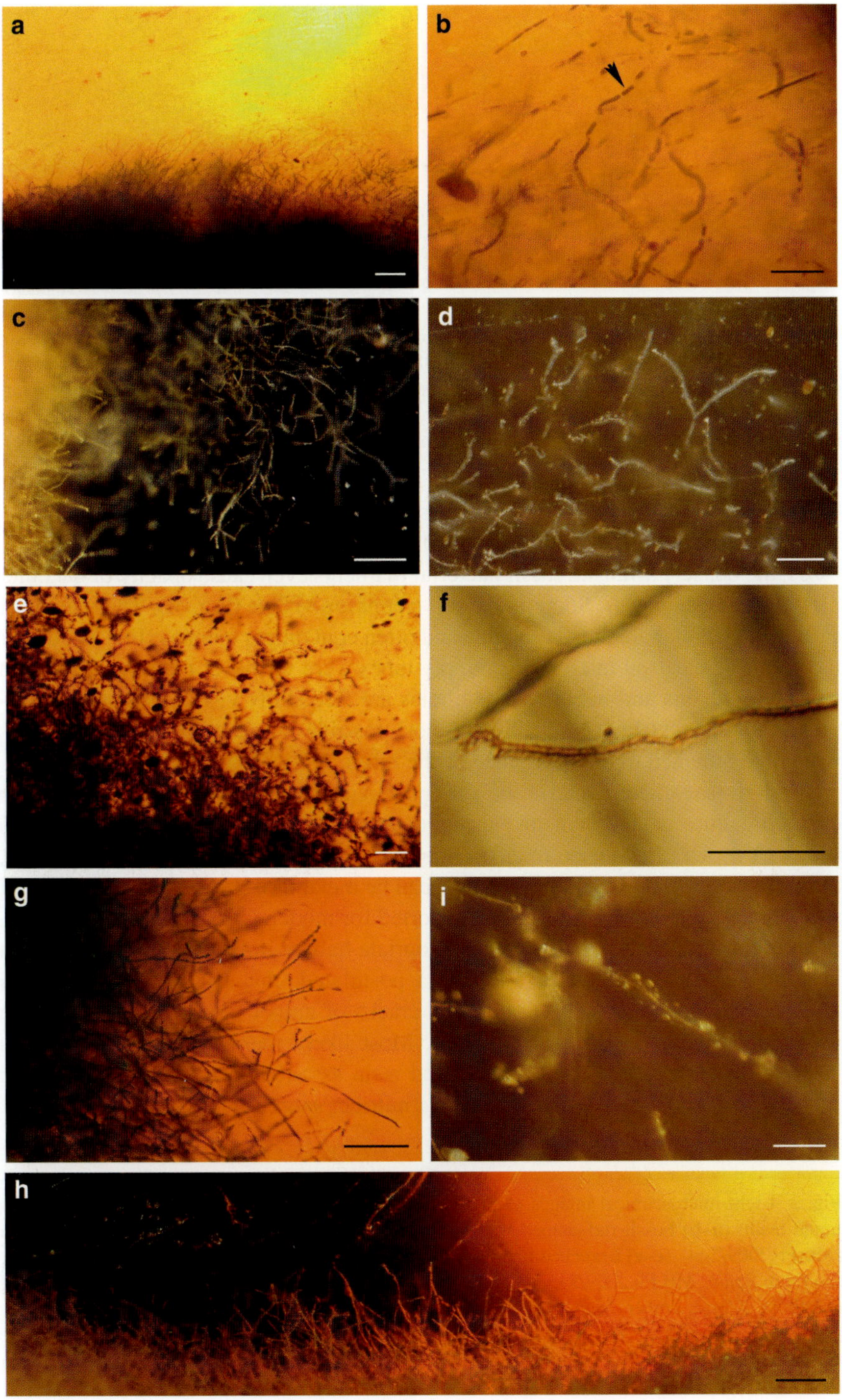
a
b
c
d
e
f
g
i
h

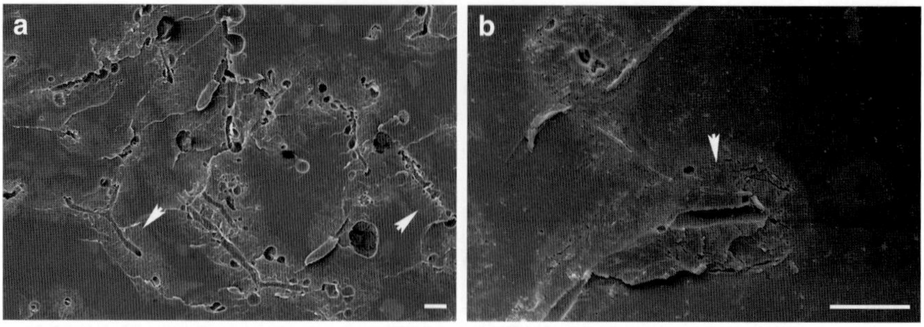

Fig. 4 Scanning electron microscope images of sheathed bacteria visible at a break surfaces of Albian amber from Santander (Spain, GZG.BST 14978): (**a**) Several branched filaments are exposed at the surface. The *arrows* indicate the surface of the fossilized sheaths that are interspersed with minute bubbles. (**b**) Detailed view. The tubular hole inside the sheath at the former location of the cell chain is well visible. Scale bars represent 5 μm

filaments and the absence of broken parts thereof suggest that these filaments grew from the surface to the inner parts of the formerly liquid resin bodies. In many cases, the densely arranged fossil bacterial filaments even cause opacity of the outer parts of the resin pieces. Sometimes, cell chains are visible (Fig. 3f). The sheaths of ca 10 μm diameter are mostly hyaline and often contain gas bubbles (Fig. 3i). Under the scanning electron microscope, empty tubes were visible at the former location of the cell chains and bubble-shaped holes are frequently visible inside the resin-preserved sheaths (Fig. 4a–b). Sometimes, possible remnants of the former cells were seen attached to the inner face of the tubes.

Modern sheathed bacteria of the genera *Leptothrix* and *Sphaerotilus* are chemoorganotrophic organisms, using a large number of carbon and nitrogen sources by tolerating a wide range of nutrient concentrations. Many carbohydrates, alcohols, organic acids and amino acids can be metabolized. Also the utilization of succinate as sole nutrient source was reported for modern sheathed bacteria. Resin, as an organic material, might be an appropriate habitat in liquid and possibly even in solid stage. Greenblatt et al. (2004) postulated, that bacteria of the genus *Micrococcus* may survive long times in cracks of amber pieces and possibly use amber compounds as a nutrient source. Most likely the Cretaceous filamentous bacteria that are preserved in amber used resin as a temporary habitat and could probably use compounds of the resin as a nutrient source.

Fig. 3 Light-microscopical images of fossil microorganisms that have grown into the formerly liquid resin: (**a–b**) Fossil rod-shaped bacteria in Lower Cretaceous Golling amber from Austria (MB.Pb.2009/344). (**c–e**) Branched filaments of sheathed bacteria in Albian amber from Santander (northern Spain; MB.Pb.2009/345 and 346). (**f**) Bacterial cell chain with only slightly developed sheath in Albian amber from Santander (GZG.BST 14979). (**g–h**) Sheathed bacteria Upper Cretaceous amber from Kuji (Japan; BMNH V.65206). Minute bubbles are sometimes visible inside the sheaths (**i**). Scale bars represent 500 μm (**a** and **h**), 50 μm (**b–c**), 100 μm (**g**), and 20 μm (**d**, **f** and **i**)

We assume that these filamentous microorganisms came into contact to the resin flows at the forest floor and used this temporary microhabitat as long as possible. Peñalver et al. (2007) described similar filamentous structures from Cretaceous amber from San Just (Spain) as "alteration crust". These authors interpret these structures as non-biological inclusions and suggest degradation structures caused by weathering of the amber pieces. According to our results, most "crusts" of Cretaceous ambers are not primarily caused by weathering but by colonization of liquid resin by filamentous microorganisms. It is, however, likely that numerous micrometer-sized tubes (see Fig. 4) that remain in the solidified resin promote subsequent physical-chemical weathering of the resin's outer parts.

Our experiments with living taxa demonstrated an active microbial colonization of liquid resin of *Cycas revoluta* and *Pinus strobus* (Fig. 2e). Growth of filaments could be observed after two or three days. Both, bacteria and fungi grew into the resins as long as it was liquid. The filaments stopped growing when the resin solidified. Interestingly, the *Cycas*-resin became darker around the hyphae (Fig. 2e) which might indicate some kind of interactions between compounds of the resin and fungal mycelia due to metabolic activity.

We found abundant fungal mycelia on recently solidified extensive resin flows of *Agathis lanceolata* from New Caledonia. In the humid forests of southeastern New Caledonia the resin flows were mostly completely overgrown by mycelia, causing a grey opaque surface of the translucent resin bodies. The hyphae did not exclusively grow at the resin surface but also penetrated the resin up to ca 3 mm deep which caused an opaque and fragile surface layer. These fungal mycelia are composed of very thin hyphae of only 1.5–3 μm diameter that are densely arranged, irregular branched and sparsely septated (Fig. 2f–h). The numerous minute tubes caused by the fungal growth are shown in the SEM images of Fig. 5a, b. Around some of these tubes, an indication of a sheath appears (see Fig. 5b) that is not visible under the transmitted-light microscope. Possibly, this structure is caused by alteration of the surrounding resin due to metabolic activity of the hyphae. These resinicolous fungi from New Caledonia are probably representatives of the Mycocaliciaceae, a group of ascomycetes whose representatives exclusively grow on resin and other exudates of vascular plants. Representatives of the genera *Chaenothecopsis*, *Mycocalicium*, *Phaeocalicium* and *Stenocybe* have been described to be resinicolous (see Tibell and Titov 1995). Rikkinen (1999, 2003) described several new recent species from temperate rain forests of western North America and one extinct species, *Chaenothecopsis bitterfeldensis*, from 24 million-year-old amber from Germany (Rikkinen and Poinar 2000). Because of their habitat specificity, it has been suggested that these fungi use liquid resin as sole nutrient source (see Rikkinen and Poinar 2000).

Also many fossil resins contain fungal hyphae which have grown into the formerly liquid resin. Mycelia often cover inclusions of arthropods. Fungi that were probably attached to the animals when embedded continued growing until the resin solidified (Fig. 6a–b). Many amber pieces contain fungal hyphae that are densely arranged particularly in outer parts. Correspondingly to filamentous bacteria, fungal growth occurred from the surface to inner parts of the amber and the

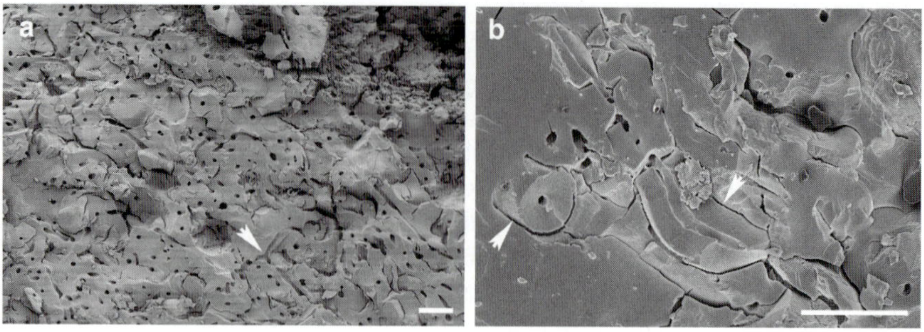

Fig. 5 Scanning electron microscope images of resinicolous fungi grown into modern resin of *Agathis lanceolata* in New Caledonia (GZG.BST 14976): (**a**) Break surface of hardened resin interspersed by minute tubes that resulted by hyphal growth. The *arrow* indicates a branching. (**b**) Detail of the filamentous structures. The *arrows* indicate the sheath-like structure around the hyphae. Scale bars represent 10 μm

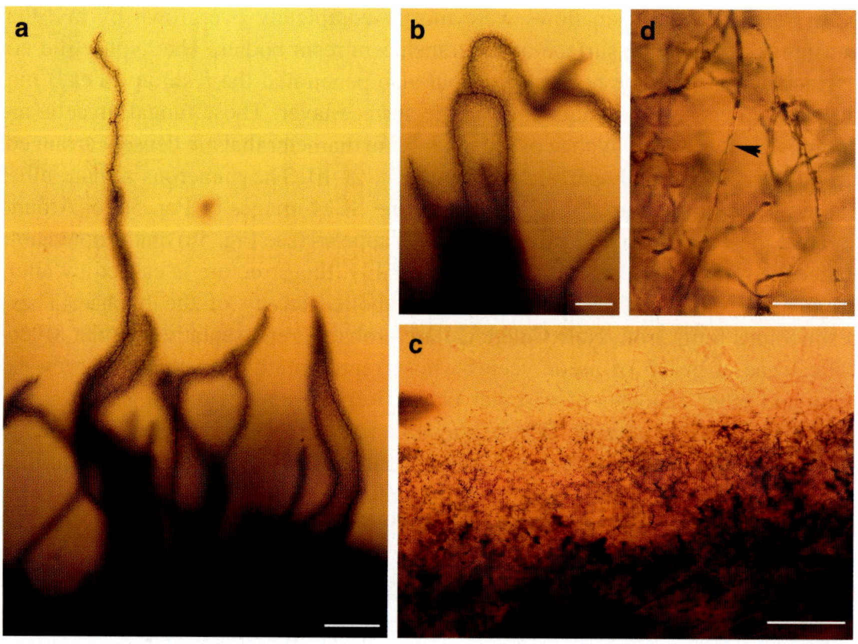

Fig. 6 Light microscopical images of fossil fungi that have grown into the formerly liquid resin: (**a–b**) Branched hyphae at the surface of an insect enclosed in Baltic amber (amber collection of the Institute of Geology and Palaeontology of the University of Hamburg, coll. no. 1805). Note the darkened amber around the filaments that resemble the structures shown in Fig. 2e from modern resin. (**c–d**) Mycelium in Cretaceous Burmese amber (BMNH In 20206). The *arrow* indicates collapsed cells. Scale bars represent 50 μm in (**a**, **b** and **d**), and 500 μm in (**c**)

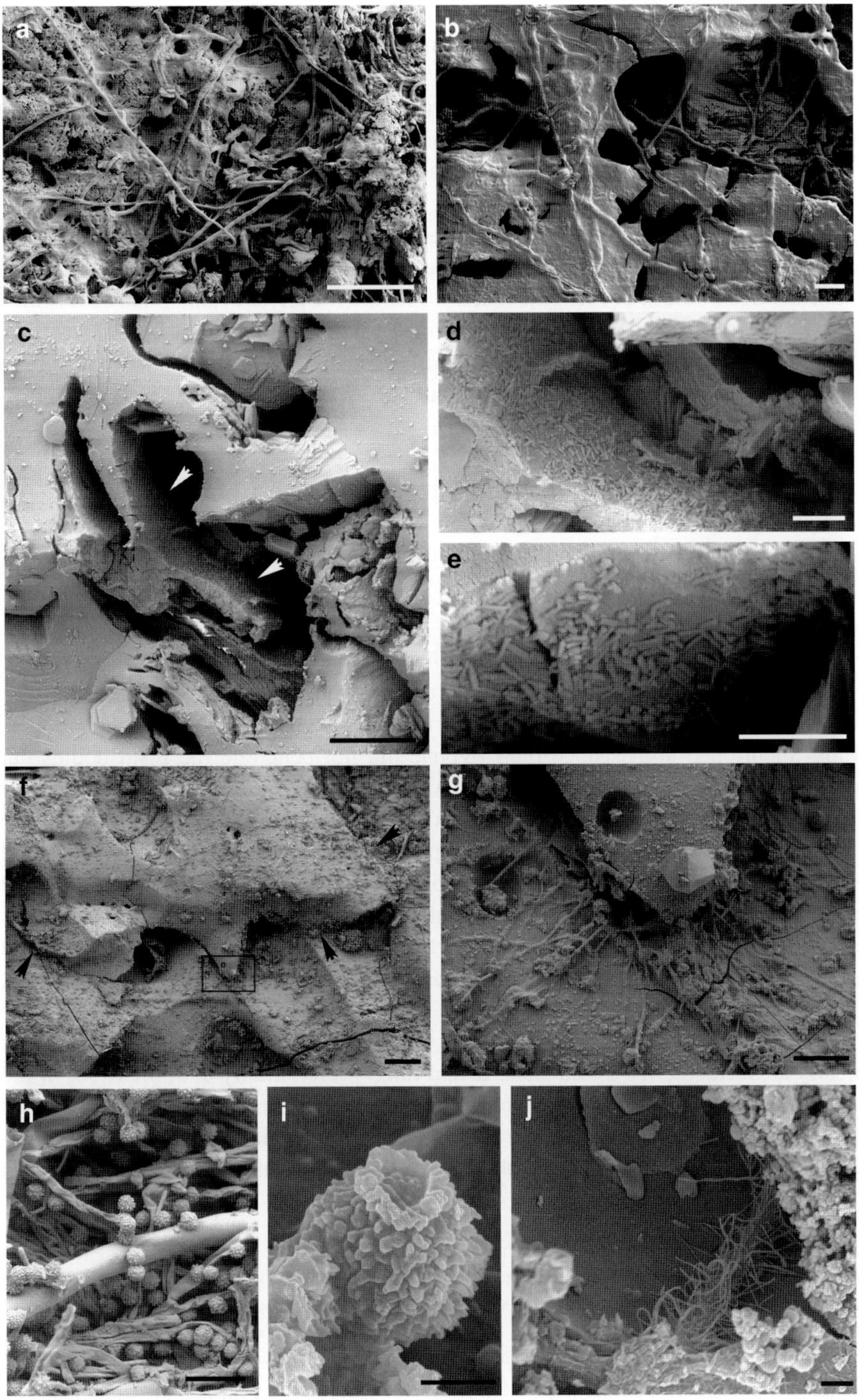

Fig. 7 Scanning electron microscope images of modern bacteria and fungi that colonize surfaces of solidified modern and fossil resins: (**a**–**b**) Hyphae at solidified resin of *Agathis ovata* (GZG.BST

hyphae are orientated in many directions (Fig. 6c–d). These fungal hyphae also derived from the ancient environment and most likely came into contact with the resin at the forest floor.

We never found cyanobacteria growing into the resin. In contrast to the filiform bacteria illustrated and discussed above, these phototrophic organisms probably did not continue to grow extensively after being embedded in the liquid resin. Current reports of abundant cyanobacteria from French Cretaceous ambers (Girard et al. 2009a) are questionable since these results are mainly based on the putative detection of phycocyanin in these resins. These authors used a probe detecting fluorescence of the wave lengths that are specific for phycocyanian. However, they did not outline how they exclude measuring of the fluorescence of the amber itself. The fact that they often got higher signals in those parts of the amber that did not contain microinclusions confirms our doubts.

5.3 Microorganisms Colonizing Surfaces of Solidified Resins

Investigating solidified modern resins in the field and ambers stored in collections, we found various examples of microbial colonization, too.

The New Caledonian resin samples of *Agathis lanceolata* and *Agathis ovata* that solidified at the forest floor or close to it at the tree trunk exhibited surfaces that were often covered by associations of bacteria, fungal mycelia and algae (Fig. 7a–b). These mycelia were distinguished from those of the resinicolous fungi mentioned above since their hyphae did not penetrate the resin bodies but overgrew their surfaces and bordering microhabitats.

Both, bacteria and fungi occurred on surfaces of Baltic amber samples of the Blumenbach collection. Hereby, fungal growth was predominantly observed. Some of these samples were strongly covered by fungal mycelia that sometimes plentiful sporulated (Fig. 7g–j). Molecular determination of the isolates revealed affiliations to the genus *Aspergillus*. Associated bacteria were mostly rod-shaped and showed affiliations to various classes, such as Betaproteobacteria, Gammaproteobacteria, Bacilli and Flavobacteria (Fig. 7c–e).

It is well known that microorganisms, particularly in form of biofilms, can cover nearly all surfaces. We found bacteria and fungi preferentially in impressions of the amber surface and in fissures. This accumulation might be caused by improved growing conditions because of the presence of inorganic and organic matter including nutrients that accumulated in deeper regions and fissures. Whether amber as an

Fig. 7 (continued) 14977), (**c–e**) Rod-shaped bacteria accumulated in deeper regions and fissures of surfaces of Baltic amber samples of the Blumenbach collection (GZG.HST 21). (**f–g**) Hyphae in cracks and minute fissures of Baltic amber samples of the Blumenbach collection (GZG.HST 19). (**h**) Sporulating *Aspergillus* sp. isolated from surface of Baltic amber of the Blumenbach collection. (**i**) Single spore of *Aspergillus* sp. (**h**) Part of a mycelium at the surface of Baltic amber of the Blumenbach collection (GZG.HST 19). Scale bars represent 50 μm in (**a** and **c**), 10 μm in (**b, d–e** and **g–h**), and 1 μm in (**i–j**) organic material can be utilized as a nutrient source for microorganisms has not been examined so far, but it is conceivable. Due to the microbial activity, recent colonization may accelerate the degradation of the amber objects. The degradation of amber is probably caused by various chemical, physical and biological processes that may act together and sometimes trigger or enforce each other. Processes of oxidation are probably most important for the degradation of amber. However, when fungal mycelia penetrate into present fissures and extent their mycelia, physical forces arise that can break up upper amber layers, thus opening new areas for oxidation processes. So far, it is unclear if also secretion of metabolic compounds can represent a damaging factor and if the microbes may utilize amber compounds as a nutrient source. This should be subject of further studies.

Usually, recent biofilms are easily distinguishable from fossil inclusions using high-magnification light-microscopy. Also, microorganisms that are situated inside fissures that extend into the amber are usually distrainable from microinclusions when tracing the refractive micrometre-sized fissures under the microscope. During the study of mid-Cretaceous amber of south-western France, Girard et al. (2008a) discovered living marine and freshwater diatoms at the amber surface and in micrometre-sized fissures. Furthermore, these authors found fungal mycelia, bacteria and cyanobacteria in fissures inside the amber samples, showing that not exclusively ambers in collections represent habitats of microorganisms.

Acknowledgements We thank Jérôme Munzinger (Nouméa) for supporting the field work in New Caledonia and David L. Dilcher (Gainesville) for making the field studies in Florida possible. We are indebted to Mike Reich (Göttingen) for amber loans from the Blumenbach collection of Baltic amber, to Wolfgang Weitschat (Hamburg) for loans of Baltic amber, to Johann Peschl (Laufen) for providing Cretaceous amber from Austria, to Markus Wilmsen (Würzburg) for Cretaceous amber samples from Spain, and to Ulf-Christian Bauer (Schliersee) for donating samples of Cedar Lake amber. Hilary Ketchum and Claire Mellish (London) kindly provided access to the amber collection of the Natural History Museum in London. We acknowledge with thanks the help of Dorothea Hause-Reitner and Andreas Reimer (Göttingen) during the SEM analyses and the help of Kerstin Schmidt (Jena) in the preparation of the manuscript. We heartily thank Joachim Reitner (Göttingen) for technical support and for his kind invitation to contribute to this special issue. This is contribution number 39 from the Courant Research Centre Geobiology that is funded by the German Initiative of Excellence.

References

Austin J, Ross AJ, Smith AB, Fortey RA, Thomas RH (1997) Problems of reproducibility – does geologically ancient DNA survive in amber-preserved insects? Proceedings Royal Society London, Series B Biological Science 264:467–621

Berkeley MJ (1848) On three species of mould detected by Dr. Thomas in the amber of East Prussia. Annual Magacine Natural History (Series 2) 2:380–383

Blunck G (1929) Bakterieneinschlüsse im Bernstein. Zentralblatt Mineralogie, Geologie, Paläntologie 11:554–555

Cano RJ, Borucki MK, Higby-Schweizer M, Poinar HN, Poinar GO Jr, Pollard KJ (1994) *Bacillus*-DNA in fossil bees: an ancient symbiosis? Applied and Environmental Microbiology 60:2164–2167

Caspary R, Klebs R (1907) Die Flora des Bernsteins und anderer fossiler Harze des ostpreußischen Tertiärs. Abhandlungen der Königlich-Preußischen Geologischen Landesanstalt 4:1–181
Cognato AI, Grimaldi DA (2009) 100 million years of morphological conservation in bark beetles (Coleoptera: Curculionidae: Scolytinae). Systematic Entomology 34:93–100
Cruickshank RD, Ko K (2003) Geology of an amber locality in the Hukawng valley, Northern Myanmar. Jouranl of Asian Earth Scieces 21:441–445
Dörfelt H, Schäfer U (2000) *Palaeozygnema spiralis*, ein Vertreter der Conjugatophyceae in mesozoischem Bernstein aus Bayern. Hoppea 61:785–793
Dörfelt H, Schmidt AR (2007) A conifer seedling with two herbicolous fungi from the Baltic amber forest. Botanical Journal of the Linnean Society 155:449–456
Galippe V (1920) Recherches sur la résistance des microzimas àlàction du temps er sur survivance dans làmbre. Comptes rendus hebdomadaires des séances de l'Académie des sciences 170:856–858
Girard V, Néraudeau D, Breton G (2008a) Contamination of amber samples by recent micro-organisms and remediation evidenced by Mid-Cretaceous amber of Charentes (SW France). Geomicrobiological Journal 26:21–20
Girard V, Schmidt AR, Saint Martin S, Struwe S, Perrichot V, Saint Martin JP, Grosheny D, Breton G, Néraudeaua D (2008b) Evidence for marine microfossils from amber. Proceedings of the National Academy of Sciences of the United States of America 105:17426–17429
Girard V, Breton G, Brient L, Néraudeau D (2009a) Sheathed prokaryotic filaments, major components of Mid-Cretaceous French amber microcoenoses. Journal of Paleolimnology 42:437–447
Girard V, Schmidt AR, Struwe S, Perrichot V, Breton G, Néraudeau D (2009b) Taphonomy and palaeoecology of mid-Cretaceous amber-preserved microorganisms from southwestern France. In: Perrichot V, Néraudeau D (eds) Studies on mid-Cretaceous amber from southwestern France. Geodiversitas 31:153–162
Greenblatt CL, Davis A, Clement BG, Kitts CL, Cox T, Cano RJ (1999) Diversity of microorganisms isolated from amber. Microbial Ecology 38:58–68
Greenblatt CL, Baum J, Klein BY, Nachson S, Cano RJ (2004) *Micrococcus luteus* – survival in amber. Microbial Ecology 48:120–127
Grimaldi DA, Engel MS, Nascimbene PC (2002) – Fossifiliferous Cretaceous amber from Myanmar (Burma): its rediscovery, biotic diversity and paleontological significance. American Museum Novitates 3361:1–72
Hofreiter M, Serre D, Poinar HN, Kuch M, Pääbo S (2001) Ancient DNA. Nature Reviews Genetics 2:353–359
Krumbiegel G, Krumbiegel B (2005) Bernstein – Fossile Harze aus aller Welt. Goldschneck Verlag, Wiebelsheim
Lambert LH, Cox T, Mitchell K, Rosselló-Mora, RA, Del Cueto C, Doge DE, Orkland P, Cano RJ (1998) *Staphylococcus succinus* sp. nov., isolated from Dominican amber. International Journal Systemaic Bacteriology 48:511–518
McKellar RC, Wolfe AP, Tappert R, Muehlenbachs K (2008) Correlation of Grassy Lake and Cedar Lake ambers using infrared spectroscopy, stable isotopes and palaeoentomology. Canadian Journal of Earth Sciences 45:1061–1082
Peñalver E, Delclòs X, Soriano C (2007) A new rich amber outcrop with palaeobiological inclusions in the Lower Cretaceous of Spain. Cretaceous Research 28:791–802
Poinar GO, Poinar R (1994) The quest for life in amber. Addison-Wesley Publishing Company, Reading, MA, 219 pp
Rikkinen J (1999) Two new species of resinicolous *Chaenothecopsis* (Mycocaliciaceae) from western North America. Bryologist 102:366–369
Rikkinen J (2003) *Chaenothecopsis nigripunctata*, a remarkable new species of resinicolous Mycocaliciaceae from western North America. Mycologia 95:98–103
Rikkinen J, Poinar GO Jr (2000) A new species of resinicolous *Chaenothecopsis* (Mycocaliciaceae, Ascomycota) from 20 million year old Bitterfeld amber, with remarks on the biology of resinicolous fungi. Mycological Research 104:7–15

Rikkinen J, Dörfeld H, Schmidt AR, Wunderlich J (2003) Sooty moulds from European Tertiary amber, with notes on the systematic position of Rosaria ("Cyanobacteria"). Mycological Research 107:251–256

Ritzkowski S (1999) Das geologische Alter der bernsteinführenden Sedimente in Sambia (Bezirk Kaliningrad), bei Bitterfeld (Sachsen-Anhalt) und bei Helmstedt (SE-Niedersachsen). In: Kosmowska-Ceranowicz B, Paner H (eds) Investigations into amber. Proceedings of the International Interdisciplinary Symposium: Baltic amber and other fossil resins, 2–6 September 1997:33–40, Gdansk, Poland

Schmidt AR, Dilcher DL (2007) Aquatic organisms as amber inclusions and examples from a modern swamp forest. Proceedings of the National Academy of Sciences of the United States of America 104:16581–16585

Schmidt AR, Schäfer U (2005) *Leptotrichites resinatus* new genus and species: a fossil sheathed bacterium in alpine Cretaceous amber. Journal of Paleontology 79:175–184

Schmidt AR, Schönborn W, Schäfer U (2004) Diverse fossil amoebae in German Mesozoic amber. Palaeontology 47:185–197

Schmidt AR, Ragazzi E, Coppellotti O, Roghi G (2006) A microworld in Triassic amber. Nature 444:835

Schönborn W, Dörfelt H, Foissner W, Krienitz L, Schäfer U (1999) A fossilized microcenosis in Triassic amber. Journal of Eukaryotic Microbiology 46:571–584

Standke G (2008) Bitterfelder Bernstein gleich Baltischer Bernstein? – Eine geologische Raum-Zeit-Betrachtung und genetische Schlußfolgerungen. In: Rascher J, Wimmer R, Krumbiegel G, Schmiedel S (eds) Bitterfelder Bernstein versus Baltischer Bernstein – Hypothesen, Fakten, Fragen. Exkursionsführer und Veröffentlichungen der Deutschen Gesellschaft für Geowissenschaften 236:11–33

Tibell L, Titov A (1995) Species of *Chaenothecopsis* and *Mycocalicium* (Caliciales) on exudate. Bryologist 98:550–560

Ting WS, Nissenbaum A (1986) Fungi in Lower Cretaceous amber from Israel. Special Publication by the Exploration and Development Research Center, Chinese Petroleum Corporation, Miaoli, Taiwan

Vávra N (2005) Bernstein und verwandte organische Minerale aus Österreich. Beiträge zur Paläontologie von Österreich 29:255–280

Waggoner BM (1993) Fossil actinomycetes and other bacteria in Eocene amber from Washington State, USA. Tertiary Research 14:155–160

Waggoner BM (1994) Fossil microorganisms from Upper Cretaceous amber of Mississippi. Review of Palaeobotany and Palynology 80:75–84

Waggoner BM (1996) Bacteria and protists from Middle Cretaceous amber of Ellsworth County, Kansas. PaleoBios 17:20–26

Willerslev E, Cooper A (2005) Ancient DNA. Proceedings Royal Society London, Series B Biological Science 272:3–16

Willerslev E, Hansen AJ, Poinar HN (2004) Isolation of nucleic acids and cultures from fossil ice and permafrost. Trends in Ecology and Evolution 19:141–147

Wilmsen M (1997) Das Oberalb und Cenoman im Nordkantabrischen Becken (Provinz Kantabrien, Nordspanien): Faziesentwicklung, Bio- und Sequenzstratigraphie. Berliner Geowissenschaftliche Abhandlungen E 23:1–167

Tolypammina gregaria Wendt 1969-*Frutexites* Assemblage and Ferromanganese Crusts: A Coupled Nutrient-Metal Interplay in the Carnian Sedimentary Condensed Record of Hallstatt Facies (Austria)

Marta Rodríguez-Martínez, Christine Heim, Klaus Simon, Thomas Zilla, and Joachim Reitner

1 Introduction

Foraminifera are the most abundant sessile organisms found on ferromanganese crusts and nodules (Greenslate 1974; Wendt 1974; Dugolinsky et al. 1977; Riemann 1983; von Stackelberg 1984; Mullineaux 1987, 1988; Verlaan 1992; Resig and Glenn 1997; Toscano and Raspini 2005). In fact, actual ferromanganese crusts and nodules share numerous similarities with their fossil counterparts. The oldest examples are described from the Devonian (Tucker 1973), and also occur widely distributed through the condensed Triassic and Lower Jurassic deposits (summarized in Jiménez-Millán and Nieto 2006) until Plio-Pleistocene records (Jenkyns 1970; Wendt 1970; Allouc 1990; Toscano and Raspini 2005).

Ferromanganese crusts and nodules are widely distributed in actual oceans, on sediment starved seafloors, slopes and seamounts. The exact mechanisms of their formation are not well understood yet, therefore abiotic, biotic as well as coupled abiotic–biotic processes are proposed. From an abiogenic point of view, Bonatti (1981) suggested four genetic types: hydrogenous, hydrothermal, diagenetic or halmyrolitic origin. On the other hand, their biogenic contribution has been linked mainly to Bacteria (Ehrlich 1963, 1968, 1971; Ghiorse and Ehrlich 1992; Wang et al. 2009) and Foraminifera. In fact, the relationship between foraminifers and their biogenic influence in the ferromanganese accretion has been widely discussed (Dugolinsky et al. 1977; Riemann 1983; Mullineaux 1987; Wendt 1974; Verlaan 1992). The most abundant foraminifers reported from ferromanganese nodules and

M. Rodríguez-Martínez (✉)
Departamento de Geología, Universidad de Alcalá, Edificio de Ciencias, N-II Km 33.6, 28871, Alcalá de Henares, Madrid, España
e-mail: marta.rodriguezm@uah.es

C. Heim, K. Simon, T. Zilla, and J. Reitner
Geoscience Centre, University of Göttingen, Goldschmidtstraße 3, 37077, Göttingen, Germany

J. Reitner et al., *Advances in Stromatolite Geobiology*, Lecture Notes in Earth Sciences 131, DOI 10.1007/978-3-642-10415-2_25, crusts by different authors are Sarcodina members and several *Tolypammina* species (Resig and Glenn 1997).

Tolypammina gregaria Wendt is an agglutinated sessile foraminifer with a surprising behaviour, forming vertical structures which were firstly described by Wendt (1969), from Carnian sedimentary condensed deposits at Feuerkogel (Austria). The vertical accreted, pillar-like structures are buildup by *Tolypammina gregaria* Wendt and *Frutexites*. *Frutexites* (Maslov 1960) is a problematic Fe-rich microfossil with shrub, colloform to microstromatolitic morphology, figured as "*Frutexites*" from Proterozoic to Recent records [Caldera lake stromatolites at Tonga, by Kazmierczak and Kempe (2006)]. The genus *Frutexites* was coined by Maslov (1960) from Ordovician stromatolites of Siberian Platform. This problematic structure has been interpreted as abiotic as well as a biologically induced precipitate.

The *Tolypammina gregaria-Frutexites* vertical structures are associated with ferromanganese crusts and surfaces. This study aims to unravel the environmental significance of *Tolypammina gregaria-Frutexites* assemblage and their interplay with ferromanganese crusts/surfaces. Therefore detailed investigations of the microfacies as well as trace and rare earth elements (REE) analysis were performed in order to be able to reconstruct the environmental conditions during deposition of the limestone and the *Tolypammina-Frutexites* assemblages and their relationship with the so called Carnian Pluvial Event.

2 Materials and Methods

After detailed field observations of the stratigraphic succession and facies changes, 15 samples ex situ were carefully selected from the type locality area for microfacies and geochemical analysis. Polished slabs, large (10 × 15 cm) and standard petrographic thin-sections were made to carry out investigation with petrographic and cathodoluminescence (CL) microscopy.

Subsequent to the microfacial studies, three samples were selected for geochemical analysis. Laser-ICP-MS measurements of trace elements and REE were performed with a Perkin Elmer SCIEX Elan DRII with a coupled laser from Lambda Physik (190 mJ, 30 kV, 10 Hz, Laser Ø 120 μm).

3 Geological Setting

Tolypammina gregaria Wendt-rich bed with ferromanganese crusts is located in the Feuerkogel peak, close to Kainisch village at Steiermark, Austria (Fig. 1). The bed, from 2 cm up to 20 cm of maximum thickness, is part of the so-called Hallstatt Limestone facies of the Northern Calcareous Alps (NCA in Fig. 1).

The Hallstatt Limestone shows two characteristic facies: proximal grey and distal red facies where condensed ammonoid beds are common. The Feuerkogel

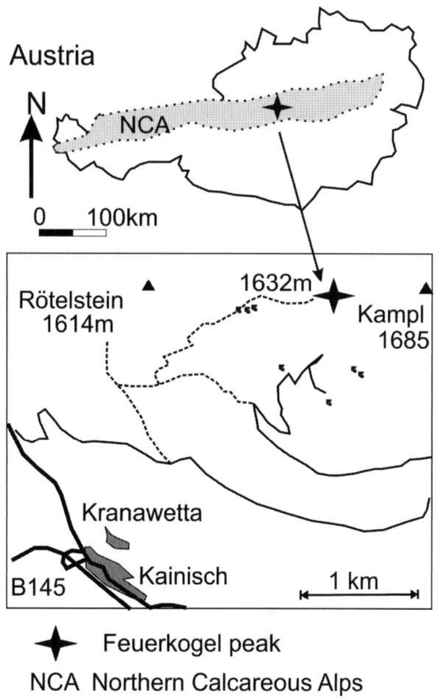

Fig. 1 Location of Feuerkogel condensed deposits, modified from Krystyn (1991)

Tolypammina-rich bed with ferromanganese crusts is interbeded between two ammonoid horizons: the "*Ellipticus* lens" sensu von Mojsisovics (1873–1902) and the "*Austriacus* lens" sensu Krystyn (1991). Thus *Tolypammina*-rich bed was deposited during the Carnian interval (top of Julian I according Fig. 17 from Krystyn op cit).

The Northern Calcareous Alps are the uppermost tectonic unit of the Eastern Alps, and form part of the Austroalpine nappe complex (Mandl 2000). The NCA is a complex E-W-trending nappe system which can be subdivided into three major tectonostratigraphic units, the Bajuvaric, the Tirolic and the Juvavic.

The NCA together with the Southern Alps and Dinarides were part of a wide epicontinental shelf at the passive continental margin of the NW Tethys during Late Triassic times. The paleogeographical reconstruction of NW Tethys by Stampfli and Borel (2002) shows two separate oceanic basins during the Carnian, the Hallstatt-Meliata Ocean in the north and the Palaeotethys in the south.

Most recent lithostratigraphic works focused on Hallstatt Limestones (Fig. 2a) have been summarized in Mandl (2000), Gawlick (2000), and Hornung and Brandner (2005). The Hallstatt Limestones, part of the Juvavic tectonostratigrapic unit, represent the sedimentary record of the pelagic realm of the so-called Hallstatt-Meliata Ocean that was deposited on salt-diapiric-induced structures (Schlager 1969). Thus, during Carnian times the NW Tethys margin was formed by the terrigenous Keuper facies at the hinterland, a wide carbonate shelf (Wetterstein Fm and Reifling Fm) which pass through the south to the basinal facies (Hallsttat Limestone and radiolarite facies Fig. 2b).

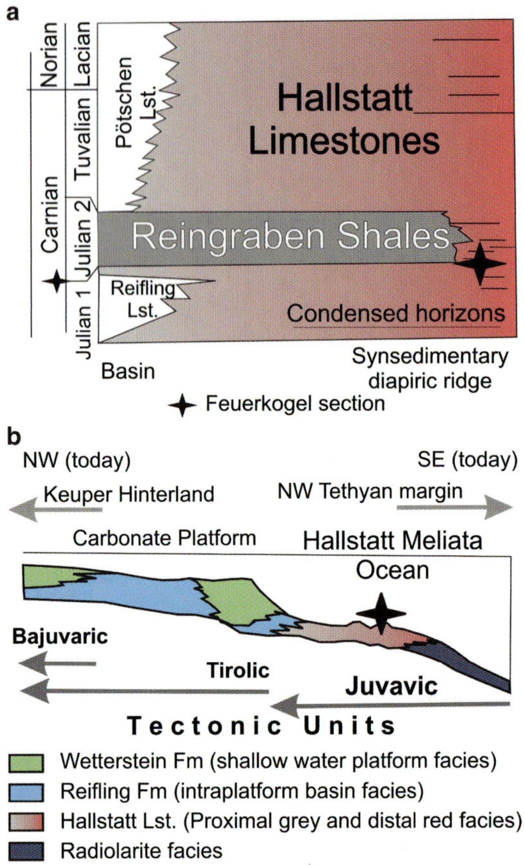

Fig. 2 (**a**) Main Carnian lithological units. Start symbol points out the lithostratigraphic situation of Feurkogel condensed deposits. (**b**) Facies model distribution of NW Tethyan margin. (**a** and **b**) modified from Hornung and Brandner (1991)

4 Previous Work

Wendt (1969) was the first author who studied in detail the *Tolypammina gregaria*-rich bed with ferromanganese crusts. He described the layer as a dense succession of dense hardgrounds colonized by numerous small, columns or pillar-like "reefs" built up by sessile foraminifers. The 95% of these pillars-like "reefs" were formed by *Tolypammina gregaria* Wendt 1969, which was also recognized by the author in other omission surfaces developed in Hallstatt facies. The rest of the subordinate foraminifer assemblage was classified as *Planiinvoluta* and *Placopsilina* ? sp. which occur concentrated in the sediment around the pillars-like "reefs" in the so-called "*Riff-Flanken*" by Wendt (op cit). The growth model proposed by Wend comprises the following steps: (1) seafloor hardground formation and later colonization by *Tolypammina gregaria* Wendt 1969; (2) hiatus development where the inorganic precipitation of ferromanganese crust (*Limonit-Krusten*) takes place onto the foraminifer; (3) new foraminifer colonization onto the previous one accompanied of a very low background sedimentation; (4) same as first and subsequent steps.

Hornung et al. (2007b) described "foraminiferal microbuildups" structures, similar as above figured by Wendt (op cit), in the C_{II} bed (Julian II in age) of the Rappoltstein section, in condensed ammonoid beds of Berchtesgaden area (southern Germany). They illustrated a similar growth model (see Fig. 5 at Hornung et al., op cit): (1) hardground surface formation; (2) growth of foraminifer "reef" generation I; (3) corrosion of "reef" generation I and impregnation by opaque pyrite; (4) subsequent growth of different "reef" generations and; (5) "drowning" beneath a bioclastic debris flow.

5 Results

At outcrop scale, the *Tolypammina*-rich bed with ferromanganese crusts show two main characters: (1) a general cm-scale, wavy morphology; and (2) a progressive increase of sub-mm, ferromanganese crusts toward the top. Both features produce the misleading appearance of a stromatolitic structure. The foraminifer pillar-like structures increase towards the top. Microfacies analysis shows a more complex vertical zonation of this sedimentary condensed record.

5.1 *Microfacies*

Two main groups of microfacies have been recognized (Figs. 3 and 4), (1) mudstones-wackestones with ferromanganese crusts/surfaces and *Tolypammina-Frutexites* structures; and (2) bioclastic wackestone to packstone with reworked cephalopods. The contact between both groups is irregular, erosional in some places, showing breakage of *Tolypammina* pillar-like structures, thus representing a discontinuity surface (DS). There is an abrupt change in fossil content as well as structures on both sides of the discontinuity surface (DS in Fig. 5). Therefore, taking the discontinuity surface as an easily identifiable reference surface, the microfacies can be grouped into the underlying and the overlying microfacies.

Two post-sedimentary fracture systems (F1 and F2 respectively) cut the microfacies (Figs. 4b and 5). F1 consists of sediment-filled fissures with "z-shape" and with a subhorizontal orientation parallel to the bedding plane, whereas F2 are vertical to oblique fractures. Fissure fillings comprise massive to graded mudstones with angular lithoclasts derived from fissure walls. In some places there are narrow cavities associated to these fissures. They show irregular contacts (dissolution wedges), with basal erosive lag filling, and cements. The succession of cements consists of a first generation of zoned radial-fibrous calcite (black core to weak orange luminescent wedge), surrounded by calcite mosaic (orange moderately luminescent) which fills the remaining porosity. F2 fractures are filled by cloudy coarse calcite mosaic (orange moderately luminescent).

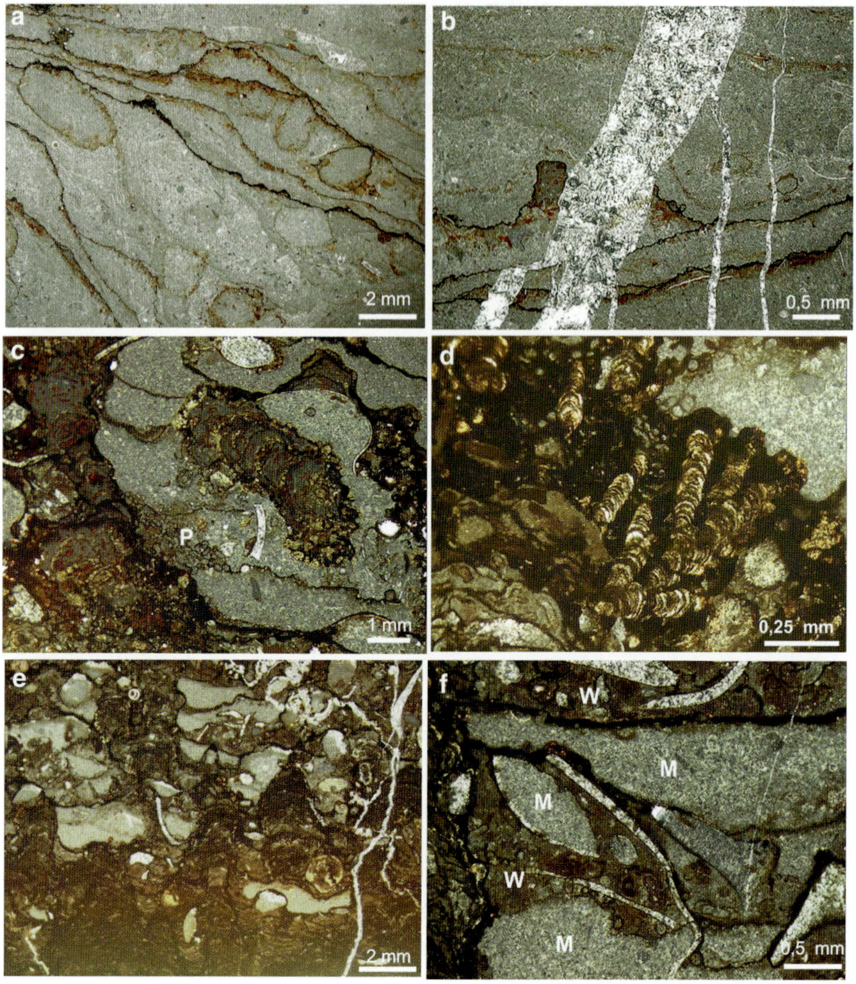

Fig. 3 (**a**) Grey grumelar to peloidal mudstone with wavy ferromanganese crusts/surfaces (microfacies A, see Fig. 5). Intraclasts, thin-shelled pelagic bivalves and small faecal pellets are common. (**b**) Detail of microfacies A. Ferromanganese crusts with microborings and attachment of *Tolypammina gregaria* Wendt crossed by F2 fracture system. (**c**) Pillar-like structures of *Tolypammina gregaria* Wendt – *Frutexites*. (P) Note the concentration in the depressed areas of *Placopsilina sp* around *Tolypammina* columns. (**d**) Detail of branched *Frutexites* growing onto *Tolypammina* buildup structures. (**e**) Microfacies C. Different growth fabrics developed by *Tolypammina*, forming massive layers (*bottom of the picture*) as well as vertical columns. (**f**) Erosive breakdown of grey, grumelar to peloidal mudstone with ferromanganese crusts/surfaces and *Tolypammina-Frutexites* buildup structures. Disrupted areas are filled by bioclastic brown wackestone (w). Note how the previously formed ferromanganese crusts protect the grey mudstones (m) from the erosion

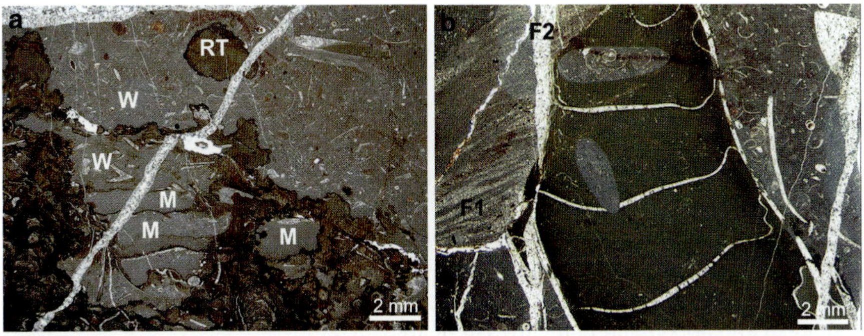

Fig. 4 (**a**) Contact between microfacies B and D (see Fig. 5). Microfacies B correspond to *Tolypammina gregaria* Wendt-*Frutexites* pillar-like structures, spaces inter-columns are grey, grumelar to peloidal mudstone with ferromanganese crusts. Microfacies D correspond to massive bioclastic wackestone to packstone with reworked cephalopods (B). The mass flow arrival of microfacies D could locally to erode the underlying microfacies (B, C-disrupted areas marked by w), breaking the *Tolypammina* columns (RT-reworked *Tolypammina*). (**b**) Detail of microfacies D, note the abraded ammonoid walls, and the presence of *Trypanites*-like borings into the ammonoid fill and septa. Graded sediment-filled fissure (F1) and oblique fractures (F2) cut the reworked ammonoid

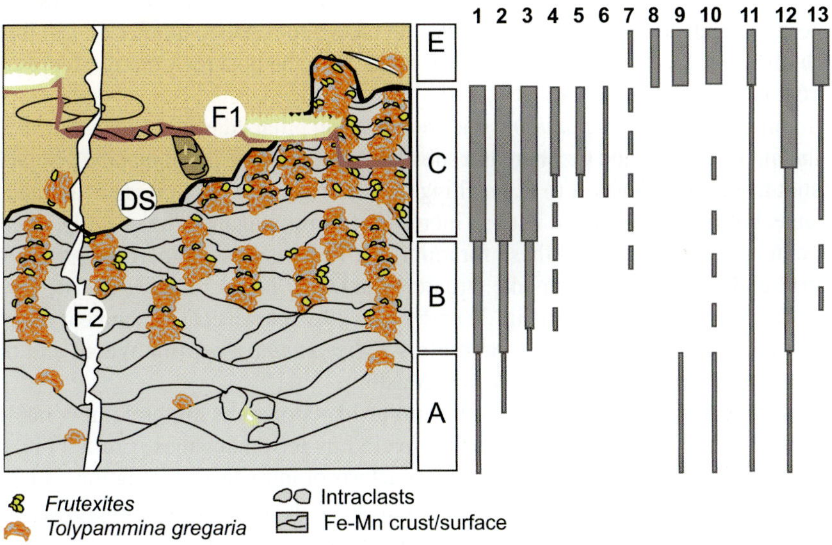

Fig. 5 Ideal microfacies sequence at Feuerkogel during Julian I–Julian II interval with the qualitative distribution of some components. A–B–C correspond to mudstone-wackestones with ferromanganese crusts/surfaces and *Tolypammina gregaria-Frutexites* buildup structures; D correspond to massive bioclastic wackestone to packstone with reworked cephalopods. F1 and F2 are two postdepositional fracture systems. DS, represent the discontinuity surface between the underlying (A–C) and the overlaying (D) microfacies. (1) ferromanganese crusts/surfaces; (2) *Tolypammina gregaria* Wendt; (3) *Frutexites*; (4) mollusc fragments; (5) big gastropods; (6) serpulids; (7) pelmatozan debris; (8) small gastropods; (9) cephalopods; (10) thin-shelled pelagic bivalves; (11) ostracods; (12) bioclastic debris; (13) infaunal foraminifer groups

5.1.1 Mudstone-Wackestones with Ferromanganese Crusts and Surfaces and *Tolypammina-Frutexites* Structures (Underlying Microfacies A–B–C, see Figs. 3 and 5)

At the base, the dominant microfacies (A) correspond to grey grumelar to peloidal mudstone with episodic development of ferromanganese crusts/surfaces (Fig. 3a). The fossil content is scarce, mainly ostracods and fragment of thin-shelled pelagic bivalves. Presence of the sessile foraminifer *Tolypammina gregaria* Wendt never exceeded 1–3% of the rock volume (Fig. 3b), whereas infaunal foraminifer groups even reach 1%. The absence of the macrofauna is significant, except for some sporadic cephalopod and pelmatozoan plates.

In microfacies A, the most noticeable features are the ferromanganese crusts/surfaces, and the presence of small faecal pellets and intraclasts. The latter appear occur floating in the surrounding matrix and present microborings and biogenic encrustations, resembling those described as "subsolution clasts" or "hardground intraclasts" (Fig. 3a).

There are two different types of ferromanganese crusts/surfaces, T1 and T2. The differences are related to variations in colour (composition), thickness, orientation, genesis, etc... Black, up to 250 μm in thickness, developed over irregular and intense microboring, hardground surfaces, with attached *Tolypammina* correspond to ferromanganese crusts (T1). Whereas the brown to reddish, thinner (<150 μm), with diffuse margins and without optical break with the host rock matrix, represent ferromanganese surfaces (T2).

In microfacies A *Tolypammina* does not develop pillar-like structures, this occur in the microfacies B and C. There is a substantial increase of *Tolypammina gregaria* Wendt from microfacies A to B, reaching values around 30–50%. It is not a gradual change, but rather associated with an increase of ferromanganese crusts (T1), which are extensively attached by these foraminifers. In fact, the volume relationship of matrix and foraminifer pillar-like structures is 60:40 in microfacies B. The foraminifer pillar-like structures (Fig. 3c) are formed by the vertical attachment of two elements: sessile agglutinated foraminifers *Tolypammina gregaria* Wendt and the probable microbial structure *Frutexites* Maslov.

In the microfacies B, the foraminifer pillar-like structures are preferably nucleated on the summit of wavy mudstones covered by ferromanganese crusts (T1), as well as using other elevated substrates as bioclasts or intraclasts. In relation with the rest of fossil content, a slight increase in bioclastic debris occurs in microfacies B (Fig. 5), but mostly concentrated around the foraminifer buildup structures. It mainly consists on mollusc fragments and a minor presence of pelmatozoan plates, ostracods and foraminiferal infaunal groups. Mollusc fragments and pelmatozoan plates show extensive development of biogenic bioerosion (microborings) filled by ferromanganese material. Thin-shelled pelagic bivalves decreases respect to microfacies A.

Frutexites does not occur in microfacies A, its first notable appearance is in microfacies B whereas C shows its maximum development. In microfacies B and C *Frutexites* appears mainly as branching-diverging columns with polymineralic convex-upward laminae (Fig. 3d). Laminae correspond to black opaque minerals (Mn-rich), brownish to reddish semitranslucent minerals (Fe-rich) and/or yellow-white carbonate minerals. The maximum width of the columns is around 60 μm and up to 750 μm in height, but normally range from 50 up to 250 μm. *Frutexites* is not exclusively associated to the foraminifer-buildup structures, also grows around bioclasts or directly over ferromanganese crusts (T1). Essentially the nucleation sites of *Frutexites* are laterally around the edges of *Tolypammina* columns, as well as at the top of them.

Thus microfacies B (Fig. 5) is a complex microfacies framework formed by wavy grey, grumelar peloidal mudstones with very low fossil content and extensive development of ferromanganese crust/surfaces (T1 and T2) which are colonized by *Tolypammina-Frutexites* buildup structures where is concentrated mostly the rest of the skeletal content.

Microfacies C are formed by patches dominated by *Tolypammina gregaria* Wendt-*Frutexites* buildup structures. In fact, the volume relationship of matrix and pillar-like structures is 30:70. Lateral distance between foraminiferal columns ranges from 2.5 mm up to 5 mm. Vertical thickness interval between successive ferromanganese crusts is around 500 μm up to 1 mm. Ferromanganese crusts (T1) are dominant.

Different growth fabrics developed by *Tolypammina gregaria* were recognized, occurring from the base to the top of these patches (Fig. 3e): (1) wide massive layers; (2) columns or pillar-like structures; and (3) dendritic, arborescent structures. It looks usually exist a vertical relationship between this growth fabric zonation and an increase of bioclastic debris. Occasionally, small articulated-shell brachiopods disrupt the massive *Tolypammina* layers as well as the pillar-like structures. The dendritic or arborescent zones are likewise prolific in branched *Frutexites*.

The quantity and diversity of foramiferal infaunal groups increase in microfacies C (Fig. 5) but is basically dominated by *Planiinvoluta*- and *Placopsilina*-like forms. *Placopsilina* concentrates near the base of *Tolypammina-Frutexites* pillar-like structures, in depressed areas, as well as in shelter areas associated to mollusc fragments. However, the distribution of *Planiinvoluta* is not only restricted to the surroundings of pillar-like structures. Big gastropods and serpulids occur also in microfacies C.

In microfacies C, the grey, grumelar to peloidal mudstone between pillar-like structures has a slightly higher skeletal content compared to the mudstones from microfacies A and B. Additionally it exhibits disrupted zones filled with massive bioclastic brown wackestone. These disrupted zones (Fig. 3f) increase from microfacies B to C where they can represent around 5% up to 10% of rock volume.

Summarizing, the sedimentary condensed deposits formed by the mudstone-wackestones with ferromanganese crusts/surfaces and *Tolypammina-Frutexites* structures show three characteristic microfacies (A, B and C in Fig. 5). It is possible reconstruct their vertical arrangement: (1) microfacies A dominates at the base; (2) B and C can occur intercalated; whereas (3) patches of microfacies C are more common toward the top.

5.1.2 Bioclastic Wackestone to Packstone with Reworked Cephalopods (Overlying Microfacies E Figs. 4 and 5)

Massive brown to reddish bioclastic wackestones to packstones overlay microfacies B and C (Figs. 4a and 5). The bioclastic content is around 35% and 45% of rock volume and is formed mainly by ostracods, benthonic foraminifers (lagenid foraminifers are dominant), articulated and disarticulated thin-shelled pelagic bivalves, small gastropods, and diverse rests of cephalopods. The fabric is chaotic, with mottled aspect where skeletal content shows a heterogeneous distribution of sizes. Thin-shelled pelagic bivalves can display imbricated fabric as well as a totally random distribution.

Cephalopods comprise orthoconic nautiolids, with geopetal filling similar as surrounding wackestone and marine calcite cements (early fibrous rim phase and later equant mosaics), and ammonoids. However, it is noticeable in this microfacies the presence of reworked ammonoids (Fig. 4b) which show imbricated orientation, microbored and abraded walls, lithological differences between its geopetal filling and the encasing rocks, local attachment by *Tolypammina*, macroborer activity cutting the lithified geopetal filling as well as septa. The boreholes may be attributed to *Trypanites* group and their infilling are equivalent to the encased lithology.

5.2 *Geochemical Analysis (Fig. 6)*

Chemical sediments in general are anticipated to reflect the chemical composition of the water from which they were precipitated. Ferromanganese nodules show an inverse REE pattern compared to a typical seawater pattern, showing a positive Ce anomaly and are slightly enriched in light REE relative to heavy REE. Interpretations must be cautious as chemical sediments/precipitates also reflect local redox conditions and maybe influenced by post depositional changes (Elderfield and Greaves 1981).

The measured REE and trace elements are given in Table 1. REE (Fig. 6a) were normalized with a PAAS standard (Post-Archaean average Australian Shale) after McLennan (1989). Anomalous behaviour of Ce, was investigated using (Ce/Ce*) ratios after Bau and Dulski (1996), calculated as [Ce/Ce* = Ce/0.5(La + Pr)] (all values PAAS normalized). It was also checked with (Pr/Pr*) ratios with [Pr/Pr* = Pre/0.5(Ce + Nd)], whether the Ce anomaly can be assigned as a Ce anomaly or is affect by high La values (Fig. 6b).

5.2.1 Matrices from Microfacies B and C

Matrix B shows a smooth and even distribution of the REE and with a negative Ce anomaly (Ce/Ce* between 0.26 and 0.51) (Table 1, Fig. 6b). Compared to matrix B, Matrix C is higher enriched in REE and shows a positive Ce anomaly (Ce/Ce*

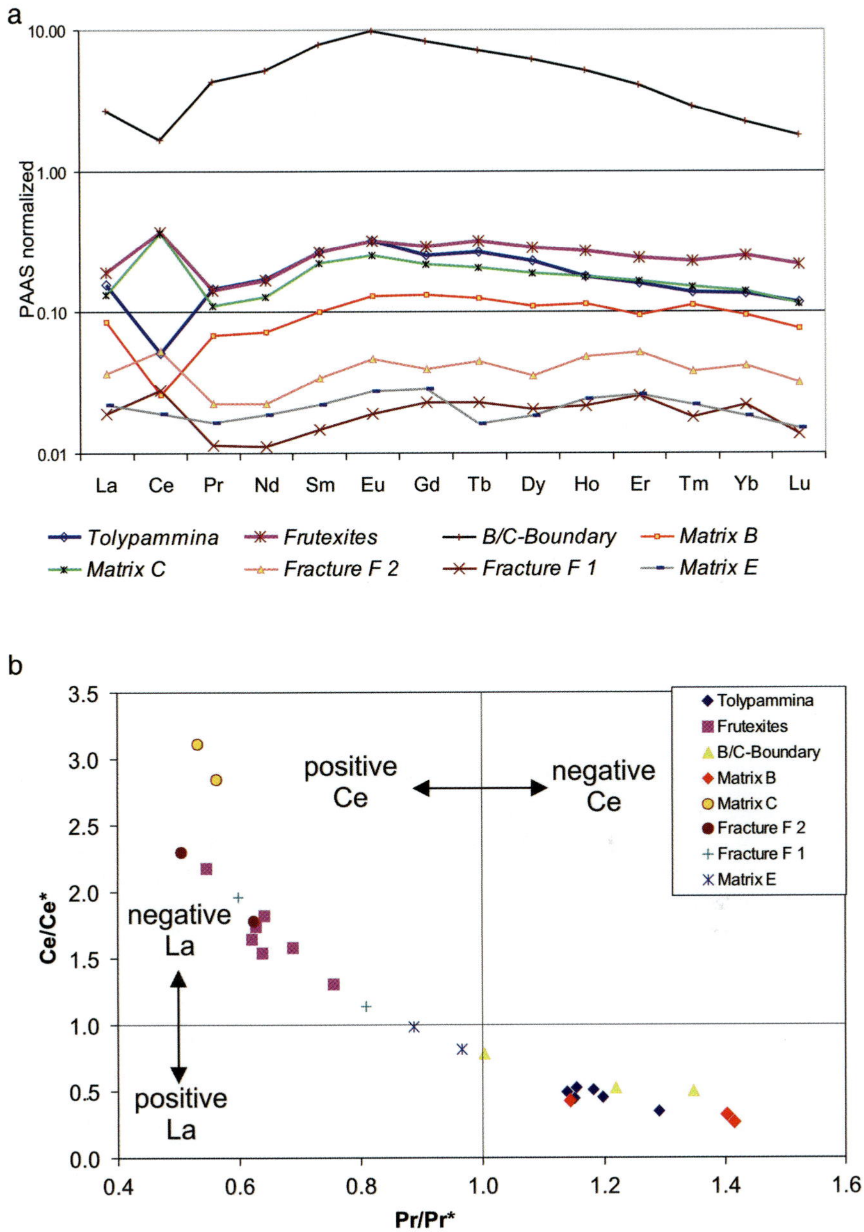

Fig. 6 (**a**) PAAS normalized REE content of *Tolypammina gregaria*, *Frutexites*, matrices from microfacies B, C and E, a distinct change of REE pattern at the B/C boundary, and the two fracture systems F1 and F2. The behaviour of the REE is generally similar, most remarkable are the occurrence of positive and negative Ce anomalies. (**b**) Plot of PAAS – normalized Pr/Pr^* $[Pr/Pr^* = Pre/0.5(Ce + Nd)]$ versus Ce/Ce^* $[Ce/Ce^* = Ce/0.5(La + Pr)]$ modified after Bau and Dulski (1996). See text for further description of the values

Table 1 Average values of REE and trace elements in parts per million (ppm) out of 1,800 measurements. See text for a detailed discussion of the element data

	Mn	Fe	Y	Zr	Nb	La	Ce	Pr	Nd	Sm	Eu	Gd	Tb	Dy	Ho	Er	Tm	Yb	Lu	Hf	Ta	Th	U	Co	Ni	Ce/Ce*	Pr/Pr*
Tolypammina	196.7	1,617.3	4.884	0.860	0.651	5.502	5.085	1.222	5.739	1.560	0.324	1.145	0.170	0.964	0.185	0.403	0.053	0.320	0.043	0.037	0.014	3.339	0.027	4.322	8.784	0.452	1.196
Tolypammina	335.1	1,674.3	4.654	0.858	0.884	6.116	6.034	1.169	5.229	1.341	0.309	1.079	0.185	0.943	0.173	0.449	0.059	0.402	0.056	0.036	0.022	3.310	0.027	3.135	8.684	0.522	1.154
Tolypammina	546.5	3,320.8	4.941	1.329	1.576	5.822	5.644	1.099	4.865	1.214	0.297	1.173	0.180	0.865	0.167	0.436	0.062	0.356	0.050	0.047	0.019	3.181	0.019	4.176	8.084	0.509	1.180
Tolypammina	180.6	992.0	5.247	0.532	0.506	5.997	4.078	1.279	5.802	1.486	0.344	1.167	0.206	1.071	0.174	0.456	0.055	0.381	0.050	0.018	0.007	3.117	0.016	5.350	10.111	0.346	1.289
Tolypammina	269.5	1,421.6	4.554	0.783	0.929	5.997	4.895	1.080	5.204	1.329	0.292	1.031	0.172	1.000	0.172	0.414	0.059	0.400	0.040	0.036	0.022	3.190	0.027	4.731	10.847	0.443	1.150
Tolypammina	351.0	2,145.0	5.497	0.975	1.156	6.166	5.971	1.274	6.034	1.577	0.380	1.391	0.205	1.116	0.203	0.475	0.067	0.412	0.052	0.033	0.014	4.021	0.027	5.182	9.263	0.494	1.138
Frutexites	1,620	13,459	3.968	5.901	6.774	5.838	16.365	0.860	3.763	0.894	0.221	0.821	0.140	0.784	0.173	0.449	0.061	0.460	0.063	0.221	0.196	3.585	0.186	4.713	10.709	1.642	0.621
Frutexites	1,483	11,225	4.803	5.848	6.950	5.873	17.259	0.994	4.332	1.071	0.262	1.072	0.155	1.026	0.200	0.494	0.071	0.472	0.061	0.235	0.268	3.334	0.195	2.384	5.654	1.578	0.689
Frutexites	2,195	18,336	6.829	10.259	12.795	7.140	25.38	1.410	6.282	1.632	0.373	1.356	0.231	1.381	0.266	0.674	0.107	0.659	0.081	0.454	0.516	5.921	0.398	1.999	4.686	1.819	0.641
Frutexites	2,036	11,989	5.431	8.509	10.007	6.847	21.583	1.126	4.775	1.174	0.316	1.058	0.180	1.034	0.214	0.593	0.071	0.559	0.077	0.301	0.297	4.243	0.239	1.686	4.170	1.742	0.628
Frutexites	1,638	12,965	4.473	5.260	6.256	6.290	16.518	0.929	4.073	0.924	0.235	0.854	0.152	0.846	0.181	0.518	0.071	0.459	0.071	0.216	0.202	3.386	0.156	1.492	5.287	1.537	0.639
Frutexites	2,898	14,375	6.239	13.460	17.603	7.306	29.578	1.256	5.714	1.471	0.345	1.369	0.248	1.346	0.269	0.689	0.093	0.709	0.093	0.588	0.606	5.340	0.566	2.234	7.843	2.177	0.546
Frutexites	1,752	9,257	6.000	6.068	7.224	7.165	18.151	1.385	6.699	1.758	0.387	1.527	0.256	1.269	0.241	0.713	0.087	0.560	0.065	0.204	0.160	3.505	0.161	3.760	4.729	1.307	0.755
B/C-Boundary	897.0	1,335.3	76.03	2.818	1.246	102.7	132.2	37.7	175.8	43.78	10.51	38.45	5.617	29.16	5.145	11.54	1.167	6.322	0.768	0.066	0.028	37.20	0.146	3.416	8.738	0.521	1.218
B/C-Boundary	348.0	2655.4	18.072	1.827	1.451	17.96	29.77	4.211	19.63	4.978	1.208	3.968	0.548	2.787	0.510	1.255	0.121	0.786	0.083	0.068	0.038	3.132	0.056	3.454	6.027	0.782	1.002
Matrix B	36.6	504.4	7.518	0.793	0.185	6.628	5.682	1.396	7.210	1.681	0.412	1.633	0.232	1.330	0.240	0.557	0.080	0.478	0.045	0.030	0.010	0.168	0.031	3.508	6.842	0.425	1.144
Matrix B	73.9	1,178.6	3.785	1.339	0.524	3.253	1.627	0.572	2.440	0.694	0.131	0.643	0.085	0.514	0.117	0.280	0.038	0.286	0.041	0.052	0.030	0.238	0.035	4.625	11.178	0.269	1.415
Matrix C	85.3	1,013.8	5.053	2.332	0.559	5.114	29.878	0.980	4.337	1.088	0.282	0.989	0.157	0.858	0.176	0.440	0.057	0.381	0.046	0.050	0.014	0.755	0.030	4.784	10.856	3.115	0.531
Matrix C	174.3	3,976.4	5.577	3.128	1.852	5.044	28.871	0.981	4.354	1.220	0.273	1.012	0.160	0.883	0.177	0.467	0.061	0.397	0.049	0.113	0.061	1.132	0.079	5.289	12.063	2.847	0.562
Matrix E	21.4	1,135.7	0.850	1.197	0.152	0.833	1.512	0.145	0.633	0.121	0.029	0.133	0.012	0.086	0.024	0.075	0.009	0.051	0.006	0.024	0.003	0.102	0.032	4.923	12.302	0.987	0.887
Matrix E	15.9	700.2	0.960	1.174	0.105	0.914	1.310	0.153	0.680	0.121	0.032	0.135	0.020	0.097	0.026	0.072	0.006	0.074	0.015	0.020	0.005	0.119	0.015	5.352	11.705	0.813	0.966
Fracture F1	95.5	87.9	0.674	0.083	0.011	0.722	2.226	0.101	0.379	0.082	0.020	0.106	0.018	0.096	0.021	0.073	0.007	0.062	0.006	0.002	0.001	0.026	0.005	0.109	0.371	1.956	0.598
Fracture F1	19.5	693.6	0.359	0.429	0.282	0.674	1.173	0.108	0.489	0.093	0.029	0.157	0.013	0.124	0.036	0.055	0.009	0.073	0.010	0.047	0.017	0.408	0.068	0.122	0.162	1.136	0.809
Fracture F2	218.6	30.8	1.301	0.049	0.002	1.374	5.458	0.214	0.951	0.110	0.030	0.196	0.043	0.200	0.038	0.144	0.008	0.162	0.024	0.000	0.002	0.005	−0.002	4.800	22.086	2.302	0.504
Fracture F2	204.3	53.7	1.374	0.061	0.004	1.414	4.248	0.198	0.762	0.190	0.051	0.184	0.035	0.165	0.048	0.148	0.016	0.118	0.014	0.006	0.000	0.024	−0.002	3.725	27.351	1.786	0.624

between 1.8 and 3.2). In both matrices the Y/Ho ratio varied between 28 and 100 and U/Th as well as Ni/Co ratios after Jones and Manning (1994), varied between 0.04–0.6 (U/Th) and 0.8–3.20 (Ni/Co) respectively.

5.2.2 Microfacies B/C-Boundary

The B/C-boundary, (Fig. 5), at the beginning of an increased colonization of *Tolypammina-Frutexites* assemblages, on the substrate, a different REE pattern is observed, with a negative Ce anomaly (Ce/Ce* between 0.4 and 0.8), a high abundance of middle REE (MREE) and a slight positive Eu anomaly. Together with the MREE enrichment, elements like Th, Y, Co and Ni occur in higher abundance (Table 1). U/Th varied between 0.01 and 0.02 and Ni/Co between 1.5 and 3.1.

5.2.3 *Tolypammina*

Tolypammina shows a negative Ce-anomaly (Ce/Ce* between 0.35 and 0.94) and a general REE pattern similar to Matrix B (Fig. 6b, Table 1). Y/Ho ratio varied between 26 and 30, U/Th 0.05–0.1 and Ni/Co varied between 1.32 and 2.9. Throughout the *Tolypammina* structures the iron/manganese (Fe/Mn) ratios remain quite stable around 5–10.

5.2.4 *Frutexites*

The *Frutexites* structures show a prominent positive Ce anomaly (Ce/Ce* between 1.3 and 2.4) and REE showing a similar distribution compared to Matrix C. The Y/Ho ratios oscillated between 18 and 25 for the *Frutexites* structures and U/Th ranged from 0.05 to 0.21 and Ni/Co varied between 1.1 and 3.8. In comparison to the surrounding limestone and the *Tolypammina* structures, *Frutexites* is up tenfold enriched in iron (Fe) and manganese (Mn) (Table 1). Interestingly, the Fe/Mn ratio of *Frutexites* is similar to those of *Tolypammina* varying between 5 and 9.

5.2.5 Matrix from Microfacies E

In Matrix E, the content of the REE is quite low, with a slight positive La anomaly (Ce/Ce* between 0.8 and 0.99) (Table 1, Fig. 6b). Y/Ho ratios ranged from 32–40, U/Th varied between 0.1 and 0.45 and Ni/Co varied between 2.1 and 2.9. Matrix E contains much lower Mn compared to Matrices B and C, thus showing Fe/Mn ratios from 42 to 54 (Table 1).

5.2.6 Fracture Systems

The fracture systems F1 and F2, show a very similar behaviour of REE, with distinct positive Ce anomalies (Ce/Ce* between 1.1 and 2.4). Y/Ho ratios ranged from 10 to 35, U/Th varied between 0 and 0.2 and Ni/Co laid between 1.3 and 7.5. The variation of the trace element content in F1 revealed two different phases within the fracture, one is enriched in Mn and Ni and depleted in Fe, the other phase is enriched in Fe, Y, Zr, Nb, Th and U. F2 shows a distinct enrichment in Mn and Ni and a depletion in Fe (Table 1).

6 Discussion

6.1 Types of Ferromanganese Crusts/Surfaces

Ferromanganese crusts and nodules are common deposits on sediment-starved basins, seamounts and slopes, mostly in and below oxygen minimum zone. They are the slowest-forming mineral deposit known, formed by heterogeneous mixtures of biogenic, authigenic and detrital particles which final composition depends on seawater chemistry (Hein et al. 1992, 1994) and supply of metals to the oxide surface (Dymond et al. 1984).

The Y/Ho ratios ranging between 18–25 and 44–100 for the ferromanganese crust and the carbonate phases respectively indicate a hydrogenous origin (Bau 1996). Further, according to U/Th and Ni/Co ratios after Jones and Manning (1994), ratios lower than 0.75 (U/Th), and 5.00 (Ni/Co) respectively, point to oxic bottom water conditions. Ratios between 0.75–1.25 (U/Th), and 5.00–7.00 (Ni/Co) represent dysoxic bottom water conditions. Therefore, with ratios not higher than 0.81 (U/Th) and 6.20 (Ni/Co), it can be stated, that the deposition of the ferromanganese crusts and the different limestone matrices occurred predominantly under oxic conditions and seldom under dysoxic conditions.

At the B/C boundary some hydrothermal influence was observed occurring as a short event as indicated by the high increase in middle REE with distinct Eu anomaly (Figs. 5 and 6). However, the Y/Ho ratios show a large variation, hampering a clear relation to a distinct hydrothermal event. Further, U/Th and Ni/Co ratio clearly indicate oxic bottom water conditions during this phase. Clear hydrothermal influence was observed in fractures F1 and F2 indicated by Y/Ho ratios of 10–15 (Bau 1996) a slight Eu anomaly (Fig. 6).

Bacteria could also concentrate Fe and Mn even from very low concentration in the seawater. Microbial activity can play an important role on the genesis of ferromanganese oxyhydroxides in several steps (Konhauser 1998) by: (1) concentration of metals by microorganisms living on water column; (2) deep transport flux as the microbes die; (3) metal release through organic matter mineralization; (4) bacterial oxidation of Mn(II) and Fe(II). In fact, the formation of microbially mediated iron and manganese minerals has been widely recognized (Nelson and Tebo 1980;

Ghiorse 1984; Konhauser 1998; Tebo et al. 2004) although their importance and exactly metal-depositing mechanisms are not well understood (Ghiorse 1984). Indeed, biologically induced minerals are governed by same equilibrium processes as their inorganic counterparts and biological as well as abiotical catalysis can occur combined in the formation of ferromanganese crusts and nodules.

The base of the Feuerkogel sedimentary condensed deposits, formed by the mudstones with ferromanganese crusts/surfaces (microfacies A, Fig. 5), records low sedimentation rates, a very scarce to absent macrofauna and a very low diverse meiofauna that points out the existence of stressed oligotrophic conditions. In such environment, during calm weather, the fluffly layer in the water/sediment interface form an important biogeochemical interface where microbial processes of Fe and Mn oxidation are most active (Strekopytov et al. 2000). Without grazer pressure, biofilms of chemolithotrophic bacteria could proliferate and promotes biologically induced epicellular ferromanganese oxyhydroxides. Indeed thin bacteria biofilms have been observed in recent manganese nodules (Burnett and Nealson 1981). Biological Mn(II) oxidation is faster than abiotic one (Tebo et al. 2004) however biological Fe(II) oxidation only prevailed acidic and microaerophilic conditions (Ghiorse 1984).

The final morphology and orientation of Feuerkogel ferromanganese crusts/surfaces (T2/T1 respectively) are related with their loci of formation as well as the with the bottom current regime. Dark ferromanganese crusts (T1), developed on microbored mudstones, were formed directly onto the wavy water/sediment interface sporadically affected by bottom currents. Thus they reproduce the seawater-sediment interface morphology. Whereas thinner, brown to reddish, ferromanganese surfaces (T2) were produced within the sediment. So their morphologies are more randomly distributed, from horizontal to vertical orientations. T2 reflects the reduction of

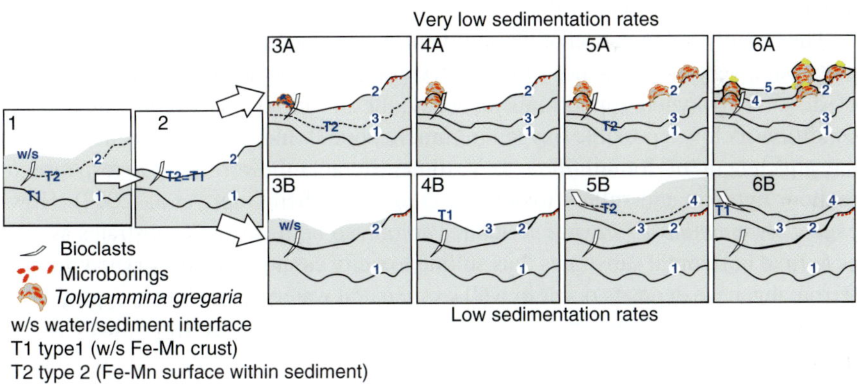

Fig. 7 Development of ferromanganese crusts onto seawater-sediment interface (T1) and ferromanganese surfaces within sediment (T2). (1) T2 cross the bioclasts and protect the underlying soft sediment from bottom currents, (2) turning into a seawater-sediment interface where a T1 is developed. Sedimentary evolution and record of the ferromanganese-rich condensed deposit formed at low (3B–6B) and at very low sedimentation rates (3A–6B). At very low sedimentation rates, *Tolypammina gregaria* Wendt can colonize the substrate. Ferromanganese crusts occur now onto seawater-sediment surface as well as surrounding *Tolypammina-Frutexites* buildup structures

Fe–Mn oxyhydroxides within sediment (by suboxic diagenesis and/or Fe and Mn reducing bacteria), followed by the upwards diffusion of Mn(II) and Fe(II) that finally are reoxidized and deposited in the suboxic-oxic sediment interface.

Both types of ferromanganese structures could act as a protecting barrier for substrate erosion (Fig. 7). The early development of ferromanganese surfaces (T2) within the still soft sediment can protect the underlying sediment of bottom currents (2 in Fig. 7). Thus, the previously formed T2 surface becomes the sediment-water interface and a new ferromanganese crust could be produced (T1 superimposed to a previous T2). The new and successive generations of ferromanganese crusts (T1) and surfaces (T2) and their relationship with biogenic encrustation allow to us to reconstruct the sedimentation rates (low and very low sedimentation rates from microfacies A to microfacies B – see 3B–6B and 3A–6A respectively in Fig. 7).

6.2 Significance of Tolypammina gregaria-Frutexites *Buildup Structures*

Thus as we have seen before, an increase of ferromanganese structures (T1 and T2) from the base to the top (from microfacies A to C, Figs. 5 and 7), could reflect a decrease of the sedimentation rate (from low to very low sedimentation rates). Both, bacterial biofilm development on water-sediment interface and very low sedimentation rates could have a positive effect on foraminiferal larval settlement, favouring the colonization by the sessile agglutinated foraminifer *Tolypammina gregaria* Wendt. The main question is which environmental parameters control/favour the different types of growth of this sessile foraminifer.

The broadly colonization of the sediment-water interface by *Tolypammina* could point out a very low presence of predatory and grazer organisms due to low values of dissolved oxygen (opportunistic behaviour in aerobic to slightly dysaerobic conditions below a possible oxygen minimum zone?). Indeed, it has been described the attachment of foraminifera to vestimentiferan tubeworms as a response to seafloor hypoxia and sulfide toxicity at cold seeps (Sen Gupta et al. 2007). However, *Tolypammina-Frutexites* structures colonized both elevated substrates as well as normal horizontal substrates thus sulfide toxicity cannot be easily argued. In fact, ferromanganese deposits occur in well oxygenated waters. In addition, agglutinated foraminifers, as a group, are much less tolerant to hypoxic conditions that calcareous assemblages, and they prevail under conditions of greater oxygenation (Gooday et al. 2000; Levin et al. 2002).

Maybe their colonial behaviour building vertical pillar-like structures could be related with their feeding behaviour. On recent manganese nodules, suspension feeders dominate the nodule summit whereas the deposit feeders concentrate near the base (Mullineaux 1989). Epibenthonic foraminifers apparently use elevated substrates under the influence of near-bottom flow where a maximum yield of food particles can be achieved (Schönfeld 2002).

The dominance of sessile *Tolypammina gregaria* in comparison with the scarcity of infaunal foraminiferal groups in microfacies B, could be also related with nutrient availability. Benthic foraminifera are not found or rare when the organic matter is scarce or absent within the sediment, and epifauna reflects better the organic flux whereas the infaunal groups are more related with the organic matter stored in the sediment (Jorissen et al. 1995; Den Dulk et al. 1998; Van der Zwaan et al. 1999).

Indeed, the different colonial growth fabrics developed by *Tolypammina gregaria* (massive layers; columns or pillar-like structures; and dendritic, arborescent structures) could be associated to both, the quantity as well as to the dominant type of organic flux input (lateral or vertical). In microfacies B (Fig. 5), the preference of *Tolypammina-Frutexites* structures to colonize the summit of wavy mudstones could respond to a dominant horizontal or lateral organic flux input in oligotrophic conditions. In oligotrophic ecosystems, the reproduction potential could play a crucial role in the competition for food (Van der Zwaan et al. 1999). Thus a massive layer growth strategy, as it is observed in microfacies C, could suggest a fast increase of biomass in response to vertical phytodetritus pulse (Graf 1989; Gooday 1988). The accretion rate of similar sessile foraminifer-rich layers has been estimated by Wendt (1974) in 1 mm/year to 1 mm/25, in recent deep-sea manganese nodules.

Microfacies B and patches of microfacies C occur intercalated although an increase of microfacies C toward the top is observed (Fig. 5). Thus the vertical increase of *Tolypammina gregaria-Frutexites* assemblage from microfacies B to C, in addition to a more diverse accompanied benthic assemblage suggest a change from oligrotrophic to slightly mesotrophic conditions associated with episodic pulses of phytodetritus.

As we see before the common nucleation sites of *Frutexites* are laterally surrounding edges of *Tolypammina gregaria* columns. *Frutexites* structures have been mainly described in marine environments from shallow to deep-water stromatolites, condensed pelagic successions and hardgrounds. *Frutexites* have been found growing directly on the seawater-sediment interface, within the micritic sediment and into marine and continental cavities and fractures. Such wider environmental distribution has been taken into consideration in their palaeoecological interpretations. Their presence in stromatolites has been linked with different types of cyanobacteria (Playford et al. 1976, 1984; Scytonematacean – Walter and Awramik 1979; Rivulariacean – Hofmann and Grotzinger 1985; *Angulocellularia* group – Riding 1991). Their growing in cavities, sheet cracks and Neptunian Dykes have been related with non-phototrophic (heterotrophic) behaviour (Tsien 1979) or cryptobiont role (Myrow and Coniglio 1991), which sometimes were linked with non-phototrophic cyanobacteria (Gischler 1996) as well as with chemosynthetic bacteria (Cavalazzi et al. 2007). The cryptobiont, negative phototactic downward growth and cryptoendopelitic behaviour of *Frutexites* were interpreted by Böhm and Brachert (1993) as a signal of oxygen deficient environments. Chafetz et al. (1998) compared Mn-rich black travertine shrubs with *Frutexites* and other structures regarded as bacterially induced products (desert varnish, deep-sea manganese nodules). Finally, the presence of *Frutexites* in condensed pelagic limestones and hardgrounds has been explained due to physicochemical origin (Wendt 1969) as well as related with iron bacteria (Reitner et al. 1995; Mamet and Préat 2006).

Frutexites have been regarded as biologically induced products (cyanobacteria and iron bacteria) in spite of their final resulting mineralogy (hematite, Fe–Mn oxyhydroxides, and carbonate and/or phosphate minerals). The final assessment depends strongly about which mineralogical composition they interpreted as primary (timing of iron mineralization) as well as its loci of growth. Thus their final mineralogy, their compositional different alternating bands, could indeed reflect microenviromental changes, as well as the same bacteria can form different minerals in different environments (Lowenstam and Weiner 1989).

In the Feuerkogel site, the preference of *Frutexites* to colonize the *Tolypammina* tests suggests certain substrate control in their development. It could be related with the foraminifer walls, rich in acid mucopolysaccharides that could favour the growth of bacteria biofilm (resulting *Frutexites* structures?) as well as promote the ferromanganese-oxide precipitation in the polymeric matrix. In Recent nodules this possibility was considered by Graham and Cooper (1959) who related the foraminiferal protein-rich exudates as a favourable substrate for growth of bacteria capable of extracting trace elements from seawater.

The role of foraminifers in the formation of ferromanganese structures has been longer and widely discussed (Greenslate 1974; Wendt 1974; Dugolinsky et al. 1977; Riemann 1983; von Stackelberg 1984; Mullineaux 1987, 1989; Mullineaux and Butman 1990; Verlaan 1992). However, in the Feuerkogel level, the development of ferromanganese crusts/surfaces took place before the extensive colonization of the substrate by *Tolypammina-Frutexites* community. But, both *Tolypammina* buildup structures, as well as ferromanganese crusts increase toward the top. Thus, probably common environmental parameters favour their increase toward the top.

6.3 Environmental Significance of Coeval Ferromanganese Crust/Surfaces and Tolypammina-Frutexites *Community During Carnian Interval*

Low sedimentation rates are favoured conditions to colonization of substrate by sessile foraminifers as well as development of ferromanganese deposits. Which could be the main parameters that control both foraminifer colonization and ferromanganese formation?

Organic flux is an important structural control in heterotrophic marine biota, and foraminifers represent more than 50% of total biomass in deep-sea environments (Snider et al. 1984; Gooday et al. 1992). Thus primary productivity in the mixed layer and vertical nutrient flux could be a critical factor in the abundance of deep water foraminifers associated to ferromanganese crusts and nodules.

We have previously seen, Foraminifera are the most abundant sessile organism found on Recent crust and nodules. In fact, Foraminifera and Bacteria are the main consumers of organic matter in deep-sea floor, where foraminifers play an important role in the fast initial processing of phytodetritus input whereas bacteria dominate the long-term carbon re-mineralization (Moodley et al. 2002). In the Feuerkogel site, the main consumers of vertical organic flux in a stressed bottom environment could have been related with the biofilms developed on water/sediment interface, *Frutexites*-induced biofilms around foraminifers, and the own *Tolypammina*. Thus, the interbedded patches of microfacies C with B, dominated by the sessile foraminifer *Tolypammina gregraria* Wendt could reflect an opportunistic behaviour to episodic phytodetritus inputs related with phytoplankton blooms. In addition, the different types of colonial growths developed by *Tolypammina gregraria* maybe reflect changes in their reproduction rate as response of the episodic inputs. Indeed, it has been noted that the response of foraminifers to phytodetritus layer deposition in deep sea environments is very fast, followed by an increasing their biomass and reproduction rate (Gooday 1988; Altenbach 1992; Moodley et al. 2000). Deep foraminifer biota from regions with episodic organic flux shows opportunistic fast-growth species which dominate the assemblages, however low seasonality and stable food input regions show higher diversity (Corliss et al. 2009). The response of actual benthic foraminifer has been specially analyzed after phytoplankton blooms related with seasonal and monsoonal periods (Schönfeld and Numberger 2007; Andersson et al. 2007; Larkin and Gooday 2009; Corliss et al. 2009).

However, high primary productivity related with phytoplankton blooms is not the only factor to explain the abundance of benthic organisms associated to actual equatorial ferromanganese nodules, because collected nodules from areas with oceanic upwelling and high biological productivity do not show high abundance of benthic foraminifers (Margolis and Burns 1976).

In addition to the primary productivity, other factors have to favour both the ferromanganese crust development and sessile colonization by foraminifers. It has been related with the own metal-ion availability. In offshore waters the bioavailability of iron micronutrient has been suggested as limiting factor of primary production (Martin and Fitzwater 1988). The phytoplankton growth can be controlled by the Fe(III) solubilities and the dissolution rates of colloidal Fe(III) phases (Wells et al. 1983; Kuma and Matsuga 1995). Although there is no clear consensus about the inorganic speciation of Fe(III) in seawater and it seem exists many unresolved questions about chemistry of iron in seawater (Wells et al. 1995; Millero et al. 1995).

The REE pattern of Feuerkogel microfacies does not show a hydrothermal overprint (except the postpositional fracture systems) thus others possible sources as atmospheric aerosols or river inputs are necessary to explain its high concentrations. As it was mentioned, the depositional setting of red Hallstatt facies was related with structural highs (tectonically-induced saline uplifts), thus the arrival of siliciclastic input due to continental run off could be not so relevant. However, the dust from desert regions is a primary source of atmospheric iron into the oceans (Donaghay et al. 1991); in fact the 80% of the iron input from the atmosphere to the oligotrophic surface waters is due to wet deposition (Uematsu et al. 1985). Thus rain and storms produce episodic inputs of iron aerosol into oceanic iron-limited waters, which favour phytoplankton growth (Hanson et al. 2001).

The ferromanganese structures with the development of *Tolypammina-Frutexites* community occur at the Feuerkogel site in the Julian I–Julian II limit, thus they coincided with the so-called Reingraben Turnover (Hornung and Brandner 2005; Hornung et al. 2007b). During Carnian, part of the marine fossil record experimented high extinction rates which has been attributed to humid climate pulse, named the Carnian Pluvial Event (Simms et al. 1994; Bellanca et al. 1995) related with a monsoonal clima (Mutti and Weissert 1995). Evidences of this humid climate pulse has been also recorded in deep-water sequences (Hornung and Brandner 2005; Rigo et al. 2007; Hornung et al. 2007a).

This humid phase related with a monsoonal climate could have produced both weathering and subsequent increase of siliciclastic input through continental run off (source of metals as well as organic matter), as well as important wet aerosol iron inputs to the offshore waters. In addition, this humid phase could also have some influence on marine currents and a slightly cooling, factors that have been linked to the growth of ferromanganese crusts (Margolis and Burns 1976; Segl et al. 1984; Puteanus and Halbach 1988; Koschinsky et al. 1997).

7 Summary

In the Feuerkogel site, the condensed Carnian sedimentary condensed Hallstatt Limestone formed by mudstone-wackestones with ferromanganese structures and *Tolypammina gregaria-Frutexites* community report to us a complex vertical evolution of environmental changes (Fig. 8):

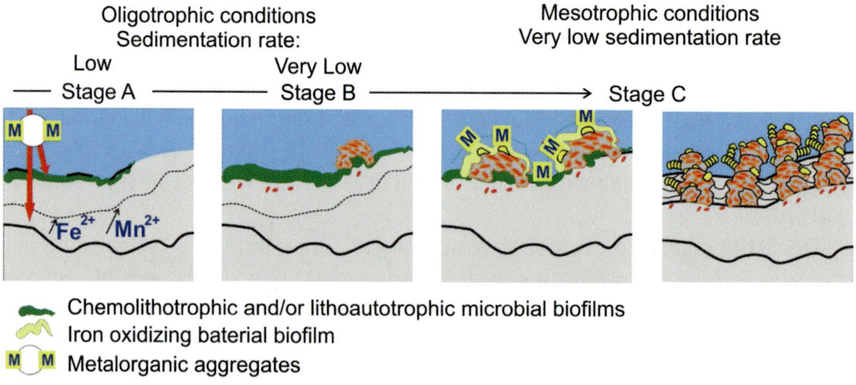

Fig. 8 Hypothetical sedimentary model of ferromanganese structures development (T1 and T2 in Fig. 7) and *Tolypammina gregaria* Wendt-*Frutexites* community: Relationship between Foraminifera and different bacteria biofilms (*Tolypammina gregaria* Wendt-*Frutexites* assemblage) coupled to nutrient and metal availability

Stage A: In deep oligotrophic conditions with low sedimentation rates, the seawater/sediment interface was colonized by microbial biofilms (lithoautotrophic and/or chemolithotrophic prokaryotes) which coupled to inorganic hydrogenous mechanisms could promote passively and/or actively organomineral ferromanganese encrustations (T1). Once within the sediment, in anaerobic conditions, the buried ferromanganese crusts as well as metal organic aggregates from water column could be used by the metabolism of other microbial populations (iron-reducing bacteria). The released metallic cations diffuse upwards and precipitate later forming ferromanganese surfaces (T2) in the suboxic–oxic boundary.

Stage B: A decrease in sedimentation rate could increase the extensive colonization of water/sediment interface by biofilms, favouring the foraminifer larval settlement of *Tolypammina gregaria* Wendt. Their colonial growth forming pillar-like structures could be related with a feeding behaviour as hypothetical suspension feeders. Its preference colonizing topographic highs and its own vertical accretion growth forming columns could increase the nutrient catchment from bottom currents in the surrounding oligotrophic conditions. Foraminifer tests rich often in acidic muchopolysaccharides could promote binding of dissolved ferric and manganic species forming passively Fe–Mn oxyhydroxides. Foraminifer tests could also serve as a favourable substrate for iron-oxidizing bacteria (maybe resulting in the biologically-induced *Frutexites* structures).

Stage C: Episodic deposition of phytodetritus layers could produce peaks of abundance and dominance of *Tolypammina gregaria-Frutexites* structures, as well as a slightly diversification of infaunal foraminiferal groups. The resulting shifts from oligotrophic to sporadic mesotrophic conditions on seafloor could be associated with changes on primary productivity in shallow waters. In off shore waters, far away from continental run off input, iron aerosol inputs during rain and storms (humid phases during a monsoonal climate: Carnian Pluvial Event) could favour these phytoplankton blooms.

Acknowledgments We thank to Jose Castillejo grant program from Universidad de Alcalá. This investigation was gratefully supported by the Deutsche Forschungsgemeinschaft Research Unit FOR 571 Geobiology of Organo- and Biofilms, publication no. 51.

References

Allouc J (1990) Quaternary crusts on slopes of Mediterranean Sea: a tentative explanation for their genesis. Marine Geology 94:205–238

Altenbach AV (1992) Short term processes and patterns in the foraminiferal response to organic flux rates. Marine Micropaleontology 19:119–129

Andersson JH, Woulds C, Shwartz M, Cowie GL, Levin LA, Soetaert K, Middelburg JJ (2007) Short-term fate of phytodetritus across the Arabian Sea oxygen minimum zone. Biogeosciences Discussion 4:2493–2523

Bau M (1996) Controls on the fractionation of isovalent trace elements in magmatic and aqueous systems: evidence from Y/Ho, Zr/Hf, and lanthanide tetrade effect. Contributions to Mineralogy and Petrology 123:323–333

Bau M, Dulski P (1996) Distribution of yttrium and rare-earth elements in the Penge and Kuruman iron-formations, Transvaal Supergroup, South Africa. Precambrian Research 79:37–55

Bellanca A, Di Stefano P, Neri R (1995) Sedimentology and isotope geochemistry of Carnian deep-water marl/limestone deposits from the Sicani Mountains, Sicily: environmental implications and evidence for planktonic source of lime mud. Palaeogeography, Palaeoclimatology, Palaeoecology 114:111–129

Böhm F, Brachert TC (1993) Deep-water stromatolites and Frutexites Maslov from the Early and Middle Jurassic of S-Germany and Austria. Facies 28:145–168

Bonatti E (1981) Metal deposits in the oceanic lithosphere. In: Emiliani C (ed) The sea, vol VII. Wiley, New York, pp 241–283

Burnett BR, Nealson KH (1981) Organic films and microorganisms associated with manganese nodules. Deep Sea Research 28A:637–645

Cavalazzi B, Barbieri R, Ori G (2007) Chemosynthetic microbialites in the Devonian carbonate mounds of Hamar Laghdad (Anti-Atlas, Morocco). Sedimentary Geology 200:73–88

Chafetz HS, Akdim B, Julia R, Reid A (1998) Mn- and Fe-rich black travertine shrubs: bacterially (and nanobacterially) induced precipitates. Journal of Sedimentary Research 68:404–412

Corliss BH, Brown CW, Sun X, Showers WJ (2009) Deep-sea benthic diversity linked to seasonality of pelagic productivity. Deep Sea Research I 56:835–841

Den Dulk M, Reichart GJ, Memon GA, Roelofs EMP, Zachariasse WJ, Van der Zwaan GJ (1998) Benthic foraminiferal response to variations in intensity of oxygen minimum zone in the northeast Arabian Sea. Marine Micropaleontology 35:43–66

Donaghay PL, Liss PS, Duce RA, Kester DR, Hanson AK, Villareal T, Tindale NW, Gifford DJ (1991) The role of episodic atmospheric nutrient inputs in the chemical and biological dynamics of the oceanic ecosystems. Oceanography 4:62–70

Dugolinsky BK, Margolis SV, Dudley WC (1977) Biogenic influence on growth of manganese nodules. Journal of Sedimentary Research 47:428–445

Dymond J, Lyle M, Finney B, Piper DZ, Murphy K, Conard R, Pisias N (1984) Ferromanganese nodules from MANOP Sites H, S, and R – Control of mineralogical and chemical composition by multiple accretionary processes. Geochimica et Cosmochimica Acta 48:931–949

Ehrlich HL (1963) Bacteriology of manganese nodules. I. Bacterial action on manganese in nodule enrichments. Applied and Environmental Microbiology 11:15–19

Ehrlich HL (1968) Bacteriology of manganese nodules. II. Manganese oxidation by cell-free extract from a manganese nodule bacterium. Applied and Environmental Microbiology 16:197–202

Ehrlich HL (1971) Bacteriology of manganese nodules. V. Effect of hydrostratic pressure on bacterial oxidation of Mn^{II} and reduction of MnO_2. Applied and Environmental Microbiology 21:306–310

Elderfield H, Greaves MJ (1981) Negative Cerium anomalies in the rare earth element pattern of oceanic ferromanganese nodules. Earth and Planetary Science Letters 55:163–170

Gawlick HJ (2000) Paläogeographie der Obertrias-Karbonatplattformen in den Nördlichen Kalkalpen. Mitteilungen der Österreichischen Geologischen Gesellschaft 44:45–95

Ghiorse WC (1984) Biology of iron- and manganese-depositing bacteria. Annual Review of Microbiology 38:515–550

Ghiorse WC, Ehrlich HL (1992) Microbial biomineralization of iron and manganese. In: Skinner HCW, Fitzpatrick RW, (eds) Biomineralization processes of iron and manganese: modern and ancient environments. Catena Verlag, Cremlingen-Destedt, pp 75–99

Gischler E (1996) Late Devonian-Early Carboniferous deep-water coral assemblages and sedimentation on a Devonian seamount: Iberg Reef, Harz Mts., Germany. Palaeogeography, Palaeoclimatology, Palaeoecology 123:297–322

Gooday AJ (1988) A response to benthic Foraminifera to the deposition of phytodetritus in the deep sea. Nature 332:70–73

Gooday AJ, Levin L, Linke P, Heeger T (1992) The role of benthic foraminifera in deep-sea food webs and carbon cycling. In: Rowe GP (ed) Deep-sea food chains and the global carbon cycling. Kluwer Academic Publisher, The Netherlands, pp 63–91

Gooday AJ, Bernhard JM, Levin LA, Suhr S (2000) Foraminifera in the Arabian Sea oxygen minimum zone and other deficient settings: taxonomic composition, diversity and relation to metazoan faunas. Deep Sea Research II 57:25–54

Graf G (1989) Benthic-pelagic coupling in a deep-sea benthic community. Nature 341:437–439

Graham JW, Cooper SC (1959) Biological origin of manganese-rich deposits of the sea floor. Nature 183:1050–1051

Greenslate J (1974) Microorganisms participate in the construction of manganese nodules. Nature 649:181–183

Hanson AK, Tindale NW, Abdel-Moati MAR (2001) An Equatorial Pacific rain event: influence on the distribution of iron and hydrogen peroxide in surface waters. Marine Chemistry 75:69–88

Hein JR, Bohrson WA, Shulz MS, Noble M, Clague DA (1992) Variations in the fine-scale composition of a central Pacific ferromanganese crust: paleooceanographic implications. Paleooceanography 7:63–77

Hein JR, Yeh H-W, Gunn SH, Gibbs AE, Wang C-H (1994) Composition of hydrothermal ironstones from central Pacific seamounts. Geochimica et Cosmochimica Acta 58:179–189

Hofmann HJ, Grotzinger JP (1985) Shelf-facies microbiotas from the Odjick and Rocknest formations (Epworth Group; 1.89 Ga), northwestern Canada. Canadian Journal of Earth Sciences 22:1781–1792

Hornung T, Brandner R (2005) Biochronostratigraphy of the Reingraben Turnover (Hallstatt Facies belt): local black shale events controlled by regional, tectonics, climatic change and plate tectonics. Facies 2005:460–479

Hornung T, Krystyn L, Brandner R (2007a) A Tethys-wide mid-Carnian (Upper Triassic) carbonate productivity crisis: evidence for the Alpine Reingraben Event from Spiti (Indian Himalaya). Journal of Asian Earth Sciences 30:285–302

Hornung T, Spatzenegger A, Joachimski MM (2007b) Multistratigraphy of condensed ammonoid beds of the Rappolstein (Berchtesgaden, southern Germany): unravelling palaeoenvironmental conditions on 'Hallstatt deep swells' during Reingraben Event (Late Lower Carnian). Facies 53:267–292

Jenkyns HC (1970) Fossil manganese nodules from the West Sicilian Jurassic. Eclogae Geologicae Helvetiae 63:741–774

Jiménez-Millán J, Nieto LM (2006) Geochemical and mineralogical evidence of tectonic and sedimentary factors controlling the origin of ferromanganese crusts associated to stratigraphic discontinuities (Betic Cordilleras, SE Spain). Chemie der Erde 68:323–336

Jones B, Manning DAC (1994) Comparison of geochemical indices used for the interpretation of palaeoredox conditions in ancient mudstones. Chemical Geology 111:111–129

Jorissen FJ, de Stigter HC, Widmark JGV (1995) A conceptual model explaining benthic foraminiferal microhabitats. Marine Micropaleontology 22:3–15

Kazmierczak J, Kempe S (2006) Modern analogues of Precambrian stromatolites from caldera lakes of Niuafo'ou Island, Tonga. Naturwissenschaften 93:119–126

Konhauser KO (1998) Diversity of bacterial iron mineralization. Earth Science Reviews 43:91–121

Koschinsky A, Stascheit A, Bau M, Halbach P (1997) Effects on phosphatization on the geochemical and mineralogical composition of marine ferromanganese crusts. Geochimica et Cosmochimica Acta 19:4079–4094

Krystyn L (1991) Die Fossillagerstätten der alpinen Trias. In: Vasicek V, Krystyn L, Golebiowski R (eds) Exkursionen im Jungpaläozoikum und Mesozoikum Österreichs, Exkusionsführer Tagung der Österreichischen Paläontologischen Gesellschaft, Wien, pp 23–78

Kuma K, Matsuga K (1995) Availability of colloidal ferric oxides to coastal marine phytoplankton. Marine Biology 122:1–11

Larkin KE, Gooday AJ (2009) Foraminiferal faunal responses to monsoon-driven changes in organic matter and oxygen availability at 140 m and 300 m water depth in the NE Arabian Sea. Deep Sea Research II 56:403–421

Levin L, Gutiérraz D, Rathburn A, Neira C, Sellanes J, Muños Gallardo V, Salamanca M (2002) Benthic processes on the Peru margin: a transect across the oxygen minimun zone during the 1997–98 El Niño. Progress in Oceanography 53:1–27

Lowenstam HA, Weiner S (1989) On biomineralization. Oxford University Press, New York

Mamet B, Préat A (2006) Iron-bacteria mediation in Phanerozoic red limestones: state of the art. Sedimentary Geology 185:147–157

Mandl GW (2000) The Alpine sector of the Tethyan shelf: examples of Triassic to Jurassic sedimentation and deformation from the Northern Calcareous Alps. Mitteilungen der Österreichischen Geologischen Gesellschaft 92:61–79

Margolis SV, Burns RG (1976) Pacific deep-sea manganese nodules: their distribution, composition and origin. Annual Review of Earth and Planetary Sciences 4:229–263

Martin JH, Fitzwater SE (1988) Iron deficiency limits phytoplankton growth in the northeast Pacific subarctic. Nature 331:341–343

Maslov VP (1960) Stromatolity (ick genezis, metodizucheniya, svjaz's fatsiyami i geologicheskoe znachenie na primere Ordovika Sibirskoj Plataformy). Trudy Instituta Geologii Akademiia nauk SSSR 41:1–188

McLennan SM (1989) Rare earth elements in sedimentary rocks: influence of provenance and sedimentary processes. In: Lipin BR McKay GA (eds) Geochemistry and Mineralogy of Rare Earth Elements. Reviews in Mineralogy 21:169–200

Millero FJ, Wensheng Y, Aicher J (1995) The speciation of Fe(II) and Fe(III) in natural waters. Marine Chemistry 50:21–39

Moodley L, Boschker HTS, Middelburg JJ, Pel R, Herman PMJ, de Deckere E, Heip C (2000) Ecological significance of benthic foraminifera: 13+C labelling experiments. Marine Ecology Progress Series 202:289–295

Moodley L, Middelbrug JJ, Boschker HTS, Duineveld GCA, Pel R, Herman PMJ, Heip CHR (2002) Bacteria and Foraminifera: key players in a short-term deep-sea benthic response to phytodetritus. Marine Ecology Progress Series 236:23–29

Mullineaux LS (1987) Organisms living on manganese nodules and crusts: distribution and abundance at three North Pacific sites. Deep Sea Research 34:165–184

Mullineaux LS (1988) Taxonomic notes on large agglutinated foraminifers encrusting manganese nodules, including the description of a new genus, *Chondrodapis* (Komokiacea). Journal of Foraminiferal Research 18:46–53

Mullineaux LS (1989) Vertical distribution of the epifauna on manganese nodules: implications for settlement and feeding. Limnology and Oceanography 34:1247–1262

Mullineaux LS, Butman CA (1990) Recruitment of encrusting benthic invertebrates in boundary-layer flows: a deep-water experiment on Cross Seamount. Limnology and Oceanography 35:409–423

Mutti M, Weissert H (1995) Triassic monsoonal climate and its signature in Ladinian-Carnian carbonate platforms (Southern Alps, Italy). Journal of Sedimentary Research B65:357–367

Myrow PM, Coniglio M (1991) Origin and diagenesis of cryptobiontic Frutexites in the Chapel island Formation (Vendian to Early Cambrian) of Southeast Newfoundland, Canada. Palaios 6:572–585

Nelson KH, Tebo B (1980) Structural features on manganese precipitating bacteria. Origins Life 10:117–126

Playford PE, Cockbain AE, Druce EC, Wray JL (1976) Devonian stromatolites from the Canning Basin, Westerns Australia. In: Walter MR (ed) Stromatolites, Developments in sedimentology 20. Elsevier, Amsterdam, pp 543–463

Playford PE, McLaren DJ, Orth C, Gilmore JS, Goodfellow WD (1984) Iridium anomaly in the Upper Devonian of the Canning Basin, Western Australia. Science 226:437–439

Puteanus D, Halbach P (1988) Correlation of Co concentration and growth rate – a method for age determination of ferromanganese crusts. Chemical Geology 69:73–85

Reitner J, Wilmsen M, Neuweiler F (1995) Cenomanian/Turonian sponge microbialite deep-water hardground community (Liencres, Northern Spain). Facies 32:203–212

Resig JM, Glenn CR (1997) Foraminifera encrusting phosphoritic hardgrounds of the Peruvian upwelling zone: taxonomy, geochemistry and distribution. Journal of Foraminiferal Research 27:133–150

Riding R (1991) Calcified Cyanobacteria. In: Riding R (ed) Calcareous algae and stromatolites, Springer, Berlin, pp 55–87

Riemann F (1983) Biological aspects of deep-sea manganese nodule formation. Oceanologica Acta 6:303–311

Rigo M, Preto N, Roghi G, Tateo F, Mietto P (2007) A rise in the Carbonate Compensation Depth of western Thetys in the Carnian (Late Triassic): deep-water evidence for the Carnian Pluvial Event. Palaeogeography, Palaeoclimatology, Palaeoecology 246:188–205

Schlager W (1969) Das Zusammenwirken von Sedimentation und Bruchtektonik in den triadischen Hallstätterkalken der Ostalpen. Geologische Rundschau 59:289–308

Schönfeld J (2002) Recent benthic foraminiferal assemblages in deep high energy environments from the Gulf of Cadiz. Marine Micropaleontology 44:141–162

Schönfeld J, Numberger L (2007) The benthic foraminiferal response to the 2004 spring bloom in the western Baltic Sea. Marine Micropaleontology 65:78–95

Segl M, Mangini A, Bonani G, Hofmann G, Nessi M, Sutter M, Wölfi W, Friedrich G, Plüger W, Wiechowski A, Beer J (1984) ^{10}Be dating of manganese crust from central North Pacific and implications for ocean paleocirculation. Nature 309:540–543

Sen Gupta BK, Smith LE, Lobegeier MK (2007) Attachment of Foraminifera to vestimentiferan tubeworms at cold seeps: refuge from seafloor hypoxia and sulfide toxicity. Marine Micropaleontology 62:1–6

Simms MJ, Johnson ALA, Ruffel A (1994) Biotic and climatic change in the Carnian (Triassic of Europe and adjacent areas). In: Fraser N, Sues HD (eds) In the shadow of the dinosaurs. Cambridge University Press, Cambridge, pp 352–365

Snider LJ, Burnett BR, Hessler RR (1984) The composition and distribution of meiofauna and nanobiota in a central North Pacific deep-sea area. Deep Sea Research 31:1225–1249

Stampfli GM, Borel GD (2002) A plate tectonic model for the Paleozoic and Mesozoic constrained by dynamic plate boundaries and restored synthetic oceanic isochrones. Earth and Planetary Science Letters 196:17–33

Strekopytov S, Dubinin A, Uspenskaya T (2000) Geochemical and mineralogical studies of Fe–Mn Nodules and crusts from the White Sea: potential role of benthic fauna in their formation. Goldschmidt 2000, Journal of Conference Abstracts 5(2):963

Tebo BM, Bargar JR, Clement BG, Dick GJ, Murray KJ, Parker D, Verity R, Webb SM (2004) Biogenic manganese oxides: properties and mechanisms of formation. Annual Review of Earth and Planetary Sciences 32:287–328

Toscano F, Raspini A (2005) Epilithozoan fauna associated with ferromanganese crustgrounds on the continental slope segment between Capri and Li Galli Islands (Bay of Salerno, Northern Tyrrehenian Sea, Italy). Facies 50:427–441

Tsien HH (1979) Paleoecology of algal-bearing facies in the Devonian (Couvirtian to Frasnian) Reef Complexes of Belgium. Palaeogeography, Palaeoclimatology, Palaeoecology 27: 103–127

Tucker ME (1973) Ferromanganese nodules from the Devonian of the Montagne Noire (S. France) and the West Germany. Geologische Rundschau 62:137–153

Uematsu M, Duce RA, Prospero JM (1985) Deposition of atmospheric mineral particles in the North Pacific Ocean. Journal of Atmospheric Chemistry 3:123–138

Van der Zwaan GJ, Duijnstee IAP, Dulk D, Ernst SR, Jannink NT, Kouwenhoven TJ (1999) Benthic foraminifers: proxies or problems? A review of paleoecological concepts. Earth Science Reviews 46:213–239

Verlaan PA (1992) Benthic recruitment and manganese crust formation on seamounts. Marine Biology 113:171–174

von Mojsisovics E (1873–1902) Das Gebirge um Hallstatt I. Abhandlungen der Geologischen Reichsanstalt 6/1:356

von Stackelberg U (1984) Significance of benthic organisms for the growth and movement of manganese nodules, Equatorial North Pacific. Geo-Marine Letters 4:37–42

Walter MR, Awramik SM (1979) *Frutexites* from stromatolites of Gunflint iron formation of Canada, and its biological affinities. Precambrian Research 9:23–33

Wang W, Schröder HC, Wiens M, Schloßmacher U, Müller WEG (2009) Manganese/polymetallic nodules: micro-structural characterization of exolithobiontic- and endolithobionthic microbial biofims by scanning electron microscopy. Micron 40:350–358

Wells ML, Zorkin NG, Lewis AG (1983) The role of colloidal chemistry in providing a source of iron to phytoplankton. Journal of Marine Research 41:731–746

Wells ML, Price NM, Bruland KW (1995) Iron chemistry in seawater and its relationship to phytoplankton: a workshop report. Marine Chemistry 48:157–158

Wendt J (1969) Foraminiferen-'Riffe' im karnischen Hallstätter Kalk des Feuerkogels (Steiermark, Österreich). Paläontologische Zeitschrift 43:177–193

Wendt J (1970) Stratigraphische Kondensation in triadischen und jurassischen Cephalopodenkalken der Tethys. Neues Jahrbuch für Geologie und Paläontologie Monatshefte 1970/7:433–448

Wendt J (1974) Encrusting organisms in deep-sea manganese nodules. In: Hsü KJ, Jenkyns HC (eds) Pelagic sediments: on land and under sea, International Association of Sedimentologists, Special Publication 1. Blackwell, Oxford, London, Edinburgh, Melbourne, pp 437–447

New Geochemical Method to Characterise Microbialites from the St. Cassian Formation, Dolomites, Northeastern Italy

Francisco Sánchez-Beristain, Nadine Schäfer, Klaus Simon, and Joachim Reitner

1 Introduction

The St. Cassian Formation (Upper Ladinian–Lower Carnian, Dolomites, Northeastern Italy) has been the subject of numerous studies "with various approaches". The St. Cassian Formation was first mentioned in the first half of the nineteenth century (e.g., Wissmann and Münster 1841). Neuweiler and Reitner (1995), Rech (1998), and Russo et al. (1998) carried out general revisions of the complex problems linked to micrite production in the Triassic platforms. The main focus of the St. Cassian researchers was on the "Cipit boulders" ("Cipitkalke", von Richthofen 1860), which represent platform-derived olistoliths that were exported into the basin (Russo et al. 1991). Towards the top of the Julian Substage (Carnian) at the base of the Heiligkreutz/Dürrenstein Formation (i.e., Alpe di Specie), biologically constructed lens bodies outcrop in alternation with marls and bioclastic calcarenites (Wendt 1982; Russo et al. 1991, 1998; Russo 2005). The investigations on Cipit Boulders have been mainly based on taxonomy or paleoecology (e.g., Dieci et al. 1968, 1974; Reitner 1987; Reitner and Engeser 1989; Reitner 1992), diagenesis and sedimentology (e.g., Russo et al. 1991, 1997), and most recently, on geobiology (Neuweiler and Reitner 1995; Russo et al. 1997, 1998; Rech 1998; Sánchez-Beristain and Reitner 2008). These carbonate boulders exhibit an extraordinary preservation, which, according to the most accepted model (e.g., Fürsich and Wendt 1977; Wendt and Fürsich 1980; Wendt 1982; Russo et al. 1991; Rech 1998), is the

F. Sánchez-Beristain (✉), N. Schäfer, and J. Reitner
Geoscience Centre, University of Göttingen, Geobiology Division, Goldschmidtstraße 3, 37077, Göttingen, Germany
e-mail: fsanche@gwdg.de

K. Simon
Geoscience Centre, University of Göttingen, Geochemistry Division, Goldschmidtstraße 1, 37077, Göttingen, Germany

J. Reitner et al., *Advances in Stromatolite Geobiology*,
Lecture Notes in Earth Sciences 131, DOI 10.1007/978-3-642-10415-2_26,

result of their rapid burial in the clay-rich basin, which prevented the main diagenetic transformation.

According to most researchers (e.g., Wendt and Fürsich 1980; Russo et al. 1991), the best preservation is found in the samples from the locality of Alpe di Specie or Seelandalpe. Another distinctive feature of these boulders is that their faunal composition mainly consists of associations of either corals or "coralline" sponges (Russo et al. 1991) or, less commonly, both. Therefore, boundstones are the more frequent facies; however, these components are neither the only nor the most abundant of the Cipit-Kalke. Russo et al. (1997), for example, identified three kinds of microbialites in the Cipit Boulders of Upper Ladinian from Punta Grohmann based on their micromorphological texture and organic matter content. For the first time, Cipit Boulders almost solely composed of microbial carbonates were found.

The purpose of this investigation was to characterise microbialites from the Alpe di Specie/Seelandalpe of the St. Cassian Formation through LA-ICP-MS and Raman Spectroscopy. Additionally, we compared the spectra of the microbialites with skeletal components in the samples to differentiate the various types.

2 Samples and Methods

A set of samples of different carbonate facies was selected for LA-ICP-MS to measure the concentration of select trace elements. Samples included a coral of the genus *Margarosmilia* (FS-SA58-K1), a "chaetetid" sponge (Cas 12-S1), three allochthonous micrite samples (Cas12-A1, Cas12-A2, and FS-SA58-A1), a late sparry cement (Cas 12-Z1), and ten samples of microbialites (Cas 12-M1, Cas 12-M2, Cas 12-M3, Cas 12-M4, FS-SA-001-M1, FS-SA-001-M2, FS-SA-001-M3, FS-SA-001-M4, FS-SA-58-M1 and FS-SA-001-M2). All samples were collected from the Alpe di Specie/Seelandalpe and they were labelled following the methodology of Kamber and Webb (2007) (each "sample" = one time-series laser ablation measurement procedure, see below for further details). The mentioned microbialites are peloidal thrombolites (Fig. 1a, b) with no visible traces of terrigenous/siliciclastic material, as was revealed by a preliminary examination of their corresponding petrographical thin sections.

Prior to the selection of the samples, we analysed their corresponding thin sections by cold cathodoluminescence microscopy using a Siemens cathode that was linked to a Zeiss microscope to assess the preservation state of the sample. Working conditions were set as the following: gun potential = 13–14 kV; current intensity = 300–350 mA. The three slabs whose corresponding thin sections displayed the best cathodoluminescence, which was based on the non-luminescence of their original skeletal aragonite, were selected. The quality of their preservation was confirmed by microprobe analysis, in which the concentrations of some elements

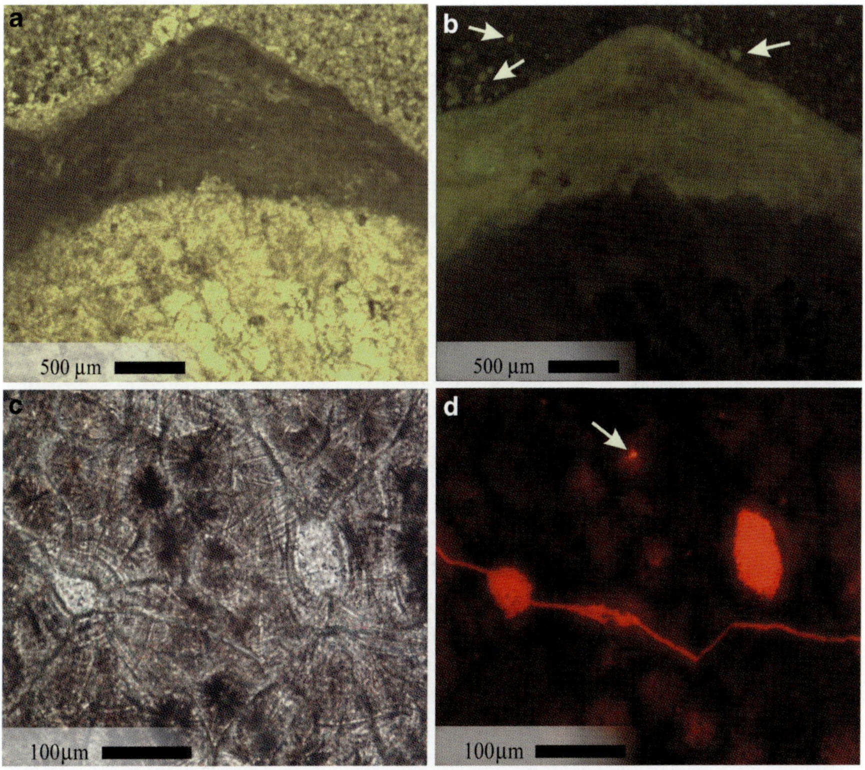

Fig. 1 (**a**) Sample FSSA 58. Clotted thrombolitic microbialite from Alpe di Specie/Seelandalpe close to Schluderbach, (Dolomites, Northeastern Italy) encrusting a coral of the genus *Margarosmilia*. Scale bar: 500 mm. (**b**) Same view in epifluorescence. Note high *yellow-green* fluorescence of the thrombolitic facies in contrast to the allomicritic-microsparitic facies (*top*) and the corallite of *Margarosmilia* (*bottom*). Note peloids in allomicrite (*arrows*), which show a fluorescence only comparable with the thrombolite. (**c**) Light microscopy view of the spherulitic microstructure of a "chaetetid" sponge. *Arrow* points to the center of the structure. (**d**) CL view, in which the center of the spherulite (*arrow*) displays a similar luminescence as the cement infillings

(Ca, Mg, Fe, Mn, Sr and S) were measured in the microbialites, skeletons, detrital micrite, and cements. Microprobe analyses were performed by Andreas Kronz at the Microprobe Laboratory of the Department of Geochemistry of the GZG Göttingen, using a JEOL JXA – 8900 RL microprobe at 15.0 kV linked to a SEM Microscope.

Additionally the microbialites were analysed by UV epifluorescence microscopy following the methodology of Neuweiler and Reitner (1995).

To perform Raman spectroscopy, we selected some additional samples from the personal collection of Joachim Reitner. The samples were thin sections in the format of 7.5 × 10 cm and were labelled with the following acronyms: JR20a, JR22, JR43, JR58, JR79, JR80, and HR66.

2.1 LA-ICP-MS

Laser ablation was performed with an Excimer 193 nm COMPEX 110 (Lambda Physik, Göttingen, Germany) equipped with an optical beam shape enhancer GEOLAS (MicroLas, Göttingen, Germany) at about 3 J/cm^2 with spot sizes and a line width of 120 μm. The ablated material was carried into the ICP-MS DRC II (Perkin-Elmer, Canada) plasma by an Argon gas flow. Kamber and Webb (2007) pioneered the detection of vital trace elements in Devonian *Renalcis* thrombolitic microbialites and associated samples via in situ time-series LA-ICP-MS. In this study, the time-series were recorded under the following conditions: 26 isotopes were measured at dwelling times of 20 ms; measurement duration was 165 s; and 402 Reading Points (RP) were taken. Measurement began after 30 s of pre-warming and ended 10 s prior to the macro end, but only measurements from 10 to 130 s (280 RPs in total) after the start of ablation were taken.

Some of the trace elements have count intensities near background levels. The quantification of a single sweep, which has a dwelling time of 20 ms for one isotope, is very erroneous; however, if five or more sweeps yield the same results, the error of the average decreases dramatically. Standardisation and quantification was done using NBS610; the internal standard sotope, ^{43}Ca, was assumed to have variable concentrations in the carbonates, as was obtained from microprobe results (390,000 ppm for the aragonite and the late cement, 360,000 ppm for the microbialite, and 350,000 ppm for the allochthonous micrite/detritus). Absolute concentrations are prone to having rather large errors, especially for those elements which have concentrations closest to background levels; however, isotopic ratios are more correct with errors less than 2%.

2.2 Raman Spectroscopy

Raman spectra were recorded using a Horiba Jobin Yvon LabRam-HR 800 UV microprobe with a focal length of 800 mm. The following were used as excitation wavelengths: (1) the 244 nm line of a frequency-doubled Ion Laser (Coherent Innova 90C FreD) with a laser power of 30 mW; and (2) the 488 nm line of an Argon Ion Laser (Melles Griot IMA 106020B0S) with a laser power of 25 mW. In the case of the 244 nm line, the laser beam was dispersed by a 2,400 lines/mm grating on a liquid nitrogen cooled CCD detector of 2,048 × 512 pixels, which yielded a spectral resolution of 1.05 cm^{-1}. The 488 nm-line laser was dispersed by a 600 line/mm grating on a CCD detector of 1,024 × 256 pixels, which yielded a spectral resolution of 0.43 cm^{-1}.

An Olympus BX41 microscope equipped with an OFR LMV-40×-UVB objective that had a numerical aperture of 0.5 focused the laser light onto the sample. To obtain an improved signal-to-noise ratio, the confocal hole diameter was set to

200 μm. The typical acquisition time varied from 30 to 120 s for a spectral range of 500–2,500 cm^{-1}. To calibrate the spectrometer, a diamond standard was used, which had its major peak at 1,332.0 cm^{-1}.

An Olympus BX41 microscope equipped with an Olympus MPlane 100× objective with a numerical aperture of 0.9 focused the laser light onto the sample. The confocal hole diameter was set to 100 μm. The acquisition time was 5 s, which yielded a spectral range of 100–2,000 cm^{-1}. Using a filter, the power of the laser was reduced to 10% of the original power at the laser exit. To calibrate the spectrometer, a silicon standard was used, which had its major peak at 520.4 cm^{-1}. All spectra were recorded and processed using LabSpecTM version 5.19.17 (Jobin-Yvon, Villeneuve d'Ascq, France).

Correctly choosing the excitation wavelength is an important part of Raman spectroscopy. Normally, using an excitation wavelength of 488 nm for carbonates (Frech et al. 1980) is sufficient, but in biogenic carbonates (Urmos et al. 1991) and microbially induced carbonates that laser excitation wavelength can cause auto-fluorescence of the sample, which masks the Raman signal. When we attempted to analyse the microbialites with a 488 nm excitation wavelength, it was only successful for the skeletal components that were present in the microbialites. Therefore, we switched to an excitation wavelength of 244 nm to avoid the fluorescence in the microbialites. It is not possible to utilise the spectrum below 500 cm^{-1} with an excitation wavelength of 244 nm because of technical limitations. Therefore, the distinction between aragonitic and calcitic phases was only possible by analysing differences in the range of 1,400–1,600 cm^{-1}.

3 Results and Discussions

3.1 *Preservation State of the Samples*

Qualitative cathodoluminescence (CL) observations did not reveal luminescence in the chaetetid sponge or in the coral; however, some anomalies were detected, particularly in the centre of some spherulites of the sponge basal skeletons. The observation of low luminescence (Fig. 1c, d) agrees with the results of Laghi et al. (1983), who found that diagenesis first occurs at the centre of the spherulites. We have observed the same phenomenon. These data are supported by microprobe analyses revealing high Sr concentrations, which are typical of aragonite and ranged from 7,293 to 8,204 ppm for the chaetetid sponges and from 6,295 to 7,296 ppm for the coral. These values are consistent with the low Mn (~10 ppm) and Fe (~70 ppm) concentrations. In the sparry cement we recorded low Sr and high Mn and Fe concentrations (~500 ppm and ~4,000 ppm, respectively). The cement also had a low-Mg calcite mineralogy, with a Mg concentration of 2.00% mol Mg.

3.2 *Magnesium*

The magnesium content in all the microbialites ranged from 12,620 to 17,846 ppm, which in principle indicates an original biological high-Mg calcite composition (Prasada Rao 1996). This result is consistent with the microprobe measurements, which ranged from 4.85 to 8.8% mol Mg and were thus beyond the accepted minimum value of 4% mol Mg (Dickson 1990; Flügel 2004). Microbialites with high-Mg calcite mineralogy have been widely reported in the literature (e.g., Reitner 1993; Zankl 1993; Schäfer 2006). The Raman spectra (see below for further explanation) further support these findings.

The Mg concentration in the allochthonous micrites was considerably higher than in the microbialites (Table 1) and may have been due to the presence of small quantities of dolomitic cement within it, which was recorded by petrographical examination. High concentrations of Mg in allochthonous components were also determined by Rech (1998) by means of ADS spectroscopy.

3.3 *Trace Elements*

3.3.1 "Non-biological" Elements and Phosphorus

In this section, we will analyse and discuss the results of trace elements without a known vital effect. Although phosphorus plays an indispensable role in all biological systems (e.g., Fraústro da Silva and Williams 1991), it is included here because of its effect on other elements that are discussed in this section.

All microbialites exhibited very high concentrations of silicon (823–4,465 ppm) and aluminium (355–1,785 ppm), except sample FSSA 58-M1, which had 6,768 ppm Si and 3,030 ppm Al, respectively. These values were not far from those of the allochthonous micrite samples, whereas the aragonitic skeletons and the sparry cement were more depleted (Table 1). Sample FSSA 58-M1 will not be further considered in the discussion of these elements because of its extremely high siliciclastic content, which is only comparable to that of allomicrites.

For the other trace elements in this section, in particular zirconium and rubidium, the well-known enrichment was observed in the allomicrite samples ($Zr > 4.50$ ppm and $Rb > 3.94$) with respect to that of microbialites ($Zr < 3.91$ ppm and $Rb < 2.86$ ppm). While Zr was depleted in both the cement and in the aragonitic skeletons, the Rb values remained high in the cement, which were comparable to those of the microbialites with lesser concentrations of this element. In the aragonitic facies, the levels of this element were negligible. Titanium exhibited features of both Zr and Rb insofar that their concentration within the skeletons was low and within the sparry cement were more similar to the lowest values of microbialites, while the allomicrite facies were close to or slightly more elevated than the highest microbialitic values (Table 1).

Table 1 LA-ICP-MS concentrations in ppm of Mg and trace elements in microbialite, allomicrite, skeletal aragonite and sparry cement facies. Concentration in the sparry cement and aragonite were normalized to a [Ca] of 390,000 ppm, whereas 360,000 and 350,000 ppm was considered for microbialite and allomicrite facies respectively. See text for explanation

Microbialites	Mg24	Al27	Si29	P31	S34	Ti47	V51	Cr53	Co59	Ni60	Cu63	Zn66	Rb85
Cas12M1	13,078.51	728.06	1,531.44	568.13	418.53	45.39	3.87	5.07	0.90	7.48	4.95	5.43	1.04
Cas12M2	12,620.07	1,612.41	3,924.30	467.78	566.30	126.19	10.53	4.85	1.75	13.49	5.01	4.39	2.42
Cas12M3	13,188.30	1,647.08	4,423.47	399.61	645.88	90.11	5.96	7.17	1.39	15.37	4.75	12.39	2.67
Cas12M4	17,714.55	1,785.89	4,465.31	215.03	1,174.16	120.37	3.42	6.91	4.96	38.24	4.76	5.03	2.86
FSSA001-M1	15,438.29	1,657.19	4,119.00	258.92	1,113.00	102.86	3.57	7.41	4.86	37.89	4.93	14.08	2.45
FSSA001-M2	13,556.99	355.39	823.59	169.66	967.30	33.68	1.40	3.79	3.40	33.57	4.14	2.16	0.42
FSSA001-M3	12,983.30	1,004.62	2,346.31	172.54	993.69	51.93	2.78	5.07	5.87	44.39	5.11	2.23	1.25
FSSA001-M4	14,889.99	1,610.47	3,566.39	240.26	881.84	93.17	2.96	5.47	4.77	35.45	4.64	10.14	2.15
FS-SA58 M1	17,846.81	3,030.08	6,768.05	2,710.11	1,009.29	150.50	6.20	15.13	1.27	6.58	4.46	8.28	3.55
FS-SA58 M2	15,170.28	369.82	1,008.55	141.69	1,296.41	23.28	2.88	9.26	1.01	8.26	4.74	7.26	0.61
r^2 (Si-element)						0.90	0.28	0.41	0.00	0.01	0.00	0.22	0.97
r^2 (Al-element)						0.89	0.25	0.44	0.00	0.01	0.00	0.18	0.93
Allochthonous micrite													
Cas 12-A1	95,113.98	1,800.27	4,178.92	1,499.32	591.97	124.50	8.83	15.05	3.71	19.30	1.64	26.19	3.94
Cas 12-A2	73,473.05	1,947.10	4,913.84	1,538.20	854.38	117.84	7.24	11.93	2.70	15.01	1.98	23.49	4.09
FSSA58-A1	49,234.56	2,828.16	6,878.54	789.48	962.16	182.63	7.85	11.35	1.68	15.75	2.39	12.74	4.92
Other facies types													
Cas 12-S1 (Chaetetid sponge)	592.44	6.06	474.83	35.69	288.84	4.31	1.17	8.68	0.49	3.01	1.59	4.31	0.24
FSSA58-K1 (*Margarosmilia*)	269.05	19.13	233.16	24.76	358.07	1.92	0.77	5.83	0.35	2.17	1.12	5.13	0.16
Cas 12-Z1 (sparry cement)	4,817.96	6.62	446.64	180.59	464.16	44.54	1.01	5.91	0.99	7.35	1.98	4.99	1.16

(continued)

Table 1 (continued)

Microbialites	Y89	Zr90	Nb93	Mo95	Sn120	Sb121	La139	Ce140	Hf178	Ta181	Th232	U238	Th/U
Cas12M1	19.45	2.47	0.20	1.04	0.67	6.85	15.43	22.18	0.03		0.13	8.84	0.02
Cas12M2	4.01	3.91	0.37	1.17	0.47	0.48	3.50	5.26	0.26		0.26	35.92	0.01
Cas12M3	5.15	3.08	0.32	1.05	2.12	0.44	4.44	7.10	0.08		0.36	24.04	0.02
Cas12M4	5.59	3.38	0.45	1.18	0.57	0.41	5.30	8.27	0.41		0.28	1.31	0.21
FSSA001-M1	4.09	3.06	0.39	1.29	0.39	0.32	3.52	5.23	0.07	0.00	0.19	1.75	0.11
FSSA001-M2	7.01	1.06	0.09	1.21	0.34	0.16	3.45	4.83	0.11	0.00	0.12	0.58	0.21
FSSA001-M3	4.11	1.57	0.16	1.50	0.31	0.19	3.89	6.07	0.14	0.00	0.12	1.47	0.08
FSSA001-M4	3.62	2.40	0.26	1.42	0.38	0.32	3.15	4.79	0.21	0.00	0.20	1.08	0.18
FS-SA58 M1	2.02	4.59	0.41	1.27	0.58	0.66	1.30	2.03	0.60		0.31	2.51	0.12
FS-SA58 M2	3.18	0.87	0.20	1.20	0.60	3.19	3.13	3.94	0.40		0.16	4.96	0.03
r^2 (Si-element)	0.19	0.83	0.75	0.00	0.06	0.20	0.15	0.12	0.25		0.65	0.02	
r^2 (Al-element)	0.17	0.82	0.68	0.01	0.02	0.16	0.13	0.11	0.29		0.14	0.01	
Allochthonous micrite													
Cas 12-A1	30.49	4.51	0.37	0.36	1.68	1.58	27.38	41.73	0.11		0.56	1.73	0.32
Cas 12-A2	30.92	7.53	0.43	0.23	0.98	0.87	28.83	48.85	0.13		0.67	2.00	0.33
FSSA58-A1	6.21	5.37	0.53	0.32	0.71	0.75	4.70	6.93	0.51		0.48	1.83	0.26
Other facies types													
Cas 12-S1 (Chaetetid sponge)	0.48	0.07	0.22	0.98	0.71	0.84	1.02	1.16	0.02		0.21	5.79	0.04
FSSA58-K1 (*Margarosmilia*)	0.14	0.44	0.16	0.81	0.47	0.34	0.14	0.15	0.52		0.21	3.20	0.07
Cas 12-Z1 (sparry cement)	2.96	0.23	0.73	5.02	0.71	0.77	2.42	2.10	2.95		2.53	0.28	9.03

A further subgroup of these elements is constituted by thorium, niobium, and hafnium. Their behaviour is well characterized, as they were most enriched within the sparry cement, whereas their concentrations in allomicritic facies were just slightly higher than in microbialites. The aragonitic skeletons exhibited concentrations similar to those of the microbialites. Hf is the only element with concentrations near background levels, and therefore had the highest average error in the time series of microbialite and allomicrite samples (see Sect. 2). Tantalum was measured only in four microbialite samples.

Webb and Kamber (2000) pioneered the determination of the genuine systematics of the distribution of Rare Earth Elements + Yttrium (REY patterns) in a set of siliciclastic-free Australian Holocene microbialites from the Great Barrier Reef. However, in all our microbialite samples, the concentrations of Si, Al, and all the trace elements that have been discussed are higher than other microbialites that have been described in the literature (Webb and Kamber 2000; Van Kranendonk et al. 2003; Nothdurft et al. 2004; Olivier and Boyet 2006). Moreover, a very strong correlation (r^2 coefficient) was found between Si and the following elements for the entire microbialite data set: Rb $= 0.97$; Ti $= 0.90$; Zr $= 0.83$; Nb $= 0.75$ and Th $= 0.67$ (Fig. 2a–e). Most of these correlations likely caused by their high siliciclastic content. Some of our samples reach Zr concentrations as high as 2% relative to the Earth's crust (Wedepohl 1995), which indicates a high degree of crustal input into the material. In addition, the concentrations of Yttrium, Lantanum, and Cerium were more enriched in the allochthonous micrite than in other phases (Table 1). In two allomicrite samples (Cas 12-A1 and Cas 12-A2), the concentration of Y was even higher than in the continental crust (compare Wedepohl 1995). In addition, La and Ce values in microbialite samples were considerably higher than those reported by several authors who measured Rare Earth Elements + Yttrium patterns (REY) for fossil and recent microbialites (e.g., Webb and Kamber 2000; Van Kranendonk et al. 2003). Although a correlation does not exist between Si and Y, La, or Ce, the increased concentrations of the latter three and their high concentrations in allomicrites and microbialites can also be interpreted in St. Cassian samples as indicators for crustal contamination. This contamination can be confirmed by a high Th/U ratio in all samples, especially in allomicrites and microbialites (Table 1). If there is a considerable siliciclastic pollution, REY patterns are of no use because they cannot reflect the partitioning coefficient of seawater (e.g., Webb and Kamber 2000; Van Kranendonk et al. 2003; Nothdurft et al. 2004). Thus it seems unlikely to find reliable REY patterns in St. Cassian microbialites because an amount of siliciclastic contamination as small as 1% effectively masks original signals (Van Kranendonk et al. 2003). Furthermore, phosphorus reaches relatively high concentrations in microbialites (up to 568 ppm) and P-free microbialites are necessary for the determination of reliable REY patterns because it is widely known that P affects their systematics considerably due to processes like phosphate coprecipitation (Lee and Byrne et al. 1996; Nothdurft et al. 2004).

Silicate weathering has proven to be an important factor for microbialite formation in some modern cryptic environments of continental islands and may be related

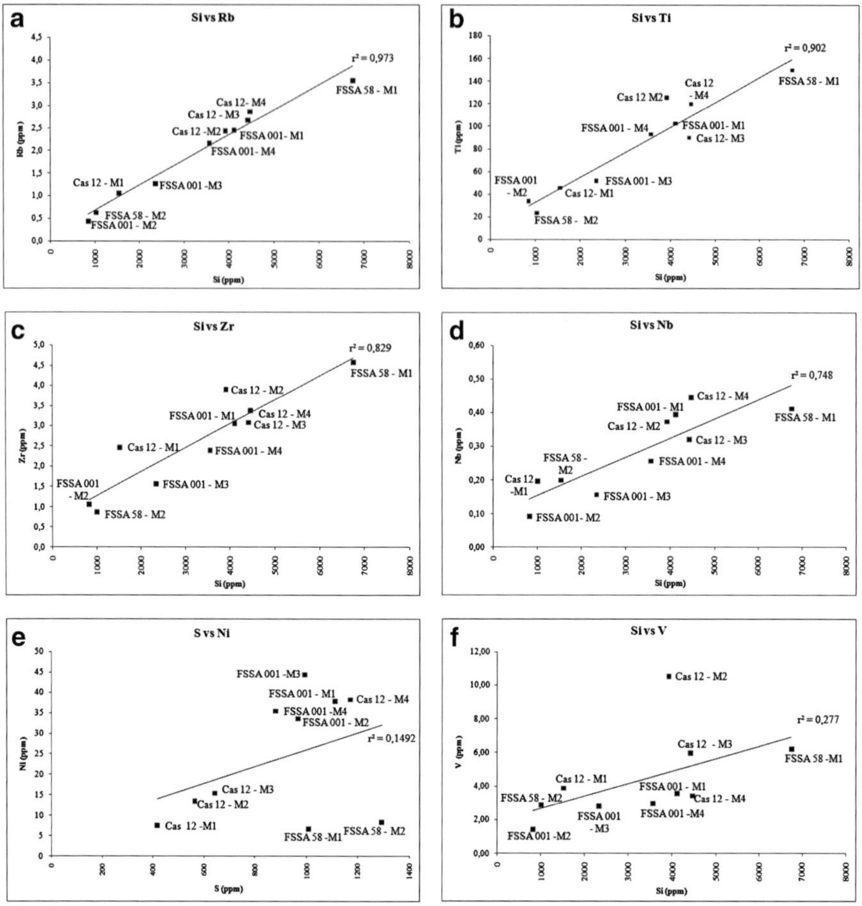

Fig. 2 (**a–e**) Correlation plots between the content of Si and various further non biological elements (Rb, Ti, Zr, Nb, Th) for microbialite samples. Note overall high values of r^2 for all elements. (**f**) Correlation plot between the content of Si vs. V among microbialites. A slight correlation exists ($r^2 = 0.27$). See text and Table 1 for a more detailed explanation

to high carbonate alkalinity (Reitner 1993). However, we believe that the main cause of the high siliciclastic content in our samples is related to the marly nature of the St. Cassian Formation, which is extensively reported in the literature (e.g., Wendt and Fürsich 1980; Russo et al. 1991).

3.4 "Biological" Elements

Elements with a possible vital effect, such as vanadium, chromium, tin, antimony, zinc, copper, molybdenum, cobalt, and nickel, have significant concentrations in many kinds of carbonates (Kamber and Webb 2007). In this section, we will discuss whether these elements reflect vital fractionation in the microbialites. Sulphur is discussed here in a similar way as P was in the non-biological elements section because its concentrations may be helpful for discerning the origin of nickel.

No significant differences were observed for V and Cr between *Margarosmilia*, chaetetids, late cements, and some of the microbialites (Table 1). Sample Cas 12-M2 appears to have a relatively high concentration of V, but a biological effect for this element cannot be confirmed in this sample because of its high Si and Al concentrations. All microbialites show a weak to medium correlation between Si and these elements, with an r2 of 0.27 for Si-V (Fig. 2f; Table 1) and of 0.40 for Si-Cr (Fig. 3a; Table 1), which could be evidence that their higher concentrations are a consequence of siliciclastic input. Furthermore, all allomicrite samples have higher concentrations than the other types of the microbialites.

The patterns of tin, antimony, and zinc are not clear. The cements were more enriched in Sn and Sb than the coral skeleton, while little difference was recordable in the chaetetid skeleton. Unfortunately, their concentrations exhibited large variations between microbialite samples (Table 1) and allomicrite was slightly enriched above all other samples, except for samples Cas 12-M3 (Sn), and Cas 12-M1, FS-SA58 M2 (Sb). Zn amounts do not have evident differences among the samples, except for allomicrite, in which the concentration is relatively enriched. Although a significant correlation does not exist between Si or Al and these elements, evidence of a vital effect for them cannot be concluded from the microbialites because only few show an enrichment relative to other samples.

Copper and molybdenum are generally more enriched in the microbialites than in cements, sponges, corals, or even allomicrites. It is unclear whether these elements have a biological role in the skeletal organisms, as both the chaetetid sponge and *Margarosmilia* have lower concentrations of all these elements than the sparry cement. These elements appear to be only weakly correlated with Si, Al, and Zr throughout the dataset (Table 1). Despite the moderate to high siliciclastic content in the microbialites, no correlation can be confirmed between Si and other elements (Fig. 3b, c). Samples Cas 12-M1 (Mo) and Cas 12-M4 (Cu) have moderate to high average error along the time series–ablation (not shown), but that does not seem to affect the distribution plots displayed in Fig. 3b, c.

It is widely known that copper plays a fundamental role in the structure of certain transmembrane enzymes, such as cytochrome c oxidase (Mathews et al. 1999) and molybdenum has been proved to be crucial in the molybdate-reducing enzymes of sulphate-reducing bacteria (Tucker et al. 1997). Therefore, we believe that the concentrations of Cu and Mo in our microbialite samples are relics of original microbial metabolism.

Cobalt concentrations were only slightly higher in allomicrites than in microbialites, but again the overall trend was the same as was observed for copper and molybdenum. The nickel concentration was significantly enriched in some of the microbialites, even more than in the allochthonous micrite. FS-SA 001-M2, the most Si-depleted microbialite sample, had a high Ni concentration of 33.57 ppm, which was more than twice the concentration of two of the allomicrite samples.

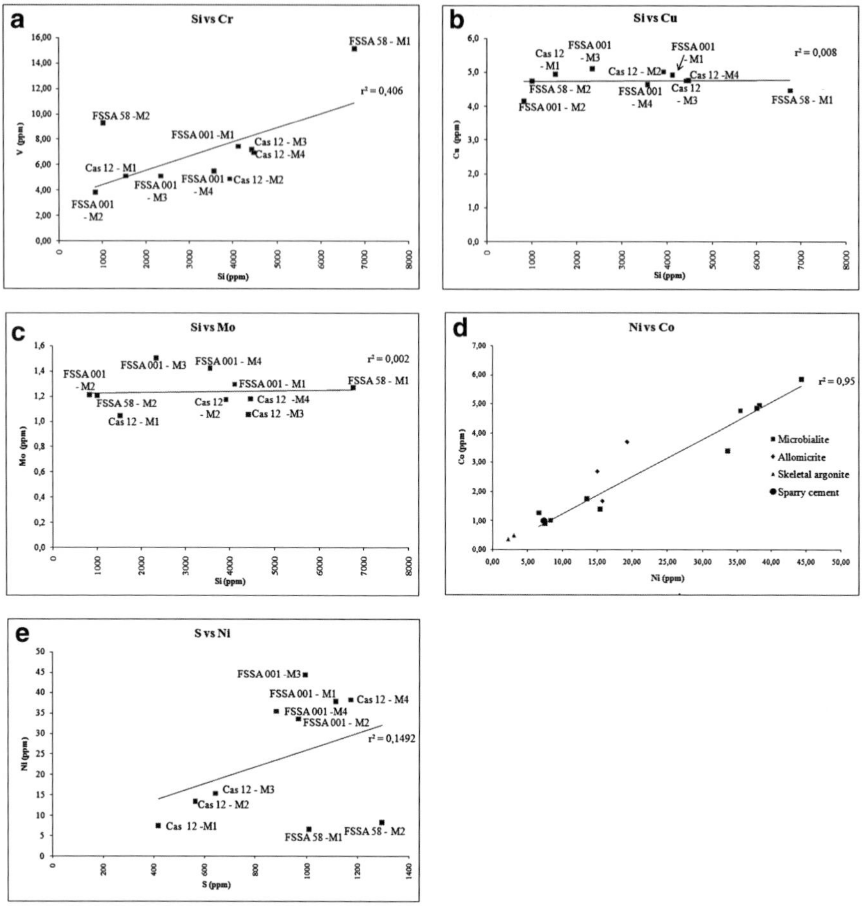

Fig. 3 (**a**) Correlation plot between the content of Si vs. Cr among microbialites. A medium correlation exists. ($r^2 = 0.40$). (**b** and **c**) Correlation plots Si vs. Cu and Mo among microbialite facies. Note the extremely low value of r^2 at both plots. Low correlation plus the higher concentration of these elements in microbialites compared to other facies permits the attribution of these concentrations to original microbial activity. (**d**) Correlation between of Ni vs. Co for all samples. Note the high correlation in microbialites, which barely changes when all samples are plotted. (**e**) Correlation between of S vs. Ni in microbialites. Only a minor correlation can be assessed, which leads to the interpretation of a non-biological origin of Ni. See Table 1 and text for further explanation

There is a strong correlation ($r^2 = 0.95$) between Ni and Co in microbialite samples (Fig. 3d), which should be a consequence of siliciclastic contamination because this correlation persists ($r^2 = 0.92$) when all samples are included. Furthermore, the Ni/Co ratio in the Earth's crust is ~2.3 (Wedepohl 1995), and in SMOW, it is around 400. In most of the measured microbialites, the allomicrites, and even the sponge this ratio is ~8.0. Hence, we cannot conclude that high Ni and Co concentrations in the microbialites are related to biological processes, but rather that they occur because of the influence of terrigenous material. Moreover, to support the hypothesis that a certain Ni concentration in microbialites is related to a biological metabolism, a correlation should exist between Ni and S in the samples (Schäfer 2006) which was not observed in the samples of this study (Fig. 3e).

4 Raman Spectroscopy

The measured components can be divided as having one of the two following structural signatures: aragonitic or calcitic. The components with an aragonitic structure are corals and sponges, whereas all the measured microbialites have a calcitic structure (Fig. 4a–c).

The most intense band in the aragonite spectra was observed at 1,086 cm^{-1}, which corresponds to the $\upsilon 1$ symmetric stretching mode of the carbonate ion. The bands with low to medium intensity in the low wavenumber region (100–300 cm^{-1}) arise from translational and rotational modes of lattice vibrations. These lattice vibrations can only be detected with the excitation wavelength of 488 nm due to technical limitations in the UV range. Furthermore, we were able to detect the $\upsilon 4$ in-plane bending mode of the carbonate ion at ~706 cm^{-1}. In the literature it is reported that this mode normally consists of a doublet with bands at 701 and 705 cm^{-1} (Frech et al. 1980; Urmos et al. 1991). In our samples this doublet can only be detected when magnifying this part of the spectra; this may be due to the relatively low accumulation time of only 5 s and the need of a filter in order to reduce sample destruction. In the $\upsilon 3$ region we observed a band at 1,464 cm^{-1}, which is in agreement with the positions reported by Frech et al. (1980) and Urmos et al. (1991).

The analysis of the structures with a calcitic signal was only possible with excitation in the UV because of the high-fluorescence background with excitation in the visible range. Therefore, we could not detect any bands in the low wavenumber region. The bands which could be assigned to calcite vary considerably, but all of them were slightly shifted to higher wavenumbers than have been reported in the literature. The strongest band, which is derived from the symmetric stretching mode ($\upsilon 1$) of the carbonate ion, appears between 1,086 and 1,091 cm^{-1}.; normally, this band is reported to appear around 1,085 cm^{-1} (Urmos et al. 1991; Rutt and Nicola 1974). The weaker bands of the $\upsilon 4$ in-plane bending and the $\upsilon 3$ antisymmetric stretching were observed between 710 and 715 cm^{-1} and between 1,419 and 1,442 cm^{-1}, respectively. Urmos et al. (1991) report these bands at 711 and 1,434 cm^{-1}, whereas Rutt and Nicola (1974) report them at 711 and 1,435 cm^{-1}. Furthermore, we detected the overtone band $2 \times \upsilon 2$ between 1,747 and 1,755 cm^{-1}. This band is reported to appear at 1,748 cm^{-1} (Urmos et al. 1991; Rutt and Nicola 1974). The shifting of the bands mentioned above strongly suggests the existence of Mg calcite. Bischoff et al. (1985) first observed the shifting of biogenic carbonates to higher wavenumbers with an increasing amount of $MgCO_3$

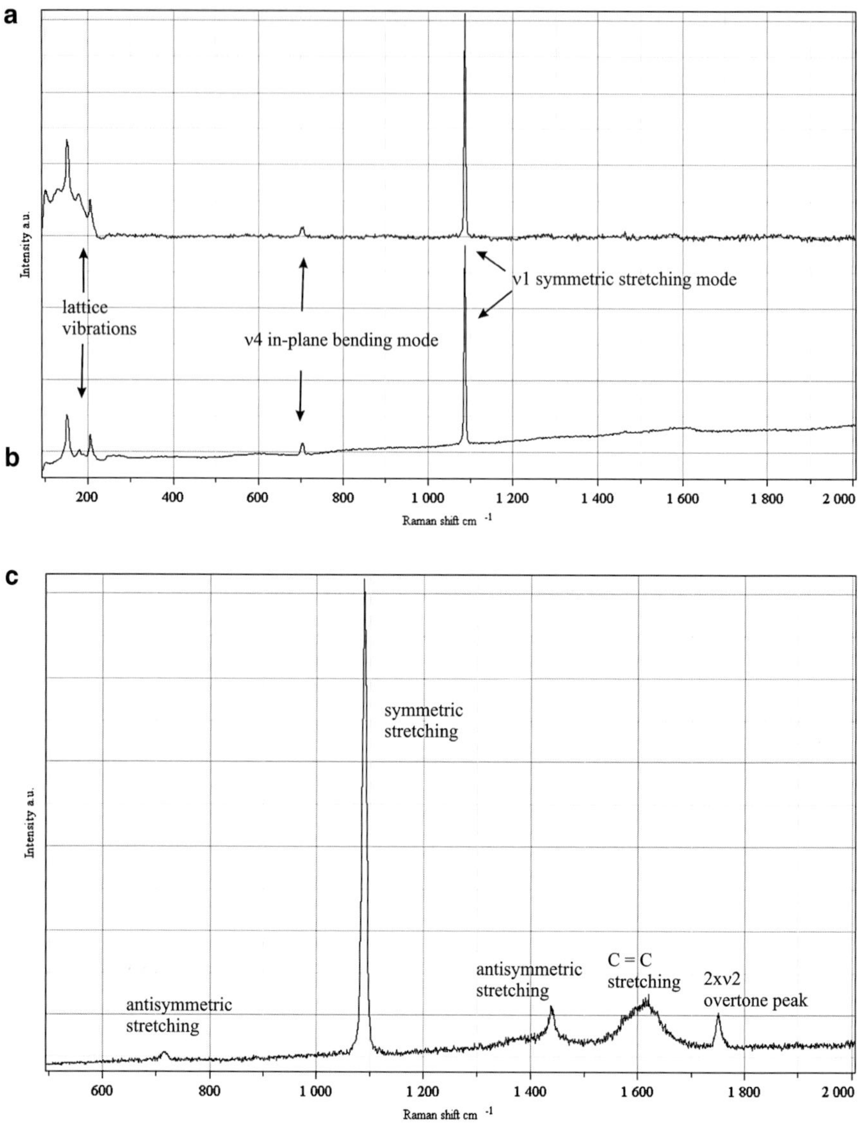

Fig. 4 Raman spectra of a coral (**a**), a chaetetid sponge (**b**) and a High-Mg Calcite Thrombolitic microbialite (**c**). In the coral and chaetetid sponge signatures, peaks linked to the effects of lattice vibrations the detection of $\upsilon 4$ in-plane bending mode and of $\upsilon 1$ symmetric stretching mode of the carbonate ion indicate that an original aragonitic mineralogy is still present. In the microbialitic signature, numerous peaks can be identified, like those related to symmetric and antisymmetric stretchings of the carbonate ion, but also C=C stretching and $2 \times \nu 2$ overtone related peaks could be observed. In particular, anti-symmetric stretching peak at 710 and 715 cm^{-1} suggests the existence of High Mg-Calcite. See text for further details

in the carbonates. The difference in the amount of shifting is likely because of the varying concentrations of $MgCO_3$ in the structures; this would be in agreement with the findings of the electron microprobe analyses.

In both the aragonitic and the calcitic spectra, an additional band between 1,603 and 1,619 cm^{-1} was observed. This band probably exists because of the C=C stretching vibrations, whereas its broad appearance suggests the beginning of the degradation of the bonding, which is often described as an amorphous carbon signature (e.g., Robertson 1986; Matsunuma 1997). Nevertheless, the existence of that peak strongly indicates that organics remain in the structure.

5 Conclusions

In addition to petrographical and epifluorescence techniques, the distinction between microbialite and allomicrite can also be recognised by the higher concentrations of selected REE and Y in the former samples.

LA-ICP-MS analysis of the Mg-content in microbialites showed a high Mg-calcite mineralogy, which was confirmed by Raman spectroscopy. The interpretation of the microbialitic spectra is in accordance with the high quality of preservation that was confirmed by CL. Furthermore, Raman spectroscopy was able to differentiate between the Mg-calcite mineralogy of the microbialites and the aragonitic mineralogy of the skeletal components. Additionally, Raman spectra revealed organic relics inside the microbialites and in the skeletal components.

LA-ICP-MS also permitted us to detect the presence of siliciclastic elements in the microbialites, which most likely indicate a marly sediment origin in the St. Cassian Formation. While most of the non-vital trace elements are markedly correlated to the content of Si, others like selected REE and Y are not; their concentrations are interpreted as being strongly affected by the content of siliciclastics and by the amount of phosphorus. Thus, it is in principle not possible to obtain a reliable microbialite-based REY pattern in our samples.

Some biologically important elements like vanadium and chromium appear to be detritus-related rather than being relics of biological activity because a medium correlation exists between them and silicon. Furthermore, allochthonous micrite more enriched with those elements. Antimony and zinc are relatively enriched in some microbialites and a strong correlation between their concentrations and Si or Al does not exist, but a biological effect cannot be assessed because their concentrations are higher in allomicrites. Copper and molybdenum are more significant, especially in the skeletons of *Margarosmilia* and in chaetetids because of their rather homogeneous distribution along the time series measurements.

Ni anomalies, which are also strongly related to Co anomalies, are probably not related to biological activity.

Acknowledgements We gratefully thank the constructive comments and reviews by Prof Dr. Franco Russo (Calabria), which considerably improved our manuscript. JFSB is supported by the German Academic Exchange Service (DAAD) with a three-year PhD research stay in Germany through the grant A/07/94459. Funding for this research comes from the Courant Research Centre in Göttingen, Germany. Dr. Andreas Kronz (Göttingen) is gratefully acknowledged for providing microprobe analyses. This is publication No. 34 of the Göttingen Courant Research Centre Geobiology (German Excellence Initiative)

References

Bischoff WD, Sharma SK, Mackenzie FT (1985) Carbonate ion disorder in synthetic and biogenic magnesian calcites: a Raman spectral study. American Mineralogist 70:581–589

Byrne RH, Liu X, Schijf J (1996) The influence of phosphate coprecipitation on rare earth distributions in natural waters. Geochimica Cosmochimica Acta 60:3341–3346

Dickson T (1990) Carbonate mineralogy and chemistry. In: Tucker ME, Wright VP (eds) Carbonate sedimentology. Blackwell Scientific Publications, Oxford, pp 284–313

Dieci G, Antonacci A, Zardini R (1968) Le spugne cassiane (Trias medio–superiore) della regione dolomítica attorno a Cortina d'Ampezzo. Bolletino della Societá Paleontologica Italiana 7:94–155

Dieci G, Russo A, Russo F (1974) Nota preliminare sulla microstrutura di spugne aragonitiche del Trias Medio–Superiore. Bolletino della Societá Paleontologica Italiana 13(1–2):99–107

Flügel E (2004) Microfacies of carbonate rocks. Springer, Berlin, Heidelberg, 976 pp

Fraústro da Silva JJR, Williams RJP (1991) The biological chemistry of the elements. Clarendon Press, Oxford, 561 pp

Frech R, Wang EC, Bates JB (1980) The IR and Raman spectra of $CaCO_3$ (aragonite). Spectrochimica Acta 36A:915–919

Fürsich FT, Wendt J (1977) Biostratinomy and palaeoecology of the Cassian Formation (Triassic) of the Southern Alps. Palaeogeography, Palaeoclimatology, Palaeoecology 22:257–323

Kamber B, Webb G (2007) Transition metal abundances in microbial carbonate: a pilot study based on in situ LA-ICP-MS analysis. Geobiology 5(4):375–389

Laghi G, Martinelli G, Russo F (1983) Localization of minor elements by EDS microanalysis in aragonitic sponges from the St. Cassian beds, Italian Dolomites. Lethaia 17:133–138

Mathews CK, Van Holde KE, Ahern KG (1999) Biochemistry. Addison Wesley, San Francisco, 1190 pp

Matsunuma S (1997) Theoretical simulation of resonance Raman bands of amorphous carbon. Thin Solid Films 306:17–22

Neuweiler F, Reitner J (1995) Epifluorescence microscopy of selected microbialites from Lower Carnian Cipit-boulders of the Cassian Formation (Seeland Alpe, Dolomites). In: Reitner J, Neuweiler F (eds) Mud Mounds: A polygenetic Spectrum of Fine-grained Carbonate Buildups. Facies 32:26–28

Nothdurft LD, Webb GE, Kamber BS (2004) Rare earth element geochemistry of Late Devonian reefal carbonates, Canning Basin, Western Australia: confirmation of a seawater REE proxy in ancient limestones. Geochimica Cosmochimica Acta 68:263–283

Olivier N, Boyet M (2006) Rare earth and trace elements of microbialites in Upper Jurassic coral- and sponge-microbialite reefs. Chemical Geology 230:105–123

Prasada Rao C (1996) Modern Carbonates tropical temperate polar. Introduction to sedimentology and geochemistry. Arts of Tasmania, University of Tasmania, Tasmania

Rech H (1998) Geobiologie der sogenanten "Cipit-Kalke" der Beckenfazies der Cassianer-Schichten, S' St. Kassian, Dolomiten. Unpublished Diploma Thesis, University of Göttingen, Göttingen, 134 pp

Reitner J (1987) A new calcitic sphinctozoan sponge belonging to the Demospongiae from the Cassian Formation (Lower Carnian; Dolomites, Northern Italy) and its phylogenetic relationship. Geobios, 20:571–589
Reitner J (1992) "Coralline Spongien". Der Versuch einer phylogenetisch-taxonomischen Analyse. Berliner Geowissenschaftliche Abhandlungen Reihe E 1:1–352
Reitner J (1993) Modern cryptic microbialite/metazoan facies from Lizard Island (Great Barrier Reef, Australia). Formation and concepts. Facies 29:3–39
Reitner J, Engeser T (1989) *Chaetosclera klipsteini* n.gen n.sp. (Halichondriida, Demospongiae), aus dem Unterkarn der Cassianer-Schichten (Dolomiten, Italien). Mitteilungen Geologisch-Paläontologisches Institut Universität Hamburg 68:159–165
Robertson J (1986) Amorphous carbon. Advances in Physics 35:317–374
Russo F (2005) Biofacies evolution in the Triassic platforms of the Dolomites, Italy. Ann Univ Stud Ferrara Mus Sci Nat vol Spec, 33–45
Russo F, Neri C, Mastandrea A, Laghi G (1991) Depositional and diagenetic history of the Alpe di Specie (Seelandalpe) fauna (Carnian,. Northeastern Dolomites). Facies 25:187–210
Russo F, Neri C, Mastandrea A, Baracca A (1997) The mud-mound nature of the Cassian platform margins of the Dolomites. A case history: the Cipit Boulders from Punta Grohmann (Sasso Piatto Massif, Northern Italy). Facies 36:25–36
Russo F, Mastandrea A, Neri C (1998) Evoluzione degli organismi costruttori nelle piattaforme triassiche delle Dolomiti (Italia). Memorie della Societá Geologica Italiana 53:479–488
Rutt HN, Nicola JH (1974) Raman spectra of carbonates of calcite structure. Journal of Physics C: Solid State Physics 7:4522–4528
Sánchez-Beristain JF, Reitner J (2008) Microbialites from the Cassian Formation: an insight. International Kalkowsky-Symposium Abstract Volume and Field Guide to Excursions, 116–117
Schäfer N (2006) Ni-Anomalien in Cold-Seep Karbonaten aus dem nordwestlichen Schwarzen Meer: Ist Ni ein kritisches Element bei der anaeroben Oxidation von Methan? Unpublished Diploma Thesis, University of Göttingen, Göttingen
Tucker MD, Barton LL, Thomson BM (1997) Reduction and immobilization of molybdenum by *Desulfovibrio desulfuricans*. Journal of Environmental Quality 26:1146–1152
Urmos J, Sharma SK, Mackenzie FT (1991) Characterization of some biogenic carbonates with Raman spectroscopy. American Mineralogist 76:641–646
Van Kranendonk M, Webb G, Kamber BS (2003) New geological and trace element evidence from 3.45 Ga stromatolitic carbonates in the Pilbara Craton: support of a marine, biogenic origin and for a reducing Archaean ocean. Geobiology 1:91–108
von Richthofen F (1860) Geognostische Beschreibung der Umgebung von Predazzo, St. Cassian und der Seisser Alpe in Südtirol, Gotha, 327 pp
Webb GE, Kamber BS (2000) Rare earth elements in Holocene reefal microbialites: a new shallow seawater proxy. Geochimica Cosmochimica Acta 64:1557–1565
Wedepohl KH (1995) The composition of the continental crust. Geochimica Cosmochimica Acta 59:1219–1232
Wendt (1982) The Cassian patch reefs (Lower Carnian, Southern Alps). Facies 6:185–202
Wendt J, Fürsich FT (1980) Facies analysis and palaeogeography of the Cassian Formation, Triassic, Southern Alps. Rivista Italiana di Paleontologia e Stratigrafía 85(3–4):1003–1028
Wissmann HL, Münster G (1841) Beitrage zur Geognosie und Petrefactenkunde des südostlichen Tirols vorzüglich der Schichten von St. Cassian. Beiträge zur Petrefaktenkunde 4:1–152, Bayreuth (Buchner)
Zankl H (1993) The origin of High-Mg-Calcite microbialites in cryptic habitats of Caribbean Coral Reefs – Their dependence on light and turbulence. Facies 29:55–60

Importance of Rare Earth Element Patterns in Discrimination Between Biotic and Abiotic Mineralization

Adriano Guido, Adelaide Mastandrea, Fabio Tosti, and Franco Russo

1 Introduction

Rare earth element (REE) patterns have been used to discuss the origin of rocks and minerals and the evolution processes that occurred during their formation (Goldschmidt 1954; Henderson 1996).

Carbonate REE patterns are similar to the REE patterns of clastic sedimentary rocks (McLennan 1989). The application of these patterns to natural water systems provided information on the incorporation of REE and the physical and chemical characteristics of the sedimentary environment (Taylor and McLennan 1988; Erel and Stopler 1993; Takahashi et al. 2000a, b; Webb and Kamber 2000; Bolhar et al. 2004). The REE signature of sediments can be used to obtain detailed information on the geochemical conditions of depositional basins.

The REE + Y carbonate patterns can also provide information on carbonatogenesis processes. Many differences in the behaviour of REE fractionation exist between microbial and skeletal carbonates. Microbial carbonates are generally formed in microenvironments that are controlled or affected by living or dead organic matter in biofilms. Microbial carbonates represent biologically induced precipitation in equilibrium with the depositional environment (Lowenstam 1981). This type of biomineralisation is reflected in the proportional fractionation of REE in the normalised patterns. During biologically controlled carbonate skeletogenesis, organisms actively manage ions to secrete the skeleton, leading to a non-proportional fractionation of REE in their pattern (Webb and Kamber 2000). This phenomenon is called the "vital effect".

A. Guido (✉), A. Mastandrea, F. Tosti, and F. Russo
Dipartimento di Scienze della Terra, Università della Calabria, Via P. Bucci Cubo 15b, 87036 Rende (CS), Italy
e-mail: aguido@unical.it

J. Reitner et al., *Advances in Stromatolite Geobiology*,
Lecture Notes in Earth Sciences 131, DOI 10.1007/978-3-642-10415-2_27,

Another important topic is to understand the REE fractionation during abiotic chemical deposition. The behaviour of REEs during evaporation and gypsum precipitation is not well understood.

The shale-normalised REE + Y patterns were used in this paper to verify the reliability of using the REE pattern as a geochemical signature of the water in which the microbialite precipitated. We tested this approach for the Messinian Calcare di Base (CDB), in which bacteria-induced precipitation has already been proven by molecular fossils and other organic markers (Guido et al. 2007) to compare the biogeochemical data to the REE + Y characteristics, we analysed the same samples studied by Guido et al. (2007). The CdB sampled in Northern Calabria lacks any diagenetic alteration, and the original mineralogy (aragonite) and microstructure are preserved.

2 Analytical Techniques

Small fragments of the carbonate samples were first hand ground in an agate mortar and then dissolved by microwave digestion using a Mars 5 microwave apparatus (CEM technologies). Then, 100 mg of the powder was dissolved in Teflon (TFM) digestion vessels using a mixture of hydrofluoric acid (2 ml HF), nitric acid (8 ml HNO_3) and perchloric acid (2 ml $HClO_4$). All reagents were of Merck "suprapur" quality. Before the mixture evaporated completely, we added 2 ml of perchloric acid to ensure the complete removal of the hydrofluoric acid. These solutions they were left to cool and then diluted to 100 ml with Millipore water.

The trace element concentrations in each sample were determined by a Perkin-Elmer Elan DRCe in the Department of Earth Science, Calabria University.

The concentration of REE, Sc, Y and Th were determined using external calibration curves, which were prepared using a Perkin Elmer "multi-element Calibration Standard 2 solution". Internal standards (indium, germanium and rhenium) were added to both the standards and the samples.

Standard reference materials (limestone from the US Geological Survey) were prepared in the same way and were used as unknowns during the analytical sequence. This procedure was suggested by Eggins et al. (1997). The measured concentrations were compared with certified values to evaluate the accuracy and precision of the analytical data.

3 REE Fractionation in Evaporites and Microbial Carbonate

REE in carbonate sediments can be used to determine the original seawater geochemistry in which the carbonates formed. The REE fractionation in the carbonates depends mainly on the following two factors: (1) the geochemistry of the water in which the carbonates mineralise and (2) the vital effect, which could perturb the equilibrium between the ion concentrations in waters and in the carbonates and cause variations in the mineralisation processes. Tanaka and Kawabe (2006) evaluated the REE partition coefficients between calcite and an aqueous solution.

REE + Y in microbial carbonates are characterised by consistent, smooth, reproducible and coherent patterns, irrespective of the total concentration (Fig. 1). The patterns are characterised by HREE enrichment, a superchondritic Y/Ho ratio, a negative Ce anomaly and positive anomalies of La_{SN}, Eu_{SN}, Gd_{SN} and Y_{SN}. These characteristics suggest that microbial carbonates incorporated REEs in equilibrium with oxygenated seawater (Bolhar et al. 2004).

The proportional incorporation of REE + Y suggests microbially-induced carbonate precipitation. This carbonate precipitation is not affected by biological fractionation, unlike the carbonate metazoan skeletogenesis (Fig. 1).

Our knowledge of REE concentrations in evaporitic sediments is very limited. Evaporites typically show parallel REE patterns when plotted against MUQ (MUd from Queensland) (Kamber et al. 2005). The relative concentrations of the LREE increase from La to Sm and, those of the middle rare earth elements (MREE) and HREE become more depleted from Sm to Lu (Playà et al. 2007). Baumer et al. (1997) observed that light REEs in gypsum show higher relative concentrations than heavy REEs because LREE are more incompatible.

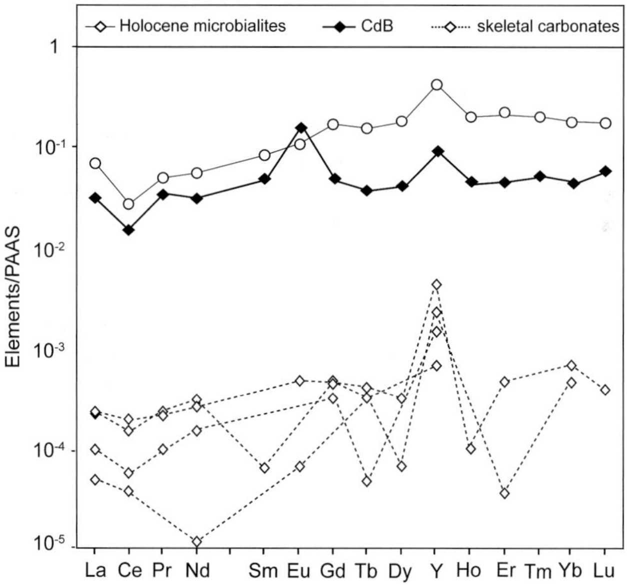

Fig. 1 PAAS-normalised REE + Y diagram showing the comparison among the average Messinian CdB composition, Holocene microbialites, and skeletal carbonates. The CdB and Holocene microbialites have similar REE + Y patterns with the same La, Ce and Y anomalies and analogous enrichment of HREE. The skeletal carbonates show a completely different trend, and in general the patterns are much more scattered [modified from Webb and Kamber (2000)]

4 Results

The data are normalised against the Post-Archean Australian Shale (PAAS, Taylor and McLennan 1985; McLennan 1989) REE + Y pattern. Y is inserted between Dy and Ho because of its ionic radius (Bau and Dulski 1996). The carbonate samples of Calcare di Base are characterised by consistent, smooth, reproducible, and coherent REE + Y patterns (Table 1; Fig. 2). The patterns show that the LREE are more depleted than the heavier REE. The HREE have a low enrichment, which is expressed by the Nd_{SN}/Yb_{SN} ratio ($Nd_{SN}/Yb_{SN} = 0.52$; SD = 0.090). The super-chondritic character is expressed as a prominent positive Y spike and by the Y/Ho ratio (Y/Ho = 57.08, SD = 3.07). The distribution of REE + Y shows a slightly positive Gd anomaly, which is expressed by the Gd/Gd* ratio [Gd/(2Tb − 1Dy) = 1.15; SD = 0.04].

The pattern also shows a consistent negative Ce_{SN} anomaly and a positive La_{SN} anomaly.

Following Bau and Dulski (1996), we calculated the La and Ce anomaly by comparing $(Ce/Ce^*)_{SN} = [Ce_{SN}/(0.5\ La_{SN} + 0.5\ Pr_{SN})]$ to $(Pr/Pr^*) = [Pr_{SN}/(0.5\ Ce_{SN} + 0.5\ Nd_{SN})]$. Our hypothesis was that no Pr_{SN} or Nd_{SN} anomalies exist. When $(Pr/Pr^*)_{SN} > 1$, a negative Ce_{SN} anomaly exists, and when $(Ce/Ce^*)_{SN} < 1$, a positive La_{SN} anomaly exists.

Table 1 REE, Y, Sc, Hf, Th contents of the Calcare di Base in ppm and the La, Ce, Eu, Gd anomalies

Elements	Samples						
	CB1	Cb2	CB3	CB4M	CB4D	CB5	Cb6
La	0.979	0.967	1.497	1.557	1.201	1.079	2.225
Ce	1.401	1.228	2.096	2.028	1.480	1.426	4.015
Pr	0.220	0.248	0.341	0.376	0.298	0.308	0.593
Nd	0.534	0.526	0.916	0.917	0.688	0.654	1.931
Sm	0.311	0.233	0.397	0.790	0.558	0.413	0.436
Eu	0.165	0.112	0.188	0.461	0.319	0.223	0.116
Gd	0.136	0.130	0.226	0.203	0.174	0.199	0.565
Tb	0.019	0.018	0.031	0.026	0.024	0.028	0.076
Dy	0.101	0.100	0.169	0.144	0.135	0.166	0.451
Y	1.801	1.415	2.788	2.575	2.356	2.497	4.4886
Ho	0.031	0.026	0.046	0.043	0.040	0.044	0.094
Er	0.092	0.080	0.142	0.131	0.122	0.143	0.282
Tm	0.017	0.014	0.024	0.027	0.025	0.024	0.039
Yb	0.090	0.080	0.138	0.144	0.129	0.139	0.241
Lu	0.022	0.018	0.034	0.067	0.046	0.032	0.039
Sc	0.789	0.894	1.294	1.526	1.229	1.002	1.504
Hf	0.073	0.061	0.097	0.134	0.094	0.070	0.137
Th	0.241	0.211	0.335	0.399	0.237	0.185	0.416
$(La/La^*)_{SN}$	1.496	1.817	1.449	1.620	1.762	1.874	1.251
$(Ce/Ce^*)_{SN}$	0.696	0.578	0.676	0.612	0.571	0.568	0.804
$(Eu/Eu^*)_{SN}$	3.367	2.898	2.843	4.001	3.810	3.344	1.267
$(Gd/Gd^*)_{SN}$	0.842	0.958	0.958	0.625	0.699	0.875	1.321

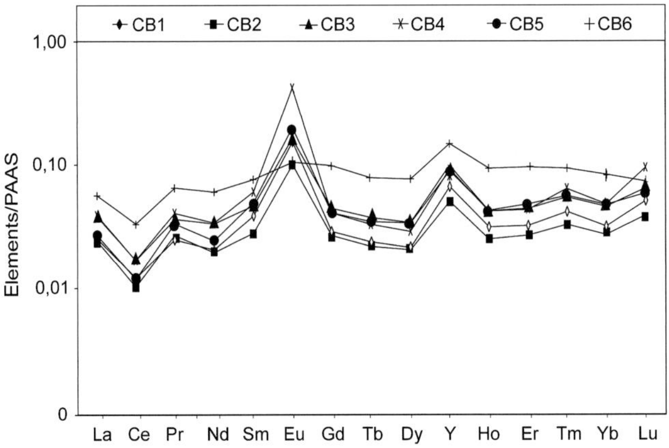

Fig. 2 REE + Y patterns of CdB normalised to PAAS

The patterns are also characterised by the presence of a strongly positive Eu anomaly, expressed as (Eu/Eu*), in which Eu* = 2/3Sm + 1/3Tb (Kamber and Webb 2001).

In our samples the Th (185–416 ppb), Hf (61–137 ppb) and Sc (788–1,525 ppb) do not obscure the original trace element signatures. The strong seawater-like signature (i.e., positive La and negative Ce anomalies and a superchondritic Y/Ho ratio) indicate a non-contaminant REE + Y pattern.

5 Interpretation and Discussion

Guido et al. (2007) demonstrated that the CdB was deposited in a marine setting and influenced by freshwater inputs. The authors attributed the thrombolitic fabric to the metabolic activity of bacteria. Bacterial signals and their metabolic mediations in the mineralisation processes have been demonstrated by molecular fossils and other organic markers.

The extraordinary preservation state of the studied carbonate sediments permits us to exclude diagenetic alterations of the trace element concentrations. This interpretation is reinforced by the presence of characteristic molecular fossils (unsaturated fatty acids and ββ hopanoids) that indicate a very low thermal stress on these sediments.

The superchondritic Y/Ho ratio and positive La and negative Ce anomalies indicate that the REE + Y patterns do not diverge from the seawater signature.

The distribution of REE + Y in CdB share the same characteristics as modern oxygenated seawater and have the following microbialitic patterns: (a) a Y/Ho ratio (>44) that is higher than chondrite (26–28) (Kamber and Webb 2001); (b) a positive La anomaly and a negative Ce anomaly (Bau and Dulski 1996; Kamber

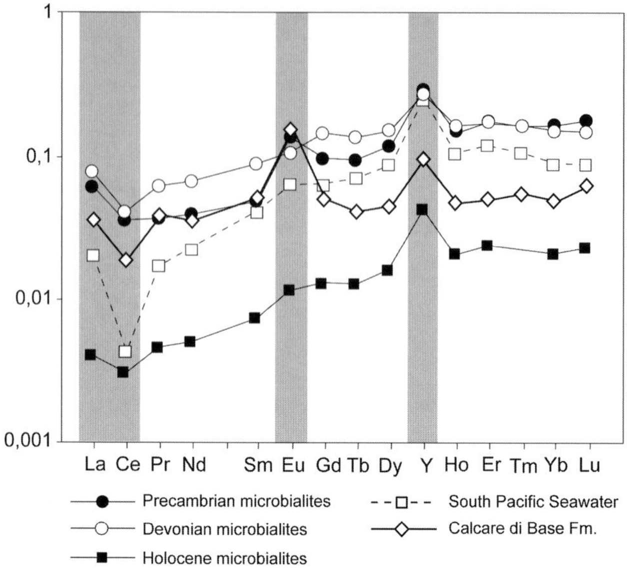

Fig. 3 PAAS-normalised REE + Y diagram comparing the average composition of South Pacific Seawater, Messinian CdB and microbialites from the Precambrian, Devonian and Holocene. All patterns and anomalies are similar [modified from Bolhar et al. (2004)]

and Webb 2001); (c) no negative Gd_{SN} anomaly, suggesting that organic compounds did not influence the incorporation of the REEs; and (d) a strong positive Eu anomaly (Fig. 3).

The La and Ce anomalies were studied in a plot of PAAS-normalised $(Pr/Pr^*)_{SN}$ $[Pr_{SN}/(0.5\ Ce_{SN} + 0.5\ Nd_{SN})]$ vs. $(Ce/Ce^*)_{SN}$ $[Ce_{SN}/(0.5\ La_{SN} + 0.5\ Pr_{SN})]$ (Bau and Dulski 1996). This plot discriminates between the La and Ce anomalies and confirms the normal marine deposition of the carbonates in this study (Fig. 4).

REE + Y dataset can be utilised as a record of the microbial origin of micrites, particularly for carbonates of doubtful genesis such as the Messinian Calcare di Base. The REE + Y pattern of the CdB is comparable to the patterns of the Archean, Devonian, and Holocene microbial carbonates (Fig. 3). The microbialites are currently the best material to use for determining the record of marine REE geochemistry.

The ability of not-microbialitic carbonates to precipitate and incorporate REE in proportion to their water environment is difficult to determine. Carbonate rarely precipitates in normal marine conditions without some biological influence.

The REE dataset in the CdB was normalised to MUQ (MUd from Queensland) to compare the distributions of this study to those of the evaporitic samples and seawater reported by Playà et al. (2007) (Fig. 5). The behaviour of REEs during evaporative concentration and gypsum precipitation is very different from the REE distribution in seawater. Most of the gypsum samples reported by Playà et al. (2007) show an increase in the light REE from La to Eu and a decrease in the heavy REE

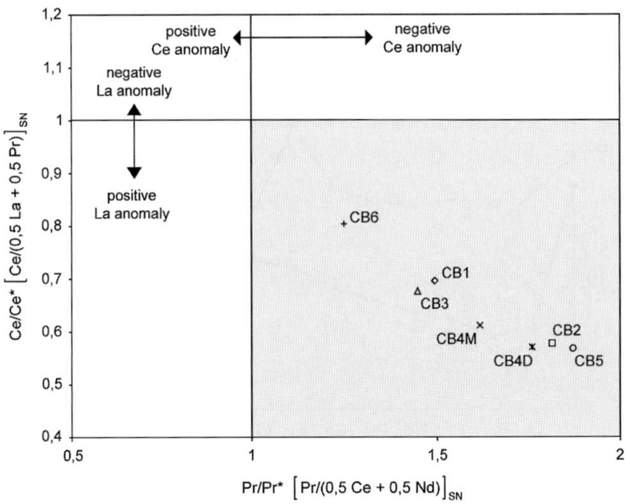

Fig. 4 Diagram of Ce/Ce* $[Ce/(0.5La + 0.5Pr)]_{SN}$ vs. Pr/Pr* $[Pr/(0.5Ce + 0.5Nd)]_{SN}$ showing the La and Ce anomalies in seawater sediments. All of the CdB samples cluster in the field of negative Ce and positive La anomalies, similar to modern open oceanic surface water

from Eu to Lu (Fig. 5c). The main differences between the CdB carbonate and evaporitic patterns are as follows: (1) the carbonates have a higher REE concentration than the evaporites; (2) the carbonates have a strong LREE depletion and HREE enrichment; (3) the evaporites are enriched in MREE. The distribution of REEs emphasises the strong difference between the CdB and the evaporitic sediments (Fig. 5a, c). The behaviour of the evaporites is the inverse of the behaviour of the microbialites and seawater.

The MUQ-normalised patterns confirm that the CdB and Holocenic microbialite REE distributions are very similar to the seawater pattern (Fig. 5a, b).

6 Conclusions

The REE + Y in the CdB are characterised by consistent, smooth, reproducible, and coherent patterns. These patterns show HREE enrichment, a superchondritic Y/Ho ratio, a negative Ce anomaly and positive anomalies of La_{SN}, Eu_{SN}, Gd_{SN} and Y_{SN}. The proportional incorporation of REE + Y suggests microbially-induced carbonate precipitation. This carbonate precipitation is not affected by biological fractionation and is comparable to the precipitation of Archean, Devonian, and Holocene microbial carbonates. This result confirms that the REE + Y dataset can be utilised as a record of the microbial origin of micrites, particularly for carbonates of doubtful genesis such as the Messinian Calcare di Base.

The MUQ-normalised pattern shows the difference between the CdB and the evaporites. The evaporite patterns are the inverse of the patterns of the CdB and

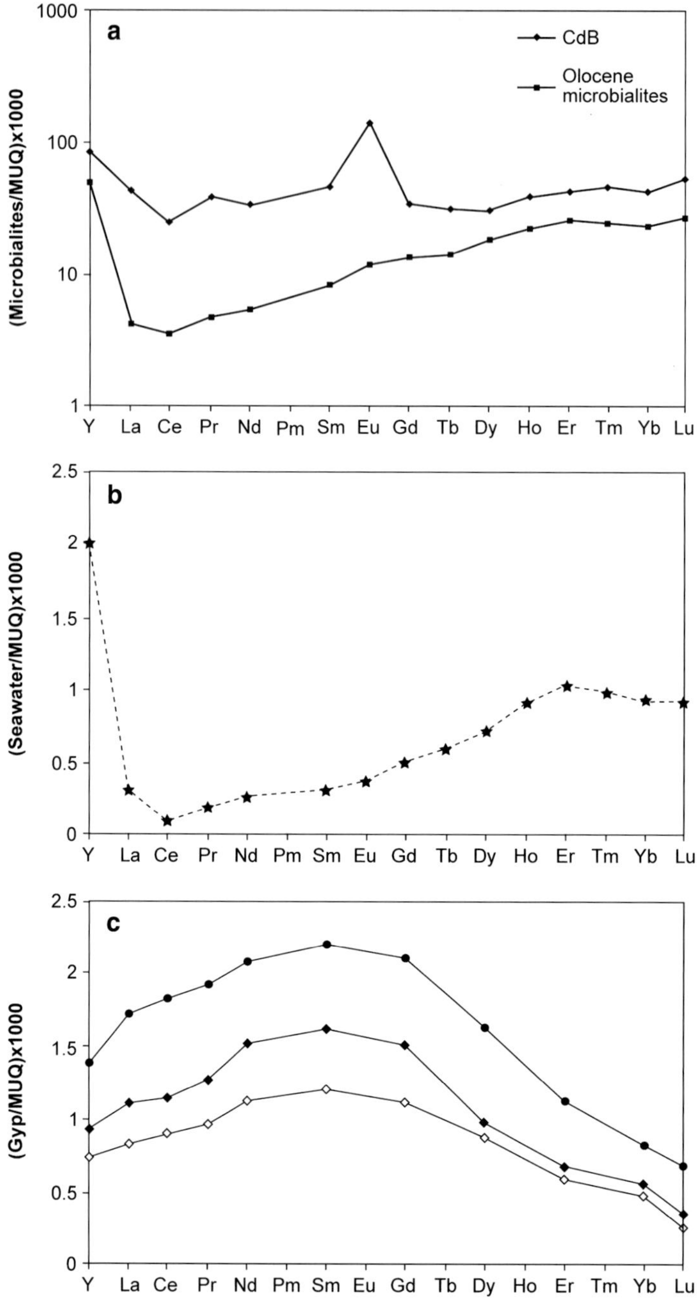

Fig. 5 MUQ-normalised patterns of: (**a**) Calcare di Base and Holocene microbialites; (**b**) seawater; and (**c**) gypsum from Carpentaria [modified from Playà et al. (2007)]. MuQ-normalised patterns confirm that the REE distributions of the CdB and Holocene microbialites are very similar to seawater and very different from gypsum

seawater: This result confirms the model for the CdB deposition that was proposed by Guido et al. (2007). This model describes biotic mineralisation in normal marine conditions rather than abiotic precipitation in an evaporitic setting.

Acknowledgments Our thanks to Prof. Dr. Joachim Reitner for the excellent organisation of the International Kalkowsky Symposium on the Geobiology of Stromatolites.
We are indebted to an anonymous reviewer for suggestions that greatly improved the manuscript. This research was funded by the Università della Calabria, PRIN 2007.

References

Bau M, Dulski P (1996) Distribution of Y and rare-earth elements in the Penge and Kuruman Iron Formations, Traansvaal Supergroup, South Africa. Precambrian Research 79:37–55

Baumer A, Blanc Ph, Cesbron F, Ohnenstetter D (1997) Cathodoluminescence of synthetic (doped with rare-earth elements) and natural anhydrites. Chemical Geology 138:73–80

Bolhar R, Kamber BS, Moorbath S, Fedo CM, Whitehouse MJ (2004) Characterisation of early Archaean chemical sediments by trace element signatures. Earth and Planetary Science Letters 222:43–60

Eggins SM, Woodhead JD, Kinsley LPJ, Mortimer GE, Sylvester P, McCulloch MT, Hergt JM, Handler MR (1997) A simple method for the precise determination of 40 trace elements in geological samples by ICPMS using enriched isotope internal standardisation. Chemical Geology 134:311–326

Erel Y, Stopler EM (1993) Modeling of rare-earth element partitioning between particles and solution in aquatic environments. Geochimica Cosmochimica Acta 57:513–518

Goldschmidt VM (1954) Geochemistry. Oxford Press, New York

Guido A, Jacob J, Gautret P, Laggoun-Défarge F, Mastandrea A, Russo F (2007) Molecular fossils and other organic markers as palaeoenvironmental indicators of the Messinian CdB formation: normal versus stressed marine deposition (Rossano Basin, Northern Calabria, Italy). Palaeogeography Palaeoclimatology Palaeoecology 255:265–283

Henderson P (1996) The rare earth elements: introduction and review. In: Jones AP (ed) Rare Earth Minerals: Chemistry, Origin and Ore Deposits. Chapman & Hall, London, 1–17

Kamber BS, Webb GE (2001) The geochemistry of late Archaean microbial carbonate: Implications for ocean chemistry and continental erosion history. Geochimica Cosmochimica Acta 65:2509–2525

Kamber BS, Greig A, Collerson KD (2005) A new estimate for the composition of weathered young upper continental crust from alluvial sediments, Queensland, Australia. Geochimica Cosmochimica Acta 69:1041–1058

Lowenstam HA (1981) Minerals formed by organisms. Science 211:1126–1131

McLennan SM (1989) Rare earth elements in sedimentary rocks: influence of provenance and sedimentary processes. In: Lipin BR McKay GA (eds) Geochemistry and Mineralogy of Rare Earth Elements. Mineralogical Society of America. Reviews in Mineralogy 21:169–200

Playà E, Cendón DI, Travé A, Chivas AR, García A (2007) Non-marine evaporites with both inherited marine and continental signatures: the Gulf of Carpentaria, Australia, at ~70 ka. Sedimentary Geology 201:267–285

Takahashi Y, Shimizu H, Usui A, Kagi H, Nomura M (2000a) Direct observation of tetravalent cerium in ferromanganese nodules and crusts by X-ray absorption near-edge structure (XANES). Geochimica Cosmochimica Acta 64:2929–2935

Takahashi Y, Shimizu H, Kagi H, Yoshida H, Usui U, Nomura M (2000b) A new method for the determination of CeIII/CeIV ratios in geological materials; application for weathering, sedimentary, and diagenetic processes. Earth and Planetary Science Letters 182:201–207

Tanaka K, Kawabe I (2006) REE abundances in ancient seawater inferred from marine limestone and experimental REE partition coefficients between calcite and aqueous solution. Geochemical Journal 40:425–435

Taylor SR, McLennan SM (1985) The Continental Crust: its Composition and Evolution. Blackwell Scientific Publications, Oxford

Taylor SR, McLennan SM, (1988) The significance of the rare earths in geochemistry and cosmochemistry. In: Gschneidner Jr KA, Eyring L (eds) Handbook on the Physics and Chemistry of Rare Earths vol. 11. Elsevier, Amsterdam, 485–578

Webb GE, Kamber BS, (2000) Rare earth elements in Holocene reefal microbialites: a shallow seawater proxy. Geochimica Cosmochimica Acta 64:1557–1565

Lower Ordovician Stromatolites from the Anhui Province of South China: Construction and Geobiological Significance

Natsuko Adachi, Yoichi Ezaki, Jianbo Liu, and Jun Cao

1 Introduction

The Cambrian through Early Ordovician was a lengthy interval when microbialites once again dominated after their decline from a peak in the Middle Proterozoic (e.g., Riding 2006a). These microbialites declined again in the Middle–Late Ordovician, although they remained locally common (e.g., Webby 2002; Shapiro 2004; Riding 2006a). The Cambrian to Early Ordovician microbialites exhibit the following characteristics: (1) abundant stromatolites, although their diversity in growth form is limited compared with typical Proterozoic stromatolites; in fact, the distinctive coniform and microdigitate forms are absent, and branched and columnar stromatolites are scarce (Awramik 1971; Riding 2000); (2) in contrast, thrombolites and dendrolites became established and diversified (Riding 2000; Shapiro 2004); and (3) calcimicrobes (e.g., *Girvanella*, *Renalcis*, and *Epiphyton*) were also distinct and became diversified during this interval (e.g., Riding 2000, 2006b; Shapiro 2004). The Proterozoic stromatolites, with the exception of silicified stromatolites, rarely preserve tubular or vesicular microfossils (e.g., Riding and Sharma 1998; Riding 2000), whereas the Cambrian to Early Ordovician stromatolites include calcimicrobial traces such as *Girvanella* (e.g., Toomey and Nitecki 1979; Cañas and Carrera 1993; Dang et al. 2009).

Previous studies of Cambrian to Early Ordovician stromatolites have examined depositional environments, environmental controls on stromatolite growth form, and morphostratigraphy (e.g., Pratt and James 1982; de Freitas and Mayr 1995;

N. Adachi (✉), J. Liu, and J. Cao
School of Earth and Space Sciences, Peking University, Haidian, Beijing 100871, PR China
e-mail: adachi@pku.edu.cn

Y. Ezaki
Department of Geosciences, Osaka City University, Sugimoto 3-3-138, Sumiyoshi-ku, Osaka 558-8585, Japan

J. Reitner et al., *Advances in Stromatolite Geobiology*,
Lecture Notes in Earth Sciences 131, DOI 10.1007/978-3-642-10415-2_28,

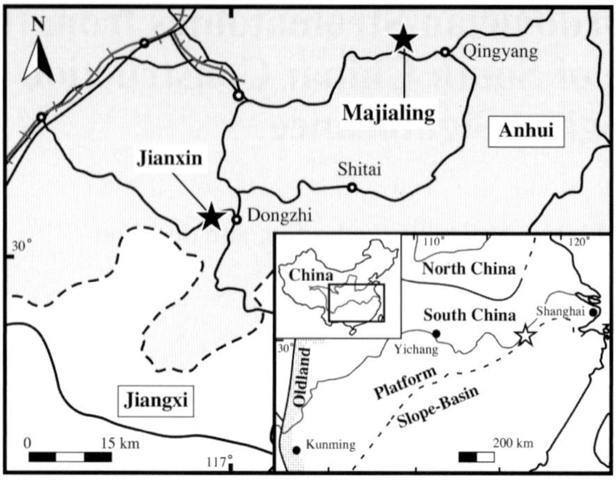

Fig. 1 Index map showing the Jianxin and Majialing sections at Anhui Province, South China. The inset map shows an Early to Middle Ordovician palaeogeographic reconstruction [modified from Zhang and Jin (2007)]. The star symbol shows the location of the larger map

Shapiro and Awramik 2000); however, little is known of their microstructures and the responsible calcimicrobes (e.g., *Girvanella*).

Stromatolite-bearing reefs associated with well-preserved calcimicrobes and metazoan (e.g., lithistid sponges, receptaculitids) can be found in the Lower Ordovician Hunghuayuan Formation (late Tremadocian–early Floian) in Anhui Province, South China (Fig. 1). Although these stromatolites show restricted morphological diversity and are small in size compared with typical Proterozoic stromatolites, the presence of abundant calcimicrobial traces provides clues to their construction. The Early Ordovician is also a particularly intriguing time, when a transition in reef type occurred between microbial- and metazoan-dominated reefs (Adachi et al. 2009).

In this paper, we first describe the megastructure of the stromatolite-bearing reefs and macrostructure (growth forms) of stromatolites. We then focus on the microstructure of the stromatolites, followed by a discussion on their construction and geobiological significance.

2 Study Area and Methods

We collected samples from two sections in the Lower Ordovician Hunghuayuan Formation in Anhui Province, South China (Fig. 1): (1) the Jianxin section (about 7 km west of Dongzhi) and (2) the Majialing section (about 7 km west of Qingyang). For microfacies analysis, we prepared polished slabs, large thin sections (7.5 × 10 cm), and small thin sections (5 × 7.5 cm). The Hunghuayuan Formation is about 70 m thick in the Jianxin section and about 100 m thick in the Majialing section. In both sections, the formation conformably overlies the middle Tremadocian Lunshan Formation and is overlain by the middle Floian Zitai Formation.

The age of the formation in both sections is interpreted to be late Tremadocian to early Floian (e.g., Adachi et al. 2009; Cao et al. 2009). Lithofacies of the formation are similar in both sections, characterised by dark gray to gray skeletal grainstones to packstones (which include abundant skeletal organisms such as pelmatozoans, trilobites, brachiopods, gastropods, cephalopods, and the problematicum *Nuia*) and mudstones. Intercalated reefal limestones include abundant lithistid sponges and receptaculitids, and microbialites, mostly stromatolites. Previous studies have examined the role of calcimicrobial activity, particularly that of *Girvanella*, in reef construction and the growth mode of reefs in the Jianxin section (Adachi et al. 2009; Cao et al. 2009).

3 Mega- and Macrostructures of Stromatolites

The examined stromatolite-bearing reefs are dome-shaped (ranging in width from several tens of centimetres to several tens of metres) and surrounded by skeletal packstones and grainstones that include abundant pelmatozoans, trilobites, and *Nuia* (Fig. 2a). Two main lithofacies (microbial bind/framestones and receptaculitid-lithistid sponge-microbial baffle/bindstones) are present within the reefs (Fig. 2b–d). The microbial bind/framestones are equivalent to the stromatolites. Distinct thrombolites have not yet been found in either section.

The macrostructure consists of nodular and columnar (sensu Walter et al. 1992) growth forms (10–40 cm wide and 10–50 cm high), with an obscure laminar texture on weathered surfaces and polished slabs (Figs. 2d, e and 3a, c). Stromatolites commonly occur side-by-side within individual beds (Fig. 2c) and are developed directly on bioclastic substrates and/or apparently upon receptaculitids and lithistid sponges. Inter-stromatolite portions are filled with bioclast-dominated sediments that contain sponge spicules, pelmatozoans, and trilobites. Tube-shaped burrows (0.2–0.5 mm in diameter) filled with coarse sparry calcite and/or dolomite are intensely developed, largely destroying the laminar textures (Fig. 3c, d).

4 Microstructures of Stromatolites

The laminae of the stromatolites are defined by irregularly repeated accumulations of *Girvanella*-rich layers and thick micritic layers containing sporadic *Girvanella* (Figs. 2e–g and 3a–b). Some of the *Girvanella*-rich layers exhibit wavy to steeply convex laminae (Fig. 3f). Bioclasts such as sponge spicules and pelmatozoans are common in the thick micritic layers, but rare in the *Girvanella*-rich layers. *Girvanella*, a non-branching filamentous calcimicrobe with a thin micritic wall (Fig. 3e) and external diameter of 10–15 μm, is the only recognisable calcimicrobe within the stromatolites. *Girvanella* generally has a randomly tangled distribution.

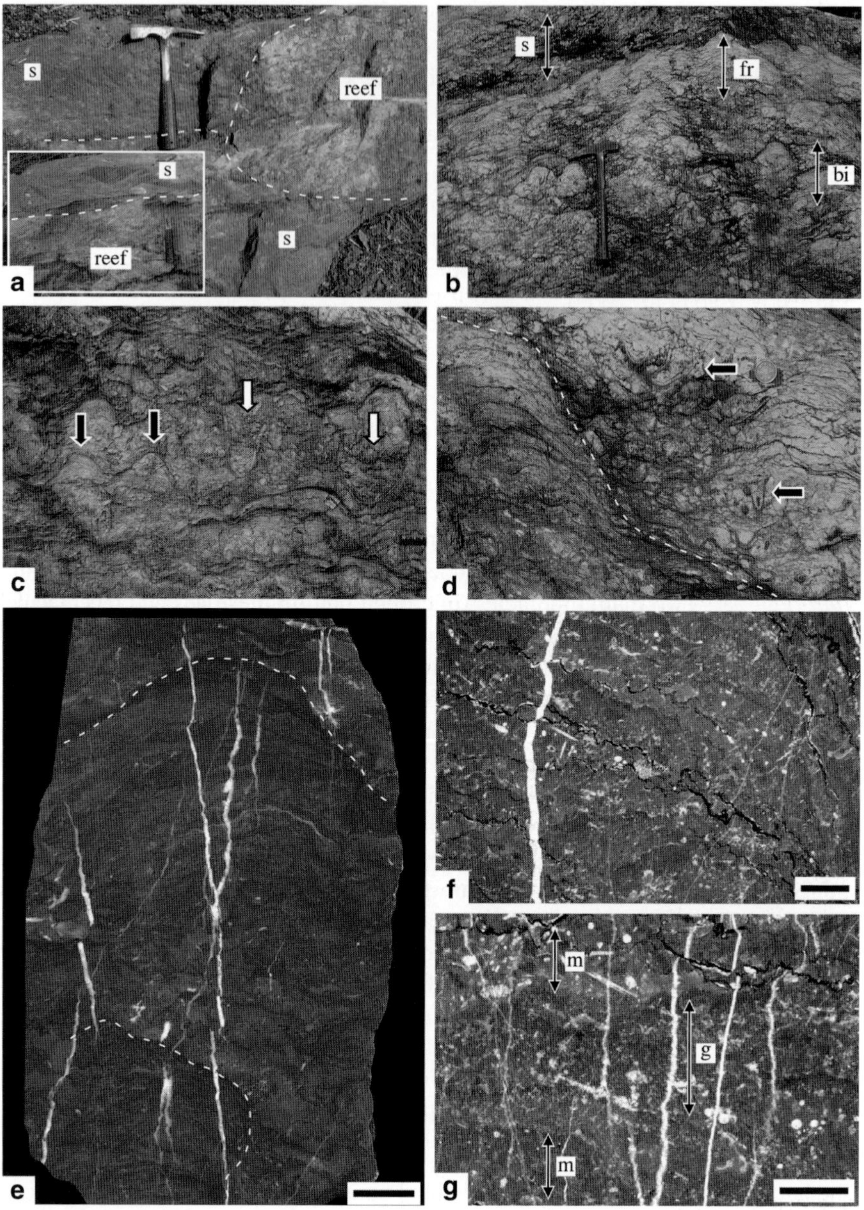

Fig. 2 Mega- and microstructures of the examined stromatolites. (**a**) Stromatolite-bearing reefs are surrounded or truncated by skeletal limestones (s). (**b**) Receptaculitid-lithistid sponge-microbial baffle/bindstones (bi) and microbial bind/framestones (fr) form reefs, which are covered by skeletal limestones (s). (**c**) Stromatolites (*black arrows*) are generally accompanied laterally by lithistid sponges and receptaculitids (*white arrows*) within the reef. (**d**) Receptaculitids (*arrows*) are abundant in inter-stromatolite spaces. (**e**) Polished surface of columnar stromatolite. Scale = 2 cm. (**f**) Enlargement of an area in (**e**), showing faint laminae within a stromatolite, as viewed under an

The *Girvanella*-rich layers contain the following textures: (1) densely packed *Girvanella*; (2) micrites with obscure *Girvanella*; (3) peloids, in and among which *Girvanella* predominates; and (4) micrites with irregular fenestral fabrics (Fig. 3e–h). *Girvanella* is obscure within the micrites because its filaments are occluded by micrite (Fig. 3f). The fenestral fabrics (0.5–1.5 mm in diameter) are filled with sparry calcite or filled in the lower parts of their voids by micrite and in the upper parts by sparry calcite, exhibiting geopetal structures. In some cases, interstices of loosely tangled *Girvanella* are filled with sparry calcite to produce fenestral-like fabrics (Fig. 3h). Individual *Girvanella*-rich layers exhibit lateral textural variations ranging from densely packed *Girvanella* to micrites with obscure *Girvanella*, peloids dominated by *Girvanella*, and micrites with fenestral fabrics.

5 Discussion: Stromatolite Construction and Geobiological Significance

Girvanella is the only recognisable calcimicrobe in the studied stromatolites, but occurs in abundance. The fact that some laminae are constructed by repeated accumulations of *Girvanella* implies that this calcimicrobe played a major (although not exclusive) role in stromatolite formation. *Girvanella* is inferred to originate from the calcification of filamentous cyanobacteria (e.g., Riding 1977), in which strands (trichomes) of cells are commonly surrounded by a protective mucilaginous sheath composed of extracellular polymeric substrates secreted by the cells (e.g., Merz-Preiß 2000; Riding 2006b).

Previous studies report that cyanobacterial calcification is closely related to photosynthetic carbon uptake (e.g., Pentecost and Riding 1986). During photosynthesis, CO_2 and HCO_3^- enter the cells and OH^- is excreted, leading to an increase in sheath pH. A sufficiently high ambient saturation state for $CaCO_3$ minerals can result in the nucleation of $CaCO_3$ crystals within the sheath. The continuation of this process impregnates the sheath with $CaCO_3$, yielding a calcified replica that preserves the sheath outline and dimensions (Riding 2006b). In addition, filamentous cyanobacteria growing on the surface of biofilm may be passively calcified to form *Girvanella*, where an increase in alkalinity results from ammonification and sulfate reduction due to the bacterial degradation of organic matter.

The decomposition of cyanobacteria themselves also leads to post-mortem calcification (e.g., Merz-Preiß 2000); however, in such a case the dead filaments are prone to bacterial attack and the sheath is progressively distorted and is unlikely to maintain its tube structure (Arp et al. 2002; Riding 2006b). Thus, in vivo

Fig. 2 (continued) optical microscope. Scale = 3 mm. (**g**) Laminae consist of the repeated accumulation of *Girvanella*-rich layers (g) and thick micritic layers (m) that contain sponge spicules. Scale = 1 mm

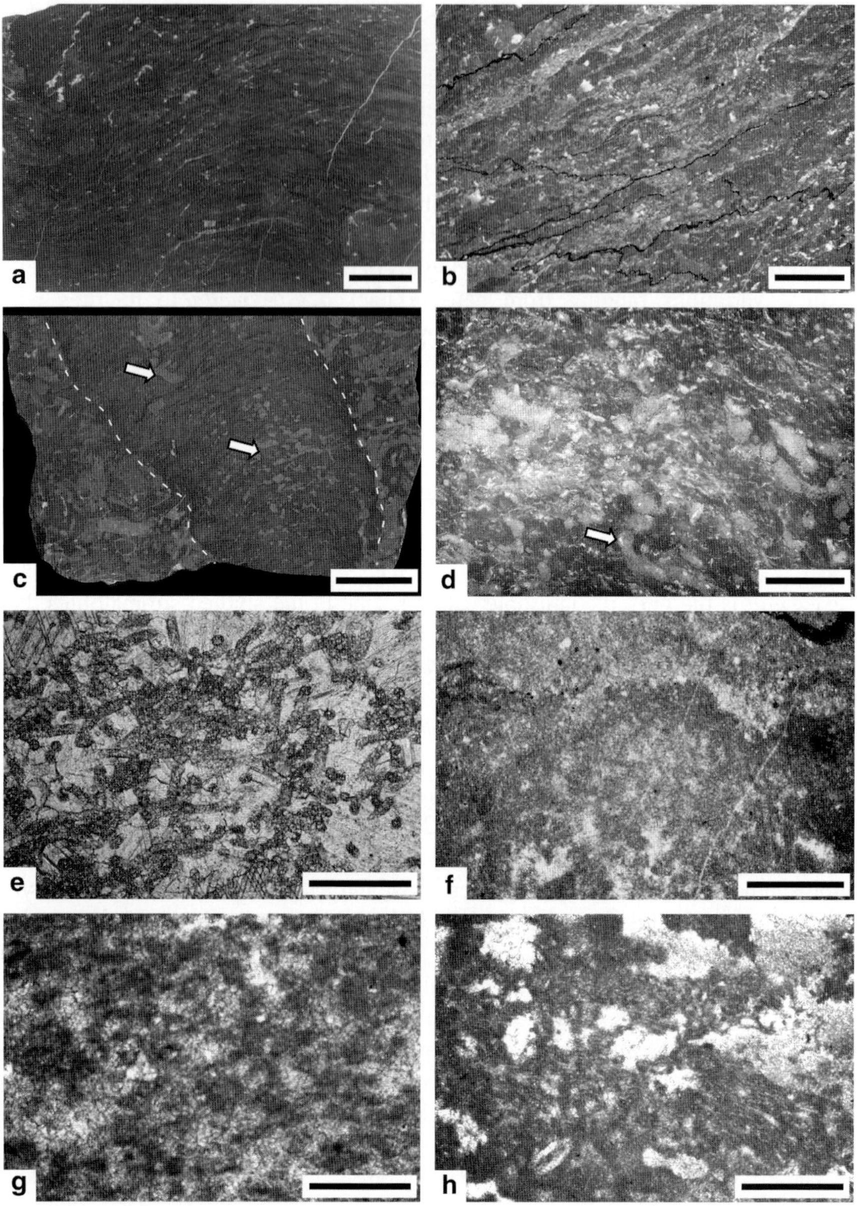

Fig. 3 Macro- and microstructures of the examined stromatolites (**a**–**d**) and features of a *Girvanella*-rich layers that are relatively dark layers (**e**–**h**). Scale = 4 mm. (**a**) Nodular stromatolites. Scale = 1 cm. (**b**) Enlargement of part of (**a**), showing *Girvanella*-rich layers that are relatively dark layers. Scale = 4 mm. (**c**) Burrows (*arrows*) are common in the stromatolites. Scale = 4 cm. (**d**) Enlargement of part of (**c**), showing tube-shaped burrows (*arrow*) that destroy the laminar textures. Scale = 3 cm. (**e**) Densely packed *Girvanella*. Scale = 0.2 mm. (**f**) Micritic texture, including obscure *Girvanella*. Scale = 0.4 mm. (**g**) Peloids are abundant in areas where *Girvanella* occur sporadically. Scale = 0.2 mm. (**h**) Fenestral-like fabric. Interstices of loosely tangled *Girvanella* are filled with sparry calcite. Scale = 0.5 mm

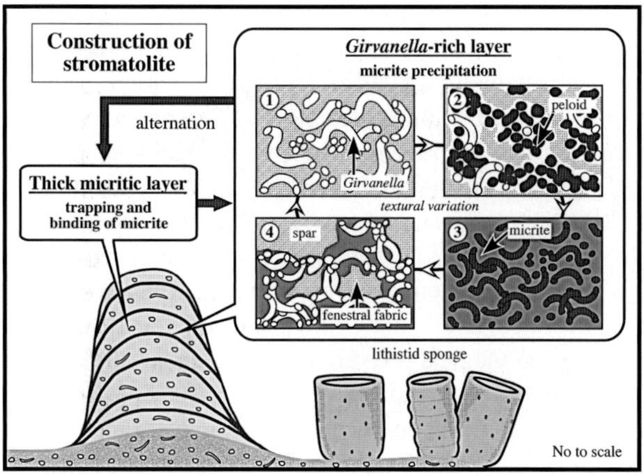

Fig. 4 Schematic model showing the construction of Lower Ordovician stromatolites. Laminae consist of alternating *Girvanella*-rich layers and thick micritic layers. *Girvanella*-rich layers exhibit varying textures ranging from (1) densely packed *Girvanella* to (2) peloids in and among which *Girvanella* predominates, (3) micrite with obscure *Girvanella*, and (4) micrite with fenestral fabrics. These different textures grade into one another within a single layer. See the text for further details

calcification of the sheath leads to the formation of *Girvanella*, whereas post-mortem degradative sheath calcification results in irregular micritic structures (e.g., Pratt 1995; Turner et al. 2000; Webb 2001; Adachi et al. 2004).

Because the textures in *Girvanella*-rich layers analysed in the present study show gradational transitions (Figs. 3e–h and 4), they are considered to reflect differences in the degree of initial calcification, and/or diagenetic alteration that affected calcimicrobe preservation. In the *Girvanella*-rich layers, *Girvanella* was produced when the sheaths of filamentous cyanobacteria were completely calcified, whereas peloids and micrite with obscure *Girvanella* were locally precipitated where the sheaths were bacterially decomposed and/or incompletely calcified. Fenestral-like fabrics are also observed, defined by the infilling by cement of interstices among tangled *Girvanella* (Fig. 4h). The dominant fenestral fabrics are possibly related to the non-calcified, void portion of a filamentous precursor (Monty 1976).

In contrast, the thick micritic layers that alternate with the *Girvanella*-rich layers contain abundant bioclasts but sporadic *Girvanella* (Figs. 2f, g and 4). It is also possible that these layers originated from micrite precipitation and decomposition, related to filamentous microbes; however, such layers were undoubtedly produced by the trapping and binding of micrite with bioclasts. In any case, the repeated changes in microbial (especially filamentous microbial) activity described above were closely linked to fluctuations in the ambient environment, resulting in the formation of the laminae that are characteristic of the stromatolites.

The examined stromatolites are considered to have formed as follows: (1) biofilms (including filamentous microbes) first encrusted the bioclastic substrates or lithistid sponges and receptaculitids, then (2) stabilised these substrates, and finally (3) grew upward and laterally, eventually forming laminated microbialites as a result of repeated changes in microbial activity (micrite precipitation or the trapping and/or binding of micrite). An additional feature unique to Lower Ordovician stromatolites is their co-occurrence with metazoans (e.g., lithistid sponges and receptaculitids) within reefs (Fig. 2c, d), although the metazoans were of low diversity before the Ordovician radiation. Such Lower Ordovician reefs are in marked contrast to subsequent shallow-marine reefs of the Middle to Late Ordovician, in which skeletal metazoans (e.g., stromatoporoids and corals) flourished to form metazoan reefs and stromatolites showed a marked declined. Even after Ordovician biodiversification, almost all reefal limestones throughout geological time are evidently the product of combined microbial and metazoan activity (e.g., Adachi et al. 2006, 2007; Adachi and Ezaki 2007); however, this Lower Ordovician co-occurrence of both stromatolites and metazoans within reefs provides critical clues regarding (1) reef palaeoecology during the period immediately prior to a great diversification of reef-building skeletal organisms and (2) the interrelationship between microbes and metazoans in the formation of age-specific reefs. More importantly, the Lower Ordovician stromatolites provide invaluable clues to the transition from microbe- to metazoan-dominated reefs and the evolution of reef ecosystems.

6 Conclusions

1. Lower Ordovician stromatolites of the Hunghuayuan Formation in Anhui Province, South China, consist of irregularly alternating accumulations of *Girvanella*-rich layers and thick micritic layers that contain sporadic bioclasts.
2. The *Girvanella*-rich layers exhibit the following textures: (1) densely packed *Girvanella*; (2) micrite with obscure *Girvanella*; (3) peloids, in and among which *Girvanella* is predominant; and (4) micrite with fenestral fabrics. These textures resulted from varying degrees of initial calcification and/or diagenetic alternation of the filamentous precursor.
3. The stromatolites are thought to have been constructed as follows: (1) biofilms including filamentous microbes became prevalent upon the bioclastic substrates or lithistid sponges, then (2) stabilised these substrates and (3) grew upward and laterally to produce laminated microbialites in response to repeated changes in microbial activity (micrite precipitation or trapping and/or binding of micrite) and fluctuations in the ambient environment.

Acknowledgements We thank anonymous reviewers and editors for their valuable comments on the manuscript. This research was supported by a Grant-in-Aid for Japan Society for the Promotion of Science Fellows and funded by a grant (17540444, 21340154) from the Scientific Research Fund of the Japan Society for the Promotion of Science, and by the National Natural Science Foundation of China (40972020), the Ministry of Science and Technology of China (2006CB806402), and the State Key Laboratory of Palaeobiology and Stratigraphy (NIGPAS20081103).

References

Adachi N, Ezaki Y (2007) Microbial impacts on the Genesis of Lower Devonian reefal limestones, Eastern Australia. Palaeoworld 16:301–310

Adachi N, Ezaki Y, Liu J (2004) The fabrics and origins of peloids immediately after the end-Permian extinction, Guizhou Province, South China. Sedimentary Geology 164:161–178

Adachi N, Ezaki Y, Pickett JW (2006) Marked accumulation patterns characteristic of Lower Devonian stromatoporoid bindstone: palaeoecological interactions between skeletal organisms and microbes. Palaeogeography, Palaeoclimatology, Palaeoecology 231:331–346

Adachi N, Ezaki Y, Pickett JW (2007) Interrelations between framework-building and encrusting skeletal organisms and microbes: more-refined growth history of Lower Devonian bindstones. Sedimentology 54:89–105

Adachi N, Ezaki Y, Liu J, Cao J (2009) Early Ordovician reef construction in Anhui Province, South China: a geobiological transition from microbial- to metazoan-dominant reefs. Sedimentary Geology 220:1–11

Arp G, Reimer A, Reitner J (2002) Calcification of cyanobacterial filaments: *Girvanella* and the origin of lower Paleozoic lime mud: comment. Geology 30:579–580

Awramik SM (1971) Precambrian columnar stromatolite diversity: reflection of metazoan appearance. Science 174:825–827

Cañas F, Carrera M (1993) Early Ordovician Microbial-sponge-receptaculitid bioherms of the Precordillera, Western Argentina. Facies 29:169–178

Cao J, Liu J, Ezaki Y, Adachi N (2009) Lower Ordovician reefs in the Honghuayuan Formation at Dongzhi, Anhui: Microbial reefs just prior to the Ordovician biodiversification. Acta Scientiarum Naturalium Universitatis Pekinensis 45:279–288 (in Chinese with English abstract)

Dang H, Liu J, Yuan X (2009) Microbialites in the Middle Cambrian Qinjianmiao Group in Xingshan, Hubei Province: implication for Palaeoenvironmental reconstruction. Acta Scientiarum Naturalium Universitatis Pekinensis 2009:289–298 (in Chinese with English abstract)

de Freitas T, Mayr U (1995) Kilometre-scale microbial buildups in a rimmed carbonate platform succession, Arctic Canada: new insight on Lower Ordovician reef facies. Bulletin of Canadian petroleum Geology 43:407–432

Merz-Preiß M (2000) Calcification in cyanobacteria. In: Riding RE, Awramik SM (eds) Microbial Sediments. Springer, Berlin, pp 50–56

Monty C (1976) The origin and development of cryptalgal fabrics. In: Walter MR (ed) Stromatolites. Developments in Sedimentology 20. Elsevier, Amsterdam, pp 193–249

Pentecost A, Riding R (1986) Calcification in cyanobacteria. In: Leadbeater BSC, Riding RE (eds) Biomineralization in Lower Plants and Animals. Systematic Association Special Volume 30. Clarendon Press, Oxford, pp 73–90

Pratt BR (1995) The origin, biota and evolution of deep-water mud-mounds: their origin and evolution. In: Monty CLV, Bosence DWJ, Bridges PH, Pratt BR (eds) Carbonate Mud-Mound. International Association of Sedimentologists, Special Publication 23. Blackwell, Oxford, pp 49–123

Pratt BR, James NP (1982) Cryptalgal-metazoan bioherms of early Ordovician age in the St George Group, western Newfoundland. Sedimentology 29:543–569

Riding R (1977) Calcified *Plectonema* (blue-green algae), a recent example of *Girvanella* from Aldabra Atoll. Palaeontology 20:33–46

Riding R (2000) Microbial carbonates: the geological record of calcified bacterial-algal mats and biofilms. Sedimentology 47:179–214

Riding R (2006a) Microbial carbonate abundance compared with fluctuations in metazoan diversity over geological time. Sedimentary Geology 185:229–238

Riding R (2006b) Cyanobacterial calcification, carbon dioxide concentrating mechanisms, and Proterozoic-Cambrian changes in atmospheric composition. Geobiology 4:299–316

Riding R, Sharma M (1998) Late Palaeoproterozoic (~1800–1600 Ma) stromatolites, Cuddapah Basin, southern India: cyanobacterial or other bacterial microfabrics? Precambrian Research 92:21–35

Shapiro R (2004) Neoproterozoic-Cambrian microbialite record. The Paleontological Society Papers 10:5–15
Shapiro R, Awramik SM (2000) Microbialite morphostratigraphy as a tool for correlating Late Cambrian–Early Ordovician sequences. Journal of Geology 108:171–180
Toomey DF, Nitecki MH (1979) Organic buildups in the Lower Ordovician (Canadian) of Texas and Oklahoma. Fieldiana Geology 2:1–181
Turner EC, James NP, Narbonne G (2000) Taphonomic control on microstructure in Early Neoproterozoic reefal stromatolites and thrombolites. Palaios 15:87–111
Walter MR, Grotzinger JP, Schopf JW (1992) Proterozoic stromatolites. In: Schopf JW, Klein C (eds) The Proterozoic Biosphere: A Multidisciplinary Study. Cambridge University Press, Cambridge, pp 253–260
Webb GE (2001) Famennian mud-mounds in the proximal fore-reef slope, Canning Basin, Western Australia. Sedimentary Geology 145:295–315
Webby BD (2002) Patterns of Ordovician reef development. In: Kiessling W, Flügel E, Golonka J (eds) Phanerozoic reef patterns. SEPM Special Publication 72:129–179
Zhan R, Jin J (2007) Ordovician–Early Silurian (Llandovery) Stratigraphy and Palaeontology of the Upper Yangtze Platform, South China, Science Press, Beijing, pp 169

Sedimentology and Palaeoecology of *Ernietta*-Bearing Ediacaran Deposits in Southern Namibia: Implications for Infaunal Vendobiont Communities

El Hafid Bououogri, Hubertus Porada, Klaus Weber, and Joachim Reitner

1 Introduction

The Ediacaran was a period of wide soft-bodied biota development in marine siliciclastic environment. The reconstruction of palaeoecology, lifestyle and environment settings of these organisms are among the challenges in characterising the earliest metazoan life on Earth. The biota colonised sea bottoms and were living as benthic communities occurring in a wide spectrum of environmental settings (e.g. Gehling 2000; Waggoner 2003; Narbonne 2005; Grazhdankin 2004; Droser et al. 2006). In the classical localities (e.g. Newfoundland, Namibia, White Sea and Australia), the fossils occur in rock successions with distinct depositional settings ranging from distributary mouth-bar of braid-delta systems to deep-water slope (Gehling 2000; Grazhdankin 2004; Narbonne 2005).

However, the preservation style of this biota in the sedimentary record as well as their lifestyle and palaeoecology are still unclear for some assemblages and genera of which data are mainly coming from isolated float samples. According to various features including among others taphonomy, ecology, functional morphology and style of preservation within the rock record, two main benthic faunas with contrasting lifestyles are distinguished: (1) epifaunal disc-shaped to sessile leaf-shaped communities, living at the water-sediment interface, generally buried as in situ biota in their habitat and preserved as epireliefs. Such types of palaeocommunities are well recognised in the Avalon-type assemblage where deep-water frondose biota are smothered in their habitat and preserved in turbiditic successions as

E.H. Bouougri
Department of Geology, Faculty of Sciences Semlalia, Cadi-Ayyad University, 40020 Marrakech, Morocco

H. Porada, K. Weber, and J. Reitner (✉)
Geowissenschaftliches Zentrum Göttingen, Universität Göttingen, Goldschmidtstrasse 3, 37077 Göttingen, Germany
e-mail: jreitne@gwdg.de

J. Reitner et al., *Advances in Stromatolite Geobiology*,
Lecture Notes in Earth Sciences 131, DOI 10.1007/978-3-642-10415-2_29,

casts on bedding surfaces (Narbonne 2005; Hofmann et al. 2008); (2) infaunal bag-shaped to canoe-shaped communities commonly preserved as three-dimensional moulds and reported in fossil-bearing rocks as reworked entities or in life position. Exclusively mouldic preservation was suggested for quilted organisms of Nama-type fossils like *Rangea*, *Namalia*, *Pteridinium* and *Ernietta* interpreted as epifaunal or infaunal biota (Jenkins et al. 1981; Jenkins 1992; Buss and Seilacher 1994; Dzik 1999; Grazhdankin and Seilacher 2002, 2005).

Although the infaunal or epifaunal lifestyle of some shallow-marine genera is still controversial (e.g. Laflamme and Narbonne 2008), it is necessary to note that various factors have been suggested to explain the diversity and distribution of the Ediacara assemblages, including biogeography, biostratigraphy, ecology and taphonomic processes (Jenkins 1992; Buss and Seilacher 1994; Gehling 1999; Waggoner 2003; Grazhdankin 2004; Narbonne 2005; Droser et al. 2006; Mapstone and McIlroy 2006). Through comparison of sedimentary features of Ediacara-bearing siliciclastic rocks and the relationship between lithofacies and biogeography, Grazhdankin (2004) documented a striking correlation between fossil association and depositional setting and distinguished three "palaeoecologic assemblages", namely, Nama, Avalon and Ediacara. These assemblages were as well recognized previously by using cluster analysis of taxonomic composition (Waggoner 2003). The three-dimensionally preserved moulds seem to be the main feature of the Nama-type assemblage and have been assigned as a type of biofacies within a "braid-delta" environment in association with sandy deposits of high energy setting (Grazhdankin 2004). The mouldic fossils preserved as endoreliefs and including *Ernietta*, *Namalia*, *Pteridinium* and *Rangea* were reported from sandy event deposits as lags, being reworked, current transported and redeposited (Jenkins 1985, 1992; Grazhdankin and Seilacher 2002, 2005; Grazhdankin 2004). By contrast, the shallow marine fossils of White Sea and Ediacara are preserved almost all in their habitat as hyporeliefs and occur within interbedded storm sandstone and mudstone of delta-front and pro-delta settings (Gehling 1999; Grazhdankin 2004).

Despite substantial progress and a large amount of studies on the structure and morphology of the three-dimensional moldic fossils (e.g. Jenkins 1992; Dzik 1999; Grazhdankin and Seilacher 2002, 2005), less is known on the major question dealing with the sedimentary setting and the precise biotope of the quilted organisms like *Ernietta*. The best direct evidence in constraining the ecological niches of this community is the preservation in life-position of body fossils in the sedimentary rock. Most of the specimens in previous studies (e.g. Jenkins et al. 1981; Dzik 1999), however, were winnowed and float specimens of which the precise style of preservation, biostratinomy and sedimentary setting is not well constrained. In this paper, we report newly found occurrences of *Ernietta* fossils from the lowermost unit of the Dabis Formation (Kanies Member) in the Nama Group. These new findings extend the stratigraphic range of Ediacara fossils in Namibia to the lowermost member of the Nama Group and provide the opportunity: (1) to constrain the sedimentary setting, preservation styles and the palaeoecology of *Ernietta* community; (2) to test the relationship between preserved three-dimensional fossils, fossil-bearing lithofacies and facies assemblage.

Ernietta is a ribbed, sack-shaped Ediacara fossil, 3–8 cm in width, mainly preserved as three-dimensional bodies. Pflug (1966) who investigated numerous specimens from Lower Nama Group sandstone and schist (Namibia), defined it as a new genus and included it in the group of Petalonamae, together with *Rangea* and *Pteridinium*. Later, *Ernietta*, represented by a single species *Ernietta plateauensis*, was attributed to the group of Vendobionta (Seilacher 1992). Up to now, *Ernietta* is only known from the Ediacaran of Namibia, where it occurs in great number over a wide area (Pflug 1966; Germs 1972), and from terminal Proterozoic siliciclastic deposits in Death Valley (southern Nevada, USA), where few specimens have been recovered (Horodyski 1991; Horodyski et al. 1994; Waggoner and Hagadorn 1997; Hagadorn et al. 2000; Corsetti and Hagadorn 2000).

In both regions *Ernietta* occurs in similar geological context. In Death Valley (see Corsetti and Hagadorn 2000), the fossils are present in the lower member of the Wood Canyon Formation which overlies, locally disconformably, the Stirling Quartzite and which consists of three parasequences, each beginning with shallow marine siliciclastic rocks that are gradationally overlain and capped by carbonate rocks. The succession is considered to record a transgressive to highstand systems tract in a shallow marine-continental braidplain transition (Hagadorn and Waggoner 2000), with *Ernietta* occurring in the lowest few metres of the first parasequence. Based on chemostratigraphic data and regional correlations, Corsetti and Hagadorn (2000) suggested a depositional age of 548–544 Ma for the upper Sterling Quartzite and the lower member of the Wood Canyon Formation.

In Namibia, *Ernietta* has been encountered in the Kuibis Subgroup at the base of the Nama Group in the southern Nama basin (Fig. 1a). This subgroup consists of two depositional sequences, each with fluvial to marginal marine sandstone overlain by subtidal carbonate (Saylor et al. 1995). The two sequences, deposited between about 555 and 550 Ma based on radiometric data (Grotzinger et al. 1995), are considered each to record low-stand to transgressive systems tracts (Saylor et al. 1995).

The fossils originally described by Pflug (1966) and referred to in the subsequent literature (see review by McCall 2006), have been recovered as isolated float samples from the Kliphoek Member of the upper sequence, just below the capping carbonate unit (Mooifontein Member).

Pflug (1966) observed *Ernietta* in steinkern preservation within a succession of shale (mudstone) with sandy to silty admixtures, in the upper part of the Kliphoek Member. He reported that in most cases ($n =$ ca. 40) the hemispheric bodies rest on bedding surfaces with the sculptured surface oriented upward, and suggested that the organisms had a benthonic-sessile lifestyle. In contrast, Jenkins (1992, p. 148, fig. 7) who studied a specific sample (GU Pf 536) from the collection of H. D. Pflug (Giessen) containing numerous three-dimensional individuals, described *Ernietta* as occurring in medium-grained, impure sandstone in which they may form groups or clusters with the sculptured, convex parts directed downward. He also considered them in life position and suggested that they "were probably buried by density currents generated as a result of wave movement and rips at times of storms". Dzik (1999, p. 519), partly referring to the same sample,

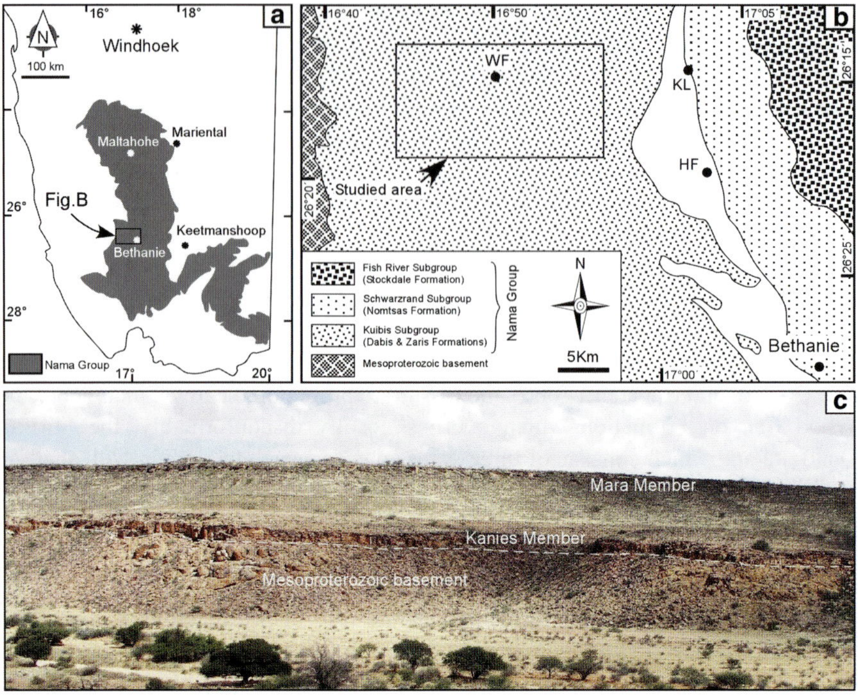

Fig. 1 (**a**) Outcrop area of the Nama Group in southern Namibia (based on Geological map of Namibia, Geological Survey of Namibia, 2003). (**b**) Simplified map showing location of the studied area. *HF* Hansburg farm, *KL* Konkiep Lapa, *WF* Wegkruip farm. (**c**) View along Hansburg River showing Mesoproterozoic basement overlain by Kanies and Mara Members

interpreted the rock matrix as "muddy sandstone" and proposed that the three-dimensional preservation of *Ernietta* resulted from "density-controlled sinking of sand-filled organic skeletons within hydrated mud layers". Seilacher (1992, p. 608) after having studied the same sample as previously Jenkins (1992) and later Dzik (1999), concluded that *Ernietta* falls into the group of "sediment stickers" which rested on the sediment surface or on a microbial mat, and whose margins grew upwards keeping pace with sand sedimentation, thus evolving into partly sand-filled, bag-like bodies; for their preservation he suggested sudden burial (obrution). This interpretation of *Ernietta's* lifestyle was resumed by Buss and Seilacher (1994) and later by Grazhdankin and Seilacher (2002) who suggested that the three-dimensional preservation of *Ernietta* and other Vendobionta (e.g., *Pteridinium*, *Namalia*, *Rangea*) might be the result of in situ preservation of "infaunal" communities of "underground vendobionts".

It has to be noted that (1) all Namibian specimens described are from float samples; (2) inferred "life positions" and life style are based on circumstantial evidence and virtually refer to one specific sample only (GU Pf 536); and (3) the sedimentological context has never been studied in detail.

The purpose of this paper is to characterize the sedimentary setting of the *Ernietta* and *Rangea*-bearing Kanies Member through facies and facies association analysis and, on the other hand, to assess the palaeoecological significance of *Ernietta* occurrences with possible implications for other occurrences of similar biota (e.g. *Rangea*, *Namalia* and *Pteridinium*). Three detailed sections were measured on Hansburg and Wegkruip farms, which show an E-W lateral variation in facies over 25 km perpendicular to the strike of the basin (Figs. 1b and 2).

2 Geological Background

The Ediacaran to Cambrian Nama Group is a fluvial and shallow marine succession of siliciclastic and carbonate sedimentary rocks, ca. 3,000 m in thickness. It was deposited in a peripheral foreland basin that developed in response to collision along the Pan-African Gariep and Damara belts and is subdivided, from base to top, into the Kuibis, Schwarzrand and Fish River Subgroups. According to paleontological and radiometric data (Germs et al. 1986; Grotzinger et al. 1995, 2000), deposition of the Kuibis and most of the Schwarzrand Subgroups occurred in the Ediacaran. The Nama basin sensu *stricto* extends north-south over more than 400 km in central to southern Namibia (Fig. 1a), and is subdivided by a flexural

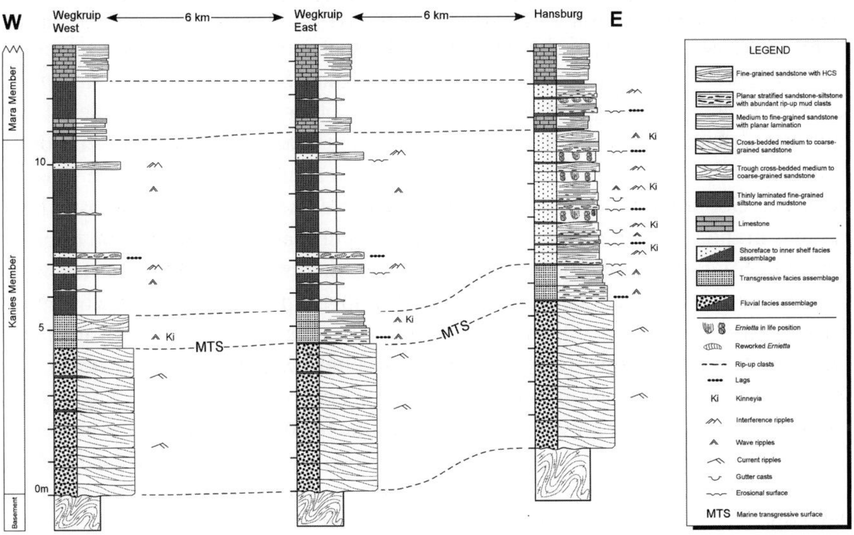

Fig. 2 Logged sections in an E-W transect through Kanies Member illustrating vertical and lateral distribution of facies associations. Note lateral continuity of the basal sandstone unit and lateral variation within the upper part of Kanies Member from sandy dominated deposits in the East (Hansburg area) to muddy dominated deposits in the west (Wegkruip area)

high (Osis ridge) into the northern Zaris and southern Witputs sub-basins (Germs 1983; Germs and Gresse 1991).

The Nama Group (Fig. 1a) is well known for its typical Ediacara fossils (~565–542 Ma), characterized by low taxonomic richness and unusual three-dimensional preservation. The fossils are commonly described in association with storm deposits and are preserved as moulds and casts at the bases, within and at the tops of sandstone beds (e.g. Jenkins 1985; Narbonne et al. 1997; Grazhdankin and Seilacher 2002). Most of the fossils were described from the Kliphoek Member of the Kuibis Subgroup (e.g., Pflug 1966; Germs 1972) and from the overlying Schwarzrand Subgroup with the youngest occurrence of *Swartpuntia* near the top of the Spitskopf Member (Narbonne et al. 1997). Occurrences of *Ernietta* as previously described and reported here, are all from the Witputs sub-basin in the south. The new locality of *Ernietta* described here, is situated on Hansburg farm (Fig. 1b), about 55 km to the northeast of the localities reported by Pflug (1966). At Hansburg farm, up to three *Ernietta* carrying horizons occur in a thin interval of cross-bedded and planar laminated sandstone, ca. 7–9 m above the base of the Nama Group and less than 1 m below the first carbonate bed. A fourth horizon is locally developed in a sandy interval situated less than 1 m above the first carbonate bed. The locality is thus stratigraphically situated mainly in the Kanies Member of the Dabis Formation (Fig. 1c), in contrast to the previously described occurrences which are in the overlying Kliphoek Member of the Zaris Formation (Pflug 1966; Germs 1995; Saylor et al. 1995).

During deposition of Kuibis and Schwarzrand strata, the Witputs sub-basin was open to the sea towards the west and received continental, terrigenous fluvial input from the east and north-east (Germs 1983; Saylor et al. 1998; Saylor 2003). Radiometrically, the age of the Kuibis Subgroup is not constrained. However, an ash layer interbedded in the overlying Schwarzrand Subgroup yielded an age of 548.1 $\pm$ 1 Ma (Grotzinger et al. 1995). On the other hand, it is considered that deposition of the Nama Group postdates the late major Ediacaran glaciation in southwestern Africa (Germs 1995). According to a south-north correlative scheme, the Kuibis Subgroup succession shows considerable thinning toward the Osis ridge, whereas in the southernmost part of the Witputs sub-basin, the succession thickens westward toward the Gariep belt (Germs 1983; Saylor et al. 1995). In this latter area, the lithostratigraphic units of the Kuibis Subgroup therefore form two depositional sequences K1 and K2, each displaying a basal siliciclastic unit (Kanies and Kliphoek Member, respectively) forming lowstand systems tracts and upper carbonate units (Mara and Mooifontein Member, respectively) forming transgressive systems tracts (Saylor et al. 1995, 1998). The siliciclastic members comprise fluvial to upper shoreface, delta or tide influenced deposits, whereas the carbonate members are made up of shallow, intertidal to subtidal deposits of an extensive carbonate platform. The transition from siliciclastic to carbonate members consists of mixed deposits which record a marine transgression and a shallow subtidal to intertidal setting (Saylor et al. 1998).

In the studied area, the Kanies Member forms the lowermost unit of the Nama Group and passes up gradationally into the carbonatic Mara Member (Fig. 1c). It consists of undeformed and flat-lying strata, up to ca. 12 m thick, exhibiting an apparent fining-upward trend and lies above the crystalline basement which belongs to the Kalahari craton. Further to the South, the Kanies Member attains a thickness of up to ca. 95 m in the depocenter near the Gariep belt (Saylor et al. 1995).

The Kanies Member in the eastern area (Hansburg farm) consists of cross-bedded sandstone at the base that is overlain by fossiliferous fine-grained sandstone interbedded with hummocky cross-stratified sandstone and mudstone. In the western area (Wegkruip farm), the upper part is mud-dominated with subordinate sandy and silty layers with HCS and scattered *Ernietta* bodies.

3 Sedimentology of Kanies Member

The Kanies Member can be traced in outcrop for several tens of kilometres along river-wall exposures and road cuts, perpendicular to the depositional strike. On Hansburg and Wegkruip farms, three vertical sections with thicknesses of up to 12 m were investigated including detailed facies analysis and control on vertical and lateral thickness and facies changes (Fig. 2). Lithologically, the Kanies Member consists of two distinct parts: (1) a lower part with thick-bedded medium to coarse-grained sandstone which forms a lateral persistent unit, (2) an upper part with thin-bedded sandstone and subordinate mudstone, that passes westward into a mud-dominated succession with subordinate thin beds of sandstone (Fig. 2). Four broad facies associations are recognised (named FA 1 to FA 4), arranged in a fining-upward succession: braided fluvial deposits, transgressive deposits, storm-influenced shoreface/subtidal lagoon deposits and inner-shelf/transitional offshore deposits. The overall stacked facies associations record a transgressive setting, whereas sedimentary characteristics indicate that the *Ernietta*-bearing lithofacies forms distinct sedimentary layers interbedded within the storm-influenced marine/lagoon deposits.

3.1 *Facies Association 1 (FA 1): Braided-Fluvial Deposits*

Facies association FA 1 dominates the lower part of Kanies Member and is characterized by thick-bedded packages of quartzose sandstone that extend over the full E-W length of the studied area without significant variation in thickness or lithology. It consists of tabular units of vertically stacked, well to moderately sorted, medium- to coarse-grained cross-bedded sandstones, interbedded with subordinate thin and lens-shaped planar laminated sandstone and mudstone layers up to 30 cm thick (Figs. 2 and 3). Sandstone beds, up to 1–5 m thick, are typically tabular with bounding surfaces dipping gently in the downflow direction. Sedimentary structures in the association include medium to large-scale cross-stratification

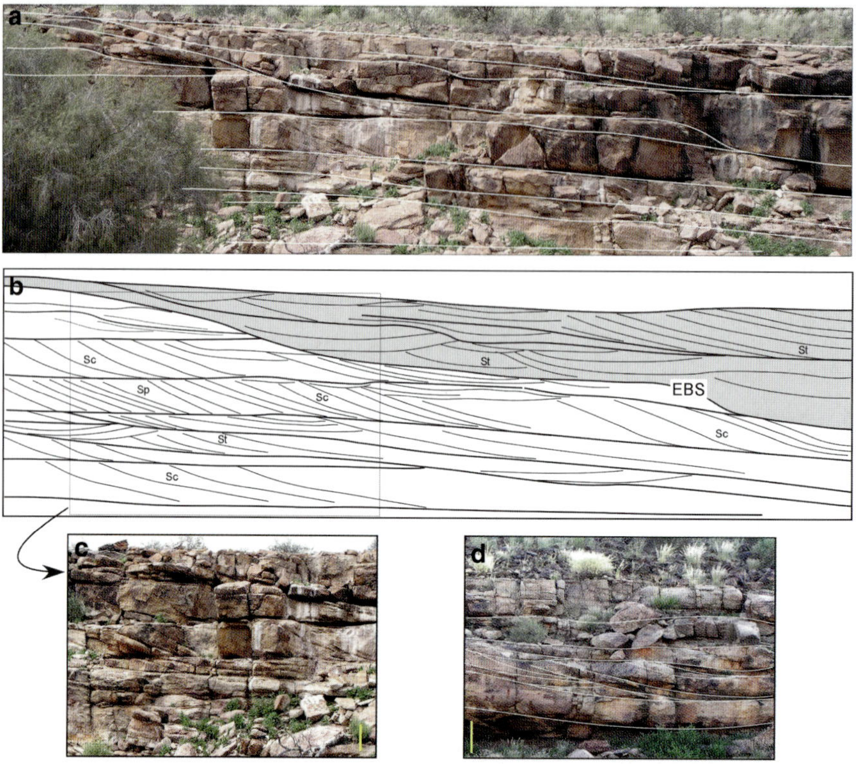

Fig. 3 Outcrop photographs and sedimentological features of facies association FA 1: (**a**) General view of the cross-bedded sandstone unit; the main bedding surfaces are marked by *white lines*; thickness of the unit is ca. 4 m. (**b**) Line drawing of bedding as seen in (**a**), illustrating details of cross-stratification characteristics [facies code from Miall (1996); see text for further details]; EBS: Erosional bounding surface. (**c**) Detail of structures in cross-bedded sandstone. (**d**) Coset with large-scale trough cross-stratification overlain by planar bedded sandstone. Scale bar is 50 cm

organized into sets and cosets with thicknesses ranging from 30 cm to 1.5 m (Fig. 3a, b). The basal surfaces of thicker sets and cosets are often marked by thin coarse-grained layers of scattered and rounded quartz granules. Small- to medium-scale sets are lenticular, whereas the cosets are either lenticular or tabular and extend laterally up to 50 m. The most common structures developed in the sandstone beds are trough (St) and concave-up (Sc) cross-bedding (Fig. 3c, d). Planar (Sp) and sigmoidal (Ss) cross-stratifications are less frequent and occur as lateral transitional structures within single cross-sets. Within some compound cross-stratified beds, the main set, often with concave-up structures (Sc), may grade into medium-scale sets of trough cross-stratification. Locally, thick cosets of cross-bedding are overlain by thin and lenticular layers of planar-laminated sandstone, up to 0.5 m thick (Fig. 3d). In section parallel to the flow direction (Fig. 3a, b), the internal architecture of the cross-bedded sandstones shows downstream dipping, large-scale concave-up erosional boundary surfaces (EB) with amplitudes of up to 3.5 m. The cross-bedding on both sides of the EB's display similar cross-stratifications with uniform foresets in dip-direction indicating unimodal paleocurrents broadly in a westward direction.

3.2 Interpretation

The tabular geometry of the cross-bedded unit together with strongly unidirectional flow (toward the west) and cross-stratification, and set/coset boundaries dipping downstream at a low angle indicate a sandy-braided fluvial depositional setting for this facies association. The most prevalent structures represented by trough and concave-up cross-stratification reflect unidirectional downstream migration and lateral accretion of large sinuous-crested dunes in relatively high-energy braid channels (e.g. Miall 1996; Bridge 2003). The architecture and internal structure of the cross-strata result from migration of bars and are similar to type-B bedforms of Røe and Hermansen (1993), interpreted as compound bars formed by local coalescence and vertical aggradation of dunes in wide and shallow channels of braided-river systems. The internal variation of cross-stratification in sets and cosets reflects current energy variations during downstream dune migration. Local occurrences of lens-shaped planar-laminated sandstone sets represent intermittent sand flat deposits which grew on top of bedforms when they reached the water surface by downstream progradation and vertical accretion. In section parallel to flow, occurrences of large erosional bounding surfaces within cross-bedded sandstones record erosional and depositional events during migration. Such type of structures is commonly associated with compound bar/dune accretion in ancient and modern braided river deposits (Miall 1985; Røe and Hermansen 1993; Sambrook Smith et al. 2006). Similar structures have been reported from compound bars in recent deposits of Saskatchewan River and form when flow over the bar-top causes bar migration into the channel talweg, with associated slip face accretion at the bar margin (Sambrook Smith et al. 2006). The similarity of the cross-stratification on both sides of the bounding surface indicates that the bedform is linked with the main downstream flow.

3.3 Facies Association 2 (FA 2): Transgressive Deposits

Facies association 2 forms a thin unit, up to 0.5 m thick, and rests directly above fluvial deposits of FA 1. The most common feature of these deposits is the abundance of well rounded and large mud-clasts which form basal concentrations. These occurrences are relatively abundant in the eastern part of the studied area and constitute a laterally continuous boundary lined with clasts marking thus an abrupt transition from FA 1 to FA 2. Two lithofacies with distinct occurrence are identified within FA 2.

3.3.1 Lithofacies 2.1: Coarse- to Medium-Grained Intraclastic Sandstone with Basal Lags

This lithofacies is widely developed in Hansburg area and consists of planar to wavy bedded granular coarse- to medium-grained intraclastic sandstone with basal lags dominantly composed of large mud-clast and quartz-granules. Internally, the mud-clasts are paving the bedding planes in association with scattered granules. Exposed bedding planes locally exhibit a densely packed fabric of the clasts. The clasts are often removed by weathering, and only casts with diameters up to 25 cm are preserved, as well as granules dominated by white well rounded quartz grains. This lithofacies pinches out westward with a gradual decrease of clast size; only thin coarse-grained sandy layers with mud-clasts of up to 5 cm in diameter were observed in the Wegkruip area.

3.3.2 Lithofacies 2.2: Trough Cross-Bedded Coarse- to Medium-Grained Sandstone

This lithofacies occurs only in the western part of the studied area (Wegkruip farm) where it overlies thin lag layers with mud-clast concentrations and abundant quartz granules, and is overlain by silty mudstone of facies association 4. It consists of coarse- to medium-grained sandstone with trough cross-bedding (Fig. 4A). The lower bedding surface is slightly erosive and marked by a thin granular horizon. This lithofacies cut into the underlying deposits and pinches out laterally with a maximum thickness of up to 60 cm. The cross-stratification is organized into stacked and lenticular sets with concave-up lower boundaries and maximum thickness of up to 10 cm. The palaeocurrent indicated by the cross-strata is uniformly toward the West.

3.3.3 Interpretation

FA 2 is interpreted to have been deposited in high-energy foreshore to shoreface positions during rapid transgression over a flat area. This is evidenced by the occurrence in lithofacies 2.1 of lag concentrations with abundant large rip-up clasts of mudstone embedded in coarse-grained material sharply overlying fluvial deposits of FA 1. Such concentrations of rip-up clasts and associated quartz granules suggest reworking of an underlying sandy substrate of fluvial deposits and constitute thus transgressive lags concentrated by winnowing during sea-level rise

a
b
M
c
2 cm
d
S
e
1
gc
2
f
b
a
c
g
1cm
h
2cm

(e.g. Kidwell 1991; Kidwell and Brenchley 1994; Zecchin 2007). The coarse grain size and the abundance of large-sized mud-clasts together with planar and wavy bedding indicate high-energy wave and storm flows acting during the transgression. Lithofacies 2.2 with trough cross-stratification results from progradation under high-energy unidirectional currents of sinuous-crested sandbodies within rip-up channels, likely produced by retreat of storm and related high energy bottom return flow in a foreshore or shoreface setting (e.g. Einsele and Seilacher 1982; Walker and Plint 1992). This is evidenced by the erosional base with granular lag and by the pinch-and-swell geometry, all together indicating a scour-and-fill regime and a rip-current channel fill. The westward directed palaeocurrents indicate basinward migration, under the action of rip-up currents, of medium-scale dunes forming thus mouth-bars of rip-up channels.

3.4 Facies Association 3 (FA 3): Storm and Wave-Dominated Shoreface Deposits

This facies association is restricted to the eastern part of the studied area (e.g. Hansburg area) and consists of medium to fine-grained sandstones which pinch out westward, and subordinate fine-grained interbeds of silty mudstone (Fig. 2). The interbeds, up to 10 cm thick, form thin discontinuous drapes that are often partially eroded by the next overlying sandstone bed. The most distinctive feature of this association is the occurrence of Ediacara fossils dominated by *Ernietta* communities, whereas other vendobionts like *Rangea* and *Pteridinium* constitute subordinate fossils. Within this facies association, two main lithofacies are distinguished: reddish to greyish, medium to fine-grained sandstones (lithofacies 3.1) and light, fossiliferous fine-grained sandstone layers with *Ernietta* (lithofacies 3.2). Both lithofacies of this association form generally lens-shaped layers interfingering laterally.

Fig. 4 Outcrop photographs and characteristics of facies associations FA 2 and 3: (**A**) Trough cross-stratification within channelised sandstone of facies association FA 2, Wegkruip-West section; scale (*pen*) is 14 cm. (**B**) Lower bedding surface of event bed in FA 3 showing lag of mud-clasts (M) and *Ernietta* bioclast (*arrow*); scale (*coin*) is 22 mm. (**C**) Lower bedding surface of storm bed showing lag with abundant mud-clasts (M) associated with winnowed *Rangea* specimen (*arrow*). (**D**) Upper bedding surface of a storm bed showing association of *Kinneyia*-type wrinkles and swales (s); scale (*coin*) is 22 mm. (**E**) Internal structures within single event beds showing hummocky cross-stratification (HCS) in (**E1**) and association of HCS and swaley cross-stratification (SCS) in (**E2**). Note rock fragment with gutter cast (gc) in upper left part of **E2**; scale (*coin*) is 22 mm for E1 and (*lenscap*) 4.5 cm for **E2**. (**F**) Bipartite sandstone bed showing lower division (**a**) with intraclastic sandstone and upper division (**b**) with large wavelength hummocky cross-stratified set. Note scoured lower boundary and partially preserved interbeds of fine-grained siltstone and mudstone (**c**); scale (*pen*) is 14 cm. (**G, H**) Outcrop photographs of *Rangea schneiderhoeni*. *Rangea* occurs rarely within *greenish-rosty* coloured laminar bedded fine grained non-mature sometimes wave rippled fine sand to siltstones in the vicinity of the *Ernietta* occurrences. Right specimen (**H**) was embedded in upright position

3.4.1 Lithofacies 3.1: Intraclastic Sandstone with Basal Lags and HCS/SCS

This lithofacies is relatively abundant in FA 3 and occurs as single beds of homogeneous composition and structure or forms bipartite sandstone beds with fining-upward trend. The beds are usually underlain and overlain by thin interbeds, up to 5 cm thick, of purple to grey mudstone which in places may completely be eroded by the overlying sandy layer.

Single layers consist of graded, medium- to fine-grained sandstone beds, up to 25 cm thick, with sharp and erosional lower surfaces. The lower surfaces display thin basal lags and occasionally gutter casts with widths and depths in the range of 48 cm (Fig. 4E2). The lags consist mainly of mud-clast concentrations, scattered quartz granules and clasts of Ediacara fauna. The mud-clasts, up to 5 cm in size, are rounded and may be aligned or densely packed. The winnowed Ediacara fossils associated with these lags encompass fragments or complete bodies of mainly *Ernietta* associated with scattered elements of *Pteridinium* and *Rangea* (Fig. 4B, C, G, H). The clasts of *Ernietta* are often broken either along the longitudunal zig-zag suture or along the transversal suture. The upper bedding surfaces are slightly undulatory with large wavelength and are covered either by interference and wave ripples or by *Kinneyia*-type wrinkle structures associated with elongate to subcircular swales (Fig. 4D). The internal stratification shows some variations, being often organized into single sets which are either planar or undulating with swales and hummocks (Fig. 4E1). In some cases, the lower part exhibits planar to gently inclined laminations grading upward into HCS. Within some event beds, large isolated swales (8 cm $\times$ 40 cm) of swaley cross-stratification (SCS) are observed in the upper part, cross-cutting trough planar-lamination or hummocky cross-stratification (Fig. 4E2). These concave-up and isolated symmetric structures seem to form part of a continuous cross-laminated set which in turn is truncated on top. The swales are symmetrically filled with laminae thickening toward the centre, which is a typical feature of SCS. Frequently, swales preserved on upper bedding surfaces are associated with *Kinneyia*-type wrinkle structure (Figs. 4D and 9c).

Bipartite sandstone beds consist of vertically arranged lower and upper divisions forming fining upward units up to 70 cm thick (Fig. 4F). The thickness of each division varies as well as the internal stratification within the upper division. The basal surface is erosional to sharp with thin basal lag concentrations of mud-clasts and scattered quartz granules, whereas the upper bedding surface is slightly undulatory with low amplitude and wavelength of up to 1 m. The lower division constitutes the main part of the bipartite sandstone beds and consists of graded and planar- to wavy-bedded fine- to medium-grained intraclastic sandstone with a large amount of rip-up mud-clasts. The internal structure is often marked by mud-clasts aligned parallel to bedding. In planar to wavy-stratified layers, most of the bedding planes are paved with large numbers of rounded mud-clasts, up to 4 cm in diameter. In some layers, mud-clasts are densely packed and form horizons up to 3 cm thick. The upper division consists of fine-grained sandstone with structures similar to those occurring in single layers and the transition with the underlying lower division is commonly sharp. In some fining upward beds, the upper division is reduced to a cm-thick layer with undulating lamination of low amplitude and wavelength up to 1 m.

3.4.2 Lithofacies 3.2: Fossiliferous Fine-Grained Sandstone with *Ernietta*

The *Ernietta*-bearing lithofacies constitutes the most distinctive feature of this association by its lithology and faunal content. The facies referred here as a biofacies occurs in at least three distinct horizons and shows lateral interfingering with surrounding tempestites. The lower bedding surface is irregular with apparent concave-upward features, while the upper bedding surface appears flat or slightly convex-up with low amplitude. The facies pinches out both along strike and basinward and shows an apparent decrease in thickness from East to West over a distance of 2 km perpendicular to shoreline. Such geometry is similar to that of lense-shaped to elongated beds. The *Ernietta*-rich lithofacies ranges in thickness from 20 to 30 cm and consists of fine-grained to very fine-grained lighter planar-bedded sandstone (Fig. 5a, b). Parts rich in *Ernietta* fossils appear moderately to poorly bedded. Interbeds of fine grained layers occur locally and consist of brown to reddish siltstone and fine sand lenses with small-scale wave ripple cross-lamination (Fig. 5b) sometimes with specimens of *Rangea* in life position (Fig. 4G, H) The *Ernietta* lithofacies appears monospecific and dominated by *Ernietta*, however scattered specimens of *Pteridinium* and *Rangea* appear locally. The internal structure shows erosional surfaces cross-cutting the primary stratification and dipping at a low angle (less than 10°) (Fig. 5a). These surfaces may locally be marked by lenses of fine-grained sandstone with gutter casts (Fig. 5b). Along these reactivation surfaces, *Ernietta* fossils form a quasi in situ concentration. Fossils are often flattened and are associated with scattered sand and mud-clasts (Fig. 5c). *Ernietta,* however, is commonly preserved as clusters of several bodies (Fig. 5d). Within such clusters, *Ernietta* fossils of variable sizes are superposed and vertically oriented (Fig. 5e, h). This preservation in life position is distinctly marked by quilts directed and tightening upward.

In thin section, the *Ernietta*-bearing lithofacies consists of quartzose mica-bearing sandstone with grain sizes ranging from 70 μm to 150 μm (mean 100 μm) and with medium sorting (Fig. 5k, l). Most of the grains are subrounded to subangular (Fig. 5k, l). The matrix surrounding the grains is abundant and forms a total of about 25%. It consists of mica (sericite?) and cryptocrystalline quartz. Abundant detrital muscovite flakes of variable size occur as lithites. They may form thin veneers parallel to bedding planes. Sediment filling moulds of *Ernietta* "bags" is more or less similar to the hosting sediment detrital mica is apparently abundant (Fig. 5l). The interface between the *Ernietta* "bags" and the hosting sediment is characterized either by small factures partly filled recent to subrecent calcrete or with parallel oriented muscovite/sericite mica flakes. The parallel oriented mica flakes could be a primary structure, may be the flakes were originally attached to the slimy surface of *Ernietta* (Fig. 5i–l).

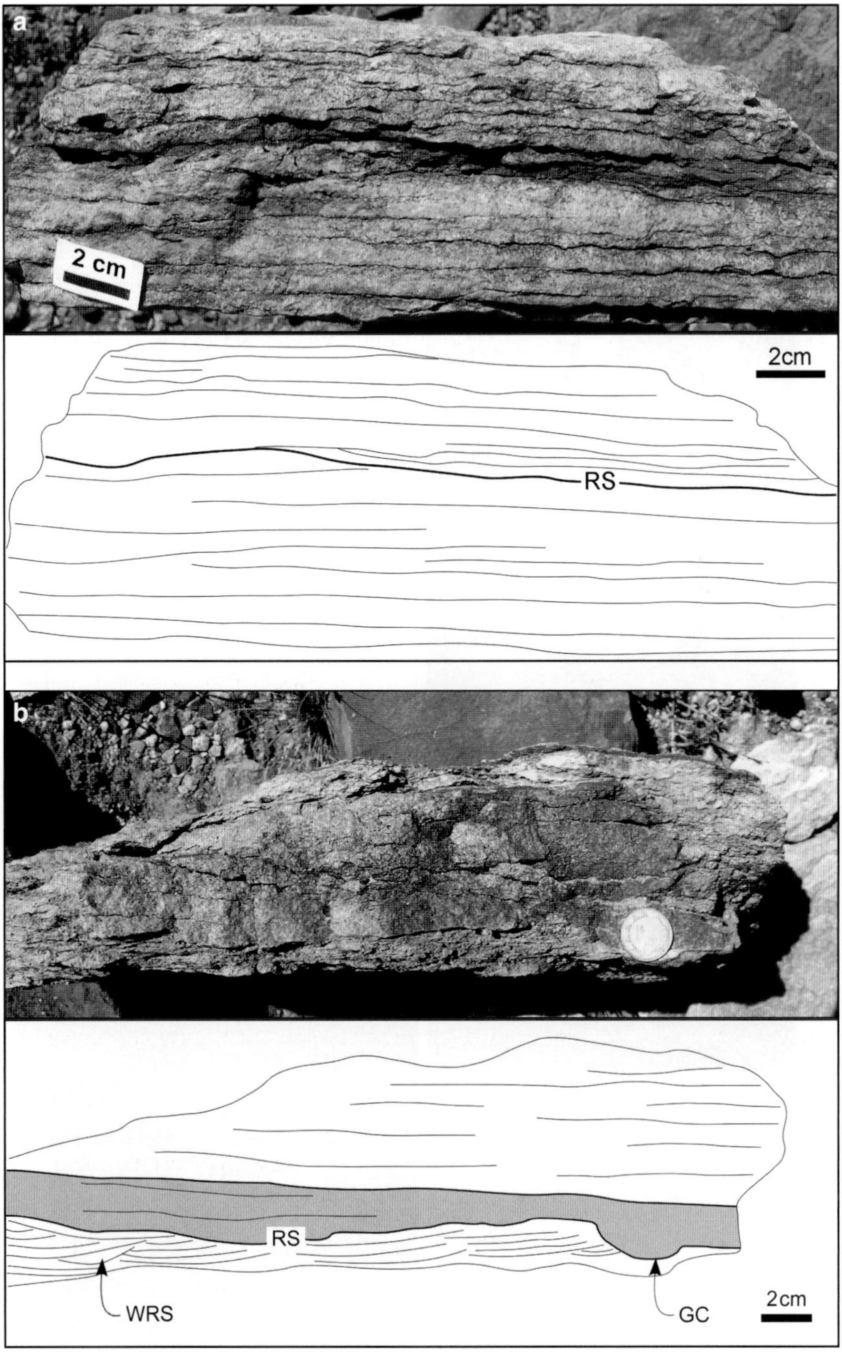
a
2 cm
2cm
RS
b
RS
WRS
GC
2cm

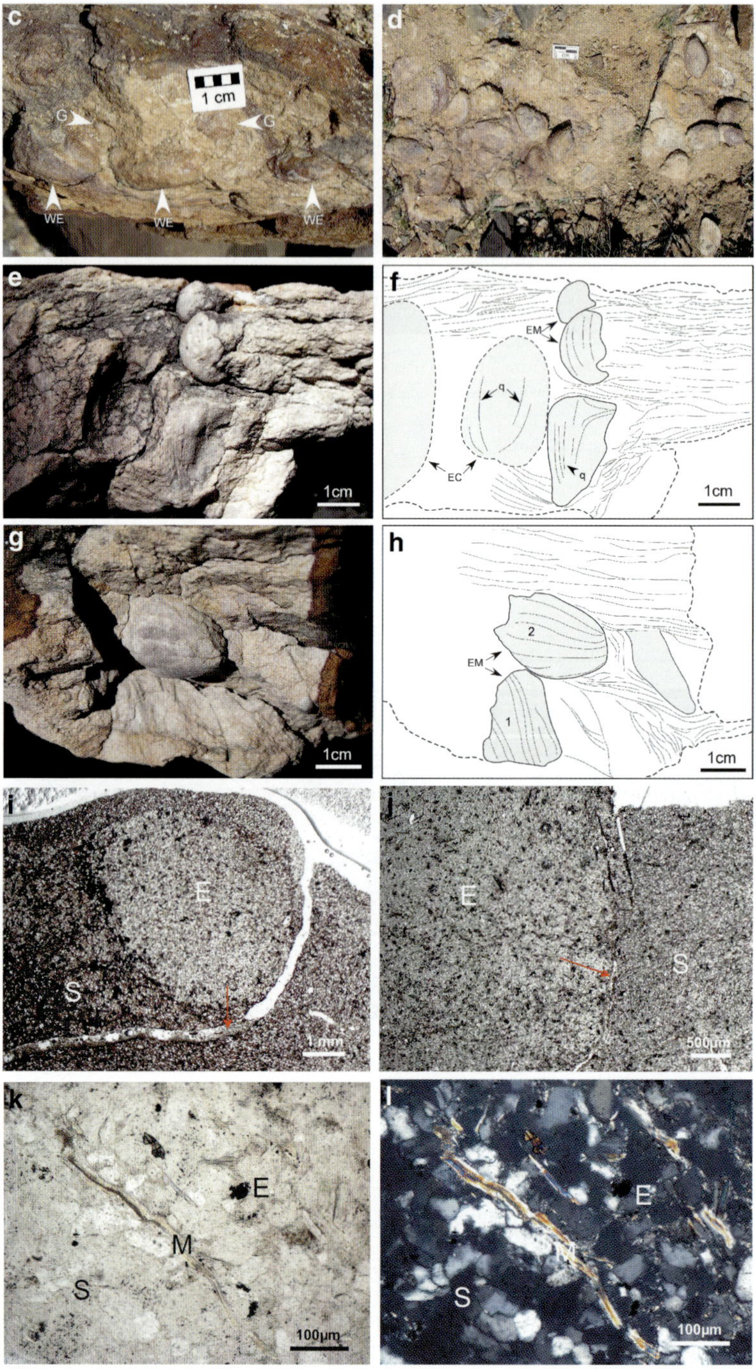
c
1 cm
G
G
WE
WE
WE
d
e
1cm
f
EM
q
q
EC
1cm
g
1cm
h
2
EM
1
1cm
i
E
S
1 mm
j
E
S
500μm
k
E
M
S
100μm
l
E
M
S
100μm

3.4.3 Interpretation

The overall sedimentary features of this facies association indicate deposition above storm wave base in a wave and storm-dominated setting with intermittent quiescent stages of low-energy permitting settling of mud suspension and growth of *Ernietta* communities (Fig. 6).

The close association in lithofacies 3.1 of fining-upward sequences, HCS, SCS, lag concentrations and erosional to sharp lower bases within sandstone beds indicates high-energy deposits related to storms (Kreisa 1981; Myrow and Southard 1996). Thin silty mudstone layers at the top of such event-strata reflect quiet-water deposition during the inter-storm periods. The low content of mudstone and the abundance of rip-up mud-clasts and intraclastic layers indicate a high-energy shoreface environment where fine-grained sediments are continuously reworked by storm- and wave-generated currents (Kreisa 1981; Walker and Plint 1992; Myrow and Southard 1996). The occurrence of interference and oscillation ripples on top of storm beds indicates deposition above fair-weather wave base were wave-generated oscillatory currents affect the seabed (Walker and Plint 1992). The intraclastic sandstones and rip-up clasts-bearing layers of the Ediacaran tempestites are analogues to shell-bearing layers in Phanerozoic tempestites (Aigner 1982; Einsele and Seilacher 1982). Each sandy bed constitutes an event bed related to a single storm event. The event beds are interpreted here as proximal tempestites containing densely packed mud-clasts or intraclasts aligned

Fig. 5 Photographs and graphic renditions illustrating main features of *Ernietta* lithofacies: (**a**) Bedding style and line drawing of *Ernietta* bearing lithofacies showing planar lamination and reactivation surface (RS). (**b**) Layer of *Ernietta* lithofacies showing interbed of wave rippled siltstone (WRS) and erosional event bed marked by reactivation surface (RS) and lens of sandstone with gutter cast (GC); scale (*coin*) is 22 mm. (**c**) Lag along a reactivation surface marked by winnowed and horizontally aligned *Ernietta* fossils (WE) and quartz granules (G). (**d**) Bedding surface showing cluster of several *Ernietta* fossils preserved in life position. Note that the transversal long axes of *Ernietta* bags are randomly oriented. (**e**) Cross-section view and (**f**) line drawing of *Ernietta* lithofacies showing cluster of vertically oriented and partly superposed *Ernietta* fossils; note differences in size of *Ernietta* and vertically oriented quilts (*EM Ernietta* moulds, *EC Ernietta* casts, *q* quilts). (**g**) Cross-section view and (**h**) line drawing showing *Ernietta* moulds (EM) in distinct styles of preservation: 1: life position with vertically oriented quilts, 2 life position but with lower part tapering obliquely in unlithified sand from left to right; note deformed lamination at end of mould and preservation of the original shape of *Ernietta* without major flattening during burial. (**i**) Cross section of a small *Ernietta* specimen. Surrounding sediment is impregnated with iron oxides. The specimen is sharply confined to the surrounding sediment by colour change and partly by a open fracture which is partly filled with caliche carbonate (calcrete) (*red arrow*) (*E Ernietta*, *S* sediment, *M* mica). (**j**) Boundary layer of another specimen of *Ernietta*. Boundary fracture is filled with caliche carbonates (calcrete) and parallel oriented bright mica (*red arrow*). The surrounding sediment exhibits a better sorting of sand grains as such ones investigated within the *Ernietta* molds. The *Ernietta* body is filled non-mature siliciclastic sediment including large mica flakes (*yellow arrow*). (**k**) Detailed magnification of (**j**) the boundary layer with aligned mica. Within the *Ernietta* mold the mica is randomly oriented. (**l**) Same specimen as (**j**) and xN. The quartz grains have an average diameter of 90–120 μm and show angular to subangular shapes

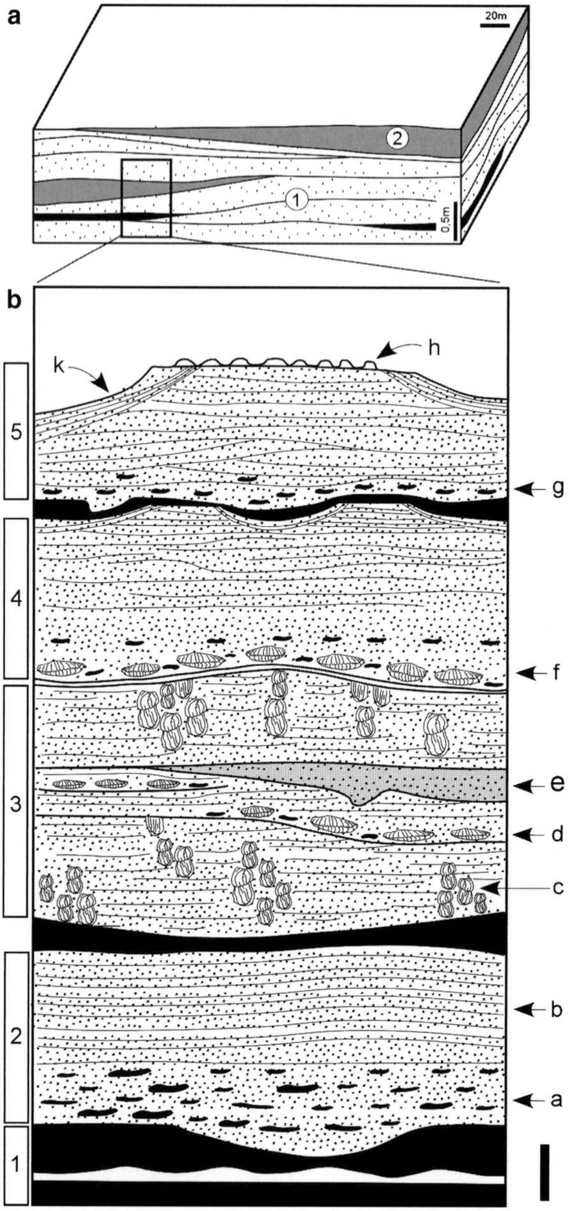

Fig. 6 Relationship of lithofacies and sedimentary features of wave and storm-dominated shoreface deposits (FA 3): (**A**) Bloc diagram showing the interfingering relationship of sandy storm deposits (1) with *Ernietta* beds (2). The *Ernietta*-bearing lithofacies occurs in shallow troughs within meter-scale undulating upper bedding surface of underlying sandy storm deposits. (**B**) Diagram summarizing sedimentary features of facies association FA3 (1) Fine-grained interbeds of laminated siltstones and silty-mudstones, wave rippled lenses; (2) Bipartite storm bed with scoured base, lower division with intraclastic sandstone (**a**) and upper division with large

parallel to bedding, which are considered as storm-generated winnowed lag deposits (Kreisa 1981; Einsele and Seilacher 1982). Bipartite beds with typically fining-upward trends and vertically arranged sedimentary features like erosive base, basal concentrations of rip-up mud clasts and clast horizons, normal graded beds, horizontal lamination of the upper plane bed and HCS/SCS, are interpreted to have resulted from waning storm-generated currents. The occurrence of SCS on the top of some storm beds indicates shallowing upward conditions and may result from aggrading hummocky beds (Walker and Plint 1992; Dumas and Arnott 2006).

The convex-up upper bedding and concave-up lower bedding surfaces as well as the pinch-and-swell structures indicate that *Ernietta* lithofacies 3.2 represents lens-shaped to elongate sandy beds, with infaunal community, aggrading in a subtidal setting. The *Ernietta* community preserved in life posture within this specific lithofacies is an evidence for an autochthonous fauna (*sensu* Kidwell et al. 1986) that lived in a shoreface setting and in fine-grained sandy matrix. Aggradation of *Ernietta* beds is considered to have started within shallow depressions of the seafloor, related to the undulatory and large wavelength of the upper bedding surface of sandy storm layers (lithofacies 3.1). This is in agreement with the interfingering relationship with tempestites of lithofacies 3.1 and with the lens-shaped geometry. Indeed, accretion and growth of *Ernietta* communities may have commenced in inherited troughs (concave-up), initiated by previous storm events and maybe rip-up currents during retreat of storm waves (Aigner 1982; Einsele and Seilacher 1982; Hequette and Hill 1995; Myrow and Southard 1996) that affected the seafloor topography. The grain size of *Ernietta* lithofacies 3.2 consisting of fine grained sand, the abundancy of muscovite flakes and the lack of structures indicating bedload transport, suggest that sediment has been transported from nearshore by suspension, and deposited in shoreface position by vertical accretion under lower flow regime conditions. Along the shoreface and subtidal zone, reworking is frequent during intermittent high-energy events, and evidence of such events is recorded as internal erosional surfaces with gutter casts in the *Ernietta* beds and by occurrences of winnowed *Ernietta* specimens redeposited as in situ lags. Growth of *Ernietta* communities and associated vendobiontan organisms likely occurred in inter-storm periods and under fair-weather conditions in a shallow subtidal setting, but the fauna intermittently became exposed to high-energy storm events as evidenced by reactivation surfaces and the intrastratal lags of winnowed fauna.

Fig. 6 (continued) wavelength HCS (**b**); (3) *Ernietta* lithofacies with clusters of *Ernietta* in life position (**c**), internal reactivation surface marked by lag of reworked *Ernietta* and mud-clasts (**d**) or by sandy lens (**e**); (4) Graded storm bed with erosional lower boundary, lag with *Ernietta* and mud-clasts (**f**), horizontal to hummocky cross-lamination and SCS, (5) Single event bed with gutter cast and intraclastic basal lag (**g**), HCS, *Kinneyia*-type wrinkles (**h**), SCS and associated swales on upper bedding surface (**k**); scale bar is 10 cm

3.5 *Facies Association 4: Inner Shelf Deposits with Distal Tempestites*

This facies association which forms a unit up to 6 m thick (Fig. 2), occurs in the western part of the studied area (e.g. Wegkruip farm) and is composed dominantly of silty mudstone with few and subordinate interbeds of fine-grained sandstone and fossiliferous micaceous siltstone with winnowed *Ernietta* specimens.

The fine-grained sediments consist of grey to purple laminated and fissile mudstone; purple interbeds likely resulted from weathering processes. In the upper part of the unit and at the transition with the overlying carbonatic Mara Member, the streaky and silty mudstones become yellowish and contain significant proportions of marl. The silty mudstones contain mm-thick streaks and lenses of very fine-grained sandstone, which show pinch-and-swell structures and occasionally represent starved ripples less than 2 cm thick (Fig. 7a). In the lower part of the silty mudstone unit, subordinate sandy interbeds form isolated layers of two types. The first type consists of fine to very fine-grained sandstone layers, up to 8 cm thick, with sharp to scoured basal surfaces and a lateral extension of tens of meters. These layers have upper bedding surfaces with large wave ripples and interference to polygonal ripples which, by contrast to FA 3, are not flat-topped. Internal structures often consist of horizontal to slightly undulating laminations. The second type consists of fossiliferous lags of very fine-grained sandstone to siltstone beds which form lenticular layers, up to 6 cm thick. In the studied area, only two occurrences have been observed. The faunal elements consist of stacked moulds and fragments of *Ernietta* which often have preserved their quilted structure (Fig. 7b). The *Ernietta* fossils, varying in size from 1 to 3 cm, are embedded in a siltstone matrix and are flattened due to differential compaction. They are associated with rounded and horizontally aligned rip-up mud-clasts, up to 2 cm in diameter.

Fig. 7 Sedimentary characteristic of facies associations FA 4: (**a**) Thinly bedded mudstone and silty mudstone with interbeds of fine-grained sandstone showing erosional base and HCS typical of storm deposits. Note occurrence of wave-rippled lenses (*arrow*) within streaky and silty mudstone; scale (*pen*) is 14 cm. (**b**) Weathered bed of fine-grained sandstone showing densely packed *Ernietta* fossils associated with mud-clasts. *Ernietta* fossils preserve quilted structure but are flattened due to compaction

3.5.1 Interpretation

The streaky and silty mudstone records low-energy deposition of fine-grained particles transported by suspension and deposited in a muddy background of inner shelf environment by settling and during fair-weather periods. The thin interbeds of sandstone with sharp to scoured basal surfaces are produced by low-density storm-induced flows and constitute distal tempestites and episodic incursions of sand during storm events (Aigner 1982; Einsele and Seilacher 1982). The occurrence of wave ripples and polygonal to interference ripples on upper bedding surfaces are related to storm generated, combined and oscillatory flows and indicate deposition near or above storm-wave base. The thin and rare occurrences of fossilferous interbeds with *Ernietta* are interpreted as distal tempestites similar to "event shell beds" of respective Phanerozoic environments (Kreisa 1981; Aigner 1982; Kidwell 1991; Myrow and Southard 1996). It is envisaged that *Ernietta* fauna-bearing banks and shoals that grew in proximal areas (e.g. lithofacies 3.3) were eroded and the fauna displaced by high-energy storm-related currents and finally deposited as fossil lags distally.

4 Vertical and Lateral Facies Distribution Within the Kanies Member

On the basis of correlation of the sections, and vertical and lateral stacking patterns of the deposits within Kanies Member, a diagram summarizing the facies distribution can be designed (Fig. 8). The facies associations are organized in a fining-upward transgressive sequence indicating landward-steepening character. The sequence is starting with the fluvial facies association FA 1 and ending with marine deposits of facies associations FA 3 and FA 4. Sedimentary data obtained from facies associations analysis and the observed vertical and lateral transitions indicate a transgressive trends punctuated by a rapid rise of sea-level and depict a basin open to the west, but without apparent differential subsidence as indicated by the lack of significant lateral change in thickness.

Within the lower part of the transgressive Kanies sequence, the fluvial facies is distinguished by sheet-like cross-bedded sandstones that extend over the whole studied area without major vertical or lateral change and with uniform palaeocurrent directed westward. However, few thin interbeds of mudstones are preserved within this sandy part, mainly in the distal section "Wegkruip west".

At the top of the braided fluvial deposits, a distinct horizon (high-energy lag layer) separates fluvial aggradation of the lower part of the transgressive unit from marine deposits of the upper part which lack major vertical trends that may indicate a continuous landward-steepening. The superposition of the marine facies on fluvial deposits is interpreted as representing a boundary of a marine transgressive surface (MTS) generated by relative sea-level rise. The lack of a gradual change into tidally or wave influenced fluvial deposits indicates a sudden transgression with a rate of sea-level rise exceeding the rate of sediment supply. Such rapid transgression might have

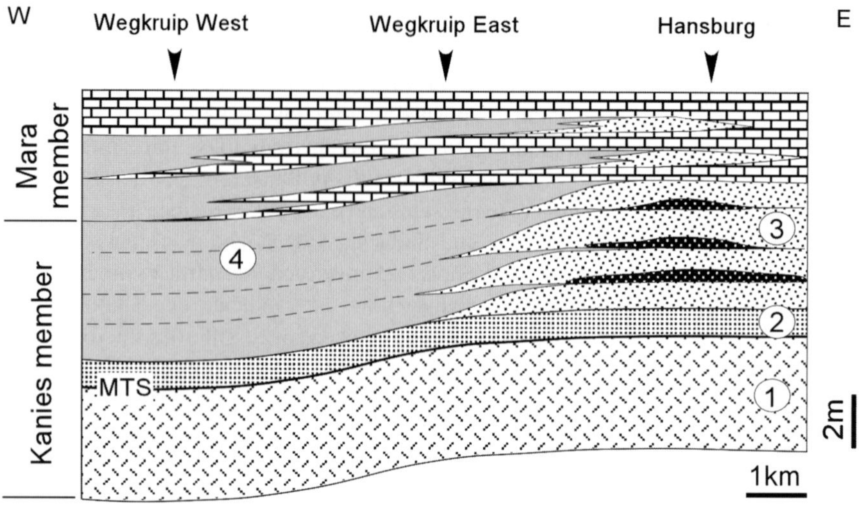

Fig. 8 Diagram summarizing vertical and lateral facies relationships within Kanies Member: (1) Fluvial facies association; *MTS* marine transgressive surface; (2) Transgressive deposits. (3) Storm deposits of facies association FA 3 with interbedded lens-shaped ridges populations of *Ernietta* (*indicated in black*); (4) Facies association FA 4 with discrete layers of distal tempestites

been accentuated by the flat-topography of the pre-existing wide braid-river plain (FA 1) which lacks narrow localized channels delimited by stabilized mud banks.

The thin transgressive interval is marked mainly by reworked sand and mud material with large quantity of rip-up clasts which was redeposited in foreshore to upper shoreface setting. In the proximal part, intense waves and storms played a major role in reworking and deposition of the coarse-grained sand and the associated lag horizons with high proportion of densely packed rip-up mud-clasts. In the distal area, the effect of geostrophic and rip-up currents induced by high-energy waves and storms is recorded as lens-shaped medium-scale 3D dunes locally developed and prograding basinward (Wegkruip sections).

A clear partitioning of the depositional setting into proximal and distal marine environment with a lateral shift of facies occurs during deposition of the upper part of Kanies transgressive sequence. This is evidenced by the basinward transition from wave and storm-dominated shoreface association (FA 3) restricted to eastern and proximal portion (Hansburg section) to muddy inner-shelf association (FA 4) with streaky silty-mudstone and subordinate interbeds of distal tempestites occurring westward in the distal area (Wegkruip sections). The establishment and growth of *Ernietta* and associated vendobiontan organisms in the proximal part constitutes the most significant change within Kanies sequence. Occurrence of *Ernietta* communities within a typical lithofacies and facies association together with their restriction to the proximal part of the shallow marine environment indicate that growth and demise of *Ernietta* infaunal communities are facies controlled, and thus *Ernietta* may be considered as an opportunistic organism.

The transition from the siliciclastic Kanies Member to the carbonatic Mara Member occurs through a mixed siliciclastic-carbonate thin interval (ca. 1.5 m thick) which forms the lowermost portion of the Mara Member (Fig. 2). In Hansburg area, this interval consists of an alternation of wave and storm-dominated shoreface deposits of FA 3 and laminated micritic to stromatolitic limestones. Both siliciclastic and carbonate facies may locally display interfingering relationships, particularly westward in Wegkruip area, where the transitional interval consists of an alternation of thinly laminated micritic limestone beds and decimetre-thick interbeds of mudstones deposited in inner shelf position. The limestone beds display thin partings and interbeds of mudstones.

The carbonate layers are resulting from an increase in carbonate productivity during ensuing regression and progradation of the Mara carbonate platform replacing thus the siliciclastic shallow shelf environment. In the proximal and shallow part of the ramp (e.g. Hansburg area), the vertical shift from shoreface siliciclastic to peritidal carbonate deposits is sharp and indicates a shallowing upward trend, with no evidence for mixing of carbonate and siliciclastic material within the limestone beds, nor evidence for interbedded siliciclastic layers and partings. By contrast, in a basinward direction (Wegkruip area), the vertical transition appears gradual as evidenced by occurrences of mudstone alternating with limestone beds. In the transitional interval of Hansburg area, scattered *Ernietta* fossils persist within siliciclastic shoreface deposits but are completely lacking in the carbonate sediments.

5 Occurrences of *Kinneyia*-Type Wrinkles Structures in Event Beds

Kinneyia-type wrinkle structures, characterized by mm-scale flat-topped winding crests and similar-sized intervening troughs and pits (see review in Porada et al. 2008), occur at several stratigraphic levels within the upper part of the Kanies Member on Hansburg farm and, sporadically, in the middle transgressive part on Wegkruip farm further to the West (see Fig. 2). In all the occurrences, the structures are developed on upper surfaces of storm layers which either are completely flattened or have preserved sets of interference ripples; in some cases, *Kinneyia* structures surround isolated (relic) interference ripple troughs or concave-up shallow scours (swales) (Fig. 9a, d).

Kinneyia structures are considered to form underneath microbial mats (e.g. Pflüger 1999; Porada et al. 2008), thus – if correct-indicating the existence of seafloor microbial mats or, in more general terms, the high potential for mat development in the area. Though, except for sporadic shrinkage cracks, further mat-related structures have not been observed in the Kanies Member on Hansburg farm as yet, it cannot be excluded that microbial mats existed throughout the upper part of the member at suitable and protected sites of low-energy hydrodynamic conditions. It may be speculated that these mats were eroded during episodic storm

Fig. 9 *Kinneyia*-type wrinkle structures developed on upper bedding surfaces of fine-grained sandy storm beds of facies association FA 3, Hansburg section: (**a**) *Kinneyia* structure with honeycomb-like pattern of flat-topped crests and intervening irregularly elongate (2–4 cm) to round pits, 1–1.5 cm in diameter. (**b**) *Kinneyia* structure with long, flat-topped, winding crests developed on flat upper bedding surface. The structure occurs in a thin top-set veneer covering the upper bedding surface. Note crest bifurcations and isolated elongate pits about 0.5–1 cm long. (**c**) *Kinneyia* structure with long, flat-topped, winding crests associated with subcircular to slightly elongate swales (S). Note the sharp and discordant limit between swales and *Kinneyia* crests. (**d**) *Kinneyia* structure with linear to honeycomb like patterns developed in a thin flat layer discordantly overlying distinct and discontinuous laminae (*indicated by arrows*) of wide and very shallow swale related to hummocky-cross stratification

events whereby mat fragments and clusters of bacteria were brought into suspension and deposited on the new sediment surface at waning storm conditions. Such "fertilization" with cyano- and other bacteria will have enhanced the rapid formation of mats in the aftermath of the storm.

6 Preservation, Growth and Controlling Factors of *Ernietta* Communities

6.1 Preservation

Within the Kanies Member, occurrences of *Ernietta* communities show two main types of preservation which, according to accompanying sedimentary features, are influenced by the frequency of changes in physical conditions in a storm and wave-dominated shoreface to inner shelf environment (see Fig. 6):

Type 1: Preservation in life position as "infaunal benthic biota" within a typical fine-grained sandy "*Ernietta* lithofacies", considered here as a biofacies (e.g. Brenchly 1990) and belonging to the "autochthonous fossil assemblages" sensu Kidwell et al. (1986). *Ernietta* occurs usually as clusters comprising individuals of variable size, likely related to various growth stages, and with vertically oriented quilts (see Fig. 5). Type 1 preservation correlates with quiet-water conditions prevailing during relatively prolonged inter-storm periods. Despite morphological variations, the fossil assemblage in this biofacies is dominated by *Ernietta*; however, scarce specimens of *Pteridinium*, *Namalia* and *Rangea* are locally found, too. In the examples shown in Fig. 10a, *Pteridinium* appears aligned along bedding planes and is preserved as "canoe-shaped" bodies with apparent symmetrical structure. A variety of preservation in life position of *Ernietta* appears on top of storm layers indicating its sand-sticking habit (Fig. 10b).

Fig. 10 Outcrop photographs illustrating various modes of preservation for *Ernietta* community in Kanies Member (**a**–**c**) and similar biota in younger strata of Schwarzrand Subgroup (**d**): (**a**) Upper bedding surface showing *Pteridinium*, with elongate and boat-shaped form, preserved in life position in *Ernietta* lithofacies. The fossil show symmetrical fusiform structure tapering at the end, it is aligned in the bedding planes indicating that it was engulfed in loose fine-grained sand, Hansburg farm. (**b**) *Ernietta* in life position sticking on upper bedding surface of a storm bed, Hansburg farm, scale (*coin*) is 22 mm. (**c**) In situ lags within *Ernietta* lithofacies formed by reworking of *Ernietta*-rich beds during intermittent high-energy event, Hansburg farm, scale (*coin*) is 22 mm. (**d**) Lower bedding surface of storm bed showing bioclast of vendobiont, probably *Pteridinium*, associated with mud-clasts (*arrows*), Büllsport farm, Schwarzrand Subgroup, Zaris sub-basin

Type 2: The taphonomic signatures of this type are consistent with reworking during high-energy events and fall within the genetic group of "sedimentologic concentration" sensu Kidwell et al. (1986). The *Ernietta* fossils are preserved as "winnowed fauna" produced during intermittent major storm events and occur in three distinct situations. The first sub-type consists of quasi in situ lags (parautochtonous fossil assemblage sensu Kidwell et al. 1986) occurring in typical "*Ernietta* lithofacies" and represented by assemblages of horizontally aligned and often unbroken fossils that underwent short transport within their biotope (Fig. 10c). The second sub-type consists of reworked and transported fauna that forms lags at the base of event beds deposited in a shoreface setting. Fossils are scarce and isolated and often broken, and may be associated with clasts of *Rangea* and *Pteridinium*. The third subtype consists of reworked fauna that is transported distally from the biotope and redeposited as thin storm lags in the mud-dominated inner shelf area.

Although sedimentary and palaeoecological settings of Erniettomorpha and associated vendobionts are not precisely constrained in other stratigraphic levels of the Nama Group, it is envisaged from the published data that similar assemblages (e.g. in the Kliphoek Member) also fall within the preservation types recognized in the Kanies Member. The preservation as winnowed fauna forming lags in sole beds appears frequent in high-energy deposits (Fig. 10d) and should be considered as occurrences associated with tempestites in corresponding wave- and storm-dominated environments.

6.2 *Growth and Controlling Factors*

According to sedimentary features of *Ernietta*-bearing deposits and their lateral and vertical relationships with surrounding sandy storm deposits (e.g. lithofacies 3.1), a model for *Ernietta*-bed growth and evolution is suggested (Fig. 11). The model considered here is based on the observation that *Ernietta* lithofacies constitutes lens-shaped elongate bodies encased within storm deposits; the bodies have a patchy distribution with lateral transitions into lens-shaped proximal storm beds with HCS and SCS. Considering taphonomic signatures and sedimentary characteristics of the facies association containing vendobionts (FA 3), several factors seem to have interplayed and controlled the accretion of *Ernietta*-rich beds, including water energy conditions, topography of the seafloor, water depth, available matrix (substrate) and rate of sediment supply.

6.2.1 Topographic Control

The lateral distribution of the lithofacies within a wave and storm dominated shoreface setting shows interfingering relationships of *Ernietta* lithofacies and sandy tempestites (Fig. 6), whereby the *Ernietta*-bearing lithofacies appears to be developed in shallow depressions as evidenced by the concave-up lower boundary

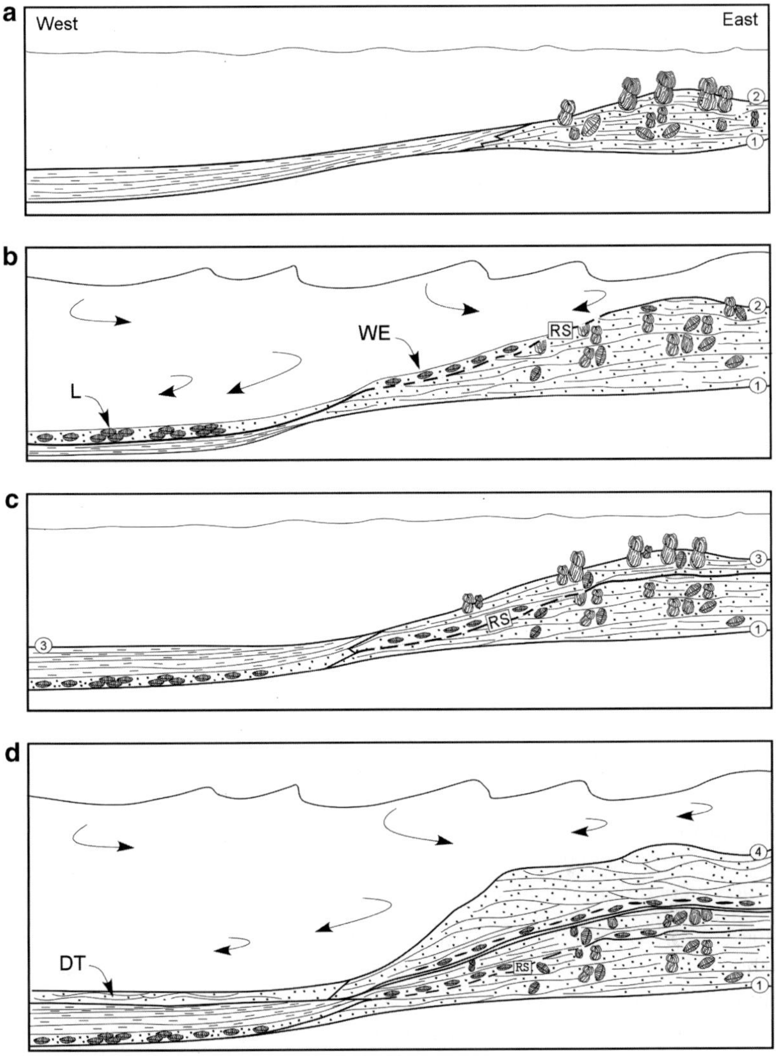

Fig. 11 Model illustrating processes recorded in *Ernietta*-rich bed (*Ernietta* lithofacies) from growth stage during low energy periods to burial stage by sandy storm deposit: (**a**) Fair-weather conditions: Growth of *Ernietta*-rich bed (1–2) within a shallow subtidal zone (shoreface), mud settling basinward in the inner shelf zone. (**b**) Intermittent minor storm event: Reworking of *Ernietta* bed, formation of internal reactivation surface (RS) and nearby redeposition of winnowed *Ernietta* (WE) fossils in front of the bed; basinward transport and redeposition of *Ernietta* as lags (L). (**c**) Low-energy stage: Reestablishment and renews growth of *Ernietta* fauna (3) and mud settling basinward. (**d**) Major storm event: *Ernietta*-rich bed partially reworked and buried by deposition of sandy storm bed (4) with lag of *Ernietta* and mud-clasts. Basinward, deposition of distal tempestites (DT)

and the pinch-and-swell geometry of the individual bodies. Such distribution and geometry indicates control by inherited topography with nucleation sites for *Ernietta* beds in troughs, which were partially filled thus leveling the seafloor. Observations in modern storm-influenced shoreface and shelf environments show the effect of storm-generated currents on the seafloor (Hequette and Hill 1995; Passchier and Kleinhans 2005; Vis-Star 2007), which create shallow, elongated to rounded meter-scale troughs in the overall lens-shaped, intraclastic, hummocky- and swaley-cross stratified storm deposit with pinch-and-swell geometry. The resulting shape of the troughs depends on the shape of sandstone body which in turn is related to water depth and predominant flow conditions. During low-energy and prolonged inter-storm periods, these depressions may have formed favourable and relatively protected sites for trapping and vertical accretion of very fine-grained sand particles and associated mica flakes, and eventually colonization by incipient *Ernietta* communities. It is difficult, in the case of Hansburg farm area, to control the exact shape of *Ernietta* beds, but the surrounding facies with hummocky and swaley cross-bedded sandstone seems likely to indicate a seafloor covered with 3D mound-like bedforms of low amplitude.

6.2.2 Control by Water Energy Conditions

It is envisaged that the change from high-energy periods dominated by storm activities to quiet-water conditions played a major role in nucleation and establishment of *Ernietta* communities. The accretion of *Ernietta*-bearing lithofacies records prolonged low-energy stages in a wave and storm-dominated shoreface setting. During growth of *Ernietta*-beds, intermittent high-energy events are recorded as reactivation surfaces and in situ lags of *Ernietta* bioclasts. The abrupt change into high-energy conditions during a major storm event, however, was the main process involved for a demise of *Ernietta* communities and their burial by the next storm deposit (Fig. 11d).

6.2.3 Water Depth Control

Within the upper part of the Kanies Member, occurrences of autochthonous *Ernietta* communities appear closely related to a specific "*Ernietta* lithofacies" within shoreface deposits (FA 3). Sedimentary characteristics including swaley-cross stratification and abundant wave-related ripples suggest deposition broadly near fair-weather wave base. By contrast, more basinward in the muddy area of the shelf (FA 4), there is no evidence for occurrence of *Ernietta* beds or for *Ernietta* preserved in life position. The only occurrences found are within thin distal tempestites as winnowed fauna. From this bathymetric distribution, it is assumed that well oxygenated zones in a shallow subtidal environment were suitable for *Ernietta* colonization.

6.2.4 Nature of Sediment and Sediment Supply

The zonality observed within the upper portion of Kanies Member shows a grain-size decrease basinward and a clear transition from a sandy dominated shoreface in the East to a mud-dominated inner shelf in the West. By considering petrographic features of the *Ernietta* lithofacies, development of *Ernietta* communities appears to have been favoured by a very fine-grained sandy matrix. The internal bedding of *Ernietta* beds dominated by horizontal to wavy lamination as well as the abundance of commonly aligned, detrital mica flakes, support vertical accretion under lower flow regime conditions. In the shoreface zone, the rate of sediment supply is mainly controlled by tide, wave and storm generated currents and by important river discharges (Johnson and Baldwin 1986; Walker and Plint 1992). Such influences were not recorded in the *Ernietta*-bearing facies association. By contrast, during low-energy, fair-weather periods when the action of strong marine currents is negligible, the rate of sediment supply is low and only very fine-grained suspended particles of sand to silt size are transported seaward from the adjacent foreshore zone by bottom currents of lower flow regime. The similarity between the sediment of *Ernietta* moulds and the surrounding matrix, together with the development and preservation of *Ernietta* in vertical life-position without major changes in the shape during burial, indicate that the organisms likely were filled progressively by sediment before complete burial.

7 Discussion: Palaeoecology of *Ernietta* and Implications for Similar Underground Vendobionts

Within the Kuibis Subgroup, development of *Ernietta* communities coincides with a major shift in sedimentation and is related to a rapid transgression event, whereas their demise coincides with a major shift to carbonate deposition during the overlying highstand systems tract of the Mara Member. A similar "invasive" pattern during rapid transgression may be invoked for the Vendobionta assemblage of the overlying siliciclastic Kliphoek Member (Grazhdankin and Seilacher 2002). Reported occurrences of underground vendobionts similar to *Ernietta* and preserved as three-dimensional moulds in Ediacaran sequences (e.g. Russia, Australia, Namibia, USA) appear, in a broad sense, associated with marine sandy environments with prevailing high-energy conditions (Jenkins 1985, 1992; Ivantsov and Grazhdankin 1997; Hagadorn and Waggoner 2000; Grazhdankin and Seilacher 2002, 2005). The absence of detailed sections and sedimentological studies of most of these occurrences does not allow a close comparison in palaeoecology of the *Ernietta* communities in the Kanies Member with similar infaunal biota. However, available sedimentological data on vendobiont-bearing successions of various localities fall within a storm-winnowing model and point broadly to wave and storm-influenced settings. Grazhdankin (2004) documented the Ediacara fauna of the Vendian Group in the White Sea area (Russia), and noted a correlation between sedimentary environment and faunal assemblage. The vendobionts consisting of three-dimensional sandstone casts including among others *Rangea* and *Onegia*, occur mainly in interstratified sandstones interpreted as fluviomarine sand-filled channels and distributary mouth-bar settings. Thereby, *Rangea* is associated with channel fill sandstone that cuts across middle-shoreface deposits, whereas *Onegia* was found at the base of channelised sandstone packages. In the Ediacara rocks of Australia, vendobionts including *Pteridinium* and *Onegia* appear to be restricted to massive fluviomarine channelised sandstone deposits (e.g. Gehling 2000). In the Kliphoek Member of the Nama Group (Namibia), vendobionts are preserved in planar and cross-bedded sandstone of a shallowly channeled coastal plain with fluvial dominance, interbedded with storm influenced lower-shoreface deposits (Saylor 2003). As noted in previous studies comparing the distribution of Ediacaran communities and considering morphological features, the infaunal vendobionts seem to be adapted to marine sandy environments of distributary mouth bar shoals and rip channels (e.g. Grazhdankin 2004; Grazhdankin and Seilacher 2002; McCall 2006). However, by comparing the specific palaeoecologic setting of *Ernietta* communities in the Kanies Member with the sedimentary settings reported, although not detailed, from similar occurrences in other successions, it appears possible to suggest similar specifically restricted niches for these vendobionts in localized troughs where incipient Vendobionta communities were able to grow during intermittent inter-storm periods. According to the sedimentary features of the *Ernietta*-bearing deposits, their modes of preservation, and their distribution in Kanies Member, *Ernietta* communities appear to be restricted to a narrow bathymetric zone in the shelf area, colonizing non-deltaic, wave and storm-dominated shoreface environments, whereas they are absent basinward in inner shelf mud-dominated environments. Such lateral change over a short distance on the Hansburg-Wegkruip shallow marine shelf indicates that growth and distribution of the biota was sensitive to water depth. On the other hand, *Ernietta* and other associated infaunal vendobionts formed and grew as "community beds" that suffered reworking during high-energy conditions. The palaeoecological model proposed here is supported by: (1) the preservation of the fauna in life position (autochtonous fauna sensu Kidwell et al. 1986) within specific fossiliferous *Ernietta* beds called here "*Ernietta* lithofacies"; (2) the geometry and lateral extent of the *Ernietta* beds encased within proximal tempestites; (3) the development and preservation of quasi *in situ* lags within the *Ernietta* beds, composed of reworked autochthonous fauna; (4) the low-diversity and monospecific feature of the beds overwhelmingly dominated by *Ernietta*. Thus, the benthic *Ernietta* communities might be considered here as the earliest precursor in the rock record of "fossil-beds", similar to those made by the autochthonous communities in the Phanerozoic. Further studies including among others sedimentary and palaeoecological setting, interaction between fauna and sediment matrix, taxonomic diversity and growth strategies in vendobiont-rich beds of the Nama Group would yield additional useful details to constrain the dynamic of infaunal Ediacaran biota like *Ernietta* and the extent of their biotope.

8 Conclusions

The siliciclastic Kanies Member in Hansburg area, southern Namibia, represents a fining-upward succession deposited during the first transgression in the Witputs sub-basin. The succession contains the oldest known Ediacaran fossils in the Nama Group represented by *Ernietta*-community. A facies analysis combined with taphonomic signatures of *Ernietta*-bearing deposits allows to constrain the paleoecological setting and the factors influencing growth of *Ernietta* with possible implications on other occurrences of similar biota.

Four facies associations (FA 1–4) ranging from fluvial to inner shelf settings are distinguished in the Kanies Member. The overall siliciclastic succession (~12 m thick) is divided by a marine flooding surface into a fluvial lower and a marine upper part. The lower part consists of braided fluvial facies association (FA 1) which forms a sheet-like unit dominated by cross-bedded sandstone. It is overlain by foreshore to shoreface high-energy transgressive deposits (FA 2) with typical coarse-grained sandstone, rich in large rip-up mud-clasts at the base marking a marine flooding surface. The overlying deposits consist in the East of wave- and storm-dominated shoreface deposits (FA 3) that grade laterally westward (basinward) into mud-dominated inner shelf deposits with interbedded distal tempestites (FA 4). Vertical and lateral distribution of the facies associations in the upper part of the Kanies Member show a clear partitioning into proximal and distal settings in a basin deepening westward.

Ernietta communities are facies related and occur commonly in typical light and fine-grained sandstone layers sometimes associated with *Rangea* encased within wave and storm-dominated shoreface deposits. They form fossil-rich beds containing moulds of *Ernietta*, most of them indicate preservation in life position as evidenced by their vertical orientation, pointing to an autochthonous benthic community. Such *Ernietta* beds were affected by intermittent storm events that produced erosional structures and quasi in situ lags containing reworked biota. During major storm events, the *Ernietta* beds were buried by tempestites with basal lags containing bioclasts of vendobionts, mostly reworked *Ernietta*. In both cases, some of reworked *Ernietta* specimens were transported distally and redeposited in an inner shelf setting. The proposed palaeoecological model emphasizes an *Ernietta* community that lived on a fine-grained sandy substrate in a stressed shoreface environment; it formed an infaunal biota engulfed in loose sediment and with vertical upward growth. Growth and distribution of *Ernietta* communities appear controlled by palaeoenvironmental factors such us inherited sea floor topography, water energy conditions, water depth, type of substrate and sediment supply. According to their in situ three-dimensional mouldic preservation and their ability to be transported from their biotope without major destruction, most of the bag-shaped *Ernietta* individuals were seemingly completely filled in situ by sediment before their final burial.

This study highlights the importance of sedimentological analysis of fossilferous Ediacaran deposits, combined with taphonomic signatures, for constraining the biotope and palaeocology of infaunal vendobionts as well as their typical preservation as three-dimensional moulds. The applicability of the model to other vendobiontan communities, especially in the Nama Group, should be tested through an integerated approach including detailed facies analysis.

Acknowledgments The authors are grateful to VW-Foundation (I/78 706) and the Göttingen Courant Center of Geobiology (German Excellence Initiative) for financial support of fieldwork in Namibia. The owners of Hansburg and Wegkruip Farms are cordially thanked for their help in accessing the outcrops. Our thanks extend also to people in Konkiep Lapa Restcamp for the hospitality and help. O. Elicki is thanked for his constructive review that improved the final manuscript, all comments and suggestions are gratefully acknowledged.

References

Aigner T (1982) Calcareous tempestites: storm-dominated stratification in Upper Muschelkalk limestones (Middle Triassic, SW Germany). In: Einsele G, Seilacher A (eds) Cyclic and Event Stratification. Springer-Verlag, Berlin, pp 180–198

Brenchly PJ (1990) Biofacies. In: Briggs DEG, Crowther PR (eds) Palaeobiology: A Synthesis. Blackwell Scientific Publication, Oxford, pp 395–400

Bridge JS (2003) Rivers and Floodplains: Forms, Processes and Sedimentary Record. Blackwell Publishing, Oxford, p 491

Buss LW, Seilacher A (1994) Thy phylum Vendobionta: a sister group of Eumetazoa? Paleobiology 20:1–4

Corsetti FA, Hagadorn JW (2000). Precambrian-Cambrian transition: Death Valley, United States. Geology 28:299–302

Droser ML, Gehling JG, Jensen S (2006) Assemblage paleoecology of the Ediacara biota: The unabridged edition? Palaeogeography, Palaeoclimatology, Palaeoecology 232:131–147

Dumas S, Arnott RWC (2006) Origin of hummocky and swaley cross-stratification – The controlling influence of unidirectional current strength and aggradation rate. Geology 34:1073–1076

Dzik J (1999) Organic membranous skeleton of the Precambrian metazoans from Namibia. Geology 27:519–522

Einsele G, Seilacher A (eds) (1982) Cyclic and Event Stratification. Springer-Verlag, Berlin

Gehling J (1999) Microbial mats in Terminal Proterozoic siliciclastic: Ediacaran death masks, Palaios 14:40–57

Gehling J (2000) Environmental interpretation and a sequence stratigraphic framework for the terminal Proterozoic Ediacara Member within the Rawnsley Quartzite, South Australia. Precambrian Research 100:65–95

Germs GJB (1972) The stratigraphy and paleontology of the lower Nama Group, South West Africa. University of Cape Town, Department of Geology, Chamber of Mines Precambrian Research Unit, Bulletin 12

Germs GJB (1983) Implications of a sedimentary facies and depositional environmental analysis of the Nama Group in South West Africa. In: Miller RM (ed) Evolution of the Damara Orogen of south West Africa. Special Publication, Geological Society of South Africa 11:89–114

Germs GJB (1995) The Neoproterozoic of southwestern Africa, with emphasis on platform stratigraphy and paleontology. Precambrian Research 73:137–151

Germs GJB, Gresse PG (1991) The foreland basin of the Damara and Gariep orogens in Namaqualand and southern Namibia: stratigraphic correlations and basin dynamics. South African Journal of Geology 94:159–169

Germs GJB, Knoll AH, Vidal G (1986) Latest Proterozoic microfossils from the Nama Group, Namibia (South West Africa). Precambrian Research 32:45–62

Grazhdankin D (2004) Patterns of distribution in the Ediacaran biotas: facies versus biogeography and evolution. Paleobiology 30:203–221

Grazhdankin D, Seilacher A (2002) Underground Vendobionta from Namibia. Palaeontology 45:57–78

Grazhdankin D, Seilacher A (2005) A re-examination of the Nama type Vendian organism Rangea schneiderhoehni. Geological Magazine 142:571–582

Grotzinger JP, Bowring SA, Saylor BZ, Kaufman AJ (1995) New biostratigraphic and geochronologic constraints on early animal evolution. Science 270:598–604

Grotzinger JP, Watters WA, Knoll AH (2000) Calcified metazoans in thrombolite-stromatolite reefs of the terminal Proterozoic Nama Group, Namibia. Paleobiology 26:334–359

Hagadorn JW, Waggoner BM (2000) Ediacaran fossils from the southwestern Great Basin, United States. Journal of Paleontology 74:349–359

Hagadorn JW, Fedo CM, Waggoner BM (2000) Early Cambrian Ediacaran-Type Fossils from California. Journal of Paleontology 74:731–740

Hequette A, Hill PR, 1995 Response of the Seabed to Storm-Generated Combined Flows on a Sandy Arctic Shoreface, Canadian Beaufort Sea. Journal of Sedimentary Research A65:461–471

Hofmann HJ, O'Brien SJ, King AF (2008) Ediacaran biota on Bonavista peninsula, Newfoundland, Canada. Journal of Paleontology 82:1–36

Horodyski RJ (1991) Late Proterozoic Megafossils from Southern Nevada. Geological Society of America Abstracts with Programs 23:163

Horodyski RJ, Gehling JG, Jensen S, Runnegar B (1994) Ediacara fauna and earliest Cambrian trace fossils in a single parasequence set, southern Nevada. Geological Society of America Abstracts with Programs 26:60

Ivantsov AY, Grazhdankin D (1997) A new representative of the Petalonamae from the upper Vendian of the Arkhangelsk Region. Paleontological Journal 31:1–16 (English Translation)

Jenkins RJF (1985) The enigmatic Ediacaran (late Precambrian) genus Rangea and related forms. Paleobiology 11:336–355

Jenkins RJF (1992) Functional and ecological aspects of Ediacaran assemblages. In: Lipps JH and Signor PW (eds) Origin and Early Evolution of the Metazoa. Plenum Press, New York, pp 131–176

Jenkins RJF, Plummer PS, Moriarty KC (1981) Late Precambrian pseudofossils from the Flinders Ranges, South Australia. Transactions of the Royal Society of South Australia 105:67–83

Johnson HD, Baldwin CT (1986) Shallow siliciclastic seas. In: Reading HG (ed) Sedimentary Environments and Facies. Blackwell, Boston, pp 229–282

Kidwell SM (1991) The stratigraphy of shell concentrations. In: Allison PA and Briggs DEG (eds) Taphonomy: Releasing the Data Locked in the Fossil Record. Plenum Press, New York, pp 211–289

Kidwell SM, Brenchley PJ (1994) Patterns in bioclastic accumulation through the Phanerozoic: changes in input or in destruction? Geology 22:1139–1143

Kidwell SM, Furish FT, Aigner T (1986) Conceptual framework for the analysis and classification of fossil concentrations. Palaios 1:228–238

Kreisa RD (1981) Storm-generated sedimentary structures in subtidal marine facies with examples from the Middle to Upper Ordovician of southwestern Virginia. Journal of Sedimentary Petrology 51:823–848

Laflamme M, Narbonne GM (2008) Ediacaran fronds. Palaeogeography, Palaeoclimatology, Palaeoecology 258:162–179

Mapstone NB, McIlroy D (2006) Ediacaran fossil preservation: Taphonomy and diagenesis of a discoid biota from the Amedeus basin, central Australia. Precambrian Research 149:126–148

McCall GJH (2006) The Vendian (Ediacaran) in the geological record: Enigmas in geology's prelude to the Cambrian explosion. Earth Science Reviews 77:1–229

Miall AD (1985) Architectural-element analysis: a new method of facies analysis applied to fluvial deposits. Earth Science Reviews 22:261–304

Miall AD (1996) The Geology of Fluvial Deposits: Sedimentary Facies, Basin Analysis and Petroleum Geology. Springer-Verlag, Berlin, p 582

Myrow PM, Southard JB (1996) Tempestite deposition, Journal of Sedimentary Research 66:875–887

Narbonne GM (2005) The Ediacara biota: Neoproterozoic origin of animals and their ecosystems. Annual Review in Earth and Planetary Sciences 33:421–442

Narbonne GM, Saylor BZ, Grotzinger JP (1997) The youngest Ediacaran fossils from Southern Africa. Journal of Paleontology 71: 953–967

Passchier S, Kleinhans MG (2005) Observations of sand waves, megaripples, and hummocks in the Dutch coastal area and their relation to currents and combined flow conditions. Journal of Geophysical Research 110:F04S15, doi:10.1029/2004JF000215

Pflug HD (1966) Neue Fossilreste aus den Nama-Schichten in Südwest-Afrika. Paläontologische Zeitschrift 40:14–25

Pflüger F (1999) Matground structures and redox facies. Palaios 14:25–39

Porada H, Chergut J, Bouougri EH (2008) Kinneyia-type wrinkle structures – Critical review and model of formation. Palaios 23:65–77

Røe S-L, Hermansen M (1993) Processes and products of large, Late Precambrian sandy rivers in northern Norway. In: Marzo M, Puigdefábreagas C (eds), Alluvial Sedimentation. Special Publication, International Association of Sedimentologists 17, pp 151–166

Sambrook Smith GH, Ashworth PJ, Best JL, Woodward J, Simpson CJ (2006) The sedimentology and alluvial architecture of the sandy braided South Saskatchewan River, Canada. Sedimentology 53:413–434

Saylor BZ (2003) Sequence stratigraphy and carbonate-siliciclastic mixing in a Terminal Proterozoic foreland basin, Urusis Formation, Nama Group, Namibia. Journal of Sedimentary Research 73:264–279

Saylor BZ, Grotzinger JP, Germs JBB (1995) Sequence stratigraphy and sedimentology of the Neoproterozoic Kuibis and Schwarzrand Subgroups (Nama Group), southwestern Namibia. Precambrian Research 73:153–171

Saylor BZ, Kaufman AJ, Grotzinger JP, Urban F (1998) A composite reference section for Terminal Proterozoic strata of southern Namibia. Journal of Sedimentary Research 68:1223–1235

Seilacher A (1992) Vendobionta and Psammocorallia: lost constructions of Precambrian evolution. Journal of the Geological Society 149:607–613

Vis-Star NC (2007) Effect of wave-topography interactions on the formation of sand ridges on the shelf. Journal of Geophysical Research 112:C06012, doi:10.1029/2006JC003844

Waggoner BM (2003) The Ediacaran biotas in space and time. Integrated Comparative Biology 43:104–113

Waggoner BM, Hagadorn JW (1997) Ediacaran fossils from western North America: Stratigraphic and biogeographic implications. Geological Society of America Abstracts with Programs 29:30

Walker RG, Plint AG (1992) Wave- and storm-dominated shallow marine systems. In: Walker RG, James NP (eds) Facies Models: Response to Sea Level Change. Geological Association of Canada, St. John's, pp 219–238

Zecchin M (2007) The architectural variability of small-scale cycles in shelf and ramp clastic systems: The controlling factors. Earth-Science Reviews 84:21–25

Biolaminated Siliciclastic Deposits

El Hafid Bouougri and Hubertus Porada

1 Introduction

"Biolaminated deposits" (Gerdes and Krumbein 1987) originate from microbial growth and activity interacting with environmental, physical and chemical processes like sedimentary deposition and mineral precipitation. The resulting "primary" sedimentary structure was named "growth bedding" by Pettijohn and Potter (1964), with clear reference to in situ formed limestones and with "stromatolitic bedding" as the most prominent example (see also Pettijohn 1975). Subsequently, it was recognized that "growth bedding" mediated by microbial mats is not restricted to carbonate environments but may occur also in peritidal siliciclastic depositional systems. Respective sedimentary deposits have been termed "siliciclastic biolaminites" (Gerdes and Krumbein 1987) which basically consist of quartz sand and interlayered dark laminae originating from monolayered microbial mats or successive generations of mats.

The process leading to siliciclastic biolaminites includes periods of non-deposition in which organisms produce and establish mat layers on a sediment surface, and periods of sedimentation in which the mat is covered by a thin layer of fine-grained sediment through which motile bacteria (or other motile organisms) can move upwards to establish a new mat atop (for details see Gerdes et al. 1991, 2000). Laboratory and field experiments (Gerdes and Klenke 2003, 2007) have shown that a new, though still thin and soft mat layer may form on the new sediment surface within few days only, whereas to establish a mature and resistant mat with fair

E.H. Bouougri (✉)
Department of Geology, Faculty of Sciences Semlalia, Cadi-Ayyad University, 40020 Marrakech, Morocco
e-mail: bouougri@ucam.ac.ma

H. Porada
Geosciences Centre, Georg-August University of Göttingen, Goldschmidt-Straße 3, 37077 Göttingen, Germany

J. Reitner et al., *Advances in Stromatolite Geobiology*,
Lecture Notes in Earth Sciences 131, DOI 10.1007/978-3-642-10415-2_30,

preservation potential, a period of non-deposition of several months may be required. The main effect of microbial mats in the process of "biosedimentary accretion" of siliciclastic biolaminites is biostabilization of the sediment surface and resistance against erosion, thus allowing deposition of a new sediment layer even from stronger currents. The accretion process may encompass (1) layers of thin to thick stacks of microbial mats which, by trapping and binding (Black 1933; see review in Schieber 2004), include sediment particles as isolated grains in their fabrics; (2) thin layers or lenses of sediment laid down during inundation or by wind action; (3) thicker layers of sediment deposited during events such as storms and sheet floods.

Siliciclastic biolaminites are basically sedimentary deposits whose accretion, however, was mediated by microbial communities (mats) on the momentary sediment surface. They are thus microbialites (Burne and Moore 1987) in the broadest sense, but deviate from "stromatolites" in two fundamental features and their combined occurrence: (1) the systematic mm-range alternation of microbial mat layers and siliciclastic, detrital sediment layers of various thickness introduced into the system by physical forces, and (2) the lack of any carbonate cementation. A comparison of siliciclastic biolaminite and stromatolite deposits and their genesis is presented in Fig. 1.

Modern occurrences of siliciclastic biolaminites have been described from the sea-marginal Gavish Sabkha (Sinai Peninsula, Red Sea) in the arid zone, and from Mellum Island (southern North Sea, Germany) in the temperate humid zone, both by Gerdes and Krumbein (1987). At both localities, sediment surfaces are reported to be permanently wetted by capillary water movement. At Gavish Sabkha, the biolaminites have formed more accidentally in isolated sandlobes along a barrier slope, whereas on Mellum Island they systematically are developed from the upper

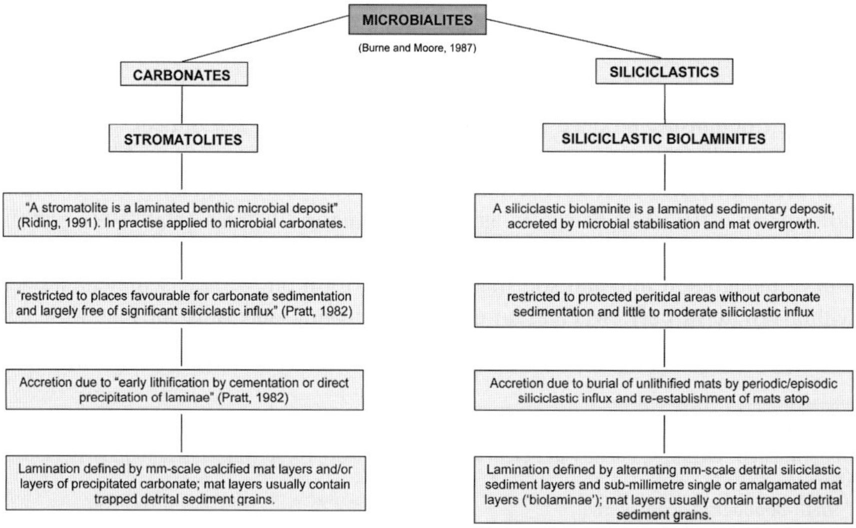

Fig. 1 Comparison of stromatolites and siliciclastic biolaminites with regard to their formation and resulting composition and structure

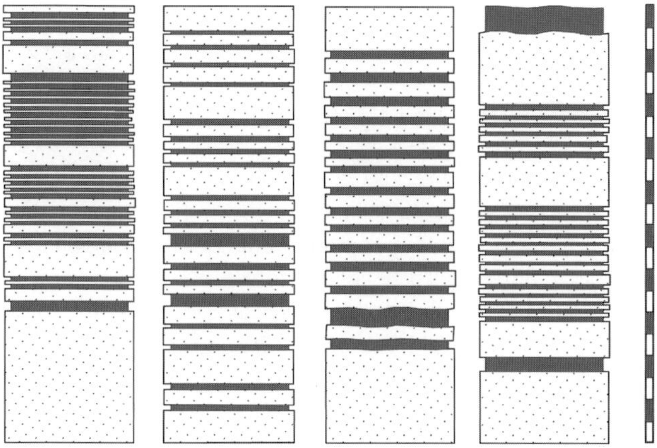

Fig. 2 Schematic presentation of siliciclastic biolaminites as developed in the upper intertidal zone of Mellum Island, southern North Sea, Germany. Note mm- to cm-scale interlayering of organic material originating from microbial mats (*grey*) and sand/silt deposits (*dotted*); thicker sediment beds result from episodic storm events. Scale bar is in cm divisions [Modified after Gerdes and Klenke (2007); Pratt (1982); Riding (1991)]

intertidal to the lower supratidal zone, occupying the so-called "versicoloured" high tidal flats ("Farbstreifensandwatt") which are open to the sea. Sediment cores recovered from the uppermost 25 cm of still unconsolidated sediment (Fig. 2) show the typical mm- to cm-scale alternation of microbial mat remains and sand to silt layers, and irregularly intercalated thicker sediment beds resulting from major wind or storm events. In detail, however, the sections differ considerably in respect of bed thicknesses and frequencies of event intercalations, depending on local, more or less sheltered situations.

The aim of this contribution is to examine a suite of macroscopic and microscopic features that may lead to recognise, in rock record, biolaminated siliciclastic deposits and to highlight the biosedimetary origin of mm-scale laminated siliciclastic deposits. For that purpose, siliciclastic biolaminites and associated structures are described from the Mediterranean coast of southern Tunisia as a further modern example, and from the Neoproterozoic of Namibia and Morocco as ancient examples.

2 Bhar Alouane, Mediterranean Coast, Southern Tunisia: Modern Siliciclastic Biolaminites and Associated Structures

Bhar Alouane is a small lagoon situated about 50 km south of Djerba Island, near the village of Zarzis, (Fig. 3). The lagoon is connected with the large Sabkha El Melah over a narrow inlet and is bordered by sand bars separating low-gradient tidal flats from the lagoon. The sand bars are intersected by narrow tidal creeks through which adjacent areas are inundated at high tides. The regularly flooded areas are characterized by well-developed microbial mats of dominant filamentous

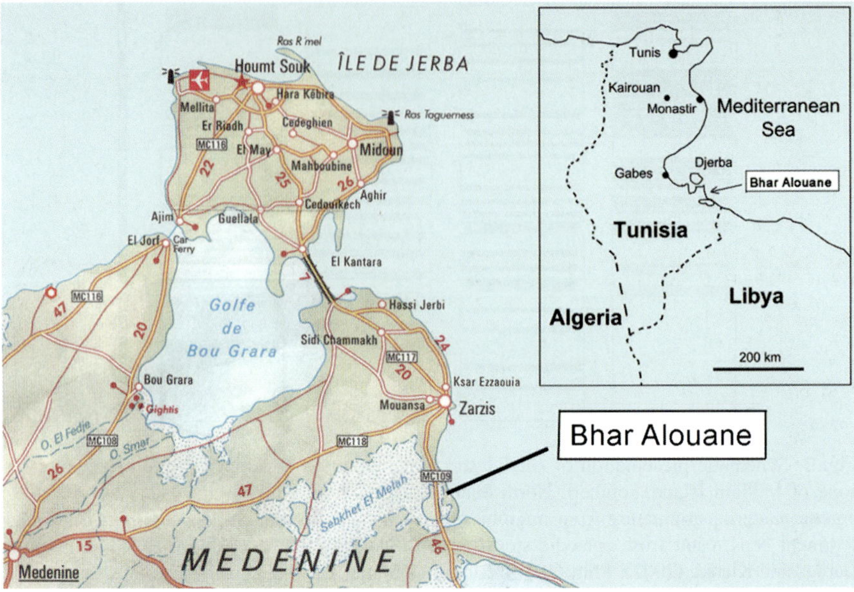

Fig. 3 Map section showing part of southern Tunisia with Bhar Alouane lagoon and tidal flats situated south of Djerba Island [Modified after Nelles Maps,1:750.000, Tunisia (2002)]

cyanobacteria, mainly *Microcoleus chthonoplastes*, and associated coccoid species (Noffke et al. 2001; Gerdes et al. 2008). Thin layers of fine-grained, sandy to silty sediment are occasionally loaded on the mats by overflowing creek waters or by wind action, whereas thicker new sediment beds overlying the mats may arise from episodic storm high-tides or sand storms. Microbes recolonizing each new sediment surface form vertically extended "biolaminated deposits" (Gerdes et al. 2008), here referred to as "siliciclastic biolaminites".

A sample cut out of the uppermost part of a succession of still unconsolidated and well laminated siliciclastic biolaminites (Fig. 4a) shows the typical mm-scale alternation of light, sediment-rich layers and greyish to dark, organic-rich layers resulting from decayed previous mats or stacks of mats which formed during periods of non-deposition. Particles in sediment-rich laminae originate mainly from settling of suspended mud but may include silt to sand-sized grains transported by wind or currents during high energy events. In this non-compacted section, microbial mat laminae appear more regular in thickness being less than 3 mm, whereas thickness of sediment-rich laminae is more variable and is up to 5 mm.

A common feature developed in the active mat on top and partly involving the underlying biolaminite succession, are dm-scale networks of polygonal cracks (Fig. 4b). Resulting "polygonal mats" have been described from various localities including the classical sites of Shark Bay (e.g. Logan et al. 1974), Trucial Coast (e.g. Kendall and Skipwith 1969) and Laguna Figueroa (e.g. Horodyski et al. 1977). In most cases, mats with polygonal networks of cracks are developed in dominantly low-energy situations on the lower intertidal zone where they are regularly flooded, or in ponded areas, but may occasionally suffer desiccation.

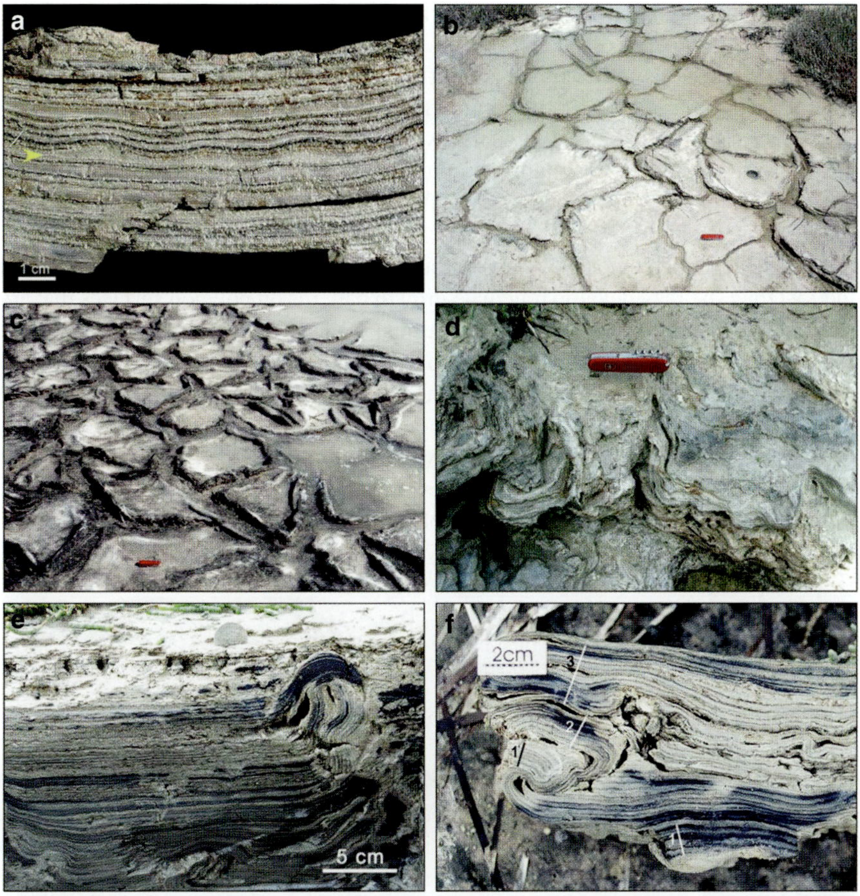

Fig. 4 Siliciclastic biolaminites and associated mat related structures as developed at Bhar Alouane tidal flat, Tunisia. (**a**) Section of siliciclastic biolaminites showing alternation of organic-rich layers originating from microbial mats (*dark*) and sediment layers deposited during storm or sand storm events. *Yellow arrow* indicates wave-rippled sediment bed. (**b**) Polygonal network of open cracks, partly deepened by ebb-flow currents. Scale (knife) is 8 cm. (**c**) Typical "polygonal mat" with upturned margins of individual polygons; note tidal water trapped in saucer-shaped polygon at *right side* of photograph. Scale (knife) is 8 cm. (**d**) Section across sediment-filled crack and adjoining upturned polygon margins. Scale (knife) is 8 cm. (**e**) Upturned margin, overgrown by new mat layers. (**f**) Complex crack margin structure resulting from progressive upturning and involution of polygon edge and concomitant mat overgrowth. Note initial upturned package of biolaminite (1) overgrown by a first generation of microbial layers (2) and subsequently involuted toward the polygon centre; overlying microbial layers (3) form the next step of overgrowth and preserve the structure

Usually, both opposing crack margins are upturned, involving 5–10 cm stacks of the underlying biolaminite succession. In this way, the mat polygons defined by the cracks attain a "saucer-shaped" geometry which allows tidal water to remain longer on the polygons than on the elevated margins (Fig. 4c). The upturned margins may be simple and almost straight, adjoining sediment trapped in the open crack

(Fig. 4d). Occasionally, when slight differences in elevation allow water to remain longer on the one side of the crack than on the other, only one of the margins is upturned whereas the other remains flat (Fig. 4e). In such cases, the crack and the upturned margin may be soon overgrown by new mat layers extending from the polygons involved. Other crack margins may develop very complex geometries (Fig. 4f) due to progressive, stepwise upturning and concomitant, steady microbial overgrowth [for details of process see Noffke et al. (2001) and Bouougri and Porada (2007a)].

Periodically, filamentous cyanobacteria, particularly *Microcoleus chthonoplastes* and *Lyngbya* sp., may develop reticluate growth patterns on polygon surfaces that are inundated first and run dry latest during a tidal cycle (Fig. 5a). The same reticulate pattern may form on saucer-shaped polygons in which water is trapped for some time (Fig. 5b). When fossilized, such growth patterns are known as "elephant skin" textures (e.g. Gehling 1999).

Towards the upper intertidal and lower supratidal zones, the mat gradually changes into a flat type which upslope develops a "pustular" surface as it becomes increasingly dominated by coccoid cyanobacteria (e.g. *Synechocystis* sp., *Synechococcus* sp.) in the top layer. Though rarely inundated, the mat even in the supratidal

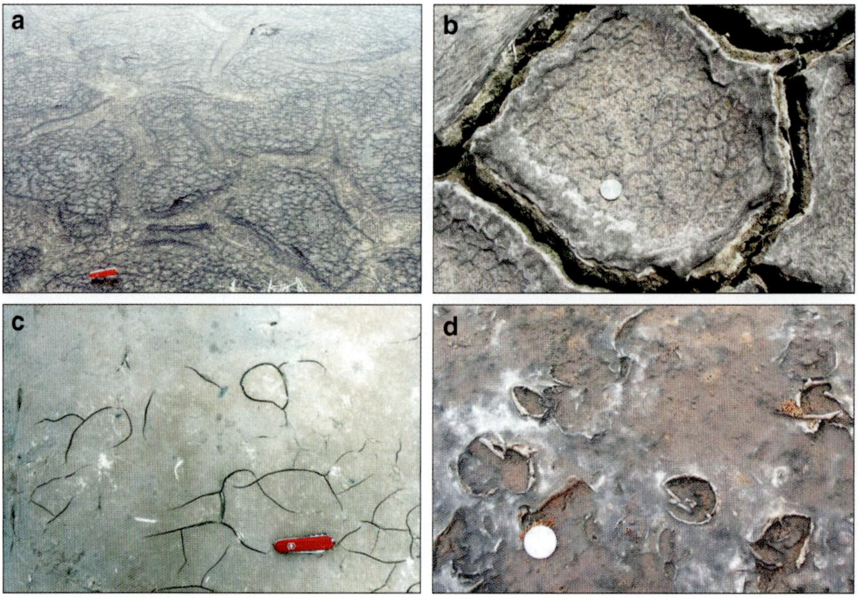

Fig. 5 Growth patterns and types of cracks developed on microbial mats. (**a**) Inundated polygons separated by wide open cracks. Note reticulate growth pattern formed on the polygons by filamentous cyanobacteria *Microcoleus chthonoplastes* and *Lyngbya* sp. Scale (knife) is 8 cm. (**b**) Reticulate growth pattern developed in centre of saucer-shaped polygon. Scale (coin) is 23 mm. (**c**) Incipient shrinkage cracks, partly forming incomplete networks in upper layer of flat mat. Scale (knife) is 8 cm. (**d**) Subcircular openings with upturned and "curled" margins developed in the uppermost, elastic layer of a thin mat dominated by coccoid cyanobacteria. Scale (coin) is 23 mm

zone is still supplied with water from below due to capillary groundwater movement and evaporative pumping (Hsü and Siegenthaler 1969; Porada et al. 2007). Nevertheless the surface layer may undergo desiccation and shrinkage and develop cracks which may be isolated and small or form incomplete networks (Fig. 5c). Incipient cracks may evolve to circular openings with curled margins when the uppermost, elastic and EPS-rich layer locally gets torn during the shrinkage process (Fig. 5d).

The flat and pustular mat types are not underlain by biolaminites which appear to be restricted to areas that regularly undergo tidal inundation, or to isolated ponds that are recharged during spring high tides.

3 Neoproterozoic Vingerbreek Member, Schwarzrand Subgroup, Nama Group, Namibia: Ancient Siliciclastic Biolaminites and Associated Structures

The Vingerbreek Member is a succession of shale, siltstone and sandstone, all deposited in tidal and low-energy shoreline depositional environments (Germs 1983), that form the upper part of the Nudaus Formation at the base of the Schwarzrand Subgroup. Its age is between ca. 548 and 545 Ma, according to U-Pb zircon dates from ash layers in the underlying Kuibis Subgroup and overlying Urusis Formation, respectively (Grotzinger et al. 1995).

In the lower part of the Vingerbreek Member, a more than 20 m thick succession of siliciclastic biolaminites with intercalated sandstone beds and heterolithic siltstone/mudstone units is preserved on Haruchas farm (Fig. 6) (Bouougri and Porada 2007b, c). Sedimentary features indicate deposition in a low-energy, shallow, intertidal position behind a protecting sand barrier. The biolaminites occur as m-scale units of thinly interbedded layers of siltstone and dark to black material of variable thickness and proportions (Fig. 7a). In thin section, the biolaminites consist of packages (0.5–1.5 mm thick) of 5–10 alternating light and dark layers, separated by mm-scale light layers of sand/silt sized sediment (Fig. 7b). The dark layers, made up of mainly sericite, Fe-oxides and scarce relics of carbonaceous material, are considered to represent former microbial mats, whereas the light layers are sediment deposited on the mats and subsequently overgrown by new mat layers.

Evidence of the participation of mats that once colonized the sedimentary surfaces and contributed to the accretion of thick stacked layers of biolaminite, include mat-related structures like "elephant skin" (reticulate growth pattern) (Fig. 7c; cf. Fig. 5a, b), spindle-shaped and lenticular shrinkage cracks (Fig. 7d), subcircular cracks with "curled" margins (Fig. 7e; cf. Fig. 5d), and Kinneyia-type wrinkle structures (Fig. 7f). Most diagnostic for the identification of biolaminites are upturned to overturned crack margins, locally combined in networks, involving stacks of biolaminites, which are overlain by continuous biolaminated deposits preserving the original sedimentation plane. Two types of crack margin structures

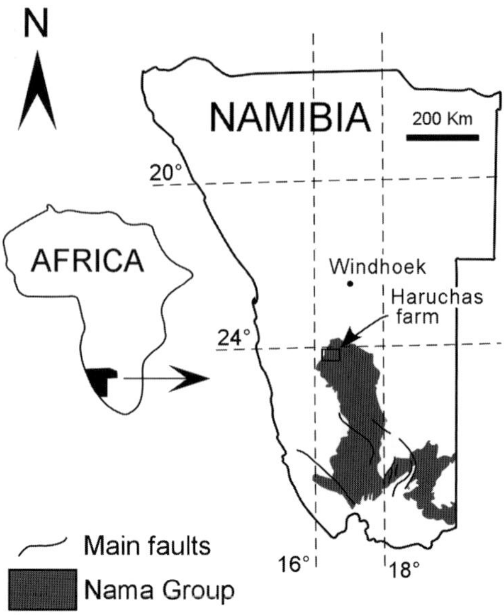

Fig. 6 Schematic map of Namibia showing distribution of Nama Group deposits in the Zaris and Witputs sub-basins, and location of Haruchas farm where siliciclastic biolaminites are preserved in the Vingerbreek Member

are observed: (1) upturned margins forming cm-scale "inverted V" structures similar to tepees, which may be symmetric or asymmetric with one margin more upturned than the other (Fig. 8a, b); on bedding surfaces they appear as straight, narrow sharp-crested folds and may occasionally combine in networks (Fig. 8c, d); (2) wide upturned to overturned margins involving thick packages of biolaminites, likely related to major cracks (Fig. 8e, f). The upturned margin structures resemble similar features observed on modern "polygonal mats" at Bhar Alouane, Tunisia, and documented by Park (1977, his Figs. 9, 11) from Holocene upper intertidal polygonal mats at Abu Dhabi, Persian Gulf, U.A.E.

4 Neoproterozoic (~0.8 Ga) Tizi n-Taghatine Group, Anti-Atlas, Morocco: Ancient Planar and Domal Siliciclastic Biolaminites and Associated Structures

In the Anti-Atlas belt, siliciclastic biolaminites of the Neoproterozoic Tizi n-Taghatine Group, are documented from Agoummy, Tirsal and Nqob sections at the southern edge of Siroua massif (Fig. 9). The Tizi n-Taghatine Group (~0.8 Ga) is a volcano-sedimentary cover, about 2 km thick, deposited on the northern continental margin of the West African craton. The biolaminated deposits occur in facies associations of peritidal and shallow subtidal lagoonal environments and form either mm- to cm-thick interbeds alternating with sand layers within a heterolithic facies, or dm- to m- thick units with mm-scale lamination (biolaminite lithofacies).

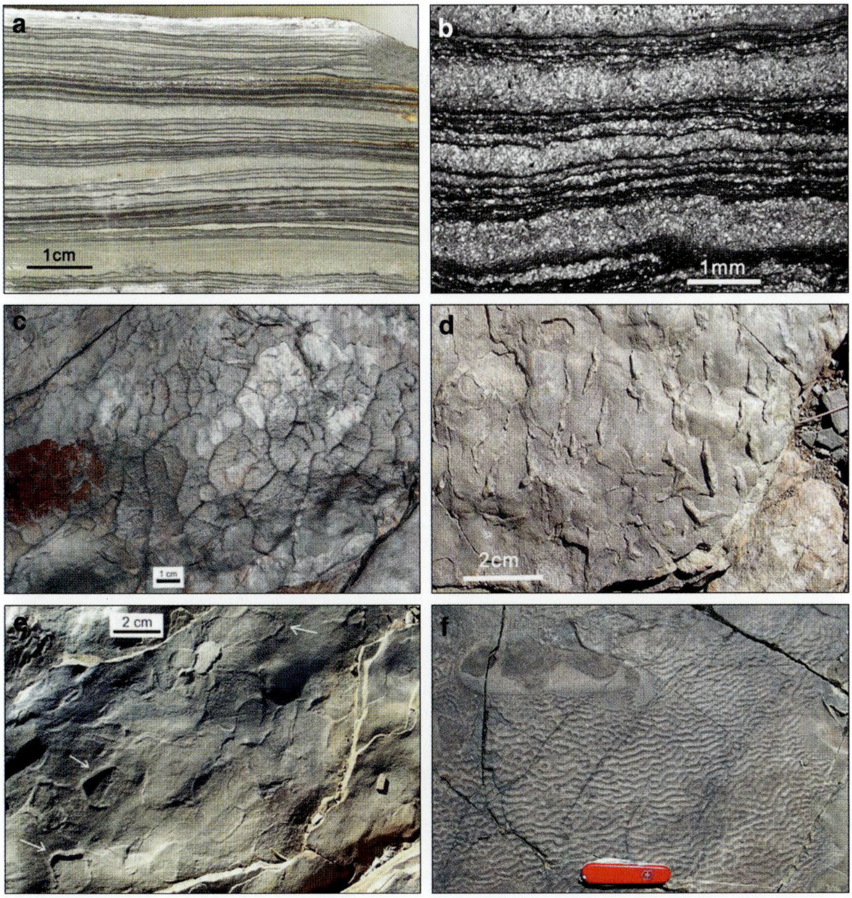

Fig. 7 Siliciclastic biolaminites and associated mat-related structures. (**a**) Younging upward succession of biolaminites with interbedded, thicker silty to sandy event layers. *Dark layers* consist of dominantly sericitic argillite and are supposed to represent previous microbial mats or stacks of mats. (**b**) Microphotograph showing alternating sedimentary layers (*light*) and microbial mat layers (*dark*). Note characteristic "wavy-crinkly" appearance of mat layers. (**c**) Upper "mud-draped" (now sericitic) bedding surface of sand layer showing reticulate pattern known as "elephant skin" texture. The pattern is considered to represent growth structures of filamentous cyanobacteria on a microbial mat that previously covered the sediment surface. (**d**) Lenticular and spindle-shaped cracks, partly forming triple junctions on upper bedding surface of sandstone layer. (**e**) Subcircular structures with elevated margins (some indicated by *arrows*) on upper bedding surface of sandstone layer. The structures are envisaged to represent subcircular openings in a thin microbial mat that had undergone desiccation and shrinkage. (**f**) Upper bedding surface of event deposit (sandstone) with typical Kinneyia-type wrinkle structures [for genetical details see Porada et al. (2008)]. Scale (knife) is 8 cm

In the heterolithic facies, the biolaminated layers may form thin multilayered veneers on sandy surfaces or form cm-thick planar laminated interbeds, commonly disrupted by shrinkage cracks of various shape and size. Several mat-related

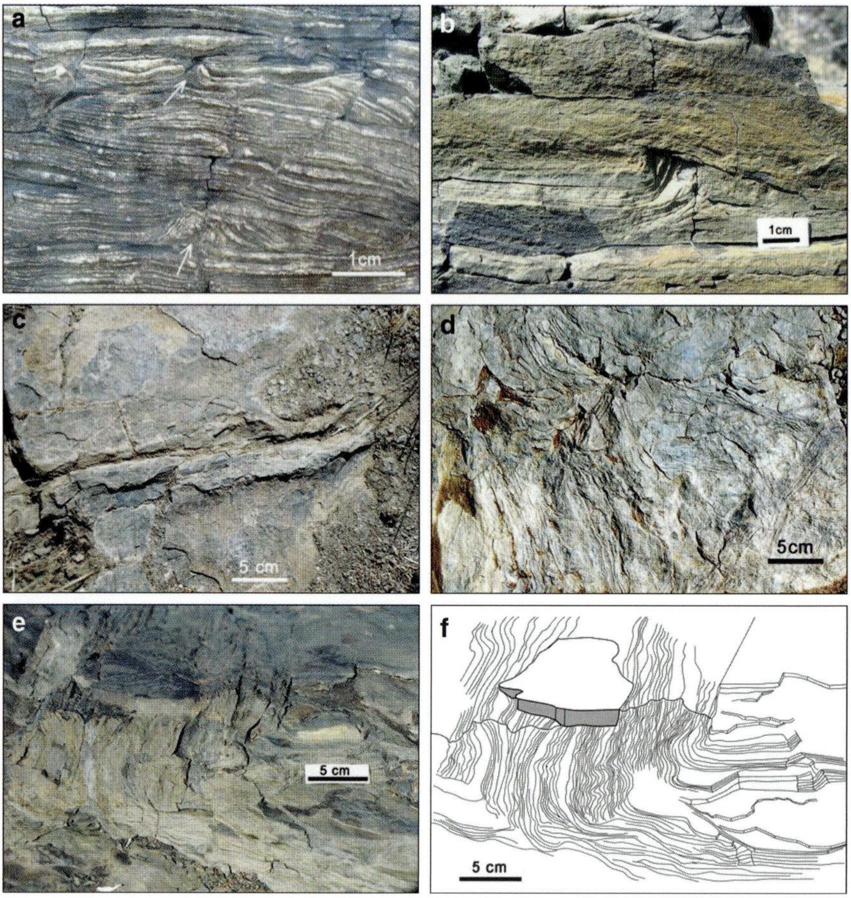

Fig. 8 Structures developed in siliciclastic biolaminites of the Vingerbreek Member, Nama Group, Namibia. (**a**) Section across package of siliciclastic biolaminites showing "upturned margin" structures similar to tepees, related to shrinkage cracks in a previous microbial mat. (**b**) Section across asymmetric "upturned margin" structure with only one margin upturned. (**c**) Upper bedding surface with shrinkage crack and upturned to overturned margins, formed in a previous thick microbial mat. (**d**) Upper bedding surface showing part of polygonal network of shrinkage cracks with upturned margins. (**e**) Upturned margin of major crack involving thick package of siliciclastic biolaminite. (**f**) Graphic rendition of (**e**)

structures have been observed in these heteroliths (e.g. Bouougri and Porada 2002, 2007d), including among others reticulate patterns and microbial shrinkage cracks (Fig. 10a, b). The crack-fills usually appear as superimposed bodies with tapering ends and overlapping relationships. They form several distinct generations indicating multiple episodes of desiccation of an organosedimentary laminated deposit. Due to high rate of compaction, the crack-fills are welded together and now appear as a single generation on the bedding surface.

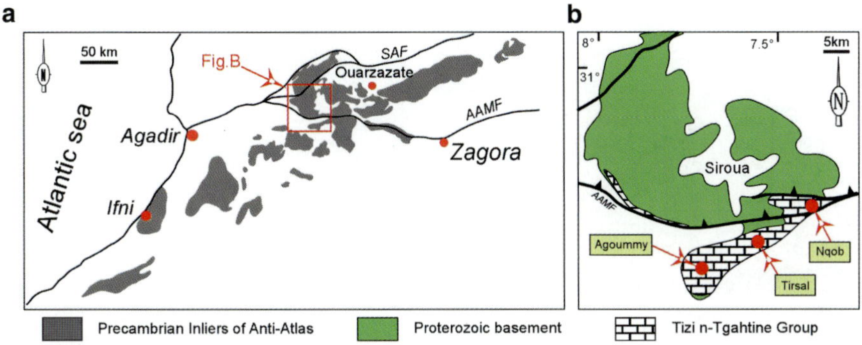

Fig. 9 (**a**) Simplified map showing distribution of the Precambrian inliers in the Anti-Atlas belt. (**b**) Outcrop distribution of the Neoproterozoic Tizi n-Taghatine Group in the Proterozoic Siroua massif. *SAF* South Atlas Fault, *AAMF* Anti-Atlas Major Fault

The thick biolaminated units form a distinct lithofacies consisting of distinctive and laterally continuous horizons, 1–3 m thick. This "biolaminite lithofacies" occurs at the top of shallowing upward peritidal parasequences or forms part of lagoon-coastal plain parasequences. It consists of thinly laminated rock bodies with mm-scale alternations of white siltstone or sandstone laminae and green sericitic laminae (Fig. 10c), and often includes some thicker, cm-scale sandy beds and lenses. The microstructure of this organo-sedimentary deposit exhibits an alternation of sand/silt sized quartz laminae and thinly laminated sericitic argillite layers considered to represent previous microbial mat layers (Fig. 10d). The quartz-rich laminae usually form layers one to four grains thick. The multilaminated fabric of the microbial layers (now sericitic argillite) contains detrital quartz as isolated "floating" grains that were trapped and bound by filamentous microbial mats, as subrounded aggregates of "coated" grains and as lenses which may represent mm-scale ripple layers.

According to their appearance in the field, two morphotypes are distinguished in the biolaminites lithofacies (Fig. 11a–b): (1) planar biolaminites characterized by a flat to wavy-crinkly laminated structure (2) domal biolaminites characterized by dm-scale domes. In many horizons, both of the structures occur together, with an upward transition from planar to domal biolaminites or inversely from domal to planar.

Domal biolaminites (Fig. 11b) appear within biolaminited units, up to 40 cm thick. They are closely related to planar forms deposited broadly in the intertidal zone and are characterised by a mound-shaped upper bedding surface. The domes, 10–30 cm in diameter and 5–20 cm high, are irregularly spaced. On bedding surfaces, the domes show subcircular to gently elongated sections and may form a single domal horizon within the biolaminated beds or vertically stacked domal structures with random distribution (Fig. 11c). The internal structures of the domes consist of mm-thick, gently to moderately convex laminae made up of fine to very fine-grained quartz sand. Lamina thickness is variable and occasionally increases outwards. At the

Fig. 10 (**a**) Close-up view of reticulate "elephant skin" pattern on upper bedding surface of thinly laminated siltstone layer. The surface shows an irregular polygonal arrangement of sharp-crested ridges indicating vertical growth of cyanobacterial filaments. Note the presence of an isolated microbial shrinkage crack crosscutting the reticulate growth pattern. (**b**) Upper surface of 2 cm thick quartzite bed exhibiting high density of sandy crack-fills. The cracks are lenticular, tri-radiate, spindle-shaped to sinuously curved and form a network of differently oriented bodies, which usually meet at high angles. (**c**) *Polished section* of ca. 3 cm thick biolaminites with mm-scale alternating sericitic (*green*) laminae and silt/sand sized quartz (*light*) laminae. Note the interbedded thicker (ca. 4 mm) lamina of fine-grained sandstone at the lowermost part of the section. (**d**) Microphotograph showing the fabric of the biolaminated deposits. Note the wavy-crinkly feature in the laminated dark layers which represent the former microbial mats

base of individual domes, the laminae along the edge tend to be progressively overturned forming thus a chevron-like structure (Fig. 11d). In many cases, the laminae in the inner part of the domes are of type SH-V sensu Logan et al. (1964) and tend to be continuous and laterally linked (LLH type) outward of the domes.

Bedding surfaces of planar biolaminites, deposited in wide intertidal to lower supratidal depressions regularly flooded during high tides, exhibit shrinkage cracks of variable size and shape, ranging from isolated, lenticular and spindle-shaped types to networks of polygonal cracks (Fig. 11e–f). Within a package of planar biolaminites, lenticular to sinuous "V"-shaped shrinkage cracks may occur at several levels as different superimposed generations indicating multiple repetitions of mat growth, shrinkage and cracking. A specific type of shrinkage structure is characterized by upturned and overgrown crack margins (Fig. 11f–g). Preserved structures record different stages of crack margin evolution, from incipient upturning into a vertical attitude or simple upturning/overgrowth to complex patterns resulting from repeated shrinkage, upturning and concomitant overgrowth

Fig. 11 Planar and domal biolaminites of the Tizi n-Taghatine Group and associated structures. (**a**) Planar biolaminites with typical mm-scale lamination. (**b**) Domal biolaminites displaying dm-scale vertically stacked domes. (**c**) Upper bedding surface showing a subcircular dome. The laminae in the inner part of the hemispheroid are overturned at the sides (*green arrow*) whereas in the outer part, the laminae are laterally continuous (*red arrow*). (**d**) Close-up view of the base of a dome showing the overturned feature of the hemispheric laminae. (**e**) Planar biolaminites cross-cut by two superposed V-shaped cracks filled with fine-grained sand. (**f**) Large polygonal upturned margin, seen in plan view, preserved in planar laminated siliciclastic biolaminites. Such thick

(Fig. 11f). Large desiccation cracks with upturned margins extending up to 4 cm upward, may cut across several layers in which small isolated and spindle-shaped cracks are preserved (Fig. 11f). Large upturned and sinuously curved crack margins commonly involve a multilayered package of sericitic-organic and siltstone layers (Fig. 11g). The large structures usually form polygonal networks and reflect the upturned margins of major polygonal cracks formed during shrinkage of thick biolaminite units. Small spindle-shaped cracks filled by sandy material constitute another aspect of desiccation with limited opening of cracks and rapid subsequent filling by sediment during tidal flood events.

Within planar biolaminites, structures developed in upturned and overgrown crack margins vary from simple to complex (Bouougri and Porada 2007d). A simple structure may develop by overgrowth of an upturned crack margin (Fig. 11h). Laminae covering the upturned margin are usually very thin and mainly of microbial origin, whereas behind the overgrown edge, lamination consists of mm-scale alternations of continuous microbial light-coloured laminae and discontinuous sedimentary brown laminae. The resulting structure implies continuous and repeated processes of shrinkage, upturning and overgrowth until the margin-structure is sealed by new and continuous mat layers. In an evolved stage, a complex structure results from shrinkage, upturning of crack margins, concomitant mat growth and deformation during burial or successive overfolding. The resulting geometry and internal organisation of laminae record multiple stages of overgrowth and crack-margin upturning, with subsequent overturning and breaking of laminae behind the margin.

5 Discussion and Conclusions

In the siliciclastic rock record, a variety of mat-related structures result from the interplay between microbial mats and a broad spectrum of processes (e.g. Schieber 2004). The resulting product of mat formation, sedimentation and microbial sediment trapping and binding (e.g. Gerdes et al. 2000; Riding 2000) is "biolamination" or "biogenic bedding" (sensu Gerdes and Klenke 2007), characterized by a typical fabric and bedding style with mm-scale lamination.

Siliciclastic biolaminites usually develop flat, parallel laminations, but excursions to more domal structures and wavy-crinkly laminations occur. At subaerial exposure, the surface mat layers of biolaminite successions may undergo desiccation and shrinkage and develop desiccation cracks of variable size and shape, ranging from

---

Fig. 11 (Continued) upturned margins are related to large and deep cracks that cut across the biolaminites and may extend up to 5 cm downwards. Note occurrence of isolated lenticular and spindle-shaped cracks (*indicated by arrows*). (**g**) Close-up view of an upturned margin showing laminated structure. *Blue arrows* indicate sericitic-organic layers; *yellow arrows* indicate siltstone layers. (**h**) Simple structure of an upturned (1) and overgrown (2) crack margin in biolaminites that underwent subaerial shrinkage. The whole structure is sealed by horizontal and continuous planar biolaminites (3). *Yellow dashed lines* indicate the initial upturned margin isolated, lenticular and spindle-shaped types to networks of dm-scale polygonal cracks with upturned and overgrown margins. Within a package of biolaminites, shrinkage cracks may be observed at several levels indicating multiple repetitions of mat growth and subaerial shrinkage during biosedimentary accretion (Fig. 12). When, at wide cracks, desiccation and shrinkage advance from the crack margins into the top portion of biolaminite succession, thick packages of biolaminite may become upturned, forming an upturned margin. Such structures constitute a diagnostic feature of peritidal biolaminites and have been identified in both recent and ancient examples.

Microbial growth, triggered by uprising groundwater in wide cracks may span the crack opening and encroach upon the margins, leading to overgrown margins. Alternating mat growth and upturning of crack margins, e.g. during tidal high and low water, respectively, may lead to complex structures along the crack margins. Depending on the rates of processes involved, resulting structures may range from rather simple to highly irregular. Complex structures reflect a non-equilibrium system of quasi-concomitant growth, destruction by desiccation, deformation by repeated upturning, and renewed overgrowth. When compared to modern examples, identification of "crack margin" related structures in ancient mm-scale laminated siliciclastic deposits may argue for: (1) the biogenicity of the deposit; (2) an

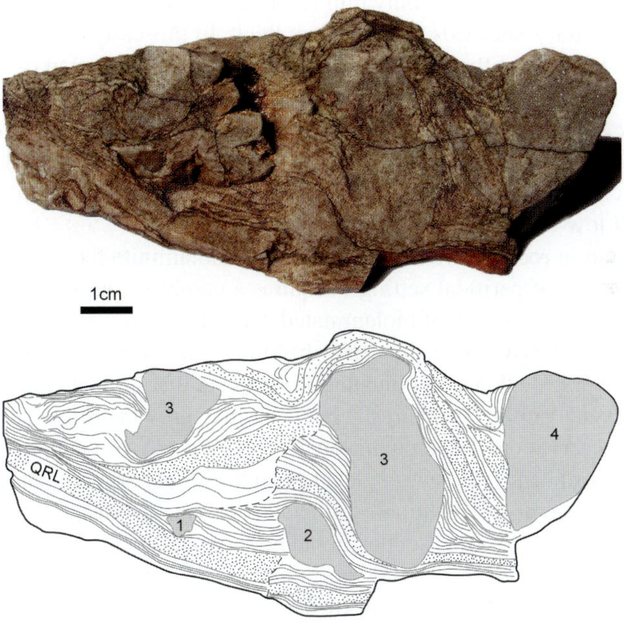

Fig. 12 (**a**) Planar biolaminites cross-cut by several generations (1–4) of V-shaped cracks. The structure indicates alternating stages of biolaminite accretion and subaerial desiccation in an intertidal setting. Depth and size of lenticular cracks are variable. Note the effect of compaction and the occurrence of comparatively thick quartz-rich laminae (QRL), less than 0.5 cm thick. Neoproterozoic of Anti-Atlas, Morocco. (**b**) Graphic rendition of (**a**)

intertidal environment, where continuous/episodic mat growth and shrinkage is closely accompanying cyclic accretion of mat-sediment doublets; (3) processes involving continuous and repeated stages of shrinkage, upturning and overgrowth until the margin structure is completely sealed by new and continuous mat layers.

In the recent example of Tunisia, topographic factors in the tidal flats appear as a significant control on biolaminite distribution and accretion. Continuous growth and accretion of siliciclastic biolaminites reflect a stable system with low rate of sediment supply and no major disturbance by high-energy events, punctuated by periodic flooding during high tides. Such conditions are met locally in shallow depressions and ponds of variable size in the intertidal zone, where they are filled periodically by sea water during high tides and where "polygonal mats" and accretion of thick biolaminites are prevailing.

In the example of the Neoproterozoic of Anti-Atlas (Morocco), dm- to m-thick planar and domal siliciclastic biolaminites, deposited in relatively protected areas, are clearly differentiated from adjacent and associated rock bodies and form thus a distinctive "biolaminite facies" within a lagoonal and pertidal facies association (Bouougri and Porada 2007d). The subtidal shallow lagoonal biolaminates are mainly domal, whereas the planar type often occurs at the base of domal forms and lacks any features indicating subaerial shrinkage. In the heterolithic facies which was deposited in an intertidal area exposed intermittently to high-energy events, biolaminites constitute only thin layers in the facies and occur as laminated organo-sedimentary interbeds up to few cm thick. In this case and according to the sedimentary features of the heteroliths, the thickness and accretion of biolaminites apparently was controlled by the frequency of periodic and aperiodic sedimentary high-energy events depositing mm- to cm-thick sand layers. A similar control was observed in the modern biolaminites of Mellum Island (Gerdes and Klenke 2007), in which accretion and thickness of biolaminites is correlated with the duration of periods with low-energy and low-rate sedimentation.

Recognition in rock records of a "siliciclastic biolaminite facies" and "biogenic bedding", mainly of peritidal settings, requires a combination of macroscopic and microscopic features typical of biolaminated deposits. The examples of siliciclastic biolaminites described above, in which bedding and accretion are driven by the interplay of mat growth and sedimentation, indicate that both ancient and modern examples reveal similar structures related to mat growth-shrinkage-overgrowth processes. Preservation and recognition in rock records of such structures may constitute a significant sedimentological tool and strongly argues for a biosedimentary origin of ancient well laminated siliciclastic deposits and for an intertidal to lower supratidal (i.e. peritidal) depositional environment periodically emerged and regularly flooded by tides or during storm events. Within facies associations containing siliciclastic biolaminites, occurrences of other "mat-related structures" constitute further evidence for the biogenicity of these deposits.

Acknowledgments The authors are very grateful to VW-Foundation (I/78 706) for financial support of this research work. Special thanks to all persons who helped us during fieldworks in Namibia, Tunisia and Morocco. Review by J. Reitner greatly improved this manuscript.

References

Black M (1933) The algal sediments of Andros Islands, Bahamas. Philosophical Transactions of the Royal Society of London. Series B 222:165–192

Bouougri EH, Porada H (2002) Mat-related sedimentary structures in Neoproterozoic peritidal passive margin deposits of the West African Craton (Anti-Atlas, Morocco). Sedimentary Geology 153:85–106

Bouougri EH, Porada H (2007a) Complex structures associated with siliciclastic biolaminites. In: Schieber J, Bose PK, Eriksson PG, Banerjee S, Sarkar S, Altermann W, Catuneanu O (eds) An Atlas of microbial mat features preserved within the siliciclastic rock record. Elsevier, Amsterdam, pp 111–115

Bouougri EH, Porada H (2007b) Siliciclastic biolaminites indicative of widespread microbial mats in the Neoproterozoic Nama Group of Namibia. Journal of African Earth Sciences 48:38–48

Bouougri EH, Porada H (2007c) Mat-related features from the Terminal Ediacaran Nudaus Formation, Nama Group, Namibia. In: Schieber J, Bose PK, Eriksson PG, Banerjee S, Sarkar S, Altermann W, Catuneanu O (eds) An Atlas of microbial mat features preserved within the siliciclastic rock record. Elsevier, Amsterdam, pp 214–221

Bouougri EH, Porada H (2007d) Mat-related features from the Neoproterozoic Tizi n-Taghatine Group, Anti-Atlas belt, Morocco. In: Schieber J, Bose PK, Eriksson PG, Banerjee S, Sarkar S, Altermann W, Catuneanu O (eds) An Atlas of microbial mat features preserved within the siliciclastic rock record. Elsevier, Amsterdam, pp 198–207

Burne RV, Moore LS (1987) Microbialites: organosedimentary deposits of benthic microbial communities. Palaios 2:241–254

Gehling JG (1999) Microbial mats in Terminal Proterozoic siliciclastics: Ediacaran death masks. Palaios 14:40–57

Gerdes G, Klenke T (2003) Geologische Bedeutung ökologischer Zeiträume in biogener Schichtung (Mikrobenmatten, potentielle Stromatolithe). Mitteilungen der Gesellschaft der Geologie- und Bergbaustudenten in Österreich 46:35–49

Gerdes G, Klenke T (2007) States of biogenic bedding as records of the interplay of ecologic time and environment: a case study of modern siliciclastic sediments, Mellum Island, southern North Sea. Senckenbergiana maritima 37:129–144

Gerdes G, Krumbein WE (1987) Biolaminated deposits. In: Bhattacharya GM, Friedmann GM, Neugebauer HJ, Seilacher A (eds) Lecture notes in earth sciences 9. Springer-Verlag, Berlin, pp 1–183

Gerdes G, Krumbein WE, Reineck HE (1991) Biolaminations-Ecological versus depositional dynamics. In: Einsele G, Ricken W, Seilacher A (eds) Cycles and events in stratigraphy. Springer Verlag, Berlin, pp 592–607

Gerdes G, Klenke T, Noffke N (2000) Microbial signatures in peritidal siliciclastic sediments: a catalogue. Sedimentology 47:279–308

Gerdes G, Porada H, Bouougri EH (2008) Bio-sedimentary structures evolving from the interaction of microbial mats, burrowing beetles and the physical environment of Tunisian coastal sabkhas. Senckenbergiana Maritima 38:45–58

Germs GBJ (1983) Implications of a sedimentary facies and depositional environmental analysis of the Nama Group in South West Africa/Namibia. Geological Society of South Africa, Special Publication 11:89–114

Grotzinger JP, Bowring SA, Saylor BZ, Kaufman AJ (1995) New biostratigraphic and geochronologic constraints on early animal evolution. Science 270:598–604

Horodyski RJ, Bloeser B, Vonder Haar S (1977) Laminated algal mats from a coastal lagoon, Laguna Mormona, Baja California, Mexico. Journal of Sedimentary Petrology 47:680–696

Hsü K, Siegenthaler C (1969) Preliminary experiments on hydrodynamic movement induced by evaporation and their bearing on the dolomite problem. Sedimentology 12:11–25

Kendall CGSC, Skipwith PAd'E (1969) Geomorphology of a recent shallow-water carbonate province, Khor Al Bazam, Trucial Coast, Southwest Persian Gulf. The Geological Society of America Bulletin 80:865–892

Logan BW, Rezak R, Ginsburg RN (1964) Classification and environmental significance of algal stromatolites. Journal of Geology 72:68–83

Logan BW, Hoffman P, Gebelein CD (1974) Algal mats, cryptalgal fabrics, and structures, Hamelin Pool, Western Australia. American Association of Petroleum Geologists Memoir 22:140–194

Noffke N, Gerdes G, Klenke T, Krumbein W (2001) Microbially induced sedimentary structures indicating climatological, hydrological and depositional conditions within Recent and Pleistocene coastal facies zones (Southern Tunesia). Facies 44:23–30

Park RK (1977) The preservation potential of some recent stromatolites. Sedimentology 24: 485–506

Pettijohn FJ (1975) Sedimentary rocks, 3rd edition. Harper & Row, New York, 628

Pettijohn FJ, Potter PE (1964) Atlas and glossary of primary sedimentary structures. Springer-Verlag, Berlin

Porada H, Bouougri EH, Ghergut J (2007) Hydraulic conditions and mat-related structures in tidal flats and coastal sabkhas. In: Schieber J, Bose PK, Eriksson PG, Banerjee S, Sarkar S, Altermann W, Catuneanu O (eds) An Atlas of microbial mat features preserved within the siliciclastic rock record. Elsevier, Amsterdam, pp 258–265

Porada H, Ghergut J, Bouougri EH (2008) Kinneyia-type wrinkle structures – critical review and model of formation. Palaios 23:63–77

Pratt BR (1982) Stromatolite decline – a reconsideration. Geology 10:512–515

Riding R (1991) Classification of microbial carbonates. In: Riding R (ed) Calcareous algae and stromatolites. Springer-Verlag, Berlin, pp 21–51

Riding R (2000) Microbial carbonates: The geological record of calcified bacterial-algal mats and biofilms. Sedimentology 47:179–214

Schieber J (2004) Microbial mats in the siliciclastic rock record: a summary of diagnostic features. In: Eriksson PG, Altermann W, Nelson D, Mueller WU, Catuneanu O, Strand K (eds) The Precambrian earth: tempos and events. Developments in Precambrian geology. Elsevier, Amsterdam, pp 663–672

Microbial Binding as a Probable Cause of Taphonomic Variability of Vendian Fossils: Carbonate Casting?

Ekaterina A. Serezhnikova

1 Introduction

The exceptional preservation of Ediacaran organisms is problematic, and the cause of the unique conservation is difficult to prove for fossil material because of restrictions imposed by the incomplete palaeontological record. Observation of the various preservation forms is particularly important and provides the basis for further reconstructions of organisms and the Precambrian community structure.

2 Short Historical Background

Studies of Precambrian organisms have repeated, to a certain extent, the following stages of the history of palaeontology: the organic origin of the first finds was doubted; the views on their systematic position were not once changed; and the usual transition along the stages of knowledge from sporadic and ambiguous finds to rival hypotheses, is observed. It appears essential to dwell on the history of taphonomic studies of Ediacaran fossils because it is responsible for the description of various taphomorphs, such as how the major part of so-called jellyfish, which is the most common representative of Late Proterozoic fossil communities, were reinterpreted as benthic organisms. The circular imprints closely resemble Solnhofen jellyfish; for instance, *Ediacaria flindersi*, which is one of the first Ediacaran finds, is similar to the Jurassic *Rhizostomites admirandus*. Subsequently it became a tradition to arrange the Ediacaran organisms within the existing phyla, to consider the majority of them as jellyfish and to consider the Late Precambrian to be the time

E.A. Serezhnikova
Borissiak Paleontological Institute, Russian Academy of Sciences, Profsoyuznaya 123, Moscow 117997, Russia
e-mail: serezhnikova@paleo.ru

J. Reitner et al., *Advances in Stromatolite Geobiology*,
Lecture Notes in Earth Sciences 131, DOI 10.1007/978-3-642-10415-2_31,

of Coelenterates (e.g., Glaessner 1984). After the first Precambrian taxa were described, it also was evident that such ancient organisms could not be reconstructed without insight into their fossilisation processes (e.g., Sprigg 1947). Wade (1968) was the first to carry out a detailed taphonomic study on fossils and to describe the schemes of their burial. Fedonkin (1985) was the first to hypothesise multiple models for Vendian taphonomy and claimed that the processes of burial and fossilisation of the organism must have been controlled by a number of biotic and abiotic factors. The most famous explanation of Ediacaran organism fossilisation was argued by Gehling (1999) who proposed that the imprints are of unique nature because of their "death mask", which appeared during early diagenesis of a fine sole layer with bacteria-deposited iron minerals. Much earlier, a similar description for botanic objects was substantiated by Krishtophovich (1957). However, Gehling (1999) reported the rare occurrence of iron mineralisation on the surface of Ediacaran fossils; framboids and isolated crystals of pyrite found on top of some Vendian imprints from the White Sea (Dzik 2003) seem to be an exception to the rule. A thin haematite/clay coating on the hyporelief surfaces of attachment disks from the Amadeus Basin, Australia, might be regarded as indirect evidence of an "iron death mask" (Mapstone and McIlroy 2006).

3 Material and Methods

In studies on taphonomic variability of palaeobotanic objects, the most reliable method has been the systematic collection of monotype samples from the same layer, and arranging them as a "monotopic series" (Meyen 1966). This method allows the establishment of relations between individual taphomorphs without distinguishing new taxa and was first applied by Harris (1932) to Triassic–Jurassic deposits in Greenland. A similar approach to Vendian material from the Khatyspyt Formation of the Siberian platform was first applied by Vodanjuk (1989) to a great number of *Hiemalora* specimens from different layers. The study on the material from the White Sea Region using this method became possible because of wide-scale excavations, layer by layer, carried out regularly by A.Yu. Ivantsov from the Laboratory of Precambrian Organisms headed by academician M.A. Fedonkin (PIN RAS). In the White Sea Region sections, various taphomorphs of identical organisms occur on the surface of the same layer, which is of particular value for palaeontological reconstructions.

4 Taphonomic Variability of Vendian Fossils

A considerable taphonomic variability of Vendian organisms was observed in representative *Hiemalora* samples from the Khatyspyt Formation of the North-Eastern Siberian Platform (Vodanjuk 1989; Serezhnikova 2005). The finds united under this generic name are also known from various other Vendian localities, but do not necessarily represent the same group. Forms like *Medusina filamentis* Sprigg 1949, *Mawsonites spriggi* Glaessner et Wade 1966, *Eoporpita medusa* Wade 1972, *Hiemalora stellaris* Fedonkin 1982, and *Hiemalora pleiomorphus* Vodanjuk 1989, comprise a problematic group of organisms with radial structures and without any other special diagnostic features. They were interpreted, at different times over the years, as jellyfish or medusoids, as remains of problematic animals, as some peculiar trace fossils, organs of attachment of some sea bottom organisms, or as remains of protozoa close to recent Xenophyophoria foraminifers. According to the results of the studies on the Siberian *Hiemalora* collection, similar remains were tentatively interpreted as attachment disks of an organism of unclear systematic position, probably on the level of sponge or coelenterate. It was proposed that when these organisms were alive, their radial structures were in the sediment piercing its different horizons (Serezhnikova 2005). Recently, feather-like imprints with diagnostic features similar to the basal part of *Hiemalora* genus were found in the Upper Precambrian Conception and Saint-Jones Groups of Newfoundland Island (Hofmann et al. 2008). Drawings of *Hiemalora* from the Siberian material together with a diagram of specimen distribution, according to the ratio of central part diameters and length of radial structures, suggest that in fossilisation only one projection, namely a conical body section, is preserved. This may account for the considerable taphonomic variability of the organisms.

Taphonomic variety of Vendian fossils has also been observed in *Fedomia* samples from the Solza River fossil locality in the White Sea Region. *Fedomia mikhaili* Serezhnikova and Ivantsov 2007, might be regarded as one of few Ediacaran taxa of sponge grade. The general morphological pattern of the organism can be reconstructed as a sack-like, thin-walled body attached at its base to the substrate; its star-shaped structures may be irregular spicules of organic nature. *Fedomia mikhaili* is mainly preserved in the three following taphomorphs: (1) flat imprint with rare negative star-shaped structures with its surface divided by a system of narrow depressions and longitudinal subparallel furrows; (2) nearly flat imprints with the surface bearing a system of negative star-shaped structures in the form of network; (3) folded imprints with star-shaped structures of positive relief. This variability of preservational forms seems to have resulted from the three-dimensional nature of *Fedomia,* which was destroyed during fossilisation. Thus, the three taphomorphs of *Fedomia* described above formed by post-mortem changes likely reflecting different stages of decay. This may imply that some rather voluminous (three-dimensional) soft bodied fossils were preserved in the form of flat imprints, i.e., as random "projections". It is quite evident that if an organism is complicated, variants of its preservation will occur. Indeed, the study of the monotopic series of the attachment disks of *Palaeophragmodictya* has shown that the three-dimensional Ediacaran organisms may be preserved as various random horizontal sections, which when combined, may give an idea of the organism's morphology (Figs. 1 to 3). Considerable differences in the structure of the imprints are observed, in spite of their apparent similarity. If the samples presented here (Fig. 3a–e) were

Fig. 1 Various preservational forms of *Palaeophragmodictya spinosa* Serezhnikova, 2007, Arkhangelsk Region, Onega Peninsula, Solza River, Upper Vendian, Verkhovka Formation: 1-PIN, N 4853/425, 2-PIN, N 4853/410, 3-PIN, N 4853/424, 4-PIN, N 4853/415, 5-PIN, N 4853/416, 6-PIN, N 4853/421. Scale 1 cm

collected as isolated finds, they may well have been considered to belong to several genera of circular organisms.

Previously, when new material was accumulated, researchers attempted to distinguish as many as possible taxa of species and genus range. Now, a different trend can be observed in the systematics, namely, a copious reduction to synonyms.

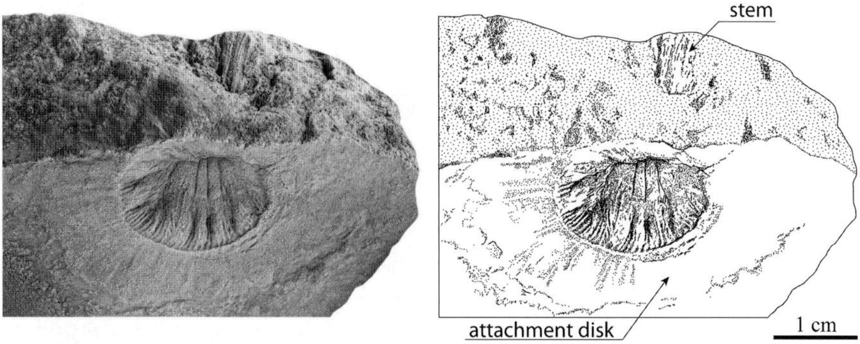

Fig. 2 *Palaeophragmodictya spinosa* Serezhnikova 2007, Arkhangelsk Region, Onega Peninsula, Solza River, Upper Vendian, Verkhovka Formation; PIN, 4853/418

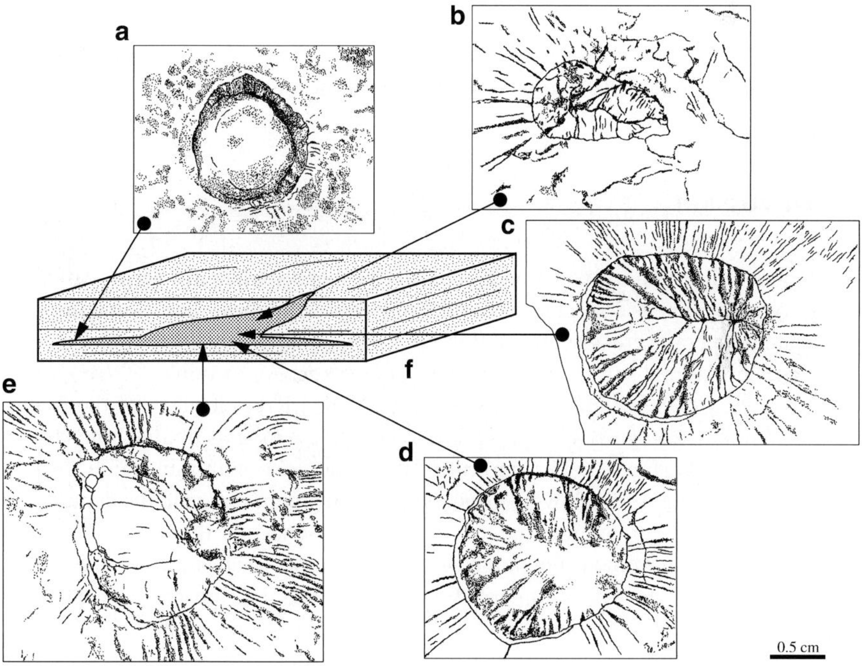

Fig. 3 Various preservational forms of *Palaeophragmodictya spinosa* Serezhnikova 2007, Arkhangelsk Region, Onega Peninsula, Solza River, Upper Vendian, Verkhovka Formation: (**a**) PIN, N 4853/425, (**b**) PIN, N 4853/412, (**c**) PIN, N 4853/416, (**d**) PIN, N 4853/415, (**e**) PIN, N 4853/424, (**f**) schematic reconstruction of the fossilised part: scale 0.5 cm [after Serezhnikova (2007), modified]

As is seen from the analysis of the *Palaeophragmodictya* monotope series from the White Sea, this unification may be justified in some cases, but generally it may be too early for application as detailed collections from the same layers are widely lacking. Furthermore, it may be expected that monotope series of different taxa may contain similar forms. As a result of the study on the imprints from the Solza locality, *Palaeophragmodictya* were suggested to be interpreted as parts of a problematic organism with attachment disks. It was considered previously to be the first Precambrian sponge represented by macro-remains, and Gehling and Rigby (1996) included it in the recent Hexactinellidae family. In spite of the fact that *Paleophragmodictya* has morphological features inherent in sponges, e.g., a net on the disk surface and spiny radial structures, the general morphology aligns them more closely to Coelenterate; although a sporadic appearance of radial elements considerably decreases the likeliness of this affinity (Serezhnikova 2007). Apparently, *Palaeophragmodictya* from the White Sea Region was the first Precambrian sedentary organism in which bilateral symmetry was detected in ontogeny (Serezhnikova 2007). The absence of a regular pattern in the appearance of radial elements somewhat reduces the likelihood of their affinity to Coelenterata, or it should be accepted that, in the Vendian, the Coelenterata had not yet acquired regular patterns that were present in later epochs. On the whole, the recognition of various taphomorphs of Precambrian organisms allows the assessment of the structure of the organisms and the possible succession of fossilisation events.

5 Discussion

The taphonomic variability of Vendian fossils has been studied in a great number of single-type specimens collected from the same depositional surface. Some imprints proved to be specific cross-sections of three-dimensional organisms preserved on an accidental horizontal surface of the fossiliferous layer. For example, *Palaeophragmodictya spinosa* Serezhnikova 2007 from the Solza River locality of the South-eastern White Sea Region were preserved as composite moulds of attachment disks and part of stems (Fig. 2). The imprint of the outer zone with thin radial furrows reflects the structure of the lower surface of the attachment disc, and the central part is a negative cast of an internal cavity. The presence of the cavity over the organism's entire lifetime is documented by the rare specimens in which the central parts have been filled with sediment (Fig. 3a). The organic substance of the inner cavity had not decayed completely at the time of lithification, and therefore various preservational forms of the central parts of the fossils developed. The specimen with a negative outer zone and mesh-like structures is unique to this locality; it appears to be an external mould of the upper surface of the attachment disc (Fig. 3a). Apparently, the variable forms of preservation which have been observed, for instance, in *Palaeophragmodictya* from the Solza River locality, give an idea of the fossilisation processes. It seems rather improbable that the various taphomorphs of the voluminous organism were preserved only due to formation of an "iron death

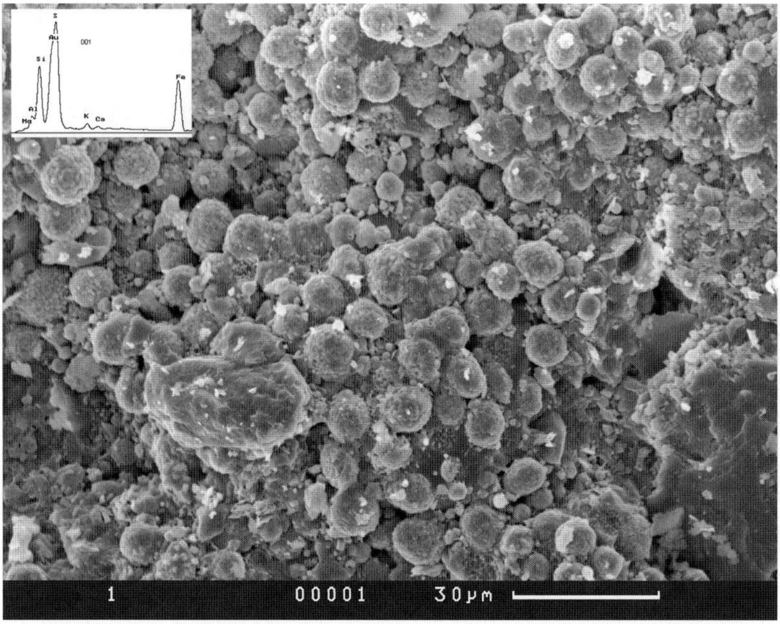

Fig. 4 SEM photograph of the surface microrelief of *Palaeophragmodictya spinosa* with pyrite framboids; qualitative elemental composition is in the upper left corner. Arkhangelsk Region, Onega Peninsula, Solza River, Upper Vendian, Verkhovka Formation

mask" on the surfaces of the collapsing bodies, though framboids and isolated crystals of pyrite do occur occasionally on the surfaces of fossil imprints (Figs. 4 and 5). Such unique fossilisation might have been also the result of a quite rapid consolidation of a thin fossiliferous layer due to cementation by carbonate. For example, carbonate accounts for 1–3% in the sandstone, but for 32% in some siltstone from the fossiliferous Neoproterozoic Arumbera Sandstone, Australia (Mapstone and McIlroy 2006), and for 10–15% in the siltstone from the Solza River locality in the Southeastern White Sea Region. The data on qualitative elemental composition of the bedding surface bearing *Palaeophragmodictya spinosa* show the presence of calcium and carbon (Fig. 6). Furthermore, it has recently been reported that calcium carbonate is found in cyanobacterial films that develop on modern beaches of the North Sea barrier island Schiermonnikoog, Netherlands (Kremer et al. 2008). It may be speculated that a similar process may have occurred in widespread microbial mat grounds of Precambrian environments (Fig. 7); but the carbonate grains might have been completely dissolved over time. According to experimental data, the dissolution of different rocks has different rates even during the same period of time (Lasaga 1983), e.g., carbonates dissolve quite rapidly in contrast to silicates. It is also well known that under anoxic conditions in modern settings, carbonate shells are destroyed over a short time interval. Hydrosulphuric contamination of the sediment, traces of which are observed in the Vendian sediments of the South-Eastern White Sea Region, seem to have been typical of

Fig. 5 SEM photograph of the surface microrelief of *Palaeophragmodictya spinosa* with prints of pyrite framboids on clay mineral; qualitative elemental composition (q.e.c.) is in the upper left corner. Arkhangelsk Region, Onega Peninsula, Solza River, Upper Vendian, Verkhovka Formation

C	O	Mg	Al	Si	S	K	Ca	Fe	Total
7,48	43,12	0,57	2,64	7,97	4,57	0,77	1,09	31,80	100,00
	33,55	0,77	3,56	11,05	7,06	1,41	1,80	40,80	100,00
9,57	30,20		4,72	13,29	4,71	3,30	1,83	32,38	100,00
	38,12		4,16	20,51	3,76	2,33	1,73	29,39	100,00
	54,10		0,94	41,94		0,80		2,22	100,00

Fig. 6 X-Ray data (EM EV050) on qualitative elemental composition of the sole surface of the layer bearing *Palaeophragmodictya spinosa* Serezhnikova 2007, Arkhangelsk Region, Onega Peninsula, Solza River, Upper Vendian, Verkhovka Formation

microbial mat grounds in Precambrian sedimentary environments (Fig. 8) (Seilacher 1999; Dzik 2003). For this reason it cannot be ruled out that the original carbonate content of the Precambrian fossiliferous layers was much higher at the time of the sedimentation and early diagenesis. Thus, the exceptional preservation of Vendian fossils could also be explained by a rapid cementation of a fossiliferous layer by microbially precipitated carbonate, though iron precipitation on the decaying tissues may have occurred as well.

Fig. 7 Prints of mat-related structures on fossiliferous bedding surface; Arkhangelsk Region, Zimnii Gory locality, Upper Vendian. The diameter of coin is about 2 cm

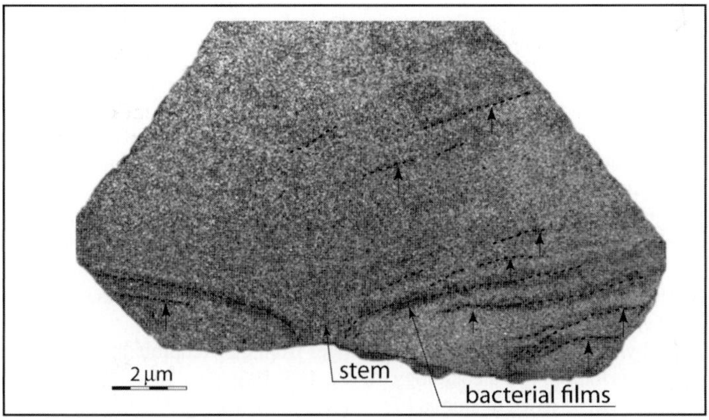

Fig. 8 Cross-section of the specimen *Palaeophragmodictya spinosa* Serezhnikova 2007, Arkhangelsk Region, Onega Peninsula, Solza River, Upper Vendian, Verkhovka Formation.(PIN, N 4853/636). *Arrows* indicate bacterial films

6 Conclusion

Detailed study of a monotopic series, with the representative samples of specimens collected from a single fossiliferous layer, has confirmed the notion of a great taphonomic variety of Vendian fossils, where the apparent diversity of fossils in some cases actually reflects taphomorph diversity rather than species diversity within a palaeocommunity. Some Vendian three-dimensional organisms such as *Hiemalora*, *Fedomia*, *Palaeophragmodictya*, etc. may be preserved as random horizontal sections on the surfaces of fossiliferous layers. Hence, the preservation variety caused by post-mortem changes appears to reflect different stages of decay of rather voluminous organisms. A quite rapid consolidation of fossiliferous layers due to carbonate cementation may be a possible explanation of such unique selective fossilisation. Thus, the recognition of various taphomorphs of Precambrian macro-organisms allows us, not only to reconstruct the morphology of the organisms, but also to hypothesise a probable scenario of their fossilisation. Obviously, it is necessary to carry out geochemical research focused on the analysis of elemental composition of the fossils and the fossil-bearing rock for more convincing proof.

Acknowledgements This work was supported in part by the Russian Foundation of Basic Research, projects nos. 08-05-00801, 08-05-90211, the Scientific School of Academicians B.S. Sokolov and M.A. Fedonkin, project no. NSH-4156.2008.5 and the Program N 18 of the Presidium of RAS. I am grateful to Academician M.A. Fedonkin and Dr. A.Yu. Ivantsov for the access to the collections and useful comments, to Dr. A.G. Ponomarenko for polite discussion, to A.V. Mazin (PIN RAS) and Yu.V. Shuvalova for the photographic works, to L.V. Zaytseva and E.A. Zhegallo (PIN RAS) for the fine-structural investigation, to N.I. Krasnova for improving the language of the manuscript and to the anonymous reviewer for critical reading and very constructive review.

References

Dzik J (2003) Anatomical information content in the Ediacaran fossils and their possible zoological affinities. Integrative and Comparative Biology 43: 114–126

Fedonkin MA (1985) Precambrian metazoans: the problems of preservation, systematics and evolution. Philosophical Transactions of the Royal Society of London, Series B 311 (1148): 27–45

Gehling JG (1999) Microbial mats in terminal Proterozoic siliciclastics: Ediacaran death masks. Palaios 14 (1): 40–57

Gehling JG, Rigby JK (1996) Long expected sponges from the Neoproterozoic Ediacara fauna of South Australia. Journal of Paleontology 70 (2): 185–195

Glaessner MF (1984) The dawn of animal life: A biohistorical study. Cambridge University Press, Cambridge, 224 pp

Harris TM (1932) The fossil flora of Scoresby Sound, East Greenland. Description of seed plants incertae sedis, together with a discussion of certain cycadophyte cuticles. Medd om Grønland 85 (5): 1–133

Hofmann HJ, O'Brien SJ, King AF (2008) Ediacaran biota on Bonavista peninsula, Newfoundland, Canada. Journal of Paleontology 82 (1): 1–36

Kremer B, Kazmierczak J, Stal LJ (2008) Calcium carbonate precipitation in cyanobacterial mats from sandy tidal flats of the North Sea. Geobiology 6: 46–56

Krishtophovich AN (1957) Palaeobotany. Gostoptekhizdat, Leningrad, 650 pp, (in Russian)

Lasaga AC (1983) Kinetics of silicate dissolution. Fourth international symposium on water-rock interaction. Misasa, Japan, pp 269–274

Mapstone NB, McIlroy D (2006) Ediacaran fossil preservation: Taphonomy and diagenesis of a discoid biota from the Amadeus Basin, central Australia. Precambrian Research 149: 126–148

Meyen SV (1966) [Cordaites from the Upper Palaeozoic of northern Eurasia (Morphology, epidermal structure, taxonomy and stratigraphic significance)]. Nauka, Moscow, 184 pp, (in Russian)

Seilacher A (1999) Biomat-related lifestyles in the Precambrian. Palaios 14: 86–93

Serezhnikova EA (2005) [New interpretation of Hiemalora imprints from the Vendian of the Olenek Uplift, northeast Siberian Platform]. Bulletin MOIP. Otdel Geologicheskii 80 (3): 26–32

Serezhnikova EA (2007) Palaeophragmodictya spinosa sp. nov., a Bilateral Benthic Organism from the Vendian of the Southeastern White Sea Region. Paleontological Journal 41 (4): 360–369 (Original Russian Text published in Paleontologicheskii Zhurnal (2007)) 4: 16–24

Serezhnikova EA, Ivantsov AYu (2007) Fedomia mikhaili – A new spicule-bearing organism of sponge grade from the Vendian (Ediacaran) of the White Sea, Russia. Palaeoworld 16: 319–324

Sprigg RC (1947) Early Cambrian? jellyfishes from the Flinders Ranges, South Australia. Transactions of the Royal Society of South Australia 71: 221–224

Vodanjuk SA (1989) [Remains of non-skeletal Metazoa from Khatyspyt formation on the Olenek Uplift]. Pozdnii dokembry and rannii paleozoi Sibiri. Actualnii voprosi stratigrafii: 61–75 [in Russian]

Wade M (1968) Preservation of soft bodied animals in Precambrian sandstones at Ediacara, South Australia. Lethaia 1 (3): 238–267

Morphology as an Indictor of Biogenicity for 3.5–3.2 Ga Fossil Stromatolites from the Pilbara Craton, Western Australia

Martin J. Van Kranendonk

1 Introduction

Stromatolites were recognised in 3.5–3.4 Ga rocks from the Pilbara Craton 30 years ago (Walter et al. 1980; Lowe 1980), but their biogenicity has been cast in doubt by recent studies of abiogenic features with gross similarity to biogenic forms (Grotzinger and Rothman 1996; Garcia-Ruiz et al. 2003) and new interpretations of geological settings that differ from original models (see below). This paper reviews the geological setting of Pilbara fossils and assesses stromatolite biogenicity, showing the critical value of stromatolite morphology when used in combination with well documented geological context.

2 Geological Setting of Proposed Fossil Localities

Proposed stromatolite fossils occur within the 3.53–3.165 Ga East Pilbara Terrane (EPT) of the Pilbara Craton, Western Australia (Fig. 1: Van Kranendonk 2007; Van Kranendonk et al. 2007a, b). The EPT consists of four unconformity-bound volcano-sedimentary groups (Pilbara Supergroup) and related ultramafic-mafic and granitic intrusive rocks (Fig. 2), all of which are unconformably overlain by rocks of the 3.07–2.94 Ga De Grey Supergroup, deformed and intruded by 3.07–2.83 Ga granitic rocks, and then unconformably overlain by the 2.78–2.63 Ga Fortescue Group (Van Kranendonk et al. 2007a). The lower three groups of the Pilbara

M.J. Van Kranendonk (✉)
Geological Survey of Western Australia, 100 Plain St., East Perth, WA 6004, Australia
Geographical Sciences, The University of Western Australia, 35 Stirling Hwy, Crawley, WA 6009, Australia
School of Earth and Environment, The University of Western Australia, 35 Stirling Hwy, Crawley, WA 6009, Australia
e-mail: vankranendonk@dmp.wa.gov.au

J. Reitner et al., *Advances in Stromatolite Geobiology*,
Lecture Notes in Earth Sciences 131, DOI 10.1007/978-3-642-10415-2_32,

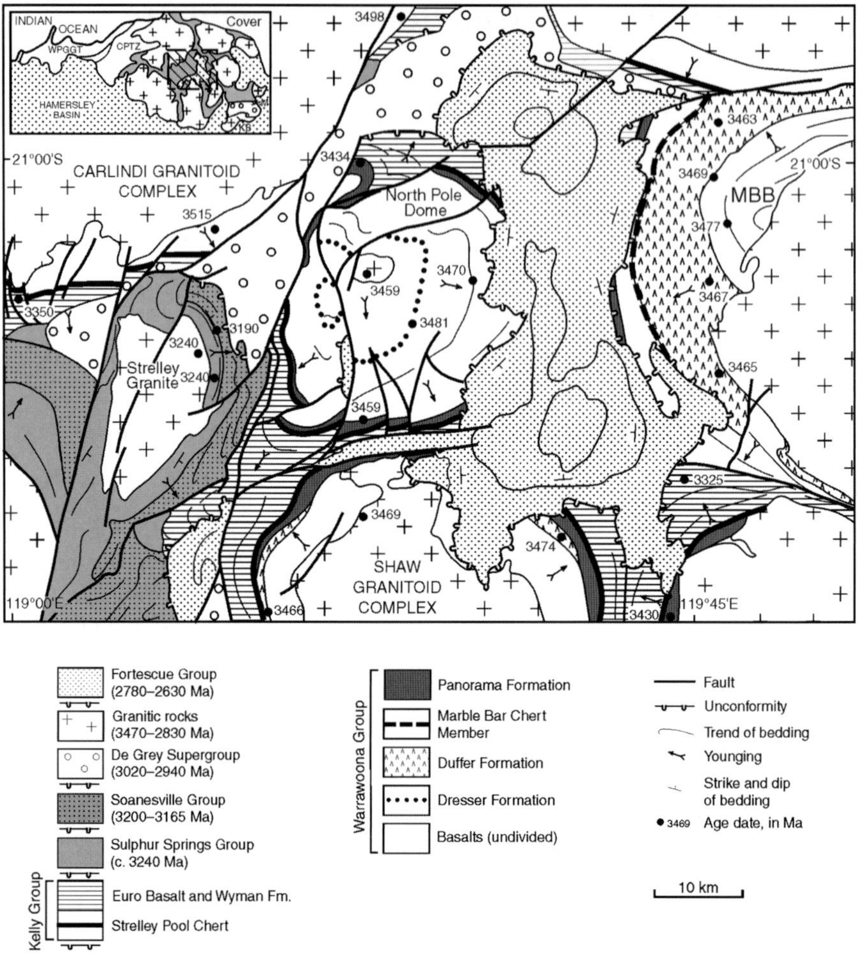

Fig. 1 Geological map of the central East Pilbara Terrane, showing main lithological units and localities mentioned in the text

Supergroup are dominantly volcanic, and were deposited from 3.53 to 3.24 Ga as the result of discrete mantle plume events, whereas the c. 3.2–3.165 Ga Soanesville Group is largely composed of clastic sedimentary rocks and is interpreted to relate to rifting of EPT margins, prior to the onset of terrane accretion at 3.07 Ga (Van Kranendonk et al. 2007a, 2010).

Strong claims for early life have been described from several stratigraphic horizons within the Pilbara Supergroup, including (Fig. 2):

1. c. 3.48 Ga Dresser Formation (diverse stromatolites, putative microfossils, highly negative $\delta^{13}C$ on carbonaceous material, sulfur isotopes) (Walter et al. 1980; Ueno et al. 2004; Van Kranendonk 2006; Philippot et al. 2007)

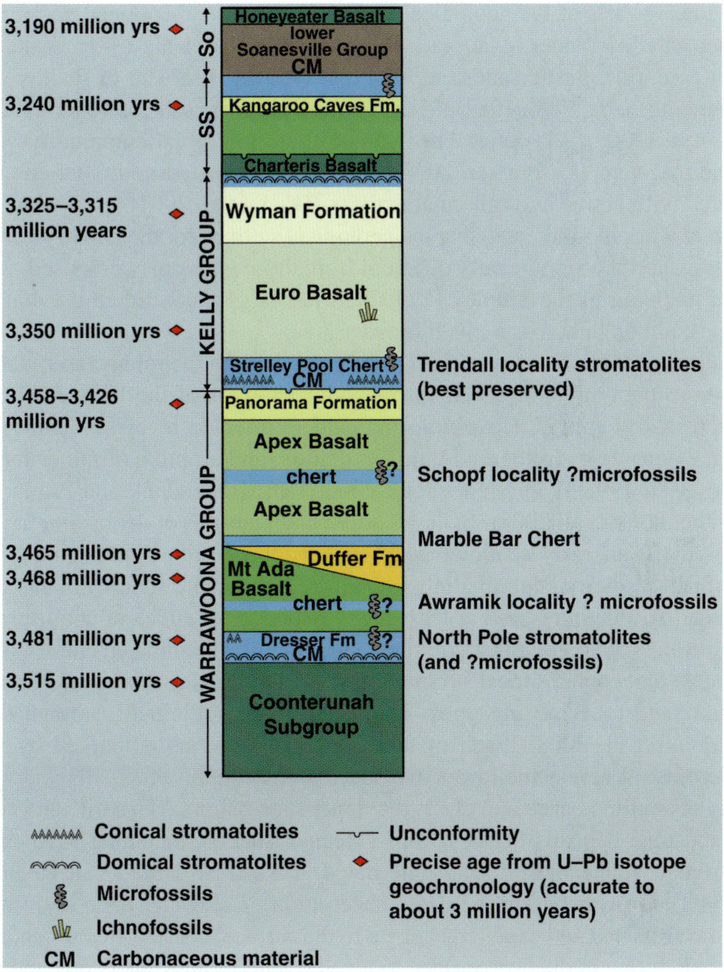

Fig. 2 Stratigraphic column of the Pilbara Supergroup, showing early life localities

2. c. 3.46 Ga Apex chert (putative microfossils, Laser-Raman spectrography) (Schopf 1993; Schopf et al. 2007)
3. c. 3.40 Ga Strelley Pool Chert (diverse stromatolites, Carbon isotope data, and organic geochemistry) (Lowe 1980; Hofmann et al. 1999; Allwood et al. 2006; Van Kranendonk 2006; Marshall et al. 2007)
4. c. 3.35 Ga Euro Basalt (ichnofossils in altered basaltic glass) (Banerjee et al. 2007; Furnes et al. 2007)
5. c. 3.24 Ga Sulphur Springs Group (putative microfossils in sulphides) (Rasmussen 2000)
6. c. 3.20 Ga Soanesville Group (organic matter, highly negative δ^{13}C on carbonaceous material, putative microfossils) (Duck et al. 2007; Glikson et al. 2008)

Previous models of the geological setting of the units mentioned above include (1) quiet, shallow water, evaporative lagoon (Groves et al. 1981; Walter et al. 1980); (2) marine clastic sandstone (Schopf 1993); (3) sabkha to shallow marine environment (Lowe 1980, 1983); (4) submarine pillow basalt (Banerjee et al. 2007; Furnes et al. 2007); (5) active chemoautotrophic microbial community within a hydrothermal vent (Rasmussen 2000); (6) sediment-hosted microbial community associated with active hydrothermal activity (Duck et al. 2007).

Current work has indicated that the geological settings for the Dresser Formation and Apex Chert are significantly different from those previously proposed, whereas detailed mapping of the Strelley Pool Chert has largely confirmed previous interpretations, although across a much broader area.

Detailed geological mapping, in combination with petrology and studies of alteration mineralogy and geochemistry, have shown that the Dresser Formation was deposited within an active volcanic caldera affected by syn-volcanic growth faulting, extensive hydrothermal fluid circulation, and eruption of minor felsic tuff (Nijman et al. 1998; Van Kranendonk 2006; Van Kranendonk and Pirajno 2004; Van Kranendonk et al. 2008). In detail, three main periods of deposition have been recognised (1) an early shallowing up succession of minor basal conglomerate, and laminated siderite, culminating with rippled carbonate sediments that contain widespread stromatolites and at least local evidence for subaerial exposure; (2) a fining upward succession of conglomerate grading up to laminated cherts (silicified carbonates) that contains local stromatolites; (3) a fining up succession of conglomerate and sandstone passing up to laminated cherts (silicified carbonates) (Van Kranendonk et al. 2008). Each of these three successions is marked by a basal unconformity. Whereas the lower two successions contain thick sills and veins of coarsely crystalline barite and chert, the upper succession lacks barite and contains thick sandstone deposits derived from volcanic sources, including a 10 cm thick unit of felsic volcaniclastic sandstone dated at $3{,}481 \pm 2$ Ma (Van Kranendonk et al. 2008). Growth faulting allowed penetration of seawater deep into the crust, where it was heated and expelled back up to the surface, causing steam-heated acid-sulphate alteration of the near-surface footwall basalts (Van Kranendonk and Pirajno 2004). Leaching of Ba, Si and Fe from footwall basalts combined with magmatic S and seawater sulphate to precipitate barite, silica, and sulphides (pyrite and sphalerite) within fossil hydrothermal veins (Fig. 3). The sedimentary units were affected by growth faulting prior to eruption of disconformably overlying high-Mg pillow basalts.

Samples with putative microfossils from the c. 3.46 Ga Apex Chert were previously interpreted to be marine sandstones (Schopf 1993). However, recent mapping and geochemical data revealed that the samples were collected from a fossil hydrothermal breccia vein in the subsurface to bedded sandstone (Brasier et al. 2002, 2005; Van Kranendonk 2006). This, in combination with more highly detailed petrography and geochemistry on the samples, has cast considerable doubt on the biogenicity of the putative microfossils (Brasier et al. 2002, 2005), although other studies continue to support a biogenic origin of these structures (Schopf et al. 2007), and it has been recognised that microfossils

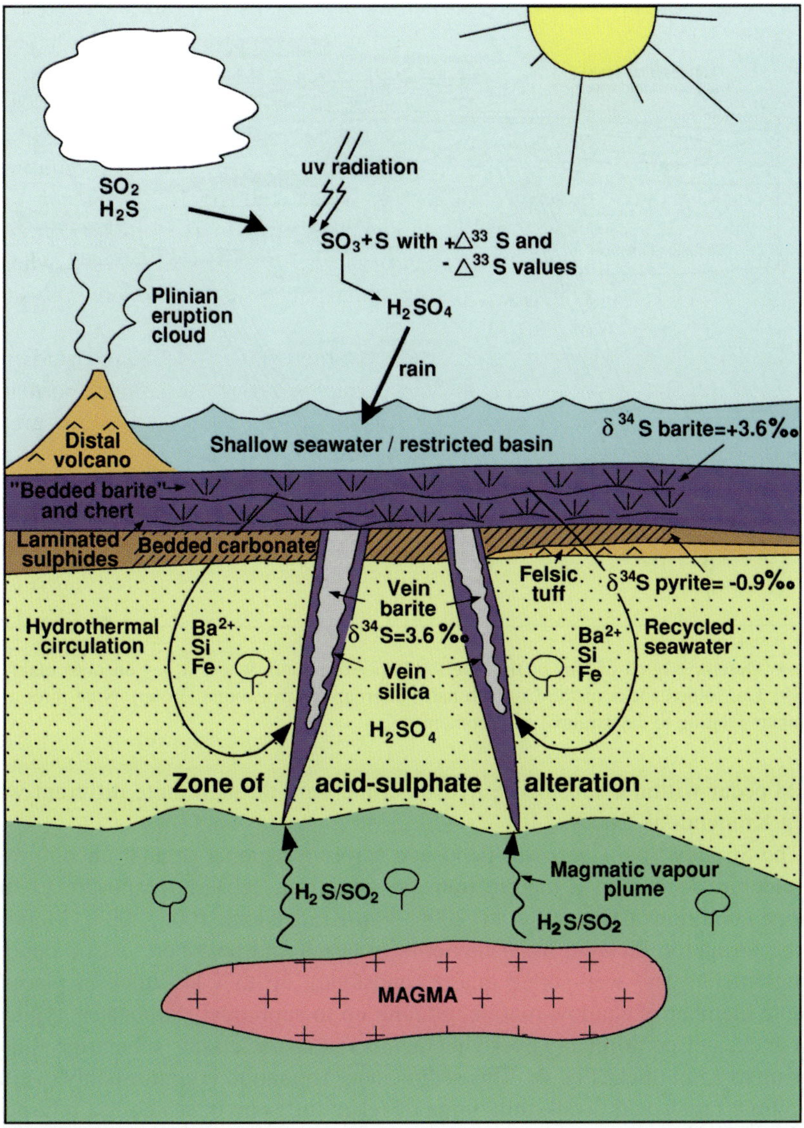

Fig. 3 Dresser Formation model of deposition within an active volcanic caldera affected by hydrothermal fluid circulation. Downward circulating seawater gets heated up and interacts with reducing fluids and volatiles from a degassing magma chamber. These fluids leach Ba, Si and Fe from footwall basalt, mix with seawater containing dissolved sulphate and are then expelled to the surface along syn-volcanic growth faults where they precipitate silica, barite and pyrite (from Van Kranendonk 2006)

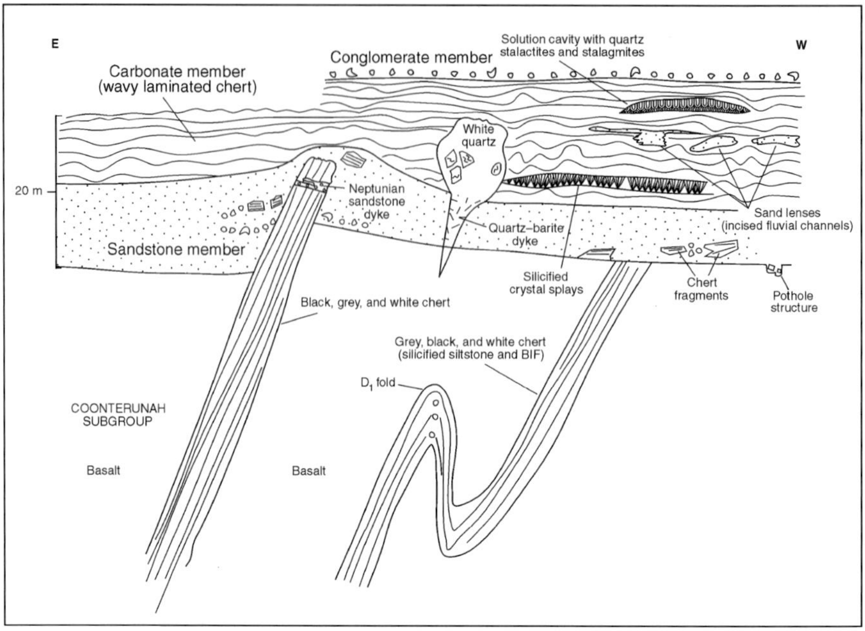

Fig. 4 Outcrop sketch of part of the Strelley Pool Formation, where it lies unconformably on c. 3.51 Ga, folded rocks of the Coonterunah Subgroup (Warrawoona Group). Note three main members relating to depositional environment

could occur within a subsurface hydrothermal vein through three different ways (Van Kranendonk 2007).

The Strelley Pool Chert was deposited across a regional angular unconformity on older rocks during a marine transgression from subaerial exposure to fully marine conditions (Allwood et al. 2006; Van Kranendonk 2006, 2007). Members of the formation include: basal fluviatile to beach and nearshore clastic deposits (conglomerate and well-sorted quartz sandstone), up to 1 km thick in places; a middle member of shallow marine, grading up to supratidal, carbonates, typically 8 m thick; a disconformably overlying member of coarse clastic, submarine alluvial fans up to 15 m thick (Fig. 4). This sedimentary formation is conformably overlain by pillow basalt with local, thin, units of felsic tuff, one of which has returned a U-Pb SHRIMP age of 3,350 ± 2 Ma (Van Kranendonk et al. 2007a).

3 Stromatolite Morphology as an Indicator of Biogenicity

Stromatolites occur at several stratigraphic levels within the Dresser Formation and have diverse forms, even within continuous horizons. Significantly, although most surface outcrops are of chert, rare outcrop and drillcore samples show that the primary material was marine carbonate (ankerite and siderite), replaced by later silica. Stromatolite forms include: coniform, columnar, domical and wrinkly mats in shallow water, rippled marine carbonate sediments (Fig. 5), suggesting possible

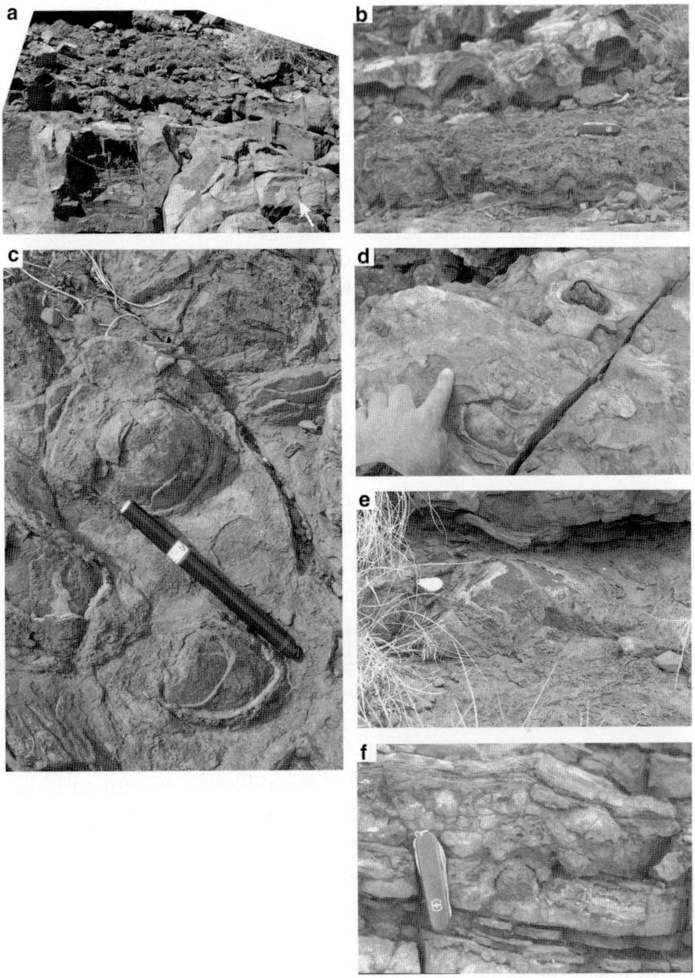

Fig. 5 Stromatolites in bedded carbonate sedimentary rocks of the c. 3.48 Ga Dresser Formation. (**a**) Wrinkly laminated stromatolite mat overlying rippled carbonate sandstone (*arrow*). (**b**) Wrinkly laminated stromatolite mat overlain by domical stromatolites. (**c**) Oblique bedding plane view of coniform stromatolites, which lie within rippled carbonate sandstone. (**d**) Oblique bedding plane view of irregular-shaped stromatolites in silicified, but finely laminated and rippled, carbonate sedimentary rock. (**e**) Oblique bedding plane view of 30-cm diameter domical stromatolite with wrinkly laminated flanks and smooth core. (**f**) Cross-sectional outcrop view of small, domical stromatolite within shallow-dipping carbonate sedimentary rocks that have been partly replaced by barite and silica: note the high amplitude of the column and the flat underlying and overlying strata

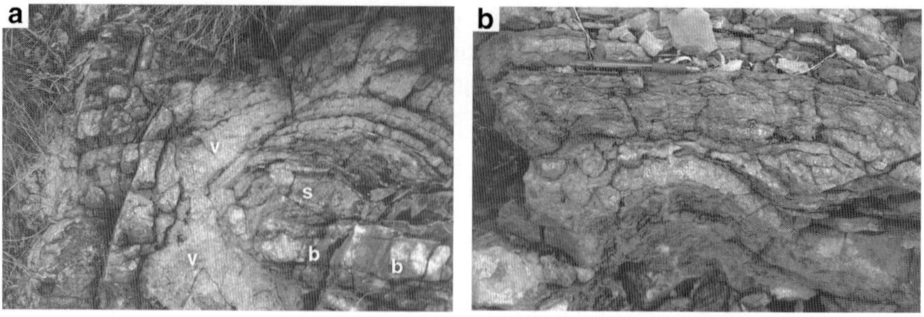

Fig. 6 Stromatolites at, and adjacent to, syn-depositional hydrothermal veins in the c. 3.48 Ga Dresser Formation, Pilbara Craton. (**a**) Low-amplitude domical stromatolite (s) at the mouth of a hydrothermal vein (v), within sedimentary rocks that have been extensively replaced by barite. Note the massive and smooth-weathering nature of vein barite (v) where it cuts bedding (**b**), vs. the more wrinkly laminated character of the barite-replaced sedimentary rocks. (**b**) Large, low-amplitude domical stromatolite, approximately 5 m along strike of (**a**), showing flat bedding overlying stromatolite and extensive replacement of primary sediment by hydrothermal barite (*light grey areas*)

photoautotrophy; domical and wrinkly laminated forms at, and downflow from, hydrothermal vents (Fig. 6), suggesting chemoautotrophy; and wrinkly laminates within some hydrothermal veins, parallel to vein walls, suggesting subsurface chemoautotrophs. Key morphological indicators of stromatolite biogenicity are coniform shapes in rippled carbonate sand (Fig. 5c), wrinkly lamination within coniform stromatolites (Fig. 7a) and laminated mats (Fig. 7b, c), microbially bound ripples in carbonate sand (Fig. 7d), irregular domical forms within 20 cm along strike of coniform varieties (Figs. 5c, d), and the presence of draped sedimentary wedges off domical stromatolite flanks, indicative of stromatolite growth during sedimentation (Van Kranendonk 2006). The occurrence of stromatolites within particulate carbonate sediment indicates the laminations are not carbonate crusts. Stromatolites are truncated by flat pebble conglomerates, indicating formation during sedimentation and not as a result of post-depositional diagenetic effects. The variation along strike from coniform to irregular shaped stromatolites within rippled sands with flat pebble conglomerates, suggests interplay between growth morphology of stromatolites and water currents in very shallow, to supratidal, conditions, in much the same way as in modern day Shark Bay (e.g. Groves et al. 1981). Thin sections show that the morphology of small (<1 cm) columnar stromatolites in pyrite-replaced carbonate sediment is independent of the diagenetic growth of barite crystals (Fig. 8), also countering a post-depositional mode of formation. These structures are characterised by very fine-scale, wavy laminations within relatively high amplitude columns (A in Fig. 8), and by laminae that either terminate at column margins (no growth wall: left of B in Fig. 8), or are bound within a smooth growth wall (C in Fig. 8). Locally, individual columns are joined by bridging laminae (D in Fig. 8). Stromatolite biogenicity is supported by sulphur

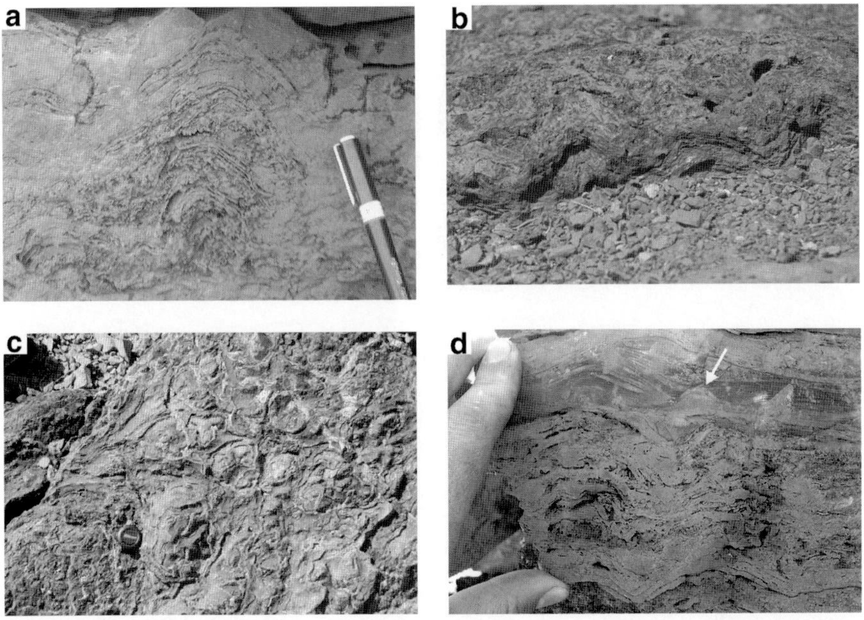

Fig. 7 Key morphological indicators of biogenicity in Dresser Formation stromatolites. (**a**) Cross-sectional outcrop view of fine-scale, wrinkly lamination within coniform stromatolite. (**b**) Cross-sectional outcrop view of wrinkly laminated stromatolite mat, directly overlying rippled carbonate sedimentary rock (width of view ~30 cm). (**c**) Oblique bedding plane view of wrinkly laminated stromatolite mat, directly overlying rippled carbonate sedimentary rock, showing numerous circular forms (lenscap is ~3 cm). (**d**) Cross-sectional outcrop view of wrinkly lamination within coniform stromatolite that is directly overlain by microbial bound rippled carbonate sedimentary rock with a small domical stromatolites (*arrow*)

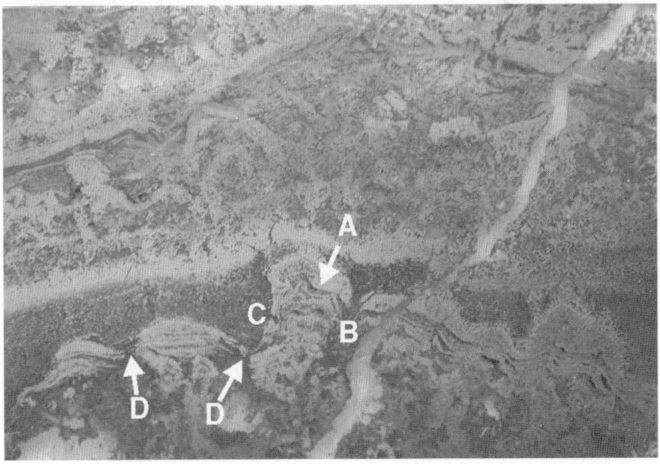

Fig. 8 Plane light thin section view of columnar, pyrite-replaced carbonate stromatolites, from 95.35 m in diamond drillhole PDP2c through the Dresser Formation (width of view ~3 cm: Van Kranendonk et al. 2008). These structures show high amplitude columnar forms in flat-bedded sedimentary rocks that have wrinkly internal laminations (A) and either ragged margins (B) or smooth growth walls (C), with bridging laminae between individual columns (D)

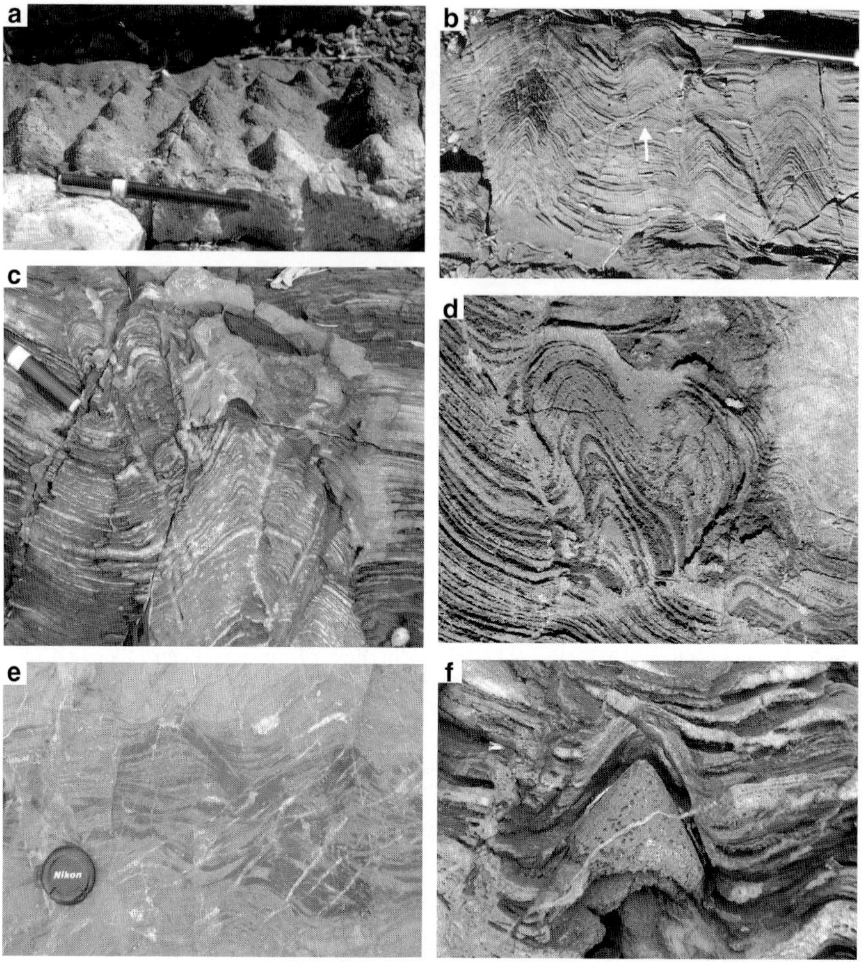

Fig. 9 Stromatolites in the marine carbonate member of the Strelley Pool Chert. (**a**) Oblique bedding plane view of relatively uniform size coniform stromatolites in silicified dolomite. (**b**) Cross-sectional outcrop view of coniform stromatolites in dolomite, showing near-isopachous laminations but varying form, including incipiently branching form on the left: note the start of coniform stromatolite from horizontally laminated dolomite at arrow. (**c**) Cross-sectional outcrop view of coniform stromatolite in partly silica-replaced dolomite, with two branching forms on left-hand side (just to right of pen). (**d**) Cross-sectional outcrop view of small, branched columnar stromatolite in partly silica-replaced dolomite (width of view ~5 cm). (**e**) Cross-sectional outcrop view of microbially-bound ripples in partly silica-replaced carbonate sandstone, which occurs along strike of coniform stromatolites (lenscap ~3 cm). (**f**) Cross-sectional outcrop view of coniform stromatolite, showing non-isopachous laminae and flat overlying bedding (width of cone base ~4 cm)

isotope data, carbon isotope data from clots of carbonaceous material in micritic carbonate, by possible microbial remains and highly negative $\delta^{13}C$ on kerogen and methane in fluid inclusions in hydrothermal veins (Ueno et al. 2001, 2004, 2006; Glikson et al. 2008; Van Zuilen et al. 2008).

Stromatolites in the Strelley Pool Chert have diverse and unique morphologies in different depositional settings: shallow marine dolomites contain conical, incipiently branching conical, and branched conical forms, columnar and branching columnar forms, and wavy laminate mats binding rippled carbonate sand (Fig. 9); supratidal carbonates with low-amplitude, locally windblown, ripples contain low amplitude domes and wrinkly laminated mats (Fig. 10); clastic rocks contain

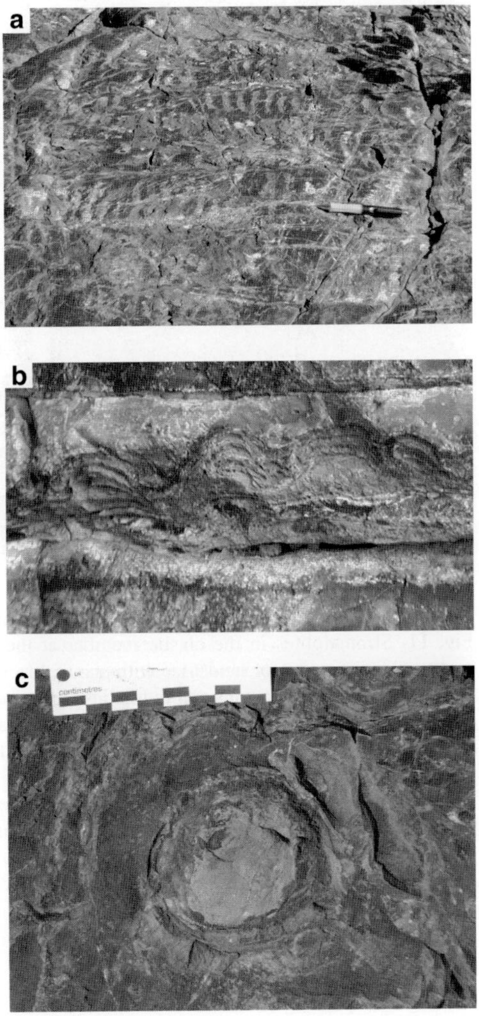

Fig. 10 Stromatolites in the supratidal carbonate member of the Strelley Pool Chert. (**a**) Oblique bedding plane view of wind-blown ripples in finely-layered silicified dolomite. (**b**) Cross-sectional outcrop view of small domical stromatolites in silicified dolomite: note horizontal overlying and underlying beds (Width of view ~12 cm). (**c**) Oblique bedding plane view of top of a small domical stromatolite in silicified, finely-laminated dolomite

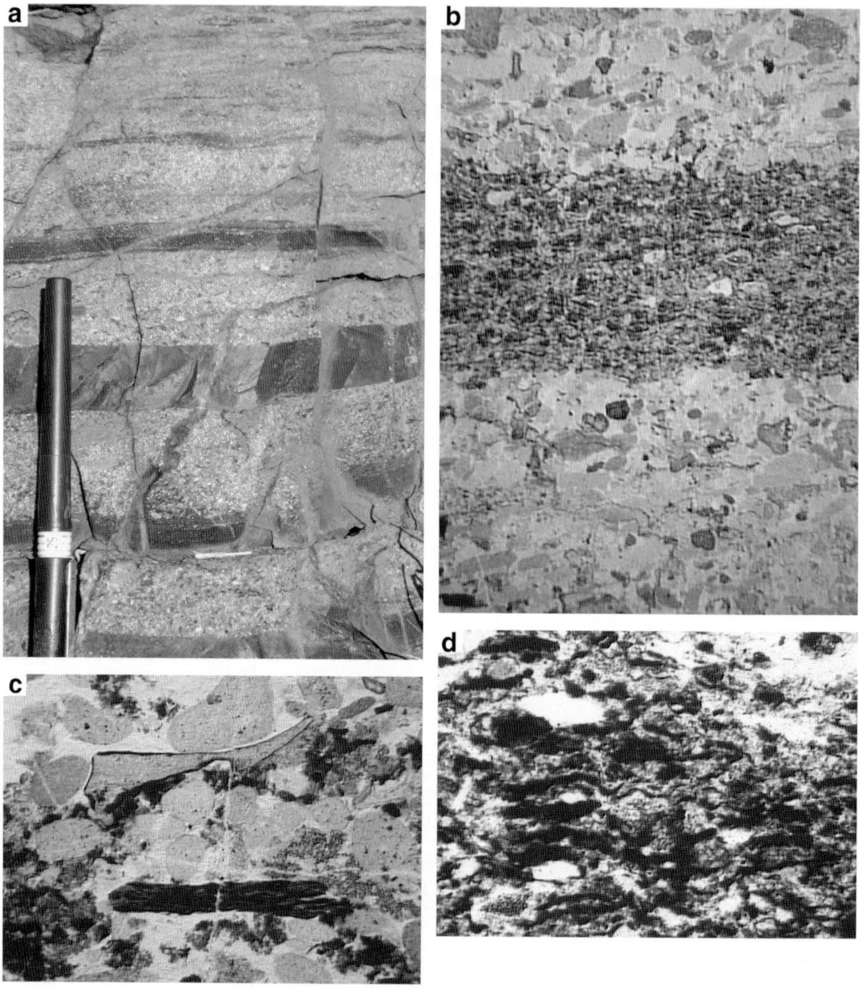

Fig. 11 Stromatolites in the clastic member at the top of the Strelley Pool Chert. (**a**) Cross-sectional outcrop view of sandstone with planar, black stromatolite mats: note trapped sand grains (*light dots*) in lower two mats and in mat just above pen, and also the cross-cutting chert vein. (**b**) Plane polarized light thin section view of planar, black stromatolite mat in sandstone from (**a**), showing the presence of clotted carbonaceous material and trapped sand grains (width of view is 2.5 cm). (**c**) Plane polarized light thin section view of clast of laminated carbonaceous material in sandstone with volcanic clasts (width of view ~5 mm). (**d**) Plane polarized light thin section view of clotty carbonaceous material, with sand grains and silica-filled gas bubbles (Width of view ~2 mm)

laminar mats with abundant clotty kerogen and clasts of laminated carbonaceous material (Fig. 11: Van Kranendonk 2007).

Key morphological indicators of biogenicity include a diverse array of macroscopic and microscopic features (Hofmann et al. 1999; Van Kranendonk et al. 2003;

Allwood et al. 2006; Van Kranendonk 2006, 2007). In marine dolomites, Van Kranendonk et al. (2003) have convincingly demonstrated that the laminae that define most of the macroscopic stromatolites, although locally isopachous (Fig. 9b), in reality show a variety of features and are not typical of seafloor crusts (Fig. 9f). Specifically, evidence for microbial binding of carbonate sediment has been observed (Fig. 9e) and it has also been observed that coniform stromatolites occur along strike of rippled sediment, indicating stromatolite growth in moving water with loose, particulate carbonate sand (Fig. 12: Van Kranendonk et al. 2003). Evidence for axial growth zones within some large coniform stromatolites has also been reported (Grey 2008).

The most compelling morphological features of biogenicity includes a high-amplitude coniform stromatolite that grew upwards through periods of alternating low energy and high energy depositional events (Fig. 13a). Another convincing piece of morphological evidence is from domical stromatolites in supratidal carbonates that show truncated laminae along a smooth growth wall (1 on Fig. 13b), onlapping laminae from one column to the next (2 on Fig. 13b), infill of the intercolumn space by particulate carbonate (3 on Fig. 13b), and truncation of stromatolite growth by a period of erosion and subsequent clastic sediment deposition (4 and sst on Fig. 13b). A final piece of evidence is the presence of different stromatolite morphologies within close proximity along strike (Fig. 13c) and different size conical forms on individual bedding planes (Fig. 13d, e).

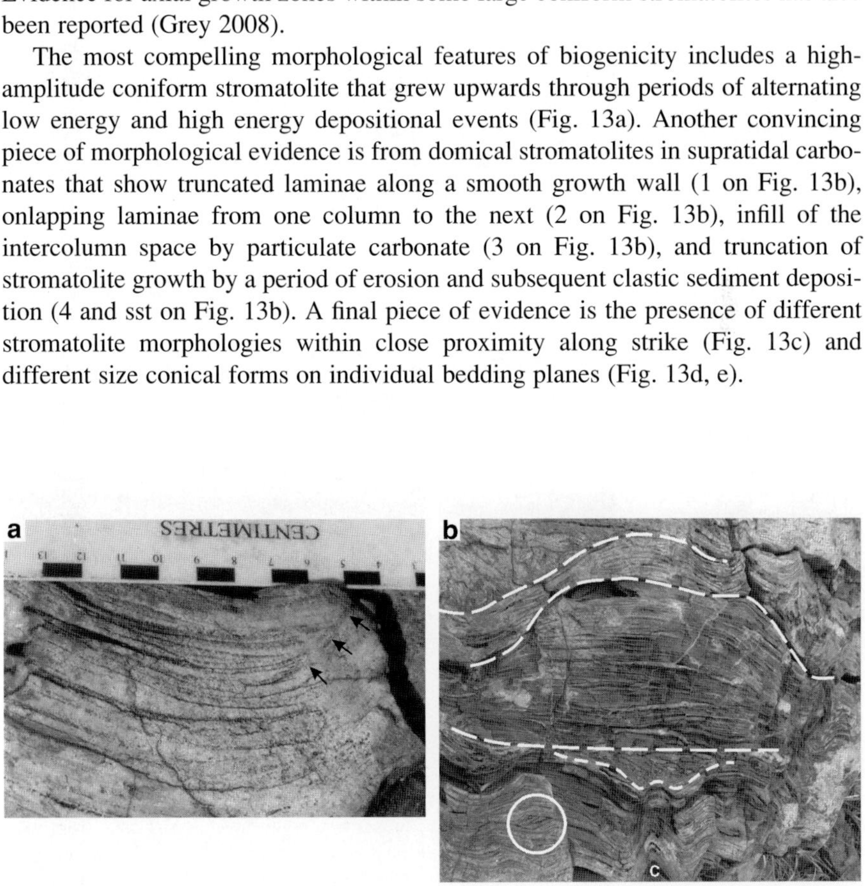

Fig. 12 Evidence for stromatolite growth during sediment accumulation in the marine carbonate member of the Strelley Pool Chert: (**a**) Cross-sectional outcrop view of silicified coniform stromatolite flank, with onlapping sediment layers (*arrows*; *bars* are in centimetres); (**b**) Oblique cross-sectional outcrop view of coniform stromatolite (**c**) in silicified carbonate sandstone with evidence of deposition under high-energy conditions, including ripples (*circle*), small channel fill of flat pebble conglomerate (*above short dashed line*), and high-amplitude sedimentary swale (top 2/3 of image, between *upper* and *lower heavy dashed lines*) with internal unconformities (*middle heavy dashed line*)

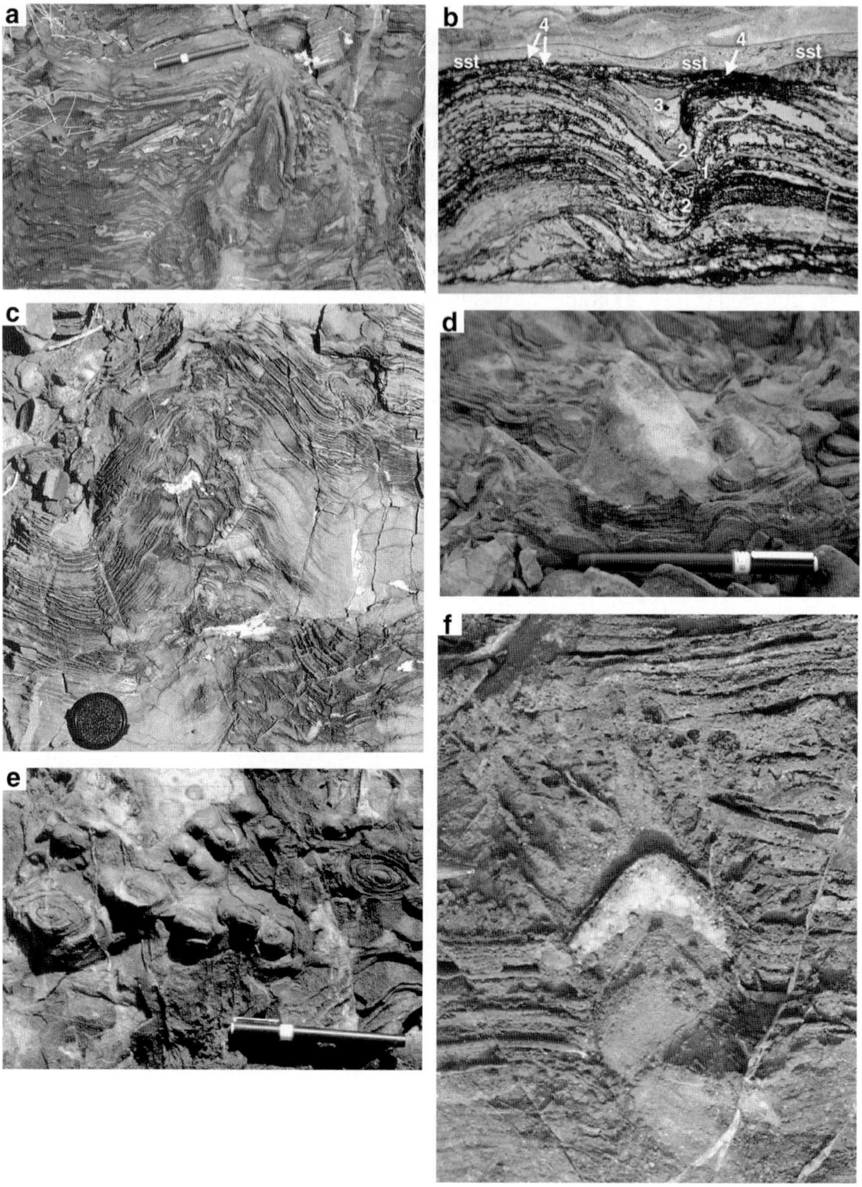

Fig. 13 Key morphological indicators of biogenicity in the Strelley Pool Chert. (**a**) Cross-sectional outcrop view of high-amplitude coniform stromatolite in dolomite, passing up from quiet water laminates, through a unit of flat-pebble conglomerate deposited during a high-energy (storm?) event, and back into quiet water laminates. (**b**) Plane polarized light thin section view of small columnar stromatolites from the supratidal carbonate member (Fig. 10b), showing smooth growth wall that truncates internal laminations in the right-side stromatolite (1), laminations of the left-side stromatolite that lap onto the growth wall of the adjacent column (2), sedimentary infill of the intercolumn space (3), and capping transgressive sand bed (sst) that truncates and erodes

Several additional features support the contention that stromatolite morphology is not the result of post-depositional diagenesis and/or growth of crystal splays, including the fact that crystal splays locally nucleate on the top of coniform stromatolites (Fig. 13f). Locally, it has been observed that horizons of crystal splays terminate the growth of stromatolites and are overlain by thinly bedded, rippled, supratidal carbonate sediments, which combines to suggest that the crystal splay beds formed during periods of subaerial exposure when stromatolite growth ceased (Fig. 14). Additional outcrop-scale evidence for biogenicity comes from the distribution of stromatolites along and across strike (Van Kranendonk 2007; Allwood et al. 2006). The best example of this is at the Trendall locality, where stromatolite size and shape varies along strike in three onlapping biostromes, from larger and more complex forms with greater inheritance (branching and incipiently branching forms) near the upslope part of the bioherm, to smaller and more simple forms downslope at the toe of the bioherms (Fig.14).

Stromatolite biogenicity is supported by highly negative carbon isotope data and organic geochemistry on kerogenous material in clastic sedimentary rocks from these outcrops (Marshall et al. 2007).

4 Summary

Convergent lines of evidence from investigations of putative fossil assemblages from several distinct horizons in the East Pilbara Terrane of the Pilbara Craton point to abundant and diverse microbial life on early Earth. A key indicator of biogenicity in all studied examples is morphology, but this must be set within a well constrained geological context.

Fig. 13 (Continued) laminae in the underlying stromatolites (4) (height of view ~2 cm). (**c**) Cross-sectional outcrop view of laminated dolomite showing large, incipiently branching stromatolite and small, branched columnar stromatolite on flank at top right (detail shown in Fig. 9d). (**d**) Highly oblique bedding plane view of large (8 cm), slightly asymmetrical columnar stromatolite in partly silicified dolomite, flanked by adjacent, smaller coniform stromatolites. (**e**) Bedding plane view of elongate coniform stromatolites with varying size. (**f**) Cross-sectional outcrop view of small coniform stromatolite capped by a radiating crystal splay

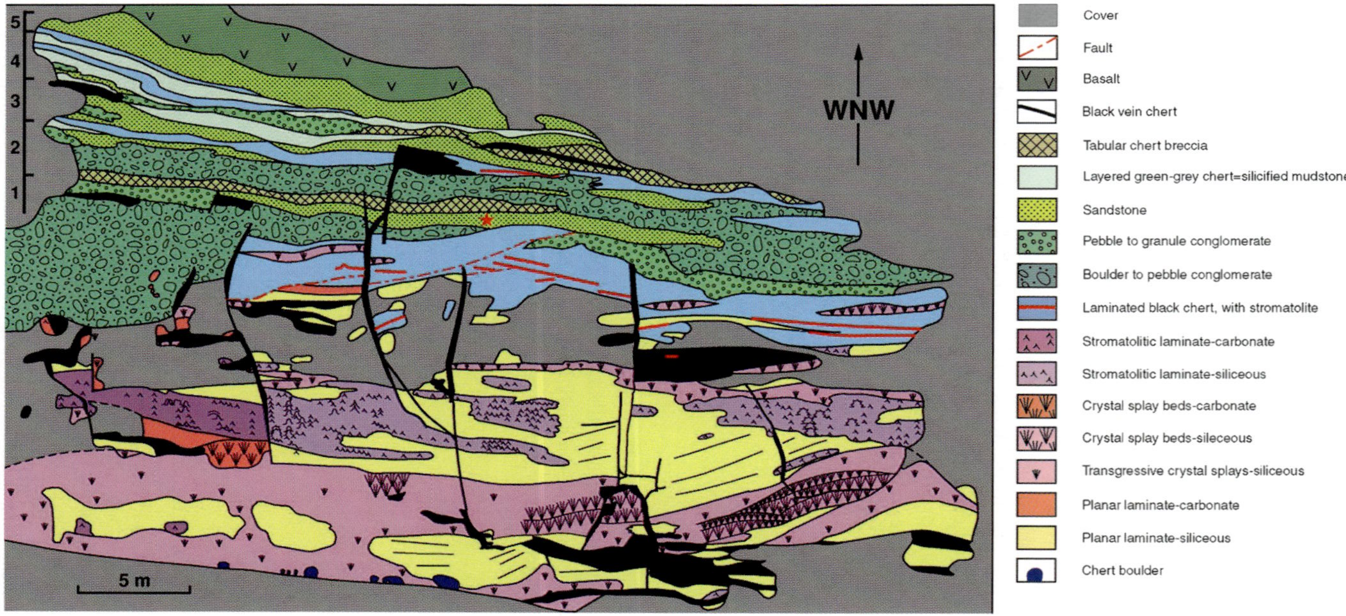

Fig. 14 Outcrop map of part of the Strelley Pool Chert, preserved in cross-section, showing carbonate-encrusted boulders of the basal clastic member, onlapping stromatolitic biostromes (*purple*) and diagenetic crystal splays within carbonate member, domical stromatolite horizons within the silicified supratidal carbonate member (*red lines in light blue unit*), and overlying clastic member comprising five fining-upward units interpreted as the depositional products of receding submarine fans: *red star* denotes occurrence of laminar stromatolites within upper clastic member, shown in Fig. 11 (from Van Kranendonk 2007)

References

Allwood AC, Walter MR, Kamber BS, Marshall CP, Burch IW (2006) Stromatolite reef from the Early Archaean era of Australia. Nature 441:714–717

Banerjee NR, Simonnetti A, Furnes H, Muehlenbachs K, Staudigel H, Heaman L, Van Kranendonk MJ (2007) Direct dating of Archean microbial ichnofossils. Geology 35:487–490

Brasier MD, Green OR, Jephcoat AP, Kleppe AK, Van Kranendonk MJ, Lindsay JF, Steele A, Grassineau N (2002) Questioning the evidence for Earth's oldest fossils. Nature 416:76–81

Brasier MD, Green OR, Lindsay JF, McLoughlin N, Steele A, Stokes C (2005) Critical testing of Earth's oldest putative fossil assemblage from the ~3.5 Ga Apex chert, Chinaman Creek, Western Australia. Precambrian Research 140:55–102

Duck LJ, Glikson M, Golding SD, Webb RE (2007) Microbial remains and other carbonaceous forms form the 3.24 Ga Sulphur Springs black smoker deposit, Western Australia. Precambrian Research 154:205–220

Furnes H, Banerjee NR, Staudigel H, Muehlenbachs K, de Wit M, Van Kranendonk M (2007) Comparing petrographic signatures of bioalteration in recent to Mesoarchean pillow lavas: Tracing subsurface life in oceanic igneous rocks. Precambrian Research 158:156–176

Garcia-Ruiz JM, Hyde ST, Carnerup AM, Christy AG, Van Kranendonk MJ, Welham NJ (2003) Self-assembled silica-carbonate structures and detection of ancient microfossils. Science 302:1194–1197

Glikson M, Duck LJ, Golding SD, Hofmann A, Bolhar R, Webb R, Baiano JCF, Sly LI (2008) Microbial remains in some earliest Earth rocks: Comparison with a potential modern analogue. Precambrian Research 164:187–200

Grey K (2008) Morphological diversity and complexity in the c. 3.40 Ga Strelley Pool Chert, Pilbara Craton – Evidence for a biogenic origin. Australian Earth Sciences Convention, Program and Abstract booklet, Perth, Western Australia

Grotzinger JP, Rothman DH (1996) An abiotic model for stromatolite morphogenesis. Nature 383:423–425

Groves DI, Dunlop JSR, Buick R (1981) An early habitat of life. Scientific American 245:64–73

Hofmann HJ, Grey K, Hickman AH, Thorpe R (1999) Origin of 3.45 Ga coniform stromatolites in the Warrawoona Group, Western Australia. The Geological Society of America Bulletin 111:1256–1262

Lowe DR (1980) Stromatolites 3,400-Myr old from the Archaean of Western Australia. Nature 284:441–443

Lowe DR (1983) Restricted shallow-water sedimentation of Early Archean stromatolitic and evaporitic strata of the Strelley Pool Chert, Pilbara Block, Western Australia. Precambrian Research 19:239–283

Marshall CP, Love GD, Snape CE, Hill AC, Allwood AC, Walter MR, Van Kranendonk MJ, Bowden SA, Sylva SP, Summons RE (2007) Structural characterization of kerogen in 3.4 Ga Archaean cherts from the Pilbara Craton, Western Australia. Precambrian Research 155:1–23

Nijman W, De Bruin K, Valkering M (1998) Growth fault control of early Archaean cherts, barite mounds, and chert-barite veins, North Pole Dome, Eastern Pilbara, Western Australia. Precambrian Research 88:25–52

Philippot P, Van Zuilen M, Lepot K, Thomazo C, Farquhar J, Van Kranendonk M (2007) Early Archean microorganisms preferred elemental sulphur, not sulphate. Science 317:1534–1537

Rasmussen B (2000) Filamentous microfossils in a 3,235-million-year-old volcanogenic massive sulphide deposit. Nature 405:676–679

Schopf JW (1993) Microfossils of the Early Archaean Apex Chert: new evidence of the antiquity of life. Science 260:640–646

Schopf JW, Kudryavtsev AB, Czaja AD, Tripathi AB (2007) Evidence of Archean life: Stromatolites and microfossils. Precambrian Research 158:141–155

Ueno Y, Isozaki Y, Yurimoto H, Maruyama S (2001) Carbon isotopic signatures of individual Archaean Microfossils(?) from Western Australia. International Geology Review 43:196–212
Ueno Y, Yoshioka H, Maruyama S, Isozaki Y (2004) Carbon isotopes and petrography in ~3.5 Ga hydrothermal silica dykes in the North Pole area, Western Australia. Geochimica et Cosmochimica Acta 68:573–589
Ueno Y, Yamada K, Yoshida N, Maruyama S, Isozaki Y (2006) Evidence from fluid inclusions for microbial methanogenesis in the early Archaean era. Nature 440:516–519
Van Kranendonk MJ (2006) Volcanic degassing, hydrothermal circulation and the flourishing of early life on Earth: new evidence from the Warrawoona Group, Pilbara Craton, Western Australia. Earth Science Reviews 74:197–240
Van Kranendonk MJ (2007) A review of the evidence for putative Paleoarchean life in the Pilbara Craton. In: Van Kranendonk MJ, Smithies RH, Bennet V (eds) Earth's Oldest Rocks. Developments in Precambrian Geology 15, Elsevier, Amsterdam, pp. 855–896
Van Kranendonk MJ, Pirajno F (2004) Geochemistry of metabasalts and hydrothermal alteration zones associated with ca. 3.45 Ga chert± barite deposits: implications for the geological setting of the Warrawoona Group, Pilbara Craton, Australia. Geochemistry: Exploration, Environment, Analysis 4:253–278
Van Kranendonk MJ, Webb GE, Kamber BS (2003) Geological and trace element evidence for a marine sedimentary environment of deposition and biogenicity of 3.45 Ga stromatolitic carbonates in the Pilbara Craton, and support for a reducing Archean ocean. Geobiology 1:91–108
Van Kranendonk MJ, Smithies RH, Hickman AH, Champion DC (2007a) Secular tectonic evolution of Archaean continental crust: interplay between horizontal and vertical processes in the formation of the Pilbara Craton, Australia. Terra Nova 19:1–38
Van Kranendonk MJ, Hickman A, Smithies RH (2007b) The East Pilbara Terrane of the Pilbara Craton, Western Australia: Formation of a continental nucleus through repeated mantle plume magmatism. In: Van Kranendonk MJ, Smithies RH, Bennet V (eds) Earth's Oldest Rocks. Developments in Precambrian Geology 15, Elsevier, Amsterdam, pp 307–337
Van Kranendonk MJ, Philippot P, Lepot K, Bodorkos S, Pirajno F (2008) Geological setting of Earth's oldest fossils in the c. 3.5 Ga Dresser Formation, Pilbara Craton, Western Australia. Precambrian Research 167:93–124
Van Kranendonk MJ, Smithies RH, Wingate MTD, Bodorkos S, Hickman AH (2010) Evidence for Mesoarchaea (~3.2 Ga) rifting of the Pilbara Craton: The missing link in an early Precambrian Wilson cycle. Precambrian Research 177:145–161
Van Zuilen M, Thomazo C, Luais B, Philippot P (2008) Photosynthesis in a 3.5 Ga old shallow marine depositional environment; clues from Carbon and Iron isotopes. Australian Earth Sciences Convention, Program and Abstract booklet, Perth, Western Australia, pp 247–248
Walter MR, Buick R, Dunlop JSR (1980) Stromatolites, 3,400–3,500 Myr old from the North Pole area, Western Australia. Nature 284:443–445

Index

J. Reitner et al., *Advances in Stromatolite Geobiology*,
Lecture Notes in Earth Sciences 131, DOI 10.1007/978-3-642-10415-2,